# Advanced Calculus

Hans Sagan
*North Carolina State University*

# Advanced Calculus

## of Real-Valued Functions of a Real Variable and Vector-Valued Functions of a Vector Variable

**Houghton Mifflin Company · Boston**
*Atlanta · Dallas · Geneva, Illinois*
*Hopewell, New Jersey · Palo Alto · London*

Printed in the United States of America

Library of Congress Catalog Card Number: 73-9399
ISBN: 0-395-17090-7

To My Wife Ingeborg

# Preface

In a calculus course, the student is expected to acquire a great number of techniques and problem-solving devices and to put them to practical use. The goal of an advanced calculus course is to put the calculus material into proper perspective and to ask the question, "Why does calculus work?" In an overwhelming number of cases, the answer is simple: "Because the real numbers have the least-upper-bound property." This text is what I hope to be an informative and entertaining documentation of this answer. At the same time, I have tried to prepare the reader for topics beyond advanced calculus such as topology, theory of functions, real variables and measure theory, functional analysis, integration on manifolds, and last but not least, what is generally referred to as applied mathematics. To avoid misunderstandings, let it be stressed that this text is not an introduction to any of these subjects but rather a preparation *for* them.

I have set myself the traditional goals of establishing the real number system as a complete ordered field, of proving Lebesgue's criterion for the existence of the Riemann integral, of establishing the inverse function theorem and the implicit function theorem, of deriving Jacobi's theorem on the transformation of multiple integrals, and of giving an introduction to line and surface integrals. There is no point in pretending that these are not difficult subjects when dealt with rigorously and in a reasonably general setting. They are and the reader cannot be spared the gory details. Jacobi's theorem and the preparatory theory, in particular, take up considerable space. This is unavoidable if one wants to establish this theorem in sufficient generality so that it is applicable in nontrivial situations and while not using ideas and techniques that require greater sophistication than the rest of this book.

The material is developed in more or less natural order proceeding from one topic to the next with few detours along the way. While some of the topics are of a supplementary nature, most stand on their own merit. Occasional departures from the straight and narrow are discussed in optional sections that are designated by an asterisk. Although these sections may be omitted without jeopardizing the continuity of the development they ought not to be ignored by the serious student of mathematics.

The first four chapters deal with real-valued functions of a real variable only. The reason for this is mostly pedagogical. It has been my experience that the student fresh out of a calculus course is overwhelmed and stunned when confronted with the concepts of limit, continuity, completeness, compactness, measure, and others in the general setting of an $n$-dimensional Euclidean space unless he has already acquired a good working knowledge of these concepts in the simple one-dimensional setting. Chapters 1 through 4 provide such a working knowledge.

I have tried to make this text as flexible as possible to accommodate a variety of needs and to fit into diverse curricula, taking into account other courses that might be available. Chapters 7 on Sequences of Functions and 14 on Infinite Series may be taken up right after Chapter 5. These topics are discussed in a one-dimensional setting and obvious generalizations to more dimensions are taken up in the exercises. The degree of generality one achieves in discussing these topics to the extent that I have done in this book in a higher dimensional setting is largely an illusion. Chapter 10 deals with the inverse function theorem and the implicit function theorem. For the convenience of the student or the instructor who does not find time to wade through all the details, I have summarized the main results in section 9.95. Similarly, the main result of Chapter 12 on transformation of multiple integrals is summarized in section 11.111. Chapter 13 on line and surface integrals, however, which may be viewed as a chapter on applications of Chapters 10 and 12, depends heavily on these chapters.

Aside from their natural order, the topics in this text may be taken up as presented in the following flow chart:

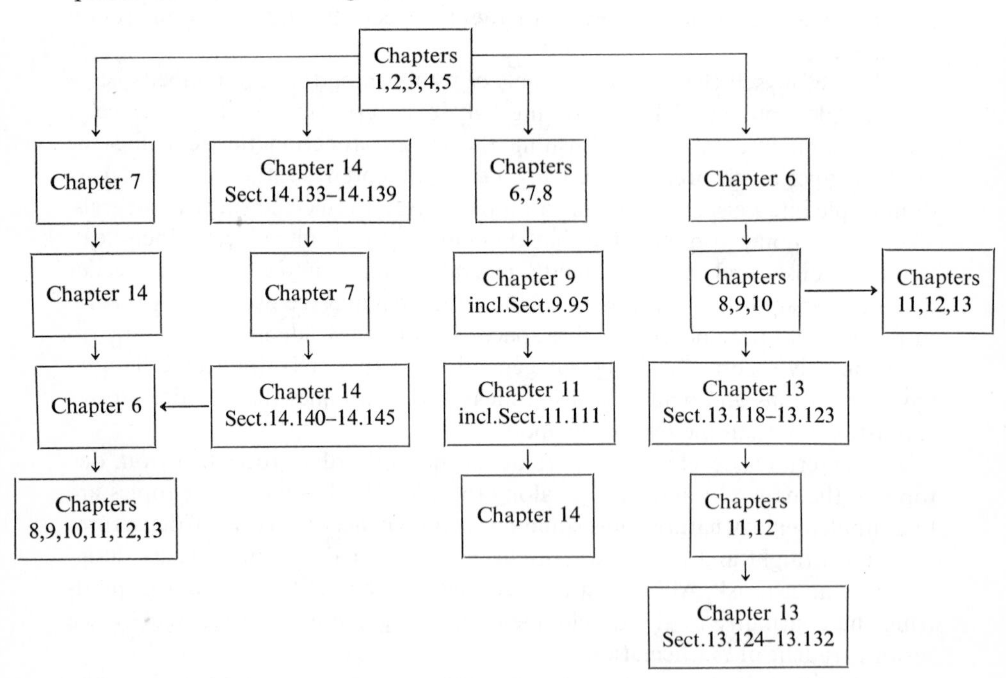


Great emphasis has been placed on detailed discussions of examples and counterexamples which are a rich source of enlightenment and contribute substantially to an understanding of mathematics. There are 380 such examples interspersed with the text. Sometimes, a displayed proof, and, in particular, a proof by contradiction, contributes little towards an understanding of a theorem. However, proofs are the tools of trade of the mathematician. Since the reader of this book is presumed to be in training to become a mathematician, I have made every effort to supply all theorems and lemmas with what I consider a valid proof by present day standards. (Some exceptions have been made in Chapter 13 where an occasional appeal to geometric insight has been substituted for what would otherwise have led to lengthy and time-consuming arguments.) Proofs are worked out in great detail to keep the reader's frustration quotient (effort required to unravel individual steps over total effort) at a minimum. When a proof is exceptionally long—and there are some such proofs—an outline of the objective precedes the formal proof.

The text is supplemented with 1204 exercises to 741 of which a hint and/or answer is provided in Appendix 2. Some of the exercises serve the purpose of helping the reader to familiarize himself with new formalisms and techniques, some supplement the theory that is developed in the main body of the text, and some complement the theory. I have not graded the exercises by degree of difficulty because I do not wish the reader to contemptuously ignore the easy ones nor do I want to scare him away from the more difficult exercises. The answers and hints in the appendix should only be used as a last resort.

A necessary and sufficient condition for a chance to succeed in a two-semester course that is based on this book is the successful completion of a calculus course and an introductory course on linear algebra. The latter may be taken concurrently with the first part of this course.

In the preparation of this text, I was inspired and assisted by the writings of many mathematicians, beginning with Archimedes of Syracuse. During the past 15 years I have used in my teaching of advanced calculus the texts by T. Apostol, R. G. Bartle, R. Courant, A. Devinatz, A. Taylor, and D. V. Widder, and I have consulted many others, particularly M. Spivak's monograph on Calculus on Manifolds. I find it very difficult to trace to its source every idea that I may have extracted from their works and incorporated in my own. Therefore, a blanket acknowledgement will have to suffice. I am also greatly indebted to my department head, N. J. Rose, who provided encouragement in word and deed and to my colleagues R. E. Chandler, W. J. Harrington, C. D. Meyer, L. Page, and R. T. Ramsay, who have class-tested my notes for the past three years and have substantially contributed to the elimination of errors and to an improvement of the presentation. Daniel Rider from the University of Wisconsin and Richard Gisselquist from the University of Wisconsin—Eau Claire have also pointed out many errors and have made very valuable suggestions for which I am grateful. I also wish to thank my many students who, through persistent questioning have led me to revise and clarify many obscure passages. Last but not least, I express my gratitude and appreciation to Marla Christopher who typed, and retyped, and retyped . . . my manuscript with patience, skill, and understanding.

Hans Sagan
*Raleigh, North Carolina*

# Contents



*optional section

*optional section

*optional section

*optional section

*optional section

# Numbers 1

## 1.1 Set Language—A Mathematical Shorthand

This section is devoted to a discussion of certain abstract concepts, namely sets, and some manipulations with these concepts, namely the elementary set operations. It is not intended as an introduction to set theory. It merely serves to introduce some terminology, the language of sets, that will prove to be helpful in our studies. A command of this language is not absolutely essential for a study of analysis but is of considerable assistance in such a study, as it provides for short and concise formulations as well as for an abstract structure for arguments which will occur repeatedly in the study of analysis. Such arguments would be awkward and overburdened with imprecise and excessive verbiage were it not for the language of sets.

We shall take a naive approach and say, in the spirit of *Georg Cantor*,† that *a set is a collection of well-distinguished objects of our conception or imagination.* (We note that this idea of a set, when carried too far, leads to grave difficulties, as we shall point out in section 3.)

The collection of coins in your pocket is a set—it may be the *empty set.*

***Definition 1.1***

The empty set $\emptyset$ is the set which contains no elements.

The collection of all matter in the universe is a set—for those who are materialistically inclined, this may well be the *universal set.*

The *universal set* is the set of all objects that are under consideration in a specific investigation.

For a number theorist the universal set may be the set of all integers; for an analyst, the universal set may be the set of all real numbers, or the set of all functions; for a geometer, the universal set may be the set of all points on a line, or the set of all lines in a plane, or what have you.

† *Georg Cantor*, 1845–1918.

In this and the next four sections, we shall discuss a variety of examples in order to illustrate the new concepts that are being introduced. To make these examples sufficiently interesting and to provide for enough variety, we shall assume, for the time being, that the reader has some intuitive notion of the real-number system, functions, graphs, coordinate systems, and other basic ideas. The concepts and operations which we are about to introduce do not, however, depend on the knowledge of, or even on the existence of these notions. In due course, we shall establish a sound foundation for the real-number system and the attendant concepts.

We use capital letters $A$, $B$, $C$, $X$, $Y$, $Z$, etc., to denote sets and lower-case letters $a$, $b$, $c$, $x$, $y$, $z$, etc., to denote elements of sets.

If *$a$ is an element of $A$*, we write

$$a \in A$$

and if *$a$ is not an element of $A$*,

$$a \notin A.$$

With the understanding that **N** *denotes the set of all natural numbers* (positive integers), we describe the set $A$ that contains the elements 2, 3, 4, 5 as follows:

(1.1) $$A = \{x \in \mathbf{N} \mid 1 < x < 6\}$$

which is to be read as "$A$ is the set of all $x$ in **N** such that $x$ is greater than 1 and less than 6." The same set may be described by

$$A = \{x \in \mathbf{N} \mid 2 \leqslant x \leqslant 5\},$$

or simply by

$$A = \{2, 3, 4, 5\}.$$

The set

(1.2) $$\mathbf{E} = \{x \in \mathbf{N} \mid x = 2y, y \in \mathbf{N}\}$$

is the set of all even numbers and

(1.3) $$\mathbf{O} = \{x \in \mathbf{N} \mid x = 2y - 1, y \in \mathbf{N}\}$$

is the set of all odd numbers.

If **R** *denotes the set of all real numbers*, then

(1.4) $$\mathfrak{I} = \{x \in \mathbf{R} \mid 0 \leqslant x \leqslant 1\}$$

denotes the set of all real numbers between 0 and 1, *endpoints included*, and

(1.5) $$\mathfrak{J} = \{x \in \mathbf{R} \mid 0 < x < 1\}$$

denotes the set of all real numbers between 0 and 1, *endpoints excluded.*

(1.6) $$F = \{x \in \mathbf{R} \mid x^2 < 0\}$$

is a complicated way of describing the empty set $\varnothing$.

***Definition 1.2***

If $X$, $Y$ are two sets, then

$$X\backslash Y = \{x \in X \mid x \notin Y\}$$

(all elements of $X$ which are not in $Y$) is called the *complement of $Y$ relative to $X$*. When $U$ is the universal set and there is no doubt about it from the context, we simply write

$$U\backslash Y = c(Y)$$

and call this set the *complement of $Y$*.

If $U = \mathbf{N}$ is the set of all natural numbers and $\mathbf{E}$ the set of all even numbers as defined in (1.2), then

$$\begin{aligned} c(\mathbf{E}) = \mathbf{N}\backslash\mathbf{E} &= \{x \in \mathbf{N} \mid x \notin \mathbf{E}\} \\ &= \{x \in \mathbf{N} \mid x \neq 2y, y \in \mathbf{N}\} \\ &= \{x \in \mathbf{N} \mid x = 2y - 1, y \in \mathbf{N}\} = \mathbf{O} \end{aligned}$$

where $\mathbf{O}$ is the set of all odd numbers in (1.3).

Similarly,

$$c(\mathbf{O}) = \mathbf{N}\backslash\mathbf{O} = \mathbf{E}.$$

If $U = \mathbf{R}$ is the set of all real numbers and if $\mathfrak{I}$ is defined as in (1.4), then

$$c(\mathfrak{I}) = \mathbf{R}\backslash\mathfrak{I} = \{x \in \mathbf{R} \mid x < 0 \text{ or } x > 1\},$$

and if $F$ is defined as in (1.6),

$$c(F) = \mathbf{R}\backslash F = \{x \in \mathbf{R} \mid x^2 \geqslant 0\} = \mathbf{R}.$$

We note that the sets $A$, $\mathbf{E}$, $\mathbf{O}$ in (1.1), (1.2), and (1.3) are contained in the set $\mathbf{N}$; that the sets $\mathfrak{I}$, $\mathfrak{J}$ in (1.4) and (1.5) are contained in $\mathbf{R}$; that $\mathbf{E}$ and $\mathbf{O}$ when combined yield $\mathbf{N}$; and that $\mathbf{E}$ and $\mathbf{O}$ have no common elements. We shall introduce some more set language that will enable us to make these statements crisp and precise:

***Definition 1.3***

$X$ is a *subset* of $Y$ ($X$ is *contained* in $Y$)

$$X \subseteq Y$$

means that if $x \in X$, then $x \in Y$. $Y$ is a *superset* of $X$ if and only if $X$ is a subset of $Y$.

$X$ is a *proper subset* of $Y$ ($X$ is *properly contained* in $Y$)

$$X \subset Y$$

means that $X$ is a subset of $Y$ and there exists at least one element $y \in Y$ such that $y \notin X$. $Y$ is a proper superset of $X$ if and only if $X$ is a proper subset of $Y$.

The sets $X$ and $Y$ are *equal*,

$$X = Y,$$

means that $X \subseteq Y$ and $Y \subseteq X$.

We have, for the sets in (1.1) to (1.6) and $\mathbf{N}$ and $\mathbf{R}$, that

$$A \subset \mathbf{N}, \quad \mathbf{E} \subset \mathbf{N}, \quad \mathbf{O} \subset \mathbf{N}, \quad \mathbf{N} \subseteq \mathbf{N}, \quad \mathfrak{J} \subset \mathfrak{I} \subset \mathbf{R}, \quad F \subset A, \quad \mathbf{R} \subseteq \mathbf{R},$$

and, of course,

$$\mathbf{N} = \mathbf{N}, \qquad \mathbf{R} = \mathbf{R}.$$

We adopt the convention that

$$\varnothing \subseteq X \tag{1.7}$$

for all sets $X$. (See also exercise 1.)

***Definition 1.4*** *(Set Union and Intersection)*

The *union* of $X$ and $Y$ (in symbols: $X \cup Y$) is the set

$$X \cup Y = \{x \mid x \in X \text{ or } x \in Y\}.$$

(Note that "or" is not exclusive, i.e., $x$ may be in $X$ as well as in $Y$.)

The *intersection* of $X$ and $Y$ (in symbols: $X \cap Y$) is the set

$$X \cap Y = \{x \mid x \in X \text{ and } x \in Y\}.$$

We have, for the sets in (1.1) to (1.5), that

$$A \cup c(A) = \mathbf{N}, \quad \mathbf{E} \cup \mathbf{O} = \mathbf{N}, \quad \mathbf{E} \cap \mathbf{O} = \varnothing, \quad \mathfrak{I} \cup \mathfrak{J} = \mathfrak{I}, \quad \mathfrak{I} \cap \mathfrak{J} = \mathfrak{J}.$$

We have, directly from the definition of a set complement and from definition 1.4 that, for every set $X$,

$$X \cup c(X) = U, \qquad X \cap c(X) = \varnothing$$

where $U$ denotes the universal set.

We call two sets $X$ and $Y$ *disjoint* if and only if $X \cap Y = \varnothing$.

In order to simplify our notation, we shall employ special symbols for certain subsets of the set $\mathbf{R}$ of real numbers, the so-called *intervals*:

$\{x \in \mathbf{R} \mid a \leqslant x \leqslant b\} = [a, b]$ is called a *closed* interval;

$\{x \in \mathbf{R} \mid a < x < b\} = (a, b)$ is called an *open* interval;

$\{x \in \mathbf{R} \mid a \leqslant x < b\} = [a, b)$ and $\{x \in \mathbf{R} \mid a < x \leqslant b\} = (a, b]$

are called *semiopen* (*semiclosed*) intervals.

We shall represent intervals geometrically as depicted in Fig. 1.1.

Occasionally we shall also employ the notation

$$\{x \in \mathbf{R} \mid x \geqslant a\} = [a, \infty),$$
$$\{x \in \mathbf{R} \mid x \leqslant a\} = (-\infty, a],$$
$$\mathbf{R} = (-\infty, \infty),$$

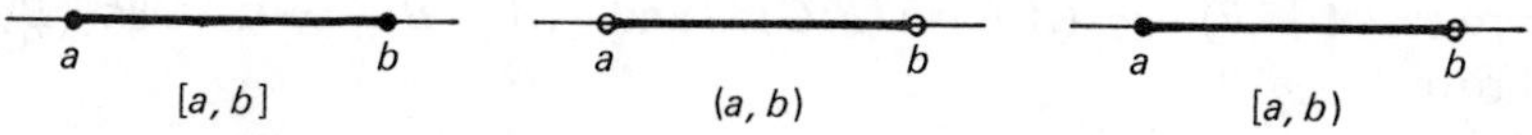

$[a, b]$ $(a, b)$ $[a, b)$

*Fig. 1.1*

and suitable modifications thereof.

## 1.1 Exercises

1. What is the alternative to (1.7)?
2. In order to show that two sets $X_1$ and $X_2$ are equal, one may proceed as follows: First, demonstrate that if $x \in X_1$, then $x \in X_2$, that is, $X_1 \subseteq X_2$. Then show the converse: if $y \in X_2$, then $y \in X_1$, that is, $X_2 \subseteq X_1$. Prove:

   (a) $c(c(A)) = A$ (b) $X \cup U = U$ (c) $X \cap U = X$
   (d) $X \cup \varnothing = X$ (e) $X \cap \varnothing = \varnothing$

3. Prove:

   (a) $X \subseteq X \cup Y, Y \subseteq X \cup Y$ (b) $X \cap Y \subseteq X, X \cap Y \subseteq Y$

4. Given $U = \mathbf{N}$ (natural numbers), $A = \mathbf{E}$ (even numbers), $B = \{1, 3, 5, 6, 8\}$, $C = \{2, 6, 10, 14\}$. Find:

$$(c(A) \cup B) \cap C$$

5. The set operations have a simple and compelling representation, as indicated in Fig. 1.2. Such a figure is called a *Venn diagram*. Draw Venn diagrams for the following set operations:

   (a) $(A \cup B) \cap C$ (b) $(A \cap B) \cup C$
   (c) $c(A) \cup B$ (d) $A \cup c(B)$

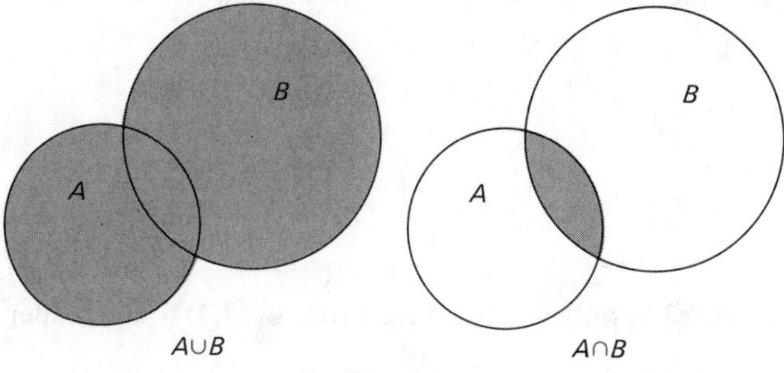


*Fig. 1.2*

6. Prove:

   (a) $A \cup B = B \cup A$; $A \cap B = B \cap A$
   (b) $(A \cup B) \cup C = A \cup (B \cup C)$; $(A \cap B) \cap C = A \cap (B \cap C)$
   (c) $A \cap (B \cup C) = (A \cap B) \cup (A \cap C)$; $A \cup (B \cap C) = (A \cup B) \cap (A \cup C)$

7. Prove: $A \backslash B = A \cap c(B)$
8. Prove: $(A \cup B)\backslash C = (A\backslash C) \cup (B\backslash C)$; and $(A \cap B)\backslash C = (A\backslash C) \cap (B\backslash C)$.
9. Prove:

   (a) If $A \cap B = \varnothing$, then $A\backslash B = A$; (b) If $A \subseteq B$, then $A\backslash B = \varnothing$;
   (c) If $A\backslash B = \varnothing$, then $A \subseteq B$.

## 1.2 Generalized Unions and Intersections; DeMorgan's Law

If $J$ denotes a given set (finite or infinite) and if with each $j \in J$ there is associated a set $X_j$, we can generalize the operation of *union* and *intersection* as follows:

$$\bigcup_{j \in J} X_j = \{x \mid x \in X_j \text{ for } \textit{some } j \in J\} \tag{2.1}$$

and

$$\bigcap_{j \in J} X_j = \{x \mid x \in X_j \text{ for } \textit{all } j \in J\}. \tag{2.2}$$

In such context, $J$ is referred to as an *index set*. Note that the *union* is the set of *all elements that are contained in at least one of the sets* and that the *intersection* is *the set of all elements that are contained in all the sets.*

Let $X_\alpha = \{x \in \mathbf{R} | \alpha/2 < x < \alpha\}$. $X_\alpha$ is well defined for all $0 < \alpha < 1$. We use $J = \{\alpha \in R \mid 0 < \alpha < 1\}$ as index set and find

$$\bigcup_{\alpha \in J} X_\alpha = (0, 1) \quad \text{and} \quad \bigcap_{\alpha \in J} X_\alpha = \varnothing.$$

The following theorem correlates complementation, union, and intersection and is an indispensible aid in streamlining proofs of many theorems.

***Theorem 2.1*** *(DeMorgan's† Law)*

If $J$ is a given index set and the sets $X_j$ are defined for all $j \in J$, then

$$c\left(\bigcup_{j \in J} X_j\right) = \bigcap_{j \in J} c(X_j), \tag{2.3}$$

$$c\left(\bigcap_{j \in J} X_j\right) = \bigcup_{j \in J} c(X_j). \tag{2.4}$$

*Proof*

We shall prove (2.3) only and leave the proof of (2.4) to the reader (exercise 1).

†*Augustus DeMorgan*, 1806–1873.

(a) Suppose that

$$x \in c\left(\bigcup_{j \in J} X_j\right),$$

that is, $x \notin \bigcup_{j \in J} X_j$. Then, $x \notin X_j$ for *all* $j \in J$. (If there were a $j_0 \in J$ such that $x \in X_{j_0}$, then $x \in \bigcup_{j \in J} X_j$.) Hence, $x \in c(X_j)$ for *all* $j \in J$ and, therefore, $x \in \bigcap_{j \in J} c(X_j)$ that is,

$$c\left(\bigcup_{j \in J} X_j\right) \subseteq \bigcap_{j \in J} c(X_j) \tag{2.5}$$

(b) Suppose that $x \in \bigcap_{j \in J} c(X_j)$. Then, $x \in c(X_j)$ for all $j \in J$, that is, $x \notin X_j$ for all $j \in J$. Hence, $x \notin \bigcup_{j \in J} X_j$ which, in turn, means that $x \in c(\bigcup_{j \in J} X_j)$ and we have

$$\bigcap_{j \in J} c(X_j) \subseteq c\left(\bigcup_{j \in J} X_j\right). \tag{2.6}$$

(2.5) and (2.6) imply set equality, and (2.3) is proved.

## 1.2 Exercises

1. Prove (2.4).
2. Given the index set $J = \mathbf{N}$ and $X_j = \{x \in \mathbf{N} \mid x \geqslant j\}$. Find:
$$\bigcup_{j \in \mathbf{N}} X_j \quad \text{and} \quad \bigcap_{j \in \mathbf{N}} X_j.$$
3. Let $\mathfrak{J}_n = \{x \in \mathbf{R} \mid 0 < x < 1/n\}$. Find:
$$\bigcup_{n \in \mathbf{N}} \mathfrak{J}_n \quad \text{and} \quad \bigcap_{n \in \mathbf{N}} \mathfrak{J}_n.$$
4. Same as in (3) for $\mathfrak{I}_n = \{x \in \mathbf{R} \mid 0 \leqslant x \leqslant 1/n\}$ used instead of $\mathfrak{J}_n$.
5. Given the sets $\mathfrak{J}_n$ in (3). Find $c(\bigcup_{n \in \mathbf{N}} \mathfrak{J}_n)$, $c(\bigcap_{n \in \mathbf{N}} \mathfrak{J}_n)$,
   (a) by DeMorgan's law, (b) by direct analysis.
6. Prove: If $A \subseteq B$, then $c(B) \subseteq c(A)$.
7. Prove: $A \backslash (B \cup C) = (A \backslash B) \cap (A \backslash C)$; $A \backslash (B \cap C) = (A \backslash B) \cup (A \backslash C)$.
8. Generalize the result of exercise 7 to the following: For a given index set $J$,
$$A \Big\backslash \bigcup_{j \in J} B_j = \bigcap_{j \in J} (A \backslash B_j), \qquad A \Big\backslash \bigcap_{j \in J} B_j = \bigcup_{j \in J} (A \backslash B_j).$$
9. Prove: $A \cap B = A \backslash (A \backslash B)$.

## 1.3 The Russell Paradox*

*Bertrand Russell*,† in his examination of the foundations of mathematics, found them to be quite shaky. The following example, due to him, has led to a thorough reexamination of these foundations.

*This section is optional and may be skipped without loss of continuity.
†*Bertrand Russell*, 1872–1970.

One would not expect a respectable set to contain itself as an element. Accordingly, we shall call *sets that do not contain themselves as elements respectable sets.* Respectable sets are certainly objects of our imagination, if not conception. We consider the *set of all respectable sets* and call it $B$. *Is $B$ a respectable set?*

Suppose that *$B$ is respectable.* Since $B$ is respectable and is the set of *all* respectable sets, it has to contain itself as an element. Hence, *$B$ cannot be respectable.*

Suppose that *$B$ is not respectable.* A set that is *not* respectable has to contain itself as an element (otherwise it would be respectable). As the set of all respectable sets, being itself an element, *$B$ has to be respectable.*

A situation in which an object (of our conception or imagination) has a certain property if it does not have that property and vice versa, is called a *paradox*. Clearly, a set $B$ such as we have attempted to construct here cannot exist, because its existence would lead to a paradox which is logically unacceptable. The lesson we learn from this experience is that in the construction of sets one has to proceed with the utmost care. We have, in the construction of $B$, abused the concept of sets by stretching it beyond its capacity.

There are some escape routes from this dilemma. One could attach safeguards to a suitably worded definition of the concept of a set, or, better still, define the concept of set not explicitly at all but rather with respect to its properties and behavior under the various set operations. The latter approach, the so-called *axiomatic approach*, is the one most widely followed and the one that provides the most comfort to mathematicians

We wish to emphasize that we shall not encounter any difficulties in our treatment because the simple sets which we shall consider behave in an acceptable manner under the elementary set operations.

## 1.3 Exercises

1. Show that the set of all sets that can be defined with less than 20 words is not a respectable set.
2. Take a blank page and inscribe the following statement on that page: "The only statement on this page is false." Is this a true statement? A false statement?
3. *Definition.* The village barber is the only man living in the village who shaves all men in the village who do not shave themselves. Does the village barber shave himself? Does he not shave himself?
4. Find a common feature of the Russell paradox, exercise 2, and exercise 3.

## 1.4 Cartesian Products and Functions

Although we shall not deal with specific functions for some time to come, it is useful to have this concept, some of its properties, and the technical language associated with it at our disposal in our discussion of the real-number system in sections 6 to 12.

The notation of a function is based on the concept of the Cartesian product:

***Definition 4.1***

The *Cartesian*† *product* $X \times Y$ of two nonempty sets $X$ and $Y$ is the *set of all ordered pairs* $(x, y)$, where $x \in X$, $y \in Y$:

$$X \times Y = \{(x, y) \mid x \in X, y \in Y\}.$$

($(x, y)$ is called an *ordered pair* whenever $(x, y) = (u, v)$ if and only if $x = u$, $y = v$. Note that, in particular, $(x, y) = (y, x)$ if and only if $x = y$.)

---

▲ *Example 1*

Let $A = \{1, 2, 3\}$, $B = \{1, 3, 5, 7\}$. Then,

$$A \times B = \{(1, 1), (1, 3), (1, 5), (1, 7), (2, 1), (2, 3), (2, 5), (2, 7), (3, 1), (3, 3), (3, 5), (3, 7)\}$$

and

$$B \times A = \{(1, 1), (1, 2), (1, 3), (3, 1), (3, 2), (3, 3), (5, 1), (5, 2), (5, 3), (7, 1), (7, 2), (7, 3)\}.$$

Note that $A \times B \neq B \times A$.

▲ *Example 2*

Let $\mathbf{R}$ denote the set of all real numbers. Then $\mathbf{R} \times \mathbf{R} = \{(x, y) \mid x \in R, y \in R\}$ is the set of all ordered pairs of real numbers. If one interprets $(x, y)$ as the Cartesian coordinates of a point in the plane, then $\mathbf{R} \times \mathbf{R}$ may be viewed as the set of all points in a *two-dimensional plane* $\mathbf{R}^2$.

---

We may consider subsets of Cartesian products, such as

$$f = \{(x, y) \in \mathbf{R} \times \mathbf{R} \mid y = x^2, x \in \mathbf{R}\} \tag{4.1}$$

The representation of this set as a set of points in a two-dimensional plane $\mathbf{R}^2$ is called the *graph of* $f$. (See Fig. 4.1.)

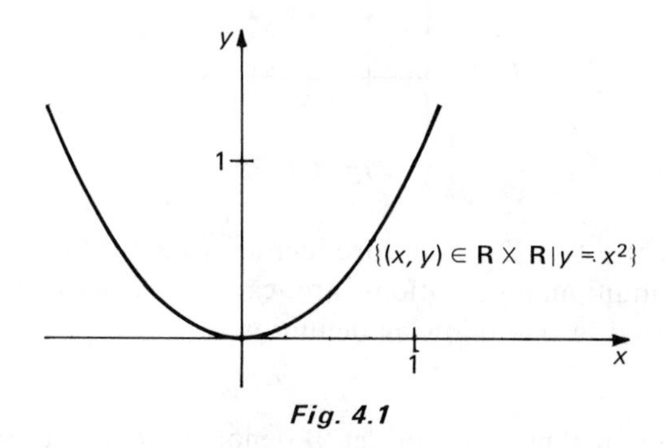


***Fig. 4.1***

† *Cartesius* is the Latinized pen name of *René Descartes*, 1596–1650.

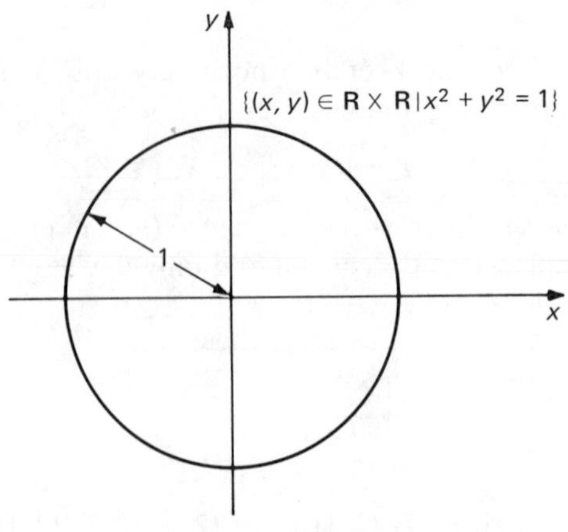


Fig. 4.2

The graph of

$$G = \{(x, y) \in \mathbf{R} \times \mathbf{R} \mid x^2 + y^2 = 1\} \tag{4.2}$$

is given in Fig. 4.2.

We have, for the sets $A$, $B$ in example 1, that $A \subset \mathbf{R}$, $B \subset \mathbf{R}$. Hence, $A \times B \subset \mathbf{R} \times \mathbf{R}$. The graph of $A \times B$ is presented in Fig. 4.3.

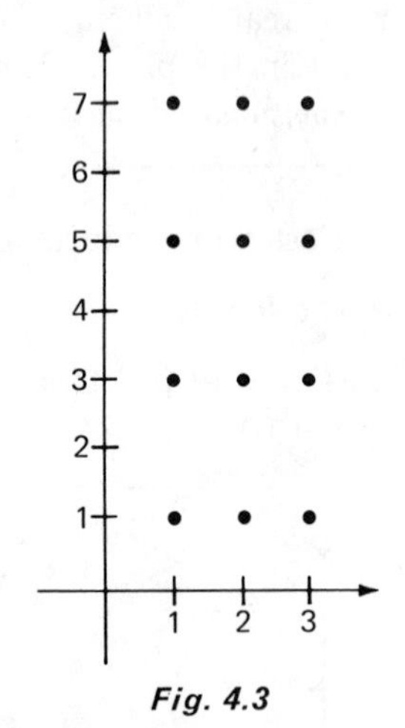


Fig. 4.3

Nonempty subsets of a Cartesian product are called *relations*. Relations that satisfy certain additional conditions are called *functions*. These additional conditions are listed in the following definition.

***Definition 4.2***

Let $X$, $Y$ denote nonempty sets and let $\mathfrak{D}$ denote a nonempty subset of $X$. We call $f$ a function from $\mathfrak{D}$ into $Y$:

$$f: \mathfrak{D} \to Y,$$

if and only if

1. $f \subseteq X \times Y, f \neq \varnothing$;
2. for each $x \in \mathfrak{D}$, there is a $y \in Y$ such that $(x, y) \in f$;
3. if $(x, y_1), (x, y_2) \in f$, then $y_1 = y_2$.

In words: A function from $\mathfrak{D} \subseteq X$ into $Y$ is a nonempty subset of ordered pairs from $X \times Y$ such that all elements $x \in \mathfrak{D}$ appear as first elements, and such that with every first element $x \in \mathfrak{D}$ in the ordered pair, there corresponds *one and only one* second element from $Y$.

$\mathfrak{D}$ is called the *domain* of $f$, $X$ is called the *supply set* (it supplies the elements for the domain), and $Y$ is called the *target set*. The set $\mathfrak{R} = \{y \in Y \mid \text{there is an } x \in \mathfrak{D} \text{ such that } (x, y) \in f\}$, the set of all such elements from $Y$ for which $(x, y) \in f$ for some $x \in \mathfrak{D}$, is called the *range* of $f$. (Occasionally, we shall employ the notation $\mathfrak{D}(f)$ and $\mathfrak{R}(f)$ for domain of $f$ and range of $f$, respectively.)

If supply set and target set of a function are the set of real numbers, we speak of a *real-valued function of a real variable.*

Whenever $(x, y) \in f$, we call $y$ the value of $f$ at $x$ and denote it by $f(x)$. The notation $f(x)$ provides a simple mechanism for the purpose of defining specific functions. For example, instead of defining a function by

(4.3) $$f = \{(x, y) \in \mathbf{R} \times \mathbf{R} \mid y = x^2, x \in [0, 1]\},$$

we shall write: Let $f$: $[0, 1] \to \mathbf{R}$ be defined by $f(x) = x^2$. The domain of this function is the interval $[0, 1]$, the supply set is $\mathbf{R}$, the target set is $\mathbf{R}$ and the range is $[0, 1]$.

Do not confuse the concepts of *function* and *value of the function.* One may view a function as a machine (e.g., a properly programmed computer). When one inserts an element $x$ from the domain of $f$ in the machine $f$ (when one plugs in $x$) and turns the crank, then the machine $f$ delivers the corresponding function value $f(x)$. (See. Fig. 4.4).

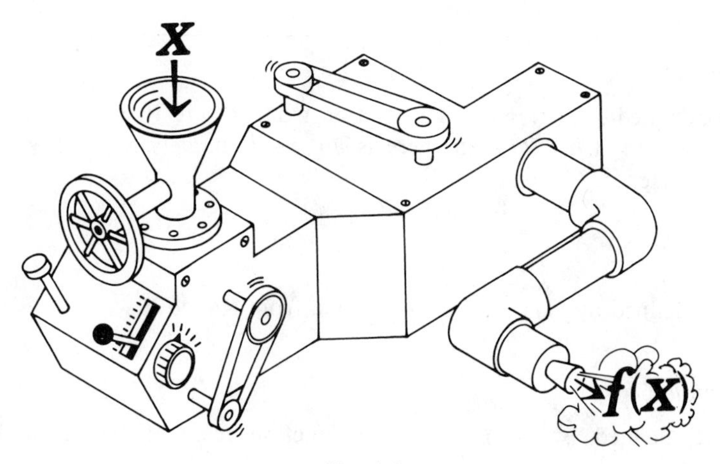


*Fig. 4.4*

Early mathematicians made less restricted use of the term function. What we now call a function was formerly called a single-valued function, as opposed to multiple-valued functions for which condition (3) of definition 4.2 is not met. (For example, $y = \pm\sqrt{x}$, $y = \sin^{-1} x$, etc.)

Observe that $G$, as defined in (4.2), is *not* a function by our definition because if $(x, y) \in G$, then also $(x, -y) \in G$ and $y \neq -y$ for all $y \neq 0$. $(y = \pm\sqrt{1 - x^2}.)$ The set $A \times B$, where $A$ and $B$ are defined in example 1, is not a function either, as the reader can easily convince himself. On the other hand, $f$ as defined in (4.3) *is* a function. First, we note that $f \subseteq \mathbf{R} \times \mathbf{R}$ and that $f \neq \varnothing$ because $(0, 0) \in f$. Next we note that for each $x \in \mathfrak{D} = [0, 1]$ there is a $y \in \mathbf{R}$, namely $y = x^2$, such that $(x, x^2) \in f$. Finally, we note that $(x, y_1), (x, y_2) \in f$ implies $y_1 = x^2, y_2 = x^2$, and hence, $y_1 = y_2$.

A seating chart defines a function from a subset of the set of chairs into the set of students. The set of all chairs is the supply set, the set of all students the target set. The domain is the set of assigned chairs, and the range is the set of students who are supposed to be in class.

Note that by definition 4.2, not all elements in the target set $Y$ have to appear as function values. When they do, we speak of an *onto function* or *surjective map.*

***Definition 4.3***

If $X$, $Y$ are two *nonempty sets*, we call $f$ a function from $\mathfrak{D} \subseteq X$ *onto* $Y$:

$$f: \mathfrak{D} \xrightarrow{\text{onto}} Y,$$

if and only if for each $y \in Y$ there is an $x \in \mathfrak{D}$ such that $(x, y) \in f$.
By the definition of range, we always have

$$f: \mathfrak{D} \xrightarrow{\text{onto}} \mathfrak{R}(f).$$

---

▲ *Example 3*

$f: \mathbf{R} \to \mathbf{R}$, defined by $f(x) = 2x$, is a function from $\mathbf{R}$ *onto* $\mathbf{R}$. It is obvious that it is a function. If $y \in \mathbf{R}$, then there is an $x \in \mathbf{R}$, namely $x = y/2$, such that $(y/2, y) \in f$. Hence, $f$ is *onto* $\mathbf{R}$.

▲ *Example 4*

$g: \mathbf{R} \to \mathbf{R}$, defined by $f(x) = x^3$, is a function from $\mathbf{R}$ *onto* $\mathbf{R}$.

▲ *Example 5*

$h: \mathbf{R} \to \mathbf{R}$, defined by $h(x) = x^2$, is a function from $\mathbf{R}$ *into* $\mathbf{R}$ and *onto* $[0, \infty) \subset \mathbf{R}$.

▲ *Example 6*

$k$: $\mathbf{N} \to \mathbf{N}$, defined by $k(x) = x$ is a function from $\mathbf{N}$ *onto* $\mathbf{N}$.

▲ *Example 7*

$l$: $\mathbf{R} \to \mathbf{R}$, defined by $l(x) = a_0 x^3 + a_1 x^2 + a_2 x + a_3$, $a_0 \neq 0$, is a function from $\mathbf{R}$ *onto* $\mathbf{R}$. Given any $b \in \mathbf{R}$, the equation $a_0 x^3 + a_1 x^2 + a_2 x + a_3 = b$ has (by the fundamental theorem of algebra and because complex roots appear in pairs) at least one real solution.

▲ *Example 8*

$m$: $\mathbf{R} \to \mathbf{R}$, defined by $m(x) = \sin x$, is a function from $\mathbf{R}$ *into* $\mathbf{R}$ and *onto* $[-1, 1]$.

---

In examples, 3, 4, and 6, every element in the range of the function appears exactly once in all the ordered pairs $(x, y)$ that make up the function. In example 5, every element in the range except 0 appears in two pairs, namely $(x, y)$ and $(-x, y)$; and in example 8, every element in the range appears in *infinitely many* pairs. Functions where every element of the range puts in exactly one appearance are of special interest to us. They are called *one-to-one functions* or *injective maps*.

***Definition 4.4***

$f$: $\mathfrak{D} \to Y$ is called a *one-to-one function* from $\mathfrak{D}$ into $Y$:

$$f: \mathfrak{D} \xrightarrow[1-1]{} Y,$$

if and only if $(x_1, y), (x_2, y) \in f$ implies $x_1 = x_2$.

The functions in examples 3, 4, and 6 are one-to-one functions; the functions in examples 5 and 8 are not. The function in example 7 may or may *not* be a one-to-one function, depending on $a_0$, $a_1$, $a_2$, $a_3$.

## 1.4 Exercises

1. Given $f$: $[-1, 1] \to \mathbf{R}$ by

$$f(x) = \begin{cases} \dfrac{1}{\sqrt{1 - x^2}} & \text{for } x \neq 1, x \neq -1, \\ 0 & \text{for } x = \pm 1. \end{cases}$$

(Note that the square-root operation is defined by $\sqrt{a^2} = |a|$ for all $a \in \mathbf{R}$—the "positive" square root.)

(a) Verify that $f$ is a function.
(b) What is the supply set of $f$ and what is the range of $f$?
(c) Is $f$ onto $\mathbf{R}$?
(d) Is $f$ one-to-one on $[-1,1]$?

2. For the function in exercise 1, find the following sets:

$$\{f(x) \mid x \in [0,1]\}; \qquad \{f(x) \mid x \in [-1,0]\};$$
$$\{f(x) \mid x \in [-1,0) \cup (0,1]\}.$$

3. State a sufficient condition on $a_0, a_1, a_2, a_3$ for the function $l$ in example 7 to be one-to-one on $\mathbf{R}$.
4. Same as in exercise 3, so that the range of $l$ consists of one element only.
5. Given $a \in \mathbf{R}$ and $f: \mathbf{R} \to \mathbf{R}$ by $f(x) = ax$. Under what condition on $a$ is $f$ onto $\mathbf{R}$?
6. Given a function $f: \mathbf{N} \to \mathbf{R}$ by $f(k) = a_k$. Such a function is called a sequence. Show: If $a_1 < a_2 < a_3 < \cdots < a_k < \ldots$, then $f$ is one-to-one on $\mathbf{N}$.
7. Let $h: \mathbf{R} \to \mathbf{R}$ be defined as in example 5. Find the following sets: $\{h(x) \mid x \in [0,1]\}$, $\{h(x) \mid x \in (0,1)\}$, $\{h(x) \mid x \in (1,2)\}$, $\{h(x) \mid x \in (-1,1)\}$.
8. Let $h: \mathbf{R} \to \mathbf{R}$ be defined as in example 5. Find a nonempty subset $A \subset \mathbf{R}$ such that $f$ is one-to-one on $A$ and onto $[0, \infty)$.
9. Let $f: \mathfrak{D} \to Y$, where $\mathfrak{D} \subseteq X$, represent a function. Let $\mathfrak{D}_1 \subset \mathfrak{D}$. We call $f_{\mathfrak{D}_1} = \{(x,y) \in f \mid x \in \mathfrak{D}_1\}$ the *restriction of $f$ to* $\mathfrak{D}_1$. Show that $f_{\mathfrak{D}_1}$ is a function from $\mathfrak{D}_1$ into $Y$.
10. Let $f: \mathbf{R} \to \mathbf{R}$ be defined by $f(x) = x^2 - x$. Sketch the graph of $f_{[0,1]}$. (See exercise 9.)
11. Let $f: \mathfrak{D} \to Y$, where $\mathfrak{D} \subseteq X$, represent a function. Let $Y_1 \subseteq Y$ such that $Y_1 \cap \mathfrak{R}(f) \neq \varnothing$. Show that $\{(x,y) \in f \mid y \in Y_1\}$ is a function. What is the domain of this function? What happens when one drops the hypothesis $Y_1 \cap \mathfrak{R}(f) \neq \varnothing$?

## 1.5 The Inverse Function

We consider the function $f: \mathfrak{D} \to Y$ where $\mathfrak{D} \subseteq X$. By definition 4.2, $f$ is a nonempty subset of $X \times Y$. Hence, $\{(y,x) \mid (x,y) \in f\}$ is a nonempty subset of $Y \times X$ and as such is a relation. This relation may or may not be a function on $Y$ or a subset thereof. If it is a function on $Y$ or on one of its subsets $\mathfrak{R}$, then we call it the *inverse function* of $f$ on $Y$ or $\mathfrak{R}$, and we call $f$ *invertible* on $Y$, or $\mathfrak{R}$.

***Definition 5.1***

If $f: \mathfrak{D} \to Y$, $\mathfrak{D} \subseteq X$, is such that $\{(y,x) \mid (x,y) \in f\}$ is a function with domain $\mathfrak{R} \subseteq Y$, then

$$f^{-1} = \{(y,x) \mid (x,y) \in f, y \in \mathfrak{R}\}$$

is called the *inverse function* of $f$ on $\mathfrak{R}$, and $f$ is called *invertible* on $\mathfrak{R}$. In this case, $f^{-1}: \mathfrak{R} \to X$.

We shall show first that, for a function to be invertible on its target set $Y$, it is necessary and sufficient that the function be one-to-one and onto. A criterion for invertibility on a suitable subset of the target set will then follow easily.

A function $f$ which is *one-to-one* from $\mathfrak{D}$ *onto* $Y$,

$$f:\mathfrak{D} \xrightarrow[1-1]{\text{onto}} Y$$

is called a *bijective map*. (A map that is surjective and injective is bijective.) We shall henceforth employ the symbol

$$f:\mathfrak{D} \leftrightarrow Y$$

to indicate that the function is one-to-one on $\mathfrak{D}$ and onto $Y$ rather than the bulky notation

$$f:\mathfrak{D} \xrightarrow[1-1]{\text{onto}} Y.$$

***Theorem 5.1***

Let $X$, $Y$ denote nonempty sets and let $\mathfrak{D} \subseteq X, \mathfrak{D} \neq \varnothing$. The function $f:\mathfrak{D} \to Y$ is invertible on $Y$ if and only if $f:\mathfrak{D} \leftrightarrow Y$. In this case, the inverse function

$$f^{-1} = \{(y, x) \mid (x, y) \in f\} : Y \to X \tag{5.1}$$

exists.

*Proof*

(a) Clearly, $f^{-1} \in Y \times X$; since $f \neq \varnothing$, we see that $f^{-1} \neq \varnothing$. Since $f$ is one-to-one, we have, from definition 4.4, that $(x_1, y), (x_2, y) \in f$ implies $x_1 = x_2$. Hence, $(y, x_1), (y, x_2) \in f^{-1}$ implies $x_1 = x_2$; that is, $f^{-1}$ is a function. Since $f$ is onto $Y$, we have, for all $y \in Y$, an $x \in \mathfrak{D}$ such that $(x, y) \in f$. Hence, for all $y \in Y$, there is an $x \in \mathfrak{D}$ such that $(y, x) \in f^{-1}$; that is, $Y$ is the domain of $f^{-1}$.

(b) Suppose that (5.1) represents a function from $Y$ into $X$. Then, for all $y \in Y$, there is an $x \in X$ such that $(y, x) \in f^{-1}$; that is, $(x, y) \in f$. Hence, $f$ is onto $Y$. Since $f^{-1}$ is a function, $(y, x_1)$, $(y, x_2) \in f^{-1}$ implies $x_1 = x_2$. Hence, $(x_1, y), (x_2, y) \in f$ implies $x_1 = x_2$; that is, $f$ is one-to-one on $\mathfrak{D}$.

By the definition of range, $f:\mathfrak{D} \xrightarrow{\text{onto}} \mathfrak{R}(f)$. Hence if $f:\mathfrak{D} \xrightarrow[1-1]{} Y$, then $f$: $\mathfrak{D} \leftrightarrow \mathfrak{R}(f)$, and we have:

***Corollary to Theorem 5.1***

If $f:\mathfrak{D} \xrightarrow[1-1]{} Y$ where $\mathfrak{D} \subseteq X$, then $f$ is invertible on its range $\mathfrak{R}(f)$ and the inverse function $f^{-1}$: $\mathfrak{R}(f) \to X$ *exists*. (See also exercise 12.)

So we see that every one-to-one function is invertible on its range. The range of $f$ becomes the domain of the inverse function $f^{-1}$, the target set of $f$ becomes the supply set of $f^{-1}$, the domain of $f$ becomes the range of $f^{-1}$, and the supply set of $f$ becomes the target set of $f^{-1}$.

▲ *Example 1*

Let $l: \mathbf{R} \to \mathbf{R}$ be given by $l(x) = ax$, $a \neq 0$. $l$ is one-to-one on $\mathbf{R}$ and onto $\mathbf{R}$, and hence, is invertible. The inverse function $l^{-1}: \mathbf{R} \to \mathbf{R}$ is given by $l^{-1}(y) = (1/a)y$.

***Theorem 5.2***

If $f: \mathfrak{D} \to Y$ is invertible on $\mathfrak{R} \subseteq Y$, then $f^{-1}: \mathfrak{R} \leftrightarrow \mathfrak{D}$.

*Proof*

Let $x \in \mathfrak{D}$. Then $(x, f(x)) \in f$ and therefore, $(f(x), x) \in f^{-1}$. Hence $f^{-1}$ is onto $\mathfrak{D}$.
If $(y_1, x), (y_2, x) \in f^{-1}$, then $(x, y_1)$, $(x, y_2) \in f$. Since $f$ is a function, $y_1 = y_2$. Hence $f^{-1}$ is one-to-one.

Throughout this book we shall make liberal use of the elementary transcendental functions (exponential function, logarithm, and the trigonometric functions and their inverses) and their basic properties in examples and exercises. We make the necessary assumption that the reader has acquired a working knowledge of these functions in a calculus course. Our informal use of these functions does not prejudice the rigorous definitions of these functions which we shall give in section 4.52 (logarithm) and sections 14.142 (exponential function) and 14.143 (trigonometric functions and their inverses). The next example deals with the exponential function and the logarithm, for the purpose of illustrating definition 5.1 and the corollary to Theorem 5.1.

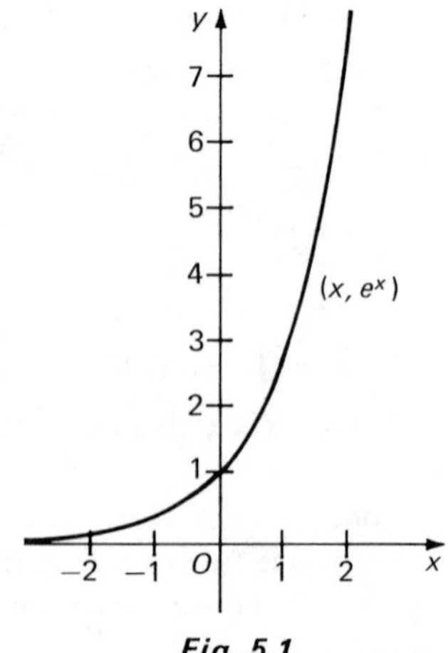


*Fig. 5.1*

▲ *Example 2*

The exponential function $E: \mathbf{R} \to \mathbf{R}$ is given by $E(x) = e^x$. The reader knows from calculus that the exponential function assumes all positive real values. Hence, the range of $E$ is $\mathbf{R}^+$ where $\mathbf{R}^+ = (0, \infty)$ denotes the set of positive real numbers. Each such value is assumed for one and only one real $x$. Hence, $E$ is one-to-one. (See Fig. 5.1.) Since $E: \mathbf{R} \to \mathbf{R}^+$ is onto and one-to-one, $E^{-1}$ exists and $E^{-1}: \mathbf{R}^+ \leftrightarrow \mathbf{R}$. The inverse function of the exponential function is the natural logarithm: $E^{-1}(y) = \log y$.

---

In example 1, we have

$$l^{-1}(l(x)) = \frac{1}{a}\, l(x) = \frac{1}{a}\, ax = x \qquad \text{for all } x \in \mathbf{R}$$

and

$$l(l^{-1}(y)) = al^{-1}(y) = a\frac{1}{a}\, y = y \quad \text{for all } y \in \mathbf{R}.$$

Similarly, we have in example 2,

$$E^{-1}(E(x)) = \log E(x) = \log e^x = x \quad \text{for all } x \in \mathbf{R}$$

and

$$E(E^{-1}(y)) = e^{E^{-1}(y)} = e^{\log y} = y \quad \text{for all } y \in \mathbf{R}^+.$$

In general:

***Theorem 5.3***

If $f: \mathfrak{D} \to Y$ is invertible on $\mathfrak{R} \subseteq Y$, then

(5.2) $f^{-1}(f(x)) = x$ for all $x \in \mathfrak{D}$ and $f(f^{-1}(y)) = y$ for all $y \in \mathfrak{R}$.

Furthermore, $f^{-1}: \mathfrak{R} \to X$ is the *only* function with this property.

*Proof*

For all $x \in \mathfrak{D}$, there is a $y \in Y$ such that $y = f(x)$ and $x = f^{-1}(y)$. Hence, $f^{-1}(f(x)) = f^{-1}(y) = x$.

For all $y \in \mathfrak{R}$, there is an $x \in X$ such that $x = f^{-1}(y)$ and $y = f(x)$. Hence $f(f^{-1}(y)) = f(x) = y$.

Suppose there is another function $g: \mathfrak{R} \to X$ with the property (5.2):

(5.3) $g(f(x)) = x$ for all $x \in \mathfrak{D}$ and $f(g(y)) = y$ for all $y \in \mathfrak{R}$.

We obtain from (5.2) and (5.3) that

$$g(y) = g(f(f^{-1}(y))) = f^{-1}(y) \quad \text{for all } y \in \mathfrak{R},$$

that is, $g = f^{-1}$ and we see that $f^{-1}$ is indeed the only function with the property (5.2).

The exponential function is *strictly increasing*. Such functions are always invertible.

***Definition 5.2***

$f: \mathfrak{D} \to \mathbf{R}$, where $\mathfrak{D} \subseteq \mathbf{R}$, is called *strictly increasing* (*decreasing*) if and only if $x_1 < x_2$ implies that $f(x_1) < f(x_2)$ ($f(x_1) > f(x_2)$).

***Theorem 5.4***

If $f: \mathfrak{D} \to \mathbf{R}$, where $\mathfrak{D} \subseteq \mathbf{R}$, is strictly increasing (decreasing), then $f^{-1}: \mathfrak{R}(f) \to \mathbf{R}$ exists and is strictly increasing (decreasing).

*Proof*

(We formulate the proof for strictly increasing functions only. The proof for strictly decreasing functions is analogous.)

If $(x_1, y), (x_2, y) \in f$, then $f(x_1) = f(x_2)$. If $x_1 \neq x_2$, then $x_1 < x_2$ or $x_1 > x_2$. Hence, $f(x_1) < f(x_2)$ or $f(x_1) > f(x_2)$, contrary to our assumption. Hence, $(x_1, y), (x_2, y) \in f$ implies that $x_1 = x_2$, that is, $f$ is one-to-one on $\mathfrak{D}$. By the Corollary to Theorem 5.1, $f^{-1}$ exists on $\mathfrak{R}(f)$.

If $y_1 < y_2$ where $y_1, y_2 \in \mathfrak{R}(f)$ and $f^{-1}(y_1) \geqslant f^{-1}(y_2)$, then, with $y_1 = f(x_1)$, $y_2 = f(x_2)$, we have $x_1 \geqslant x_2$ while $f(x_1) < f(x_2)$, contrary to our assumption that $f$ is strictly increasing. Hence, $f^{-1}(y_1) < f^{-1}(y_2)$, that is, $f^{-1}$ is also strictly increasing.

---

▲ *Example 3*

The function $f: \mathbf{R} \to \mathbf{R}$, defined by $f(x) = x^3$, is strictly increasing. Hence, $f^{-1}$ exists and is found to be given by $f^{-1}(y) = \sqrt[3]{y}$. (See Fig. 5.2.)

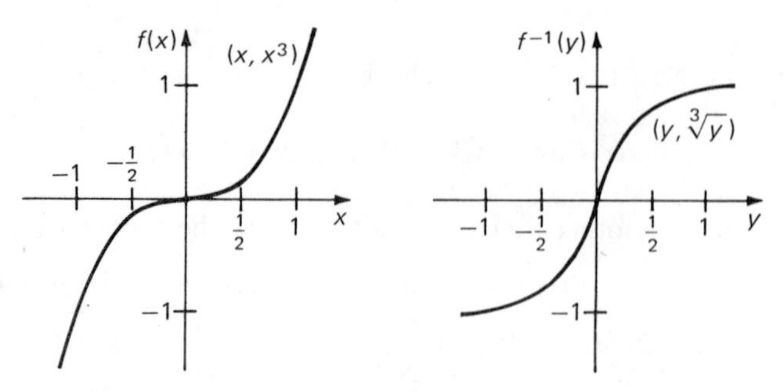


*Fig. 5.2*

▲ *Example 4*

Let $A, B$ denote finite sets which contain the same number $n$ of elements. We label the elements of $A$ by $a_k$ and the elements of $B$ by $b_k$. Then

$$A = \{a_1, a_2, \ldots, a_n\}, \qquad B = \{b_1, b_2, \ldots, b_n\}.$$

The function $f: A \to B$, defined by

$$f(a_k) = b_k, \qquad k = 1, 2, \ldots, n,$$

is one-to-one and onto.

---

Clearly, such a one-to-one and onto function can always be constructed when two finite sets have the same number of elements. We call such sets *equivalent*. In general:

***Definition 5.3***

Two sets $X, Y$ are called equivalent (*or* are said to be in *one-to-one correspondence*),

$$X \sim Y,$$

if and only if there exists a function $f: X \leftrightarrow Y$.

---

▲ *Example 5*

If two sets are equivalent, there may be more than one function with the required property. In example 4, the function $g$: $A \to B$, defined by

$$g(a_k) = b_{n-k}, \qquad k = 1, 2, \ldots, n$$

is also one-to-one and onto.

▲ *Example 6*

We have seen in example 2 that $E$: $\mathbf{R} \leftrightarrow \mathbf{R}^+$. Hence, $\mathbf{R} \sim \mathbf{R}^+$.

▲ *Example 7*

Let $\mathbf{O}$ denote the set of odd positive integers. The function $f: \mathbf{N} \to \mathbf{O}$, defined by

$$f(x) = 2x - 1$$

is one-to-one and onto. Hence, $\mathbf{N} \sim \mathbf{O}$.

---

## 1.5 Exercises

1. Given $f: \mathbf{R}^+ \to \mathbf{R}$ by $f(x) = x^2$. Show that $f: \mathbf{R}^+ \leftrightarrow \mathbf{R}^+$ and find $f^{-1}(y)$.
2. Given $f: \mathbf{R}^+ \to \mathbf{R}$ by $f(x) = 1/x$. Show that $f: \mathbf{R}^+ \leftrightarrow \mathbf{R}^+$ and find $f^{-1}(y)$.

3. Given $f: \mathbf{R} \to \mathbf{R}$ by $f(x) = x^3$. Show that $f: \mathbf{R} \leftrightarrow \mathbf{R}$ and find $f^{-1}(y)$.
4. Given $f: \mathbf{R} \to \mathbf{R}$ by $f(x) = x^3 - x$. Find three subsets $\mathfrak{D}_1, \mathfrak{D}_2, \mathfrak{D}_3$ of $\mathbf{R}$ such that $\mathfrak{D}_1 \cup \mathfrak{D}_2 \cup \mathfrak{D}_3 = \mathbf{R}$, $\mathfrak{D}_i \cap \mathfrak{D}_k = \varnothing$ for $i \neq k$, and such that $f: \mathfrak{D}_i \xrightarrow[1-1]{} \mathbf{R}$, $i = 1, 2, 3$.
5. Given the sets $A = \{0, 1, 2\}$, $B = \{-5, -7, -9\}$. Find three different functions $f_1, f_2, f_3$ such that $f_i : A \leftrightarrow B$ for $i = 1, 2, 3$.
6. Let $A, B$ denote finite sets. Prove: $A \sim B$ if and only if $A$ and $B$ have the same number of elements.
7. Show that $\mathbf{R}^+ \sim \mathbf{R}^-$ where $\mathbf{R}^-$ denotes the set of negative real numbers.
8. Prove: (a) $A \sim A$; (b) If $A \sim B$ then $B \sim A$; (c) If $A \sim B$, $B \sim C$, then $A \sim C$.
9. Given the sets $(0, 1)$ and $\mathbf{R}^+$. Show that $(0, 1) \sim \mathbf{R}^+$.
10. Show that $(0, 1) \sim \mathbf{R}$.
11. Let $f: \mathfrak{D} \to Y$ where $\mathfrak{D} \subseteq X$. Show: If $\mathfrak{D}_1 \subset \mathfrak{D}$, if $f_{\mathfrak{D}_1}(x) = f(x)$ for all $x \in \mathfrak{D}_1$ and $f_{\mathfrak{D}_1}: \mathfrak{D}_1 \xrightarrow[1-1]{} Y$, then $f_{\mathfrak{D}_1}$, the *restriction of $f$* to $\mathfrak{D}_1$, is invertible on $\{f(x) \mid x \in \mathfrak{D}_1\}$. (See also exercise 9, section 1.4.)

## 1.6 The Field of Rational Numbers

*Leopold Kronecker*† said that God created the natural numbers and that everything else is the work of man. Be that as it may, we shall take the natural numbers for granted and take it from there.

We denote the set of all natural numbers by $\mathbf{N}$:

$$\mathbf{N} = \{1, 2, 3, \ldots\}.$$

We know that the sum and the product of natural numbers are again natural numbers:

$$\text{if } a, b \in \mathbf{N}, \quad \text{then } a + b, a \cdot b \in \mathbf{N};$$

and that addition and multiplication are *commutative*:

$$a + b = b + a, \qquad a \cdot b = b \cdot a;$$

*associative*:

$$(a + b) + c = a + (b + c), (ab)c = a(bc);$$

and *distributive*:

$$a(b + c) = ab + ac.$$

Subtraction, the inverse operation to addition, cannot be carried out freely within $\mathbf{N}$ (e.g., $5 - 7$ is not a natural number). In order to free ourselves from any restrictions, we adjoin new elements to $\mathbf{N}$, namely 0 and the negative integers, and thus arrive at the set of all integers $\mathbf{Z}$:

$$\mathbf{Z} = \{0, 1, -1, 2, -2, 3, -3, \ldots\}.$$

† *Leopold Kronecker*, 1823–1891.

We require that addition and multiplication within **Z** be governed by the commutative, associative, and distributive laws and, guided by experience, we impose the following rules on subtraction: 0 is the only element such that

$$a + 0 = a \quad \text{for all } a \in \mathbf{Z}$$

and for each $a \in \mathbf{Z}$ there is a unique element $(-a)$ in **Z** such that

$$a + (-a) = 0.$$

(We refer to 0 as the *additive identity* in **Z** and to $(-a)$ as the *additive inverse of a*.)

The system **Z** which we have obtained is closed under addition, multiplication, and subtraction; i.e., these operations can be carried out freely within **Z** and will always result in an element from **Z**.

Division, the inverse operation to multiplication, cannot be carried out freely in **Z**. (While $6/-2 = -3 \in \mathbf{Z}$, $7/2 \notin \mathbf{Z}$.) Again, we adjoin new elements to **Z**, namely, all the results than can be obtained from carrying out the four elementary operations with elements from **Z** except for division by 0. This leads to the set of *rational numbers* **Q**:

$$\mathbf{Q} = \left\{ \frac{m}{n} \,\middle|\, m \in \mathbf{Z}, n \in \mathbf{N}, (m, n) = 1 \right\}$$

where $(m, n) = 1$ means that $m, n$ are *relatively prime* (have no common factor other than 1 and $-1$). We require that all the rules which we have imposed up to now govern the operations in **Q**. This is the case if we define addition and multiplication of fractions in the customary manner by $m_1/n_1 + m_2/n_2 = a/b$, where $an_1n_2 = b(m_1n_2 + n_1m_2)$ and $(a, b) = 1$, and by $(m_1/n_1) \cdot (m_2/n_2) = c/d$, where $cn_1n_2 = dm_1m_2$ and $(c, d) = 1$. Then, 1 is the only element in **Q** for which

$$a \cdot 1 = 1 \cdot a = a \quad \text{for all } a \in \mathbf{Q},$$

and for all $a \in \mathbf{Q}$, $a \neq 0$, there is a unique $(1/a) \in \mathbf{Q}$ such that

$$a \cdot \frac{1}{a} = 1.$$

(We call 1 the *multiplicative identity* in **Q** and $(1/a)$ the *multiplicative inverse* of $a$.)

**Q** is *closed* under the four elementary operations of addition, multiplication, subtraction, and division; i.e., these operations can be carried out freely within **Q** (except division by 0) without ever arriving at a result that is not in **Q**. A collection of elements with this property is called a *field*:

***Definition 6.1***

A set **F** of elements, denoted by $a, b, c, \ldots$, is called a *field* if and only if two operations, called addition $(+)$ and multiplication $(\cdot)$, are defined such that

1. If $a, b \in \mathbf{F}$, then $a + b, a \cdot b \in \mathbf{F}$;
2. $a + b = b + a$, $a \cdot b = b \cdot a$ for all $a, b \in \mathbf{F}$;

3. $(a + b) + c = a + (b + c)$, $(a \cdot b) \cdot c = a \cdot (b \cdot c)$ for all $a, b, c \in \mathbf{F}$;
4. $a \cdot (b + c) = a \cdot b + a \cdot c$, for all $a, b, c \in \mathbf{F}$;
5. There is a unique *additive identity* in $\mathbf{F}$, call it 0, such that

$$a + 0 = a \quad \text{for all } a \in \mathbf{F};$$

6. Each $a \in \mathbf{F}$ has a unique *additive inverse* in $\mathbf{F}$, call it $(-a)$, such that

$$a + (-a) = 0;$$

7. There is a unique *multiplicative identity* in $\mathbf{F}$ different from the additive identity, call it 1, such that

$$a \cdot 1 = a \quad \text{for all } a \in \mathbf{F};$$

8. Each $a \in \mathbf{F}$, $a \neq 0$, has a unique *multiplicative inverse* in $\mathbf{F}$, call it $(1/a)$, such that

$$a \cdot \frac{1}{a} = 1.$$

Examples of fields abound in mathematics. We have seen that $\mathbf{Q}$ is a field and is, as a matter of fact, the "smallest" field that contains $\mathbf{N}$ as a subset. Many more examples of fields are treated in exercises 2, 3, 5, 6, and 7.

The rules of arithmetic which we are accustomed to can all be derived from the field axioms (1) to (8). This task is to be carried out in exercises 8 through 12.

Henceforth we shall adopt the standard convention of writing $ab$ for $a \cdot b$.

## 1.6 Exercises

1. Show that $\sqrt{2}$ is not a rational number.
2. Show that the set $\mathbf{Q}(\sqrt{2}) = \{a + \sqrt{2}\,b \mid a, b \in \mathbf{Q}\}$ is a field. (Addition and multiplication are defined in the usual manner.)
3. Show that the set $\mathbf{Q}(i) = \{x + iy \mid x, y \in \mathbf{Q}, i^2 = -1\}$ is a field if addition and multiplication are defined by:

$$(x_1 + iy_1) + (x_2 + iy_2) = x_1 + x_2 + i(y_1 + y_2),$$
$$(x_1 + iy_1)(x_2 + iy_2) = x_1x_2 - y_1y_2 + i(x_1y_2 + x_2y_1).$$

4. Show that $\sqrt{p}$, where $p$ is a prime number, is not a rational number.
5. Show that $\mathbf{Q}(\sqrt{p}) = \{a + \sqrt{p}\,b \mid a, b \in \mathbf{Q}\}$ is a field if $p$ is a given prime number.
6. Show that the set $L = \{0, 1\}$ is a field, if addition and multiplication are defined by:

| + | 0 | 1 |
|---|---|---|
| 0 | 0 | 1 |
| 1 | 1 | 0 |

| | 0 | 1 |
|---|---|---|
| 0 | 0 | 0 |
| 1 | 0 | 1 |

7. Same as in exercise 6 for $M = \{0, -1, 1\}$ with

| $+$ | $0$ | $-1$ | $1$ |
|---|---|---|---|
| $0$ | $0$ | $-1$ | $1$ |
| $-1$ | $-1$ | $1$ | $0$ |
| $1$ | $1$ | $0$ | $-1$ |

| $\cdot$ | $0$ | $-1$ | $1$ |
|---|---|---|---|
| $0$ | $0$ | $0$ | $0$ |
| $-1$ | $0$ | $1$ | $-1$ |
| $1$ | $0$ | $-1$ | $1$ |

8. Derive the following rules from the field axioms (1) to (8):

(a) $(a+b)c = ac + bc$, $a, b, c \in \mathbf{F}$.
(b) If $a + x = a$, $a \in \mathbf{F}$, then $x = 0$.
(c) If $a + b = 0$, $a, b \in \mathbf{F}$, then $a = -b$.
(d) If $ab = 1$, $a, b \in \mathbf{F}$, $a \neq 0$, then $b = 1/a$.
(e) $a \cdot 0 = 0$
(f) $(-1)a = -a$ (g) $-(a+b) = -a - b$
(h) $-(-1) = 1$ (i) $(-1)(-1) = 1$

9. Use the field axioms, and whatever results of exercise 8 prove useful, to show that each of the equations

$$a + x = b \quad \text{and} \quad cx = d,$$

where $a, b, c, d \in \mathbf{F}$, and $c \neq 0$, has a unique solution.

10. Use the field axioms, and whatever results from exercises 8 and 9 prove useful, to show that:

(a) If $a \in \mathbf{F}$, $a \neq 0$, then $a = 1/(1/a)$.
(b) If $ab = 0$, $a, b \in \mathbf{F}$, then $a = 0$ or $b = 0$.
(c) $(-a)(-b) = ab$ for all $a, b \in \mathbf{F}$.

11. Define $a^2 = aa$ and, inductively, $a^{n+1} = aa^n$, $n \in \mathbf{N}$. Prove that $a^{m+n} = a^m a^n$ for all $a \in \mathbf{F}$, $m, n \in \mathbf{N}$.
12. Generalize the result in exercise 11 to $m, n \in \mathbf{Z}$ with the understanding that $a^{-n} = 1/a^n$, $n \in \mathbf{N}$.

## 1.7 Ordered Fields

It is our goal to establish a unique characterization of the set of real numbers $\mathbf{R}$. The real numbers, as we know them, are certainly a field but we know with equal certainty that they are not uniquely characterized by the field axioms. As we have seen, the rationals $\mathbf{Q}$ are a field and so are many other collections of numbers or other abstract elements for which addition and multiplication are suitably defined. (See section 1.6, exercises 2, 3, 5, 6, 7.)

In our search for additional properties that might be useful for a characterization of $\mathbf{R}$, we let experience be our guide. We know that any two real numbers are either equal or one is larger than the other one. Fields where this holds true are called *ordered fields*.

***Definition 7.1***

**F** is an ordered field if and only if

1. **F** is a field.
2. There exists a nonempty subset $P \subset \mathbf{F}$, called a *positive class*, with the following properties
   (a) If $a, b \in P$, then $a + b$, $ab \in P$.
   (b) If $a \in \mathbf{F}$, then one of the following is true: $a = 0$, or $a \in P$, or $-a \in P$.

We have as an immediate consequence of this definition that, whenever $a \in \mathbf{F}$ and $a \neq 0$, where **F** denotes an ordered field, then

$$a^2 \in P. \tag{7.1}$$

This may be seen as follows: If $a \in P$, then, by 2(a) of Definition 7.1, $a \cdot a = a^2 \in P$. If $a \notin P$, then, by 2(b) of definition 7.1, $(-a) \in P$. Since $(-a)(-a) = a \cdot a$ (see exercise 10(c), section 1.6), we have again $a^2 \in P$.

The requirement that a field be ordered eliminates a number of fields from consideration.

---

▲ *Example 1*

The field $\mathbf{Q}(i)$ of exercise 3, section 1.6, is *not* ordered. Since $1 \in \mathbf{Q}(i)$, we have from (7.1) that $1 = 1^2 \in P$. On the other hand, $i \in \mathbf{Q}(i)$ and, again by (7.1), $i^2 = -1 \in P$, which contradicts the preceding result that $1 \in P$.

▲ *Example 2*

The set $L = \{0, 1\}$ of exercise 6, section 1.6, is not an ordered field either: By 2(b) of definition 7.1 we have $\{0\} \cap P = \emptyset$. Hence, if $L$ contains a positive class at all, we have to have $P = \{1\}$. By 2(a) of definition 7.1, $1 \in P$ implies $1 + 1 \in P$, but by our definition of addition, we have $1 + 1 = 0$ and 2(a) is violated.

---

However, the set **Q** of rationals is an ordered field, and we still have not made any progress towards a unique characterization of **R**, except that we have uncovered another important property; namely, that an order relationship can be defined in **Q** (and **R**):

***Definition 7.2***

Let $a, b \in \mathbf{F}$ where **F** is an *ordered field.* Then

$$a < b$$

if and only if $b - a \in P$, and

$$a \leqslant b$$

if and only if either $b - a \in P$ or $b - a = 0$. ($b > a$ is equivalent to $a < b$.)

An immediate consequence of this definition is that $a \in P$ if and only if $a > 0$, and the *positive class P* stands revealed as the *class of all positive numbers* in **F** —if **F** is a field of numbers.

We obtain, as another consequence from definition 7.2, that

$$a \geqslant b, c \geqslant d \quad \text{implies that} \quad a + c \geqslant b + d \tag{7.2}$$

because $a \geqslant b$, $c \geqslant d$ means, by definition 7.2, that $a - b \geqslant 0$, $c - d \geqslant 0$. Hence, by 2(a) of definition 7.1,

$$a - b + c - d \geqslant 0,$$

or equivalently,

$$(a + c) - (b + d) \geqslant 0$$

from which (7.2) follows readily. The other familiar rules for operating with inequalities are left for exercises 4, 9, 10, 11, 12.

***Definition 7.3***

If **F** is an ordered field, then for all $a \in \mathbf{F}$,

$$|a| = \begin{cases} a & \text{if } a \geqslant 0 \ (a \in P \text{ or } a = 0), \\ -a & \text{if } a < 0 \ (-a \in P) \end{cases}$$

is called the *absolute value of a*. (Intuitively, the reader may view the absolute value of $a$ as the "distance" from $a$ to 0.)

We obtain immediately

$$\pm a \leqslant |a|. \tag{7.3}$$

This can be seen as follows: If $a > 0$, then $|a| = a$; if $a = 0$, then $|a| = 0$; if $a < 0$, then $-a > 0$ and $|a| - a > 0$, by 2(a) of definition 7.1. Hence, $a \leqslant |a|$. A symmetric argument applies to the case $-a \leqslant |a|$. (See exercise 1.)

The fundamental properties of the absolute value are given in the following theorem:

***Theorem 7.1***

1. $|a| \geqslant 0$ for all $a \in \mathbf{F}$
2. $|a| = 0$ if and only if $a = 0$
3. $|ab| = |a|\,|b|$
4. $|a \pm b| \leqslant |a| + |b|$ (Triangle inequality)

*Proof*

1: follows directly from definition 7.3.

2: If $a = 0$, then, by definition 7.3, $|a| = 0$. If $a \neq 0$, then, by definition 7.3, $|a| > 0$. Hence, $|a| = 0$ implies $a = 0$.

3: If $a > 0$, $b > 0$, then, by 2(a) of definition 7.1, $ab > 0$ and $|ab| = ab = |a|\,|b|$. If $a > 0$, $b < 0$, then $-b > 0$ and $a(-b) > 0$. Hence, $|ab| = a(-b) = |a|\,|b|$. If $a = 0$, then $ab = 0$ and $|ab| = 0 = 0 \cdot b = 0 \cdot |b| = |a|\,|b|$.

4: If $a + b > 0$, then $|a + b| = a + b \leqslant |a| + |b|$, by (7.2) and (7.3). If $a + b < 0$, then $|a + b| = -a - b \leqslant |a| + |b|$, by (7.2) and (7.3). If $a + b = 0$, then $|a + b| = 0 \leqslant |a| + |b|$, because $|a| + |b| \geqslant 0$ for all $a, b \in \mathbf{F}$. The case $|a - b| \leqslant |a| + |b|$ is left for exercise 2.

***Corollary to Theorem 7.1***

$$||a| - |b|| \leqslant |a + b| \tag{7.4}$$

*Proof*

Let $x = -a$, $y = b + a$. Then, $x + y = b$ and we have, from (4) of Theorem 7.1, that

$$|b| = |x + y| \leqslant |x| + |y| = |-a| + |a + b| = |a| + |a + b|$$

if we also note that $|-a| = |(-1)a| = |-1|\,|a| = |a|$. Hence, $|b| - |a| \leqslant |a + b|$. Interchange of $a$ and $b$ yields $|a| - |b| \leqslant |a + b|$ and (7.4) follows readily. The case $||a| - |b|| \leqslant |a - b|$ is left for exercise 3.

We shall now show that $\mathbf{Q}$ is the "smallest" ordered field in the sense that every ordered field $\mathbf{F}$ contains $\mathbf{Q}$ as a subfield. By *subfield* of $\mathbf{F}$ we mean a subset of $\mathbf{F}$ which is itself a field under the same operations as $\mathbf{F}$.

***Theorem 7.2***

$\mathbf{Q}$ is the smallest ordered field i.e., every ordered field $\mathbf{F}$ contains $\mathbf{Q}$ as a subfield.

*Proof*

By field axiom (7) of definition 6.1, $\mathbf{F}$ has to contain at least two distinct elements, the additive identity, 0, and the multiplicative identity, 1.

It follows from the field postulates that 0 and 1 have to behave under addition and multiplication as follows:

| + | 0 | 1 |
|---|---|---|
| 0 | 0 | 1 |
| 1 | 1 | ? |

| · | 0 | 1 |
|---|---|---|
| 0 | 0 | 0 |
| 1 | 0 | 1 |

(See also exercise 8(e), section 1.6.)

If $1 + 1 = 1$, then we do not have a field, since then 1 does not have an additive inverse. Hence, $1 + 1 \neq 1$.

If $1 + 1 = 0$ we do have a field, as we have seen in exercise 6, section 1.6, but not an ordered field as we have seen in example 2. Hence, $1 + 1 \neq 0$.

If $\mathbf{F}$ is an ordered field we have from (7.1) that $1 = 1^2 \in P$. Hence, if $k \in P$, then $k + 1 \in P$. By letting $1 + 1 = 2$, $2 + 1 = 3, \ldots$, we obtain $\mathbf{N} \subseteq \mathbf{F}$ and, by field axiom (6), also $\mathbf{Z} \subseteq \mathbf{F}$. Consider $m/n$, where $m \in \mathbf{Z}$, $n \in \mathbf{N}$, $(m, n) = 1$. Since $n \in \mathbf{N} \subseteq \mathbf{F}$, we have, from field axiom (8) that $1/n \in \mathbf{F}$. Since $m \in \mathbf{Z}$, we have, from field axiom (1), that $m(1/n) = m/n \in \mathbf{F}$. Hence, $\mathbf{Q} \subseteq \mathbf{F}$. Since $\mathbf{Q}$ itself is an ordered field, we see that $\mathbf{Q}$ is the "smallest" ordered field.

Note that if we had denoted the additive and multiplicative identities by other symbols, such as $\theta$ and $e$, and introduced other symbols for $e + e$, $e + e + e, \ldots$, we still would have obtained $\mathbf{Q}$, except in a new and awkward notation.

## 1.7 Exercises

1. Prove: $-a \leqslant |a|$ for all $a \in \mathbf{F}$.
2. Prove: $|a - b| \leqslant |a| + |b|$.
3. Prove: $||a| - |b|| \leqslant |a - b|$.
4. Let $\mathbf{F}$ represent an ordered field. Prove, for $a, b, c, d \in \mathbf{F}$:

   (a) If $a > b$, $b > c$, then $a > c$.
   (b) If $a, b \in \mathbf{F}$, then either $a > b$ or $a = b$, or $a < b$.
   (c) If $a \geqslant b$ and $b \geqslant a$, then $a = b$.
   (d) If $a > b$, then $a + c > b + c$ for all $c \in \mathbf{F}$.
   (e) If $a > b$ and $c > d$, then $a + c > b + d$.
   (f) If $a > b$ and $c > 0$, then $ac > bc$.
   (g) If $a > b$ and $c < 0$, then $ac < bc$.
   (h) If $a > 0$, then $1/a > 0$.
   (i) If $a < 0$, then $1/a < 0$.
   (j) If $ab > 0$, then either $a > 0$, $b > 0$ or $a < 0$, $b < 0$.
   (k) If $a^2 + b^2 = 0$, then $a = b = 0$.

5. Show: If $a_1, a_2, \ldots, a_k \in \mathbf{F}$, where $\mathbf{F}$ is an ordered field, then
   $$|a_1 + a_2 + \cdots + a_k| \leqslant |a_1| + |a_2| + \cdots + |a_k|.$$
6. Show: An ordered field has no smallest positive element.
7. If $\mathbf{F}$ is an ordered field, show that if $a < b$, then $a < (a + b)/2 < b$.
8. Show: If $\mathbf{F}$ is an ordered field and $a < b$, then there are infinitely many elements $x$ with $a < x < b$.
9. Show:

   (a) $|-a| = |a|$.
   (b) If $b \geqslant 0$, then $|a| \leqslant b$ if and only if $-b \leqslant a \leqslant b$.
   (c) $-|a| \leqslant a \leqslant |a|$.

10. Show: If $a > 0$, then $(1 + a)^n \geqslant 1 + na$, $n \in \mathbf{N}$.
11. Show: If $c > 1$, then $c^m \geqslant c^n$ for $m \geqslant n$, $m, n \in \mathbf{N}$ and if $0 < c \leqslant 1$, then $0 < c^m \leqslant c^n$ for $m \geqslant n$, $m, n \in \mathbf{N}$.

12. Show: If $a < b$, and $a > 0$, then $a^n < b^n$ for $n \in \mathbf{N}$, and vice versa.
13. An ordered field $\mathbf{F}$ is called *Archimedean* if, for any $x \in \mathbf{F}$, there exists an $n \in \mathbf{N} \subset \mathbf{F}$ such that $n > x$. Prove: The ordered field $\mathbf{Q}$ of rationals is Archimedean.
14. Given the collection

    $\mathbf{Q}(x) = \{f(x)/g(x) \mid f(x) = a_0 x^n + a_1 x^{n-1} + \cdots + a_n, g(x) = b_0 x^m + b_1 x^{m-1} + \cdots + b_m$, where $n, m \in \mathbf{N}$, $a_i, b_j \in \mathbf{Q}$, and $(b_0, b_1, \ldots, b_m) \neq (0, 0, \ldots, 0)$; $f(x)$, $g(x)$ have no linear factor in common$\}$,

    (a) Show that $\mathbf{Q}(x)$ is a field.
    (b) Show that $\mathbf{Q} \in \mathbf{Q}(x)$.
    (c) Show that $\mathbf{Q}(x)$ is an ordered field with the positive class containing all elements $f(x)/g(x)$ for which $a_0 \cdot b_0 > 0$ and only those elements.
    (d) Show that $\mathbf{Q}(x)$ is *not* Archimedean.

## 1.8 The Real Numbers as a Complete Ordered Field

When we plunged from $\mathbf{Z}$ into $\mathbf{Q}$ we adjoined, in effect, the solutions $x$ of all linear equations $ax = b$, $a,b \in \mathbf{Z}$, $a \neq 0$, to $\mathbf{Z}$—except the ones that were already in $\mathbf{Z}$. This might mislead us to expect that the real numbers can be obtained by a similar process, namely, by adjoining to $\mathbf{Q}$ the solutions of all possible algebraic equations $a_0 x^n + a_1 x^{n-1} + \cdots + a_{n-1}x + a_n = 0$, where $a_i \in \mathbf{Z}$ (or $a_i \in \mathbf{Q}$, for that matter—see exercise 1), and $n \in \mathbf{N}$, and where not all of the coefficients $a_0, a_1, \ldots, a_{n-1}$ vanish. (See also exercise 2, section 1.6.) Such a process does not yield enough and, at the same time, yields too much. We shall see later (exercise 10, section 1.13) that this process leads to a subset of the set of complex numbers.

Since the rational numbers already form an ordered field, the field axioms and the order axioms do not suffice to characterize the real numbers. It appears that some important, but seemingly elusive, property of the real numbers is not captured in these axioms.

A study of the set

$$A = \{1, 1.01, 1.01001, 1.010010001, 1.01001000100001, 1.0100100010000100001, \ldots\}$$

of rational numbers will help us to flush out this elusive property. Clearly, the rational number $u_1 = 2$ is an *upper bound* of $A$, that is, $a < u_1$ for all $a \in A$. Also, $u_2 = 1.02 \in \mathbf{Q}$ is an upper bound of $A$, $a < u_2$ for all $a \in A$, and $u_1 > u_2$. Likewise, $u_3 = 1.01002 \in \mathbf{Q}$ has the property that $a < u_3$ for all $a \in A$, and we have $u_1 > u_2 > u_3$. We continue in this manner and obtain rational numbers $u_4 = 1.010010002$, $u_5 = 1.01001000100002$, ... such that for all $j \in \mathbf{N}$, $a < u_j$ for all $a \in A$, and $u_1 > u_2 > u_3 > u_4 > u_5 > \cdots$. The question arises: Is there a *smallest* rational number $u$ such that $a < u$ for all $a \in A$? There is not. The successive construction of increasingly smaller upper bounds leads to the infinite decimal

$$1.010010001000010000010000001000000010000000001\ldots$$

which does *not* represent a rational number because all rational numbers can be represented by a finite decimal or a repeating decimal. (See exercise 6.)

It turns out that one obtains the set of all real numbers if one adjoins to **Q** all *least upper bounds* of all sets of rational numbers which are bounded above, and that the resulting set is *complete* in the sense that the least upper bounds of all bounded sets of real numbers are again real numbers which have already been included.

In order to develop this process with precision, let us first introduce the appropriate terminology:

***Definition 8.1***

A subset $A$ of an ordered field **F** is *bounded above* if and only if there exists an element $u \in \mathbf{F}$ such that $x \leqslant u$ for all $x \in A$. $u$ is called an *upper bound of A*. ($A$ is *bounded below* if and only if there exists an element $l \in \mathbf{F}$ such that $x \geqslant l$ for all $x \in A$. $l$ is called a *lower bound of A*.)

---

▲ *Example 1*

The set $\mathbf{N} = \{1, 2, 3, \ldots\} \subset \mathbf{Q}$ is *not* bounded above but is bounded below by 1 (or by 0, or by $-27$, etc ...).

▲ *Example 2*

The set $\{1, \frac{1}{2}, \frac{1}{3}, \frac{1}{4}, \ldots\} \subset \mathbf{Q}$ is bounded above by 1 (or by 2, or 54/37, etc ...), and is bounded below by 0 (or $-3$, or $-153/19$, etc ...).

▲ *Example 3*

The set $\{1, 1.4, 1.41, 1.414, 1.4142, 1.41421, 1.414214, \ldots\} \subset \mathbf{Q}$ is bounded above by 2 (or 1.5, or 1.42, etc ...) and bounded below by 1 (or 0, or $-123$, etc ...).

▲ *Example 4*

The set $\{3, 3.1, 3.14, 3.141, 3.1415, 3.14159, 3.141592, \ldots\} \subset \mathbf{Q}$ is bounded above by 3.15 (or 3.1416, or 3.141593, etc ...).

---

In example 1, the lower bound 1 is the greatest lower bound. In example 2, the upper bound 1 is the least upper bound, and 0 appears to be the greatest lower bound; in example 3 and 4, the bounds 1 and 3 are the greatest lower bounds, respectively, but we do not really know what the least upper bounds are, although we have strong suspicions.

*Definition 8.2*

If $A \subset \mathbf{F}$ is bounded above, $M \in \mathbf{F}$ is the *least upper bound* of $A$; in symbols,

$$M = \sup A,$$

if and only if:
1. $M$ is an upper bound of $A$,
2. $M \leqslant u$ for *all* upper bounds $u$ of $A$.

(*sup* in the above notation is an abbreviation for *supremum*.) ($m \in \mathbf{F}$ is the *greatest lower bound* of $A$,

$$m = \inf A,$$

if and only if $m$ is a lower bound of $A$ and $l \leqslant m$ for all lower bounds $l$ of $A$; —*inf* is an abbreviation for *infimum*.)

*Theorem 8.1*

The least upper bound and the greatest lower bound of a subset of an ordered field are unique, provided they exist.

*Proof*

If $M_1$, $M_2$ both are least upper bounds of $A$, then, by (2) of definition 8.2, $M_1 \leqslant M_2$ as well as $M_2 \leqslant M_1$. Hence $M_1 = M_2$. An analogous proof establishes the uniqueness of the greatest lower bound.

*Definition 8.3*

An ordered field $\mathbf{F}$ is called *complete* if and only if every nonempty subset that is bounded above has a least upper bound in $\mathbf{F}$.

Occasionally, this is phrased as follows: $\mathbf{F}$ is *complete* if and only if it has the *least-upper-bound property*. (Clearly, if $\mathbf{F}$ is complete, then every subset that is bounded below has a greatest lower bound, and vice versa.)

If we require that our ordered field of numbers has the least-upper-bound property, we catch the real numbers in body and in spirit:

*Definition 8.4*

A complete ordered field is denoted by $\mathbf{R}$ and *the elements of a complete ordered field are called real numbers.*

This definition is meaningful only if there *is* such a thing as a complete ordered field. The existence of a complete ordered field can be proved in a number of ways. One such proof is given in section 1.11.

Even if the question as to the existence of a complete ordered field is settled, the question as to its uniqueness is still to be studied. We shall explain in section

1.11 to what extent a complete ordered field is to be considered as unique. Uniqueness certainly cannot mean that elements have a unique representation. Take, for example, the set **Q** of rationals. **Q** may be viewed as the collection of all fractions $m/n$ where $m \in \mathbf{Z}$, $n \in \mathbf{N}$, $(m, n) = 1$. ($m$ and $n$, themselves may be represented in the customary manner in the decimal system, or may be represented in the binary system, or ternary system, or, for that matter, by Roman numerals I, II, III, IV, V, VI, VII, ...). **Q** may also be viewed as the collection of all repeating decimals, or repeating binaries, or repeating ternaries (see exercises 5, 6, 7, 11) or, for that matter, as a collection of points on the number line where the point $m/n$ is located at a distance that is $|m|$ times one $n$th of the unit distance from 0, to the right if $m > 0$ and to the left if $m < 0$. Still, we speak in every one of these instances of the rational numbers, no matter how they manifest themselves. What makes these objects rational numbers is not their formal appearance but their intrinsic properties that are captured in the field axioms and order axioms.

By the same token, $\sqrt{2}$ is a real number, no matter how it presents itself: as an infinite decimal, or an infinite binary, or an infinite ternary, or as the length of the hypotenuse of a right triangle with the two sides forming the right angle having unit length, or, for that matter, as the symbol $\sqrt{2}$ itself (or $\sqrt{\text{II}}$, $\sqrt{e+e}$, etc...).

In section 1.10 we shall show that if a complete ordered field **R** exists, then there is a one-to-one correspondence between **R** and the set of all decimals. Our discussion of existence and uniqueness of **R** in section 1.11 will be based on the representation of elements of **R** by decimals.

## 1.8 Exercises

1. Show that the algebraic equation $b_0 x^n + b_1 x^{n-1} + \cdots + b_{n-1}x + b_n = 0$, $b_k \in \mathbf{Q}$, is equivalent to an algebraic equation $a_0 x^n + a_1 x^{n-1} + \cdots + a_{n-1}x + a_n = 0$, $a_k \in \mathbf{Z}$, in the sense that every solution of the one equation is also a solution of the other equation, and vice versa.
2. Show: If $x^2 < 4$, then $x < 2$.
3. Let $A = \{x \in \mathbf{Q} | x^2 \leqslant 2\}$. Show that 3/2 and 71/50 are upper bounds of $A$.
4. Show: If $S \subset \mathbf{F}$ and if $u \in S$ is an upper bound of $S$, then $u = \sup S$.
5. Show:

   (a) Every finite decimal represents a rational number.
   (b) Every repeating infinite decimal represents a rational number.

6. Show:

   (a) A rational number $m/n$ where $(m, n) = 1$, can be represented by a finite decimal if and only if $n = 2^j 5^k$ for some $j, k \in \{0, 1, 2, \ldots\}$.
   (b) Every rational number $m/n$ where $n = Np$, $N \in \mathbf{Z}$, $p$ prime and $p \neq 2$, $p \neq 5$, can be represented by an infinite repeating decimal.
7. Show: There is a one-to-one correspondence between **Q** and all repeating infinite decimals.

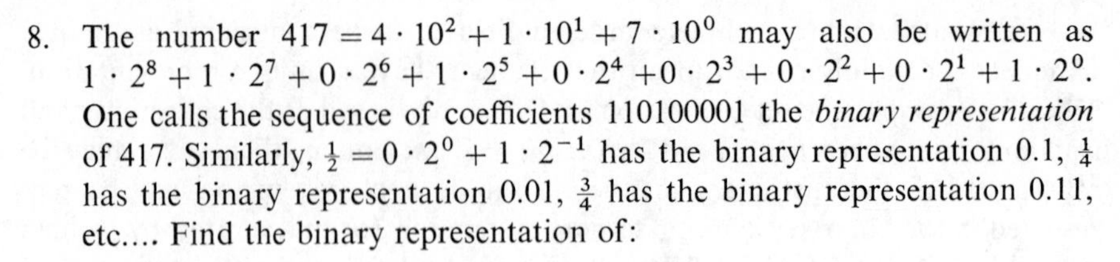

8. The number $417 = 4 \cdot 10^2 + 1 \cdot 10^1 + 7 \cdot 10^0$ may also be written as $1 \cdot 2^8 + 1 \cdot 2^7 + 0 \cdot 2^6 + 1 \cdot 2^5 + 0 \cdot 2^4 + 0 \cdot 2^3 + 0 \cdot 2^2 + 0 \cdot 2^1 + 1 \cdot 2^0$. One calls the sequence of coefficients 110100001 the *binary representation* of 417. Similarly, $\frac{1}{2} = 0 \cdot 2^0 + 1 \cdot 2^{-1}$ has the binary representation 0.1, $\frac{1}{4}$ has the binary representation 0.01, $\frac{3}{4}$ has the binary representation 0.11, etc.... Find the binary representation of:

   (a) 513, (b) $17\frac{3}{8}$ (c) 0.125

9. Write the following binary representations (binaries) as decimal representations (decimals):

   (a) 1111 (b) 101.1
   (c) 0.11001 (d) $0.\overline{01}$

10. Show:

    (a) A rational number $m/n$ where $(m, n) = 1$, has a finite binary representation if and only if $n = 2^k$ for some $k \in \{0, 1, 2, 3, \ldots\}$.
    (b) A rational number $m/n$ has a repeating binary representation if $n = N \cdot p$ where $p \neq 2$ is a prime number.

11. Show: There is a one-to-one correspondence between **Q** and all repeating infinite binaries.
12. 1022102 is called the ternary representation of the number $1 \cdot 3^6 + 0 \cdot 3^5 + 2 \cdot 3^4 + 2 \cdot 3^3 + 1 \cdot 3^2 + 0 \cdot 3^1 + 2 \cdot 3^0$. Find the ternary representation of:

    (a) 321 (b) $15\frac{26}{27}$
    (c) $0.\overline{2592}$ (d) $\frac{2}{3}$

13. Show: Every rational number with a finite decimal expansion has an infinite ternary expansion, and every number with a finite ternary expansion has an infinite decimal expansion.
14. Same as in exercise 13 with *binary* substituted for *decimal*.
15. Show: If $n \in \mathbf{N}$, then the ternary representation of $n^2$ has either the digit 0 or the digit 1 in its unit place. (For example, $4^2 = 1 \cdot 3^2 + 2 \cdot 3^1 + 1 \cdot 3^0$.)
16. Use the result of exercise 15 to show that $\sqrt{2}$ is not rational.
17. Show: If **F** is a complete ordered field, then every subset $A$ of **F** which is bounded below has a greatest lower bound, and vice versa.

## 1.9 Properties of the Least Upper Bound

We need the following characterization of the least upper bound for our future work.

***Theorem 9.1***

An upper bound $M$ of $A \subset \mathbf{F}$, where **F** is an ordered field, is the least upper bound of $A$ if and only if for every $\varepsilon > 0$ there is an element $x \in A$ such that $M - \varepsilon < x \leqslant M$. (See Fig. 9.1.) The lower bound $m$ is the greatest lower

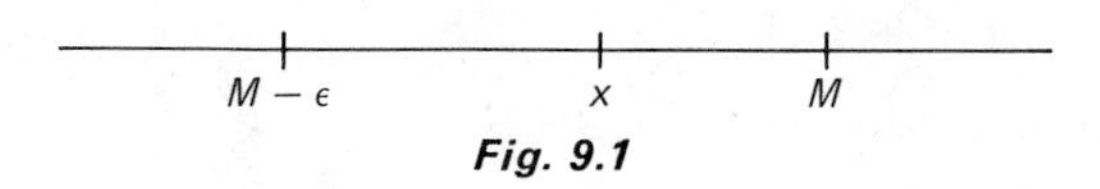


*Fig. 9.1*

bound of $A$ if and only if for every $\varepsilon > 0$ there is an element $y \in A$ such that $m \leqslant y < m + \varepsilon$.

*Proof*

We shall prove only the statement pertaining to the least upper bound and leave the proof of the second statement to the reader. (See exercise 1.)

(a) Suppose that $M = \sup A$ but that the statement of the theorem is not true, i.e., there is an $\varepsilon_0 > 0$ such that $x \leqslant M - \varepsilon_0$ for all $x \in A$. Since $M - \varepsilon_0 < M$, it follows that $M$ is *not* the least upper bound of $A$. (See (2) of definition 8.2.)

(b) Suppose that $M$ is an upper bound of $A$, that is,

$$x \leqslant M \qquad \text{for all } x \in A,$$

and that for every $\varepsilon > 0$, there is an element $x \in A$ such that

$$M - \varepsilon < x \leqslant M.$$

If $M$ is not the least upper bound of $A$, then there is a $M_0 < M$ such that $x_0 \leqslant M$ for all $x \in A$. Let $\varepsilon = M - M_0 > 0$. Then, there is an element $x \in A$ such that

$$M - (M - M_0) < x \leqslant M,$$

and we obtain $x > M_0$, contrary to our assumption that $M_0$ is an upper bound of $A$.

The condition in Theorem 9.1 may or may not be trivially satisfied, as we shall see in the following examples:

---

▲ *Example 1*

Let $A = \{1, 2, 3\}$. Then, $\sup A = 3$, and since $3 \in A$, we have, for every $\varepsilon > 0$ that $3 - \varepsilon < 3 \leqslant 3$.

▲ *Example 2*

Let $B = \{\frac{1}{2}, \frac{2}{3}, \frac{3}{4}, \frac{4}{5}, \ldots, n/(n+1), \ldots\}$. Then $\sup B = 1$ but $1 \notin B$. Still, for every $\varepsilon > 0$, there is a $n/(n+1) \in B$ such that $1 - \varepsilon < n/(n+1) \leqslant 1$, or, equivalently, $1 - n/(n+1) = 1/(n+1) < \varepsilon$.

---

One would expect to find, in a set of integers that is bounded above, a largest integer. This is indeed so, as we shall see from the following theorem.

*Theorem 9.2*

If $A \subset \mathbf{Z}$ is bounded above, then $\sup A \in A$.

*Proof* (by contradiction)

Since $\mathbf{Z} \subset \mathbf{R}$ and since $\mathbf{R}$ is complete, $\sup A \in \mathbf{R}$ exists. Suppose that $\sup A = M \notin A$. Then $x < M$ for all $x \in A$, and we have, from Theorem 9.1, that for $\varepsilon_1 = 1$ there is an $x' \in A$ such that $M - 1 < x' < M$. Again, by Theorem 9.1, we have for $\varepsilon_2 = M - x' > 0$, an $x'' \in A$ such that

$$M - (M - x') = x' < x'' < M.$$

Thus we obtain two distinct integers $x'$, $x''$ which differ by less than 1, which is impossible. (See also Fig. 9.2.) Hence, $\sup A \in A$.

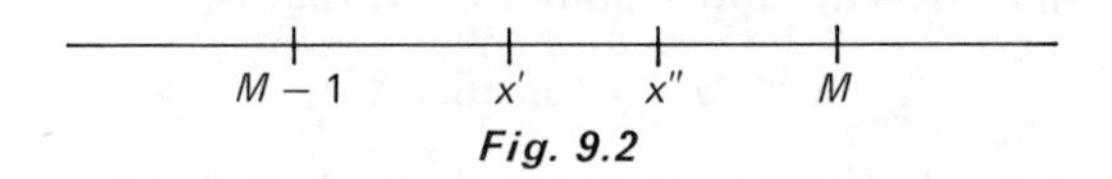


*Fig. 9.2*

*Definition 9.1*

An ordered field $\mathbf{F}$ is called *Archimedean* if and only if, for every $x \in \mathbf{F}$ there is an $n \in \mathbf{N}$ such that $n > x$.

(Note, that by Theorem 7.2, $\mathbf{N} \subset \mathbf{Q} \subseteq \mathbf{F}$. See also exercise 13, section 1.7.) We obtain from Theorem 9.2:

*Theorem 9.3*

A complete ordered field $\mathbf{R}$ is Archimedean.

*Proof* (by contradiction)

Suppose that there is an $x \in \mathbf{R}$ such that $n \leqslant x$ for all $n \in \mathbf{N}$. By Theorem 9.2, $\sup \{n \in \mathbf{N} \mid n \leqslant x\} \in N$, that is, there exists an $n_0 \in \mathbf{N}$ such that $n \leqslant n_0$ for all $n \in \mathbf{N}$. But $n_0 + 1 \in \mathbf{N}$ and $n_0 + 1 > n_0$, contrary to our assumption.

(Note that an ordered field that is not complete need not be Archimedean. See exercise 14(d), section 1.7.)

---

▲ *Example 3*

If $\mathbf{F}$ is an ordered field with the Archimedean property (not necessarily complete), then for every $\varepsilon > 0$, $\varepsilon \in \mathbf{F}$, there is some $n \in \mathbf{N}$ such that

$$\frac{1}{2^n} < \varepsilon. \tag{9.1}$$

This may be seen as follows: By the Archimedean property, there is some $n \in \mathbf{N}$ such that $n > 1/\varepsilon$. By induction, $2^n > n$ for all $n \in \mathbf{N}$. Hence, $2^n > n > 1/\varepsilon$, and (9.1) follows readily.

---

We state for the record, in order to facilitate later references:

***Theorem 9.4***

If $\mathbf{F}$ is an ordered field and if $A \subset \mathbf{F}$ contains finitely many (distinct) elements only, then $\sup A$ exists and $\sup A \in A$.

The proof is left to the reader.

Note that Theorem 9.4 implies, in particular, that a set of integers that is *bounded above by an integer* possesses a least upper bound, and that the least upper bound is an element of the set. This result does *not* depend on the completeness of $\mathbf{R}$. Note the subtle difference between this statement and Theorem 9.2. (See also exercise 5.)

The following lemma concerning some manipulations of least upper bound and greatest lower bound will be needed in Chapter 4.

***Lemma 9.1***

If $A \subset \mathbf{R}$ is bounded above and below and if $\lambda > 0$, then

$$\sup\{\lambda a \mid a \in A\} = \lambda \sup A, \qquad \inf\{\lambda a \mid a \in A\} = \lambda \inf A. \tag{9.2}$$

If $\Gamma$ is a given index set and if $A = \{a_\gamma \mid \gamma \in \Gamma\} \subset \mathbf{R}$, $B = \{b_\gamma \mid \gamma \in \Gamma\} \subset \mathbf{R}$ are bounded above and below, then

$$\sup\{a_\gamma + b_\gamma \mid \gamma \in \Gamma\} \leqslant \sup A + \sup B, \tag{9.3}$$

$$\inf\{a_\gamma + b_\gamma \mid \gamma \in \Gamma\} \geqslant \inf A + \inf B. \tag{9.4}$$

If $\Gamma$, $\Delta$ are given index sets and if $A = \{a_\gamma \mid \gamma \in \Gamma\} \subset \mathbf{R}$, $B = \{b_\delta \mid \delta \in \Delta\} \subset \mathbf{R}$ are bounded above and below, then

$$\sup\{a_\gamma + b_\delta \mid \gamma \in \Gamma, \delta \in \Delta\} = \sup A + \sup B, \tag{9.5}$$

$$\inf\{a_\gamma + b_\delta \mid \gamma \in \Gamma, \delta \in \Delta\} = \inf A + \inf B. \tag{9.6}$$

*Proof*

We shall prove only (9.4) and leave the rest to the reader (exercises 7, 8, 9, 10). Suppose that

$$\inf\{a_\gamma + b_\gamma \mid \gamma \in \Gamma\} < \inf A + \inf B.$$

Then, there is an $\varepsilon > 0$ such that

$$\inf\{a_\gamma + b_\gamma \mid \gamma \in \Gamma\} + \varepsilon = \inf A + \inf B. \tag{9.7}$$

By Theorem 9.1, there is an element $a_{\gamma_0} + b_{\gamma_0} \in \{a_\gamma + b_\gamma \mid \gamma \in \Gamma\}$ such that

$$a_{\gamma_0} + b_{\gamma_0} < \inf \{a_\gamma + b_\gamma \mid \gamma \in \Gamma\} + \frac{\varepsilon}{2},$$

and we obtain from (9.7) that $a_{\gamma_0} + b_{\gamma_0} < \inf A + \inf B$. On the other hand, $a_\gamma \geqslant \inf A$, $b_\gamma \geqslant \inf B$ for all $\gamma \in \Gamma$ and, in particular, for $\gamma = \gamma_0$. Hence, $a_{\gamma_0} + b_{\gamma_0} \geqslant \inf A + \inf B$, and we have a contradiction.

## 1.9 Exercises

1. Prove: A lower bound $m$ of $A \subset \mathbf{F}$, where $\mathbf{F}$ is a complete ordered field, is the greatest lower bound of $A$ if and only if for every $\varepsilon > 0$, there is an element $x \in A$ such that $m \leqslant x < m + \varepsilon$,
2. The ordered pair $(A, B)$ of nonempty subsets $A$, $B$ of an ordered field $\mathbf{F}$ defines a *Dedekind*† *cut* in $\mathbf{F}$ if

   1. $a \in A$, $b \in B$ implies $a < b$,
   2. $A \cap B = \varnothing$, $A \cup B = \mathbf{F}$.

   Prove: $\mathbf{F}$ is complete if and only if for every Dedekind cut $(A, B)$ in $\mathbf{F}$, there exists an $x \in \mathbf{F}$ such that $x \geqslant a$ for all $a \in A$ and $x \leqslant b$ for all $b \in B$.
3. Let $a \in \mathbf{R}^+$ and $a_i = \sup \{x \in \mathbf{Z} \mid 10^i a_0 + 10^{i-1} a_1 + \cdots + 10 a_{i-1} + x \leqslant 10^i a\}$, where $i = 0, 1, 2, \ldots$ . Show that $10^i a_0 + 10^{i-1} a_1 + \cdots + 10 a_{i-1} + a_i + 1 > 10^i a$.
4. Prove Theorem 9.4 by induction.
5. Let $A$ denote a set of integers which is bounded above by an integer $n_0$. Show, without using the completeness of $\mathbf{R}$, that $\sup A$ exists and $\sup A \in A$.
6. Show: If $\mathbf{F}$ is a complete ordered field, and if $x \in \mathbf{F}$, $x \neq 0$, then there is an $n \in \mathbf{N}$ such that $1/n < |x|$.
7. Prove (9.2).
8. Prove (9.3).
9. Prove (9.5) and (9.6).
10. Let $A = \{a_\gamma \mid \gamma \in \Gamma\} \subset \mathbf{R}$ represent a set which is bounded above *and* below. Show that

$$\inf \{-a_\gamma \mid \gamma \in \Gamma\} = -\sup A, \qquad \sup \{-a_\gamma \mid \gamma \in \Gamma\} = -\inf A.$$

11. Show that an ordered Archimedean field need not be complete.

## 1.10 Representation of the Real Numbers by Decimals*

A *decimal* is a formal expression $\pm a_0 \,.\, a_1 a_2 a_3 \ldots$ where $a_0 \in \{0, 1, 2, 3, \ldots\}$, where $a_j \in \{0, 1, 2, 3, 4, 5, 6, 7, 8, 9\}$, and where infinitely many digits 9 do not occur consecutively. (We do admit, however, infinitely many consecutive

† *Richard Dedekind*, 1831–1916.

* The reader may omit this and the next section if he is already familiar with the fact that a complete ordered field exists and may be represented by the set of decimals, or if he is willing to accept this fact on faith.

digits 0, or 1, or 2, ... , or 8. If infinitely many consecutive digits 0 occur, we speak of a finite decimal. When infinitely many consecutive digits occur, we indicate this by a dot above the first one in the chain, such as $3.46777... = 3.46\dot{7}$.) For the time being, we consider the decimals merely as symbols without any quantitative meaning.

Two decimals $a_0 . a_1 a_2 a_3 \ldots$ and $b_0 . b_1 b_2 b_3 \ldots$ are equal if and only if $a_j = b_j$ for all $j \in \{0, 1, 2, 3, \ldots\}$.

We assume that a complete ordered field **R** of real numbers exists, and we then note that, by Theorem 7.2, **R** contains the rationals **Q** as a subfield. We shall demonstrate that, under this assumption, **R** and the set **R*** of all decimals are equivalent sets by constructing a *one-to-one* function $f$ from **R*** *onto* **R**. In the next section we shall introduce an order relation in **R*** and show that **R*** is complete. We shall then define field operations in **R***, and it will develop that **R*** is a complete ordered field and that the above mentioned function $f$ preserves the order relation and the field operations. This, in turn, implies that there are real numbers and that they are uniquely determined, but for their representation, by the field postulates, the order postulate, and the completeness postulate.

The construction of a function $f: \mathbf{R}^* \leftrightarrow \mathbf{R}$ is based on the following lemma:

***Lemma 10.1***

Let $a_0 \in \{0, 1, 2, 3, \ldots\}$ and let $a_j \in \{0, 1, 2, 3, 4, 5, 6, 7, 8, 9\}$ for all $j \in \mathbf{N}$. If $k \in \mathbf{N}$ is fixed, then

$$a_0 + \frac{a_1}{10} + \frac{a_2}{10^2} + \cdots + \frac{a_{k-1}}{10^{k-1}} + \frac{a_k}{10^k} + \cdots + \frac{a_l}{10^l} \leqslant a_0 + \frac{a_1}{10} + \frac{a_2}{10^2} + \cdots + \frac{a_k + 1}{10^k} \tag{10.1}$$

for all $l \geqslant k$.

***Proof***

The inequality is trivial for $l = k$. If $l > k$, we have

$$\frac{a_{k+1}}{10^{k+1}} + \cdots + \frac{a_l}{10^l} \leqslant \frac{9}{10^{k+1}}\left(1 + \frac{1}{10} + \cdots + \frac{1}{10^{l-k-1}}\right)$$
$$= \frac{9}{10^{k+1}} \cdot \frac{10}{9}\left(1 - \frac{1}{10^{l-k}}\right) < \frac{1}{10^k},$$

and (10.1) follows readily.

***Theorem 10.1***

If **R** is a complete ordered field, then

$$f(a_0 . a_1 a_2 a_3 \ldots) = \sup\left\{a_0, a_0 + \frac{a_1}{10}, a_0 + \frac{a_1}{10} + \frac{a_2}{10^2}, a_0 + \frac{a_1}{10} + \frac{a_2}{10^2} + \frac{a_3}{10^3}, \ldots\right\} \tag{10.2}$$

$$f(-a_0 . a_1 a_2 a_3 \ldots) = -f(a_0 . a_1 a_2 a_3 \ldots),$$

where $a_0 \in \{0, 1, 2, 3, \ldots\}$, where $a_j \in \{0, 1, 2, 3, 4, 5, 6, 7, 8, 9\}$, and where infinitely many digits 9 do not occur consecutively, defines a function from the set $\mathbf{R}^*$ of all decimals into $\mathbf{R}$. This function is *one-to-one* on $\mathbf{R}^*$ and *onto* $\mathbf{R}$. Hence, $\mathbf{R}^*$ and $\mathbf{R}$ are equivalent sets.

Note that for any finite decimal $a_0 . a_1 a_2 \ldots a_k \dot{0} \in \mathbf{R}^*$, $f(a_0 . a_1 a_2 \ldots a_k \dot{0}) = a_0 + a_1/10 + a_2/10^2 + \cdots + a_k/10^k$.

*Proof*

We call

$$\{a_0 . a_1 a_2 a_3 \cdots \mid a_0 \in \{0, 1, 2, 3, \ldots\},\quad a_j \in \{0, 1, 2, 3, 4, 5, 6, 7, 8, 9\},$$
$$\text{infinitely many digits 9 do not occur consecutively}\}$$

the set of *nonnegative decimals*. Its complement with respect to $\mathbf{R}^*$ is called the set of *negative decimals*. That subset of nonnegative decimals where at least one digit is not zero is called the set of *positive decimals*. Observe that a decimal is either positive, or zero $(0.\dot{0})$, or negative. It suffices to show that $f$ is *one-to-one* on the set of nonnegative decimals *onto* the nonnegative reals. The rest follows by a symmetric argument.

(a) First we show that $f$ is a function with domain $\mathbf{R}^*$. By (10.1), $\{a_0, a_0 + a_1/10, a_0 + a_1/10 + a_2/10^2, \ldots\} \subset \mathbf{Q} \subset \mathbf{R}$ is bounded. Since $\mathbf{R}$ is assumed to be complete, the least upper bound $a$ exists and is unique. Since $a_0 + a_1/10 + \cdots + a_k/10^k \geqslant 0$, it follows that $a \geqslant 0$. Similarly, one can show that $f$ maps the negative decimals uniquely into the negative reals. Hence, $f$ is a function with domain $\mathbf{R}^*$ and target set $\mathbf{R}$.

(b) Next we show that $f$ is *onto* the reals $\mathbf{R}$. Again, we need only consider the nonnegative reals. We represent the set of nonnegative reals as a union of half-open disjoint intervals

$$\bigcup_{k=0}^{\infty} [k, k+1). \tag{10.3}$$

(See also exercises 1 and 2.)

If $a \geqslant 0$, then there is a unique $a_0 \in \{0, 1, 2, 3, \ldots\}$ such that $a \in [a_0, a_0 + 1)$. If $a = a_0$, we let $a_1 = a_2 = a_3 = \cdots = 0$, and obtain an element $a_0 . \dot{0} \in \mathbf{R}^*$ for which $f(a_0 . \dot{0}) = \sup\ \{a_0,\ a_0 + 0,\ a_0 + 0 + 0,\ \ldots\} = a_0 = a$. If $a > a_0$, we represent $[a_0, a_0 + 1)$ as a union of disjoint half-open intervals

$$[a_0, a_0 + 1) = \bigcup_{k=0}^{9} \left[a_0 + \frac{k}{10}, a_0 + \frac{k+1}{10}\right),$$

and note that there is a unique $a_1 \in \{0, 1, 2, 3, 4, 5, 6, 7, 8, 9\}$ such that $a \in [a_0 + a_1/10, a_0 + (a_1 + 1)/10)$. If $a = a_0 + (a_1/10)$, we let $a_2 = a_3 = a_4 = \cdots = 0$ and obtain an element $a_0 . a_1\dot{0} \in \mathbf{R}^*$ for which $f(a_0 . a_1\dot{0}) = \sup\ \{a_0, a_0 + a_1/10, a_0 + a_1/10 + 0, \ldots\} = a_0 + a_1/10 = a$. If $a > a_0 + (a_1/10)$, we consider

$$\left[a_0 + \frac{a_1}{10}, a_0 + \frac{a_1 + 1}{10}\right) = \bigcup_{k=0}^{9}\left[a_0 + \frac{a_1}{10} + \frac{k}{10^2}, a_0 + \frac{a_1}{10} + \frac{k+1}{10^2}\right),$$

and proceed as before. (See also Fig. 10.1.)

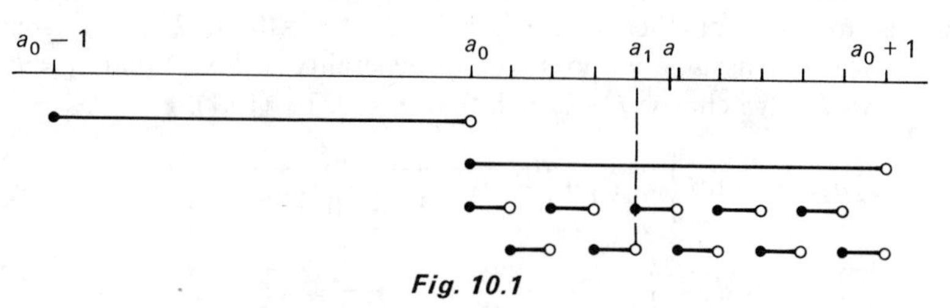


*Fig. 10.1*

This process either breaks up after finitely many steps or it does not break up. If it breaks up after $k$ steps, we obtain an element $a_0 . a_1 a_2 \ldots a_k \dot{0} \in \mathbf{R}^*$ for which $f(a_0 . a_1 a_2 \ldots a_k \dot{0}) = a_0 + a_1/10 + a_2/10^2 + \cdots + a_k/10^k = a$. If it does not break up, we obtain an infinite decimal $a_0 . a_1 a_2 a_3 \ldots$, where infinitely many digits 9 cannot occur consecutively, for, suppose to the contrary, that $a_{k-1} \neq 9$ but $a_k = a_{k+1} = \cdots = 9$ for some $k \in \{1, 2, 3, \cdots\}$. Then, for all integers $j \geqslant k$,

$$a_0 + \cdots + \frac{a_{k-1}}{10^{k-1}} + \frac{9}{10^k} + \cdots + \frac{9}{10^j} < a < a_0 + \cdots + \frac{(a_{k-1} + 1)}{10^{k-1}}.$$

Since $a_0 + \cdots + (a_{k-1} + 1)/10^{k-1} = \sup_{(j \geqslant k)} \{a_0 + \cdots + a_{k-1}/10^{k-1} + 9/10^k + \cdots + 9/10^j\}$ because

$$\frac{9}{10^k} + \cdots + \frac{9}{10^j} = \frac{1}{10^{k-1}}\left(1 - \frac{1}{10^{j-k+1}}\right)$$

and Theorem 9.1, and since $\sup_{(j \geqslant k)} \{a_0 + \cdots + a_{k-1}/10^{k-1} + 9/10^k + \cdots + 9/10^j\} \leqslant a$, we obtain $a \geqslant a_0 + \cdots + (a_{k-1} + 1)/(10^{k-1})$ contrary to the fact that, by construction,

$$a \in [a_0 + \cdots + a_{k-1}/10^{k-1}, a_0 + \cdots + (a_{k-1} + 1)/10^{k-1}).$$

In order to clinch our argument that $f$ is onto $\mathbf{R}$, we still have to show that, whenever our construction leads to an infinite decimal $a_0 . a_1 a_2 a_3 \ldots$, then

$$a = \sup \{a_0, a_0 + a_1/10, a_0 + a_1/10 + a_2/10^2, \ldots\}. \tag{10.4}$$

By construction,

$$a \in [a_0 + a_1/10 + \cdots + a_j/10^j, a_0 + a_1/10 + \cdots + (a_j + 1)/10^j) \text{ for all } j \in \mathbf{N}.$$

For any $\varepsilon > 0$, we can find a $j \in \mathbf{N}$ such that $(1/10^j) < \varepsilon$ and hence,

$$a - \varepsilon < a_0 + \frac{a_1}{10} + \cdots + \frac{a_j}{10^j} \leqslant a.$$

(10.4) follows readily from Theorem 9.1.

(c) Finally, we have to show that $f$ is one-to-one on $\mathbf{R}^*$. In order to show that $f(a_0 \,.\, a_1a_2a_3 \ldots) = f(b_0 \,.\, b_1b_2b_3 \ldots)$ implies $a_0 \,.\, a_1a_2a_3 \ldots = b_0 \,.\, b_1b_2b_3 \cdots$ (see definition 4.4), we assume, to the contrary, that $a_0 \,.\, a_1a_2a_3 \ldots \neq b_0 \,.\, b_1b_2b_3 \ldots$, and show that the corresponding function values cannot be equal. If the two decimals are unequal, then, there is a smallest $k \in \{0, 1, 2, 3, \ldots\}$ such that $a_k \neq b_k$. We may assume without loss of generality (w.l.o.g.) that $a_k < b_k$, that is, $a_k + 1 \leqslant b_k$. We choose $j > k$ such that $a_j < 9$. By (10.1),

$$
\begin{aligned}
f(a_0 \,.\, a_1a_2a_3 \ldots) &= \sup\left\{a_0, a_0 + \frac{a_1}{10}, a_0 + \frac{a_1}{10} + \frac{a_2}{10^2}, \ldots\right\} \\
&\leqslant a_0 + \frac{a_1}{10} + \cdots + \frac{a_k}{10^k} + \cdots + \frac{9}{10^j} \\
&\leqslant a_0 + \frac{a_1}{10} + \cdots + \frac{a_k}{10^k} + \frac{9}{10^{k+1}}\left(1 + \cdots + \frac{1}{10^{j-k-1}}\right) \\
&= a_0 + \frac{a_1}{10} + \cdots + \frac{a_k}{10^k} + \frac{9}{10^{k+1}}\frac{10}{9}\left(1 - \frac{1}{10^{j-k}}\right) \\
&< a_0 + \frac{a_1}{10} + \cdots + \frac{a_k}{10^k} + \frac{1}{10^k} \\
&\leqslant b_0 + \frac{b_1}{10} + \cdots + \frac{b_k}{10^k} \leqslant \sup\left\{b_0, b_0 + \frac{b_1}{10}, b_0 + \frac{b_1}{10} + \frac{b_2}{10^2}, \ldots\right\} \\
&= f(b_0 \,.\, b_1b_2b_3 \ldots);
\end{aligned}
$$

that is, $f(a_0 \,.\, a_1a_2a_3 \ldots) < f(b_0 \,.\, b_1b_2b_3 \ldots)$, contrary to our assumption. Since $f\colon \mathbf{R}^* \leftrightarrow \mathbf{R}$, we have, from definition 5.3, that $\mathbf{R}^*$ and $\mathbf{R}$ are equivalent.

Theorem 10.1 implies that every element in a complete ordered field may be represented by a unique decimal, and that every decimal represents a unique element of a complete ordered field, provided, of course, that a complete ordered field exists. In the next section we shall show that $\mathbf{R}^*$, with order relation and field operations properly defined, is a complete ordered field.

If $\mathbf{D}^*$ denotes the set $\{a_0 \cdot a_1a_2 \ldots a_k\}$ of all finite decimals, and if $\mathbf{D}$ denotes the set of all rational numbers that can be represented in the form $\{a_0 + a_1/10 + \cdots + a_k/10^k\}$, then the restriction of $f$ to the domain $\mathbf{D}^*$ is one-to-one on $\mathbf{D}^*$ and onto $\mathbf{D}$. This can be established *without* having to assume the completeness of $\mathbf{R}$ as the reader can easily convince himself by checking every pertinent step in the above proof.

## 1.10 Exercises

1. Show that (10.3) is a valid representation of the set of nonnegative real numbers.
2. Show that the intervals in (10.3) are disjoint.

3. Show that the function $f$ of Theorem 10.1 maps all negative decimals one-to-one onto the set of negative reals.
4. A *binary* is a formal expression $\pm b_0 \,.\, b_1 b_2 b_3 \ldots$ where $b_0 \in \{0, 1, 2, 3, \ldots\}$, where $b_j \in \{0, 1\}$ for all $j \in \mathbf{N}$, and where infinitely many digits 1 do not occur consecutively. The integral part $b_0$ is represented by $\beta_n \beta_{n-1} \ldots \beta_1 \beta_0$ if and only if $b_0 = \beta_n 2^n + \beta_{n-1} 2^{n-1} + \cdots + \beta_1 2 + \beta_0$, $\beta_j \in \{0, 1\}$. The collection of all binaries is denoted by $\mathbf{B}^*$. Show that the function $f$, defined by
$$f(b_0 \,.\, b_1 b_2 b_3 \ldots) = \sup \left\{ b_0, b_0 + \frac{b_1}{2}, b_0 + \frac{b_1}{2} + \frac{b_2}{2^2}, \ldots \right\},$$
$$f(-b_0 \,.\, b_1 b_2 b_3 \ldots) = -f(b_0 \,.\, b_1 b_2 b_3 \ldots),$$
is one-to-one on $\mathbf{B}^*$ and onto $\mathbf{R}$, provided that $\mathbf{R}$ is a complete ordered field. Deduce from this that $\mathbf{B}^*$ and $\mathbf{R}^*$ are equivalent.
5. A *ternary* is a formal expression $\pm t_0 \,.\, t_1 t_2 t_3 \ldots$ where $t_0 \in \{0, 1, 2, 3, \ldots\}$, where $t_j \in \{0, 1, 2\}$ for all $j \in \mathbf{N}$, and where infinitely many digits 2 do not occur consecutively. The integral part $t_0$ is represented by $\tau_n \tau_{n-1} \ldots \tau_1 \tau_0$ if and only if $t_0 = \tau_n 3^n + \tau_{n-1} 3^{n-1} + \cdots + \tau_1 3 + \tau_0$, $(\tau_j \in \{0, 1, 2,\})$. The collection of all ternaries is denoted by $\mathbf{T}^*$. Show: If $\mathbf{R}$ is a complete ordered field, then the sets $\mathbf{B}^*$ of exercise 4 and $\mathbf{T}^*$ are equivalent.
6. Prove: If $\mathbf{R}$ is a complete ordered field, then the set $\mathfrak{I} = \{x \in \mathbf{R} \mid 0 \leqslant x \leqslant 1\}$, and the set of decimals that consists of $1.\dot{0}$ and all decimals for which $a_0 = 0$, are equivalent.
7. Same as in exercise 6 for the set of binaries that consist of $1.\dot{0}$ and all binaries for which $b_0 = 0$.
8. Prove: If $\mathbf{R}$ is a complete ordered field. then the set $\mathfrak{J} = \{x \in \mathbf{R} \mid 0 < x < 1\}$, and the set of binaries for which $b_0 = 0$ and at least one $b_j \neq 0$, $j \in \mathbf{N}$, are equivalent.
9. Same as in exercise 8, for ternaries with $t_0 = 0$ and at least one $t_j \neq 0$.

## 1.11 Existence and Uniqueness of a Complete Ordered Field*

We shall demonstrate in this section that the set $\mathbf{R}^*$ of all decimals, with an order relation and field operations properly defined, is a complete ordered field. This will establish the existence of complete ordered fields. We shall then proceed to show that, if $\mathbf{R}$ is any complete ordered field, then the function $f: \mathbf{R}^* \leftrightarrow \mathbf{R}$ of Theorem 10.1 preserves the order relation and the field operations. This means that any complete ordered field is, but for the representation of its elements, identical with $\mathbf{R}^*$.

In section 1.7 we defined an order relation in terms of the field operations. (See definition 7.1.) Here, we shall define an order relation first. After we have also defined the field operations, we shall see that the order relation satisfies definition 7.1.

*See footnote on p. 36.

***Definition 11.1***

$$a_0 \,.\, a_1 a_2 a_3 ... < b_0 \,.\, b_1 b_2 b_3 ...$$

if and only if there is a $k \in \{0, 1, 2, 3, ...\}$ such that

$$a_0 = b_0, a_1 = b_1, ..., a_{k-1} = b_{k-1}, \text{ but } a_k < b_k.$$

$$-a_0 \,.\, a_1 a_2 a_3 ... < -b_0 \,.\, b_1 b_2 b_3 ...$$

if and only if $a_0 \,.\, a_1 a_2 a_3 ... > b_0 \,.\, b_1 b_2 b_3 ...$

$$-a_0 \,.\, a_1 a_2 a_3 ... < b_0 \,.\, b_1 b_2 b_3 ....$$

(For example, $3.1415... < 3.1421...$, $-1.2137... < -1.2136...$, $-0.39 < 1.4\dot{6}$.) It is easily seen that for any $a^*, b^*, c^* \in \mathbf{R}^*$,

(11.1) $\qquad a^* < b^*, b^* < c^* \quad \text{implies} \quad a^* < c^*.$

(See exercise 1.)

On the basis of definition 11.1, we can show:

***Lemma 11.1***

Every subset of $\mathbf{R}^*$ which is bounded above has a least upper bound.

(We advise the reader at this point to check definition 8.2 of least upper bound.)

***Proof***

We shall restrict ourselves to subsets with nonnegative elements. Let

$$A^* = \{a_0^\alpha \,.\, a_1^\alpha a_2^\alpha a_3^\alpha ... \mid \alpha \in J\} \subset \mathbf{R}^*,$$

where $J$ is a given index set, represent a set that is bounded above. Then, there is a $u_0 \,.\, u_1 u_2 u_3 ... \in \mathbf{R}^*$ such that $a_0^\alpha \,.\, a_1^\alpha a_2^\alpha a_3^\alpha ... \leqslant u_0 \,.\, u_1 u_2 u_3 ...$ for all $\alpha \in J$.
Then $A_0^* = \{a_0^\alpha \mid \alpha \in J\}$ is bounded above and, by Theorem 9.4, $a_0 = \sup A_0^* \in A_0^*$. Hence, $J_1 = \{\alpha \in J \mid a_0^\alpha = a_0\}$ is not empty. The set $A_1^* = \{a_1^\alpha \mid \alpha \in J_1\}$ is bounded above by 9 and, again by Theorem 9.4, $a_1 = \sup A_1^* \in A_1^* \subseteq \{0, 1, 2, 3, 4, 5, 6, 7, 8, 9\}$. Hence, $J_2 = \{\alpha \in J_1 \mid a_1^\alpha = a_1\}$ is not empty. Next, we consider $A_2^* = \{a_2^\alpha \mid \alpha \in J_2\}$, find $a_2 = \sup A_2^* \in A_2^* \subseteq \{0, 1, 2, 3, 4, 5, 6, 7, 8, 9\}$, and obtain in turn that $J_3 = \{\alpha \in J_2 \mid a_2^\alpha = a_2\}$ is not empty. We proceed in this manner, arriving at $a_0, a_1, a_2, a_3, ....$

We consider

(11.2) $\qquad a^* = a_0 \,.\, a_1 a_2 a_3 ...,$

where we assume for the moment that infinitely many digits 9 do not occur consecutively. (This could happen, for example, if $A^* = \{9, 9.9, 9.99, 9.999, ...\}$.)

By construction, $a^*$ as defined in (11.2) is an upper bound of $A^*$. Suppose that $b^* = b_0 \,.\, b_1 b_2 b_3 ...$ is also an upper bound of $A^*$ and that $b^* < a^*$. Then there is a $k \in \{0, 1, 2, 3...\}$ such that $b_0 = a_0$, $b_1 = a_1$, ..., $b_{k-1} = a_{k-1}$ but

$b_k < a_k$. By construction of $a^*$, there is an $\alpha \in J_{k+1}$ such that $a_0^\alpha = a_0$, $a_1^\alpha = a_1, \ldots, a_{k-1}^\alpha = a_{k-1}$, $a_k^\alpha = a_k$. Hence, $b^* < a_0^\alpha . a_1^\alpha a_2^\alpha a_3^\alpha \ldots$ and cannot be an upper bound of $A^*$. Hence, $a^*$ is the least upper bound of $A^*$.

If infinitely many digits 9 occur consecutively in (11.2), we replace the digit immediately preceding the chain of nines, say $a_k$, by $a_k + 1$ and show in a similar manner that $a_0 . a_1 a_2 \ldots a_{k-1}(a_k + 1)$ is the least upper bound of $A^*$. We leave the details of this argument to the reader. We also leave it to the reader to prove the theorem for sets that contain negative and nonnegative elements and for sets that contain negative elements only.

We define addition and multiplication of finite decimals in the usual manner and note that these operations are commutative, associative, and that multiplication is distributive over addition. It is easy to see that the function $f$ of Theorem 10.1, when restricted to the set $\mathbf{D}^*$ of finite decimals, preserves the order relation. This means that we can manipulate inequalities of finite decimals according to the rules we are accustomed to.

As a next step, we define addition and multiplication of infinite decimals. We let experience be our guide. If we wish to express $\pi + e$, where $\pi = 3.141592653\ldots$, $e = 2.718281828\ldots$ we realize, of course, that we can never write this down in its entirety but that we can come as close as we please by considering the successive approximations $3 + 2 = 5, 3.1 + 2.7 = 5.8, 3.14 + 2.71 = 5.85, \ldots$. We proceed similarly with multiplication. This suggests the following definitions of addition and multiplication in $\mathbf{R}^*$:

$$a^* + b^* = \sup \{a_0 + b_0,\ (a_0 . a_1 + b_0 . b_1),\ (a_0 . a_1 a_2 + b_0 . b_1 b_2),\ \ldots\}, \tag{11.3}$$

$$a^* b^* = \sup \{a_0 b_0,\ (a_0 . a_1)(b_0 . b_1),\ (a_0 . a_1 a_2)(b_0 . b_1 b_2),\ \ldots\}. \tag{11.4}$$

These least upper bounds exist, because $a_0 . a_1 a_2 \ldots a_k \leqslant a^* < a_0 + 1$, $b_0 . b_1 b_2 \ldots b_k \leqslant b^* < b_0 + 1$ and hence, $a_0 . a_1 a_2 \ldots a_k + b_0 . b_1 b_2 \ldots b_k$ and $(a_0 . a_1 a_2 \ldots a_k) \cdot (b_0 . b_1 b_2 \ldots b_k)$ are bounded above by $a_0 + b_0 + 2$ and $(a_0 + 1)(b_0 + 1)$ respectively.

Although the above definitions are perfectly all right, it is more practical to work with the following equivalent, and only seemingly more general, definitions:

***Definition 11.2***

Let $\mathbf{D}^*$ denote the set of all finite decimals; let $A = \{\alpha^* \in \mathbf{D}^* \mid \alpha^* < a^*\}$, $B = \{\beta^* \in \mathbf{D}^* \mid \beta^* < b^*\}$, and let

$$A \oplus B = \{\alpha^* + \beta^* \mid \alpha^* \in A,\quad \beta^* \in B\},\quad A \odot B = \{\alpha^* \beta^* \mid \alpha^* \in A,\quad \beta^* \in B\}.$$

Then

$$a^* + b^* = \sup A \oplus B,\quad a^* b^* = \sup A \odot B. \tag{11.5}$$

For negative decimals and a mixture of negative and nonnegative decimals, these definitions have to be suitably modified.

We observe that when $a^* > 0.\dot{0}$, $b^* > 0.\dot{0}$, then $a^* + b^* > 0.\dot{0}$ and $a^*b^* > 0.\dot{0}$. Since we also have that a decimal is either positive, or zero, or negative, definition 11.1 of order relation is equivalent to definition 7.1.

Addition and multiplication, as defined above, are clearly commutative because this is true for finite decimals. To show that they are also associative, and that multiplication is distributive over addition, is tedious and time-consuming. The reader will get a taste of it in exercise 8, where the associativity of addition is to be shown. For the rest, we shall take these properties for granted.†

Since $a^* + 0.\dot{0} = \sup\{a_0, a_0 . a_1, a_0 . a_1a_2, \ldots\} = a^*$ and $(a^*)(1.\dot{0}) = \sup\{a_0, a_0 . a_1, a_0 . a_1a_2, \ldots\} = a^*$, we see that $0.\dot{0}$ and $1.\dot{0}$ are the additive and multiplicative identities, respectively, in $\mathbf{R}^*$. The additive inverse of $a_0 . a_1a_2a_3 \ldots$ is clearly $-a_0 . a_1a_2a_3 \ldots$. In order to establish the existence of a multiplicative inverse, one may proceed as follows:

Let $a^* > 0.\dot{0}$ and consider the set of all finite decimals $\beta^*$ such that $a^*\beta^* < 1.\dot{0}$. It is clear that there are such finite decimals. (If $a^* = 19.7396\ldots$, take $\beta^* = 0.01$. Then, $a^*\beta^* = \sup\{0, 0, (19.73)(0.01), (19.739)(0.010), \ldots\} = 0.197396\ldots$). It is easy to see that the set of all such elements $\beta^*$ is bounded above (see exercise 10). Hence,

$$b^* = \sup\{\beta^* \in \mathbf{D}^* \mid a^*\beta^* < 1.\dot{0}\} \tag{11.6}$$

exists. It develops that $a^*b^* = 1.\dot{0}$ and that $b^*$ is the only element with this property. (See exercises 11 and 12.) Hence, $b^*$ is the multiplicative inverse of $a^*$.

Lemma 11.1 and what we have demonstrated and hinted at in the foregoing discussion may now be summarized in the following fundamental theorem:

***Theorem 11.1***

The set $\mathbf{R}^*$ of all decimals with an order relation and field operations as in definitions 11.1 and 11.2, is a *complete ordered field.*

Note that the results of this section are *not* based on the assumption that $\mathbf{R}$ is a complete ordered field. Had they been, our results would be meaningless. Now that we know that there exists a complete ordered field, namely $\mathbf{R}^*$, we can say on the basis of Theorem 10.1 that every complete ordered field is equivalent to $\mathbf{R}^*$ in the sense of set equivalence. Theorem 10.1 does not give any information as to how the algebraic structures in $\mathbf{R}$ and $\mathbf{R}^*$ are related to each other. It turns out that the function $f: \mathbf{R}^* \leftrightarrow \mathbf{R}$ of Theorem 10.1 preserves the order relation and the field operations, as the reader has probably expected all along. We establish these facts in the following two lemmas:

***Lemma 11.2***

If $\mathbf{R}$ is a complete ordered field and if $f: \mathbf{R}^* \leftrightarrow \mathbf{R}$ is the function of Theorem 10.1, then $a^* < b^*$ if and only if $f(a^*) < f(b^*)$.

† See also J. F. Ritt, *Theory of Functions* (New York: King's Crown Press, 1946), pp. 3–8.

*Proof*

(a) We have seen in part (c) of the proof of Theorem 10.1 that, in the light of definition 11.1 of order relation, $a_0 . a_1 a_2 a_3 \ldots < b_0 . b_1 b_2 b_3 \ldots$ implies $f(a_0 . a_1 a_2 a_3 \ldots) < f(b_0 . b_1 b_2 b_3 \ldots)$.

(b) The converse follows from part (a) and the fact that $f$ is one-to-one.

***Lemma 11.3***

If $\mathbf{R}$ is a complete ordered field and if $f: \mathbf{R}^* \leftrightarrow \mathbf{R}$ is the function of Theorem 10.1, then, for any $a^*, b^* \in \mathbf{R}^*$,

$$f(a^*) + f(b^*) = f(c^*) \quad \text{if and only if} \quad a^* + b^* = c^*, \tag{11.7}$$

$$f(a^*)f(b^*) = f(d^*) \quad \text{if and only if} \quad a^*b^* = d^*. \tag{11.8}$$

*Proof*

We shall prove (11.7) only for nonnegative decimals, and leave the rest to the reader. From

$$a_0 + \cdots + \frac{a_k}{10^k} \leqslant a_0 + \cdots + \frac{a_k}{10^k} + \frac{a_{k+1}}{10^{k+1}}$$

and an analogous relation for $b$, we have

$$\sup\left\{a_0 + b_0, a_0 + \frac{a_1}{10} + b_0 + \frac{b_1}{10}, \ldots\right\} = \sup\left\{a_0, a_0 + \frac{a_1}{10}, \ldots\right\} + \sup\left\{b_0, b_0 + \frac{b_1}{10}, \ldots\right\}. \tag{11.9}$$

Since $f$ preserves the order relation, the image in $\mathbf{R}$ of any bounded subset of $\mathbf{R}^*$ is a bounded subset of $\mathbf{R}$, and the least upper bound in $\mathbf{R}$ is the image of the least upper bound in $\mathbf{R}^*$, and vice versa. Hence, if

$$a^* + b^* = \sup\{a_0 + b_0, \quad a_0 . a_1 + b_0 . b_1, \quad \ldots\},$$

then

$$f(a^* + b^*) = \sup\left\{a_0 + b_0, a_0 + \frac{a_1}{10} + b_0 + \frac{b_1}{10}, \ldots\right\}.$$

By (11.9), $f(a^* + b^*) = f(a^*) + f(b^*)$ and (11.7) obtains readily. The converse is shown easily, using again (11.9).

A function $f$ which is one-to-one and onto, and which preserves the order relation and the field operations is called an *order-isomorphism*. Theorems 10.1, 11.1, and Lemmas 11.2 and 11.3 yield:

***Theorem 11.2***

There are real numbers and they are uniquely determined but for an order-isomorphism.

## 1.11 Exercises

1. Using only definition 11.1, show that $a^* < b^*$, $b^* < c^*$ implies $a^* < c^*$, where $a^*, b^*, c^* \in \mathbf{R}^*$.
2. Let $a^*$ denote a nonnegative decimal and let $A$ denote the set of all finite decimals $\alpha^*$ for which $\alpha^* < a^*$. Show that $a^* = \sup A$.
3. Show that the definition of addition in (11.3) and in definition 11.2 are equivalent.
4. Show that the definition of multiplication in (11.4) and in definition 11.2 are equivalent.
5. Let $c^* \geqslant a^*, d^* \geqslant b^*$ where $a^*, b^*, c^*, d^*$ are nonnegative decimals. Show that $c^* + d^* \geqslant a^* + b^*$.
6. Show that the result of exercise 5 is also true if $a^* = b^* = 0.\dot{0}$ and if $a^* = 0.\dot{0}$, $b^* > 0.\dot{0}$.
7. Let $a^* \geqslant b^*$ where $a^*, b^*$ denote nonnegative decimals. Show that $a^*c^* \geqslant b^*c^*$ for $c^* > 0.\dot{0}$.
8. Let $a^*, b^*, c^*$ denote nonnegative decimals. Show that addition is associative, that is, $a^* + (b^* + c^*) = (a^* + b^*) + c^*$.
9. Given $0.\dot{0} < a^* < 1.\dot{0}$. Show that there is a finite decimal $\delta^* > 1.\dot{0}$ such that $a^*\delta^* < 1.\dot{0}$.
10. Let $a^*$ denote a nonnegative decimal. Show that the set $\{\beta^* \in \mathbf{D}^* \mid a^*\beta^* < 1\}$ is bounded above.
11. Let $a^*$ denote a nonnegative decimal and let $b^*$ be as defined in (11.6). Show that $a^*b^* = 1.\dot{0}$.
12. Prove that the multiplicative inverse is unique.
13. Prove that, for nonnegative decimals $a^*, b^*$,

$$f(a^*b^*) = f(a^*)f(b^*),$$

and deduce (11.8) from this result.

## 1.12 The Hierarchy of Numbers

The complete ordered field $\mathbf{R}$ of real numbers contains the set $\mathbf{Q}$ of rational numbers as a *proper* subset, because there is at least one element in $\mathbf{R}$, namely $\sqrt{2}$, which is not in $\mathbf{Q}$. All elements in $\mathbf{R}$ which are not in $\mathbf{Q}$ (that is, which are not rational) are called *irrational.* We denote the set of all irrational numbers by $\mathbf{I}$. $\mathbf{Q}$ contains the set $\mathbf{Z}$ of integers as a proper subset. The set $\mathbf{Q}\setminus\mathbf{Z}$ is the set of *proper fractions* $\{m/n \mid m \in \mathbf{Z}, n \in \mathbf{N}, n \neq 1, (m, n) = 1\}$. We have

$$\mathbf{R} = \mathbf{Q} \cup \mathbf{I} \quad \text{where} \quad \mathbf{Q} \cap \mathbf{I} = \varnothing,$$

$$\mathbf{Q} = \mathbf{Z} \cup (\mathbf{Q}\setminus\mathbf{Z}) \quad \text{where} \quad \mathbf{Z} \cap (\mathbf{Q}\setminus\mathbf{Z}) = \varnothing.$$

The set $\mathbf{I}$ of irrational numbers itself is the union of the set of *algebraic irrational numbers* $\mathbf{I}_a$ and the set of *transcendental numbers* $\mathbf{I}_t$:

***Definition 12.1***

$x_0 \in \mathbf{R}$ is called *algebraic irrational* if and only if there is an integer $n \geqslant 2$ and integers $a_0, a_1, a_2, \ldots, a_n$ such that

$$a_0 x_0^n + a_1 x_0^{n-1} + \cdots + a_{n-1} x_0 + a_n = 0$$

is satisfied but there is *no* linear equation

$$b_0 x + b_1 = 0, \qquad b_0, b_1 \in Z,$$

of which $x_0$ is a solution.

$x_0 \in R$ is called *transcendental* if and only if $x_0 \notin \mathbf{Q} \cup \mathbf{I}_a$, that is, if and only if $x_0$ is irrational but *not* algebraic irrational.

For example, $x_0 = 2$ satisfies the equation

$$x_0^2 - 4 = 0$$

but also satisfies the linear equation

$$x_0 - 2 = 0.$$

Hence, 2 is *not* algebraic irrational. On the other hand, $x_0 = \sqrt{2}$ *is* algebraic irrational because

$$x_0^2 - 2 = 0$$

but $b_0 x + b_1 = 0$ is *not* satisfied by $x_0 = \sqrt{2}$, no matter how $b_0, b_1, b_0 \neq 0$, are chosen. (See also exercise 1, section 1.6.)

That there are transcendental numbers will be shown in a somewhat roundabout way in section 1.13. As a matter of fact, we shall see that there are, in a certain technical sense, more transcendental numbers than any other real numbers (exercise 9, section 1.13), even though not many of these are popularly known. Let us just mention here that $e$ is a transcendental number as *Charles Hermite*† showed in 1873, and that $\pi$ is also a transcendental number as *C. L. F. Lindeman*‡ showed in 1882. Also, that the Hilbert number $\sqrt{2}^{\sqrt{2}}$ is transcendental follows from a theorem by *A. Gelfond* and *Th. Schneider* that was proved in 1934; there are many others. (Incidentally, there are numbers of which it is not even known as yet whether they are rational or irrational. A case in point is Euler's constant

$$\gamma = \lim_{n \to \infty} \left( 1 + \frac{1}{2} + \cdots + \frac{1}{n} - \log n \right) = 0.57721566 \ldots).$$

† *Charles Hermite*, 1822–1901.
‡ *C. L. F. Lindeman*, 1852–1939.

The following diagram represents a summary of our discussion:

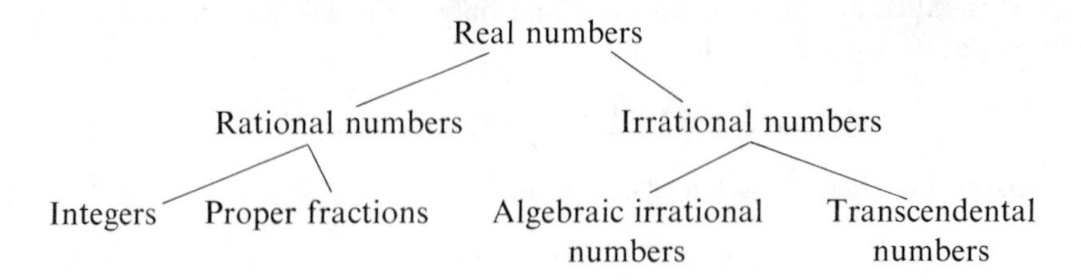


## 1.12 Exercises

1. Let $p > 0$ denote a prime number. Show that $\sqrt{p}$ is algebraic irrational.
2. A number in **R** is called *algebraic* if it is either rational or algebraic irrational. Prove: A number is algebraic if and only if it satisfies some algebraic equation $a_0 x^n + a_1 x^{n-1} + \cdots + a_{n-1}x + a_n = 0$ with $n \geqslant 1$ and $a_0, a_1, \ldots, a_n \in \mathbf{Z}$.
3. A theorem by Gelfond and Schneider states: If $\alpha, \beta$ are algebraic numbers, where $\alpha \neq 0$, $\alpha \neq 1$ and $\beta \notin \mathbf{Q}$, then $\alpha^\beta$ is a transcendental number. Deduce from this theorem that $\log_{10} 2$ is either rational or transcendental.

## 1.13 Cardinal Numbers

> Principle of the Conservation of Ignorance: "A false notion, once arrived at, is not easily dislodged." (Georg Cantor)

In this section, we shall introduce Cantor's concept of the cardinality of sets, which is indispensable to modern analysis. This concept will enable us to compare infinite sets with each other, and make meaningful statements about certain infinite sets being larger than certain other infinite sets. Upon superficial examination, one would think that there are more rational numbers than there are integers, that there are more integers than positive integers, more positive integers than even positive integers, etc. These are some of the false notions which we shall try to dislodge.

The cardinal number of a given finite set is the number of elements that are contained in the set.

If $A = \{1, 2, 3\}$, then, the cardinal number $\overline{\overline{A}}$ of $A$ is given by $\overline{\overline{A}} = 3$.

If two finite sets $A, B$ have the same cardinal number (i.e., the same number of elements), then there exists a function $f: A \leftrightarrow B$, and $A, B$ are called equivalent. (See definition 5.3 and example 4, section 1.5.) Two finite sets are equivalent if and only if they have the same cardinal number. (See exercise 6, section 1.5.)

If $A = \{1, 2, 3\}$, $B = \{\frac{1}{2}, \frac{1}{4}, \frac{1}{8}\}$, then $\overline{\overline{A}} = \overline{\overline{B}}$, and a function $f: A \leftrightarrow B$ is given by $f(x) = 1/2^x$.

If one wishes to compare infinite sets in regard to their size, counting elements is clearly out of the question. The above discussion suggests the following definition:

***Definition 13.1***

The two sets $A, B$ (finite or infinite, no matter) have the same cardinal number,

$$\bar{\bar{A}} = \bar{\bar{B}}$$

if and only if $A$ and $B$ are equivalent, i.e., if and only if there exists a function $f: A \leftrightarrow B$. (See also definition 5.3.) (For finite sets, this definition is a theorem; see exercise 6, section 1.5.)

In view of this definition we can now give precise meaning to the terms finite set and infinite set: A set is *finite* if and only if it has cardinal number $n$ for some $n \in \mathbf{N}$, that is, if and only if it is equivalent to $\{1, 2, 3, \dots, n\}$ for some $n \in \mathbf{N}$. A set is *infinite* if and only if it is not finite.

Assuming that every living person has a head (which is sometimes doubtful), and that no living person has more than one head (which is a pity), we can say that the set of living persons and the set of all heads have the same cardinality, without having to count either.

A cardinal number that is not finite is called a *transfinite cardinal number.*

The simplest and most common infinite set that comes to mind is the set $\mathbf{N}$ of all positive integers. We assign to this set the transfinite cardinal number $\aleph_0$:

$$\bar{\bar{\mathbf{N}}} = \aleph_0.$$

(Read: aleph nought.) $\aleph$ is the first, or last, letter in the Hebrew alphabet, depending on whether you read from right to left or from left to right.

If $\mathbf{E} = \{2, 4, 6, \dots\}$ is the set of all even positive integers, then we see, with some surprise, that

$$\mathbf{E} \sim \mathbf{N}, \qquad \text{that is,} \qquad \bar{\bar{\mathbf{E}}} = \bar{\bar{\mathbf{N}}}.$$

The one-to-one correspondence is given by the function $f: \mathbf{N} \leftrightarrow \mathbf{E}$, where $f(x) = 2x$. (Naively, one would expect $\mathbf{E}$ to have only half as many elements as $\mathbf{N}$. That this is not so is part of the mysteries of the infinite and the human mind.)

Similarly, we see that

$$\mathbf{Z} \sim \mathbf{N}, \qquad \text{that is,} \qquad \bar{\bar{\mathbf{Z}}} = \bar{\bar{\mathbf{N}}},$$

because the function $f: \mathbf{Z} \to \mathbf{N}$, given by

$$f(x) = \left\{ \begin{array}{ll} 2x & \text{if } x > 0, \\ -2x + 1 & \text{if } x \leqslant 0, \end{array} \right\},$$

is one-to-one and onto.

This is getting exasperating. Aren't there any sets with cardinality other than $\aleph_0$? No infinite set that is equivalent to a subset of a set of cardinality $\aleph_0$ can have cardinality other than $\aleph_0$:

***Theorem 13.1***

If $A$ is an infinite set and if $A \sim B \subseteq \mathbf{N}$, then $\overline{\overline{A}} = \aleph_0$.

(Note that we have written $\mathbf{N}$ instead of a general set of cardinality $\aleph_0$. We may do this w.l.o.g. because, by definition, all sets of cardinality $\aleph_0$ are equivalent to $\mathbf{N}$.)

*Proof*

We check off the elements in $\mathbf{N} = \{1, 2, 3, \ldots\}$ until we reach the first element that is in $B$, say $n_1$. We proceed in this manner until we reach the next element that is in $B$, say $n_2$, etc.... This process never breaks up because $B$ is equivalent to $A$ and $A$ is infinite. (No infinite set can be equivalent to a finite set. See exercise 1.) We arrive at the set

$$B = \{n_1, n_2, n_3, \ldots\},$$

and see that $B \sim \mathbf{N}$ since $f: \mathbf{N} \to B$, given by $f(x) = n_x$, is one-to-one and onto. Hence, $\overline{\overline{B}} = \aleph_0$ and since $A \sim B$, also $\overline{\overline{A}} = \aleph_0$.

We have for finite sets $A, B$ that $\overline{\overline{A}} \leqslant \overline{\overline{B}}$ if and only if $A \sim C \subseteq B$. (See exercise 2.) In view of this, we give the following definition:

***Definition 13.2***

$\overline{\overline{A}} \leqslant \overline{\overline{B}}$ if and only if $A \sim C \subseteq B$. $\overline{\overline{A}} < \overline{\overline{B}}$ if and only if $\overline{\overline{A}} \leqslant \overline{\overline{B}}$ and $\overline{\overline{A}} \neq \overline{\overline{B}}$.

In view of this definition, and because every infinite set contains a set of cardinality $\aleph_0$ as a subset, we have:

***Corollary to Theorem 13.1***

$\aleph_0$ is the smallest transfinite cardinal number.

*Proof*

The only subsets of $\mathbf{N}$ that do not have cardinality $\aleph_0$ are, by Theorem 13.1, the finite sets.

In the light of this result we have to search for infinite sets with cardinality other than $\aleph_0$ among sets which contain a proper subset that is equivalent to $\mathbf{N}$. $\mathbf{Z}$, as we have seen, won't do, because $\overline{\overline{\mathbf{Z}}} = \aleph_0$. Hence, we have to look among sets that contain $\mathbf{Z}$ as a proper subset. $\mathbf{Q}$ comes to mind. We shall return to this set after some preliminary investigations.

In order to simplify and streamline the terminology, we give the following definition:

***Definition 13.3***

A set is called *denumerable* if and only if it has cardinality $\aleph_0$.

***Theorem 13.2***

The union $S = \bigcup_{k=1}^{\infty} S_k$ of denumerably many denumerable, mutually disjoint sets $S_1, S_2, S_3, \ldots$ is again denumerable:

If $\overline{\overline{S}}_k = \aleph_0$ for all $k \in \mathbf{N}$ and if $S_i \cap S_j = \varnothing$ for $i \neq j$, then $\overline{\overline{S}} = \overline{\overline{\bigcup_{k=1}^{\infty} S_k}} = \aleph_0$.

*Proof*

We denote the elements of $S_k$ by

$$S_k = \{a_1^{(k)}, a_2^{(k)}, a_3^{(k)}, \ldots\}.$$

(We can do this because, by hypothesis, $S_k \sim \mathbf{N}$.) All element of $S = \bigcup_{k=1}^{\infty} S_k$ and only such elements are contained in the following scheme:

$$\begin{array}{ccccc} a_1^{(1)} & a_2^{(1)} & a_3^{(1)} & \cdots \\ a_1^{(2)} & a_2^{(2)} & a_3^{(2)} & \cdots \\ a_1^{(3)} & a_2^{(3)} & a_3^{(3)} & \cdots \\ \vdots & & & \end{array}$$

($a_i^{(k)}$ is contained in the $k$th row and $i$th column.) We demonstrate the denumerability of this set by arranging all its elements in a row, as indicated by arrows in the above scheme:

$$\{a_1^{(1)}, a_1^{(2)}, a_2^{(1)}, a_1^{(3)}, a_2^{(2)}, a_3^{(1)}, a_1^{(4)}, \ldots\}.$$

There is a first, second, third, etc... element in this arrangement and all elements from $S$ are present. Hence, $S$ is denumerable, that is, $\overline{\overline{S}} = \aleph_0$. (The reader may convince himself that this arrangement defines the function $f: S \leftrightarrow \mathbf{N}$ which is given by

$$f(a_i^{(k)}) = \frac{(k+i-2)(k+i-1)}{2} + i.$$

See also exercise 12 where another one-to-one function from $S$ onto $\mathbf{N}$ is discussed.)

***Corollary to Theorem 13.2***

$$\overline{\overline{\mathbf{Q}}} = \aleph_0.$$

*Proof*

We have $\mathbf{Q} = \{m/n \mid m \in \mathbf{Z}, n \in \mathbf{N}, (m,n) = 1\} \subset \{m/n \mid m \in \mathbf{Z}, n \in \mathbf{N}\}$. The latter set is the set of all formal fractions. Since $\mathbf{Z}$ is denumerable, we recognize the set of all formal fractions as the union of denumerably many denumerable sets. (There are denumerably many numerators and with every numerator there go denumerably many denominators.) By Theorem 13.2, such a set has cardinality $\aleph_0$. Since $\mathbf{Q}$ is an infinite set and is equivalent to a subset of the set of formal fractions, we have from Theorem 13.1 that $\overline{\overline{\mathbf{Q}}} = \aleph_0$.

We are in for still another disappointment if we consider the set of algebraic irrational numbers.

***Theorem 13.3***

The set of all algebraic irrational numbers has cardinality $\aleph_0$.

*Proof*

Let $\mathbf{I}_a$ denote the set of all algebraic irrational numbers. First, we realize that $\mathbf{I}_a$ is infinite because there are infinitely many primes and $\{\sqrt{2}, \sqrt{3}, \sqrt{5}, \sqrt{7}, \sqrt{11}, \sqrt{13}, \ldots\} \subset \mathbf{I}_a$. Next, we shall demonstrate that $\mathbf{I}_a \sim B \subseteq \mathbf{N}$. It will then follow from Theorem 13.1 that $\overline{\overline{\mathbf{I}}}_a = \aleph_0$.

We consider the set of *all algebraic equations*

$$(13.1) \qquad a_0 x^n + a_1 x^{n-1} + \cdots + a_{n-1}x + a_n = 0, a_i \in \mathbf{Z}, n = 2, 3, 4, \ldots.$$

We call $h = n + |a_0| + |a_1| + \cdots + |a_n|$ the *height of the algebraic equation* (13.1). There are *no* algebraic equations with $n \geqslant 2$ of height 1 and 2. There are two algebraic equations of height 3, namely $x^2 = 0$ and $-x^2 = 0$; there are six algebraic equations of height 4, namely

$$\begin{array}{ccc} x^2 + 1 = 0, & x^2 - 1 = 0, & -x^2 + 1 = 0, \\ -x^2 - 1 = 0, & 2x^2 = 0, & -2x^2 = 0. \end{array}$$

The point is that for every given height, there are only *finitely* many algebraic equations. Each algebraic equation has, in turn, only finitely many real roots. Hence, for every given height $h \in \mathbf{N}, h \geqslant 3$, we have finitely many roots $x_1^{(h)}, x_2^{(h)}, \ldots x_{n_h}^{(h)}$. Hence, all roots of all algebraic equations of order $\geqslant 2$ are contained in the scheme

$$\begin{array}{cccc} x_1^{(3)}, & x_2^{(3)}, & \ldots, & x_{n3}^{(3)} \\ x_1^{(4)}, & x_2^{(4)}, & \ldots, & x_{n4}^{(4)} \\ x_1^{(5)}, & x_2^{(5)}, & \ldots, & x_{n5}^{(5)} \\ \vdots & & & \end{array}$$

The set of all these roots is denumerable:

$$x_1^{(3)}, \; x_2^{(3)}, \; \ldots, \; x_{n_3}^{(3)}, \; x_1^{(4)}, \; x_2^{(4)}, \; \ldots, \; x_{n4}^{(4)}, \; x_1^{(5)}, \; x_2^{(5)}, \; \ldots, \; x_{n5}^{(5)}, \; x_1^{(6)}, \; \ldots$$

because $f: \{x_i^{(k)}\} \to \mathbf{N}$, as given by

$$f(x_i^{(k)}) = n_3 + n_4 + \cdots + n_{k-1} + i,$$

where $k = 3, 4, 5 \ldots, \quad i = 1, 2, \ldots, n_k, \; n_2 = 0$, is onto and one-to-one. Since $\mathbf{I}_a$ is an infinite subset of the set of all roots of all algebraic equations of order $\geqslant 2$ (see definition 12.1), we have from Theorem 13.1 that $\overline{\overline{\mathbf{I}}}_a = \aleph_0$.

As a last desperate effort, we investigate $\mathbf{R}$ itself. Here we strike pay dirt:

***Theorem 13.4***

$$\overline{\overline{\mathbf{R}}} > \aleph_0.$$

*Proof*

By Theorem 10.1 there is a one-to-one correspondence between $\mathbf{R}$ and the set of all decimals $\mathbf{R}^*$: $\mathbf{R} \sim \mathbf{R}^*$. Hence, it will suffice to show that $\overline{\overline{\mathbf{R}^*}} > \aleph_0$. We assume, to the contrary, that $\mathbf{R}^*$ is denumerable, i.e., there is a scheme

$$\begin{array}{l} a_1 \,.\, a_{11}\, a_{12}\, a_{13}\, a_{14} \cdots \\ a_2 \,.\, a_{21}\, a_{22}\, a_{23}\, a_{24} \cdots \\ a_3 \,.\, a_{31}\, a_{32}\, a_{33}\, a_{34} \cdots \\ a_4 \,.\, a_{41}\, a_{42}\, a_{43}\, a_{44} \cdots \\ \vdots \end{array} \tag{13.2}$$

where $a_i \in \mathbf{Z}$, $a_{ik} \in \{0, 1, 2, 3, 4, 5, 6, 7, 8, 9\}$, which contains *all* elements of $\mathbf{R}^*$. We consider $0 \,.\, b_1 b_2 b_3 b_4 \ldots \in \mathbf{R}^*$ where

$$b_k = \begin{cases} 0 \text{ if } a_{kk} \neq 0 \\ 1 \text{ if } a_{kk} = 0 \end{cases}$$

and note that

$$0 \,.\, b_1 b_2 b_3 b_4 \cdots \neq a_k \,.\, a_{k1} a_{k2} a_{k3} a_{k4} \cdots$$

for all $k \in \mathbf{N}$ since $b_k \neq a_{kk}$. Hence, the scheme in (13.2) does *not* contain all elements from $\mathbf{R}^*$ and we have arrived at a contradiction to our assumption that $\overline{\overline{\mathbf{R}^*}} = \aleph_0$. Hence, $\overline{\overline{\mathbf{R}^*}} \neq \aleph_0$ and, since $\mathbf{N} \subset \mathbf{R}^*$, we have, from definition 13.2, that $\overline{\overline{\mathbf{R}}} = \overline{\overline{\mathbf{R}^*}} > \aleph_0$.

***Definition 13.4***

The cardinal number of $\mathbf{R}$ is called $\aleph : \overline{\overline{\mathbf{R}}} = \aleph$. ($\aleph$ is often referred to as the *cardinality of the continuum*, $\mathbf{R}$ being the continuum.)

We have, finally, succeeded in discovering a set of cardinality greater than $\aleph_0$. Since $\overline{\overline{\mathbf{Q}}} = \aleph_0$, $\overline{\overline{\mathbf{I}_a}} = \aleph_0$, we have $\overline{\overline{\mathbf{Q} \cup \mathbf{I}_a}} = \aleph_0$. However, $\overline{\overline{\mathbf{R}}} = \aleph > \aleph_0$. Hence:

***Theorem 13.5***

There are transcendental numbers.

In exercise 9 the reader is asked to show that the cardinality of the set of transcendental numbers is actually $\aleph$, i.e., that there are "as many" transcendental numbers as there are real numbers. Other sets of cardinality $\aleph$ are also discussed in the exercises (exercise 4, 5, 6). In exercises 13–16 we develop a method for the construction of sets of increasingly higher cardinality.

## 1.13 Exercises

1. Show: If $A$ is finite and $B$ is infinite, then $A \sim B$ is not possible.
2. Show: If $A, B$ are finite sets, then $\bar{\bar{A}} \leqslant \bar{\bar{B}}$ if and only if $A \sim C \subseteq B$.
3. Show: A set $A$ is infinite if and only if it is equivalent to one of its proper subsets.
4. Show that $(0, 1) = \{x \in \mathbf{R} \mid 0 < x < 1\}$ has cardinality $\aleph$.
5. Show that $(a, b) \sim (0, 1) \sim \mathbf{R}$.
6. Show that $[0, 1] \sim [0, 1)$.
7. Show: If $C \subset A$, $A \cap B = \varnothing$, and $\bar{\bar{B}} = \bar{\bar{C}} = \aleph_0$, then $\overline{\overline{A \cup B}} = \bar{\bar{A}}$.
8. Show: If $t$ is a transcendental number, then $t^n$ is also a transcendental number for every $n \in \mathbf{N}$.
9. Show: $\bar{\bar{\mathbf{I}}}_t = \aleph$, where $\mathbf{I}_t$ denotes the set of transcendental numbers.
10. The set $\mathbf{C}$ of complex numbers is defined by

$$\mathbf{C} = \{x + iy \mid x, y \in \mathbf{R}, i^2 = -1\}.$$

Show that

$$\mathbf{C}_a = \{z \mid a_0 z^n + a_1 z^{n-1} + \cdots + a_{n-1} z + a_n = 0, \quad a_i \in \mathbf{Z}, n \in \mathbf{N}\}$$

is a proper subset of $\mathbf{C}$.

11. Show: If $t \in \mathbf{I}_t$, then $(1/t) \in \mathbf{I}_t$ and if $a \in \mathbf{I}_a$, then $(1/a) \in \mathbf{I}_a$.
12. Show that $f(a_i^{(k)}) = (2j - 1)2^{k-1}$ is one-to-one from $S$ onto $\mathbf{N}$, where $S$ is defined in Theorem 13.2.
13. The set of all subsets (including the empty set) of $A_n = \{a_1, a_2, \ldots, a_n\}$ is called the power set of $A_n$ and is denoted by $2^{A_n}$. Let $A_3 = \{a_1, a_2, a_3\}$ and find $2^{A_3}$.
14. A subset, say $\{a_1, a_2, a_5\}$, of $\{a_1, a_2, a_3, a_4, a_5, a_6\}$ may be characterized by $\{1, 1, 0, 0, 1, 0\}$, where 0 in the $j$th place signifies that $a_j$ is not in the set and where 1 in the $k$th place signifies that $a_k$ is in it. Use this device to show that

$$\overline{\overline{2^{A_n}}} = 2^n,$$

where $A_n = \{a_1, a_2, \ldots, a_n\}$.

15. Let $A$ denote a set (finite or infinite) and show that there is a set $B$ such that $A \sim B \subseteq 2^A$, where $2^A$ is the power set (set of all subsets) of $A$.
16. Suppose that $A \sim 2^A$. Then there is a function $f: A \leftrightarrow 2^A$ with values $f(x) = S_x \in 2^A$ for $x \in A$. Consider

$$T = \{x \in A \mid x \notin S_x\} \in 2^A.$$

Then $T = S_{x_1}$ for some $x_1 \in A$. Show that if $x_1 \in T$, then $x_1 \notin T$, and if $x_1 \notin T$, then $x_1 \in T$. Deduce from this result and exercise 15 that $\overline{\overline{2^A}} > \bar{\bar{A}}$.

17. Use the technique of exercise 14 to show that $2^{\mathbf{N}}$ is equivalent to the set of all sequences of 0's and 1's.
18. The set of all formal binaries $0 . a_1 a_2 a_3 \ldots$, $a_j \in \{0, 1\}$ is the union of the set of all formal binaries where infinitely many 1's do not occur consecu-

tively and the set of all formal binaries where infinitely many 1's do occur consecutively. These two sets are disjoint. The one is equivalent to $\mathbf{R}$ and the other to $\mathbf{N}$. (Why?) Using the fact that $\mathbf{R} \sim \mathbf{I}_t$ (see exercise 9) and that $\mathbf{N} \sim \mathbf{I}_a \cup \mathbf{Q}$ (see Theorems 13.2 and 13.3), derive, from the result in exercise 17, that $2^{\mathbf{N}} \sim \mathbf{R}$, that is, $\overline{\overline{2^{\mathbf{N}}}} = \aleph$.

19. Use the result of exercise 16 to show that there are at least denumerably many cardinal numbers.
20. Let $C$ denote the set of all cardinal numbers. For each $\lambda \in C$ pick a set $S_\lambda$ with cardinality $\lambda$, and consider $P = \bigcup_{\lambda \in C} S_\lambda$. Show that $P$ cannot be a set.
21. Use the result of exercise 16 to show that the set of all sets cannot be a set.
22. Let $F$ denote the set of all functions $f: \mathbf{R} \to \mathbf{R}$. Show that $\overline{\overline{F}} > \aleph$.

## 1.14 The Cantor Set*

In this section we shall discuss a most remarkable set. Its most distinguished feature is the fact that it has as many elements as there are real numbers but that its representative points occupy, in the geometric sense, *no* space on the number line. This in itself is peculiar enough to whet one's appetite for a closer scrutiny of this set. As a fringe benefit, such a study raises an almost ulimited number of rewarding problems that are suitable for exercises.

Let us consider the following sequence of sets, each of which is the union of open intervals:

$$\begin{aligned} D_0 &= \varnothing, \\ D_1 &= (\tfrac{1}{3}, \tfrac{2}{3}), \\ D_2 &= (\tfrac{1}{9}, \tfrac{2}{9}) \cup (\tfrac{7}{9}, \tfrac{8}{9}), \\ D_3 &= (\tfrac{1}{27}, \tfrac{2}{27}) \cup (\tfrac{7}{27}, \tfrac{8}{27}) \cup (\tfrac{19}{27}, \tfrac{20}{27}) \cup (\tfrac{25}{27}, \tfrac{26}{27}), \\ &\vdots \end{aligned} \tag{14.1}$$

(See Fig. 14.1.)

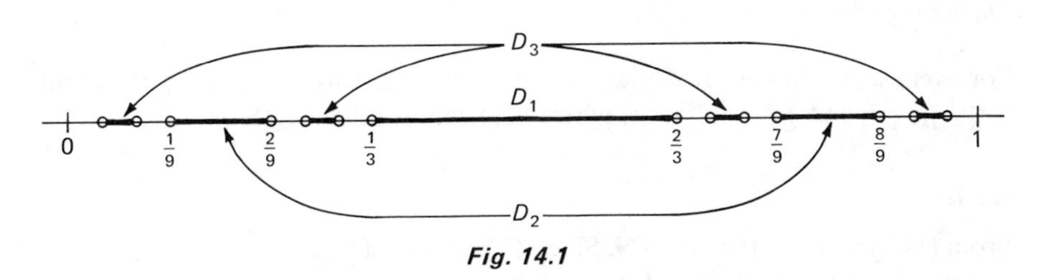


*Fig. 14.1*

We see that $D_n$ is the union of open intervals, each of which is the middle third of the intervals that are left after the points of $D_{n-1}$ have been removed from $\mathfrak{J} = [0, 1]$.

***Definition 14.1***

The set

$$C = \left\{x \in \mathfrak{J} \mid x \notin \bigcup_{k=0}^{\infty} D_k\right\} = [0, 1] \Big| \bigcup_{k=0}^{\infty} D_k$$

is called the *Cantor set*. (This set is often referred to as the "*set of the excluded middle thirds*"—for obvious reasons.)

If we identify $[0, 1]$ with the universal set, we may also write

$$C = c\left(\bigcup_{k=0}^{\infty} D_k\right).$$

By DeMorgan's law (Theorem 2.1), we have

$$c\left(\bigcup_{k=0}^{\infty} D_k\right) = \bigcap_{k=0}^{\infty} c(D_k).$$

If we denote $c(D_k) = C_k$, where

$C_0 = [0, 1]$,
$C_1 = [0, \frac{1}{3}] \cup [\frac{2}{3}, 1]$,
$C_2 = [0, \frac{1}{9}] \cup [\frac{2}{9}, \frac{1}{3}] \cup [\frac{2}{3}, \frac{7}{9}] \cup [\frac{8}{9}, 1]$,
$C_3 = [0, \frac{1}{27}] \cup [\frac{2}{27}, \frac{1}{9}] \cup [\frac{2}{9}, \frac{7}{27}] \cup [\frac{8}{27}, \frac{1}{3}] \cup [\frac{2}{3}, \frac{19}{27}] \cup [\frac{20}{27}, \frac{7}{9}] \cup [\frac{8}{9}, \frac{25}{27}] \cup [\frac{26}{27}, 1]$,
$\vdots$

we have the following representation of the Cantor set:

$$C = \bigcap_{k=0}^{\infty} C_k. \tag{14.2}$$

In a geometric sense, the Cantor set does not occupy any space in $[0, 1]$:

***Theorem 14.1***

For every $\varepsilon > 0$, there are finitely many closed intervals $\mathfrak{J}_1, \mathfrak{J}_2, \ldots, \mathfrak{J}_n$ such that $C \subseteq \bigcup_{k=1}^{n} \mathfrak{J}_k$ but $\sum_{k=1}^{n} l(\mathfrak{J}_k) < \varepsilon$, where $l(\mathfrak{J}_k)$ denotes the length of $\mathfrak{J}_k$.

*Proof*

From (14.2), $C \subset C_j$ for all $j \in \mathbf{N}$. Since $C_j$ consists of $2^j$ closed, nonoverlapping intervals $\mathfrak{J}_k$ of length $l(\mathfrak{J}_k) = (1/3^j)$ each, we have:

$$C \subset \bigcup_{k=1}^{2^j} \mathfrak{J}_k \text{ where } \sum_{k=1}^{2^j} l(\mathfrak{J}_k) = \sum_{k=1}^{2^j} \frac{1}{3^j} = \left(\frac{2}{3}\right)^j < \varepsilon$$

for sufficiently large $j$. Let $n = 2^j$, and our theorem is proved.

At this point, the reader might begin to doubt that $C$ contains any points at all. This is not so: $0, 1, \frac{1}{3}, \frac{2}{3}, \frac{1}{9}, \frac{2}{9}, \ldots$ are all contained in $C$. As a matter of fact, there are as many points in $C$ as there are in $\mathbf{R}$.

***Theorem 14.2***

$$\bar{\bar{C}} = \aleph.$$

*Proof*

We represent each number in $[0, 1]$ as a ternary $0.a_1a_2a_3\ldots$, where $a_i \in \{0, 1, 2\}$ and where 1 is to be represented by $0.\dot{2}$. If $x = 1/3^k$ for some $k \in \mathbf{N}$, then the ternary representation of $x$ is finite. If the last nonvanishing digit is 1, we replace it by 0 and all following 0's by 2's (for example, $1 = 0.\dot{2}$, $\frac{1}{3} = 0.0\dot{2}$, ...) and if the last nonvanishing digit is 2, we leave it alone (for example, $\frac{2}{3} = 0.2$, $\frac{2}{9} = 0.02$, ...). Of course, we leave 0 as it is, namely, $0 = 0.\dot{0}$. By this process we achieve a unique representation of all reals in $[0, 1]$ by a ternary that either terminates with a 2 or is infinite, and every such ternary represents a real number in $[0, 1]$.

By definition, $C_1 = [0, 1] \backslash D_1$; that is, to obtain the elements of $C_1$, we remove from $[0, 1]$ all numbers whose ternary representation starts with 0.1. ($\frac{1}{3}$ is *not* removed because $\frac{1}{3} = 0.0\dot{2}$, by our agreement.) Similarly, since $C_2 = [0, 1] \backslash (D_1 \cup D_2)$, we obtain the numbers in $C_2$ by removing from what is left all numbers whose ternary representation starts with 0.01 or 0.21. We proceed in this manner, and arrive at the conclusion that the elements of $C$ have a ternary representation that contains the digits 0 and 2 only, and that every number so represented is a member of the Cantor set. (See exercise 3.) Hence, the Cantor set is equivalent to the set of all sequences of 0's and 2's, and this set, in turn, is equivalent to the set of all sequences of 0's and 1's. The latter set has cardinality $\aleph$ (see exercise 18, section 1.13) and our theorem is proved.

## 1.14 Exercises

1. Show: The Cantor set does *not* contain any open interval $(a, b)$.
2. Show by induction: $C_k$ consists of $2^k$ open intervals of length $(1/3^k)$ each.
3. Show: If $x$ is represented by $0.a_1a_2a_3\ldots$, where $a_i \in \{0, 2\}$, then $x \in C$.

# Functions 2

## 2.15 Limit of a Sequence

A function $f: \mathbf{N} \to \mathbf{R}$ is called a *sequence*, or, more precisely, an *infinite sequence of real numbers*. It is customary to denote an infinite sequence by

$$\{a_1, a_2, a_3, \ldots\}, \qquad \text{or} \qquad \{a_k\}_1^\infty, \qquad \text{or simply by } \{a_k\},$$

rather than by "$f: \mathbf{N} \to \mathbf{R}$ where $f$ is given by $f(k) = a_k$". Examples of sequences abound in mathematics:

$$\{k\} = \{1, 2, 3, \ldots\},$$

$$\left\{\frac{1}{k}\right\} = \left\{1, \frac{1}{2}, \frac{1}{3}, \ldots\right\},$$

$$\{1^k\} = \{1, 1, 1, \ldots\},$$

$$\{(-1)^k\} = \{-1, 1, -1, \ldots\},$$

$$\{2k - 1\} = \{1, 3, 5, \ldots\},$$

$$\left\{\frac{1}{k^2}\right\} = \left\{1, \frac{1}{4}, \frac{1}{9}, \ldots\right\},$$

all of these are examples of sequences. When all but finitely many elements in $\{a_k\}$ are equal, we call $\{a_k\}$ a *constant sequence*: $\{a_k\}$ is a *constant sequence* if and only if there is an $n \in \mathbf{N}$ such that $a_{n+1} = a_{n+2} = a_{n+3} = \cdots$.

The sequences which are of special interest and upon which we shall focus our attention are the convergent sequences:

***Definition 15.1***

The sequence $\{a_k\}$ *converges to the limit* $a \in \mathbf{R}$:

$$\lim_{k \to \infty} a_k = a \qquad \text{or} \qquad \lim_{\infty} a_k = a,$$

if and only if for every $\varepsilon > 0$ there is an $N(\varepsilon) \in \mathbf{N}$ such that

$$|a_k - a| < \varepsilon \qquad \text{for all } k > N(\varepsilon).$$

A sequence that converges to a limit is called a *convergent sequence*. (See also Fig. 15.1.)

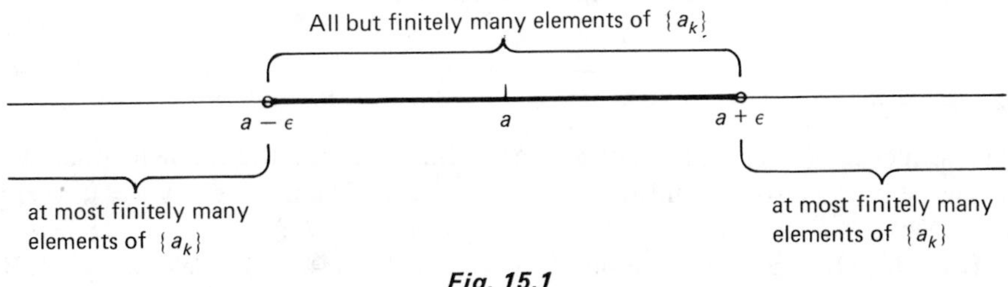


***Fig. 15.1***

---

▲ *Example 1*

The sequence $\{1/k\}$ converges to the limit $a = 0$ $(\lim_{\infty}(1/k) = 0)$: We have, for every $\varepsilon > 0$, that

$$\left|\frac{1}{k} - 0\right| = \frac{1}{k} < \varepsilon \qquad \text{for all } k > \left[\frac{1}{\varepsilon}\right] = N(\varepsilon).$$

($[x]$ denotes the smallest integer that is greater than or equal to $x$.)

▲ *Example 2*

The sequence $\{1^k\}$ converges to the limit $a = 1$ $(\lim_{\infty} 1^k = 1)$: We have, for every $\varepsilon > 0$, that $|1^k - 1| = 0 < \varepsilon$ for all $k \geqslant 1 = N(\varepsilon)$. This is a *constant sequence*. *A constant sequence always converges.*

▲ *Example 3*

The sequence $\{(-1)^k/k^2\}$ converges to the limit $a = 0$: We have, for every $\varepsilon > 0$, that $|(-1)^k/k^2 - 0| = 1/k^2 < \varepsilon$ for all $k > N(\varepsilon) = [1/\sqrt{\varepsilon}]$.

---

Definition 15.1 of *convergence* and *limit* is due to *Karl Weierstrass*.† This definition establishes limit and convergence *conceptually* but is not a *practical* test for convergence as it requires the *a priori* knowledge of the limit. If the limit of a sequence is already known, then one need not test the convergence any more, as the existence of the limit implies convergence of the sequence. Still, this definition has some limited practical use in cases where one can make an educated guess as to what the limit might be, as we have done in examples 1, 2, and 3, or for the computation of limits of sequences in terms of the known limits of simpler sequences, as we shall demonstrate in section 2.16.

In section 5.62 we shall develop a criterion (necessary and sufficient condition) for the convergence of a sequence that does not require any knowledge of a

†*Karl Weierstrass*, 1815–1897.

limit but is purely based on the behavior of the elements of the sequence as $k$ becomes large.

One may use definition 15.1 to demonstrate the nonconvergence of a sequence, as we shall see from the following two examples:

---

▲ *Example 4*

The sequence $\{(-1)^k\}$ does not converge. Suppose, for the moment, that the sequence does converge and that $\lim_\infty (-1)^k = a$. Then, for every $\varepsilon > 0$, there is an $N(\varepsilon) \in \mathbf{N}$ such that $|(-1)^k - a| < \varepsilon$ for all $k > N(\varepsilon)$.

Let $\varepsilon = \frac{1}{2}$. If $a \geqslant 0$, then for odd $k$, $|(-1)^k - a| = 1 + a < \frac{1}{2}$ yields $a < -\frac{1}{2}$. If $a < 0$, then for even $k$, $|(-1)^k - a| = 1 - a < \frac{1}{2}$, yields $a > \frac{1}{2}$. Hence, there is no $N(\frac{1}{2})$ with the required property, and therefore, $\{(-1)^k\}$ does not converge.

▲ *Example 5*

The sequence $\{k\}$ does *not* converge because, no matter what $a$ we choose, $|k - a| \geqslant k - |a| > 1$ for all $k > [|a|]$.

---

A sequence that does not converge is said to *diverge*. A nonconvergent sequence is called a *divergent sequence*. Specifically:

***Definition 15.2***

The sequence $\{a_k\}$ diverges to $\infty$ $(-\infty)$, $\lim_\infty a_k = \infty$ $(-\infty)$, if and only if, for all $A > 0$, there is an $N(A) \in \mathbf{N}$ such that

$$a_k > A (a_k < -A) \qquad \text{for all } k > N(A).$$

---

▲ *Example 6*

The sequence $\{k\}$ diverges to $\infty$ because, for every $A > 0$, we have $k > A$ for all $k > [A] = N(A)$. (By the same token, $\{-k\}$ diverges to $-\infty$.)

---

## 2.15 Exercises

1. Investigate the following sequences for convergence or divergence:

(a) $\left\{\frac{(-1)^k}{k}\right\}$ (b) $\left\{1 - \frac{(-1)^k}{k}\right\}$

(c) $\{1, \frac{1}{2}, \frac{1}{3}, \frac{2}{3}, \frac{1}{4}, \frac{3}{4}, \frac{1}{5}, \frac{4}{5}, \frac{1}{6}, \frac{5}{6}, \ldots\}$ (d) $\left\{\frac{k}{2k - 1}\right\}$

(e) $\left\{\frac{k^2}{k + 1}\right\}$ (f) $\left\{\frac{1}{k} + (-1)^k\right\}$

(g) $\left\{\frac{1}{1+ak}\right\}$, $a > 0$.

2. Show that $(1 + a)^n > 1 + na$ for $a > 0$ and all integers $n > 1$.
3. Show that $\lim_\infty c^k = 0$ for any given $c \in (0, 1)$.
4. Given $\{d_k\} = \{a_0, a_0 . a_1, a_0 . a_1a_2, a_0 . a_1a_2a_3, \ldots\}$. Prove: $\lim_\infty d_k = a_0 . a_1a_2a_3\ldots$
5. Show: If $\lim_\infty a_k = b$, then, for every $\varepsilon > 0$, there is an $N(\varepsilon) \in \mathbf{N}$ such that $|a_k - a_l| < \varepsilon$ for all $k, l > N(\varepsilon)$.

## 2.16 Three Fundamental Properties of Convergent Sequences

Limit and convergence, as we have defined them in the preceding section, have the following fundamental properties: The limit of a sequence is unique, a convergent sequence is bounded, and any subsequence of a convergent sequence converges to the same limit. We shall take up these properties one at a time.

***Theorem 16.1***

If $\{a_k\}$ converges, then the *limit is uniquely determined.*

*Proof* (by contradiction)

Suppose that $\lim_\infty a_k = a$ and $\lim_\infty a_k = b$, where $a \neq b$. We have, from definition 15.1, that, for every $\varepsilon > 0$ and in particular for $\varepsilon = \frac{1}{4}|b - a| > 0$, there exists an $N_a(\varepsilon) \in \mathbf{N}$ such that

$$|a_k - a| < \varepsilon \qquad \text{for all } k > N_a(\varepsilon)$$

and an $N_b(\varepsilon) \in \mathbf{N}$ such that

$$|a_k - b| < \varepsilon \qquad \text{for all } k > N_b(\varepsilon).$$

Let $N(\varepsilon) = \max(N_a(\varepsilon), N_b(\varepsilon))$. Then,

$$|a - b| \leqslant |a - a_k| + |a_k - b| < \tfrac{1}{4}|b - a| + \tfrac{1}{4}|b - a| = \tfrac{1}{2}|b - a|$$

for all $k > N(\varepsilon)$, which implies $|a - b| = 0$, contrary to our assumption that $a \neq b$.

***Theorem 16.2***

If $\{a_k\}$ converges to the limit $a \in \mathbf{R}$, then $\{a_k\}$ is *bounded*, i.e., there is an $M > 0$ such that $|a_k| \leqslant M$ for all $k \in \mathbf{N}$.

*Proof*

By definition 15.1, we have, for every $\varepsilon > 0$ and in particular for $\varepsilon = 1$, an $N(1) \in \mathbf{N}$ such that $|a_k - a| < 1$ for all $k > N(1)$. Hence,

$$|a_k| - |a| \leqslant |a_k - a| < 1 \qquad \text{for all } k > N(1),$$

and therefore,

$$|a_k| < 1 + |a| \qquad \text{for all } k > N(1).$$

Let $M = \max\{|a_1|, |a_2|, \ldots, |a_{N(1)}|, 1 + |a|\}$. Then $|a_k| \leqslant M$ for all $k \in \mathbf{N}$.

In order to establish the third property we have to define what we mean by subsequence:

***Definition 16.1***

The sequence $\{a_{n_k}\} = \{a_{n_1}, a_{n_2}, a_{n_3}, \ldots\}$ is a *subsequence* of $\{a_k\}$ if and only if $n_1, n_2, n_3, \ldots$ are positive integers, with $n_1 < n_2 < n_3 < \cdots$

---

▲ *Example 1*

$\{1, 3, 5, \ldots\}$ is a subsequence of $\{1, 2, 3, \ldots\}$. So is $\{28, 29, 30, \ldots\}$.

▲ *Example 2*

$$\left\{1, \frac{1}{2}, \frac{1}{3}, \frac{1}{5}, \frac{1}{7}, \frac{1}{11}, \frac{1}{13}, \frac{1}{17}, \frac{1}{19}, \frac{1}{23}, \ldots\right\}$$

is a subsequence of $\{1/k\}$ because $\{1, 2, 3, 5, 7, 11, 13, 17, 19, 23, \ldots\} = \{p\}$, the sequence of all prime numbers, has the required property.

---

Note that, in accordance with our definition, every sequence is a subsequence of itself.

***Theorem 16.3***

If $\lim_\infty a_k = a$ and if $\{a_{n_k}\}$ is a *subsequence* of $\{a_k\}$, then $\lim_\infty a_{n_k} = a$.

*Proof*

First, we show, by induction, that

$$n_k \geqslant k \qquad \text{for all } k \in \mathbf{N}. \tag{16.1}$$

For $k = 1$, this is trivially satisfied because $n_1$ is a positive integer; hence, $n_1 \geqslant 1$. We assume that $n_k \geqslant k$. Since $n_{k+1} > n_k$ and hence, $n_{k+1} \geqslant n_k + 1$, we have $n_{k+1} \geqslant n_k + 1 \geqslant k + 1$, and (16.1) stands proved.

Since $\lim_\infty a_k = a$, we have, from definition 15.1 that, for every $\varepsilon > 0$, there is an $N(\varepsilon) \in \mathbf{N}$ such that $|a_k - a| < \varepsilon$ for all $k > N(\varepsilon)$. Hence

$$|a_{n_k} - a| < \varepsilon \qquad \text{for all } n_k > N(\varepsilon).$$

Since $n_k \geqslant k$, we have also

$$|a_{n_k} - a| < \varepsilon \qquad \text{for all } k > N(\varepsilon),$$

that is, $\lim_\infty a_{n_k} = a$.

The following result will prove useful on a number of occasions:

***Corollary to Theorem 16.3***

$\lim_\infty a_k = a$ if and only if for every fixed $n$, $\lim_{k\to\infty} a_{n+k} = a$.

*Proof*

$\{a_{n+k}\}$ is a subsequence of $\{a_k\}$ and $\lim_\infty a_k = a$ implies $\lim_{k\to\infty} a_{n+k} = a$. The converse is trivial.

This corollary says, in effect, that when investigating the convergence of a sequence, we may just as well ignore a finite number (e.g., a million or so) of terms of the sequence.

## 2.16 Exercises

1. Show: If $\{a_k\}$ converges, then $\{a_{2k+1}\}$ converges, and both sequences have the same limit.
2. Show: The two sequences $\{\sqrt{7}, \pi^2, \log 2, e^{\sqrt{3}}, -1/e^\pi, \frac{1}{6}, \frac{1}{7}, \frac{1}{8}, \frac{1}{9}, \ldots\}$ and $\{17, 31, 11, \frac{1}{3}, \frac{1}{6}, \frac{1}{9}, \frac{1}{12}, \ldots\}$ have the same limit.
3. Prove: If $\lim_\infty a_k = a$, then $\lim_\infty (a_k + 3) = a + 3$.
4. Suppose $\lim_\infty a_{n_k} = a$ and $\{a_{n_k}\}$ is a subsequence of $\{a_k\}$. Is it always true that $\{a_k\}$ converges? To the same limit?

## 2.17 Manipulation of Sequences

We shall discuss in this section a number of theorems that prove helpful in the computation of limits of sequences in terms of known limits of simpler sequences.

***Theorem 17.1***

If $\lim_\infty a_k = a$, $\lim_\infty b_k = b$, then,

(a) $\lim_\infty (a_k + b_k) = a + b$, and (b) $\lim_\infty a_k b_k = ab$;

and, if $b_k \neq 0$ for all $k \in \mathbf{N}$ and $b \neq 0$, then

(c) $\lim \dfrac{a_k}{b_k} = \dfrac{a}{b}$.

*Proof*

We leave the proofs of (a) and (b) for exercises 1 and 2. In order to prove (c), we first note that the elements of $\{b_k\}$ are bounded away from 0:

$$|b_k| \geqslant m \qquad \text{for some } m > 0. \tag{17.1}$$

(See exercise 3.) In order to show that $\lim_\infty a_k/b_k = a/b$, we have to demonstrate that $|(a_k/b_k) - (a/b)|$ can be made arbitrarily small for sufficiently large $k$. We have, in view of (17.1):

$$\begin{aligned}\left|\frac{a_k}{b_k} - \frac{a}{b}\right| &= \left|\frac{a_k b - ab_k}{b_k b}\right| = \frac{1}{|b_k||b|}\,|a_k b - ab_k| \\ &\leqslant \frac{1}{m|b|}\,|a_k b - ab_k| \leqslant \frac{1}{m|b|}\,(|a_k b - ab| + |ab - ab_k|) \\ &= \frac{1}{m|b|}\,|b||a_k - a| + \frac{1}{m|b|}\,|a||b - b_k|.\end{aligned}$$

Since $\{a_k\}$, $\{b_k\}$ converge to $a$ and $b$, respectively, we have, for every $\varepsilon > 0$, an $N_a(\varepsilon) \in \mathbf{N}$ and an $N_b(\varepsilon) \in \mathbf{N}$ such that

$$|a_k - a| < \frac{\varepsilon m}{2} \quad \text{for all } k > N_a(\varepsilon), \qquad |b_k - b| < \frac{\varepsilon |b| m}{2|a|} \quad \text{for all } k > N_b(\varepsilon).$$

Hence,

$$\left|\frac{a_k}{b_k} - \frac{a}{b}\right| < \varepsilon \qquad \text{for all } k > N(\varepsilon) = \max(N_a(\varepsilon), N_b(\varepsilon)),$$

that is, $\lim_\infty (a_k/b_k) = a/b$.

***Corollary to Theorem 17.1***

If $\lim_\infty a_k = a$, and if $\lambda \in \mathbf{R}$, then $\lim_\infty \lambda a_k = \lambda a$. (See exercise 4.)

---

▲ *Example 1*

$$\left\{\frac{1 + (-1)^k}{k+1}\right\} = \left\{\frac{1}{k+1} + \frac{(-1)^k}{k+1}\right\}.$$

Since $\lim_\infty 1/(k+1) = 0$ and $\lim_\infty (-1)^k/(k+1) = 0$, we have

$$\lim_\infty \left\{\frac{1 + (-1)^k}{k+1}\right\} = 0 + 0 = 0.$$

▲ *Example 2*

$$\left\{\frac{k^2}{k^2 + 3k + 2}\right\} = \left\{\frac{k}{k+1} \cdot \frac{k}{k+2}\right\}.$$

Since $\lim_\infty k/(k+1) = 1$ and $\lim_\infty k/(k+2) = 1$, we have

$$\lim_\infty \frac{k^2}{k^2+3k+2} = 1 \cdot 1 = 1.$$

▲ *Example 3*

$$\left\{\frac{k^2+5}{3k^2-1}\right\} = \left\{\frac{1+\dfrac{5}{k^2}}{3-\dfrac{1}{k^2}}\right\}.$$

Since $\{1/k^2\}$ is a subsequence of $\{1/k\}$ and $\lim_\infty 1/k = 0$, we have $\lim_\infty 1/k^2 = 0$. By the corollary to Theorem 17.1, $\lim_\infty 5/k^2 = 5 \lim_\infty 1/k^2 = 0$. By (a) of Theorem 17.1, $\lim_\infty (1 + 5/k^2) = 1$ and $\lim_\infty (3 - 1/k^2) = 3$. Since $b_k = 3 - 1/k^2 \neq 0$ for all $k \in \mathbf{N}$, and since $b = 3 \neq 0$, we have from (c) of Theorem 17.1, that

$$\lim_\infty \frac{k^2+5}{3k^2-1} = \lim_\infty \left\{\frac{1+\dfrac{5}{k^2}}{3-\dfrac{1}{k^2}}\right\} = \frac{\lim_\infty \left\{1+\dfrac{5}{k^2}\right\}}{\lim_\infty \left\{3-\dfrac{1}{k^2}\right\}} = \frac{1}{3}.$$

---

We note that it is *not* possible to draw any conclusions about the convergence of $\{a_k\}$, $\{b_k\}$ from the fact that $\lim_\infty (a_k \pm b_k)$ exists, or that $\lim_\infty \{a_k b_k\}$ or $\lim_\infty \{a_k/b_k\}$ exists, as we shall demonstrate in the following examples:

---

▲ *Example 4*

Consider $\{(-1)^{k+1} + (-1)^k\}$. Since $(-1)^{k+1} + (-1)^k = 0$ for all $k \in \mathbf{N}$, we have

$$\lim_\infty ((-1)^{k+1} + (-1)^k) = 0.$$

However, neither $\lim_\infty (-1)^{k+1}$ nor $\lim_\infty (-1)^k$ exists.

▲ *Example 5*

Consider

$$\{a_k\} = \{k + (-1)^k k\} \equiv \{0, 4, 0, 8, 0, 12, \ldots\}$$

and

$$\{b_k\} = \{k + (-1)^{k+1} k\} \equiv \{2, 0, 6, 0, 10, 0, \ldots\}.$$

Then, $\{a_k b_k\} = \{0, 0, 0, \ldots\}$ and $\lim_\infty a_k b_k = 0$, but neither $\lim_\infty a_k$ nor $\lim_\infty b_k$ exists.

▲ *Example 6*

$\lim_\infty (k+2)/(k^2+2) = 0$ and $k^2+2 \neq 0$ for all $k \in \mathbf{N}$ but neither $\lim_\infty (k+2)$ nor $\lim_\infty (k^2+2)$ exists.

---

If the elements of two sequences $\{a_k\}$ and $\{b_k\}$ satisfy the relation $a_k \leqslant b_k$ for all $k \in \mathbf{N}$, we say that the sequence $\{a_k\}$ is *dominated* by the sequence $\{b_k\}$. The sequence $\{b_k\}$ is called the *dominating* sequence. If both sequences converge, it is reasonable to expect that the limit of the dominating sequence *cannot be less* than the limit of the dominated sequence. This is indeed so:

***Theorem 17.2***

If $a_k \leqslant b_k$ for all $k \in \mathbf{N}$, and if $\lim_\infty \{a_k\} = a$ and $\lim_\infty \{b_k\} = b$ exist, then $a \leqslant b$.

*Proof* (by contradiction)

Suppose that $a > b$ and choose $\varepsilon = (a-b)/2 > 0$. Then there is an $N(\varepsilon) \in \mathbf{N}$ such that

$$|a_k - a| < \frac{a-b}{2} \qquad \text{for all } k > N(\varepsilon)$$

or, equivalently, $-(a-b)/2 < a_k - a < (a-b)/2 \quad$ for all $k > N(\varepsilon)$.
Hence, $b_k \geqslant a_k > (a+b)/2$, and therefore,

$$b_k - b > \frac{a-b}{2} \qquad \text{for all } k > N(\varepsilon).$$

This means that $\{b_k\}$ cannot converge to $b$, contrary to our assumption.

The result in the following corollary follows directly from Theorem 17.2:

***Corollary to Theorem 17.2***

If $a_k \geqslant b$ for all $k \in \mathbf{N}$ and if $\lim_\infty a_k = a$ exists, then $a \geqslant b$.

A sequence that converges to 0 is called a *null sequence* (*Null* is the German word for *zero*). If a sequence of nonnegative terms is dominated by a null sequence, then it is itself a null sequence:

***Lemma 17.1***

If $0 \leqslant a_k \leqslant b_k$ for all $k \in \mathbf{N}$ and if $\lim_\infty b_k = 0$, then $\lim_\infty a_k = 0$.

*Proof*

$0 \leqslant a_k - 0 \leqslant b_k - 0$ for all $k \in \mathbf{N}$ implies $\lim_\infty a_k = 0$, in view of the assumed convergence of $\{b_k\}$ to the limit zero.

***Corollary to Lemma 17.1***

If $a_k \leqslant b_k \leqslant c_k$ for all $k \in \mathbf{N}$, and if $\lim_\infty a_k = \lim_\infty c_k = a$, then $\lim_\infty b_k$ exists and $\lim_\infty b_k = a$.

*Proof*

Since $0 \leqslant b_k - a_k \leqslant c_k - a_k$, since $\lim_\infty (c_k - a_k) = 0$, the result follows from Lemma 17.1, and $b_k = a_k + (b_k - a_k)$.

---

▲ *Example 7*

We shall illustrate the use of some of the manipulative rules and results of this section by studying the sequence $\{\sqrt[k]{c}\}$ for $c > 0$. The result of our deliberations will be that

(17.2) $$\lim_\infty \sqrt[k]{c} = 1, \qquad c > 0.$$

First, we consider the case where $c = 1$. Then $\sqrt[k]{c} = 1$ and we have $\lim_\infty \sqrt[k]{c} = \lim_\infty 1 = 1$.
Next, we look at $c > 1$. We put

$$\sqrt[k]{c} = 1 + \varepsilon_k,$$

and observe that $\varepsilon_k > 0$. (See exercise 5.) By exercise 2, section 2.15, we have

$$c = (1 + \varepsilon_k)^k \geqslant 1 + k\varepsilon_k,$$

and hence,

$$0 < \varepsilon_k \leqslant \frac{c-1}{k} \qquad \text{where } c - 1 > 0.$$

Since $\lim_\infty (c-1)/k = (c-1) \lim_\infty 1/k = 0$, we obtain, from Lemma 17.1, that $\lim_\infty \varepsilon_k = 0$. Hence, by Theorem 17.1,

$$\lim_\infty \sqrt[k]{c} = \lim_\infty (1 + \varepsilon_k) = \lim_\infty 1 + \lim_\infty \varepsilon_k = 1 + 0 = 1.$$

Finally, we consider the case $0 < c < 1$. We let $b = 1/c$, and note that $b > 1$. Hence, by the preceding result, $\lim_\infty \sqrt[k]{b} = 1$. Since $\sqrt[k]{c} = 1/\sqrt[k]{b}$ and $\sqrt[k]{b} \neq 0$ for all $k \in \mathbf{N}$, $1 \neq 0$, we have from Theorem 17.1(c), that $\lim_\infty \sqrt[k]{c} = \lim_\infty 1/\sqrt[k]{b} = 1/\lim_\infty \sqrt[k]{b} = 1$.

---

We proved the theorems of this section on the basis of the validity of certain relations that were assumed to hold for all $k \in \mathbf{N}$. These assumptions can be weakened in view of the Corollary to Theorem 16.3, to the extent that the validity of these relations is required only for *all but finitely many* $k \in \mathbf{N}$. (See exercises 6 through 10.)

## 2.17 Exercises

1. Prove Theorem 17.1(a).
2. Prove Theorem 17.1(b).
3. Given $\{b_k\}$. Show: If $b_k \neq 0$ for all $k \in \mathbf{N}$ and $\lim_\infty b_k = b \neq 0$, then there exists an $m > 0$ such that $|b_k| \geqslant m$ for all $k \in \mathbf{N}$.
4. Prove the Corollary to Theorem 17.1.
5. Show: If $\sqrt[k]{c} = 1 + \varepsilon_k$ and $c > 1$, then $\varepsilon_k > 0$.
6. Prove: If $\lim_\infty a_k = a$, $\lim_\infty b_k = b$, then $\lim_\infty s_k = a \pm b$, if $s_k = a_k \pm b_k$ *for all but finitely many* $k \in \mathbf{N}$.
7. Prove: If $\lim_\infty a_k = a$, $\lim_\infty b_k = b$, then $\lim_\infty p_k = ab$ if $a_k b_k = p_k$ *for all but finitely many* $k \in \mathbf{N}$.
8. Prove: If $\lim_\infty a_k = a$, $\lim_\infty b_k \neq 0$, then $\lim_\infty q_k = a/b$ if $b_k \neq 0$ *for all but finitely many* $k \in \mathbf{N}$, and

$$q_k = \begin{cases} \dfrac{a_k}{b_k} & \text{if } b_k \neq 0, \\ 0 & \text{if } b_k = 0. \end{cases}$$

9. Prove Theorem 17.2 under the weaker condition that $a_k \leqslant b_k$ *for all but finitely many* $k \in \mathbf{N}$.
10. Prove the Corollary to Theorem 17.2 under the weaker condition that $a_k \geqslant b$ for all but finitely many $k \in \mathbf{N}$.
11. Show that

$$(1 + a)^n \geqslant 1 + na + \left(\frac{n(n-1)}{2}\right) a^2,$$

where $a > 0$, for all $n \geqslant 1$.
12. Show that $\lim_\infty \sqrt[k]{k} = 1$.
13. Given $\lim_\infty a_k = 0$ and $|b_k| \leqslant M$ for all $k \in \mathbf{N}$ and some $M \geqslant 0$. Prove that $\lim_\infty a_k b_k = 0$.
14. Find $\lim_\infty \sin k/k$.

## 2.18 Monotonic Sequences

We have seen in section 2.16 that a convergent sequence is bounded. Clearly, the converse is not necessarily true, as the example $\{(-1)^k\}$ shows. However, if we impose a certain additional, admittedly strong, condition on bounded sequences, then we do obtain convergence. The additional condition is that the sequence be *monotonic*. Monotonic sequences are either *increasing* or *decreasing*.

***Definition 18.1***

The sequence $\{a_k\}$ is *increasing* (*decreasing*) if and only if $a_k \leqslant a_{k+1}$ ($a_k \geqslant a_{k+1}$) for all $k \in \mathbf{N}$.

We have

***Theorem 18.1***

If the sequence $\{a_k\}$ is *bounded above* and *increasing*, then $\lim_\infty a_k$ exists and $\lim_\infty a_k = \sup \{a_k | k \in \mathbf{N}.\}$ (An analogous theorem holds for *decreasing* sequences $\{b_k\}$ with $\lim_\infty b_k = \inf \{b_k | k \in \mathbf{N}.\}$)

*Proof*

Since $\mathbf{R}$ is complete and since $\{a_k\}$ is bounded, $\sup \{a_k | k \in \mathbf{N}\} = a$ exists. By Theorem 9.1, we have, for every $\varepsilon > 0$, an element $a_{N(\varepsilon)} \in \{a_k\}$ such that

$$a - \varepsilon < a_{N(\varepsilon)} \leqslant a. \tag{18.1}$$

Since $\{a_k\}$ is *increasing*, we have

$$a_{N(\varepsilon)} \leqslant a_n \qquad \text{for all } n > N(\varepsilon), \tag{18.2}$$

and simultaneously,

$$a_n \leqslant a \qquad \text{for all } n \in \mathbf{N}. \tag{18.3}$$

From (18.1), (18.2), and (18.3) we obtain readily that

$$|a_n - a| < \varepsilon \qquad \text{for all } n > N(\varepsilon),$$

that is, $\lim_\infty a_k = a$.

We have also seen, in section 2.16, that every subsequence of a convergent sequence converges to the same limit. Again, the converse is not necessarily true, unless we impose the additional condition of monotonicity, as we shall now demonstrate:

***Theorem 18.2***

If the sequence $\{a_k\}$ is increasing and if some subsequence $\{a_{n_k}\}$ of $\{a_k\}$ converges and $\lim_{k \to \infty} a_{n_k} = a$, than $\{a_k\}$ itself converges, and $\lim_\infty a_k = a$.

*Proof*

Since $\{a_{n_k}\}$ is also increasing, we have from Theorems 16.2 and 18.1 that

$$a = \sup\{a_{n_k} | k \in \mathbf{N}\}.$$

Hence, we have, for every $\varepsilon > 0$, some $N(\varepsilon) \in \mathbf{N}$ such that

$$a - \varepsilon < a_{n_{N(\varepsilon)}} \leqslant a.$$

By (16.1), $a_k \leqslant a_{n_k}$ for all $k \in \mathbf{N}$. Hence, $a_k \leqslant a$ for all $k \in \mathbf{N}$, and $a_{n_{N(\varepsilon)}} \leqslant a_k \leqslant a$ for all $k > n_{N(\varepsilon)}$. Therefore,

$$|a_k - a| < \varepsilon \qquad \text{for all } k > n_{N(\varepsilon)},$$

that is, $\lim_\infty a_k = a$. (See also Fig. 18.1.)

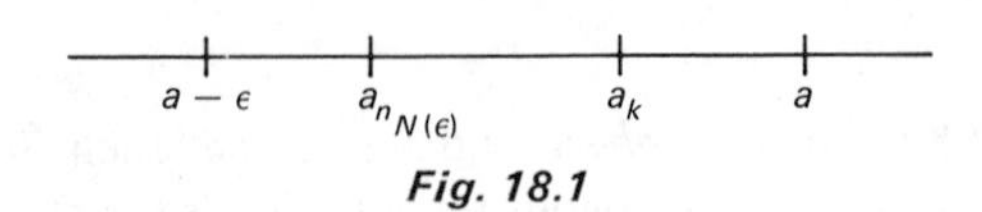


*Fig. 18.1*

## 2.18 Exercises

1. Given the sequence $\{a_k\} = \{1 + q + q^2 + \cdots + q^{k-1}\}$, where $0 < q < 1$. Show that $\lim_\infty \{a_k\}$ exists.
2. Given the sequence $\{a_k\}$ constructed from:

$$\begin{aligned} a_1 &= \sqrt{2}, \\ a_2 &= \sqrt{2 + \sqrt{2}}, \\ &\vdots \\ a_k &= \sqrt{2 + a_{k-1}}. \end{aligned}$$

   Show that $\{a_k\}$ converges.
3. Prove: If $\{b_k\}$ is *decreasing* and $b_k \geqslant -M$ for some $M > 0$ and all $k \in \mathbf{N}$, then $\lim_\infty b_k = \inf\{b_k \mid k \in \mathbf{N}\}$.
4. Prove: If $\{b_k\}$ is *decreasing* and if $\lim_\infty b_{n_k} = b$ exists, where $\{b_{n_k}\}$ is a subsequence of $\{b_k\}$, then $\lim_\infty \{b_k\} = b$.
5. Given two sequences $\{a_k\}$, $\{b_k\}$, where $\{a_k\}$ is *increasing* and $\{b_k\}$ is *decreasing*. Show: If $0 \leqslant b_k - a_k \leqslant 1/2^k$ for all $k \in \mathbf{N}$, then $\lim_\infty a_k = \lim_\infty b_k$.
6. Let $a_1 = \sqrt{1}$, $a_k = \sqrt{1 + a_{k-1}}$. Show that $\tau = \lim_\infty a_k$ exists and find $\tau$. ($\tau$ is called the *golden ratio*.)

## 2.19 The Geometric Series

Infinite series will be discussed systematically in Chapter 14 which may be taken up immediately following Chapter 7. In the present section, we shall discuss some special aspects of this subject which will be needed in Chapter 4.

### *Definition 19.1*

The *infinite series* $\sum_{k=1}^\infty a_k$ generated by the sequence $\{a_k\}$ is the sequence $\{s_n\}$ of *partial sums* $s_n = a_1 + a_2 + \cdots + a_n$. ($s_n$ is called the $n$th partial sum.) The infinite series $\sum_{k=1}^\infty a_n$ *converges* if and only if the sequence $\{s_n\}$ converges.

Furthermore, $\lim_\infty s_n$ is called the *sum of the infinite series*. We often write, for convenience,

$$\sum_{k=1}^{\infty} a_k = \lim_{\infty} s_n.$$

In such context, $\sum_{k=1}^\infty a_k$ is viewed as the sum of the infinite series.

▲ *Example 1* (Geometric series)

The sequence $\{q^{k-1}\}$, where $|q| < 1$, generates the infinite series $\sum_{k=1}^{\infty} q^{k-1}$, the partial sums of which are given by $s_n = 1 + q + \cdots + q^{n-1} = (1 - q^n)/(1 - q)$. We have, from Theorem 17.1, that

(19.1) $$\lim_{\infty} s_n = \lim_{\infty} \frac{1 - q^n}{1 - q} = \frac{\lim_{\infty}(1 - q^n)}{\lim_{\infty}(1 - q)} = \frac{1}{1 - q}(1 - \lim_{\infty} q^n).$$

We have seen in exercise 3, section 2.15, that

(19.2) $$\lim_{\infty} \{q^n\} = 0 \qquad \text{if } 0 < q < 1.$$

If $q = 0$, this result is trivially true. If $-1 < q < 0$, we have $|q^n - 0| = |q^n| = |q|^n < \varepsilon$ for all $n > N(\varepsilon)$, because $\lim_{\infty} \{|q|^n\} = 0$ by (19.2). Hence,

(19.3) $$\lim_{\infty} q^n = 0 \qquad \text{if } -1 < q < 1$$

and we obtain

$$\lim_{\infty} s_n = \frac{1}{1 - q},$$

or, by definition 19.1,

(19.4) $$\sum_{k=1}^{\infty} q^{k-1} = \frac{1}{1 - q} \qquad \text{if } |q| < 1.$$

This infinite series, generated by the sequence $\{q^k\}$, is called the *geometric series* and we have just shown that the geometric series converges for $|q| < 1$.

▲ *Example 2*

The sequence $\{1/k^{\alpha}\}$ generates an infinite series, the partial sums of which are given by

$$s_n = 1 + \frac{1}{2^{\alpha}} + \cdots + \frac{1}{n^{\alpha}}.$$

We note that $\{s_n\}$ is an increasing sequence. If we can show that a subsequence converges, then it follows from Theorem 18.2 that $\{s_n\}$ itself converges.

We assume that $\alpha > 1$ (convergence does not take place for $\alpha \leqslant 1$), and pick a subsequence

$$\{s_{n_k}\} \qquad \text{where } n_k = 2^k - 1.$$

We have, for $k = 1$,

$$s_{n_1} = s_1 = 1,$$

for $k = 2$,

$$s_{n_2} = s_3 = 1 + \left(\frac{1}{2^\alpha} + \frac{1}{3^\alpha}\right) < 1 + \frac{1}{2^\alpha} + \frac{1}{2^\alpha} = 1 + \frac{1}{2^{\alpha-1}},$$

(note that $\alpha - 1 > 0$). For $k = 3$,

$$s_{n_3} = s_7 = 1 + \left(\frac{1}{2^\alpha} + \frac{1}{3^\alpha}\right) + \left(\frac{1}{4^\alpha} + \frac{1}{5^\alpha} + \frac{1}{6^\alpha} + \frac{1}{7^\alpha}\right) < 1 + \frac{1}{2^{\alpha-1}} + 4 \cdot \frac{1}{4^\alpha}$$

$$= 1 + \frac{1}{2^{\alpha-1}} + \frac{1}{4^{\alpha-1}}.$$

We obtain, by induction (see exercise 1), that

(19.5)

$$s_{n_k} = s_{2^k-1} = 1 + \frac{1}{2^\alpha} + \frac{1}{3^\alpha} + \cdots + \frac{1}{(2^k-1)^\alpha} < 1 + \frac{1}{2^{\alpha-1}} + \frac{1}{4^{\alpha-1}} + \cdots + \frac{1}{(2^{k-1})^{\alpha-1}}$$

$$= 1 + \frac{1}{2^{\alpha-1}} + \left(\frac{1}{2^{\alpha-1}}\right)^2 + \cdots + \left(\frac{1}{2^{\alpha-1}}\right)^{k-1} = \frac{1 - \left(\dfrac{1}{2^{\alpha-1}}\right)^k}{1 - \dfrac{1}{2^{\alpha-1}}}.$$

Since $0 < (1/2^{\alpha-1}) < 1$, we have

$$\frac{1 - \left(\dfrac{1}{2^{\alpha-1}}\right)^k}{1 - \dfrac{1}{2^{\alpha-1}}} < \frac{1}{1 - \dfrac{1}{2^{\alpha-1}}} = \frac{2^{\alpha-1}}{2^{\alpha-1} - 1},$$

and hence,

$$0 < s_{n_k} < \frac{2^{\alpha-1}}{2^{\alpha-1} - 1} \qquad \text{for all } k \in \mathbf{N}. \tag{19.6}$$

Now we know that $\{s_{n_k}\}$ is bounded. Since $\{s_{n_k}\}$ is also increasing, we obtain, from Theorem 18.1, that $\lim_\infty s_{n_k} = \sup\{s_{n_k} \mid k \in \mathbf{N}\}$ exists and, from Theorem 18.2, that

$$\lim_\infty s_n = \sup\{s_{n_k} \mid k \in \mathbf{N}\}.$$

We have, from (19.6), that $\sup\{s_{n_k} \mid k \in \mathbf{N}) \leqslant 2^{\alpha-1}/2^{\alpha-1} - 1$ and hence,

$$\sum_{k=1}^{\infty} \frac{1}{k^\alpha} = \lim_\infty s_n \leqslant \frac{2^{\alpha-1}}{2^{\alpha-1} - 1}. \tag{19.7}$$

---

## 2.19 Exercises

1. Given $s_n = 1 + 1/2^\alpha + \cdots + 1/n^\alpha$ and $n_k = 2^k - 1$. Show by induction, that

$$0 < s_{n_k} < 1 + \frac{1}{2^{\alpha-1}} + \frac{1}{4^{\alpha-1}} + \cdots + \frac{1}{(2^{k-1})^{\alpha-1}}.$$

2. Prove: The infinite series generated by the sequence $\{k/2^{k+1}\}$ converges.
3. Given the infinite series in exercise 2. Show that $\sum_{k=1}^{\infty} k/2^{k+1} < 3$.
4. Interpret the infinite decimal $d_0 . d_1 d_2 d_3 \ldots$ as the limit of the sequence of partial sums

$$\left\{d_0,\quad d_0 + \frac{d_1}{10},\quad d_0 + \frac{d_1}{10} + \frac{d_2}{10^2},\quad \ldots\right\}.$$

Show that $0.\dot{9} = 1$.

5. Same as in exercise 4, for infinite binaries and $0.\dot{1} = 1$.
6. Same as in exercise 4, for infinite ternaries and $0.\dot{2} = 1$.

## 2.20 The Number Line

In this section we free ourselves from the constraints of rigor and logical reasoning, and let intuition and geometric insight serve as our guides.

A horizontal line with two points marked on it, the one on the left labeled 0, the one on the right labeled 1, is called a *number line*. The geometric distance between 0 and 1 is called *unit distance*. *Positive* integers $n$ are represented by points $n$ unit-distances to the right of 0, and *negative* integers $m$ by points $|m|$ unit-distances to the left of 0. Rational numbers $m/n$ are represented by points $|m|$ times one $|n|$th the unit-distance away from zero, to the right if $m/n > 0$, to the left if $m/n < 0$. (For the construction of 7/3, see Fig. 20.1.)

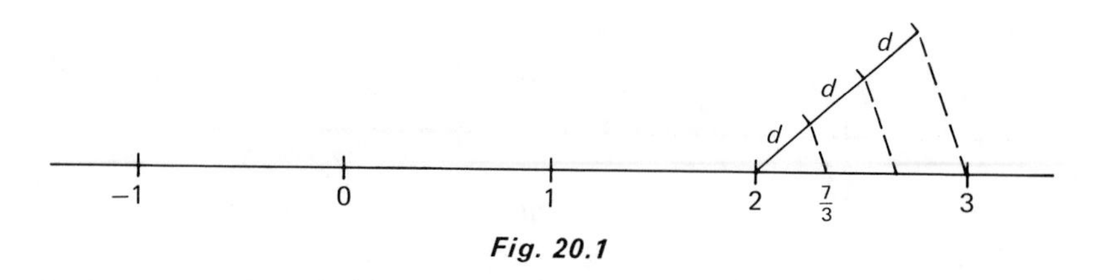


*Fig. 20.1*

In this manner we establish a one-to-one correspondence between **Q** and certain points on the number line, the *rational points*. Between any two rational points, no matter how close, there is *another* rational point, corresponding to the fact that, if $a < b$ are rational numbers, then $(a + b)/2$ is again rational, and

$$a < \frac{a+b}{2} < b.$$

Still, the *rational* points cannot possibly fill the number line, because there is a point at distance $\sqrt{2}$ from 0 (see Fig. 20.2), and $\sqrt{2}$ is *not* a rational number; hence, this location on the number line cannot be occupied by a rational point. If we consider that the same is true for all the points at distances $r\sqrt{2}$, $r \in \mathbf{Q}$ (and many others), we see that the collection of rational points on the number line represents what one might describe as a one-dimensional sieve—and a very crude one at that.

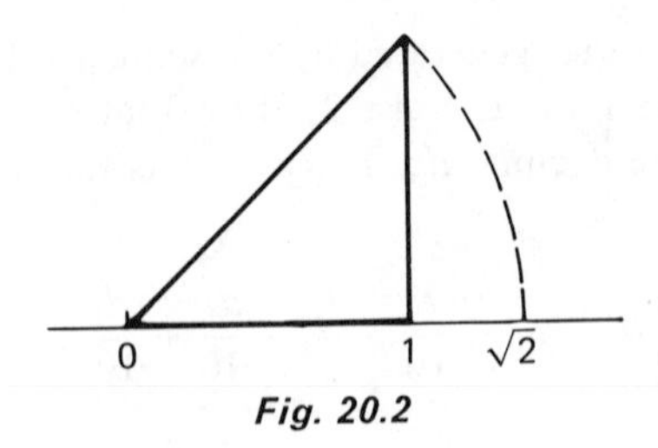


*Fig. 20.2*

There is no obvious way of filling the gaps, as we have no clear geometric insight into the "atomic" structure of the line. We overcome this difficulty by forcing the line to conform to our mental image of it by means of an axiom.

Let $A$, $B$ denote two points on the line, $B$ to the right of $A$. $[A, B]$, the set of all points between $A$ and $B$, $A$, $B$ included, is called a *closed interval* on the line. Suppose we have a sequence $\{[A_k, B_k]\}$ of intervals with the property that

$$[A_1, B_1] \supseteq [A_2, B_2] \supseteq [A_3, B_3] \supseteq \cdots \supseteq [A_k, B_k] \supseteq \cdots.$$

This is called a *nested sequence* of closed intervals. It is intuitively clear that there is at least one point $P$ on the line which lies in *all intervals* of the nested sequence. The existence of such a point is called the *nested-interval property of the line:*

*For every nested sequence of closed intervals on the number line, there exists at least one point on the number line that lies in all intervals of the nested sequence.* (See Fig. 20.3)

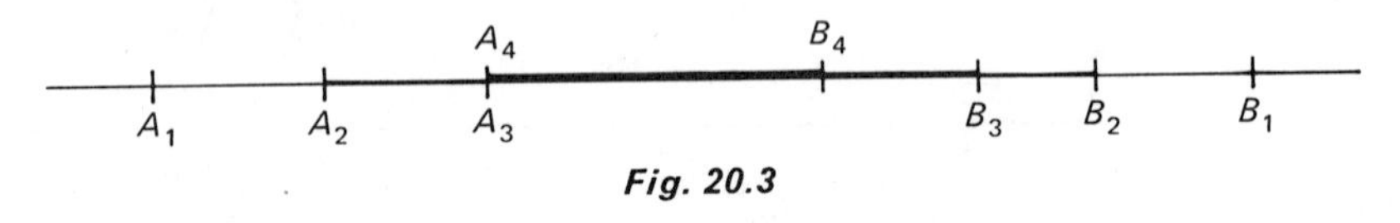


*Fig. 20.3*

We have picked this property of the number line above other equivalent properties not merely because of its intuitive plausibility. We have chosen it because it leads us, in conjunction with our understanding of the number line as a representation of the set of real numbers, to an important property of a complete ordered field, the nested-interval property. The nested-interval property is equivalent to the least-upper-bound property and is, in many instances, easier to handle.

To formulate the nested-interval property for **R** and to demonstrate its equivalence with the least-upper-bound property of **R** is the objective of the next section.

## 2.20 Exercises

1. Let the points $A_k$ represent the rational numbers 0.1, 0.10, 0.101, 0.1010, 0.10101, 0.101010, ... , and $B_k$ represent the rational numbers 0.2, 0.11, 0.102, 0.1011, 0.10102, 0.101011, ....

(a) Show: $\{[A_k, B_k]\}$ is a nested sequence of closed intervals.
(b) Show: $\lim_\infty l([A_k, B_k]) = 0$, where $l([A_k, B_k]) = |a_k - b_k|$, $a_k$, $b_k$ being the rational numbers represented by $A_k$, $B_k$.
(c) Is there a rational point $P \in [A_k, B_k]$ for all $k$?

2. Given a nested sequence $\{[A_k, B_k]\}$ of closed intervals, where $\lim_\infty l[A_k, B_k] = 0$. It is possible for two rational points $P$, $Q$, where $P \neq Q$, to lie in all intervals $[A_k, B_k]$?

## 2.21 The Nested-Interval Property of R

We shall demonstrate in this section that the nested-interval property, properly translated into the language of ordered fields, characterizes the complete ordered field **R**.

***Definition 21.1***

The sequence $\{\mathfrak{J}_k\} = \{\{x \in \mathbf{R} \mid a_k \leqslant x \leqslant b_k\}\}$, where $a_k, b_k \in \mathbf{R}$, is called a *nested sequence of closed, nonempty intervals* if and only if

1. $a_k \leqslant a_{k+1}, \qquad b_k \geqslant b_{k+1},$
2. $a_k \leqslant b_k$

for all $k \in \mathbf{N}$, that is, $\mathfrak{J}_n \neq \varnothing$ and $\mathfrak{J}_k \supseteq \mathfrak{J}_{k+1}$ for all $k \in \mathbf{N}$.

We note that (1) and (2) imply that

$$a_k \leqslant b_j \qquad \text{for all } j, k \in \mathbf{N}, \tag{21.1}$$

because we have, for all $n \in \mathbf{N}$, that

$$a_1 \leqslant a_2 \leqslant \cdots \leqslant a_n \leqslant b_n \leqslant b_{n-1} \leqslant \cdots \leqslant b_1.$$

***Definition 21.2***

An ordered field **F** is said to have the *nested-interval property* if and only if for every nested sequence $\{\mathfrak{J}_k\}$ of closed, nonempty intervals, there is at least one $x \in \mathbf{F}$ such that $x \in \mathfrak{J}_k$ for all $k \in \mathbf{N}$.

The following theorem establishes the equivalence of the *least-upper-bound property* with the *nested-interval property* of an ordered Archimedean field, and enables us, as a consequence, to characterize complete ordered fields by the Archimedean property and the nested-interval property instead of the least-upper-bound property.

***Theorem 21.1***

The ordered field **R** is complete (has the least-upper-bound property) if and only if it is Archimedean and has the nested-interval property.

*Proof*

(a) We assume that $\mathbf{R}$ is complete and consider the nested sequence $\{\mathfrak{J}_k\} = \{[a_k, b_k]\}$ of closed nonempty intervals. We have, from (21.1), that $A = \{a_1, a_2, a_3, \ldots\}$ is bounded above by $b_1$ and hence,

$$x = \sup A$$

exists and $x \geqslant a_i$ for all $i \in \mathbf{N}$. Suppose that $x > b_k$ for some $k \in \mathbf{N}$. Then, by Theorem 9.1, we have, for $\varepsilon = x - b_k > 0$, an element $a_j \in A$ such that

$$x - (x - b_k) < a_j \leqslant x,$$

and hence, $b_k < a_j$, contrary to (21.1). Therefore, $a_k \leqslant x \leqslant b_k$ for all $k \in \mathbf{N}$. Hence, $\mathbf{R}$ has the nested-interval property. That $\mathbf{R}$ is also Archimedean follows from Theorem 9.3.

(b) We assume that $\mathbf{R}$ is Archimedean and has the nested-interval property, and intend to show that, if $A \subset \mathbf{R}$ is bounded above, then $\sup A$ exists. Let $b_1$ be an upper bound of $A$, and choose $a_1 \in \mathbf{R}$ such that $a_1$ is not an upper bound of $A$. Then, $\mathfrak{J}_1 = [a_1, b_1]$ contains at least one element of $A$. Let $m_1 = (a_1 + b_1)/2$. If $m_1$ is an upper bound of $A$, we let $a_2 = a_1$ $b_2 = m_1$ and recognize that $\mathfrak{J}_2 = [a_2, b_2]$ contains at least one element of $A$. If $m_1$ is not an upper bound of $A$, we let $a_2 = m_1$, $b_2 = b_1$, and recognize again that $\mathfrak{J}_2 = [a_2, b_2]$ contains at least one element of $A$. In either case, $\mathfrak{J}_1 \supseteq \mathfrak{J}_2$. (See also Fig. 21.1.) We proceed

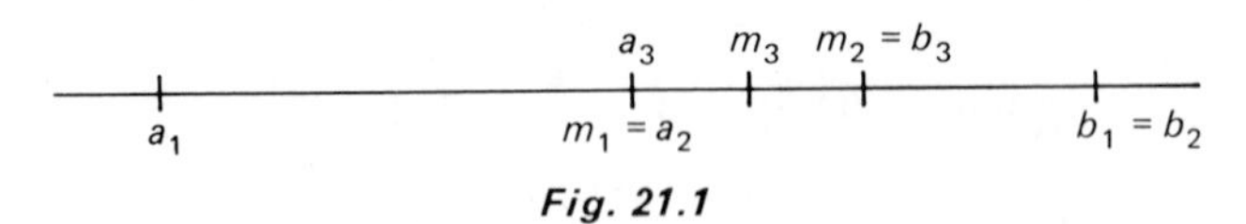


*Fig. 21.1*

with this construction and obtain a nested sequence of closed, nonempty intervals $\{\mathfrak{J}_k\}$, where each $\mathfrak{J}_k$ contains at least one element of $A$. By the nested-interval property, there exists at least one $x \in \mathbf{R}$ such that $x \in \mathfrak{J}_k$ for all $k \in \mathbf{N}$.

By construction, $l(\mathfrak{J}_k) = (b_1 - a_1)/2^{k-1}$ where $l(\mathfrak{J}_n) = |b_n - a_n|$. Since $\mathbf{R}$ is Archimedean, $\lim_\infty l(\mathfrak{J}_k) = 0$. (See example 3, section 1.9.) Suppose there are at least two distinct elements $x, y \in \mathbf{R}$, $x \neq y$, so that $x, y \in \mathfrak{J}_k$ for all $k \in \mathbf{N}$. Then, $|x - y| = \delta > 0$. On the other hand, we have, from

$$a_k \leqslant x \leqslant b_k, \qquad -b_k \leqslant -y \leqslant -a_k,$$

that $|x - y| \leqslant |a_k - b_k| = l(\mathfrak{J}_k)$. Since $\lim_\infty l(\mathfrak{J}_k) = 0$, we have, for every $\varepsilon > 0$ and in particular for $\varepsilon = \delta/2$, an $N(\varepsilon) \in \mathbf{N}$ such that $l(\mathfrak{J}_k) = |a_k - b_k| < \delta/2$ for all $k > N(\varepsilon)$, and we obtain

$$\delta = |x - y| \leqslant |a_k - b_k| < \frac{\delta}{2},$$

which is only possible if $\delta = 0$, that is, $x = y$, contrary to our assumption. Hence, $x \in \mathfrak{J}_k$ for all $k \in \mathbf{N}$ is uniquely determined.

It remains to be shown that $x$ is the least upper bound of $A$. We do this in two steps:

First, we show that $x$ is an upper bound of $A$. Suppose there is an element $x \in A$ such that $a > x$. By construction, $\lim_\infty b_k = x$. (See exercise 2.) Hence, for $\varepsilon = a - x > 0$, there is a $k \in \mathbf{N}$ such that $0 \leqslant b_k - x < a - x$, that is, $a > b_k$, contrary to the fact that $b_k$ is an upper bound of $A$. Hence, $x$ is an upper bound of $A$.

Secondly, we show that $x$ is the *least* upper bound of $A$. By construction, $\lim_\infty a_k = x$. (See exercise 2.) Hence, for every $\varepsilon > 0$, we have $0 \leqslant x - a_k < \varepsilon$ for all $k > N(\varepsilon)$ for some $N(\varepsilon)$. Since $a_k$ is not an upper bound of $A$, we have, for every $a_k$, an element $a \in A$ such that $a_k < a \leqslant x$. Therefore,

$$x - \varepsilon < a_k < a \leqslant x,$$

that is, $x$ is the least upper bound of $A$.

Arguing as in part (b) of this proof, we obtain the following important result:

***Corollary to Theorem 21.1***

If $\mathbf{R}$ is a complete ordered field and if $\{\mathfrak{J}_k\}$ is a nested sequence of closed, nonempty intervals such that

$$\lim_\infty l(\mathfrak{J}_k) = 0,$$

then *there is exactly one element* $x \in \mathbf{R}$ such that $x \in \mathfrak{J}_k$ for all $k \in \mathbf{N}$.

To review briefly: An ordered field need not be Archimedean (exercise 14, section 1.7); an ordered Archimedean field need not be complete (exercise 13, section 1.7); but an ordered Archimedean field with the nested-interval property is complete. Conversely, a complete ordered field is Archimedean and has the nested-interval property.

## 2.21 Exercises

1. Given a bounded set $A \subset \mathbf{R}$ with infinitely many elements: $-m \leqslant x \leqslant M$ for all $x \in A$, where $m, M > 0$. Construct a nested sequence of closed, nonempty intervals $\{\mathfrak{J}_k\}$ where $\lim_\infty l(\mathfrak{J}_k) = 0$ such that each $\mathfrak{J}_k$ contains infinitely many elements of $A$.
2. Let $\{a_k\}$ and $\{b_k\}$ denote the sequences that are defined in the proof of Theorem 21.1, and let $x \in [a_k, b_k]$ for all $k \in \mathbf{N}$. Show that $\lim_\infty a_k = \lim_\infty b_k = x$.
3. Prove: If an ordered Archimedean field $\mathbf{F}$ has the property that, for every nested sequence of closed, nonempty intervals $\{\mathfrak{J}_k\}$ where $\lim_\infty l(\mathfrak{J}_k) = 0$, there is a unique point $x \in \mathbf{F}$ such that $x \in \mathfrak{J}_k$ for all $k \in \mathbf{N}$, then $\mathbf{F}$ has the nested-interval property, and vice versa.
4. Show: There are nested sequences of *open*, nonempty intervals $\{\mathfrak{J}_k\}$, such that $\bigcap_{k=1}^\infty \mathfrak{J}_k = \varnothing$.

## 2.22 The Location of the Rationals and Irrationals Relative to the Reals

If $a$, $b$ are rational numbers, then $(a + b)/2$ is also a rational number and lies between $a$ and $b$. We shall demonstrate in this section that there is a rational number between any two real numbers, and that there is an irrational number between any two real numbers.

***Theorem 22.1***

For any $a, b \in \mathbf{R}$, with $a < b$, there is a rational number $r$ such that $a < r < b$.

*Proof*

Let $a, b \in \mathbf{R}$, $a < b$, where we assume w.l.o.g. that $a \geqslant 0$. Then, there exists some $n \in \mathbf{N}$ such that

$$n > \frac{1}{b - a}$$

(Archimedean property, Theorem 9.3), and hence,

$$\frac{1}{n} < b - a. \tag{22.1}$$

If $a = 0$, then $0 < 1/n < b$, and our theorem is proved. If $a > 0$, then we can take a suitable multiple of $1/n$ such that $a < l(1/n) < b$. (See Fig. 22.1 and exercise 8.)

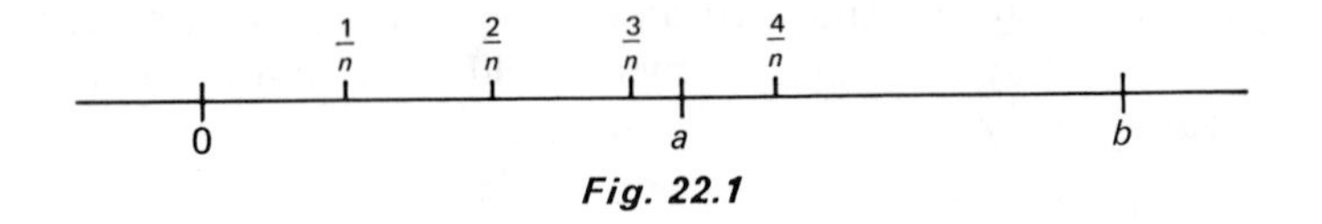


***Fig. 22.1***

***Theorem 22.2***

For any $a, b \in \mathbf{R}$, $a < b$, there is an irrational number $x$ such that $a < x < b$.

*Proof*

Let $a, b \in \mathbf{R}$, $a < b$. By Theorem 22.1, there is a $q_1 \in \mathbf{Q}$ such that $a < q_1 < b$, and a $q_2 \in \mathbf{Q}$ such that $q_1 < q_2 < b$. Hence,

$$a < q_1 < q_2 < b, \qquad q_1, q_2 \in \mathbf{Q}.$$

Let

$$x = q_1 + \frac{q_2 - q_1}{\sqrt{2}}.$$

Then $x > q_1$ since $q_2 - q_1 > 0$ and $\sqrt{2} > 0$. Since $\sqrt{2} > 1$,

$$x = q_1 + \frac{q_2 - q_1}{\sqrt{2}} < q_1 + (q_2 - q_1) = q_2.$$

Hence,

$$a < q_1 < x < q_2 < b.$$

Our theorem is proved if we can show that $x \in \mathbf{I}$. Suppose, to the contrary, that $x \in \mathbf{Q}$. Then,

$$\frac{q_2 - q_1}{x - q_1} = \sqrt{2}$$

is rational, which is absurd. Hence, $x$ is irrational, and the theorem is proved.

An important consequence of Theorems 22.1 and 22.2 is the following one:

***Theorem 22.3***

For every $x \in \mathbf{R}$, there is a nonconstant sequence $\{r_k\}$ of rational numbers and a nonconstant sequence $\{i_k\}$ of irrational numbers such that

$$\lim_{\infty} r_k = x, \qquad \lim_{\infty} i_k = x.$$

(See exercises 2, 3, 4, 5.)

## 2.22 Exercises

1. Show that $x + r$ is irrational if $x \in \mathbf{I}$ and $r \in \mathbf{Q}$.
2. Prove: If $x \in \mathbf{Q}$, then $\lim_{\infty} (x + 1/k) = x$ where $x + 1/k \in \mathbf{Q}$.
3. Prove: If $x \in \mathbf{I}$, then $\lim_{\infty} (x + 1/k) = x$ where $x + 1/k \in \mathbf{I}$.
4. Prove: If $x \in \mathbf{Q}$, then, there exist irrational numbers $i_1, i_2, i_3, \ldots$ such that $\lim_{\infty} (x + i_k) = x$ where $x + i_k \in \mathbf{I}$.
5. Prove: If $x \in \mathbf{I}$, then there exist rational numbers $r_1, r_2, r_3, \ldots$ such that $\lim_{\infty} r_k = x$.
6. Given the function $f: [0, 1] \to \mathbf{R}$ defined by:
$$f(x) = \begin{cases} 1 & \text{for } x \in [0, 1] \cap \mathbf{I}, \\ 0 & \text{for } x \in [0, 1] \cap \mathbf{Q}. \end{cases}$$
Let $a \in [0, 1]$. Find two sequences $\{x_k\}$ and $\{x_k'\}$ such that $\lim_{\infty} x_k = \lim_{\infty} x_k' = a$ and $\lim_{\infty} f(x_k) = 1$ but $\lim_{\infty} f(x_k') = 0$.
7. Show: There is a rational number (irrational number) between any two rational numbers and between any two irrational numbers.
8. Show that $l$, in the proof of Theorem 22.1, may be taken to be $l = \inf\{k \in \mathbf{N} \mid k/n > b\} - 1$.

## 2.23 Accumulation Point

In the next section, when we discuss the limit of a function $f$, we shall investigate what happens to the values $f(x)$ as $x$ approaches a certain value, say $x = a$. While it is not relevant whether or not $f$ is defined at $x = a$, it will matter that $f$ be defined on a sufficiently rich point set around $a$. In order to formulate this idea with precision, we introduce the concept of *accumulation point*. In order to do this concisely and without excessive verbiage, we first introduce the concepts of *neighborhood* and *deleted neighborhood:*

***Definition 23.1***

$N_\delta(a) = \{x \in \mathbf{R} \mid |x - a| < \delta\}$ is called a *$\delta$-neighborhood of $a \in \mathbf{R}$* and $N'_\delta(a) = \{x \in \mathbf{R} \mid 0 < |x - a| < \delta\}$ is called a *deleted $\delta$-neighborhood of $a \in \mathbf{R}$*. (See Fig. 23.1.)

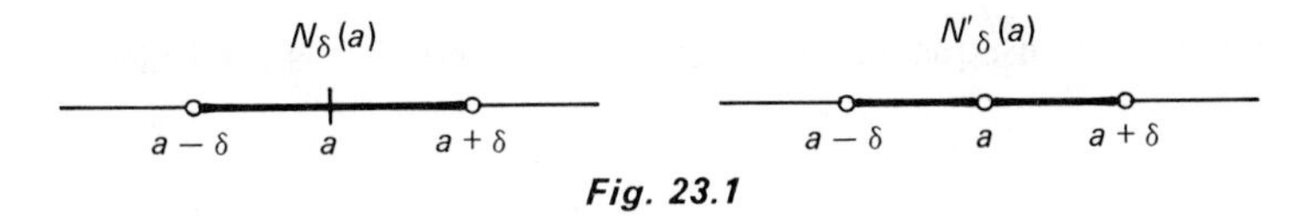


***Fig. 23.1***

Observe that $N'_\delta(a) = N_\delta(a) \backslash \{a\}$; that is, a *deleted neighborhood* of $a$ is obtained from a *neighborhood* of $a$ by *deleting* the point $a$.

***Definition 23.2***

$a \in \mathbf{R}$ is called an *accumulation point of $S \subseteq \mathbf{R}$* if and only if every deleted $\delta$-neighborhood $N'_\delta(a)$ contains a point of $S$.

---

▲ *Example 1*

Every $a \in [0, 1]$ is an accumulation point of $S = (0, 1)$.

▲ *Example 2*

$a = 0$ is an accumulation point of $S = \{1, \frac{1}{2}, \frac{1}{3}, \frac{1}{4}, \ldots\}$.

▲ *Example 3*

$S = \{1, 2, 3, \ldots\}$ does not have any accumulation points. If $a = n \in \mathbf{N}$, then $N'_{1/2}(a)$ does not contain any element of $S$. If $a \neq n$ for all $n \in \mathbf{N}$, then $N'_\delta(a)$, where

$$\delta = \tfrac{1}{2} \inf\{|a - x| \mid x \in S\},$$

does not contain any element of $S$.

▲ *Example 4*

Every $a \in [0, 1]$ is an accumulation point of $S = \{x \in \mathbf{Q} \mid 0 < x < 1\}$, because between any two real numbers is a rational number; see Theorem 22.1.

---

All these results can be stated more simply in terms of the *derived set*:

***Definition 23.3***

The *derived set* $A'$ of $A$ is the set of all accumulation points of $A$.

---

▲ *Example 5*

We have, from examples 1 through 4, that $(0, 1)' = [0, 1]$, $\{1, \frac{1}{2}, \frac{1}{3}, \frac{1}{4}, \ldots\}' = \{0\}$ $\{1, 2, 3, \ldots\}' = \varnothing$, $\{x \in \mathbf{Q} \mid 0 < x < 1\}' = [0, 1]$.

---

We have seen in all these examples that not only is there one point of $S$ in any deleted neighborhood of an accumulation point, but there are infinitely many points from $S$ in any such deleted neighborhood. As a matter of fact, one can actually form a sequence, from elements of $S$, that converges to $a$:

***Theorem 23.1***

$a \in \mathbf{R}$ is an accumulation point of $S \subseteq \mathbf{R}$ if and only if there exists a sequence $\{x_k\}$ where $x_k \in S\backslash\{a\}$ such that $\lim_\infty x_k = a$.

*Proof*

(a) If there is such a sequence, then infinitely many elements of $S$ lie in every deleted $\delta$-neighborhood of $a$. Hence, there is one element of $S$ in every deleted $\delta$-neighborhood of $a$, and $a$ is an accumulation point of $S$.

(b) Suppose that $A$ is an accumulation point of $S$. Then, for every $k \in \mathbf{N}$, there is an $x_k \in S$ such that $x_k \in N'_{1/k}(a)$. Hence, $0 < |a - x_k| < 1/k$ for all $k \in \mathbf{N}$, and we have $\lim_\infty x_k = a$, where $x_k \in S\backslash\{a\}$.

We shall frequently use this theorem in the following weaker form:

***Corollary to Theorem 23.1***

$a \in \mathbf{R}$ is an accumulation point of $S \subseteq \mathbf{R}$ if and only if every deleted $\delta$-neighborhood of $a$ contains infinitely many elements of $S$.

By Theorems 22.1 and 22.2, there is a rational number and an irrational number between any two real numbers. This fact is often expressed by saying that the rational numbers and the irrational numbers are *everywhere dense* in the set of real numbers:

***Definition 23.4***

The set $A \subseteq X$ is *everywhere dense* in $X$ if and only if $A' \cup A = X$.

***Theorem 23.2***

**Q** and **I** are everywhere dense in **R**.

*Proof*

By Theorem 22.1, there is a rational number between any two real numbers. Hence, every real number is an accumulation point of **Q**; that is, $\mathbf{Q}' = \mathbf{R}$. By the same token we obtain from Theorem 22.2 that $\mathbf{I}' = \mathbf{R}$.

If $S \subseteq \mathbf{R}$ is a nonempty set and if $a \in S$, then $a$ is either an accumulation point of $S$ or it is *not* an accumulation point of $S$. If it is an accumulation point of $S$, then every $N'_\delta(a)$ contains at least one point of $S$. Hence, if $a$ is not an accumulation point of $S$, then there exists an $N'_\delta(a)$ which does not contain any points from $S$. Such a point is called an *isolated point* of $S$.

***Definition 23.5***

The point $a \in S \subseteq \mathbf{R}$ is called an *isolated point* of $S$ if and only if there exists a $\delta > 0$ such that $N'_\delta(a) \cap S = \varnothing$.

We have from the discussion that preceded this definition:

***Lemma 23.1***

If $S \subseteq \mathbf{R}$ denotes a nonempty set and if $a \in S$, then $a$ is either an accumulation point of $S$ or an isolated point of $S$. (See also exercises 8 and 9.)

---

▲ *Example 6*

All points of $S = \{0, 1, \frac{1}{2}, \frac{1}{3}, \frac{1}{4}, \ldots\}$ except the point 0 are isolated points of $S$. 0 is an accumulation point of $S$.

▲ *Example 7*

$2, -2$ are isolated points of $S = [0, 1] \cup \{-2, 2\}$. All other points of $S$ are accumulation points of $S$.

▲ *Example 8*

**Q** has no isolated points.

---

## 2.23 Exercises

1. Prove: If $\lim_\infty x_k = a$, and $\{x_k\}$ is not a constant sequence, then $a$ is an accumulation point of $S = \{x_1, x_2, x_3, \ldots\}$.

2. Find $A'$ if:

   (a) $A = \{1, \frac{2}{3}, \frac{3}{4}, \frac{4}{5}, \frac{5}{6}, \ldots\}$

   (b) $A = \left\{\frac{1}{k} + (-1)^k \mid k \in \mathbf{N}\right\}$

   (c) $A = \left\{n + \frac{1}{k} \middle| k, n \in \mathbf{N}\right\}$

   (d) $A = \left\{k - \frac{1}{k} \middle| k \in \mathbf{N}\right\}$

3. Prove: If the sequence $\{x_k\}$ has two distinct accumulation points, then $\{x_k\}$ does not converge but contains at least two convergent subsequences, one converging to the one accumulation point, the other one converging to the other accumulation point.
4. Show: If $\lim_\infty x_k = a$ where $x_k \neq a$ for all $k \in \mathbf{N}$, then $\{x_k\}$ is not a constant sequence and does not contain a constant subsequence.
5. Show that $\mathbf{R}$ is everywhere dense in $\mathbf{R}$.
6. Show that $[0, 1] \cap \mathbf{Q}$ is everywhere dense in $[0, 1]$.
7. Show that $[0, 1] \cap \mathbf{I}$ is everywhere dense in $[0, 1]$.
8. Show that the Cantor set (section 1.14) has no isolated points.
9. Prove by contradiction: If $a \in \mathbf{R}$ is an accumulation point of $S \subseteq \mathbf{R}$, then every deleted $\delta$-neighborhood of $a$ contains infinitely many points of $S$.
10. Prove: $(A \cup B)' = A' \cup B'$.

## 2.24 Limit of a Function

We consider a function $f : \mathfrak{D} \to \mathbf{R}$ where $\mathfrak{D} \subseteq \mathbf{R}$, and assume that $a \in \mathbf{R}$ is an accumulation point of $\mathfrak{D}$. ($a$ need not be an element of $\mathfrak{D}$.) By Theorem 23.1 we may pick a nonconstant sequence $\{x_k\}$ from $\mathfrak{D}$ such that $\lim_\infty x_k = a$. Since

$$x_k \in \mathfrak{D} \qquad \text{for all } k \in \mathbf{N},$$

$f(x_k)$ is defined for all $k \in \mathbf{N}$. The sequence $\{f(x_k)\}$ may or may not converge. If it does converge, we pick another nonconstant sequence $\{x_k'\}$ from $\mathfrak{D}$ which converges to $a$, and consider $\{f(x_k')\}$. Again, $\{f(x_k')\}$ may or may not converge. If it does converge, it may or may not converge to the same limit as $\{f(x_k)\}$. If it converges to the same limit, we pick still another sequence $\{x_k''\}$ from $\mathfrak{D}$ which converges to $a$, consider $\{f(x_k'')\}$, etc... If $\{f(x_k)\}$ converges to the same limit $b$ for all possible sequences from $\mathfrak{D}$ which converge to $a$, then we say that $f$ has the limit $b$ as $x$ approaches $a$.

---

▲ *Example 1*

Let $f : [0, 1] \to \mathbf{R}$ be given by

$$f(x) = \begin{cases} 1 & \text{if } x \in [0, 1] \cap \mathbf{I}, \\ 0 & \text{if } x \in [0, 1] \cap \mathbf{Q}. \end{cases}$$

$a = 0$ is an accumulation point of $[0, 1]$. Let $x_k = 1/k$. Then $x_k \in [0, 1]$, $x_k \neq 0$, for all $k \in \mathbf{N}$ and $\lim_\infty x_k = 0$. We have $f(x_k) = 0$ for all $k \in \mathbf{N}$ and hence $\lim_\infty f(x_k) = 0$. On the other hand, if we let $x_k' = \sqrt{2}/2k$, we also have $x_k' \in [0, 1]$, $x_k' \neq 0$ for all $k \in \mathbf{N}$ and $\lim_\infty x_k' = 0$ but $\lim_\infty f(x_k') = 1$ because $f(x_k') = f(\sqrt{2}/2k) = 1$ for all $k \in \mathbf{N}$. (See also exercise 1.)

▲ *Example 2*

Let $f : [0, 2] \to \mathbf{R}$ be given by

$$f(x) = \begin{cases} -1 & \text{if } 0 \leqslant x < 1, \\ 0 & \text{if } x = 1, \\ 1 & \text{if } 1 < x \leqslant 2. \end{cases}$$

(See Fig. 24.1) $a = 1$ is an accumulation point of $[0, 2]$. The sequence $\{x_k\} = k/(k+1)$ satisfies all our requirements and, in particular, $\lim_\infty k/(k+1) = 1$. Since $f\{x_k\} = -1$, for all $k \in \mathbf{N}$, we have $\lim_\infty f(x_k) = -1$. On the other hand, we have, for $x_k' = (k+1)/k$ where $\lim_\infty (k+1)/k = 1$, that $\lim_\infty f(x_k') = 1$ because $f((k+1)/k) = 1$. (See also exercise 2.)

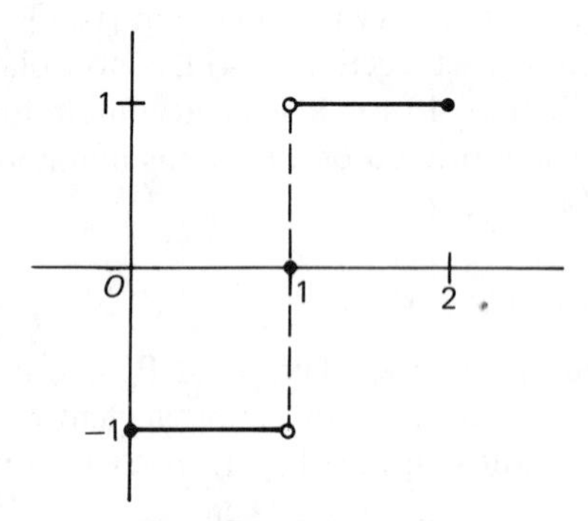


*Fig. 24.1*

▲ *Example 3*

Let $f : [-1, 0) \cup (0, 1] \to \mathbf{R}$ be given by $f(x) = x^2$. (See Fig. 24.2.) $a = 0$ is an accumulation point of $[-1, 0) \cup (0, 1]$. We pick *any* nonconstant sequence $\{x_k\}$ such that $x_k \in [-1, 0) \cup (0, 1]$, and $\lim_\infty x_k = 0$ and obtain $\lim_\infty f(x_k) = \lim_\infty x_k^2 = (\lim_\infty x_k)(\lim_\infty x_k) = 0$. Since $\lim_\infty f(x_k) = 0$ no matter what sequence $\{x_k\}$ we choose, as long as it satisfies our requirements, we say that $f(x)$ approaches the limit 0 as $x$ approaches 0.

▲ *Example 4*

Let $f : [0, 1] \to \mathbf{R}$ be given by

$$f(x) = \begin{cases} x & \text{if } x \in [0, 1] \cap \mathbf{I}, \\ 1 - x & \text{if } x \in [0, 1] \cap \mathbf{Q}. \end{cases}$$

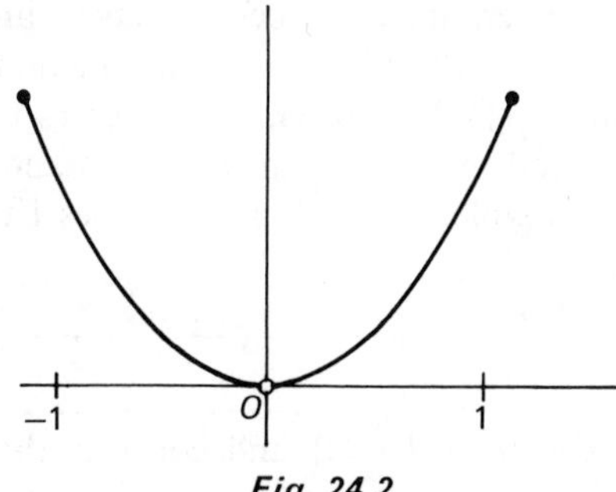


*Fig. 24.2*

$a = \frac{1}{2}$ is an accumulation point of $[0, 1]$. We pick any sequence $\{x_k\}$, $x_k \neq \frac{1}{2}$, $x_k \in [0, 1]$, such that $\lim_\infty x_k = \frac{1}{2}$, and obtain:

$$|f(x_k) - \tfrac{1}{2}| = \left.\begin{cases} |x_k - \frac{1}{2}| & \text{if } x_k \in [0, 1] \cap \mathbf{I} \\ |1 - x_k - \frac{1}{2}| & \text{if } x_k \in [0, 1] \cap \mathbf{Q} \end{cases}\right\} = |x_k - \tfrac{1}{2}| < \varepsilon$$

for all $k > N(\varepsilon)$ since $\lim_\infty x_k = \frac{1}{2}$. Hence, $\lim_\infty f(x_k) = \frac{1}{2}$ for any such sequence. Again we say that $f(x)$ approaches the limit $\frac{1}{2}$ as $x$ approaches $\frac{1}{2}$.

If $a \neq \frac{1}{2}$, but $a \in [0, 1]$, then $a$ is still an accumulation point of $[0, 1]$. In this case, we can pick sequences such that $\{f(x_k)\}$ converges to $(1 - a)$, or to $a$, or does not converge at all. (See exercises 3 and 4.)

---

***Definition 24.1***

If $f : \mathfrak{D} \to \mathbf{R}$, $\mathfrak{D} \subseteq \mathbf{R}$, and $a \in \mathbf{R}$ is an accumulation point of $\mathfrak{D}$, $f$ is said to approach the limit $b \in \mathbf{R}$ as $x$ approaches $a$:

$$\lim_{x \to a} f(x) = b, \qquad \text{or} \qquad \lim_a f(x) = b$$

if and only if $\lim_\infty f(x_k) = b$ for *all* nonconstant sequences $\{x_k\}$ in $\mathfrak{D} \backslash \{a\}$ which converge to $a$.

We obtain immediately from Theorem 16.1:

***Theorem 24.1***

If $\lim_a f(x) = b$ exists, then it is unique. (See also exercise 8.)

---

▲ *Example 5*

Let $f : (0, 1] \to \mathbf{R}$ be given by $f(x) = x \sin(1/x)$. $a = 0$ is an accumulation point of $(0, 1]$. Choose any sequence $\{x_k\}$ such that $x_k \in (0, 1]$ for all $k \in \mathbf{N}$ and $\lim_\infty x_k = 0$. Then $\lim_\infty x_k \sin(1/x_k) = 0$ because $\lim_\infty x_k = 0$, and $|\sin(1/x_k)| \leqslant 1$ for all $k \in \mathbf{N}$. (See exercise 13, section 2.17.) Hence, $\lim_0 x \sin(1/x) = 0$.

---

In order to show that a given function does not have a limit as $x$ approaches a given accumulation point $a$ of the domain of the function, it suffices either to find a nonconstant sequence $\{x_k\}$ in $\mathfrak{D}$ which converges to $a$ such that $\{f(x_k)\}$ does not converge, or to find two such sequences $\{x_k\}$ and $\{x'_k\}$ for which $\lim_\infty f(x_k) \neq \lim_\infty f(x'_k)$. (See examples 1 and 2 and exercises 1 through 5.)

---

▲ *Example 6*

Let $S = \{x \in \mathbf{R} \mid x = 2/(\pi + 2k\pi),\ k \in \mathbf{N}\}$ and consider the function $f : \mathfrak{D} \to \mathbf{R}$, $\mathfrak{D} = \mathbf{R} \backslash S$, given by

$$f(x) = x \tan \frac{1}{x}.$$

(Observe that this function is well defined on $\mathfrak{D}$ because $1/x \neq \pi/2 + k\pi$ for all $x \in \mathfrak{D}$.)

The point $a = 0$ is an accumulation point of $\mathfrak{D}$ since $\{1/k \mid k \in \mathbf{N}\} \subset \mathfrak{D}$. Does $\lim_0 f(x)$ exist? First, we consider the sequence

$$\{x_k\} = \left\{\frac{1}{\frac{\pi}{4} + k\pi}\right\} \subset \mathfrak{D} \qquad x_k \neq 0, \qquad \lim_\infty x_k = 0.$$

Then

$$\lim_\infty f(x_k) = \lim_\infty \frac{1}{\frac{\pi}{4} + k\pi} \tan\left(\frac{\pi}{4} + k\pi\right) = \lim_\infty \frac{1}{\frac{\pi}{4} + k\pi} = 0.$$

Next, we construct a sequence $\{x'_k\}$, $x'_k \in \mathfrak{D}$, $x'_k \neq 0$ such that $\lim_\infty x'_k = 0$ but $x'_k \tan 1/x'_k = 1$ for all $k \in \mathbf{N}$. This is possible: Let $1/x = z$. Then, the equation $1/z \tan z = 1$, or equivalently,

$$\tan z = z$$

has a solution $z_1 \in (\pi/2, 3\pi/2)$, a solution $z_2 \in (\pi/2 + \pi, 3\pi/2 + \pi)$, etc... In general, $z_k \in (\pi/2 + (k-1)\pi, 3\pi/2 + (k-1)\pi)$. For $x'_k = 1/z_k$, $\lim_\infty x'_k = 0$ while $\lim_\infty f(x'_k) = x'_k \tan(1/x'_k) = 1$. (See Fig. 24.3.) Hence $\lim_0 x \tan(1/x)$ does not exist.

---

## 2.24 Exercises

1. Given the function $f$ of example 1. Find a sequence $\{x_k\}$, where $x_k \neq 0$, $x_k \in [0, 1]$ for all $k \in \mathbf{N}$, and $\lim_\infty x_k = 0$, such that $\lim_\infty f(x_k)$ does *not* exist.
2. Same as in exercise 1, for the function $f$ in example 2 and $a = 1$.
3. Consider the function $f$ in example 4 and $a \neq \frac{1}{2}$. Pick a sequence $\{x_k\}$ with

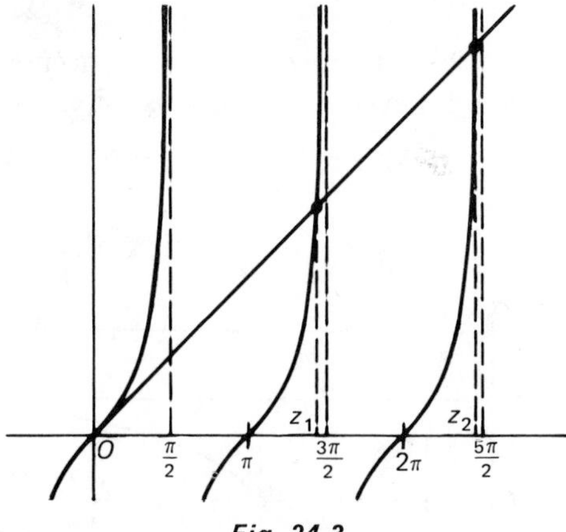


*Fig. 24.3*

all the required properties such that $\lim_\infty f(x_k) = 1 - a$, and another sequence $\{x_k'\}$ such that $\lim_\infty f(x_k') = a$.

4. Same as in exercise 1, for the function in example 4 and $a \neq \frac{1}{2}$.
5. Given $f : (0, 1] \to \mathbf{R}$ by $f(x) = 1/x$. Show that $\lim_0 f(x)$ does not exist.
6. Show that $\lim_0 1/(1 + e^{1/x})$ does not exist.
7. Find $\lim_0 x^a \sin(1/x)$ for a given $a > 0$. (See exercise 13, section 2.17.)
8. Prove Theorem 24.1.
9. $f(a + 0) = \lim_{a+0} f(x)$ is called the right-sided limit of $f$ as $x$ approaches $a$ from the right if and only if, for every nonconstant sequence $\{x_k\}$ in $\mathfrak{D}$ for which $x_k > a$ and which converges to $a$, $\lim_\infty f(x_k) = f(a + 0)$. Similarly, $f(a - 0) = \lim_{a-0} f(x)$ if and only if, for every nonconstant sequence $\{x_k\}$ in $\mathfrak{D}$ for which $x_k < a$ and which converges to $a$, $\lim_\infty f(x_k) = f(a - 0)$. Show: $\lim_a f(x)$ exists if and only if $f(a - 0) = f(a + 0)$.
10. Given the function $f : \mathbf{R} \to \mathbf{R}$ by

$$f(x) = \begin{cases} x^2 & \text{for } x < 1 \\ 0 & \text{for } x = 1 \\ -x + 3 & \text{for } x > 1 \end{cases}$$

Find $f(1 + 0)$ and $f(1 - 0)$.

11. Let $f : [0, 1] \to \mathbf{R}$. Show: If $\lim_0 f(x)$ exists, then $\lim_0 f(x) = f(0 + 0)$; and if $\lim_1 f(x)$ exists, then $\lim_1 f(x) = f(1 - 0)$.

## 2.25 Limit of a Function—Neighborhood Definition

When we defined $\lim_a f(x) = b$, we required, in essence, that $f(x)$ be close to $b$ whenever $x$ is close to $a$. Pursuit of this idea leads to an alternative characterization of the limit of a function in which the somewhat ambiguous-sounding reference to "all nonconstant sequences in $\mathfrak{D}$ which converge to $a$" does not appear, namely:

***Theorem 25.1***

Let $f: \mathfrak{D} \to \mathbf{R}$ where $\mathfrak{D} \subseteq \mathbf{R}$, and let $a \in \mathbf{R}$ denote an accumulation point of $\mathfrak{D}$. $\lim_a f(x) = b$ if and only if, for every $\varepsilon > 0$, there is a $\delta(\varepsilon) > 0$ such that

$$|f(x) - b| < \varepsilon \qquad \text{for all } x \in \mathfrak{D} \quad \text{for which } 0 < |x - a| < \delta(\varepsilon).$$

(See Fig. 25.1.)

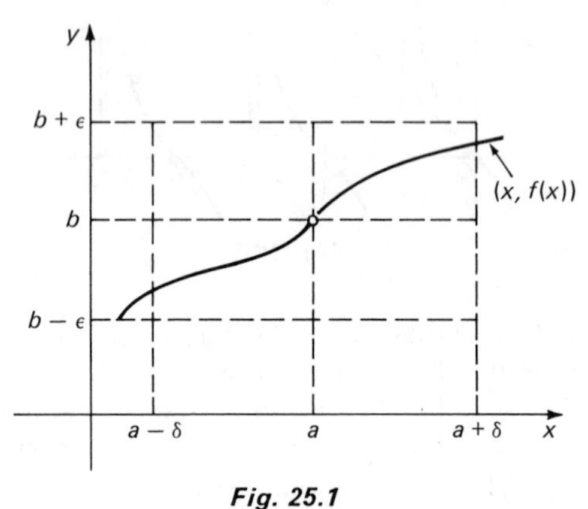


***Fig. 25.1***

In view of the definition of neighborhoods and deleted neighborhoods in definition 23.1, we may state this theorem in the following equivalent form.

***Theorem 25.1(a)***

$\lim_a f(x) = b$ if and only if, for every $\varepsilon > 0$, there is a $\delta(\varepsilon) > 0$ such that

$$f(x) \in N_\varepsilon(b) \qquad \text{for all } x \in N'_{\delta(\varepsilon)}(a) \cap \mathfrak{D}.$$

*Proof of Theorem 25.1:*

(a) We assume that the condition of Theorem 25.1 is satisfied: For every $\varepsilon > 0$, there is a $\delta(\varepsilon)$ such that $|f(x) - b| < \varepsilon$ for all $x \in \mathfrak{D}$ for which $0 < |x - a| < \delta(\varepsilon)$. Let $\{x_k\}$ denote any sequence where $x_k \in \mathfrak{D}$, $x_k \neq a$ for all $k \in \mathbf{N}$, and $\lim_\infty x_k = a$. Then, for every $\delta > 0$, there is an $N(\delta) \in \mathbf{N}$ such that $|x_k - a| < \delta$ for all $k > N(\delta)$. Since $x_k \neq a$ for all $k \in \mathbf{N}$, we have $0 < |x_k - a| < \delta$ for all $k > N(\delta)$. We identify $\delta$ with $\delta(\varepsilon)$ and obtain $|f(x_k) - b| < \varepsilon$ for all $k > N(\delta(\varepsilon))$, that is, $\lim_\infty f(x_k) = b$. Since this is true for every such sequence, the sufficiency of the condition in Theorem 25.1 is established.

(b) We prove the converse by contradiction, assuming that the condition in Theorem 25.1 is not satisfied.

If there is an $\varepsilon > 0$ such that, for all $\delta > 0$, there are points $x \in \mathfrak{D}$ for which $0 < |x - a| < \delta$ but $|f(x) - b| \geqslant \varepsilon$, we choose a sequence $\{\delta_k\}$ where $\delta_k > 0$ and $\lim_\infty \delta_k = 0$ and pick for each $\delta_k$ an $x_k \in \mathfrak{D}$, such that $0 < |x_k - a| < \delta_k$ but $|f(x_k) - b| \geqslant \varepsilon$. By construction, $\lim_\infty x_k = a$ but $\lim_\infty f(x_k) \neq b$ or does not exist. Hence, there exists a sequence $\{x_k\}$ with all the required properties for which $\lim_\infty f(x_k)$ is either different from $b$ or does not exist, and the condition in definition 24.1 is not satisfied either.

The following theorem on manipulation of limits can be proved either with reference to definition 24.1 or on the basis of Theorem 25.1.

***Theorem 25.2***

If $a \in \mathbf{R}$ is an accumulation point of $\mathfrak{D} \subseteq \mathbf{R}$ and $f, g : \mathfrak{D} \to \mathbf{R}$, where $\mathfrak{D} \subseteq \mathbf{R}$, and if $\lim_a f(x) = b$, $\lim_a g(x) = c$, then

$$\lim_a (f(x) \pm g(x)) = b \pm c$$

$$\lim_a (f(x)g(x)) = bc.$$

The proof is left to the reader. (See exercises 3 and 4.)

In order to prove a similar theorem about the limit of a quotient of two functions, we need the following lemma.

***Lemma 25.1***

If $a \in \mathbf{R}$ is an accumulation point of $\mathfrak{D} \subseteq \mathbf{R}$, if $f : \mathfrak{D} \to \mathbf{R}$, and if $\lim_a f(x) = b > 0$ then there exists a deleted neighborhood $N'_\delta(a)$ of $a$ such that $f(x) > 0$ for all $x \in N'_\delta(a) \cap \mathfrak{D}$.

***Proof***

By Theorem 25.1 we obtain, for $\varepsilon = (b/2) > 0$, a $\delta(\varepsilon) > 0$ such that $-b/2 < f(x) - b < b/2$ or equivalently, $b/2 < f(x) < 3b/2$ for all $x \in \mathfrak{D}$ for which $0 < |x - a| < \delta(\varepsilon)$, that is, for all $x \in N'_{\delta(\varepsilon)}(a) \cap \mathfrak{D}$.

***Theorem 25.3***

If $a \in \mathbf{R}$ is an accumulation point of $\mathfrak{D}$, and if $f, g : \mathfrak{D} \to \mathbf{R}$, where $\mathfrak{D} \subseteq \mathbf{R}$, and if $\lim_a f(x) = b$ and $\lim_a g(x) = c \neq 0$, then $\lim_a (f(x)/g(x)) = b/c$.

That $f/g$ is defined on some $N'_\delta(a) \cap \mathfrak{D}$ and that $a$ is an accumulation point of this set follow from Lemma 25.1. The remaining part of the proof is straightforward, and is left for the exercises (exercise 4).

Before closing this section, let us briefly discuss the case where the values of a function grow beyond all bounds as $x$ approaches an accumulation point of the domain of the function. In such a case, we speak of an *improper limit* of the function.

***Definition 25.1***

Let $f: \mathfrak{D} \to \mathbf{R}$ where $\mathfrak{D} \subseteq \mathbf{R}$, and let $a$ denote an accumulation point of $\mathfrak{D}$. Then,

$$\lim_{a} f(x) = \infty$$

if and only if, for every $M > 0$, there is a $\delta(M) > 0$ such that $f(x) > M$ for all $x \in N'_{\delta(M)}(a) \cap \mathfrak{D}$ ($\lim_a f(x) = -\infty$ if $f(x) < -M$). (See also exercise 6.)

---

▲ *Example 1*

Let $f: \mathbf{R} \to \mathbf{R}$ be defined by $f(0) = 0$ and $f(x) = 1/x^2$ for $x \neq 0$. Then

$$\lim_0 f(x) = \infty .$$

▲ *Example 2*

Let $f: (-\infty, 0) \to \mathbf{R}$ be defined by $f(x) = 1/x$. Then, $\lim_0 f(x) = -\infty$.

---

## 2.25 Exercises

1. Show that Theorem 25.1 and 25.1a are equivalent.
2. Prove Theorem 25.2 on the basis of definition 24.1.
3. Prove Theorem 25.2 on the basis of Theorem 25.1, or 25.1a.
4. Prove Theorem 25.3.
5. Let $f: (0, 1) \to \mathbf{R}$ be defined by $f(x) = \sin(1/x)$. Show that $\lim_0 f(x)$ does not exist.
6. Prove: $\lim_a f(x) = \infty$ if and only if for every sequence $\{x_k\}$ for which $x_k \in \mathfrak{D}(f)$, $x_k \neq a$ for all $k \in \mathbf{N}$ and $\lim_\infty x_k = a$, $\{f(x_k)\}$ diverges to $\infty$.

## 2.26 Continuous Functions

It is a part of the mathematical folklore that we pick up in calculus that a function is continuous at a point if and only if the limit of the function is equal to the value of the function. If we were to follow this narrow view, many functions would be relegated to a mathematical limbo, being neither continuous nor discontinuous. A case in point is the function $f: \{0\} \to \mathbf{R}$ that is given by $f(x) = 1$. (See Fig. 26.1.)

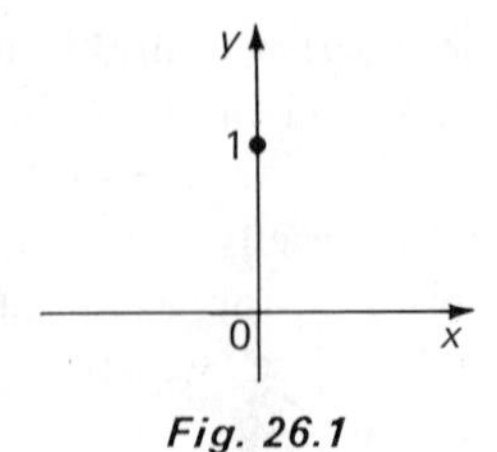


*Fig. 26.1*

Since the domain of this function, consisting of the point 0 only, has no accumulation point, it is senseless to talk about $\lim_0 f(x)$. However, this function does not even have a chance to be not continuous, and not calling it continuous would be penalizing it for the paucity of its domain.

We shall give here a definition of continuity that is general enough to embrace cases such as the one we have just discussed, and still is compatible with the familiar definition that is given in calculus for the case where the point under consideration is an accumulation point of the domain of the function.

***Definition 26.1***

Let $f: \mathfrak{D} \to \mathbf{R}$, $\mathfrak{D} \subseteq \mathbf{R}$, and $a \in \mathfrak{D}$. $f$ is continuous at $a$ if and only if, for every $\varepsilon > 0$, there is a $\delta(\varepsilon) > 0$ such that

(26.1) $\quad |f(x) - f(a)| < \varepsilon \quad$ for all $x \in \mathfrak{D}$ for which $|x - a| < \delta(\varepsilon)$,

or equivalently,

(26.2) $\quad f(x) \in N_\varepsilon(f(a)) \quad$ for all $x \in N_{\delta(\varepsilon)}(a) \cap \mathfrak{D}$.

We observe two formal differences between definition 26.1 and Theorem 25.1: Instead of $b$, we now have $f(a)$, and instead of $N_\delta'(a)$, we now have $N_\delta(a)$. We also observe that now $a \in \mathfrak{D}$, while in Theorem 25.1 (or definition 26.1) $a$ need not be in $\mathfrak{D}$. These observations lead immediately to:

***Theorem 26.1***

If $f: \mathfrak{D} \to \mathbf{R}$, $\mathfrak{D} \subseteq \mathbf{R}$, and $a \in \mathfrak{D}$ is an accumulation point of $\mathfrak{D}$, then $f$ is continuous at $a$ if and only if $\lim_a f(x) = f(a)$.

(For proof, see exercise 1.)

---

▲ *Example 1*

We consider the function $f: \{0\} \to \mathbf{R}$, given by $f(x) = 1$, which we discussed at the beginning of this section. This function is continuous at $0 \in \{0\} = \mathfrak{D}$: $f(x) - f(0) = 1 - 1$ for all $x \in \mathfrak{D}$. We have $|f(x) - f(0)| < \varepsilon$ for all $x \in N_\delta(0) \cap \mathfrak{D}$ for any $\varepsilon > 0$, $\delta > 0$. Theorem 26.1 does not apply to this case because 0 is *not* an accumulation point of $\{0\}$. (For similar examples, see exercises 2 and 3.)

---

In the above example, 0 is an isolated point of the domain of $f$ (see definition 23.5). A function is always continuous at an isolated point:

***Theorem 26.2***

If $f: \mathfrak{D} \to \mathbf{R}$, $\mathfrak{D} \subseteq \mathbf{R}$, and $a \in \mathfrak{D}$ is an isolated point of $\mathfrak{D}$, then $f$ is continuous at $a$.

*Proof*

By definition 23.5, there is a $\delta > 0$ such that $N'_\delta(a) \cap \mathfrak{D} = \varnothing$. Hence we have, for any $\varepsilon > 0$, that $f(x) \in N_\varepsilon(f(a))$ for all $x \in N_\delta(a) \cap \mathfrak{D} = \{a\}$.

By Lemma 23.1, any point $a \in \mathfrak{D}$ is either an accumulation point of $\mathfrak{D}$ or an isolated point. Therefore, Theorems 26.1 and 26.2 take care of all contingencies. In the first case, the familar test for continuity applies, and in the latter case, the function is trivially continuous.

---

▲ *Example 2*

Let $f: \mathbf{N} \to \mathbf{R}$ be defined by $f(x) = 1/x$. Since every point of $\mathbf{N}$ is an isolated point of $\mathbf{N}$, $f$ is continuous at every point of $\mathbf{N}$.

▲ *Example 3*

Consider $f: \mathbf{R} \to \mathbf{R}$, given by $f(x) = x^n$. $f$ is continuous for all $a \in \mathbf{R}$: Since $a \in \mathbf{R}$ is an accumulation point of $\mathbf{R}$, we may apply Theorem 26.1. We have, for *any* sequence $\{x_k\}$ for which $x_k \neq a$ and $x_k \in \mathbf{R}$ for all $k \in \mathbf{N}$ and $\lim_\infty x_k = a$, that $\lim_\infty f(x_k) = \lim_\infty x_k^n = a^n$. Hence, $\lim_a f(x) = a^n$. Since $f(a) = a^n$, we have $\lim_a f(x) = f(a)$; that is, $f$ is continuous at $a$.

▲ *Example 4*

The function $f: \mathbf{R} \to \mathbf{R}$, defined by

$$f(x) = \begin{cases} 0 & \text{for } x \neq 0, \\ 1 & \text{for } x = 0, \end{cases}$$

is *not* continuous at 0. We have $\lim_0 f(x) = 0$ but $f(0) = 1$. A discontinuity of this type is called a *removable discontinuity*. (See Fig. 26.2 and exercise 13.)

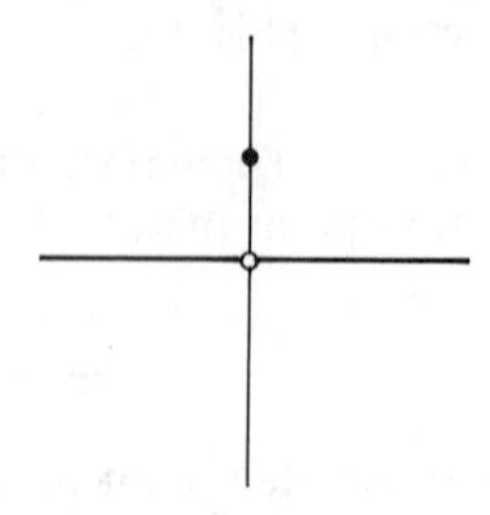

*Fig. 26.2*

▲ *Example 5*

The function $g : \mathbf{R} \to \mathbf{R}$, defined by

$$g(x) = \begin{cases} 0 & \text{for } x < 0, \\ 1 & \text{for } x \geqslant 0, \end{cases}$$

is *not* continuous at $x = 0$. $\lim_0 g(x)$ does not exist. A discontinuity of this type is called a *jump discontinuity*. (See Fig. 26.3 and exercise 14.)

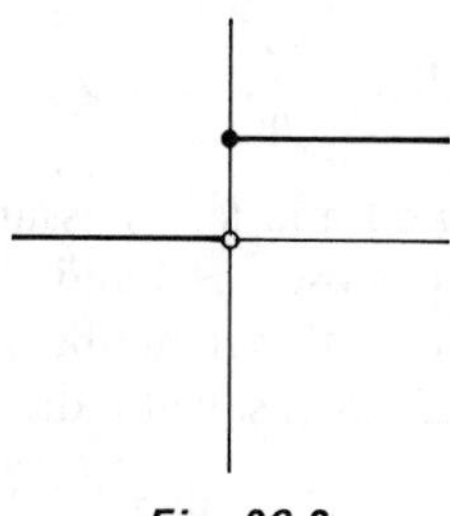

*Fig. 26.3*

▲ *Example 6*

The function $h : \mathbf{R} \to \mathbf{R}$, defined by

$$h(x) = \begin{cases} 0 & \text{for } x = 0, \\ \dfrac{1}{x} & \text{for } x \neq 0, \end{cases}$$

is *not* continuous at $x = 0$. $\lim_0 h(x)$ does not exist. A discontinuity of this type is called an *infinite discontinuity*. (See Fig. 26.4 and exercise 15.)

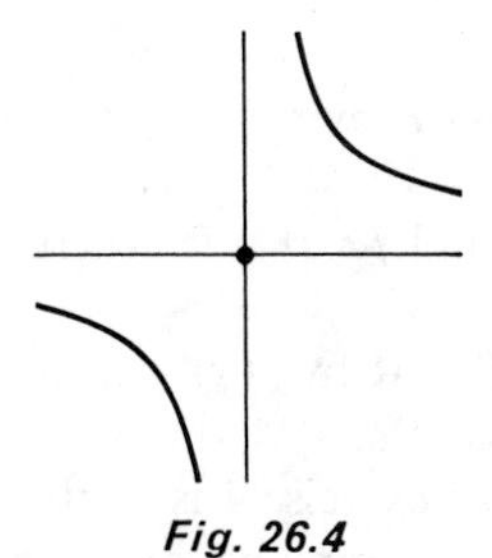

*Fig. 26.4*

---

***Definition 26.2***

$f : \mathfrak{D} \to \mathbf{R}$, where $\mathfrak{D} \subseteq \mathbf{R}$, is called continuous on $A \subseteq \mathfrak{D}$ if and only if $f$ is continuous for *all* $a \in A$.

▲ *Example 7*

The function of example 3 is continuous for all $a \in \mathbf{R}$. Hence it is continuous on $\mathbf{R}$.

▲ *Example 8*

The function $f : \mathbf{R} \to \mathbf{R}$, given by

$$f(x) = \begin{cases} 0 & \text{for all } x \in \mathbf{I}, \\ \dfrac{1}{n} & \text{for all } x = \dfrac{m}{n}, \quad m \in \mathbf{Z}, \quad n \in \mathbf{N}, \quad (m, n) = 1, \end{cases}$$

is continuous on $\mathbf{I} \subset \mathbf{R}$: Let $a \in \mathbf{I}$ and let us assume w.l.o.g. that $a > 1$. Then, $a \in (k, k+1)$ for some $k \in \mathbf{N}$. Choose $\varepsilon > 0$ and choose $n_0 \in \mathbf{N}$ such that $(1/n_0) < \varepsilon$. There are only finitely many rational numbers $m/n$ in $(k, k+1)$ with $n < n_0$ and hence, one of them is closest to $a$, say at a distance $\delta$. Since

$$|f(x) - f(a)| = \begin{cases} |0 - 0| = 0 & \text{if } x \in \mathbf{I} \\ \left|\dfrac{1}{n} - 0\right| = \dfrac{1}{n} & \text{if } x = \dfrac{m}{n} \end{cases}$$

and since $1/n < 1/n_0 < \varepsilon$ for all $m/n \in N_\delta(a) \cap \mathbf{Q}$, we see that $|f(x) - f(a)| < \varepsilon$ for all $x \in N_\delta(a)$, that is, $f$ is continuous at $a$. Since $a$ is any irrational number, $f$ is continuous on $\mathbf{I}$. $f$ is not continuous at any rational number because one can find in every neighborhood of $m/n$ an irrational number $\alpha$ such that

$$f\left(\frac{m}{n}\right) - f(\alpha) = \frac{1}{n}.$$

## 2.26 Exercises

1. Prove Theorem 26.1.
2. Given the function $f : \mathbf{N} \to \mathbf{R}$ by $f(x) = 1/(\sin x)$. Show that $f$ is continuous on $\mathbf{N}$.
3. Let $\mathfrak{D} = \{1, \frac{1}{2}, \frac{1}{3}, \frac{1}{4}, \ldots\}$ and $f : \mathfrak{D} \to \mathbf{R}$ where $f(x) = 1/x$. Is $f$ continuous on $\mathfrak{D}$?
4. Given the function $f : \mathbf{R} \to \mathbf{R}$ by $f(x) = x^n$ and $a \in \mathbf{R}$. Prove, by definition 26.1, that $f$ is continuous at $a$.
5. Show that the function of exercise 4 is continuous on $\mathbf{R}$.
6. Show: If $f : [a, b] \to \mathbf{R}$ is continuous on $[a, b]$ and if $\eta \in \mathbf{R}$, then the function $F$, defined by $F(x) = f(x) - \eta$ is also continuous on $[a, b]$.
7. Show that the function $f : [0, 1] \to \mathbf{R}$, defined by

$$f(x) = \begin{cases} 0 & \text{if } x = 0 \\ \dfrac{1}{x} & \text{if } x \neq 0 \end{cases}$$

is *not* continuous on [0, 1].

8. Same as in exercise 7, for the function $f: [0, 1] \to \mathbf{R}$ that is given by

$$f(x) = \begin{cases} 0 & \text{if } x = 0 \\ \sin \dfrac{1}{x} & \text{if } x \neq 0 \end{cases}$$

9. Show that the function $f: \mathbf{R}^+ \to \mathbf{R}$ that is defined by $f(x) = 1/x$ is continuous on $\mathbf{R}^+$.
10. Given the function $f: [0, 1] \to \mathbf{R}$ by

$$f(x) = \begin{cases} 0 & \text{if } x = 0 \\ x \sin \dfrac{1}{x} & \text{if } x \neq 0 \end{cases}$$

Show that $f$ is continuous on [0, 1].

11. Given $f: (a, b) \to \mathbf{R}$. Show: If $f$ is continuous on $(a, b)$, and if $\lim_a f(x)$ and $\lim_b f(x)$ exist, then there is a function $F: [a, b] \to \mathbf{R}$ which is continuous on $[a, b]$ and is such that $F(x) = f(x)$ for all $x \in (a, b)$.
12. Prove: If $f: \mathcal{A} \to \mathbf{R}$ is continuous at $a \in \mathcal{A}$, then the function $|f|: \mathcal{A} \to \mathbf{R}$, defined by $|f|(x) = |f(x)|$, is also continuous at $a$.
13. Let $f: \mathfrak{D} \to \mathbf{R}$ where $\mathfrak{D} \subseteq \mathbf{R}$, and let $a \in \mathfrak{D}$ denote an accumulation point of $\mathfrak{D}$. $f$ is said to have a *removable discontinuity* at $a$ if and only if $\lim_a f(x)$ exists but $\lim_a f(x) \neq f(a)$.

    (a) Show that the function $f$ in example 4 has a removable discontinuity at 0 by this definition.
    (b) Explain the term "removable discontinuity."

14. Let $f: \mathfrak{D} \to \mathbf{R}$ where $\mathfrak{D} \subseteq \mathbf{R}$, and let $a \in \mathfrak{D}$ denote an accumulation point of $\mathfrak{D}$. Then $f$ is said to have a *jump discontinuity* at $a$ if and only if $\lim_{a-0} f(x)$ and $\lim_{a+0} f(x)$ both exist but $\lim_{a-0} f(x) \neq \lim_{a+0} f(x)$. Show that the function $g$ of example 5 has a jump discontinuity at 0 by this definition.
15. Let $f: \mathfrak{D} \to \mathbf{R}$ where $\mathfrak{D} \subseteq \mathbf{R}$, and let $a \in \mathfrak{D}$ denote an accumulation point of $\mathfrak{D}$. Then $f$ is said to have an *infinite discontinuity* at $a$ if and only if $\lim_{a-0} f(x) = \pm\infty$ or $\lim_{a+0} f(x) = \pm\infty$ or both. Show that the function $h$ of example 6 has an infinite discontinuity at 0 by this definition.
16. Show: If $f: \mathcal{A} \to \mathbf{R}$ is continuous on $\mathcal{A}$, then the function $g: \mathbf{R} \to \mathbf{R}$, which is defined by $g(x) = f(x)$ for $x \in \mathcal{A}$, $g(x) = 0$ for $x \notin \mathcal{A}$, need not be continuous on $\mathcal{A}$.

## 2.27 Properties of Continuous Functions

The intuitive meaning of continuity of a function at an accumulation point $a$ of its domain is that the change of the function must remain small provided that the change of $x$ is restricted to a sufficiently small neighborhood of $a$. Therefore, the following lemma, reflecting a very important property of continuity, is easy to understand.

***Lemma 27.1***

If $f: \mathfrak{D} \to \mathbf{R}$, where $\mathfrak{D} \subseteq \mathbf{R}$, is continuous at $a \in \mathfrak{D}$, and if $f(a) \neq 0$, then there exists a $\delta > 0$ such that $f(x) \neq 0$ for all $x \in N_\delta(a) \cap \mathfrak{D}$. Specifically, if $f(a) > 0$ ($< 0$), then $f(x) > 0$ ($< 0$) for all $x \in N_\delta(a) \cap \mathfrak{D}$.

This lemma is trivial if $a$ is an isolated point of $\mathfrak{D}$ and follows from Lemma 25.1 and Theorem 26.1 if $a$ is an accumulation point of $\mathfrak{D}$.

Note that, in order for Lemma 27.1 to hold, $f$ need be continuous at $a$ only and nowhere else. A case of a function which is continuous at one point only is discussed in the following example:

---

▲ *Example 1*

Let $f: \mathbf{R} \to \mathbf{R}$ be defined by

$$f(x) = \begin{cases} x & \text{if } x \in \mathbf{I}, \\ 1 - x & \text{if } x \in \mathbf{Q}. \end{cases}$$

We have seen in example 4, section 2.24 that $\lim_{1/2} f(x) = \frac{1}{2}$. Since $f(\frac{1}{2}) = \frac{1}{2}$, $f$ is continuous at $\frac{1}{2}$. Since $\lim_a f(x)$ does not exist for $a \neq \frac{1}{2}$ (see exercises 3, 4, section 2.24), $f$ is not continuous anywhere but at $\frac{1}{2}$. Still, there is a $N_\delta(\frac{1}{2})$, for example, $N_{1/2}(\frac{1}{2})$ such that $f(x) > 0$ for all $x \in N_{1/2}(\frac{1}{2})$.

---

The intuitive meaning of continuity is also reflected in the so-called *intermediate-value property* of continuous functions, meaning that a function which is continuous on an interval assumes every value between any two of its values. In order to prove this, we establish first the following lemma.

***Lemma 27.2***

If $f: [a, b] \to \mathbf{R}$ is continuous on $[a, b]$ and if $f(a) < 0$, $f(b) > 0$, then there is a point $\xi \in (a, b)$ such that $f(\xi) = 0$. (See Fig. 27.1.)

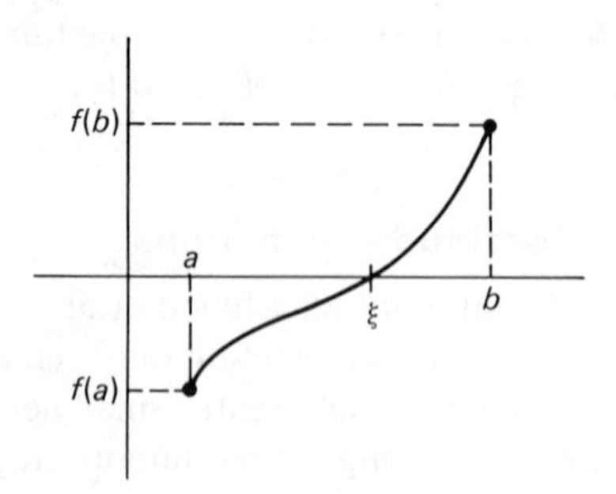


*Fig. 27.1*

*Proof*

The set $A = \{x \in [a, b] \mid f(x) < 0\}$ is not empty, since $a \in A$. $A$ is bounded above by $b$; hence, sup $A = \xi$ exists and $\xi \in [a, b]$. By Theorem 9.1, we have, for every $k \in \mathbf{N}$, an element $x_k \in A$ such that $\xi - 1/k < x_k \leqslant \xi$.

Hence, $|\xi - x_k| < 1/k$ and, consequently, $\lim_\infty x_k = \xi$. $\xi \in [a, b]$ is an accumulation point of $[a, b]$, and since $f$ is continuous on $[a, b]$, we have from Theorem 26.1 that $\lim_\infty f(x_k) = f(\xi)$. Since $f(x_k) < 0$ for all $k \in \mathbf{N}$, we have $\lim_\infty f(x_k) \leqslant 0$ (see the Corollary to Theorem 17.2), and hence, $f(\xi) \leqslant 0$.

Suppose that $f(\xi) < 0$. Then, $\xi < b$ because $f(b) > 0$. By Lemma 27.1,

$$f(x) < 0 \qquad \text{for all } x \in N_\delta(\xi) \cap [a, b]$$

for some $\delta > 0$. Since $\xi \in [a, b)$, there are, by Lemma 27.1, points $x$, to the right of $\xi$, for which $f(x) < 0$, contrary to the definition of $\xi$. Hence, $f(\xi) = 0$ and $\xi > a$.

The intermediate-value property of continuous functions is now a simple consequence of Lemma 27.2:

***Theorem 27.1*** *(Intermediate-Value Theorem)*

If $f: [a, b] \to \mathbf{R}$ is continuous on $[a, b]$, $f(a) \neq f(b)$, and if $\eta$ is any value between $f(a)$ and $f(b)$, then there is a $\xi \in (a, b)$ such that $f(\xi) = \eta$. (See Fig. 27.2.)

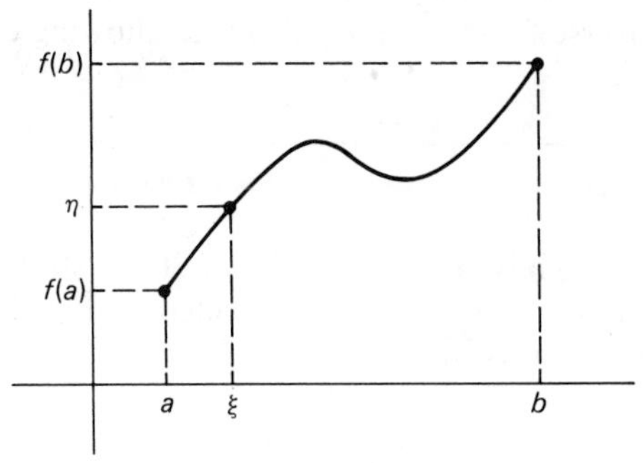


***Fig. 27.2***

*Proof*

We assume w.l.o.g. that $f(a) < f(b)$ and $f(a) < \eta < f(b)$. The function $F$: $[a, b] \to \mathbf{R}$, defined by $F(x) = f(x) - \eta$, satisfies the hypothesis of Lemma 27.2. Hence, there is a $\xi \in (a, b)$ such that $F(\xi) = 0$; that is, $f(\xi) = \eta$. (See also exercise 6, section 2.26.)

The following theorem on the inverse of a continuous, strictly increasing function, very important in itself, also serves well to illustrate the intermediate-value property:

***Theorem 27.2***

If $f:(a, b) \to \mathbf{R}$ is continuous and strictly increasing (decreasing), then $f^{-1}$: $\mathcal{R}(f) \to \mathbf{R}$ is also continuous.

*Proof*

By Theorem 5.4, $f^{-1}: \mathcal{R}(f) \to \mathbf{R}$ exists and is strictly increasing (decreasing). We assume that $f$ is strictly increasing, and demonstrate that $f$ maps open intervals in its domain onto open intervals. Let $\alpha, \beta \in (a, b)$ and $\alpha < \beta$. Then, by definition 5.2, $f(\alpha) < f(\beta)$. By the intermediate-value theorem there is, for every $\eta \in (f(\alpha), f(\beta))$, a $\xi \in (\alpha, \beta)$ such that $f(\xi) = \eta$. Since $f$ is strictly increasing and hence, one-to-one, there is exactly one such $\xi$. Conversely, for every $\xi \in (\alpha, \beta)$, $f(\xi) \in (f(\alpha), f(\beta))$. Hence, $f: (\alpha, \beta) \leftrightarrow (f(\alpha), f(\beta))$.

If $c \in (a, b)$ and if $N_\varepsilon(c) \subseteq (a, b)$, then $f: N_\varepsilon(c) \leftrightarrow (f(c - \varepsilon), f(c + \varepsilon))$. Therefore, there is a $\delta(\varepsilon) > 0$ with $N_{\delta(\varepsilon)}(f(x_0)) \in (f(c - \varepsilon), f(c + \varepsilon))$ such that $x \in N_\varepsilon(c)$ for all $f(x) \in N_{\delta(\varepsilon)}(f(c))$. With $f(c) = d$ and $f(x) = y$ we see that, for every $\varepsilon > 0$, there is a $\delta(\varepsilon) > 0$ such that $f^{-1}(y) \in N_\varepsilon(f^{-1}(d))$ for all $y \in N_{\delta(\varepsilon)}(d)$; that is, $f^{-1}$ is continuous at $d$. Since there is, for every $d \in \mathcal{R}(f)$, a $c \in (a, b)$ such that $f(c) = d$, we see that $f^{-1}$ is continuous in $\mathcal{R}(f)$.

In section 6.69 where we shall reopen this question in a more general setting, we shall see that the assumption in Theorem 27.2 that $f$ is continuous may be dispensed with. That $f$ is strictly increasing suffices already to guarantee the continuity of the inverse function. (See also the following example.)

---

▲ *Example 2*

The function whose graph is depicted in Fig. 27.3(a) is strictly increasing but not continuous. The graph of its inverse is depicted in Fig. 27.3(b). Clearly, this function is continuous on its domain.

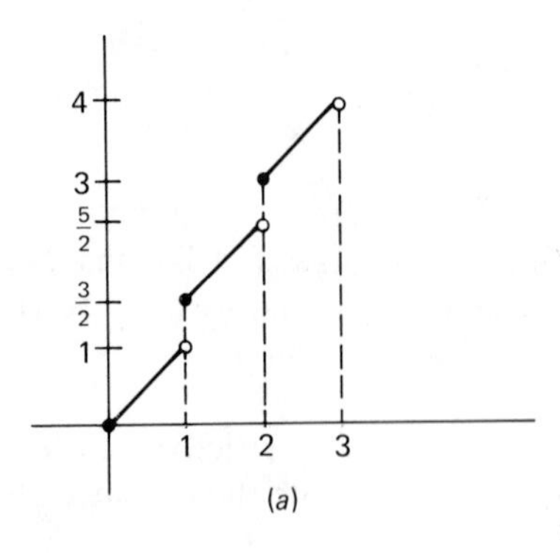


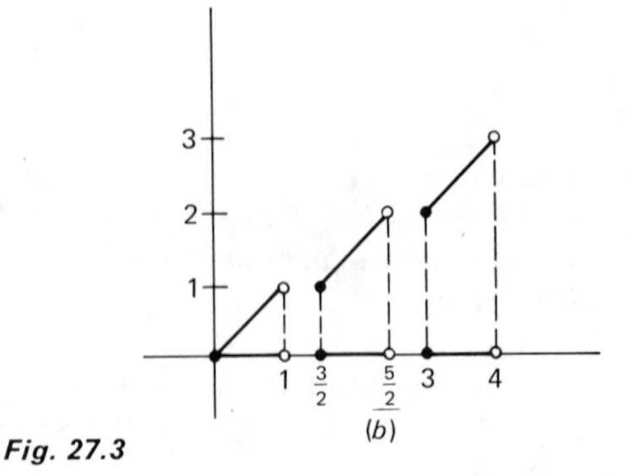


*Fig. 27.3*

---

Sums and products of continuous functions are again continuous functions: If $f, g : \mathfrak{D} \to \mathbf{R}$ are given functions, then their sum $f + g$ and their product $fg$ are defined by

$$(f + g)(x) = f(x) + g(x)$$
$$(fg)(x) = f(x)g(x)$$

for all $x \in \mathfrak{D}$ (ordinatewise addition and multiplication). We have:

***Lemma 27.3***

If $f, g : \mathfrak{D} \to \mathbf{R}$ are continuous at $a \in \mathfrak{D}$, then $f + g$ and $fg$ are continuous at $a$.

*Proof*

If $a$ is an accumulation point of $\mathfrak{D}$, the lemma follows from Theorem 25.2. If $a$ is an isolated point of $\mathfrak{D}$, the lemma is trivially true.

We obtain from Lemma 27.3 and definition 26.2:

***Theorem 27.3***

If $f, g$: $\mathfrak{D} \to \mathbf{R}$ are continuous on $A \subseteq \mathfrak{D}$, then $f + g$ and $fg$ are also continuous on $A$.

## 2.27 Exercises

1. Let $f$: $\mathfrak{D} \to \mathbf{R}$ where $\mathfrak{D} \subseteq \mathbf{R}$, and let $a \in \mathfrak{D}$ denote an accumulation point of $\mathfrak{D}$. Let

$$\mathfrak{D}_1 = \{x \in \mathfrak{D} \mid f(x) = 0\}$$

   and let $a \in \mathfrak{D}_1'$, Show: If $f(a) \neq 0$, then $f$ cannot be continuous at $a$.
2. Show: If $f : [a, b] \to \mathbf{R}$ is continuous and if $f(x) = 0$ for all $x \in \mathbf{Q}$, then $f(x) = 0$ for all $x \in [a, b]$.
3. Prove: $f : \mathfrak{D} \to \mathbf{R}$ is continuous at $a \in \mathfrak{D}$ if and only if, for all $A > f(x_0)$, there exists a $\delta(A) > 0$ such that $A > f(x)$ for all $x \in N_{\delta(A)}(a) \cap \mathfrak{D}$, and, for all $B < f(x_0)$ there exists a $\delta(B) > 0$ such that $B < f(x)$ for all $x \in N_{\delta(B)}(a) \cap \mathfrak{D}$.
4. Derive Lemma 27.1 from the result of exercise 3. 
5. Let $f : \mathfrak{D} \to \mathbf{R}$ denote a continuous function, let $a, b \in \mathfrak{D}$, and assume that $f(a) < 0$, $f(b) > 0$. Construct an example to show that Lemma 27.2 need not apply if $\mathfrak{D}$ is not an interval.
6. Prove: If $f : \mathfrak{D} \to \mathbf{R}$ is continuous at $a \in \mathfrak{D}$, then, $-f : \mathfrak{D} \to \mathbf{R}$, defined by

$$(-f)(x) = -f(x),$$

   is continuous at $a$.

7. Prove: If $f, g : \mathfrak{D} \to \mathbf{R}$ are continuous at $a \in \mathfrak{D}$ and if $g(a) \neq 0$, then there is a $\delta > 0$ such that the function $f/g$, defined by
$$\frac{f}{g}(x) = \frac{f(x)}{g(x)},$$
is defined for all $x \in N_\delta(a) \cap \mathfrak{D}$.
8. Prove: If $f, g : \mathfrak{D} \to \mathbf{R}$ are continuous at $a \in \mathfrak{D}$ and if $g(a) \neq 0$, then $f/g$ is continuous at $a$.
9. Show that $f : \mathbf{R} \to \mathbf{R}$, given by $f(x) = a_0 x^n + a_1 x^{n-1} + a_1 x^{n-2} + \cdots + a_{n-1} x + a_n, a_i \in \mathbf{R}$, is continuous on $\mathbf{R}$.
10. Let $f : \mathbf{R} \to \mathbf{R}$ be defined by $f(x) = x^{2n+1} + a_1 x^{2n} + \cdots + a_{2n} x + a_{2n+1}$. Show that $f(x) = 0$ has at least one real solution.
11. Assuming the continuity of the cosine function, show that $\cos x = x$ has at least one solution in $[0, 1]$.
12. Assuming that the temperature changes continuously along the equator, show that there are, at any given time, antipodal points on the equator with the same temperature.
13. Let $f$: $[a, b] \to \mathbf{R}$. Show: If $f(x) \neq (f(a) + f(b))/2$ for all $x \in [a, b]$, then $f$ is not continuous on $[a, b]$.

## 2.28 Some Elementary Point Set Topology

"God created the open sets, everything else is the work of topologists."

Neighborhoods have played an integral part in our discussions of limits and continuity of functions. A neighborhood is a special case of an *open set*. Open sets play a fundamental role in mathematical analysis, as we shall see in later developments. We shall define open sets in terms of *interior points*:

***Definition 28.1***

The point $a \in A \subseteq \mathbf{R}$ is an *interior point* of $A$ if and only if there exists a $\delta > 0$ such that $N_\delta(a) \subseteq A$.

---

▲ *Example 1*

$\mathfrak{J} = (0, 1) \subset \mathbf{R}$ consists of interior points only.

▲ *Example 2*

$B = \{1, \frac{1}{2}, \frac{1}{3}, \frac{1}{4}, \ldots\} \subset \mathbf{R}$ has no interior points.

▲ *Example 3*

$\mathfrak{I} = [0, 1] \subset \mathbf{R}$ has all $x \in (0, 1)$ as interior points. The endpoints 0, 1 are *not* interior points.

▲ *Example 4*

Let $D = \{0, 1, \frac{1}{2}, \frac{1}{3}, \frac{1}{4}, \ldots\} \subset \mathbf{R}$ and $E = c(D)$. All $x \in E$ are interior points of $E$: if $x \in E$, then $x \neq 0$. Either $x < 0$ and hence $N_\delta(x) \subset E$ for $\delta = |x|/2$, or $x > 0$. If $x > 0$, the $1/(n+1) < x < 1/n$ for some $n \in \mathbf{N}$ and $N_\delta(x) \subset E$ for $\delta = \frac{1}{2}$ min $(x - 1/(n+1), 1/n - x)$.

▲ *Example 5*

**R** consists of interior points only.

---

***Definition 28.2***

The set $A \subseteq \mathbf{R}$ is called an *open set* if and only if all points of $A$ are interior points.

---

▲ *Example 6*

**R** *is an open set and the empty set* $\varnothing$ *is an open set.* (The empty set contains no points and hence, all its points are interior points.)

▲ *Example 7*

Every open interval $(a, b)$ is an open set.

▲ *Example 8*

A closed interval $[a, b]$ is *not* an open set. Any neighborhood of $a$ (and $b$) contains points not in $[a, b]$, and there goes the neighborhood!

▲ *Example 9*

The set $E = c(D)$ of example 4 is an open set.

---

While the lid of a Bavarian beer mug is either open or closed, sets that are not open are not necessarily closed. Sets that are not open are, without further qualifications, of little interest in mathematics; the *complements* of open sets, however, are of interest, and these are the ones which we shall call *closed sets*.

***Definition 28.3***

The set $A \subseteq \mathbf{R}$ is called a *closed set* if and only if $c(A) = \mathbf{R} \backslash A$ is open.

---

▲ *Example 10*

The set $E = c(D) \subset \mathbf{R}$ of example 4 is open. Hence, $D = c(E)$ is closed.

▲ *Example 11*

$\mathfrak{J} = [0, 1] \subset \mathbf{R}$ is closed because $c(\mathfrak{J}) = \{x \in \mathbf{R} \mid x < 0 \text{ or } x > 1\}$ consists of interior points only and is therefore open.

▲ *Example 12*

$B = \{1, \frac{1}{2}, \frac{1}{3}, \frac{1}{4}, \cdots\} \subset \mathbf{R}$ is not closed because its complement $c(B)$ is not open: $0 \in c(B)$ is not an interior point because, for every $\delta > 0$, there is an element $1/n \in B$, $1/n \notin c(B)$, such that $1/n \in N_\delta(0)$. $B$ is not open either, as we have seen in example 2.

▲ *Example 13*

The *empty set* $\varnothing$ *is closed* because $c(\varnothing) = \mathbf{R}$ is open and $\mathbf{R}$ *is closed* because $c(\mathbf{R}) = \varnothing$ is open. (See example 6.)

---

We have seen in example 6 that $\mathbf{R}$ and $\varnothing$ are open, and in example 13 that $\mathbf{R}$ and $\varnothing$ are closed. While this strikes us as somewhat peculiar, we are not overly concerned because $\mathbf{R}$, as the universal set, and $\varnothing$, as the empty set, are exceptional sets. However, if this were true for sets other than $\varnothing$ and $\mathbf{R}$, we would have reasons to be concerned. Fortunately, $\varnothing$ *and* $\mathbf{R}$ *are the only subsets of* $\mathbf{R}$ *which are both open and closed.* We shall prove this in a more general setting in section 6.72. For those who can't wait, a proof is outlined in exercise 10.

We have recognized the sets $D$ of example 10, $\mathfrak{J}$ of example 11, and $\varnothing$ and $\mathbf{R}$ of example 13 as closed sets. It strikes us immediately that the one feature these sets have in common is that they contain all their accumulation points. This is characteristic of closed sets:

***Theorem 28.1***

The set $A \subseteq \mathbf{R}$ is closed if and only if $A' \subseteq A$; that is, $A$ contains all its accumulation points.

*Proof*

(a) Suppose that $A$ is closed and that $x \notin A$; that is, $x \in c(A)$. By definition 28.3, $c(A)$ is open; i.e., there is some $\delta > 0$ such that $N_\delta(x) \subset c(A)$. Hence, $x$

is not an accumulation point of $A$. Since this is true for all $x \notin A$, it follows that $A$ contains all its accumulation points.

(b) Suppose that $A' \subseteq A$. If $x \notin A$, then $x \notin A'$ and there is a $\delta > 0$ such that $N_\delta(x)$ contains no point from $A$; that is, $N_\delta(x) \subset c(A)$. This means that $x$ is an interior point of $c(A)$. Since this is true for all $x \notin A$, it follows that $c(A)$ is open and hence, $A$ is closed.

Observe that closed sets may or may not contain interior points. While $[0, 1]$ contains infinitely many interior points, $\{0, \frac{1}{2}, \frac{1}{3}, \frac{1}{4}, \cdots\}$ contains no interior points. Every finite set is closed but contains no interior points. (See exercise 2.)

## 2.28 Exercises

1. Prove: If $x$ is an interior point of $A \subseteq \mathbf{R}$, then $x$ is an accumulation point of $A$.
2. Prove: Every finite set $A \subset \mathbf{R}$ is closed.
3. Show that $A \cup A'$ is closed, where $A$ is any subset of $\mathbf{R}$.
4. Given any set $B \subset \mathbf{R}$. Is $B'$ closed?
5. Show by Theorem 28.2 that $\mathbf{R}$ is closed.
6. Show by Theorem 28.2 that $\varnothing$ is closed.
7. Is $(a, b]$ closed? Open?
8. Prove: Every set that contains all but finitely many real numbers is open.
9. Let $A \subset \mathbf{R}$ be bounded above. Then $a_0 = \sup A$ exists. Show: If $a_0 \notin A'$, then $a_0 \in A$ but the converse is not necessarily true.
10. Let $A \subset \mathbf{R}$, where $A \neq \varnothing$ and $A \neq \mathbf{R}$. Assume that $A$ is open as well as closed and derive a contradiction.
11. Show: An open subset of $\mathbf{R}$ does not contain any isolated points.

## 2.29 Unions and Intersections of Open Sets and Closed Sets

Let us consider some unions and intersections of open sets and closed sets, respectively:

---

▲ *Example 1*

$A_k = (1/k, 1 - 1/k)$, $k = 3, 4, 5, \ldots$, is open, and

$$\bigcup_{k=3}^{\infty} A_k = (0, 1), \qquad \bigcap_{k=3}^{\infty} A_k = (\tfrac{1}{3}, \tfrac{2}{3})$$

are also open.

▲ *Example 2*

$B_k = (-1/k, 1 + 1/k)$, $k \in \mathbf{N}$, is open, and $\bigcup_{k \in \mathbf{N}} B_k = (-1, 2)$ is open, while $\bigcap_{k \in \mathbf{N}} B_k = [0, 1]$ is closed.

▲ *Example 3*

$C_k = [-1/k, 1 + 1/k]$, $k \in \mathbf{N}$ is closed, and

$$\bigcup_{k \in \mathbf{N}} C_k = [-1, 2], \qquad \bigcap_{k \in \mathbf{N}} C_k = [0, 1]$$

are also closed.

▲ *Example 4*

$D_k = [1/k, 1 - 1/k]$, $k = 3, 4, 5 \ldots$, is closed, and $\bigcap_{k=3}^{\infty} D_k = [\frac{1}{3}, \frac{2}{3}]$ is closed, while $\bigcup_{k=3}^{\infty} D_k = (0, 1)$ is open.

---

The following pattern seems to emerge: The union of open sets is open and the intersection of closed sets is closed, while the union of closed sets may or may not be closed and the intersection of open sets may or may not be open. We have:

***Theorem 29.1***

The union of any collection of open sets is open and the intersection of any collection of closed sets is closed.

*Proof*

Let $A$ denote an index set and let $\{\mathcal{O}_\alpha \mid \alpha \in A\}$ denote a collection of open sets. (Note that the index set need not be denumerable.) If

$$x \in \bigcup_{\alpha \in A} \mathcal{O}_\alpha,$$

then $x \in \mathcal{O}_{\alpha_0}$ for some $\alpha_0 \in A$. Since $\mathcal{O}_{\alpha_0}$ is open, there is a $\delta > 0$ such that

$$N_\delta(x) \in \mathcal{O}_{\alpha_0} \subseteq \bigcup_{\alpha \in A} \mathcal{O}_\alpha.$$

Hence, $x$ is an interior point of $\bigcup_{\alpha \in A} \mathcal{O}_\alpha$. This is true for all $x$ in the union, and hence, the union is open.

Let $B$ denote an index set and let $\{\mathcal{C}_\beta \mid \beta \in B\}$ denote a collection of closed sets. By De Morgan's law (Theorem 2.1),

$$c \bigcap_{\beta \in B} \mathcal{C}_\beta = \bigcup_{\beta \in B} c(\mathcal{C}_\beta).$$

Since $\mathcal{C}_\beta$ is closed, $c(\mathcal{C}_\beta)$ is open, and by our preceding result, $\bigcup_{\beta \in B} c(\mathcal{C}_\beta)$ is open. Hence, $\bigcap_{\beta \in B} \mathcal{C}_\beta = c \bigcup_{\beta \in B} c(\mathcal{C}_\beta)$ is closed.

---

▲ *Example 5*

The Cantor set $C$, as defined in section 1.14, is a closed set. (See exercises 2 and 3.)

---

We have seen in examples that the intersection of infinitely many open sets is not necessarily open, and that the union of infinitely many closed sets is not necessarily closed. However, if we take union and intersection of finitely many sets only, then the property of openness and closedness is preserved.

***Theorem 29.2***

The intersection of finitely many open sets is open and the union of finitely many closed sets is closed. (See exercise 1.)

While the union of open intervals is an open set, it is not necessarily an open interval, as the following example shows:

---

▲ *Example 6*

Let $\mathfrak{J}_1 = (0, 1)$ and $\mathfrak{J}_2 = (2, 3)$. Then $\mathfrak{J}_1 \cup \mathfrak{J}_2 = \{x \in \mathbf{R} \mid 0 < x < 1,\ 2 < x < 3\}$, which is *not* an open interval.

---

However, it is true that the intersection of closed intervals is always a closed interval, or a point, or the empty set. Without going into detail, let us outline how one can arrive at this conclusion:

A closed interval may be characterized as a closed and bounded set, with the property that, with any two elements $x$, $y$ in the set, all points between $x$ and $y$ lie in the set. (Convexity property. See exercises 4 and 5.) Let $\mathfrak{J}_\alpha$, $\alpha \in A$, denote closed intervals. Either $\bigcap_{\alpha \in A} \mathfrak{J}_\alpha = \varnothing$, or $\bigcap_{\alpha \in A} \mathfrak{J}_\alpha$ contains exactly one point, or $\bigcap_{\alpha \in A} \mathfrak{J}_\alpha$ contains more than one point. In the latter case, $x, y \in \bigcap_{\alpha \in A} \mathfrak{J}_\alpha$ where $x \neq y$. Then $x, y \in \mathfrak{J}_\alpha$ for all $\alpha \in A$ and hence, all points between $x$ and $y$ lie in $\mathfrak{J}_\alpha$ for all $\alpha \in A$, and hence in $\bigcap_{\alpha \in A} \mathfrak{J}_\alpha$. Therefore, $\bigcap_{\alpha \in A} \mathfrak{J}_\alpha$ is a closed interval. (It is bounded because each $\mathfrak{J}_\alpha$ is bounded, and closed because the intersection of closed sets is closed.)

The following two examples illustrate some interesting consequences of Lemma 27.1.

---

▲ *Example 7*

If $f:(a, b) \to \mathbf{R}$ is continuous on $(a, b)$, then the set

$$A = \{x \in (a, b) \mid f(x) \neq 0\}$$

is an *open* set. Whenever $f(x_0) \neq 0$ for some $x_0 \in (a, b)$, then, by Lemma 27.1, there exists a $\delta > 0$ such that

$$f(x) \neq 0 \qquad \text{for all } x \in N_\delta(x_0) \cap (a, b).$$

Since the intersection of two open sets is open, $x_0$ is an interior point of $A$ and hence, $A$ is open.

▲ *Example 8*

If $f: [a, b] \to \mathbf{R}$ is continuous on $[a, b]$, then the set

$$B = \{x \in [a, b] \mid f(x) = 0\}$$

is a *closed* set. We extend the definition of $f$ to $\mathbf{R}$ as follows:

$$f_1(x) = \begin{cases} f(x) & \text{for all } x \in [a, b] \\ f(a) & \text{for all } x < a \\ f(b) & \text{for all } x > b \end{cases}$$

(see Fig. 29.1) and see that $f_1: \mathbf{R} \to \mathbf{R}$ is continuous on $\mathbf{R}$.

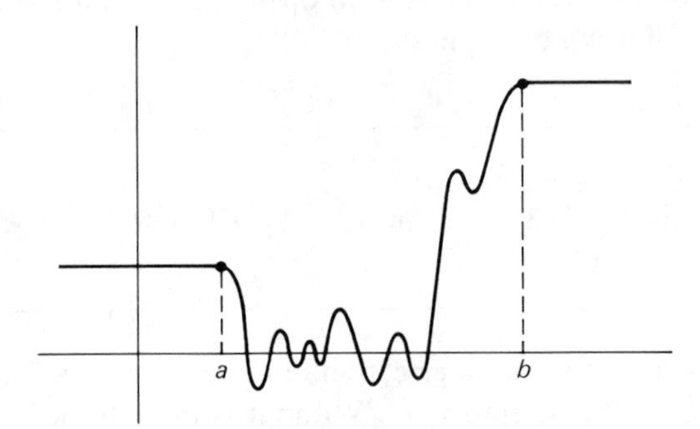


*Fig. 29.1*

As in example 7, we see that $C = \{x \in \mathbf{R} \mid f_1(x) \neq 0\}$ is open. Hence, $c(C)$ is closed. Since $B = [a, b] \backslash C = [a, b] \cap c(C)$, and since the intersection of two closed sets is closed we see that $B$ is closed.

If $f$ is defined on an open interval, the above result need not be true. Let $f: (0, 1) \to \mathbf{R}$ be defined by

$$f(x) = x \sin \frac{1}{x}.$$

This function is continuous for all $x \in (0, 1)$ (see exercise 10, section 2.26), and vanishes in $(0, 1)$ for $x = 1/\pi, 1/2\pi, 1/3\pi, \ldots$. Hence,

$$B = \left\{x \in (0,1) \,\middle|\, x \sin \frac{1}{x} = 0\right\} = \left\{\frac{1}{k\pi} \,\middle|\, k \in \mathbf{N}\right\}$$

is *not* closed. If one extends the definition of $f$ to $[0, 1]$ such that the extended function is continuous on $[0, 1]$, then, by necessity, $f(0) = 0$ because 0 is an accumulation point of $B$. (For the sufficiency of this condition see exercise 10, section 2.26.)

---

## 2.29 Exercises

1. Prove Theorem 29.2.
2. Prove that the Cantor set $C$ is closed, by using the representation in definition 14.1.

3. Prove that the Cantor set $C$ is closed by using the representation (14.2).
4. Given the closed interval $[a, b]$. Show: If $x, y \in [a, b]$ and $x < y$, then $z = x + t(y - x) \in [a, b]$ for all $t \in [0, 1]$. (This means that all points between $x$ and $y$ lie in $[a, b]$.)
5. Given a closed and bounded set $S \subset \mathbf{R}$ with the property that with $x, y \in S$, $x < y$, all $z = x + t(y - x) \in S$ for all $t \in [0, 1]$. Show that $S$ is a closed interval.
6. Show: If $A$ is a given index set and $\mathfrak{J}_\alpha, \alpha \in A$, is an open interval, and if $\mathfrak{J}_\alpha \cap \mathfrak{J}_\beta = \varnothing$ for $\alpha \neq \beta$, then $A$ is at most denumerable. (In words: A collection of disjoint open intervals is at most denumerable.)
7. Show: $A \subseteq \mathbf{R}$ is open if and only if it is the union of $\delta$-neighborhoods.
8. Show: Any open subset of $\mathbf{R}$ is the union of at most denumerably many disjoint open intervals.
9. Let $f: (a, b) \to \mathbf{R}$ denote a continuous function. Show that the set $\{x \in (a, b) | f(x) > 0\}$ is open.

## 2.30 The Bolzano–Weierstrass Theorem

If a subset of $\mathbf{R}$ with infinitely many elements is bounded, then it is reasonable to expect that such a set has at least one accumulation point (that need not be a member of the set), because there simply is not enough room for infinitely many elements to keep their distance from each other. This is, in essence, the content of a celebrated theorem by Bolzano† and Weierstrass:

***Theorem 30.1*** *(Theorem of Bolzano–Weierstrass)*

If $S \subset \mathbf{R}$ contains infinitely many elements and if $S$ is bounded, then $S$ has at least one accumulation point that need not lie in $S$.

*Proof*

By hypothesis, $S$ is bounded. Hence, there is an $M > 0$ such that $S \subseteq [-M, M]$. Since infinitely many elements of $S$ lie in $[-M, M]$, infinitely many elements of $S$ have to lie in $[-M, 0]$ or in $[0, M]$, or in both. We assume w.l.o.g. that infinitely many elements lie in $[0, M]$ and denote $0 = a_1, M = b_1$, and $[0, M] = [a_1, b_1] = \mathfrak{J}_1$. Then, infinitely many elements of $S$ have to lie in $[a_1, (a_1 + b_1)/2]$ or in $[(a_1 + b_1)/2, b_1]$, or in both. We assume w.l.o.g. that infinitely many elements lie in $[(a_1 + b_1)/2, b_1]$, and denote $(a_1 + b_1)/2 = a_2$, $b_1 = b_2$, $[(a_1 + b_1)/2, b_1] = [a_2, b_2] = \mathfrak{J}_2$. We proceed in this manner and obtain a nested sequence of closed, nonempty intervals $\{\mathfrak{J}_k\}$ where, by construction, $\lim_\infty l(\mathfrak{J}_k) = 0$. Also, by construction, each $\mathfrak{J}_k$ contains infinitely many elements of $S$. By the nested-interval property of $\mathbf{R}$ (Theorem 21.1), there exists an element $a \in \mathbf{R}$, such that $a \in \mathfrak{J}_k$ for all $k \in \mathbf{N}$. We can find, for every $\delta > 0$, a

† *Bernard Bolzano*, 1781–1848.

$k \in \mathbf{N}$ such that $\mathfrak{J}_k \in N_\delta(a)$ because $\lim_\infty l(\mathfrak{J}_k) = 0$ and $a \in \mathfrak{J}_k$. Hence, $N'_\delta(a)$ contains infinitely many elements of $S$ and $a$ is, by the Corollary to Theorem 23.1, an accumulation point of $S$.

We obtain as a consequence of the Bolzano–Weierstrass theorem:

***Theorem 30.2***

If $S \subset \mathbf{R}$ is bounded, then one can select from every infinite subset of $S$ a non-constant convergent sequence. (The limit of the sequence need not be in $S$.)

*Proof*

By Theorem 30.1, every infinite subset $S_1$ of $S$, being bounded, has at least one accumulation point $a \in \mathbf{R}$. By Theorem 23.1, there is at least one sequence $\{x_k\}$, where $x_k \in S_1 \backslash \{a\}$ for all $k \in \mathbf{N}$, such that $\lim_\infty x_k = a$.

A set with the property that one can select from every one of its infinite subsets a convergent sequence is often called *Bolzano-compact* or *B-compact*. If the limits of all these convergent sequences also lie in the set, then the set is often referred to as *compact in itself*. (See exercises 1, 2, 3.) We shall not make any use of this particular terminology in our further developments.

## 2.30 Exercises

1. Prove: If $S \subset \mathbf{R}$ is bounded, then $S$ is Bolzano-compact.
2. Prove: If $S \subset \mathbf{R}$ is bounded and closed, then $S$ is compact in itself.
3. Prove: An open set other than the empty set cannot be compact in itself.
4. Prove: For a bounded set $S$ not to have an accumulation point, it is necessary and sufficient that $S$ be finite.
5. Prove: If $S \subset \mathbf{R}$ is bounded and closed and contains infinitely many elements then $S$ contains at least one accumulation point.

## 2.31 The Heine–Borel Property

We consider the continuous function $f: \mathbf{R} \to \mathbf{R}$ which is defined by $f(x) = x^2$. (See also example 3, section 2.26.) $f$ maps the closed interval $[-1, 1]$ onto a closed interval, namely $[0, 1]$ but does not map the open interval $(-1, 1)$ onto an open interval: $(\{f(x) \mid x \in (-1, 1)\} = [0, 1))$. $f$ also maps the closed set $A = \{0, 1, \frac{1}{2}, \frac{1}{3}, \ldots\}$ onto a closed set, namely $\{0, 1, \frac{1}{4}, \frac{1}{9}, \ldots\}$.

On the other hand, if we consider the continuous function $g: \mathbf{R}^+ \to \mathbf{R}$ that is given by $g(x) = (1/x)$ (see also exercise 9, section 2.26), we see that the closed set $\mathbf{N} = \{1, 2, 3, \ldots\}$ is *not* mapped onto a closed set, but rather onto

$$\{1, \tfrac{1}{2}, \tfrac{1}{3}, \tfrac{1}{4}, \ldots\}$$

which is neither open nor closed. The difference between $[-1, 1]$ and $\mathbf{N}$, or between $A$ and $\mathbf{N}$, for that matter, is that $[-1, 1]$ and $A$ are also bounded while $\mathbf{N}$ is not bounded.

It develops that continuous functions map closed and bounded sets onto closed and bounded sets; this leads us to a celebrated theorem of Weierstrass according to which a continuous function assumes its maximum value and minimum value on a closed interval. This theorem, in turn, has far reaching consequences and important applications. We shall, therefore, subject the closed and bounded sets to further scrutiny.

We shall characterize these sets in a manner that is adaptable to generalizations in topology and functional analysis and, what is even more important at the moment, is practical and illuminating in the proofs of a number of important theorems, namely by the so called *Heine–Borel property.*

In order to explain the Heine–Borel† property, we need the concept of an *open cover*:

***Definition 31.1***

The collection of open sets $\Omega = \{\Omega_\alpha \mid \alpha \in A\}$, $\Omega_\alpha \subset \mathbf{R}$, where $A$ is a given index set, is an *open cover* of $S \subseteq \mathbf{R}$ if and only if

$$S \subseteq \bigcup_{\alpha \in A} \Omega_\alpha.$$

---

▲ *Example 1*

If $\Omega_\alpha = \{x \in \mathbf{R} \mid \alpha/2 < x < \alpha\}$, then $\Omega = \{\Omega_\alpha \mid \alpha \in (0, 1)\}$ is an open cover of $(0, 1)$ because

$$(0, 1) \subseteq \bigcup_{\alpha \in (0, 1)} \Omega_\alpha = (0, 1).$$

▲ *Example 2*

If $\Omega_k = \{x \in \mathbf{R} \mid 1/k < x < 1 - 1/k\}$, then $\Omega = \{\Omega_k \mid k \in \{3, 4, 5, \ldots\}\}$ is an open cover of $(0, 1)$ because

$$(0, 1) \subseteq \bigcup_{k=3}^{\infty} \Omega_k = (0, 1).$$

▲ *Example 3*

If $\Omega_k = \{x \in \mathbf{R} \mid -1/k < x < 1 + 1/k\}$, then $\Omega = \{\Omega_k \mid k \in \mathbf{N}\}$ is an open cover of $[0, 1]$ because

$$[0, 1] \subseteq \bigcup_{k \in \mathbf{N}} \Omega_k = (-1, 2).$$

---

† *Edward Heine*, 1821–1881; *Émile Borel*, 1871–1938.

In example 3, we can select finitely many sets from $\Omega$; as a matter of fact, one, namely $\Omega_2 = (-\frac{1}{2}, \frac{3}{2})$, which is already an open cover of $[0, 1]$.

This is *not* possible in example 2. Suppose there were finitely many sets $\Omega_{k_1}$, $\Omega_{k_2}, \ldots, \Omega_{k_n} \in \Omega$ such that $\{\Omega_{k_1}, \Omega_{k_2}, \ldots, \Omega_{k_n}\}$ is an open cover of $(0, 1)$. Let $k_0 = \max\{k_1, k_2, \ldots, k_n\}$. Then

$$\bigcup_{i=1}^{n} \Omega_{k_i} = \Omega_{k_0} = \left(\frac{1}{k_0}, 1 - \frac{1}{k_0}\right)$$

but $(0, 1) \not\subseteq (1/k_0, 1 - 1/k_0)$.

In example 3 we could find a *finite subcover* of $[0, 1]$; in example 2, a *finite subcover* of $(0, 1)$ does not exist.

***Definition 31.2***

If $\Omega = \{\Omega_\alpha \mid \alpha \in A\}$ is an open cover of $S \subseteq \mathbf{R}$ and if $\Omega' = \{\Omega_{\alpha_1}, \Omega_{\alpha_2}, \ldots, \Omega_{\alpha_n}\} \subset \Omega$ for some $\alpha_1, \alpha_2, \ldots, \alpha_n \in A$ is also an open cover of $S$, $\Omega'$ is called a *finite subcover* of $S$.

***Definition 31.3***

A set $S \subseteq \mathbf{R}$ has the *Heine–Borel property* if and only if one can select, from *every* open cover $\Omega$ of $S$, a finite subcover $\Omega'$ of $S$.

---

▲ *Example 4*

$(0, 1)$ does *not* have the Heine–Borel property, because there is at least one open cover of $(0,1)$, namely $\Omega = \{\Omega_k \mid k \in \{3, 4, 5, \ldots\}\}$, where $\Omega_k = \{x \in \mathbf{R} \mid 1/k < x < 1 - 1/k\}$, from which a finite subcover cannot be selected.

---

## 2.31 Exercises

1. Show that $S = [0, 1] \cap \mathbf{I}$ does *not* have the Heine–Borel property, where $\mathbf{I}$ denotes the set of irrationals.
2. Show that $S = [0, 1] \cap \mathbf{Q}$ does not have the Heine–Borel property.
3. Show that $\mathbf{R}$ does not have the Heine–Borel property.
4. Show that $\mathbf{N}$ does not have the Heine–Borel property.
5. Show: Every finite set has the Heine–Borel property.
6. Let $S \subseteq \mathbf{R}$ and let $\Omega = \{\Omega_\alpha \mid \alpha \in A\}$ represent an open cover of $S$. Show: $\Omega$ contains a *denumerable* subcover of $S$.

## 2.32 Compact Sets

***Definition 32.1***

The set $K \subseteq \mathbf{R}$ is called *compact* if and only if it has the Heine–Borel property.

We are now ready to characterize the bounded and closed subsets of **R** by the Heine–Borel property:

***Theorem 32.1*** *(Heine-Borel Theorem)*

The set $K \subseteq \mathbf{R}$ is *compact* (i.e., has the Heine–Borel property) if and only if $K$ is *bounded and closed.*

*Proof*

(a) We assume that $K$ is compact. If $\Omega_k = (-k, k)$, then $\Omega = \{\Omega_k \mid k \in \mathbf{N}\}$ is an open cover of $K$, because $K \subseteq \mathbf{R} = \bigcup_{k \in \mathbf{N}} \Omega_k$. Since $K$ is compact, there exists a finite subcover $\Omega' = \{\Omega_{k_1}, \Omega_{k_2}, \ldots, \Omega_{k_n}\} \subset \Omega$ of $K$. Let $k_0 = \max\{k_1, k_2, \ldots, k_n\}$. Then $\bigcup_{i=1}^n \Omega_{k_i} = (-k_0, k_0)$ and, since $K \subseteq \bigcup_{i=1}^n \Omega_{k_i}$, we have $K \subseteq (-k_0, k_0)$; that is, *$K$ is bounded.*

In order to show that $K$ is closed, we shall demonstrate that $c(K)$ is open. Let $a \notin K$, that is, $a \in c(K)$, and let $\Omega_k = \{x \in \mathbf{R} \mid |x - a| > 1/k\}$. Then $\bigcup_{k \in \mathbf{N}} \Omega_k = \mathbf{R} \backslash \{a\}$ and $\Omega = \{\Omega_k \mid k \in \mathbf{N}\}$ emerges as an open cover of $K$. Since $K$ is compact, there exists a finite subcover $\Omega' = \{\Omega_{k_1}, \Omega_{k_2}, \ldots, \Omega_{k_n}\} \subset \Omega$ of $K$. Let $k_0 = \max\{k_1, k_2, \ldots, k_n\}$. Then, $\bigcup_{i=1}^n \Omega_{k_i} = \{x \in \mathbf{R} \mid |x - a| > 1/k_0\}$ and $K \subseteq \{x \in \mathbf{R} \mid |x - a| > (1/k_0)\}$. Hence

$$N_{(1/k_0)}(a) = \left\{x \in \mathbf{R} \,\middle|\, |x - a| < \frac{1}{k_0}\right\} \subseteq c(K);$$

that is, $a$ is an interior point of $c(K)$. This is true of all points in $c(K)$. Hence, $c(K)$ is open and *$K$ is closed.*

(b) We assume that $K$ is bounded and closed, and demonstrate that $K$ is compact. It suffices, for this purpose, to show that a closed and bounded interval is compact: Since $K$ is bounded, there is an $M > 0$ such that $K \subseteq [-M, M]$. If $\Omega = \{\Omega_\alpha \mid \alpha \in A\}$ is any open cover of $K$, then $\{\Omega, c(K)\}$ is an open cover of $[-M, M]$ because $c(K) \cup \left(\bigcup_{\alpha \in A} \Omega_\alpha\right) = \mathbf{R}$ and $c(K)$ is open, $K$ being closed. If $[-M, M]$ is compact, then finitely many sets from $\{\Omega, c(K)\}$, say $c(K), \Omega_{\alpha_1}, \Omega_{\alpha_2}, \ldots, \Omega_{\alpha_n}, \alpha_j \in A$, cover $[-M, M]$, and hence, $\Omega_{\alpha_1}, \Omega_{\alpha_2}, \ldots, \Omega_{\alpha_n}$ is a finite subcover of $K$. Hence, $K$ is also compact.

We shall establish the compactness of $[-M, M]$ by contradiction. We assume that there is an open cover $\Omega = \{\Omega_\beta \mid \beta \in B\}$ of $[-M, M]$ such that $[-M, M]$ cannot be covered by finitely many sets from $\Omega$. Then, $[-M, 0]$, or $[0, M]$, or both, cannot be covered by finitely many sets from $\Omega$. We assume w.l.o.g. that $[0, M]$ cannot be covered by finitely many sets from $\Omega$. Then $[0, M/2]$, or $[M/2, M]$, or both, cannot be covered by finitely many sets from $\Omega$. We continue this process *ad infinitum.* (A termination of this process stands in contradiction to our assumption that $[-M, M]$ cannot be covered by finitely many sets from $\Omega$.)

This process generates a nested sequence of nonempty, closed intervals $\{\mathfrak{J}_k\}$ where $\lim_\infty l(\mathfrak{J}_k) = 0$. (See also Fig. 32.1.) Since **R** has the nested-interval property, we obtain, from the Corollary to Theorem 21.1, that there is exactly

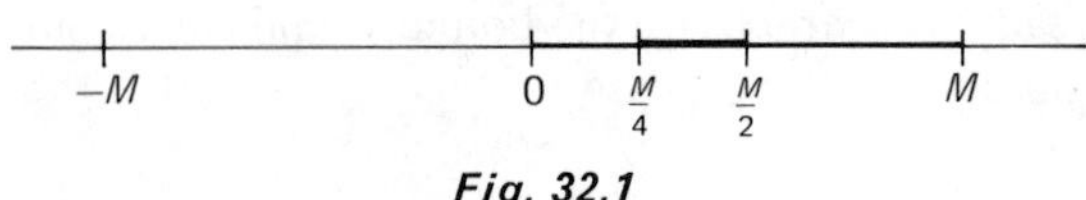


*Fig. 32.1*

one $x \in \mathbf{R}$ such that $x \in \mathfrak{J}_k$ for all $k \in \mathbf{N}$. By construction, $\mathfrak{J}_k \subseteq [-M, M]$ for all $k \in \mathbf{N}$. Hence, $x \in [-M, M]$. Therefore, there is a set $\Omega_\beta \in \Omega$ such that $x \in \Omega_\beta$. Since $\Omega_\beta$ is open, there is a $\delta > 0$ such that $N_\delta(x) \subseteq \Omega_\beta$. Because $\lim_\infty l(\mathfrak{J}_k) = 0$, we may choose a $k \in \mathbf{N}$ so large that $\mathfrak{J}_k \subseteq N_\delta(x) \subseteq \Omega_\beta$. Hence, $\mathfrak{J}_k$ may be covered by *one* set from $\Omega$, contrary to our assumption that none of the intervals $\mathfrak{J}_k$ can be covered by finitely many sets from $\Omega$. Hence, $[-M, M]$ is compact and, consequently, $K$ is compact.

The next theorem will show that the compact sets are what we called *compact in itself* in section 2.30.

### *Theorem 32.2*

$K \subseteq \mathbf{R}$ is compact if and only if one can select from every infinite subset of $K$ a nonconstant sequence that converges to an element of $K$.

*Proof*

(a) Suppose that $K$ is compact. Then, by Theorem 32.1, $K$ is bounded and, by Theorem 30.2, one can select from every infinite subset of $K$ a convergent sequence. The limit of a sequence is an accumulation point of the sequence (see exercise 1, section 2.23), and hence, of $K$. Since $K$ is also closed, the limit of the sequence lies in $K$.

(b) In order to prove the first part of the converse, we assume w.l.o.g. that $K$ is not bounded above. Then, for every $k \in \mathbf{N}$, there is a $x_k \in K$ such that $x_k > k$. Then $\{x_k\}$ diverges to $\infty$ (see definition 15.2), and every subsequence diverges to $\infty$. (See exercise 6.) Hence, *$K$ has to be bounded.*

Suppose $a \in \mathbf{R}$ is an accumulation point of $K$. Then, by Theorem 23.1, we can find a sequence $\{x_k\}$, $x_k \in K \backslash \{a\}$ for all $k \in \mathbf{N}$, such that $\lim_\infty x_k = a$ and $a \in K$ by hypothesis. Hence, *$K$ is also closed.* Therefore, by Theorem 32.1, $K$ is compact.

## 2.32 Exercises

1. Show: Every closed interval $[a, b]$ is compact.
2. Show: $\mathbf{R}$ is not compact.
3. Show: $\mathbf{Q}$ is not compact.
4. Show: If $A \subset \mathbf{R}$ is bounded, then $A \cup A'$ is compact.
5. Show: Every finite set is compact.
6. Show: Every subsequence of a sequence that diverges to $\infty$, diverges to $\infty$.
7. Let $f: [a, b] \to \mathbf{R}$ denote a continuous function. Show that $\{x \in [a, b] \mid f(x) = 0\}$ is compact.

## 2.33 Uniformly Continuous Functions

The function $f: (0, 1) \to \mathbf{R}$ which is given by $f(x) = 1/x$ is continuous on (0, 1). (See exercise 9, section 2.26). For every $a \in (0, 1)$, and every $\varepsilon > 0$, we can find a $\delta(\varepsilon)$ such that

$$\left|\frac{1}{x} - \frac{1}{a}\right| < \varepsilon \qquad \text{for all } x \in N_{\delta(\varepsilon)}(a) \cap (0, 1).$$

Cumbersome but elementary manipulations yield $\delta(\varepsilon) = \varepsilon a^2/(1 + \varepsilon a)$. That this is the best we can do can be seen by taking

$$x = a - \frac{\varepsilon a^2}{1 + \varepsilon a} = \frac{a}{1 + \varepsilon a} \in (0, 1) \qquad \text{where } |x - a| = \delta(\varepsilon).$$

Then, $1/x - 1/a = \varepsilon$ and *not* $< \varepsilon$.

We observe that $\delta(\varepsilon)$ depends not only on $\varepsilon$ but also on the choice of $a$. If we keep $\varepsilon$ fixed and let $a$ approach zero, we see that $\delta(\varepsilon)$ also approaches zero. The point we wish to make is this: While it is possible to find for each particular $a \in (0, 1)$ and every given $\varepsilon > 0$ a $\delta(\varepsilon) > 0$ with the required property, it is *not* possible to find a $\delta(\varepsilon)$ that will work for *all* $a \in (0, 1)$.

This is possible, however, for the function $g : (0, 1) \to \mathbf{R}$, given by $g(x) = x^2$. This function is also continuous on (0, 1)—see example 3, section 2.26—and we can find, for every $\varepsilon > 0$ and each $a \in (0,1)$, a $\delta(\varepsilon) > 0$ such that

$$|x^2 - a^2| < \varepsilon \qquad \text{for all } x \in N_{\delta(\varepsilon)}(a) \cap (0, 1).$$

A simple computation shows that $\delta(\varepsilon) = \varepsilon/2$ will do. We see that $\delta(\varepsilon)$ is independent of $a$ and hence, the same $\delta(\varepsilon)$ will ensure that $|x^2 - a^2| < \varepsilon$ as long as $a, x \in (0, 1)$ and $|a - x| < \delta(\varepsilon)$. For this reason we call $g$ *uniformly* continuous on (0, 1).

***Definition 33.1***

The function $f : \mathfrak{D} \to \mathbf{R}$, $\mathfrak{D} \subseteq \mathbf{R}$, is *uniformly continuous on* $\mathfrak{D}$ if and only if for every $\varepsilon > 0$, there is a $\delta(\varepsilon) > 0$ such that

$$|f(x_1) - f(x_2)| < \varepsilon \qquad \text{for all } x_1, x_2 \in \mathfrak{D} \quad \text{for which } |x_1 - x_2| < \delta(\varepsilon).$$

Obviously, if $f: \mathfrak{D} \to \mathbf{R}$ is uniformly continuous on $\mathfrak{D}$, then $f$ is also continuous on $\mathfrak{D}$ but the converse is not true, as the example at the beginning of the section shows.

A function that is continuous on an open interval may or may not be uniformly continuous on that open interval (as exemplified by the functions $f$ and $g$ which we discussed at the beginning of this section). However, a function that is continuous on a closed interval is uniformly continuous on that closed interval. More generally:

***Theorem 33.1***

If $\mathfrak{K} \subset \mathbf{R}$ is *compact* and if $f: \mathfrak{K} \to \mathbf{R}$ is continuous on $\mathfrak{K}$, then $f$ is uniformly continuous on $\mathfrak{K}$.

*Proof*

By hypothesis, $f$ is continuous on $\mathcal{K}$; that is, if $a \in \mathcal{K}$ and $\varepsilon > 0$, then there is a $\delta(\varepsilon, a) > 0$ such that

$$(33.1) \qquad |f(a) - f(x)| < \frac{\varepsilon}{2} \qquad \text{for all } x \in N_{\delta(\varepsilon, a)}(a) \cap \mathcal{K}.$$

If $\Omega_a = N_{\delta(\varepsilon, a)/2}(a)$, then $\Omega = \{\Omega_a \mid a \in \mathcal{K}\}$ is an open cover of $\mathcal{K}$. Since $\mathcal{K}$ is compact, there exists a finite subcover $\Omega' = \{\Omega_{a_1}, \Omega_{a_2}, \ldots, \Omega_{a_n}\}$, $a_j \in \mathcal{K}$, of $\mathcal{K}$. Let

$$(33.2) \qquad \delta(\varepsilon) = \tfrac{1}{2} \min \{\delta(\varepsilon, a_1), \delta(\varepsilon, a_2), \ldots, \delta(\varepsilon, a_n)\}.$$

If $x_1 \in \mathcal{K}$, then $x_1 \in \Omega_{a_j}$ for some $j \in \{1, 2, \ldots, n\}$. Hence,

$$(33.3) \qquad |x_1 - a_j| < \tfrac{1}{2}\,\delta(\varepsilon, a_j).$$

If $x_2 \in \mathcal{K}$ such that

$$(33.4) \qquad |x_1 - x_2| < \delta(\varepsilon),$$

then, by (33.3) and (33.4),

$$|x_2 - a_j| \leqslant |x_2 - x_1| + |x_1 - a_j| < \delta(\varepsilon) + \tfrac{1}{2}\,\delta(\varepsilon, a_j) < \delta(\varepsilon, a_j),$$

and we have by (33.1) that

$$|f(x_1) - f(x_2)| \leqslant |f(x_1) - f(a_j)| + |f(a_j) - f(x_2)| < \frac{\varepsilon}{2} + \frac{\varepsilon}{2} = \varepsilon.$$

Thus, we see that we can find, for every $\varepsilon > 0$, a $\delta(\varepsilon) > 0$, given by (33.2), such that the condition of definition 33.1 is satisfied.

If $f$ is continuous on a closed interval $[a, b]$, then $f$ is uniformly continuous on $[a, b]$ because $[a, b]$ is a compact set.

---

▲ *Example 1*

$f: \mathbf{R} \to \mathbf{R}$, given by $f(x) = x^n$, is continuous on $\mathbf{R}$. (Example 3, section 2.26.) Hence, $f$ is uniformly continuous on any compact subset of $\mathbf{R}$, for example, on $[a, b]$, where $a, b \in \mathbf{R}$.

▲ *Example 2*

$f: \mathbf{R}^+ \to \mathbf{R}$, given by $f(x) = 1/x$ is continuous on $\mathbf{R}^+$. Hence, $f$ is uniformly continuous on every closed interval $[a, b]$ as long as $a > 0$.

▲ *Example 3*

$f: [0, 1] \to \mathbf{R}$, given by $f(0) = 0$, $f(x) = 1/x$ for $x \in (0, 1]$ is not continuous on

[0, 1]. Since uniform continuity implies continuity, $f$ cannot be uniformly continuous on [0, 1].

---

If we negate the defining property of uniform continuity in definition 32.1, we obtain: The function $f: \mathfrak{D} \to \mathbf{R}$ is *not* uniformly continuous on $\mathfrak{D}$ if and only if there exists an $\varepsilon > 0$ such that, for all $\delta > 0$, there are points $x_1, x_2 \in \mathfrak{D}$ for which $|x_1 - x_2| < \delta$ and $|f(x_1) - f(x_2)| \geqslant \varepsilon$. This condition is satisfied for all $\delta > 0$ if and only if it is satisfied for all $\delta_k = 1/k$, $k \in \mathbf{N}$. (See also exercise 1.) Hence:

***Theorem 33.2***

The function $f: \mathfrak{D} \to \mathbf{R}$ is *not* uniformly continuous on $\mathfrak{D}$ if and only if there exist an $\varepsilon > 0$ and two sequences $\{x_k^{(1)}\}$ and $\{x_k^{(2)}\}$ such that $|x_k^{(1)} - x_k^{(2)}| < 1/k$ and $|f(x_k^{(1)}) - f(x_k^{(2)})| \geqslant \varepsilon$. (See also exercise 1.)

This theorem may be used to prove Theorem 33.1 by contradiction. (See exercise 2.)

---

▲ *Example 4*

The function $f: (0, 1) \to \mathbf{R}$, given by $f(x) = 1/x$ is not uniformly continuous on $(0, 1)$: Let $x_k^{(1)} = 1/k, x_k^{(2)} = 1/(k+1)$. Then $|x_k^{(1)} - x_k^{(2)}| = 1/k(k+1) < 1/k$ and $|f(x_k^{(1)}) - f(x_k^{(2)})| = 1 = \varepsilon$.

▲ *Example 5*

The function $f: \mathbf{R} \to \mathbf{R}$, given by $f(x) = x^2$, is not uniformly continuous on $\mathbf{R}$: Let $x_k^{(1)} = k$, $x_k^{(2)} = k + 1/2k$. Then, $|x_k^{(1)} - x_k^{(2)}| = 1/2k < 1/k$ and $|f(x_k^{(1)}) - f(x_k^{(2)})| = |k^2 - k^2 - 1 - 1/4k^2| = 1 + 1/4k^2 > 1 = \varepsilon$.

---

## 2.33 Exercises

1. Show: For all $\delta > 0$, there are points $x_1, x_2$ for which $|x_1 - x_2| < \delta$ and
$$|f(x_1) - f(x_2)| \geqslant \varepsilon$$
if and only if, for all $k \in \mathbf{N}$, there are points $x_1, x_2$ for which $|x_1 - x_2| < 1/k$ and
$$|f(x_1) - f(x_2)| \geqslant \varepsilon.$$
2. Prove Theorem 33.1 by contradiction.
3. Show: $f: \mathbf{R} \to \mathbf{R}$ given by $f(x) = x$ is *uniformly continuous* on $R$.
4. Given that $\sin = \{(x, y) \in \mathbf{R} \times \mathbf{R} \mid y = \sin x\}$ is continuous on $[0, 2\pi]$. Show that sin is *uniformly* continuous on $\mathbf{R}$.
5. Generalize the result of exercise 4 as follows: If $f: \mathbf{R} \to \mathbf{R}$ is continuous on $\mathbf{R}$ and if $f(x + 1) = f(x)$ for all $x \in \mathbf{R}$, then $f$ is uniformly continuous on $\mathbf{R}$.

## 2.34 The Continuous Image of a Compact Set

We are now ready to return to the problem which we discussed in the beginning of section 2.31 and which motivated our discussion of compact sets. We have:

***Theorem 34.1***

If $f: \mathcal{K} \to R$ is continuous on $\mathcal{K} \subset \mathbf{R}$ and if $\mathcal{K}$ is compact, then $f$ maps $\mathcal{K}$ onto a compact set, that is, $\{f(x) \mid x \in \mathcal{K}\}$ is compact.

*Proof*

The proof we shall present here utilizes the fact that a compact set is bounded and closed. Another proof that makes explicit use of the Heine–Borel property of compact sets will be given in section 6.70, where this theorem will be discussed in a more general setting.

In order to simplify the notation, we use the symbol $f(\mathcal{K})$ to denote the range of $f: f(\mathcal{K}) = \{f(x) \mid x \in \mathcal{K}\}$. We assume that $f(\mathcal{K})$ is not bounded and suppose w.l.o.g. that it is not bounded above. Then, for each $k \in \mathbf{N}$, there is an element $x_k \in \mathcal{K}$ such that $f(x_k) > k$. Since $\mathcal{K}$ is compact, we obtain, from Theorem 32.2, that $\{x_k\}$ contains a subsequence $\{x_{n_k}\}$ such that $\lim_\infty x_{n_k} = a \in \mathcal{K}$. $f$ is continuous at $a$; hence, for $\varepsilon = 1$, there is a $\delta(1)$ such that $|f(x)| - |f(a)| \leqslant |f(x) - f(a)| < 1$ for all $x \in N_{\delta(1)}(a) \cap \mathcal{K}$. Hence,

$$|f(x)| \leqslant |f(a)| + 1 \qquad \text{for all } x \in N_{\delta(1)}(a) \cap \mathcal{K}.$$

Since $\lim_\infty x_{n_k} = a$, we have $x_{n_k} \in N_{\delta(1)}(a)$ for all $n_k > N(\delta)$; that is,

$$|f(x_{n_k})| \leqslant |f(a)| + 1 \qquad \text{for all } n_k > N(\delta). \tag{34.1}$$

On the other hand, we have, from our assumption, that

$$f(x_{n_k}) > n_k \qquad \text{for } all \ n_k$$

which stands in contradiction to (34.1). Hence, $f(\mathcal{K})$ *is bounded.*

Suppose that $b$ is an accumulation point of $f(\mathcal{K})$. (If $f(\mathcal{K})$ does not have any accumulation points, then $f(\mathcal{K})$ is trivially compact, since it is bounded.) Then, for every $k \in \mathbf{N}$, there is some $f(x_k)$, $x_k \in \mathcal{K}$, such that

$$|f(x_k) - b| < \frac{1}{k}\cdot \tag{34.2}$$

By Theorem 32.2, $\{x_k\}$ has a subsequence $\{x_{n_k}\}$ such that $\lim_\infty x_{n_k} = a \in \mathcal{K}$ and we have from (34.2) that $\lim_\infty f(x_{n_k}) = b$. Since $f$ is continuous at $a$, $\lim_\infty f(x_{n_k}) = f(a) = b$, that is, $b \in f(\mathcal{K})$. Since this is true for every accumulation point $b$ of $f(\mathcal{K})$, $f(\mathcal{K})$ *is closed.*

Since $f(\mathcal{K}) \subset \mathbf{R}$ is bounded and closed, it is compact and the theorem is proved.

That a continuous function assumes its maximum (and minimum) value on a compact set is based on the fact that sup $\mathcal{K}$ and inf $\mathcal{K}$ of a compact set $\mathcal{K}$ are contained in $\mathcal{K}$.

***Theorem 34.2***

If $\mathcal{K} \subset \mathbf{R}$ is compact, then $\sup \mathcal{K} \in \mathcal{K}$, $\inf \mathcal{K} \in \mathcal{K}$.

*Proof*

Since $\mathcal{K}$ is bounded, $\sup \mathcal{K}$ exists.

By Theorem 9.1, we have, for every $\delta > 0$, an element $a \in \mathcal{K}$ such that $M - \delta < a \leqslant M$; that is, either $a = M$ and the theorem is proved or $a \in N'_\delta(M)$. Since this is true for every $\delta > 0$, $M \in \mathcal{K}$ or $M$ is an accumulation point of $\mathcal{K}$. Since $\mathcal{K}$ is closed, $M \in \mathcal{K}$ after all. (The proof that $\inf \mathcal{K} \in \mathcal{K}$ is analogous. See exercise 1.)

***Definition 34.1***

If $f: \mathcal{D} \to \mathbf{R}$, $\mathcal{D} \subseteq \mathbf{R}$ and there is an $x_0 \in \mathcal{D}$ such that $f(x_0) \geqslant f(x)$ for all $x \in \mathcal{D}$, then $f(x_0)$ is called the *maximum* of $f$ on $\mathcal{D}$:

$$f(x_0) = \max_{\mathcal{D}} f(x).$$

(If $f(x_1) \leqslant f(x)$, $x_1 \in \mathcal{D}$ for all $x \in \mathcal{D}$, then $f(x_1)$ is called the minimum of $f$ on $\mathcal{D}: f(x_1) = \min_{\mathcal{D}} f(x)$.)

***Theorem 34.3***

If $f$: $\mathcal{K} \to \mathbf{R}$ is continuous and $\mathcal{K} \subset \mathbf{R}$ is compact, then $f$ assumes its *maximum* $\max_{\mathcal{K}} f(x)$ and *minimum* $\min_{\mathcal{K}} f(x)$ on $\mathcal{K}$.

*Proof*

By Theorem 34.1, $f(\mathcal{K})$ is compact. By Theorem 34.2, $\sup f(\mathcal{K}) \in f(\mathcal{K})$; i.e., there exists an $x_0 \in \mathcal{K}$ such that $f(x_0) \geqslant f(x)$ for all $x \in \mathcal{K}$. (The proof of the existence of the minimum is analogous. See exercise 2.)

---

▲ *Example 1*

If the domain of the function is *not* compact, a continuous function need not assume its maximum, or minimum, or both:

Let $f: (0, 1) \to \mathbf{R}$ be given by $f(x) = x$. (See Fig. 34.1.) Then $\sup \{f(x) \mid x \in (0, 1)\} = 1$ and $\inf \{f(x) \mid x \in (0, 1)\} = 0$, but $f(x) \neq 1$ and $f(x) \neq 0$ for all $x \in (0, 1)$.

▲ *Example 2*

If the domain is compact but $f$ is not continuous, then $f$ need not assume its maximum, or minimum, or both.

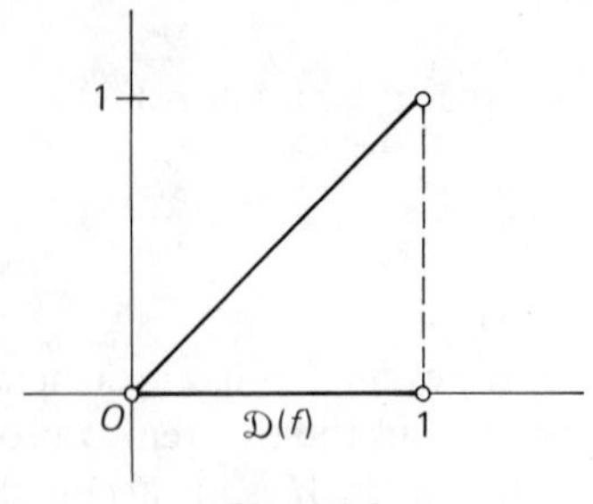


*Fig. 34.1*

Let $f\colon [-1, 1] \to \mathbf{R}$ be given by

$$f(x) = \begin{cases} 0 & \text{for } x = -1, 1 \\ x^2 & \text{for } x \in (-1, 0) \cup (0, 1) \\ -1 & \text{for } x = 0 \end{cases}$$

(See Fig. 34.2.) This function does *not* assume its maximum on $[-1, 1]$ but it does assume its minimum. (See exercises 4 and 5.)

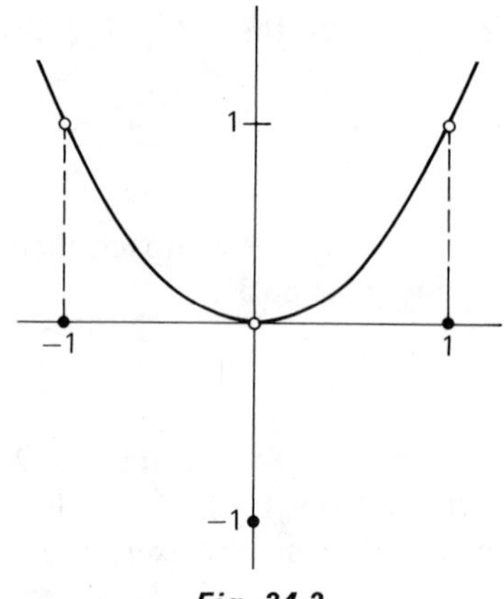


*Fig. 34.2*

---

***Corollary to Theorem 34.3***

If $f\colon [a, b] \to \mathbf{R}$ is continuous, then the range of $f$ is the closed interval $[m, M]$ where $m = \min_{[a,b]} f(x)$, $M = \max_{[a,b]} f(x)$ if $m < M$ and the point $\{m\}$ if $m = M$. (See exercise 8.)

## 2.34 Exercises

1. Prove: If $K \subset \mathbf{R}$ is compact, then $\inf K \in K$.
2. Prove: If $\mathcal{K} \subset \mathbf{R}$ is compact and $f\colon \mathcal{K} \to \mathbf{R}$ is continuous on $\mathcal{K}$, then $f$ assumes its minimum on $\mathcal{K}$.

3. $f: \mathfrak{D} \to \mathbf{R}$ is called *upper semicontinuous* at $a \in \mathfrak{D}$ if and only if, for every $\varepsilon > 0$, there is a $\delta(\varepsilon) > 0$ such that $f(x) - f(a) < \varepsilon$ for all $x \in N_{\delta(\varepsilon)}(a) \cap \mathfrak{D}$ and *lower semicontinuous* if $f(a) - f(x) < \varepsilon$ for all $x \in N_{\delta(\varepsilon)}(a) \cap \mathfrak{D}$. Show that the function $f$ of example 2 is *lower semicontinuous* on $[-1, 1]$.
4. Show: If $\mathcal{K} \subset \mathbf{R}$ is compact and $f: \mathcal{K} \to \mathbf{R}$ is upper *semicontinuous* on $\mathcal{K}$, then $f(\mathcal{K})$ is bounded above. (Same for lower semicontinuous, $f(\mathcal{K})$ bounded below.) (See exercise 3.)
5. Show: If $f: \mathcal{K} \to \mathbf{R}$ is upper (lower) semicontinuous on $\mathcal{K}$ and if $\mathcal{K}$ is compact, then $f$ assumes its maximum (minimum) on $\mathcal{K}$.
6. Given $f: [a, b] \to \mathbf{R}$ and $a < x_1 < x_2 < \cdots < x_{n-1} < b$. Show: If $f$ is continuous on $[a, b]$, then
$$\max_{[x_{k-1}, x_k]} f(x), \qquad k = 1, 2, \ldots, n; \quad x_0 = a, x_n = b,$$
exists.
7. Show that Theorem 34.2 reduces to Theorem 9.4 if $\mathcal{K}$ has finitely many elements.
8. Prove the Corollary to Theorem 34.3.
9. Given $f: [a, b] \to \mathbf{R}$. Show: If $f$ maps the closed interval $[\alpha, \beta] \subseteq [a, b]$ onto the open interval $(\gamma, \delta) \subseteq \mathbf{R}$, then $f$ is not continuous on $[a, b]$.

# 3 The Derivative

### 3.35 Definition of the Derivative

We consider a function $f\colon \mathfrak{D} \to \mathbf{R}$ and assume that $a \in \mathfrak{D}$ is an accumulation point of $\mathfrak{D}$. Then

$$\frac{f(x) - f(a)}{x - a} \tag{35.1}$$

is defined for all $x \in \mathfrak{D}\backslash\{a\}$, and $a$ is an accumulation point of $\mathfrak{D}\backslash\{a\}$. Thus, the stage is set for considering the limit of (35.1) as $x$ approaches $a$. (See definition 24.1.)

***Definition 35.1***

If $f\colon \mathfrak{D} \to \mathbf{R}$, $\mathfrak{D} \subseteq \mathbf{R}$, and $a \in \mathfrak{D}$ is an accumulation of $\mathfrak{D}$, and if

$$\lim_{a} \frac{f(x) - f(a)}{x - a}$$

exists, the function $f$ is called *differentiable* at $a$ and

$$f'(a) = \lim_{a} \frac{f(x) - f(a)}{x - a} \tag{35.2}$$

is called the *derivative* of $f$ at $a$.

We observe that the derivative is unique, if its exists. (See Theorem 24.1.)

Ordinarily, the domains of functions that we shall discuss are open intervals or closed intervals. The reason for defining the derivative of a function at an accumulation point of its domain is not merely to stretch this concept as far as it will go, but rather to avoid a bothersome distinction between an interior point of an interval and an endpoint of a closed interval. (See also exercises 5 and 6.) The one obvious feature such points have in common is that they are both accumulation points. Therefore, it is just as well to define the derivative at accumulation points, especially since the concept of the limit of a function is general enough to cover such a contingency.

▲ *Example 1*

0 is an accumulation point of $\mathfrak{D} = \{0, 1, \frac{1}{2}, \frac{1}{3}, \frac{1}{4}, \ldots\}$. Let $f : \mathfrak{D} \to \mathbf{R}$ be defined by $f(x) = x$. Then

$$f'(0) = \lim_0 \frac{f(x) - f(0)}{x - 0} = \lim_\infty \frac{x_k - 0}{x_k - 0} = 1$$

for all sequences $\{x_k\}$, where $x_k \in \mathfrak{D}\backslash\{0\}$ for all $k \in \mathbf{N}$, and $\lim_\infty x_k = 0$.

▲ *Example 2*

0 is an accumulation point of $\mathbf{R}$. We consider $f : \mathbf{R} \to \mathbf{R}$, given by $f(x) = |x|$. Since $\lim_0 (f(x) - f(0)/(x - 0) = \lim_\infty (x_k - 0)/(x_k - 0) = 1$ for all sequences $\{x_k\}$ for which $x_k > 0$ and $\lim_\infty x_k = 0$, and since $\lim_0 (f(x) - f(0))/(x - 0) = \lim_\infty (0 + x_k')/(0 - x_k') = -1$ for all sequences $\{x_k'\}$ for which $x_k' < 0$ and $\lim_\infty x_k' = 0$, we see that $f$ is not differentiable at 0.

▲ *Example 3*

Let $f$: $\mathbf{R} \to \mathbf{R}$ be given by

$$f(x) = \begin{cases} 0 & \text{if } x = 0 \\ x^2 \sin \dfrac{1}{x} & \text{if } x \neq 0. \end{cases}$$

0 is an accumulation point of $\mathbf{R}$ and we obtain

$$f'(0) = \lim_0 \frac{f(x) - f(0)}{x - 0} = \lim_0 \frac{x^2 \sin \dfrac{1}{x}}{x} = \lim_0 x \sin \frac{1}{x} = 0.$$

(See also example 5, section 2.24.)

The physical interpretation of the derivative as the velocity of the motion of a point that is described by $f$ as a function of the time is assumed to be known, and so is the geometric interpretation of the derivative at an interior point as the slope of the tangent line to the graph of the function.

We also assume that the reader is familiar with the differentiation formulas of the elementary functions (polynomials, algebraic functions, trigonometric functions, exponential functions, hyperbolic functions, and their inverses). All these formulas which are taught in calculus are correct, even though some of the derivations that are presented there may be lacking in mathematical rigor and may be given under assumptions that are difficult to justify. Rigorous derivations of most of these differentiation formulas will be given in sections 14.142 and 14.143. In this chapter, our discussion will concentrate on more general and theoretical aspects of the derivative that are not given proper attention in a calculus course.

In order to obtain a characterization of the derivative that will be useful for later generalization to vector-valued functions of a vector variable, we observe that, by Theorem 25.1, Eq. (35.2) is equivalent to the statement that, for every $\varepsilon > 0$, there is a $\delta(\varepsilon) > 0$ such that

$$\left| \frac{f(x) - f(a)}{x - a} - f'(a) \right| < \varepsilon \tag{35.3}$$

for all $x \in N'_{\delta(\varepsilon)}(a) \cap \mathfrak{D}$.

We define a function $\alpha_a : \mathfrak{D} \backslash \{a\} \to \mathbf{R}$ by

$$\alpha_a(x) = \frac{f(x) - f(a)}{x - a} - f'(a), \tag{35.4}$$

and see from (35.3) that $\lim_a \alpha_a(x) = 0$. With this in mind, we may characterize the derivative as follows.

***Theorem 35.1***

The function $f : \mathfrak{D} \to \mathbf{R}$, $\mathfrak{D} \subseteq \mathbf{R}$, is differentiable at $a \in \mathfrak{D}$, where $a$ is an accumulation point of $\mathfrak{D}$, if and only if there exists a number $f'(a) \in \mathbf{R}$, a $\delta > 0$, and a function $\alpha_a : \mathfrak{D} \backslash \{a\} \to \mathbf{R}$ with the property that $\lim_a \alpha_a(x) = 0$, such that

$$f(x) - f(a) = f'(a)(x - a) + \alpha_a(x)(x - a) \tag{35.5}$$

for all $x \in N'_\delta(a) \cap \mathfrak{D}$.

*Proof*

(a) If (35.5) is true, we obtain, via (35.4), that $f'(a) = \lim_a (f(x) - f(a))/(x - a)$ exists.

(b) If the derivative exists, then (35.3) is true and hence, $\alpha_a$ in (35.4) has the required property.

The function in example 2 is continuous at 0 but does not have a derivative at 0. Hence, continuity does not imply differentiability. The converse, however, is true.

***Theorem 35.2***

Let $f : \mathfrak{D} \to \mathbf{R}$, $\mathfrak{D} \subseteq \mathbf{R}$, and let $a \in \mathfrak{D}$ be an accumulation point of $\mathfrak{D}$. If $f'(a)$ exists, then $f$ is continuous at $a$.

*Proof*

By Theorem 35.1, we have a function $\alpha_a : \mathfrak{D} \backslash \{a\} \to \mathbf{R}$ such that

$$f(x) - f(a) = f'(a)(x - a) + \alpha_a(x)(x - a),$$

where $\lim_a \alpha_a(x) = 0$. Since $\lim_a (f'(a) + \alpha_a(x)) = f'(a)$ and $\lim_a (x - a) = 0$, we have $\lim_a f(x) = f(a)$. Hence, $f$ is continuous at $a$.

***Definition 35.2***

If all points of $\mathfrak{D} \subseteq \mathbf{R}$ are accumulation points of $\mathfrak{D}$†, and if $f: \mathfrak{D} \to \mathbf{R}$ is differentiable for all $a \in \mathfrak{D}$, then $f$ is called *differentiable on* $\mathfrak{D}$.

***Definition 35.3***

If $f: \mathfrak{D} \to \mathbf{R}$, $\mathfrak{D} \subseteq \mathbf{R}$, then

(35.6) $\quad f \in C(\mathfrak{D})$ if and only if $f$ is continuous on $\mathfrak{D}$,

(35.7) $\quad f \in C'(\mathfrak{D})$ if and only if $f$ is differentiable on $\mathfrak{D}$.

With this notation, we can state the following

***Corollary to Theorem 35.2***

$$\text{If } f \in C'(\mathfrak{D}), \qquad \text{then } f \in C(\mathfrak{D}).$$

(In particular, we have that $f \in C'(a, b)$ implies $f \in C(a, b)$; $f \in C'[a, b]$ implies $f \in C[a, b]$, and $f \in C'(a, b]$ implies $f \in C(a, b]$.)

## 3.35 Exercises

1. Prove: If $f: \mathfrak{D} \to \mathbf{R}$.is defined by $f(x) = C$, then $f'(a) = 0$ for all interior points $a \in \mathfrak{D}$. Is the converse also true?
2. Show: If $f \in C'[a, b]$, then $f$ is *uniformly* continuous on $[a, b]$.
3. Prove: If $f'(a)$, $g'(a)$ exist, then $f + g$ and $fg$ are differentiable at $a$, and

$$(f + g)'(a) = f'(a) + g'(a),$$
$$(fg)'(a) = f'(a)g(a) + f(a)g'(a).$$

4. Prove: If $f'(a)$, $g'(a)$ exist and $g(a) \neq 0$, then $f/g$ is differentiable at $a$, and

$$\left(\frac{f}{g}\right)'(a) = \frac{f'(a)g(a) - f(a)g'(a)}{g^2(a)}.$$

5. $f'_+(a) = \lim_{a+0} (f(x) - f(a))/(x - a)$ is called the *right-sided derivative* of $f$ at $a$ (see also exercise 9, section 2.24.), and

$$f'_-(a) = \lim_{a-0} (f(x) - f(a))/(x - a)$$

is called the *left-sided derivative* of $f$ at $a$. Prove: $f'(a)$ exists if and only if $f'_-(a)$ and $f'_+(a)$ both exist and are equal.

†Note that this does *not* mean that $\mathfrak{D}$ contains all its accumulation points.

6. If $\mathfrak{D}(f) = [a, b]$ and if $f \in C'[a, b]$, then $f'(a)$ is, according to definition 35.1, in effect the right-sided derivative of $f$ at $a$. Why? (Same for $f'(b)$ and left-sided derivative.)
7. As in exercise 9, section 2.24, $f'(a+0) = \lim_{a+0} f'(x)$ and $f'(a-0) = \lim_{a-0} f'(a)$. Consider $f: [0, 1] \to \mathbf{R}$, given by $f(0) = 0$, $f(x) = x^2 \sin(1/x)$, and show that $f'_+(0) \neq f'(0+0)$.
8. Prove, using Theorem 35.1, that $f'(a)$ is unique, provided it exists.
9. Use definition 35.1 to find $f'(a)$, where $a \in \mathbf{R}$ for the functions $f: \mathbf{R} \to \mathbf{R}$ which are defined below:

   (a) $f(x) = x$ (b) $f(x) = x^2$
   (c) $f(x) = x^n$, $n \in \mathbf{N}$ (d) $f(x) = \sin x$
   (e) $f(x) = \cos x$

10. Use the results of exercises 3, 4, and 9 to find $f'(a)$ where:

    (a) $f(x) = a_0 x^n + a_1 x^{n-1} + \cdots + a_{n-1} x + a_n$, $a \in \mathbf{R}$
    (b) $f(x) = \tan x$, $a \in (-\pi/2, \pi/2)$
    (c) $f(x) = \cot x$, $a \in (0, \pi)$

11. Given $f: \mathbf{R} \to \mathbf{R}$ by

$$f(x) = \begin{cases} x^2 & \text{for } x \in \mathbf{Q}, \\ 0 & \text{for } x \in \mathbf{I}. \end{cases}$$

    Find all points $a \in \mathbf{R}$ for which $f'(a)$ exists and find $f'(a)$.

## 3.36 Differentiation of a Composite Function

(A much more general version of the material in this section will be presented in sections 6.67 and 9.92.)

A composite function is exactly what the name indicates: If $f: \mathfrak{D}(f) \to \mathbf{R}$ and $g: \mathfrak{R}(f) \to \mathbf{R}$ where $\mathfrak{D}(f) \subseteq \mathbf{R}$, then,

$$h = g \circ f: \mathfrak{D}(f) \to \mathbf{R},$$

defined by

$$h(x) = g(f(x)) \quad \text{for all} \quad x \in \mathfrak{D}(f),$$

is called the *composite function* of $g$ and $f$, or the *composition* of $g$ and $f$.

First we note that $h$ is well-defined for all $x \in \mathfrak{D}(f)$ because $\mathfrak{D}(g) = \mathfrak{R}(f)$. That $h$ is a function follows from the fact that both $f$ and $g$ are functions, and hence, whenever $(x, z_1)$, $(x, z_2) \in h$, then $(f(x), z_1)$, $(f(x), z_2) \in g$, and, therefore, $z_1 = z_2$. The other conditions of definition 4.2 are obviously satisfied.

---

▲ *Example 1*

Let $f: \mathbf{R}\backslash\{0\} \to \mathbf{R}$ be defined by $f(x) = x^2$ and $g: \mathbf{R}^+ \to \mathbf{R}$ by $g(y) = 1/\sqrt{y}$. Then, $\mathfrak{R}(f) = \mathbf{R}^+ = \mathfrak{D}(g)$ and the composite function $h = g \circ f: \mathbf{R}\backslash\{0\} \to \mathbf{R}$ is defined by

$$(g \circ f)(x) = \frac{1}{\sqrt{f(x)}} = \frac{1}{|x|}.$$

▲ *Example 2*

Let $f: \mathbf{R} \to \mathbf{R}$ be defined by $f(x) = \sin x$ and $g: [-1, 1] \to \mathbf{R}$ by $g(y) = \sqrt{1 - y^2}$. Since $\mathfrak{R}(f) = [-1, 1] = \mathfrak{D}(g)$, we have

$$(g \circ f)(x) = \sqrt{1 - \sin^2 x} = |\cos x|$$

for all $x \in \mathbf{R}$.

---

In the following theorem we state the differentiation rule for composite functions, widely referred to as the *chain rule*.

***Theorem 36.1***

If $f: \mathfrak{D}(f) \to \mathbf{R}$, $g: \mathfrak{R}(f) \to \mathbf{R}$ (where $\mathfrak{D}(f) \subseteq \mathbf{R}$), if $a$ is an accumulation point of $\mathfrak{D}(f)$, if $f(a) = b$ is an accumulation point of $\mathfrak{R}(f)$, and if $f'(a)$, $g'(b)$ exist, then $h = g \circ f: \mathfrak{D}(f) \to \mathbf{R}$ is differentiable at $a$, and

$$(g \circ f)'(a) = g'(f(a))f'(a).$$

*Proof*

Since $f'(a)$, $g'(b)$ exist, we have, from Theorem 35.1, that, for some $\delta_1, \delta_2 > 0$,

$$f(x) - f(a) = f'(a)(x - a) + \alpha_a(x)(x - a), \tag{36.1}$$

$$g(y) - g(b) = g'(b)(y - b) + \beta_b(y)(y - b), \tag{36.2}$$

for all $x \in N'_{\delta_1}(a) \cap \mathfrak{D}(f)$ and all $y \in N'_{\delta_2}(b) \cap \mathfrak{D}(g)$, where $\lim_a \alpha_a(x) = 0$, $\lim_b \beta_b(y) = 0$.

Since $g(b) = g(f(a)) = h(a)$ and, for $y = f(x)$, $g(y) = g(f(x)) = h(x)$, we obtain from (36.1) and (36.2) that

$$\begin{aligned} h(x) - h(a) = g(y) - g(b) &= g'(b)(y - b) + \beta_b(y)(y - b) \\ &= g'(b)(f(x) - f(a)) + \beta_b(y)(f(x) - f(a)) \\ &= (g'(b) + \beta_b(y))(f'(a)(x - a) + \alpha_a(x)(x - a)) \\ &= g'(b)f'(a)(x - a) + [g'(b)\alpha_a(x) + f'(a)\beta_b(y) + \alpha_a(x)\beta_b(y)](x - a). \end{aligned}$$

Let

$$\gamma_a(a) = g'(f(a))\alpha_a(x) + f'(a)\beta_b(f(x)) + \alpha_a(x)\beta_b(f(x)). \tag{36.3}$$

Then, $\lim_a \gamma_a(x) = 0$ (see exercise 1), and we obtain from Theorem 35.1 that $h'(a)$ exists and

$$h'(a) = g'(f(a))f'(a).$$

▲ *Example 3*

Let $f:\mathbf{R}\to\mathbf{R}$ be defined by $f(x)=x^3$, and $g:[0,\infty)\to\mathbf{R}$ by $g(y)=\sqrt{y}$. Since $\mathfrak{R}(f)=\mathbf{R}$ but $\mathfrak{D}(g)=[0,\infty)$, we have $\mathfrak{R}(f)\neq\mathfrak{D}(g)$, and we cannot apply our formalism immediately. However, if we consider the restriction $f_1$ of $f$ to the domain $\mathfrak{D}(f_1)=[0,\infty)$, then $\mathfrak{R}(f_1)=[0,\infty)=\mathfrak{D}(g)$, and we obtain

$$(36.4)\qquad (g\circ f_1)(x)=\sqrt{x^3}\qquad \text{for all}\quad x\in[0,\infty).$$

Since $f_1'(x)=3x^2$ and $g'(y)=1/2\sqrt{y}$ for all $y\in\mathbf{R}^+$, we obtain, from Theorem 36.1,

$$(36.5)\qquad (g\circ f_1)'(x)=\frac{1}{2\sqrt{x^3}}\,3x^2=\frac{3}{2}\sqrt{x}\qquad \text{for all}\quad x\in\mathbf{R}^+.$$

The chain rule does *not* apply to $x=0$ since the derivative of $g$ at $f(0)=0$ does not exist. However, (36.5) is valid at $x=0$, as one can see by differentiating (36.4) directly.

---

The composition of functions where the range of $f$ and the domain of $g$ are not identical but do overlap, such as in example 3, will be discussed in full generality in section 6.67 and the differentiation of such functions will be discussed in section 9.92.

If $f:(a,b)\to\mathbf{R}$ is invertible, then $f^{-1}\circ f:(a,b)\leftrightarrow(a,b)$ is the identity function; that is, $f^{-1}\circ f(x)=x$ for all $x\in(a,b)$. Hence, $(f^{-1}\circ f)'(x)=1$. If $f$ is differentiable at $c$ and $f^{-1}$ is differentiable at $f(c)$, we obtain, in view of the chain rule, that $(f^{-1}\circ f)'(c)=(f^{-1})'(f(c))f'(c)$ and hence, $(f^{-1})'(f(c))=1/(f'(c))$ or, equivalently, $(f^{-1})'(d)=1/f'(f^{-1}(d))$ where $d=f(c)$. In most instances, $f^{-1}$ is not explicitly known and hence, its differentiability cannot be established by a direct examination of $f^{-1}$. Still, it is often possible to establish the differentiability of $f^{-1}$ and find its derivative from known properties of $f$ as, for example, in the following theorem.

***Theorem 36.2***

If $f\in C'(a,b)$ is strictly increasing, then $(f^{-1})'$ exists for all $d\in\mathfrak{R}(f)$ for which $'(f^{-1}(d))\neq 0$, and is given by

$$(f^{-1})'(d)=\frac{1}{f'(f^{-1}(d))}.$$

*Proof*

By Theorem 27.2, $f^{-1}:\mathfrak{R}(f)\to\mathbf{R}$ exists and is continuous. Hence, $\lim_{y\to d}f^{-1}(y)=f^{-1}(d)$; that is, $\lim_{y\to d}x=c$ where $f(x)=y$, $f(c)=d$. Therefore,

$$(f^{-1})'(d)=\lim_{y\to d}\frac{f^{-1}(y)-f^{-1}(d)}{y-d}=\lim_{x\to c}\frac{1}{\dfrac{f(x)-f(c)}{x-c}}=\frac{1}{f'(c)}=\frac{1}{f'(f^{-1}(d))}.$$

▲ *Example 4*

The exponential function $E: \mathbf{R} \to \mathbf{R}$, given by $E(x) = e^x$, is strictly increasing, and $E'(x) = e^x > 0$. Hence,

$$(E^{-1})'(y) = \frac{1}{E(E^{-1}(y))}.$$

Since $E^{-1}(y) = \log y$ and $E(E^{-1}(y)) = e^{\log y} = y$, we have

$$(\log)'(y) = \frac{1}{y}.$$

(See also example 2, section 1.5.)

## 3.36 Exercises

1. Show that $\lim_a \gamma_a(x) = 0$ where $\gamma_a$ is defined in (36.3).
2. Consider $f: \{x \in \mathbf{R} \mid x = 0 \text{ or } \sin 1/x \geqslant 0\} \to \mathbf{R}$, defined by

$$f(x) = \begin{cases} 0 & \text{for } x = 0, \\ x^2 \sin \dfrac{1}{x} & \text{for } x \neq 0, \end{cases}$$

   and $g$: $[0, \infty) \to \mathbf{R}$, defined by $g(y) = \sqrt{y}$.
   (a) Show that $\mathcal{R}(f) = \mathcal{D}(g)$.
   (b) Show that $x = 0$ is an accumulation point of $\mathcal{D}(f)$ and $f(0) = 0$ is an accumulation point of $\mathcal{D}(g)$.
   (c) Does $(g \circ f)'(0)$ exist?
   (d) Does $(g \circ f)'(x)$ exist for $x \in \mathcal{D}(f)\backslash\{0\}$?
   (e) If the answer to (c) and/or (d) is *yes*, find the derivative.
3. Let $f: \mathcal{A} \to \mathcal{B}$, $g: \mathcal{B} \to \mathcal{C}$, $h: \mathcal{C} \to \mathcal{D}$, where $\mathcal{A}, \mathcal{B}, \mathcal{C}, \mathcal{D} \subseteq \mathbf{R}$.
   (a) Define the composite function $h \circ g \circ f: \mathcal{A} \to \mathcal{D}$.
   (b) Impose suitable conditions and derive the following differentiation rule for $h \circ g \circ f$:

$$(h \circ g \circ f)'(x) = h'(g(f(x)))g'(f(x))f'(x).$$

4. Use the result of exercise 3 to differentiate the function $f: \mathbf{R} \to \mathbf{R}$ which is defined by

$$f(x) = (\sin \sqrt{x^2 + 1})^3.$$

5. Same as in exercise 4, for the function $f: \mathbf{R} \to \mathbf{R}$, defined by

$$f(x) = \sqrt{x^2 + \sin(x^2)}.$$

6. Let $f: \mathbf{R} \to \mathbf{R}$ be defined by $f(x) = x$. Find $(f^{-1})'(y)$ for all $y \in \mathbf{R}$.
7. Let $f: (0, \infty) \to \mathbf{R}$ be defined by $f(x) = x^2$. Find $(f^{-1})'(y)$ for all $y \in (0, \infty)$.
8. Let $f: (-\pi/2, \pi/2) \to \mathbf{R}$ be defined by $f(x) = \sin x$. Find $(f^{-1})'(y)$ for all $y \in (-1, 1)$.

### 3.37 Mean-Value Theorems

***Definition 37.1***

If $f:\mathfrak{D}\to\mathbf{R}$, $\mathfrak{D}\subseteq\mathbf{R}$, then $f$ assumes a *relative maximum* at $a\in\mathfrak{D}$ if and only if there is a $\delta>0$ such that $f(a)\geqslant f(x)$ for all $x\in N_\delta(a)\cap\mathfrak{D}$. ($f$ assumes a *relative minimum* at $a\in\mathfrak{D}$ if and only if $f(a)\leqslant f(x)$ for all $x\in N_\delta(a)\cap\mathfrak{D}$.)

---

▲ *Example 1*

The function $f:\mathbf{R}\to\mathbf{R}$, given by $f(x)=|x|$, assumes a *relative minimum* at $a=0$. (See Fig. 37.1.)

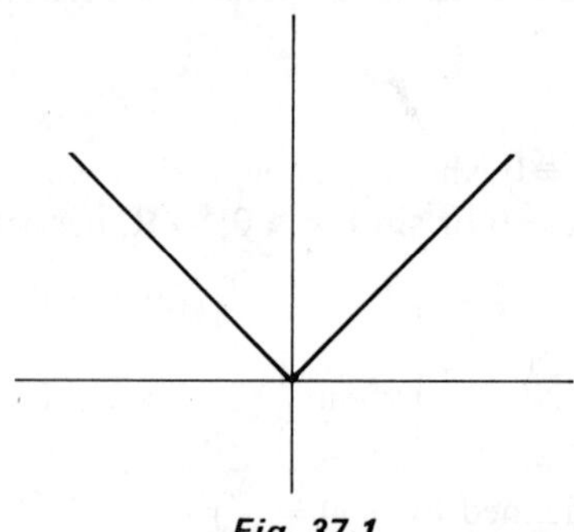

***Fig. 37.1***

▲ *Example 2*

The function $g:\mathbf{R}\to\mathbf{R}$, given by $g(0)=1/10$, $g(x)=x^2$ for $x\neq 0$, assumes a *relative maximum* at $a=0$. (See Fig. 37.2)

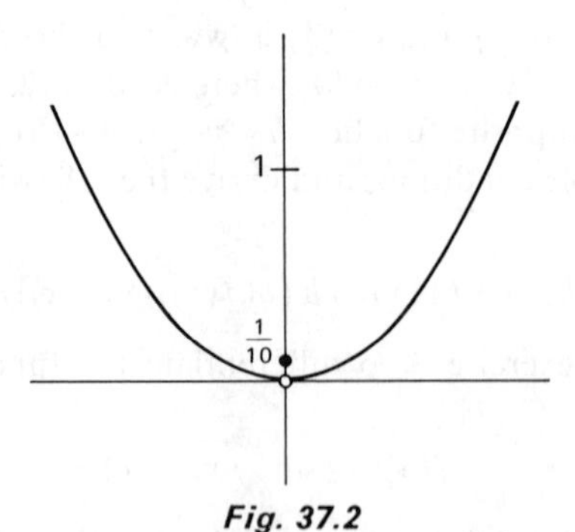


***Fig. 37.2***

▲ *Example 3*

The function $h:\mathbf{R}\to\mathbf{R}$, given by

$$h(x)=\begin{cases}x & \text{for } x\leqslant 1\\ 1 & \text{for } 1<x<2\\ 3-x & \text{for } 2\leqslant x\end{cases}$$

assumes a *relative maximum* for all $a\in[1,2]$. (See Fig. 37.3.)

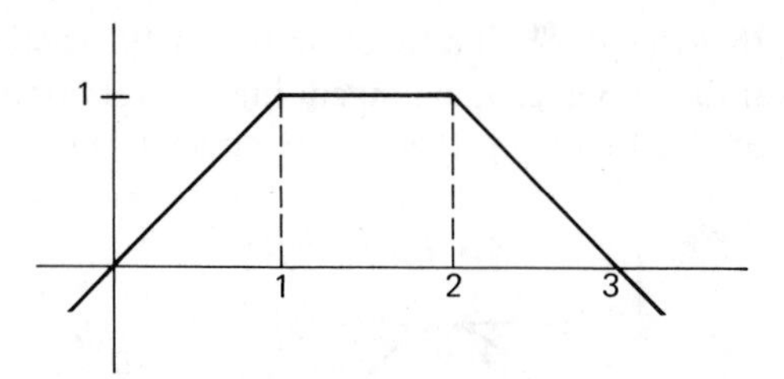


*Fig. 37.3*

---

Suppose that $f$ assumes a *relative maximum* at $a \in \mathfrak{D}$, where *a is an interior point of* $\mathfrak{D}$. Then,

$$\frac{f(x) - f(a)}{x - a} \begin{cases} \geqslant 0 & \text{if } x < a, \\ \leqslant 0 & \text{if } x > a. \end{cases} \tag{37.1}$$

If $f'(a)$ exists, then the limit of (37.1) exists and is the same, regardless of whether $x$ approaches $a$ from the left ($x < a$) or from the right ($x > a$). Since

$$f'(a) = \lim_a \frac{f(x) - f(a)}{x - a} \begin{cases} \geqslant 0 & \text{if } x \text{ approaches } a \text{ from the left,} \\ \leqslant 0 & \text{if } x \text{ approaches } a \text{ from the right,} \end{cases}$$

we conclude that, by necessity, $f'(a) = 0$, and we have the following theorem:

***Theorem 37.1***

If $f: \mathfrak{D} \to \mathbf{R}$, $\mathfrak{D} \subseteq \mathbf{R}$, assumes a *relative maximum* at $a \in \mathfrak{D}$, where $a$ is an interior point of $\mathfrak{D}$, and if $f'(a)$ exists, then, by necessity, $f'(a) = 0$. (If $f$ assumes a *relative minimum* at $a$, then, $f'(a) = 0$.)

Note that we could not have reached this conclusion, had we assumed only that $a$ was an accumulation point of $\mathfrak{D}$. Then it might not have been possible to approach $a$ from both sides. (See exercise 1.)

---

▲ *Example 4*

If $f: \mathbf{R} \to \mathbf{R}$ is given by $f(x) = x^3$, then $f'(0) = 0$ but $f$ does *not* assume a relative maximum or minimum at 0. This demonstrates that Theorem 37.1 is *strictly* a necessary condition.

▲ *Example 5*

Theorem 37.1 is not applicable to the functions $f$ and $g$ in examples 1 and 2 because $f'(0)$, $g'(0)$ do *not* exist. It *is* applicable to the function $h$ in example 3, but only for the points $a \in (1, 2)$. At $a = 1$ and $a = 2$, the derivative does not exist but the function $h$ still assumes relative maxima at these points.

---

If one observes that the graph of a function that is differentiable at every point of an open interval $(a, b)$ possesses a tangent line at every point of that interval, then the validity of the following theorem is strongly indicated by Fig. 37.4.

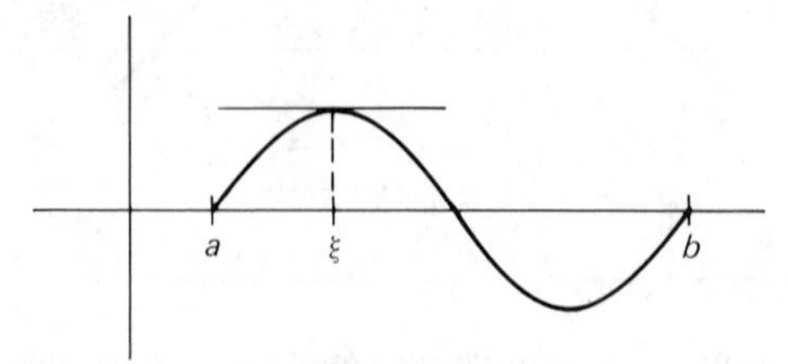


*Fig. 37.4*

***Theorem 37.2*** *(Theorem of Rolle)*†

If $f \in C[a, b]$ and $f \in C'(a, b)$, and if $f(a) = f(b) = 0$, then there exists at least one $\xi \in (a, b)$ such that $f'(\xi) = 0$.

*Proof*

(a) If $f(x) = 0$ for all $x \in (a, b)$, then $f'(x) = 0$ for all $x \in (a, b)$, and the theorem is trivially true.

(b) If $f$ does *not* vanish for all $x \in (a, b)$, then there is at least one $c \in (a, b)$, so that $f(c) \neq 0$. We may assume w.l.o.g. that $f(c) > 0$. Since $f \in C[a, b]$ and $[a, b]$ is compact, $f$ assumes its maximum for some $\xi \in [a, b]$ (Theorem 34.3) and $f(\xi) = \max_{[a,b]} f(x) \geqslant f(c) > 0$. Hence, $\xi \in (a, b)$. Since $f(\xi)$ is a relative maximum and since $f'(\xi)$ exists, we have, from Theorem 37.1, that $f'(\xi) = 0$. (See also exercise 2.)

We obtain from Rolle's theorem:

***Theorem 37.3*** *(Mean-Value Theorem of the Differential Calculus)*

If $f \in C[a, b]$, $f \in C'(a, b)$, then there exists at least one $\xi \in (a, b)$ such that

$$f'(\xi) = \frac{f(b) - f(a)}{b - a}. \tag{37.2}$$

*Proof*

The function $g : [a, b] \to \mathbf{R}$, defined by

$$g(x) = f(x) - f(a) - \frac{x - a}{b - a}(f(b) - f(a)),$$

satisfies all hypotheses of Theorem 37.2. Hence, there is a $\xi \in (a, b)$ such that $g'(\xi) = 0$. Equation (37.2) follows readily. (See also Fig. 37.5.)

†*Michel Rolle*, 1652–1719.

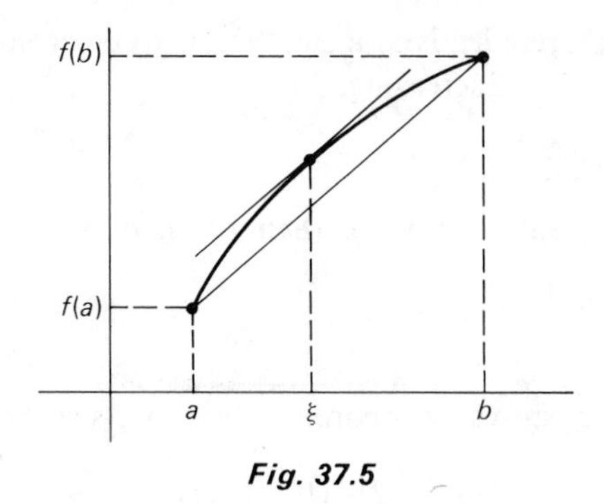


*Fig. 37.5*

***Corollary 1 to Theorem 37.3***

If $f : \mathfrak{J} \to \mathbf{R}$ and $f'(x) = 0$ for all $x \in \mathfrak{J}$, where $\mathfrak{J}$ denotes an interval (open, or closed, or neither), then $f(x) =$constant for all $x \in \mathfrak{J}$.

*Proof*

If $a, b \in \mathfrak{J}$, then, by the Corollary to Theorem 35.2, $f \in C[a, b]$ and, by hypothesis, $f \in C'(a, b)$. Hence, by Theorem 37.3, there is a $\xi \in (a, b)$ such that $f'(\xi) = (f(b) - f(a))/(b - a)$. Since $f'(\xi) = 0$, by hypothesis, we have $f(a) = f(b)$. Since this is true for every $a, b \in \mathfrak{J}$, $f(x) =$ constant for all $x \in \mathfrak{J}$.

***Corollary 2 to Theorem 37.3***

If $f, g : \mathfrak{J} \to \mathbf{R}$, where $\mathfrak{J}$ denotes an interval, and if $f'(x) = g'(x)$ for all $x \in \mathfrak{J}$, then

$$f(x) = g(x) + C \qquad \text{for all } x \in \mathfrak{J}$$

for some constant $C$.

*Proof*

The function $h = f - g$ satisfies the hypotheses of Corollary 1. Hence, $f(x) - g(x) =$ constant for all $x \in \mathfrak{J}$.

---

▲ *Example 6*

The function $f : (-1, 0) \cup (0, 1) \to \mathbf{R}$, defined by

$$f(x) = \begin{cases} 0 & \text{if } x \in (-1, 0) \\ 1 & \text{if } x \in (0, 1) \end{cases}$$

has the property that $f'(x) = 0$ for all $x \in (-1, 0) \cup (0, 1)$. Corollary 2 does *not* apply because the domain is not an interval, and we see that the function is *not* constant on its domain.

---

The mean-value theorem leads to a simple test to determine whether a differentiable function is strictly increasing.

***Lemma 37.1***

If $f'(x) > 0$ $(<0)$ for all $x \in (a, b)$, then $f: (a, b) \to \mathbf{R}$ is strictly increasing (decreasing).

*Proof*

Let $x_1 < x_2$. By the mean-value theorem, there exists a $\xi \in (x_1, x_2)$ such that

$$\frac{f(x_2) - f(x_1)}{x_2 - x_1} = f'(\xi) > 0.$$

Hence, $f(x_2) > f(x_1)$.

This lemma, in conjunction with Theorem 36.2, is very important for establishing the differentiability of inverse functions.

***Corollary to Lemma 37.1***

If $f'(x) > 0$ for all $x \in (a, b)$, then $(f^{-1})'(y) = 1/f'(f^{-1}(y))$ for all $y \in \mathfrak{R}(f)$.

---

▲ *Example 7*

Let $f: \mathbf{R} \to \mathbf{R}$ be defined by

$$f(x) = 137x^{205} - 19x^{39} + x^{38} - 47x^{27} + 1035x^{19} + 2x^2 + x + 13.$$

Since $f'(0) = 1$ and since $f'$ is continuous, we see that $f'(x) > 0$ in $N_\delta(0)$ for some $\delta > 0$. Hence, the restriction of $f$ to $N_\delta(0)$ is invertible and $(f^{-1})'(13) = 1$. (Note that $f(0) = 13$.) We could find this derivative without explicit knowledge of the inverse $f^{-1}$. To find the inverse would be a rather formidable task, not made any easier by Galois' Theorem† that the general algebraic equation of degree $n \geqslant 5$ cannot be solved by radicals.

---

In Chapter 14, extensive use will be made of the above corollary in finding the derivatives of inverse functions of the elementary functions, which, in turn, will lead to explicit representations of these inverse functions.

The following, more general version of the mean-value theorem has a number of practical applications, some of which will be discussed in the sequel.

***Theorem 37.4*** *(Cauchy's‡ Mean-Value Theorem)*

If $f, g \in C[a, b]$ and $f, g \in C'(a, b)$, then there exists a point $\xi \in (a, b)$ such that

$$(f(b) - f(a))g'(\xi) = (g(b) - g(c))f'(\xi). \tag{37.3}$$

†*Évariste Galois*, 1811–1832. For Galois' Theorem, see I. N. Herstein, *Topics in Algebra* (Waltham, Mass.: Blaisdell Publishing Company, 1964), p. 214.

‡ *Augustin-Louis Cauchy*, 1789–1857.

*Proof*

If $g(b) - g(a) = 0$, we obtain, from Theorem 37.3, a $\xi \in (a, b)$ such that $g'(\xi) = 0$ and (37.3) is satisfied.

If $g(b) - g(a) \neq 0$, we consider the function $h : [a, b] \to \mathbf{R}$ defined by

$$h(x) = f(x) - f(a) - \frac{f(b) - f(a)}{g(b) - g(a)}(g(x) - g(a)),$$

which satisfies the hypotheses of Theorem 37.2 (Theorem of Rolle). Application of Theorem 37.2 yields the desired result. (Note that Theorem 37.3 is a special case of Theorem 37.4 for $g(x) = x$.)

Two practical applications of Cauchy's mean-value theorem will be discussed in the following two examples:

---

▲ *Example 8* (*Rule of l'Hôpital*† *for indeterminate forms* 0/0)

We consider two functions $f, g : \mathfrak{D} \to \mathbf{R}$. Suppose that, for some $a \in \mathfrak{D}$, $\delta > 0$, $f, g \in C(N_\delta(a))$, $f, g \in C'(N'_\delta(a))$, and $f(a) = g(a) = 0$. Formal application of Theorem 25.3 to $\lim_a (f(x)/g(x))$ yields the result $\lim_a (f(x)/g(x)) = 0/0$, which is called an *indeterminate form*. Still, if $g'(x) \neq 0$ in $N'_\delta(a)$, and if

$$\lim_a (f'(x)/g'(x))$$

exists, then $\lim_a (f(x)/g(x))$ also exists and is given by

$$\lim_a \frac{f(x)}{g(x)} = \lim_a \frac{f'(x)}{g'(x)}. \tag{37.4}$$

This may be seen as follows: By Theorem 37.4, we have, for every $x \in N_\delta(a)$, that

$$f(x)g'(\xi) = g(x)f'(\xi).$$

Then $g(x) \neq 0$ for all $x \in N'_\delta(a)$. Otherwise we would obtain, from Theorem 37.3, that $g(x) = g'(\xi)(x - a) = 0$ for some $\xi \in N'_\delta(a)$, contrary to our assumption that $g'(x) \neq 0$ in $N'_\delta(a)$. Hence,

$$\frac{f(x)}{g(x)} = \frac{f'(\xi)}{g'(\xi)}$$

for some $\xi \in (a, x)$ or $\xi \in (x, a)$. Equation (37.4) follows readily.

Equation (37.4) obviously applies to $\lim_0 (\sin x)/x$, and we obtain

$$\lim_0 \frac{\sin x}{x} = \lim_0 \frac{\cos x}{1} = 1.$$

†*Guillaume François Marquis de l'Hôpital*, 1661–1704.

▲ *Example 9* *(Rule of l'Hôpital for indeterminate forms $\infty/\infty$)*

An analogous rule holds for $\lim_a (f(x)/g(x))$ for the case where $\lim_a f(x) = \pm\infty$ $\lim_a g(x) = \pm\infty$. (See definition 25.1.) In such a case, formal application of Theorem 25.3 would yield $\lim_a (f(x)/g(x)) = \pm\infty/\infty$, an indeterminate form. However, if $f, g \in C'(N'_\delta(a))$ for some $\delta > 0$, if $g'(x) \neq 0$ in $N'_\delta(a)$, and if $\lim_a (f'(x)/g'(x))$ exists, then

$$\lim_a \frac{f(x)}{g(x)} = \lim_a \frac{f'(x)}{g'(x)}. \tag{37.5}$$

The proof of this statement again makes use of Cauchy's mean-value theorem. Let $\lim_a (f'(x)/g'(x)) = L$. Then for every $\varepsilon > 0$, there is a $\delta_1 > 0$, $\delta_1 < \delta$, such that

$$\left|\frac{f'(x)}{g'(x)} - L\right| < \varepsilon \qquad \text{for all} \quad x \in N'_{\delta_1}(a).$$

For any $x \in (a - \delta_1, a)$ or $(a, a + \delta_1)$, we have, from Theorem 37.4, that

$$\frac{f(x) - f(a - \delta_1)}{g(x) - g(a - \delta_1)} = \frac{f'(\xi)}{g'(\xi)} \qquad \text{for some } \xi \in (a - \delta_1, x) \quad \text{or} \quad \xi \in (x, a - \delta_1).$$

(That $g(x) - g(a - \delta_1) \neq 0$ follows from $g'(x) \neq 0$ in $N'_\delta(a)$ and Theorem 37.3.) Hence,

$$\left|\frac{f(x) - f(a - \delta_1)}{g(x) - g(a - \delta_1)} - L\right| = \left|\frac{f(x)}{g(x)} \cdot \frac{1 - \dfrac{f(a - \delta_1)}{f(x)}}{1 - \dfrac{g(a - \delta_1)}{g(x)}} - L\right| < \varepsilon$$

for all $x \in (a - \delta_1, a)$ or $x \in (a, a + \delta_1)$. As $x$ approaches $a$,

$$\lim_a \frac{1 - \dfrac{f(a - \delta_1)}{f(x)}}{1 - \dfrac{g(a - \delta_1)}{g(x)}} = 1$$

and (37.5) follows readily.

Equation (37.5) obviously applies to $\lim_0 x \log x^2 = \lim_0 \dfrac{\log x^2}{1/x}$, and we obtain

$$\lim_0 \frac{\log x^2}{\dfrac{1}{x}} = \lim_0 \frac{\dfrac{2}{x}}{-\dfrac{1}{x^2}} = \lim_0 (-2x) = 0.$$

## 3.37 Exercises

1. Suppose $f: \mathfrak{D} \to \mathbf{R}$, and $a \in \mathfrak{D}$ is an *accumulation point* of $\mathfrak{D}$. Suppose, further, that $f(a) \geqslant f(x)$ for all $x \in N_\delta(a) \cap \mathfrak{D}$ for some $\delta > 0$ and that $f'(a)$ exists. Show that $f'(a)$ need not vanish.
2. Show: If $f: [a, b] \to \mathbf{R}$, $\xi \in (a, b)$ and $f(\xi) \geqslant f(x)$ for all $x \in [a, b]$, then $f(\xi)$ is a relative maximum, by definition 37.1.
3. Apply Rolle's Theorem to $f(x) = x^m(1 - x)^n$, $x \in [0, 1]$, $m, n \in \mathbf{N}$. What is $\xi$?
4. Given $f \in C'[a, b]$ and $[f'(x)]^2 = 0$ for all $x \in (a, b)$ and $f(a) = 0$. Show that $f(x) = 0$ for all $x \in [a, b]$.
5. Given $f \in C'(a, b)$ and $\xi \in (a, b)$. Show: If $\lim_\xi f'(x) = l$, then $f'(\xi) = l$.
6. If $f': \mathfrak{D} \to \mathbf{R}$ and if $a \in \mathfrak{D}$ is an accumulation point of $\mathfrak{D}$, then

$$f''(a) = \lim_a \frac{f'(x) - f'(a)}{x - a}$$

is called the *second* (iterated) *derivative* of $f$ at $a$, provided that limit exists. If $f''$ exists for all $x \in \mathfrak{D}$, we denote this by $f \in C''(\mathfrak{D})$. Show: If $f' \in C[a, b]$ and $f \in C''(a, b)$, then

$$f(b) - f(a) = f'(a)(b - a) + f''(\xi)\frac{(b - a)^2}{2}$$

for some $\xi \in (a, b)$.

7. Define the $n$th derivative of a function inductively in terms of the $(n - 1)$th derivative, and prove: If $f: [a, b] \to \mathbf{R}$ where $f^{(n-1)} \in C[a, b]$, $f^{(n)}$ exists in $(a, b)$, then there exists a $\xi \in (a, b)$ such that

$$\begin{aligned} f(b) - f(a) &\\ = f'(a)(b - a) &+ f''(a)\frac{(b - a)^2}{2!} + \cdots + f^{(n-1)}(a)\frac{(b - a)^{n-1}}{(n - 1)!} \\ + f^{(n)}(\xi)&\frac{(b - a)^n}{n!}. \end{aligned}$$

This is *Taylor's formula*.

8. Use the rule of l'Hôpital (example 8) to find the following limits:

(a) $\displaystyle\lim_0 \frac{x^n}{e^x - 1}$, $n \in \mathbf{N}$ (b) $\displaystyle\lim_0 \frac{x}{\tan x}$

(c) $\displaystyle\lim_1 \frac{\log x}{x - 1}$ (d) $\displaystyle\lim_1 \frac{\sin (x - 1)}{(\log (2x - 1))}$

9. Apply the rule of l'Hôpital repeatedly to find the following limits:

(a) $\displaystyle\lim_0 \frac{\sin^2 x}{x^2}$ (b) $\displaystyle\lim_0 \frac{e^x - x - 1}{x^2}$

(c) $\displaystyle\lim_0 \frac{\sin (2x^3)}{\sin^3 (2x)}$ (d) $\displaystyle\lim_0 \frac{\mathrm{Tan}^{-1}(x^2)}{x \sin x}$

10. Use the rule of l'Hôpital (example 9) to find the following limits:

(a) $\lim_{0} \dfrac{\log(\sin^2 x)}{\cot x}$ (b) $\lim_{0} \dfrac{\log(\cot x)}{\cot x}$

11. Prove: If $f$ has a continuous second-order derivative in $(a, b)$, and if $c \in (a, b)$, $f'(c) = 0$, $f''(c) > 0$, $(f''(c) < 0)$, then $f$ assumes a relative minimum (maximum) at $c$.
12. Let $f : [a, b] \to \mathbf{R}$. Prove: If $f$ assumes a relative maximum (minimum) at $a$ and if $f'(a)$ exists, then by necessity, $f'(a) \leqslant 0$ ($\geqslant 0$); and if $f$ assumes a relative maximum (minimum) at $b$ and if $f'(b)$ exists, then, by necessity, $f'(b) \geqslant 0$ ($\leqslant 0$).
13. Let $f(x) = 17x^{12} - 124x^9 + 16x^3 - 129x^2 + x - 1$. Find $(f^{-1})'(-1)$.
14. Let $f(x) = \cos x$, $x \in (0, \pi)$. Find $(f^{-1})'(y)$ for $y \in (-1, 1)$.

## 3.38 The Intermediate-Value Property of Derivatives

If $f : \mathfrak{D} \to \mathbf{R}$ and if $f \in C'(\mathfrak{D})$, then $f'$ is also a function from $\mathfrak{D}$ into $\mathbf{R}$. $f'$ need not be a continuous function. Take, for example, $f : \mathbf{R} \to \mathbf{R}$, given by

$$f(x) = \begin{cases} 0 & \text{if } x = 0 \\ x^2 \sin \dfrac{1}{x} & \text{if } x \neq 0. \end{cases}$$

We have seen in example 3, section 3.35, that $f'(0) = 0$, and we obtain for $x \neq 0$, that

$$f'(x) = 2x \sin \frac{1}{x} - \cos \frac{1}{x}.$$

The function $f' : \mathbf{R} \to \mathbf{R}$, given by

$$f'(x) = \begin{cases} 0 & \text{if } x = 0 \\ 2x \sin \dfrac{1}{x} - \cos \dfrac{1}{x} & \text{if } x \neq 0 \end{cases} \tag{38.1}$$

is *not* continuous at 0 because $\lim_0 f'(x)$ does not even exist. Still, being a derivative, $f'$ has a property (discovered by *Gaston Darboux*†) which it shares with continuous functions, namely, the intermediate-value property.

***Theorem 38.1***

If $f \in C'[a, b]$ and $\eta$ is any number between $f'(a)$ and $f'(b)$, then there exists at least one $\xi \in [a, b]$ such that $f'(\xi) = \eta$.

†*Gaston Darboux*, 1842–1917.

*Proof*

(a) If $f(a) = f'(b)$, then $\eta = f'(a)$ and $\xi = a$.

(b) We may assume w.l.o.g. that $f'(a) < \eta < f'(b)$. Then, no matter what $(f(b) - f(a))/(b - a)$ is, $\eta$ is either equal to $(f(b) - f(a))/(b - a)$, or between $f'(a)$ and $(f(b) - f(a))/(b - a)$, or between $(f(b) - f(a))/(b - a)$ and $f'(b)$.

The functions $\phi, \psi : [a, b] \to \mathbf{R}$, defined by

$$\phi(x) = \begin{cases} f'(a) & \text{for } x = a, \\ \dfrac{f(x) - f(a)}{x - a} & \text{for } 0 < x \leqslant b, \end{cases}$$

$$\psi(x) = \begin{cases} \dfrac{f(b) - f(x)}{b - x} & \text{for } a \leqslant x < b, \\ f'(b) & \text{for } x = b, \end{cases}$$

are continuous on $[a, b]$, because $\lim_a \phi(x) = f'(a) = \phi(a)$ and $\lim_b \psi(x) = f'(b) = \psi(b)$.

If $\eta = (f(b) - f(a))/(b - a)$, then we have, from Theorem 37.3, that $f'(\xi) = \eta$ for some $\xi \in (a, b)$, and our theorem is proved.

If $\eta$ is between $f'(a)$ and $(f(b) - f(a))/(b - a)$, then we have, from the intermediate-value theorem for continuous functions (Theorem 27.1) that there is a $c \in (a, b)$ such that $\phi(c) = \eta$; that is,

$$\frac{f(c) - f(a)}{c - a} = \eta.$$

By Theorem 37.3, there is a $\xi \in (a, c) \subset (a, b)$ such that $f'(\xi) =$

$$(f(c) - f(a))/(c - a)$$

and hence, $f'(\xi) = \eta$.

If $\eta$ is between $(f(b) - f(a))/(b - a)$ and $f'(b)$, we argue as before but use the function $\psi$ instead. (See exercise 2.)

---

▲ *Example 1*

Theorem 38.1 is useful when one wishes to demonstrate that a given function cannot be a derivative. Take $g : \mathbf{R} \to \mathbf{R}$, given by

$$g(x) = \begin{cases} 0 & \text{if } x < 0, \\ 1 & \text{if } x \geqslant 0. \end{cases}$$

Clearly, there is no function $f : \mathbf{R} \to \mathbf{R}$ of which $g$ is the derivative, because $g$ does *not* have the intermediate-value property. ($g$ does *not* assume the value $\frac{1}{2}$ which is between $g(-127{,}384) = 0$ and $g(\pi) = 1$.)

---

### 3.38 Exercises

1. Show that the function $f'$ defined in (38.1) is not continuous at 0.
2. Carry out the proof of Theorem 38.1 for the case where $\eta$ lies between $(f(b) - f(a))/(b - a)$ and $f'(b)$.
3. Show that the function $f' : \mathbf{R} \to \mathbf{R}$, which is defined in (38.1), assumes the values $\frac{1}{2}$ and $-\frac{1}{2}$ for infinitely many distinct values of $x$.
4. Let $f \in C'[a, b]$ and $f'(x) \neq 0$ for all $x \in [a, b]$. Show that $x_1 \neq x_2$, $x_1$, $x_2 \in [a, b]$, implies $f(x_1) \neq f(x_2)$.

### 3.39 The Differential*

***Definition 39.1***

A function $l : \mathbf{R} \to \mathbf{R}$ is called *linear* if and only if

1. $l$ is *additive*; that is, $l(x + y) = l(x) + l(y)$ for all $x, y \in \mathbf{R}$;
2. $l$ is *homogeneous*; that is, $l(\lambda x) = \lambda l(x)$ for all $x \in \mathbf{R}$ and any given $\lambda \in \mathbf{R}$.

Since $x = x \cdot 1$, the general form of a linear function $l : \mathbf{R} \to \mathbf{R}$ is given by

$$l(x) = l(x \cdot 1) = xl(1),$$

or, if $l(1) = m$, by

$$l(x) = mx \qquad \text{for all } x \in \mathbf{R}.$$

We view the derivative $f'(a)$ of a function $f$ at a point $a$ as a *linear function* $f'(a) : \mathbf{R} \to \mathbf{R}$ that is defined by

$$f'(a)(x) = f'(a)x.$$

With this in mind, we may reformulate Theorem 35.1 as follows:

***Theorem 39.1***

$f : \mathfrak{D} \to \mathbf{R}$, $\mathfrak{D} \subseteq \mathbf{R}$, is differentiable at $a \in \mathfrak{D}$, where $a$ is an accumulation point of $\mathfrak{D}$ and $f'(a)$ is its derivative at that point, if and only if there exists a linear function $l_a : \mathbf{R} \to \mathbf{R}$ and a function $\alpha_a : \mathfrak{D}\backslash\{a\} \to \mathbf{R}$ with $\lim_a \alpha_a(x) = 0$, such that

$$f(x) - f(a) = l_a(x - a) + \alpha_a(x)(x - a) \tag{39.1}$$

for all $x \in N'_\delta(a) \cap \mathfrak{D}$ for some $\delta > 0$ and $l_a$ is given by $l_a(h) = f'(a)h$ for all $h \in \mathbf{R}$.

*Proof*

(a) If $f$ is differentiable at $a$ and $f'(a)$ is its derivative, we have from Theorem 35.1 that

$$f(x) - f(a) = f'(a)(x - a) + \alpha_a(x)(x - a)$$

*The reader may omit this section at this time and read it before studying Chapter 9.

for all $x \in N'_\delta(a) \cap \mathfrak{D}$ for some $\delta > 0$, where $\lim_a \alpha_a(x) = 0$. We define a function $l_a : \mathbf{R} \to \mathbf{R}$ by $l_a(h) = f'(a)h$ and note that it is a linear function. Hence, we have $f'(a)(x - a) = l_a(x - a)$ and (39.1) is established.

(b) If (39.1) holds, we obtain, in view of the homogeneity of $l_a$, that $l_a(x - a) = (x - a)l_a(1)$, and hence,

$$\frac{f(x) - f(a)}{x - a} = l_a(1) + \alpha_a(x).$$

We take the limit as $x$ approaches $a$ and obtain that $f'(a)$ exists and $f'(a)h = l_a(h)$.

Since $\lim_a \alpha_a(x) = 0$, we have, for every $\varepsilon > 0$, a $\delta(\varepsilon) > 0$ (which is not necessarily the same as the $\delta$ of Theorem 39.1) such that $|\alpha_a(x)| < \varepsilon$ for all $x \in N'_{\delta(\varepsilon)}(a) \cap \mathfrak{D}$. If we also observe that the left side in (39.1) vanishes for $x = a$, we may state Theorem 39.1 as follows:

***Theorem 39.2***

$f : \mathfrak{D} \to \mathbf{R}$, $\mathfrak{D} \subseteq \mathbf{R}$, is differentiable at $a \in \mathfrak{D}$, where $a$ is an accumulation point of $\mathfrak{D}$ and $f'(a)$ is its derivative if and only if there is a linear function $l_a : \mathbf{R} \to \mathbf{R}$ and, for every $\varepsilon > 0$, a $\delta(\varepsilon) > 0$ such that

(39.2) $$|f(x) - f(a) - l_a(x - a)| \leqslant \varepsilon |x - a|$$

for all $x \in N_{\delta(\varepsilon)}(a) \cap \mathfrak{D}$ and $l_a(h) = f'(a)h$ for all $h \in \mathbf{R}$.

***Definition 39.2***

If $f : \mathfrak{D} \to \mathbf{R}$, $\mathfrak{D} \subseteq \mathbf{R}$, is differentiable for all $x \in \mathcal{A} \subseteq \mathfrak{D}$, then the function $df : \mathcal{A} \times \mathbf{R} \to \mathbf{R}$, which is defined by

$$df(x, h) = f'(x)h, \qquad x \in \mathcal{A}, \quad h \in \mathbf{R},$$

is called the *differential* of $f$ at $x$ with increment $h$. (The differential is *not* an infinitesimally small quantity, as the engineers would have it, and whatever that might mean.)

The term $\alpha_a(x)(x - a)$ in the representation of $f(x) - f(a)$ in (39.1) is a product of two terms, each of which approaches zero as $x$ approaches $a$. Since one of the two terms approaches zero linearly (of first order), the product is said to be *small of higher than first order*. Then $l_a(x - a)$ in (39.1) is an approximation to $f(x) - f(a)$, where terms that are small of higher than first order, are neglected. Such an approximation is called a *linear approximation*. Since $l_a(x - a) = df(a, x - a) = f'(a)(x - a)$, we say that $f'(a)(x - a)$ is a linear approximation to $f(x) - f(a)$.

If we substitute, in (39.1), $df(a, x - a)$ for $l_a(x - a)$, we obtain

(39.3) $$f(x) - f(a) = df(a, x - a) + \alpha_a(x)(x - a).$$

This together with $\tan \alpha = f'(a)$ yields a compelling geometric interpretation of the differential, as depicted in Fig. 39.1.

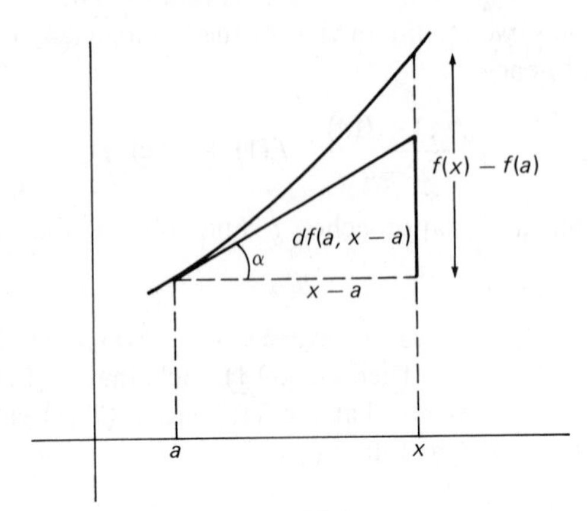


*Fig. 39.1*

### 3.39 Exercises

1. Show: If $h : \mathbf{R} \to \mathbf{R}$ is homogeneous; that is, if $h(\lambda x) = \lambda h(x)$ for every $\lambda \in \mathbf{R}$ and all $x \in \mathbf{R}$, then $h$ is also additive; that is, $h(x + y) = h(x) + h(y)$ for all $x, y \in \mathbf{R}$.
2. (a) Given $f : \mathbf{R} \to \mathbf{R}$ by $f(x) = x^3$. Find $df(1, h)$, $df(a, x - a)$, and $df(0, 1)$.
   (b) Given $f : \mathbf{R} \to \mathbf{R}$ by $f(x) = \sin x$. Find $df(x, h)$, $df(0, h)$, and $df(\pi, h)$.
3. Find a linear approximation to $\sin 31°$.
4. Find a linear approximation to $\sqrt{4.03}$.
5. Find a linear approximation to $(10.009032)^2$.
6. Let $f \in C[a, b]$, $f \in C'(a, b)$. Show: There is a $\xi \in (a, b)$ such that $f(b) - f(a) = df(\xi, b - a)$.

# 4 The Riemann Integral

## 4.40 Area Measure

In this and the following two sections we sketch an axiomatic development of area measure and, in doing this, provide a motivation for the introduction of the integral as an instrument for measuring *areas* of regions other than polygonal regions.

We shall not attempt to develop the concept of area measure in a manner that meets very exacting standards of rigor, economy, and elegance. We merely wish to outline how one might go about such a task and, in doing this, impress upon the reader that area measure is not presented to us by some mathematical deity but is conceived by man (mathematicians, to be sure) to serve certain practical needs and possibly provide intellectual fulfillment.

We may wish to reflect for a moment on the concept of length measure. A length measure, which assigns nonnegative numbers to certain points sets (on a line, or on a curve) serves the purpose of giving precise meaning to practical statements such as "Susan's legs are longer than Sheila's," or "a frankfurter is twice as long as a pork sausage," etc. For example, any length measure that is worth its salt assigns the number $(b - a)$ to the point set $[a, b] = \{x \in \mathbf{R} \mid a \leqslant x \leqslant b\}$ and, as a matter of fact, also to the point set $(a, b) = \{x \in \mathbf{R} \mid a < x < b\}$. Whether or not a length measure can be sensibly assigned to point sets such as $[0, 1] \cap \mathbf{Q}$ we are in no position to say at the moment. (This depends to a large extent on how the length measure is defined. See sections 4.48 and example 4, section 11.105.)

As is the case with length measure, it is the purpose of area measure to assign nonnegative numbers to point sets in the two-dimensional plane in a manner that statements such as "a given point set has smaller area than another given point set" (meaning: it takes less paint to paint the one wall than it takes to paint the other wall), or "a given point set has twice the area of another given point set," take on precise meaning.

Before we can formulate some reasonable demands as to what properties an area measure should have, we have to introduce some elementary concepts of the topology of the two-dimensional plane. We do this with the understanding that there is a one-to-one correspondence between the set of all ordered pairs of numbers $(x, y) \in \mathbf{R} \times \mathbf{R}$ and all points in the two-dimensional plane $\mathbf{R}^2$ where the ordered pair $(x_1, x_2)$ is interpreted as the Cartesian coordinates of the point it corresponds with. (See also section 2.20.)

***Definition 40.1***

The *open disk* $N_\delta(a_1, a_2) = \{(x_1, x_2) \in \mathbf{R} \times \mathbf{R} \mid (x_1 - a_1)^2 + (x_2 - a_2)^2 < \delta^2\}$ is called a *$\delta$-neighborhood of the point* $(a_1, a_2)$.

***Definition 40.2***

$(a_1, a_2)$ is an *interior point* of $A \subseteq \mathbf{R}^2$ if and only if there is a $\delta > 0$ such that $N_\delta(a_1, a_2) \subseteq A$. $A^\circ$ denotes the set of interior points of $A$ and is called the *interior of A*.

---

▲ *Example 1*

The point set $A = \{(x_1, x_2) \in \mathbf{R} \times \mathbf{R} \mid x_1 = x_2\}$ does not have any interior points. (See Fig. 40.1.)

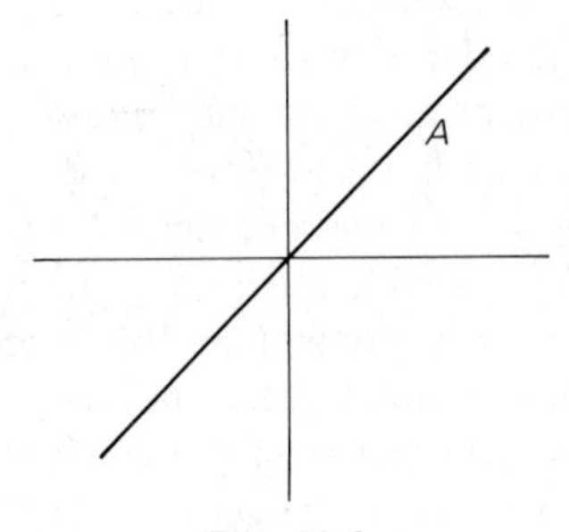


***Fig. 40.1***

▲ *Example 2*

If $B = \{(x_1, x_2) \in \mathbf{R} \times \mathbf{R} \mid x_1^2 + x_2^2 \leqslant 1\}$, then $B^\circ = \{(x_1, x_2) \in \mathbf{R} \times \mathbf{R} \mid x_1^2 + x_2^2 < 1\}$. (See Fig. 40.2.)

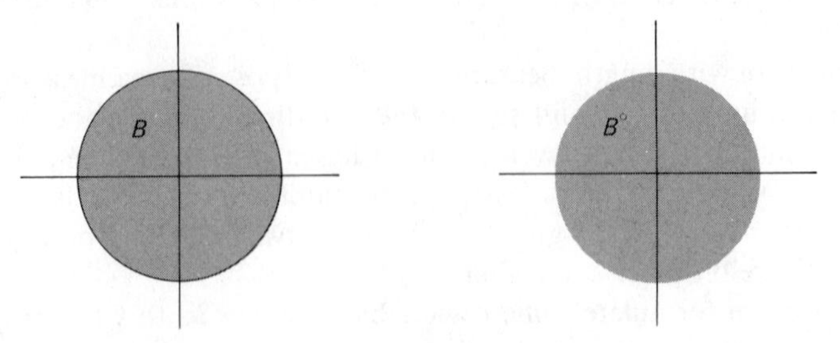


***Fig. 40.2***

---

It is not at all clear that every point set in the plane can be assigned an area measure in a meaningful manner. Take, for example, the point set $\{(x_1, x_2) \in \mathbf{R} \times \mathbf{R} \mid x_1 > 0, x_2 < 1/x_1\}$. We shall eliminate such point sets from our consideration by restricting ourselves to *bounded* point sets.

***Definition 40.3***

A point set $A \subseteq \mathbf{R}^2$ is called *bounded*, if and only if there exists an $R > 0$ such that $A \subseteq N_R(0, 0) = \{(x_1, x_2) \in \mathbf{R} \times \mathbf{R} \mid x_1^2 + x_2^2 < R^2\}$. (See Fig. 40.3.)

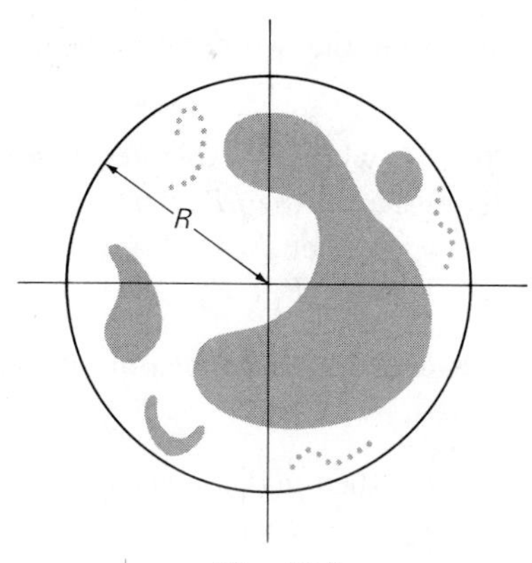


*Fig. 40.3*

We shall not go into the question of whether or not all bounded point sets can be measured in a meaningful and practical manner, but shall be content with demonstrating that there is a nonempty collection of bounded point sets $\mathfrak{I}$ in the plane for which area measure can be sensibly defined. The next section is devoted to a characterization of this collection $\mathfrak{I}$.

## 4.40 Exercises

1. Given the set $A = \{(x, 0) \in \mathbf{R} \times \mathbf{R} \mid x = 1/k, k \in N\}$. Find $A^\circ$.
2. Same as in exercise 1, for the set $A = \{(0, y) \in \mathbf{R} \times \mathbf{R} \mid 0 < y < 1\}$.
3. Same as in exercise 1, for the set $A = \{(x, y) \in \mathbf{R} \times \mathbf{R} \mid x^2 + y^2 < 1\} \cup \{(0, 1)\}$.
4. Same as in exercise 1, for the set $A = \{(x, y) \in \mathbf{R} \times \mathbf{R} \mid x < 0\} \cup \{(x, y) \in \mathbf{R} \times \mathbf{R} \mid x = 0, y > 0\}$.

## 4.41 Properties of Area Measure

Practical considerations dictate that squares belong to the class $\mathfrak{I}$ of point sets that have area measure. If a certain point set is in $\mathfrak{I}$ and another point set is congruent to it, then surely, this other set is also in $\mathfrak{I}$ and should be assigned the same area measure. (Two point sets are congruent if they can be brought to coincidence by means of a translation and an orthogonal transformation.)

If one removes a point set with area measure from another point set with area measure, then whatever is left over is expected to have area measure. Also, the union of point sets with area measure and nonoverlapping interiors will have area measure which is given by the sum of the area measures of its parts. Finally, one needs some unit of area measure and it is practical to choose as such a unit the square with side length 1, the so-called unit square. Let us now translate these loosely formulated demands into mathematical language:

***Definition 41.1***

If $\mathfrak{I}$ is a collection of point sets in the two-dimensional plane $\mathbf{R}^2$ with the properties that:

($\mathfrak{I}_1$) $U_a \in \mathfrak{I}$ for all $a \in (0, \infty)$, where $U_a$ denotes a square of side length $a$;
($\mathfrak{I}_2$) If $P_1 \in \mathfrak{I}$, $P_1 \sim P_2$ (congruent), then $P_2 \in \mathfrak{I}$;
($\mathfrak{I}_3$) If $P_1, P_2 \in \mathfrak{I}$, $P_1^\circ \cap P_2^\circ = \varnothing$, then $P_1 \cup P_2 \in \mathfrak{I}$;
($\mathfrak{I}_4$) If $P_1, P_2 \in \mathfrak{I}$, $P_1 \subseteq P_2$, then $P_2 \backslash P_1 \in \mathfrak{I}$;

then a function $\mu : \mathfrak{I} \to \mathbf{R}$ is called an *area measure* on $\mathfrak{I}$ if and only if

($\mu_1$) $\mu(P) \geqslant 0$ for all $P \in \mathfrak{I}$;
($\mu_2$) If $P_1, P_2 \in \mathfrak{I}$ and $P_1 \sim P_2$, then $\mu(P_1) = \mu(P_2)$;
($\mu_3$) If $P_1, P_2 \in \mathfrak{I}$ and $P_1^\circ \cap P_2^\circ = \varnothing$, then $\mu(P_1 \cup P_2) = \mu(P_1) + \mu(P_2)$;
($\mu_4$) $\mu(U_1) = 1$.

An area measure makes sense only when it does not assign to a contained point set a larger number than to the containing point set. There was no need to postulate this property, as it follows already from the postulates in definition 41.1:

***Theorem 41.1***

If $P_1, P_2 \in \mathfrak{I}$, and $P_1 \subseteq P_2$, then $\mu(P_1) \leqslant \mu(P_2)$. (The area measure is *monotonic*.)

***Proof***

By ($\mathfrak{I}_4$), $P_2 \backslash P_1 \in \mathfrak{I}$. Since $P_1 \subseteq P_2$, we have $P_2 = (P_2 \backslash P_1) \cup P_1$. Since $(P_2 \backslash P_1) \cap P_1 = \varnothing$, then $(P_2 \backslash P_1)^\circ \cap P_1^\circ = \varnothing$, and we obtain from ($\mu_3$) that $\mu(P_2) = \mu((P_2 \backslash P_1) \cup P_1) = \mu(P_2 \backslash P_1) + \mu(P_1)$. By ($\mu_1$), $\mu(P_2 \backslash P_1) \geqslant 0$, and hence, $\mu(P_1) \leqslant \mu(P_2)$.

If a point set with nonempty interior has area measure, we expect this area measure to be positive. This is indeed true.

***Theorem 41.2***

If $P \in \mathfrak{I}$ and $P^\circ \neq \varnothing$, then $\mu(P) > 0$.

*Proof*

Let $(x_1^\circ, x_2^\circ) \in P^\circ$. Then, by definition 40.2, there is a $N_\delta(x_1^\circ, x_2^\circ) \subset P^\circ$ for some $\delta > 0$. We choose $n$ so large that the square $U_{1/n} \subset N_\delta(x_1^\circ, x_2^\circ)$. Then we have from $(\mathfrak{S}_1)$, $(\mu_1)$, and Theorem 41.1 that $0 \leqslant \mu(U_{1/n}) \leqslant \mu(P)$. We may view $U_1$ as the union of $n^2$ squares $U_{1/n}$ of nonoverlapping interiors and obtain, from $(\mu_3)$ and $(\mu_4)$, that

$$1 = \mu(U_1) = n^2 \mu(U_{1/n}),$$

and hence, $\mu(U_{1/n}) = (1/n^2)$ and $\mu(P) \geqslant (1/n^2) > 0$ follows readily.

By contrast, a point set with no interior points has area measure zero, if it has area measure at all.

***Theorem 41.3***

If $P \in \mathfrak{S}$ and $P^\circ = \varnothing$, then $\mu(P) = 0$.

*Proof*

Since $P^\circ \cap P^\circ = \varnothing \cap \varnothing = \varnothing$ and $P \cup P = P$, we have, from $(\mu_3)$, that $\mu(P) = \mu(P \cup P) = \mu(P) + \mu(P) = 2\mu(P)$. Hence, $\mu(P) = 0$.

When measuring area, only the interior matters. This is reflected in the next theorem.

***Theorem 41.4***

If $P, P^\circ \in \mathfrak{S}$, then $\mu(P) = \mu(P^\circ)$.

*Proof*

By $(\mathfrak{S}_4)$, $B = P \backslash P^\circ \in \mathfrak{S}$. By construction $B^\circ = \varnothing$. Since $P = B \cup P^\circ$ and $B^\circ \cap P^\circ = \varnothing \cap P^\circ = \varnothing$, we have, from $(\mu_3)$ and Theorem 41.3, that

$$\mu(P) = \mu(B \cup P^\circ) = \mu(B) + \mu(P^\circ) = \mu(P^\circ).$$

## 4.41 Exercises

1. Show, by $(\mu_3)$ and $(\mu_4)$ of definition 41.1, that $\mu(U_r) = r^2$, where $r$ is rational.
2. Given a rectangle $\square$ of dimensions $a$, $b$. Show: If $\square \in \mathfrak{S}$, and if $a, b \in \mathbf{Q}$, then $\mu(\square) = ab$.
3. Show that it is possible to have $P, Q \in \mathfrak{S}$, $P \subset Q$, but $\mu(P) = \mu(Q)$.
4. Prove: If $P_1, P_2 \in \mathfrak{S}$, and $P_1 \subseteq P_2$, then $\mu(P_2 \backslash P_1) = \mu(P_2) - \mu(P_1)$.

## 4.42 Area Measure of Rectangles*

Although we have not yet demonstrated the existence of $\mathcal{T}$, we shall demonstrate in this section that, if $\mathcal{T}$ contains all bounded rectangles (including line segments and points, which are to be viewed as degenerate rectangles), then the area measure of a rectangle of dimensions $a$ and $b$ is, by necessity, given by $ab$. This is not a startling revelation. It is quite easy to see that such an area measure satisfies all the relevant axioms in definition 41.1. However, our discussion will show that this is the only possible way to define the area measure of a rectangle if one wishes to stay within the framework of definition 41.1, which, in turn, seems a reasonable and sensible way to define area measure.

We note that a rectangle $\square_{a,b}$ is uniquely determined by its base $a$ and width $b$. Hence, its area measure, if it exists at all, is a function of $a$ and $b$:

$$\mu(\square_{a,b}) = f(a,b). \tag{42.1}$$

It is our task to determine the function $f: [0, \infty) \times [0, \infty) \to \mathbf{R}$ in the spirit of definition 41.1.

In order to study the dependence of $f$ on the first argument, we shall consider all bounded rectangles of base $x$ and a given height $b$ and denote

$$f(x, b) = \phi(x). \tag{42.2}$$

We obtain, from $(\mu_3)$, with reference to Fig. 42.1, that

$$\phi(x_1 + x_2) = \phi(x_1) + \phi(x_2), \qquad x_1 > 0, \quad x_2 > 0.$$

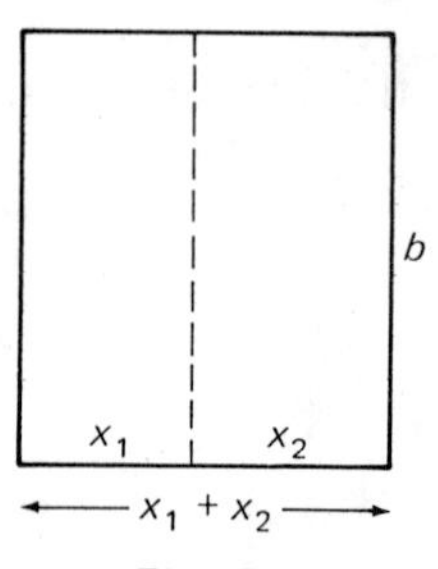


*Fig. 42.1*

By Theorem 41.3 and in view of our assumption that line segments are in $\mathcal{T}$, we have $\phi(0) = 0$. Hence, we can extend the validity of this formula to

$$\phi(x_1 + x_2) = \phi(x_1) + \phi(x_2) \qquad \text{for all} \quad x_1 \geqslant 0, \quad x_2 \geqslant 0. \tag{42.3}$$

We have, in view of $(\mu_3)$, that

$$\phi(1) = \phi\left(\frac{1}{n}\right) + \phi\left(\frac{1}{n}\right) + \cdots + \phi\left(\frac{1}{n}\right),$$

and hence, $\phi(1/n) = \phi(1)/n$, $n \in \mathbf{N}$. We can find, for every $\varepsilon > 0$, some $m \in \mathbf{N}$ such that $\phi(1)/m < \varepsilon$, and a $\delta > 0$ such that $\delta < 1/m$. Then we obtain, from Theorem 41.1, that

$$0 \leqslant \phi(x) \leqslant \phi(\delta) \leqslant \phi\left(\frac{1}{m}\right) = \frac{\phi(1)}{m} < \varepsilon$$

for all $0 \leqslant x < \delta$. Since $\phi(0) = 0$, we have, for every $\varepsilon > 0$, a $\delta > 0$ such that

$$|\phi(0) - \phi(x)| < \varepsilon \qquad \text{for all} \quad 0 \leqslant x - 0 < \delta;$$

that is, $\phi$ is continuous at 0. Since $\phi$ is also additive by (42.3), it follows that $\phi$ is continuous on $[0, \infty)$. (See exercise 9.) This, in turn, implies that $\phi$ is homogeneous; that is,

$$\phi(\lambda x) = \lambda\phi(x)$$

for all $\lambda \in [0, \infty)$ and all $x \in [0, \infty)$. (See exercises 8 through 11.) Hence, we obtain for (42.2),

$$f(x, b) = \phi(x) = x\phi(1) = xf(1, b). \tag{42.4}$$

Since a rectangle of base 1 and width $b$ is congruent to a rectangle of base $b$ and width 1, we obtain, from $(\mu_2)$,

$$f(1, b) = f(b, 1).$$

If we denote $f(y, 1) = \psi(y)$, we obtain, as before, $\psi(y) = y\psi(1)$ for all $y \in [0, \infty)$, and (42.4) yields

$$f(a, b) = af(1, b) = af(b, 1) = a\psi(b) = ab\psi(1).$$

$\psi(1)$, by definition, is the area measure of $U_1$, and hence, $\psi(1) = 1$. So we arrive at the desired result that

$$\mu(\square_{a,b}) = ab. \tag{42.5}$$

So far, we have shown that, if $\mathfrak{T}$ contains all bounded rectangles, then (42.5) is the area measure of a rectangle. With reference to Fig. 42.2, and in view of

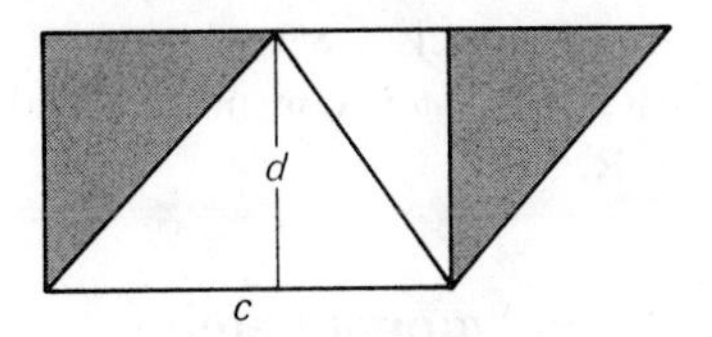


*Fig. 42.2*

$(\mu_2)$ and $(\mu_3)$, we can extend the area measure to triangles, under the provision that $\mathfrak{T}$ contains all bounded triangles; and we then obtain, by necessity, that:

$$\mu(\Delta_{c,d}) = \tfrac{1}{2}cd. \tag{42.6}$$

By means of $(\mu_3)$ we can extend the definition of $\mu$ to all bounded point sets that can be triangulated, i.e., decomposed into finitely many triangles of nonoverlapping interiors. If $P$ is such a point set and if $T_1, T_2, \ldots, T_n$ are its triangular components, then we have, from $(\mu_3)$ that

$$\mu(P) = \mu(T_1) + \mu(T_2) + \cdots + \mu(T_n). \tag{42.7}$$

The collection $\mathfrak{T}$ of all bounded point sets that can be triangulated satisfies axioms $(\mathfrak{T}_1)$ to $(\mathfrak{T}_4)$ of definition 41.1 (see exercises 1 to 5) and $\mu$, as defined in (42.7) with $\mu(T_j)$ defined as in (42.6), satisfies $(\mu_1)$ to $(\mu_4)$ of definition 41.1. (See exercises 6 and 7.) We have thus found a nonempty set $\mathfrak{T}$ and an area measure $\mu$ on $\mathfrak{T}$.

## 4.42 Exercises

1. Show: Every quadrangle, pentagon, and hexagon can be triangulated.
2. Show: If $P_1$, $P_2$ can be triangulated and $P_1^\circ \cap P_2^\circ = \varnothing$, then $P_1 \cup P_2$ can be triangulated.
3. Show: If $T_1$, $T_2$ are triangles, then $T_1 \cap T_2$ is either empty, or a point, or a line segment, or a triangle, or a quadrilateral, or a pentagon, or a hexagon.
4. Show: If $P_1$, $P_2$ can be triangulated, and if $P_1 \subseteq P_2$, then $P_2 \backslash P_1$ can be triangulated.
5. Show: The collection $\mathfrak{T}$ of all bounded point sets that can be triangulated satisfies $(\mathfrak{T}_1)$ to $(\mathfrak{T}_4)$ of definition 41.1.
6. Let $\mathfrak{T}$ denote the collection of exercise 5. Show: If $\mu$ is defined on $\mathfrak{T}$ as in (42.7), then $\mu$ is a function; i.e., for each $P \in \mathfrak{T}$ there is exactly one $p \in \mathbf{R}$ such that $\mu(P) = p$.
7. Show that $\mu$, as defined by (42.7) on the collection of all point sets that can be triangulated, satisfies $(\mu_1)$ to $(\mu_4)$ of definition 41.1. (See also exercise 6.)
8. Show: If $\phi : [0, \infty) \to \mathbf{R}$ is additive, then $\phi(0) = 0$ and $\phi(nx) = n\phi(x)$ for any $n \in \mathbf{N}$ and all $x \in [0, \infty)$.
9. Show: If $\phi : [0, \infty) \to \mathbf{R}$ is additive, that is, if $\phi(x + y) = \phi(x) + \phi(y)$ for all $x, y \in [0, \infty)$, and if $\phi$ is continuous at 0, then $\phi$ is continuous on $[0, \infty)$.
10. Show, for the function of exercises 8 and 9, that $\phi(rx) = r\phi(x)$ for any positive rational $r$ and all $x \in [0, \infty)$.
11. Show: If $\phi : [0, \infty) \to \mathbf{R}$ is additive and continuous, then it is also positive homogeneous; that is, $\phi(\lambda x) = \lambda\phi(x)$ for any $\lambda \in [0, \infty)$ and all $x \in [0, \infty)$.
12. Generalize the results of exercises 8 to 11 as follows: If $\phi(x + y) = \phi(x) + \phi(y)$ for all $x, y \in \mathbf{R}$, and if $\phi$ is continuous at 0, then $\phi(\lambda x) = \lambda\phi(x)$ for any $\lambda \in \mathbf{R}$ and all $x \in \mathbf{R}$.

## 4.43 Approximation by Polygonal Regions

Based on our knowledge of the area measure of polygonal regions that can be triangulated as the sum of the area measures of their triangular components, we shall now investigate the possibility of assigning area measure to other than polygonal regions. If a circular disk, for example, has an area measure at all, then, by Theorem 41.1, its area measure has to be smaller than the area measure of every polygonal region that contains the disk and larger than the area measure of every polygonal region that is contained in the disk.

By taking a sequence of circumscribed regular polygonal regions with 4, 8, 16, ... vertices

$$A_4 \supset A_8 \supset A_{16} \supset \cdots \supset A_{2^k} \supset \cdots,$$

and a sequence of inscribed regular polygonal regions with the corresponding number of vertices,

$$a_4 \subset a_8 \subset a_{16} \subset \cdots \subset a_{2^k} \subset \cdots$$

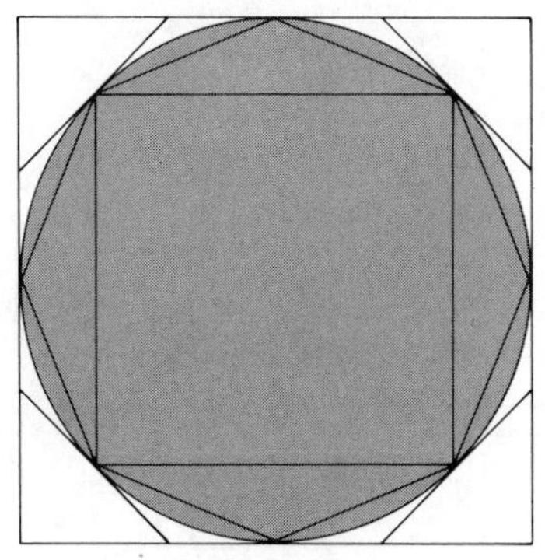

*Fig. 43.1*

(see Fig. 43.1), and by computing the area measure of these polygonal regions by triangulation, we obtain two monotonic sequences

$$\mu(A_4) \geqslant \mu(A_8) \geqslant \mu(A_{16}) \geqslant \cdots \geqslant \mu(A_{2^k}) \geqslant \cdots$$
$$\mu(a_4) \leqslant \mu(a_8) \leqslant \mu(a_{16}) \leqslant \cdots \leqslant \mu(a_{2^k}) \leqslant \cdots$$

(see also Theorem 41.1). The first sequence is bounded below by, say $\mu(a_4)$, and the second sequence is bounded above by, say $\mu(A_4)$. Hence, by Theorem 18.1, both sequences converge. It turns out that $\lim_\infty \mu(a_{2^k}) = \lim_\infty \mu(A_{2^k})$. (See exercises 1–6.) In view of this result we can argue that, if the circular disk has an area measure at all, then its area measure is given by either of these two limits.

Archimedes† of Syracuse has, among his many spectacular exploits as an engineer and physicist, utilized a similar idea to devise an algorithm for the computation of $\pi$. He argued that the length $2\pi$ of the circumference of the unit disk is shorter than the length of the circumference of any *circumscribed* regular polygon and longer than the length of the circumference of any *inscribed* regular polygon. He then established recursion formulas for the computation of the length of the circumference of the inscribed and circumscribed regular polygons with $2n$ vertices in terms of the ones with $n$ vertices. (See exercises 8, 9.) Starting with $n = 6$ and working against overwhelming odds in the form of a cumbersome and grossly inadequate number system, he found that the length $\pi$ of half the circumference of the unit disk lies between $3\frac{10}{71}$ and $3\frac{1}{7}$ (see exercise 10).

We shall return to a discussion of the area measure of planar regions in Chapters 11 and 12, where we deal with this matter in a more general setting. In this chapter we deal only with planar regions of the type

$$\{(x, y) \in \mathbf{R} \times \mathbf{R} \mid a \leqslant x \leqslant b, 0 \leqslant y \leqslant f(x)\},$$

where $f: [a,b] \to \mathbf{R}$ is a bounded function and where we assume, for the moment, that $f(x) \geqslant 0$ for all $x \in [a, b]$. (See Fig. 43.2.)

†'Ἀρχιμήδης, 287–212 B.C.

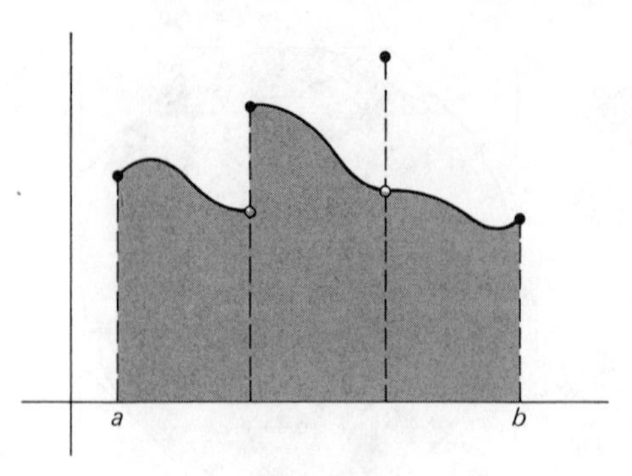


*Fig. 43.2*

In order to find out if such a point set has area measure at all, and if so, what its area measure is, we consider circumscribed and inscribed polygonal regions as depicted in Fig. 43.3 and Fig. 43.4. We shall argue that if the point set in Fig. 43.2 has an area measure, then this area measure is greater than, or equal to, the area measure of all possible polygonal regions of the type depicted in Fig. 43.3 and less than, or equal to, the area measure of all possible polygonal regions of the type depicted in Fig. 43.4. The area measures of these polygonal regions, in turn, can be found as a sum of the area measures of their finitely many rectangular parts.

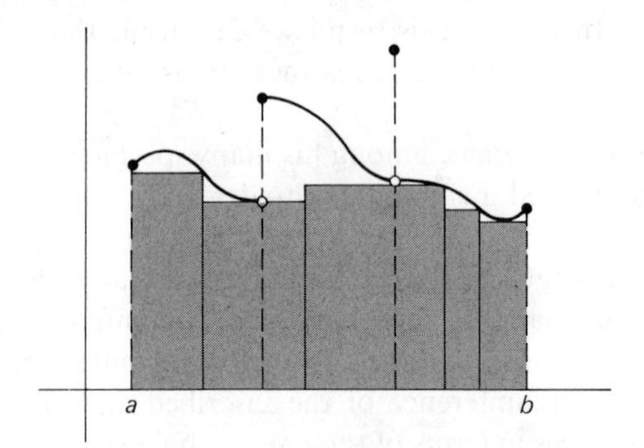


*Fig. 43.3*

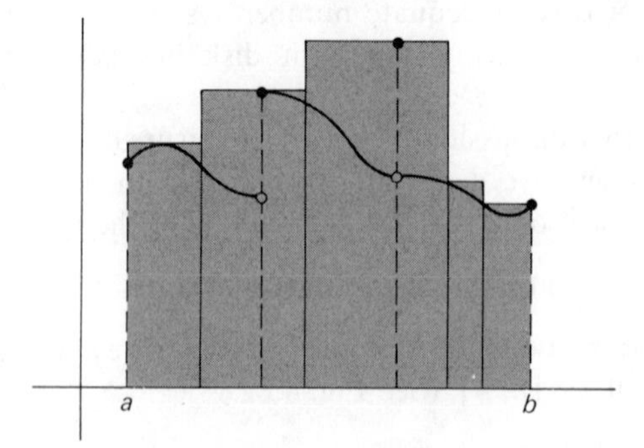


*Fig. 43.4*

In our forthcoming discussion, we shall drop the restriction that $f(x) \geqslant 0$ for all $x \in [a, b]$, but keep in mind that our results will be suitable for a development of area measure in case $f(x) \geqslant 0$ does hold true for all $x \in [a, b]$.

## 4.43 Exercises

1. Given a disk of radius 1. Find the area measure $\mu(a_{2^n})$ of an inscribed regular polygon $a_{2^n}$ with $2^n$ vertices.
2. Same as in exercise 1, for the circumscribed regular polygon $A_{2^n}$ with $2^n$ vertices.
3. Show that $\mu(a_{2^n})/\mu(A_{2^n}) = \cos^2(180/2^n)$.
4. Given the sequence $\{a_k\}$ where $a_1 = \sqrt{2}$, $a_k = \sqrt{2 + a_{k-1}}$. Show that $\lim_\infty a_k = 2$.
5. Show that $\lim_\infty \cos^2(180/2^n) = 1$.
6. Show that $\lim_\infty \mu(a_{2^n}) = \lim_\infty \mu(A_{2^n})$.
7. If the unit disk has an area measure $\mu(\bigcirc)$, then
$$\mu(a_{2^n}) \leqslant \mu(\bigcirc) \leqslant \mu(A_{2^n}).$$
Evaluate $\mu(a_{2^n})$ and $\mu(A_{2^n})$ for $n = 4, 5, 7$, and show that $\mu(\bigcirc) = 3.141\ldots$.
8. Let $c_n$ denote the length of the circumference of a regular polygon with $n$ vertices which is inscribed in a circle of radius 1, and let $C_n$ denote the length of the circumference of the circumscribed regular polygon with $n$ vertices. Show, by elementary geometrical methods, that
$$C_{2n} = \frac{2c_n C_n}{c_n + C_n}, \qquad c_{2n} = \sqrt{c_n C_{2n}}.$$
9. Show that the sequences $\{c_{2^n \cdot 6}\}$ and $\{C_{2^n \cdot 6}\}$ converge to the same limit, where the elements $c_{2^n \cdot 6}$, $C_{2^n \cdot 6}$ are recursively defined in exercise 8.
10. Find $\frac{1}{2}C_{96}$, $\frac{1}{2}c_{96}$ to two decimal places.
11. Consider the unit square and construct two sequences of polygonal regions, one inscribed, the other circumscribed, with perimeters $\{l_n\}$ and $\{L_n\}$, such that $\lim_\infty l_n = \lim_\infty L_n \neq 4$, 4 being the perimeter of the unit square. Where does this leave Archimedes?

## 4.44 Upper and Lower Sums

We shall now carry out the program that was outlined in the preceding section. Towards this end, we partition the interval $[a, b]$ into a number of subintervals.

***Definition 44.1***

$P = \{x_0, x_1, x_2, \ldots, x_{n-1}, x_n\}$ is called a *partition* of $[a, b]$ if and only if $a = x_0 < x_1 < x_2 < \cdots < x_{n-1} < x_n = b$.

$P'$ is a *refinement* of $P : P' \supseteq P$, if and only if $P'$ is a partition of $[a, b]$ and all elements of $P$ are also elements of $P'$.

---

▲ *Example*

$P = \{0, 1/n, 2/n, \dots, (n-1)/1, 1\}$ is a partition of $[0, 1]$. $P$ itself, as well as $P' = \{0, 1/2n, 1/n, \dots, (n-1)/n, 1\}$ are examples of refinements of $P$.

---

***Lemma 44.1***

If $P_1$, $P_2$ are any two partitions of $[a, b]$, then $P = P_1 \cup P_2$ is a refinement of $P_1$ and a refinement of $P_2$.

(This follows immediately from definition 44.1 and $A \cup B \supseteq A, \supseteq B$.)

We consider a function $f: [a, b] \to \mathbf{R}$ which is bounded on $[a, b]$, i.e., there is an $M > 0$ such that

$$|f(x)| \leqslant M \qquad \text{for all} \quad x \in [a, b].$$

Since $f$ is bounded on $[a, b]$, $f$ is bounded on any subset of $[a, b]$, and we have, for any partition $P = \{x_0, x_1, \dots, x_n\}$ of $[a, b]$, that

$$\sup_{[x_{k-1}, x_k]} f(x) = M_k(f) \leqslant M$$

(44.1)

$$\inf_{[x_{k-1}, x_k]} f(x) = m_k(f) \geqslant -M$$

exist for all $k = 1, 2, \dots, n$.

Note that if $f \in C[a, b]$, then, by Theorem 34.3,

$$M_k(f) = \max_{[x_{k-1}, x_k]} f(x), \qquad m_k(f) = \min_{[x_{k-1}, x_k]} f(x).$$

If $f(x) \geqslant 0$ for all $x \in [a, b]$, then $M_k(f)$ and $m_k(f)$ are the heights of the rectangles making up the circumscribed and inscribed polygonal regions in Figs. 43.3 and 43.4.

In the sequel we shall use the following abbreviating notation:

$$\Delta x_k = x_k - x_{k-1}, k = 1, 2, \dots n.$$

***Definition 44.2***

For any partition $P = \{x_0, x_1, \dots, x_n\}$ of $[a, b]$,

$$\underline{S}(f; P) = \sum_{k=1}^{n} m_k(f)\, \Delta x_k$$

is called a *lower sum* of $f$ on $[a, b]$, and

$$\bar{S}(f; P) = \sum_{k=1}^{n} M_k(f)\, \Delta x_n$$

is called an *upper sum* of $f$ on $[a, b]$.

If $f(x) \geqslant 0$ for all $x \in [a, b]$, then the lower sum represents the area measure of the inscribed polygonal region in Fig. 43.3 and the upper sum represents the area measure of the circumscribed polygonal region in Fig. 43.4.

Since

$$m_k(f) = \inf_{[x_{k-1}, x_k]} f(x) \leqslant \sup_{[x_{k-1}, x_k]} f(x) = M_k(f),$$

we have immediately that

(44.2) $\underline{S}(f;P) \leqslant \bar{S}(f;P)$ for every partition $P$ of $[a,b]$.

Geometric intuition suggests that a lower sum cannot decrease and an upper sum cannot increase if the partition is refined. This is indeed the case.

***Lemma 44.2***

If $P_2 \supseteq P_1$ ($P_2$ is a refinement of $P_1$), then

(44.3) $$\underline{S}(f;P_1) \leqslant \underline{S}(f;P_2),$$

(44.4) $$\bar{S}(f;P_1) \geqslant \bar{S}(f;P_2).$$

*Proof*

We shall prove (44.3) only. (44.4) follows by symmetric reasoning.

If $P_2 \supseteq P_1$, then either $P_2 = P_1$ and the statement holds trivially, or else $P_2 \supset P_1$, that is, $P_2$ contains at least one more point than $P_1$. We may assume w.l.o.g. that $P_2$ contains precisely one more point, say $\bar{x}$, than $P_1$. (If $P_2$ contains more such points, then the following argument is to be repeated until the set of new points is exhausted.)

If $\bar{x} \notin P_1$, then there is a $j \in \{1, 2, \ldots, n\}$ such that

$$x_{j-1} < \bar{x} < x_j,$$

and we have

(44.5) $$\inf_{[x_{j-1}, \bar{x}]} f(x) \geqslant m_j(f), \qquad \inf_{[\bar{x}, x_j]} f(x) \geqslant m_j(f).$$

(See also exercise 1.) Since all the other subintervals are the same for $P_1$ and for $P_2$, we obtain, in view of (44.5),

$$\begin{aligned}
\underline{S}(f;P_1) &= \sum_{k=1}^{n} m_k(f)\,\Delta x_k \\
&= \sum_{k=1}^{j-1} m_k(f)\,\Delta x_k + m_j(f)(\bar{x} - x_{j-1}) + m_j(f)(x_j - \bar{x}) \\
&\quad + \sum_{k=j+1}^{n} m_k(f)\,\Delta x_k \\
&\leqslant \sum_{k=1}^{j-1} m_k(f)\,\Delta x_k + \inf_{[x_{j-1}, \bar{x}]} f(x)(\bar{x} - x_{j-1}) + \inf_{[\bar{x}, x_j]} f(x)(x_j - \bar{x}) \\
&\quad + \sum_{k=j+1}^{n} m_k(f)\Delta x_k \\
&= \underline{S}(f;P_2),
\end{aligned}$$

and (44.3) is proved.

The inequality (44.2) not only is true if the same partition is used on both sides, but is true for *any two partitions*, as we can easily prove by means of Lemma 44.2.

***Lemma 44.3***

If $P_1$, $P_2$ are any two partitions of $[a, b]$, then

$$\underline{S}(f; P_1) \leqslant \bar{S}(f; P_2).$$

*Proof*

Let $P = P_1 \cup P_2$. Then, by Lemma 44.1, $P$ is a refinement of $P_1$ and of $P_2$. Hence, we have, from Lemma 44.2 and (44.2), that

$$\underline{S}(f; P_1) \leqslant \underline{S}(f; P) \leqslant \bar{S}(f; P) \leqslant \bar{S}(f; P_2).$$

## 4.44 Exercises

1. Show: If $\mathcal{A} \subseteq \mathcal{B} \subseteq \mathbf{R}$ and if $f : \mathcal{B} \to \mathbf{R}$ is bounded on $\mathcal{B}$, then
$$\inf_{(\mathcal{A})} f(x) \geqslant \inf_{(\mathcal{B})} f(x), \qquad \sup_{(\mathcal{A})} f(x) \leqslant \sup_{(\mathcal{B})} f(x).$$
2. Given the function $f : [0, 2] \to \mathbf{R}$ by
$$f(x) = \begin{cases} 1 & \text{for } 0 \leqslant x < 1, \\ x + 1 & \text{for } 1 \leqslant x < 2, \\ 4 & \text{for } x = 2, \end{cases}$$
and the partition $P = \{0, \frac{1}{2}, 1, \frac{4}{3}, 2\}$. Find $\underline{S}(f; P)$ and $\bar{S}(f; P)$.
3. Given the function $f : [0, 1] \to \mathbf{R}$ by $f(x) = x$ and the partition $P_n = \{0, 1/n, 2/n, \ldots, (n-1)/n, 1\}$. Find $\underline{S}(f; P_n)$ and $\bar{S}(f; P_n)$. Show that $\lim_\infty \underline{S}(f; P_n) = \lim_\infty \bar{S}(f; P_n)$. $(1 + 2 + \cdots + (n-1) = (n-1)n/2.)$
4. Same as in exercise 3, for $f(x) = x^2$. $(1 + 2^2 + 3^2 + \cdots + (n-1)^2 = \frac{1}{6}n(n-1)(2n-1).)$
5. Given the function $f : \mathbf{R} \to \mathbf{R}$ which is defined by $f(x) = x$ for $x \in [0, 1)$ and $f(x + 1) = f(x)$ for all $x \in \mathbf{R}$. Consider the partition $P = \{0, 1, 2, \ldots, 10\}$ of $[0, 10]$, and evaluate $\bar{S}(f; P)$.

## 4.45 The Riemann Integral

In this section, we are going to show that if $f : [a, b] \to \mathbf{R}$ is bounded, then $\sup_{(P)} \underline{S}(f; P)$ and $\inf_{(P)} \bar{S}(f; P)$ exist (the notation $(P)$ indicates "to be taken over all possible partitions of $[a, b]$") and $\sup_{(P)} \underline{S}(f; P) \leqslant \inf_{(P)} \bar{S}(f; P)$. If $\sup_{(P)} \underline{S}(f; P) = \inf_{(P)} \bar{S}(f; P)$, then we shall say that $f$ is Riemann-integrable† on $[a, b]$, and define the Riemann integral of $f$ on $[a, b]$ to be the $\sup_{(P)} \underline{S}(f; P)$ (or, equivalently, the $\inf_{(P)} \bar{S}(f; P)$).

†*Bernhard Riemann*, 1826–1866.

***Lemma 45.1***

If $f:[a, b] \to \mathbf{R}$ is bounded on $[a, b]$, then $\sup_{(P)}\underline{S}(f; P)$ and $\inf_{(P)}\bar{S}(f; P)$ exist.

*Proof*

We obtain, from Lemma 44.3 for a fixed partition $P_0$, that

$$\underline{S}(f;P) \leqslant \bar{S}(f;P_0) \tag{45.1}$$

for all partitions $P$ of $[a, b]$. Hence, the set $\{\underline{S}(f; P)\}$ is bounded above and, in view of the least-upper-bound property of $\mathbf{R}$, $\sup_{(P)}\underline{S}(f; P)$ exists. (A similar argument yields the existence of $\inf_{(P)}\bar{S}(f; P)$.)

***Lemma 45.2***

$$\sup_{(P)} \underline{S}(f;P) \leqslant \inf_{(P)} \bar{S}(f;P). \tag{45.2}$$

*Proof*

From (45.1), $\sup_{(P)}\underline{S}(f; P) \leqslant \bar{S}(f; P_0)$. Since this is true for every partition $P_0$ of $[a, b]$, (45.2) follows readily.

***Definition 45.1*** *(The Riemann Integral)*

The function $f:[a,b] \to \mathbf{R}$, which is presumed to be bounded on $[a,b]$, is called *Riemann-integrable* on $[a,b]$ if and only if

$$\sup_{(P)} \underline{S}(f;P) = \inf_{(P)} \bar{S}(f;P).$$

This number is called the *Riemann integral* of $f$ on $[a,b]$ and is denoted by $\int_a^b f$ or by $\int_a^b f(x)\,dx$:

$$\int_a^b f = \int_a^b f(x)\,dx = \sup_{(P)} \underline{S}(f;P) = \inf_{(P)} \bar{S}(f;P).$$

The notation $\int_a^b f$ contains all the relevant information for all our theoretical investigations and will be employed in such context. At certain occasions, when we deal with more practical aspects of Riemann integration, we shall use the more elaborate notation $\int_a^b f(x)\,dx$ with which the reader is familiar from calculus.

---

▲ *Example 1*

The function $f:[0, 1] \to \mathbf{R}$, given by $f(x) = 1$, is Riemann-integrable on $[0, 1]$, since $\underline{S}(f;P) = 1$ and $\bar{S}(f;P) = 1$ for every partition $P$ of $[0, 1]$, and hence,

$$\int_0^1 dx = \sup_{(P)} \underline{S}(f;P) = \inf_{(P)} \bar{S}(f;P) = 1.$$

▲ *Example 2*

The Dirichlet† function $\delta : [0, 1] \to \mathbf{R}$, given by

$$\delta(x) = \begin{cases} 0 & \text{if } x \in [0,1] \cap \mathbf{Q}, \\ 1 & \text{if } x \in [0,1] \cap \mathbf{I}, \end{cases}$$

is *not* Riemann-integrable on $[0, 1]$. If $P$ is any partition of $[0, 1]$, then every subinterval contains rational as well as irrational numbers, and we have

$$M_k(\delta) = 1, m_k(\delta) = 0 \qquad \text{for all} \quad k = 1, 2, \ldots, n.$$

Hence, $\underline{S}(\delta; P) = 0$ and $\bar{S}(\delta; P) = 1$ for all partitions $P$ of $[0, 1]$, and therefore

$$\sup_{(P)} \underline{S}(\delta; P) = 0 < 1 = \inf_{(P)} \bar{S}(\delta; P).$$

---

Let us assume that $f : [a, b] \to \mathbf{R}$ is Riemann-integrable and that $f(x) \geqslant 0$ for all $x \in [a, b]$. Since

$$\underline{S}(f; P) \leqslant \int_a^b f \leqslant \bar{S}(f; P)$$

for all partitions $P$ of $[a, b]$ and since $\int_a^b f$ is the only number with this property, we can say: If the point set $A = \{(x, y) \mid a \leqslant x \leqslant b,\ 0 \leqslant y \leqslant f(x)\}$ has area measure $\mu(A)$, then, by necessity,

$$\mu(A) = \int_a^b f.$$

(See also section 4.43.)

## 4.45 Exercises

1. Suppose that the bounded function $f : [a, b] \to \mathbf{R}$ is Riemann-integrable on $[a, b]$. Show: There exists a sequence $\{P_k\}$ of partitions of $[a, b]$ such that $\lim_\infty \underline{S}(f; P_k) = \int_a^b f$ and a sequence $\{P_k'\}$ such that $\lim_\infty \bar{S}(f; P_k') = \int_a^b f$.
2. Given $h : [a, b] \to \mathbf{R}$, where $h(x) \geqslant 0$ for all $x \in [a, b]$. Prove: If $h$ is Riemann-integrable on $[a, b]$, then $\int_a^b h \geqslant 0$.
3. Given $h : [a, b] \to \mathbf{R}$, where $h(x) \geqslant 0$ for all $x \in [a, b] \setminus \{c\}$, where $c \in (a, b)$ and where $h(c) = -1$. Prove: If $h$ is integrable on $[a, b]$, then $\int_a^b h \geqslant 0$.
4. Show, without reference to upper sums, that $\sup_{(P)} \underline{S}(f; P)$ exists, provided that $f$ is bounded on $[a, b]$.
5. $\sup_{(P)} \underline{S}(f; P) = \underline{\int_a^b} f$ is called the *lower Darboux* integral of $f$ on $[a, b]$, and $\inf_{(P)} \bar{S}(f; P) = \overline{\int_a^b} f$ the *upper Darboux* integral of $f$ on $[a, b]$. Prove: If $f : [a, b] \to \mathbf{R}$ is bounded, then its upper and lower Darboux integrals exist and $\underline{\int_a^b} f \leqslant \overline{\int_a^b} f$.
6. Formulate definition 45.1 in terms of upper and lower Darboux integrals.

†*Peter Gustav Lejeune Dirichlet*, 1805–1859.

7. If $P = \{x_0, x_1, \ldots, x_n\}$ is a partition of $[a, b]$, and if $x_{k-1} \leqslant \xi_k \leqslant x_k$ for all $k = 1, 2, \ldots, n$, then $\Xi = \{\xi_1, \xi_2, \ldots, \xi_n\}$ is called a *set of intermediate values* of $P$. $S(f; P, \Xi) = \sum_{k=1}^{n} f(\xi_k)\, \Delta x_k$ is called an *intermediate sum* of $f$ for the partition $P$ and the set of intermediate values $\Xi$. Show: $\underline{S}(f; P) \leqslant S(f; P, \Xi) \leqslant \bar{S}(f; P)$ for all partitions $P$ of $[a, b]$ and all sets of intermediate values $\Xi$.
8. Show: For any partition $P$ of $[a, b]$ and every $\varepsilon > 0$, there are sets of intermediate values $\Xi'$ and $\Xi''$ such that $|\bar{S}(f; P) - S(f; P, \Xi')| < \varepsilon$ and $|\underline{S}(f; P) - S(f; P, \Xi'')| < \varepsilon$, assuming that $f$ is bounded on $[a, b]$.

## 4.46 The Riemann Criterion for Integrability

From definition 45.1 of the Riemann integral, it is reasonable to expect that the existence of the integral will depend on the possibility of bringing upper and lower sums as close together as one pleases through a judicious choice of the partition. This is, as a matter of fact, a necessary and sufficient condition for the existence of the integral.

***Definition 46.1***

The bounded function $f : [a, b] \to \mathbf{R}$ satisfies the *Riemann condition* on $[a, b]$ if and only if, for every $\varepsilon > 0$, there is a partition $P_\varepsilon$ of $[a, b]$ such that

$$\bar{S}(f; P_\varepsilon) - \underline{S}(f; P_\varepsilon) < \varepsilon.$$

We obtain immediately from Lemma 44.3:

***Lemma 46.1***

$$\bar{S}(f; P_\varepsilon) - \underline{S}(f; P_\varepsilon) < \varepsilon$$

if and only if

$$\bar{S}(f; P) - \underline{S}(f; P) < \varepsilon$$

for all $P \supseteq P_\varepsilon$. (See exercise 1.)

We shall now show that the Riemann condition is a necessary and sufficient condition for integrability.

***Theorem 46.1*** *(Riemann Criterion for Integrability)*

The bounded function $f : [a, b] \to \mathbf{R}$ is Riemann integrable on $[a, b]$ if and only if it satisfies the Riemann condition.

*Proof*

(a) We assume that $f$ is integrable on $[a, b]$; that is,

$$\sup_{(P)} \underline{S}(f; P) = \inf_{(P)} \bar{S}(f; P).$$

By Theorem 9.1, we have, for every $\varepsilon > 0$, partitions $P_1$, $P_2$ of $[a, b]$ such that

$$\sup_{(P)} \underline{S}(f; P) - \frac{\varepsilon}{2} < \underline{S}(f; P_1),$$

$$\bar{S}(f; P_2) < \inf_{(P)} \bar{S}(f; P) + \frac{\varepsilon}{2}.$$

Hence,

$$\bar{S}(f; P_2) - \underline{S}(f; P_1) < \inf_{(P)} \bar{S}(f; P) + \frac{\varepsilon}{2} - \sup_{(P)} \underline{S}(f; P) + \frac{\varepsilon}{2} = \varepsilon.$$

Let $P_\varepsilon = P_1 \cup P_2$. Then we have, from Lemmas 44.1 and 44.3, that

$$\bar{S}(f; P_\varepsilon) - \underline{S}(f; P_\varepsilon) < \bar{S}(f; P_2) - \underline{S}(f; P_1) < \varepsilon,$$

i.e., the Riemann condition is satisfied.

(b) We now show, by contradiction, that the Riemann condition implies integrability.

By Lemma 45.2, $\sup_{(P)} \underline{S}(f; P) \leqslant \inf_{(P)} \bar{S}(f; P)$. Suppose that $f$ is not Riemann-integrable on $[a, b]$, that is,

$$\sup_{(P)} \underline{S}(f; P) < \inf_{(P)} \bar{S}(f; P).$$

Then, there is an $\varepsilon > 0$ such that

$$\sup_{(P)} \underline{S}(f; P) + \varepsilon = \inf_{(P)} \bar{S}(f; P). \tag{46.1}$$

Since $\underline{S}(f; P) \leqslant \sup_{(P)} \underline{S}(f; P)$, $\inf_{(P)} \bar{S}(f; P) \leqslant \bar{S}(f; P)$ for all $P$, we obtain from (46.1) that

$$\underline{S}(f; P) + \varepsilon \leqslant \sup_{(P)} \underline{S}(f; P) + \varepsilon = \inf_{(P)} \bar{S}(f; P) \leqslant \bar{S}(f; P),$$

that is,

$$\bar{S}(f; P) - \underline{S}(f; P) \geqslant \varepsilon$$

for *all* $P$; i.e., the Riemann condition cannot be satisfied, and we have a contradiction to our assumption.

---

▲ *Example 1*

We have, for the Dirichlet function $\delta : [0, 1] \to \mathbf{R}$ of example 2, section 4.45, that

$$\bar{S}(\delta; P) - \underline{S}(\delta; P) = 1 \qquad \text{for all} \quad P,$$

i.e., $\delta$ does *not* satisfy the Riemann condition. This was to be expected, of course, because we recognized already in section 4.45 that $\delta$ is not integrable.

▲ *Example 2*

Let $f : [0, 1] \to \mathbf{R}$ be given by $f(x) = x$. Let

$$P_n = \left\{0, \frac{1}{n}, \frac{2}{n}, \dots, \frac{n-1}{n}, 1\right\}.$$

Then

$$\underline{S}(f; P_n) = \frac{n-1}{2n}, \qquad \bar{S}(f; P_n) = \frac{n+1}{2n}$$

(see also exercise 3, section 4.44). Hence,

$$\bar{S}(f; P_n) - \underline{S}(f; P_n) = \frac{1}{n} < \varepsilon$$

if $n$ is sufficiently large, and we see that $f$ is integrable on $[0, 1]$.

---

If the bounded function $f : [a, b] \to \mathbf{R}$ is integrable, we have, from the Riemann condition, and from

$$\underline{S}(f; P) \leqslant \sup_{(P)} \underline{S}(f; P) = \int_a^b f = \inf_{(P)} \bar{S}(f; P) \leqslant \bar{S}(f; P),$$

that, for every $\varepsilon > 0$, there is a partition $P_\varepsilon$ such that

$$(46.2) \qquad 0 \leqslant \int_a^b f - \underline{S}(f; P_\varepsilon) < \varepsilon \qquad \text{and} \qquad 0 \leqslant \bar{S}(f; P_\varepsilon) - \int_a^b f < \varepsilon.$$

Hence, lower and upper sums may be used to approximate the integral.

If $P = \{x_0, x_1, \dots, x_n\}$ is a partition of $P$ and if $\xi_j \in [x_{j-1}, x_j]$, $j = 1, 2, \dots, n$, then $\Xi = \{\xi_1, \xi_2, \dots, \xi_n\}$ is called a set of *intermediate values* of $P$, and

$$(46.3) \qquad S(f; P, \Xi) = \sum_{k=1}^{n} f(\xi_k)\, \Delta x_k$$

is called an *intermediate sum* of $f$ for the partition $P$ and the set $\Xi$ of intermediate values. (See also exercise 7, section 4.45.) Since

$$m_k(f) \leqslant f(x) \leqslant M_k(f) \qquad \text{for all} \quad x \in [x_{k-1}, x_k], \qquad k = 1, 2, \dots, n,$$

we obtain immediately that

$$(46.4) \qquad \underline{S}(f; P) \leqslant S(f; P, \Xi) \leqslant \bar{S}(f; P)$$

for all partitions $P$ and all sets $\Xi$ of intermediate values of $P$. We have, from (46.2) and (46.4), that if $f : [a, b] \to \mathbf{R}$ is integrable, then, for every $\varepsilon > 0$, there is a partition $P_\varepsilon$, such that

$$(46.5) \qquad \left| \int_a^b f - S(f; P_\varepsilon, \Xi) \right| < \varepsilon$$

for all sets $\Xi$ of intermediate values of $P_\varepsilon$. The converse is also true. (See exercises 4 through 7.)

## 4.46 Exercises

1. Prove Lemma 46.1.
2. Let $f : [0, 1] \to \mathbf{R}$ represent a continuous function such that $f(x_1) \leqslant f(x_2)$ whenever $x_1 \leqslant x_2$. Show that $(1/2n)\ [f(0) + 2f(1/n) + \cdots + 2f((n-1)/n) + f(1)]$ is an intermediate sum for the partition $P = \{0, 1/n, 2/n, \ldots, (n-1)/n, 1\}$ of $[0, 1]$.
3. Given $f : [0, 1] \to \mathbf{R}$ by $f(x) = x^2$. Use Theorem 46.1 to show that $f$ is integrable.
4. Show: If there is a number $I \in \mathbf{R}$ such that, for every $\varepsilon > 0$, there is a partition $P_\varepsilon$ such that

$$(*) \qquad |S(f; P, \Xi) - I| < \varepsilon \qquad \text{for all} \quad P \supseteq P_\varepsilon$$

   and all $\Xi$ of $P$, then $f$ satisfies the Riemann condition.
5. Show: If the bounded function $f$ satisfies the Riemann condition on $[a, b]$, then there is a real number $I$ such that (*) is satisfied.
6. Use the results of exercises 4 and 5 to establish the following theorem: The bounded function $f : [a, b] \to \mathbf{R}$ is Riemann-integrable on $[a, b]$ if and only if there is a real number $I$ such that, for every $\varepsilon > 0$, there is a partition $P_\varepsilon$ of $[a, b]$ such that

$$|S(f; P, \Xi) - I| < \varepsilon$$

   for all $P \supseteq P_\varepsilon$ and all $\Xi$ of $P$.
7. Show that $I = \int_a^b f$, where $I$ is defined in exercise 6.

The concept of the Riemann–Stieltjes† integral is developed in the following exercises.

8. Let $f, g : [a, b] \to \mathbf{R}$ represent bounded functions. If $P$ is a partition of $[a, b]$ and $\Xi$ any set of intermediate values of $P$, then

$$S(f, g; P, \Xi) = \sum_{k=1}^{n} f(\xi_k)[g(x_k) - g(x_{k-1})]$$

   is called a *Riemann–Stieltjes* intermediate sum of $f$ with respect to $g$. $f$ is integrable with respect to $g$ if and only if there exists a real number $I$ such that, for every $\varepsilon > 0$, there is a partition $P_\varepsilon$ such that

$$|S(f, g; P, \Xi) - I| < \varepsilon$$

   for all $P \supseteq P_\varepsilon$ and all $\Xi$ of $P$. $I = \int_a^b f\,dg$ is called the *Riemann–Stieltjes integral* of $f$ with respect to $g$ on $[a, b]$. Show: If $g(x) = x$ for all $x \in [a, b]$ and if $f$ is Riemann-integrable on $[a, b]$, then $\int_a^b f\,dg = \int_a^b f$, where the latter is the Riemann integral.
9. Let $g : [a, b] \to \mathbf{R}$ be given by

$$g(x) = \begin{cases} 0 & \text{for } x = a, \\ 1 & \text{for } x \neq a, \end{cases}$$

   and let $f \in C[a, b]$. Show that $\int_a^b f\,dg = f(a)$.

†*Thomas Joannes Stieltjes*, 1856–1894.

10. Show: if $f:[a,b]\to\mathbf{R}$ is integrable with respect to $g:[a,b]\to\mathbf{R}$, then $\int_a^b f\,dg$ is unique.

## 4.47 Integration of Continuous Functions and Monotonic Functions

It is our ultimate aim to characterize integrable functions. Two classes of functions, the integrability of which is easy to demonstrate, are the *continuous functions* and the so-called *monotonic functions.* As we shall see in section 4.49, these are by no means the only kind of integrable functions.

***Theorem 47.1***

If $f:[a,b]\to\mathbf{R}$ is continuous on $[a,b]$, then $f$ is Riemann-integrable on $[a,b]$.

*Proof*

Since $f$ is continuous on the compact set $[a,b]$, it is, by Theorem 33.1, *uniformly* continuous on $[a,b]$, i.e., for every $\varepsilon>0$, there is a $\delta(\varepsilon)>0$ such that

$$|f(x)-f(y)|<\frac{\varepsilon}{b-a} \tag{47.1}$$

for all $x, y\in[a,b]$ for which $|x-y|<\delta(\varepsilon)$.

By Theorem 34.3, $f$ assumes its maximum and minimum on any closed subinterval of $[a,b]$. Hence, if $P$ is any partition of $[a,b]$, there are points $\xi_k'$, $\xi_k''\in[x_{k-1},x_k]$, where $k=1,2,\ldots,n$, such that

$$f(\xi_k')=M_k(f),\qquad f(\xi_k'')=m_k(f).$$

We choose a partition $P_\varepsilon$ such that

$$\max_{(k=1,2,\ldots,n)}(x_k-x_{k-1})<\delta(\varepsilon).$$

$\max_{(k=1,2,\ldots,n)}(x_k-x_{k-1})=\|P_\varepsilon\|$ is called the *norm* of the partition $P_\varepsilon$. Then, by (47.1)

$$|f(\xi_k')-f(\xi_k'')|<\frac{\varepsilon}{b-a},$$

since $\xi_k', \xi_k''\in[x_k,x_{k-1}]$, and hence $|\xi_k'-\xi_k''|<\delta(\varepsilon)$. Therefore,

$$\begin{aligned}\bar{S}(f;P_\varepsilon)-\underline{S}(f;P_\varepsilon)&=\sum_{k=1}^{n}(f(\xi_k')-f(\xi_k''))\,\Delta x_k\\&\leqslant\sum_{k=1}^{n}|f(\xi_k')-f(\xi_k'')|\,\Delta x_n<\frac{\varepsilon}{b-a}(b-a)=\varepsilon,\end{aligned}$$

i.e., $f$ satisfies the Riemann condition and is, according to Theorem 46.1, integrable on $[a,b]$.

▲ *Example 1*

That Theorem 47.1 is only a sufficient condition, and by no means necessary, can be demonstrated by exhibiting a function that is not continuous and still is integrable. An example of such a function is $f : [0, 1] \to \mathbf{R}$, which is given by

$$f(x) = \begin{cases} 0 & \text{for } 0 \leqslant x < \frac{1}{2} \\ 1 & \text{for } \frac{1}{2} \leqslant x \leqslant 1 \end{cases}$$

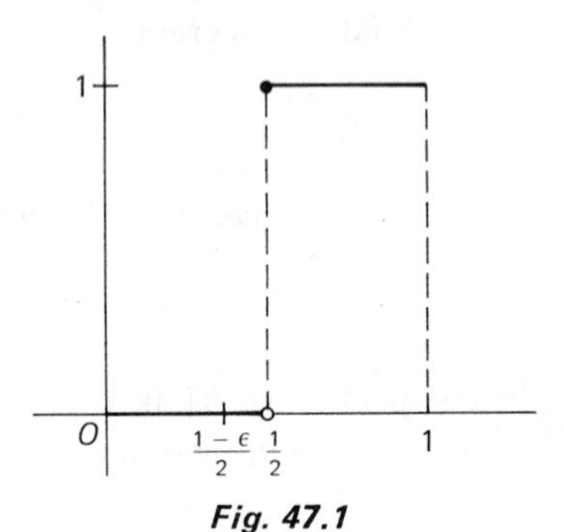


*Fig. 47.1*

(see Fig. 47.1). For any $\varepsilon > 0$, we choose a partition $P_\varepsilon$ that contains $(1 - \varepsilon)/2$, $\frac{1}{2}$ as consecutive points. Then, no matter what the other points of $P_\varepsilon$ might be, we have

$$\bar{S}(f; P) - \underline{S}(f: P) = \frac{\varepsilon}{2} < \varepsilon,$$

and hence, $f$ is Riemann-integrable.

We can do much better than that. We can show, without much effort, that at least some functions with infinitely many discontinuities are integrable. Such functions may be found among the *increasing* and *decreasing* (*monotonic*) functions, which we shall deal with in the next theorem.

**Definition 47.1**

$f : [a, b] \to \mathbf{R}$ is called *increasing* (*decreasing*) if and only if $x_1 < x_2$ implies $f(x_1) \leqslant f(x_2)$ ($f(x_1) \geqslant f(x_2)$).

(Note the subtle distinction between increasing (decreasing) functions and *strictly* increasing (decreasing) functions. See definition 5.2.)

**Theorem 47.2**

If $f : [a, b] \to \mathbf{R}$ is increasing (decreasing), then $f$ is Riemann-integrable on $[a, b]$.

*Proof*

We may assume without loss of generality that $f(a) < f(b)$. (If $f(a) = f(b)$, then $f(x) = f(a)$ for all $x \in [a, b]$ and $\int_a^b f = f(a)(b-a)$. (See exercise 2.) We choose an $\varepsilon > 0$ and, subsequently, a partition $P_\varepsilon$ such that

$$\|P_\varepsilon\| = \max_{k \in \{1, 2, \ldots, n\}} (x_k - x_{k-1}) < \frac{\varepsilon}{f(b) - f(a)}.$$

Since $f$ is increasing,

$$M_k(f) = \max_{[x_{k-1}, x_k]} f(x) = f(x_k),$$

$$m_k(f) = \min_{[x_{k-1}, x_k]} f(x) = f(x_{k-1}).$$

(See exercise 3.) Hence,

$$\bar{S}(f; P_\varepsilon) - \underline{S}(f; P_\varepsilon)$$

$$= \sum_{k=1}^{n} (f(x_k) - f(x_{k-1}))\, \Delta x_k$$

$$< \frac{\varepsilon}{f(b) - f(a)} (f(x_1) - f(a) + f(x_2) - f(x_1) + \cdots + f(b) - f(x_{k-1}))$$

$$= \frac{\varepsilon}{f(b) - f(a)} (f(b) - f(a)) = \varepsilon,$$

i.e., the Riemann condition is satisfied, and $f$ is integrable.

---

▲ *Example 2*

Theorem 47.2 applies, for example, to the following function $f: [0, 1] \to \mathbf{R}$ with infinitely many discontinuities at $\frac{1}{2}, \frac{2}{3}, \frac{3}{4}, \frac{4}{5}, \ldots$, defined by

$$f(x) = \begin{cases} 0 & \text{for } 0 \leqslant x < \frac{1}{2}, \\ 1 & \text{for } \frac{1}{2} \leqslant x < \frac{2}{3}, \\ 1 + \frac{1}{2} & \text{for } \frac{2}{3} \leqslant x < \frac{3}{4}, \\ 1 + \frac{1}{2} + \frac{1}{4} & \text{for } \frac{3}{4} \leqslant x < \frac{4}{5}, \\ \vdots & \\ 1 + \dfrac{1}{2} + \cdots + \dfrac{1}{2n} & \text{for } \dfrac{n+1}{n+2} \leqslant x < \dfrac{n+2}{n+3}, \\ \vdots & \\ 2 & \text{for } x = 1. \end{cases}$$

This function is increasing and hence, integrable. (See exercise 4.)

---

## 4.47 Exercises

1. Find $\int_0^1 f$ where $f$ is the function in example 1.
2. Show: If $f(x) = f(a)$ for all $x \in [a, b]$, then $\int_a^b f = f(a)(b - a)$.
3. Show: If $f$ is increasing on $[a, b]$ and if $[x_1, x_2] \subseteq [a, b]$, then $f$ assumes its minimum and maximum on $[x_1, x_2]$ and $\min_{[x_1, x_2]} f(x) = f(x_1)$, $\max_{[x_1, x_2]} f(x) = f(x_2)$.
4. Show that the function $f$ of example 2 is increasing, and hence, integrable.
5. Prove Theorem 47.2 for decreasing functions.
6. Let $f \in C[a, b]$ and let $g : [a, b] \to \mathbf{R}$ be increasing on $[a, b]$. Let

$$\bar{S}(f, g; P) = \sum_{k=1}^{n} M_k(f)(g(x_k) - g(x_{k-1})),$$

$$\underline{S}(f, g; P) = \sum_{k=1}^{n} m_k(f)(g(x_k) - g(x_{k-1})).$$

Show:

(a) $\underline{S}(f, g; P_1) \leqslant \underline{S}(f, g; P_2)$ for all $P_2 \supseteq P_1$
(b) $\bar{S}(f, g; P_1) \geqslant \bar{S}(f, g; P_2)$ for all $P_2 \supseteq P_1$
(c) $\underline{S}(f, g; P) \leqslant S(f, g; P, \Xi) \leqslant \bar{S}(f, g, P)$ for all $\Xi$ of $P$
(d) $\underline{S}(f, g; P_1) \leqslant \bar{S}(f, g; P_2)$ for any $P_1, P_2$
(e) $\sup_{(P)} \underline{S}(f, g; P) \leqslant \inf_{(P)} \bar{S}(f, g; P)$.

(See also exercise 8, section 4.46).

7. Prove: If $f \in C[a, b]$ and $g : [a, b] \to \mathbf{R}$ is increasing on $[a, b]$, then $\int_a^b f\, dg$ exists. (Use the result of exercise 6. See also exercise 8 of section 4.46.)
8. Derive Theorem 47.1 from the result of exercise 7.
9. Let $a_k \in \mathbf{R}$ $k \in \mathbf{N}$. Write $\sum_{k=1}^{n} a_k$ as a Riemann–Stieltjes integral.
10. Let $f : [a, b] \to \mathbf{R}$ represent a bounded function. Show: If there is a real number $I$ such that, for every $\varepsilon > 0$, there is a $\delta(\varepsilon) > 0$ such that

$$|S(f; P, \Xi) - I| < \varepsilon$$

for all partitions $P$ for which $\|P\| = \max_{(k)} (x_k - x_{k-1}) < \delta(\varepsilon)$ and all $\Xi$ of $P$, then $f$ is Riemann-integrable on $[a, b]$ and $I = \int_a^b f$.

11. Prove the converse of the theorem in exercise 10; namely, that if $f$ is integrable on $[a, b]$, then, for every $\varepsilon > 0$, there is a $\delta(\varepsilon) > 0$ such that

$$(*) \qquad |S(f; P, \Xi) - I| < \varepsilon$$

for all $P$ with $\|P\| < \delta(\varepsilon)$ and all $\Xi$ of $P$.

12. Prove: $f : [a, b] \to \mathbf{R}$ is Riemann-integrable on $[a, b]$ if and only if there is an $I \in \mathbf{R}$ such that, for every $\varepsilon > 0$, there is a $\delta(\varepsilon) > 0$ such that

$$|S(f; P, \Xi) - I| < \varepsilon$$

for all $P$ for which $\|P\| < \delta(\varepsilon)$ and all $\Xi$ of $P$.

13. Prove: If $f : [a, b] \to \mathbf{R}$ is Riemann-integrable and if

$$\underline{S}_n(f) = \frac{1}{2^n} \sum_{k=1}^{2^n} m_k(f),$$

where

$$m_k(f) = \inf \left\{ f(x) \,\middle|\, a + \frac{k-1}{2^n}(b-a) \leqslant x \leqslant a + \frac{k}{2^n}(b-a) \right\},$$

then $\lim_\infty \underline{S}_n(f) = \int_a^b f$.

14. Use the above result to show that $\int_0^1 x^2 dx = \frac{1}{3}$.

## 4.48 Sets of Lebesgue Measure Zero

"Almost everywhere" means "everywhere except almost nowhere." To explain what we mean by "almost nowhere" is the aim of this section. Our purpose in introducing this concept is to characterize Riemann-integrable functions by their intrinsic properties rather than by reference to certain properties of upper and lower sums that are associated with them. We shall see, in section 4.49, that a bounded function is Riemann-integrable if and only if it is continuous "almost everywhere."

When we consider the set of all real numbers, say between 0 and 1, each occupying its natural position on the number line, we recognize that this set takes up space. As a matter of fact, it takes up the entire interval (0, 1) of length 1. On the other hand, if we consider the point set $S_1 = \{1\}$, then we see that we can find, for every $\varepsilon > 0$, no matter how small, an open interval $\mathfrak{J}_1 = (1 - \varepsilon/3, 1 + \varepsilon/3)$ such that $S_1 \subseteq \mathfrak{J}_1$ and $l(\mathfrak{J}_1) < \varepsilon$ (where $l(\mathfrak{J}_1)$ denotes the length of $\mathfrak{J}_1$). Since we can choose $\varepsilon$ as small as we please, we can say, loosely speaking, that $S_1$ does not take up any space on the number line.

Similarly, if we consider

$$S_n = \{x_1 x_2, \ldots, x_n\} \subset \mathbf{R}$$

and choose any $\varepsilon > 0$, we can construct open intervals

$$\mathfrak{J}_1 = \left(x_1 - \frac{\varepsilon}{3n}, x_1 + \frac{\varepsilon}{3n}\right), \quad \ldots, \quad \mathfrak{J}_n = \left(x_n - \frac{\varepsilon}{3n}, x_n + \frac{\varepsilon}{3n}\right)$$

such that $S_n \subseteq \bigcup_{k=1}^n \mathfrak{J}_k$ and $\sum_{k=1}^n l(\mathfrak{J}_k) = \dfrac{2\varepsilon}{3} < \varepsilon$.

$S_1$ and $S_n$ are sets that do not take up space on the number line. We call them *sets of Lebesgue measure zero*.† In general:

***Definition 48.1***

The point set $S \subset \mathbf{R}$ has *Lebesgue measure zero*:

$$\lambda(S) = 0,$$

†*Henri Lebesgue*, 1875–1941.

if and only if one can find, for every $\varepsilon > 0$, a sequence of *open* intervals $\{\mathfrak{J}_k\}$ such that

(a) $$S \subseteq \bigcup_{k=1}^{\infty} \mathfrak{J}_k,$$

(b) $$\sum_{k=1}^{\infty} l(\mathfrak{J}_k) < \varepsilon.$$

Clearly, *any subset of a set of Lebesgue measure zero has Lebesgue measure zero.* (See exercise 1.)

We have seen that every finite subset of **R** has Lebesgue measure zero. A more interesting, powerful, and profitable result is the following one:

***Lemma 48.1***

Every denumerable subset of **R** has Lebesgue measure zero.

*Proof*

Since $S$ is denumerable, we may represent it by $S = \{x_1, x_2, x_3, \ldots\}$. Let $\mathfrak{J}_k = (x_k - \varepsilon/2^{k+2}, x_k + \varepsilon/2^{k+2})$. Then $S \subset \bigcup_{k=1}^{\infty} \mathfrak{J}_k$ and

$$\textstyle\sum_{k=1}^{\infty} l(\mathfrak{J}_k) = \varepsilon \sum_{k=1}^{\infty} 1/2^{1+k} = \varepsilon/2 < \varepsilon.$$

Hence, $\lambda(S) = 0$.

---

▲ *Example 1*

The set of rational numbers **Q** is denumerable. Hence, $\lambda(\mathbf{Q}) = 0$.

---

Clearly, a union of finitely many sets of Lebesgue measure zero has again Lebesgue measure zero (see exercise 2). We have more generally:

***Lemma 48.2***

If each of the sets $S_1, S_2, S_3, \ldots \subset \mathbf{R}$ has Lebesgue measure zero, then

$$S = \bigcup_{k \in \mathbf{N}} S_k$$

has Lebesgue measure zero. In words: The union of denumerably many sets of Lebesgue measure zero has Lebesgue measure zero.

*Proof*

Since $\lambda(S_k) = 0$, we can find a sequence of open intervals $\{\mathfrak{J}_i^k\}$ such that $S_k \subseteq \bigcup_{i=1}^{\infty} \mathfrak{J}_i^k$ and $\sum_{i=1}^{\infty} l(\mathfrak{J}_i^k) < \varepsilon/2^{k+1}$.
Clearly,

(48.1) $$S = \bigcup_{k=1}^{\infty} S_k \subseteq \bigcup_{i,k \in \mathbf{N}} \mathfrak{J}_i^k.$$

It remains to be shown that the set of open intervals $\Omega = \{\mathfrak{J}_i^k \mid i \in \mathbf{N},$ can be arranged in a sequence, and that the sum of the lengths of these i does not exceed $\varepsilon$. We use for $\Omega$ the same arrangement as for the set $\{a_i^{(k)} \mid i \in \mathbf{N},$ $k \in \mathbf{N}\}$ in the proof of Theorem 13.2, namely

$$(48.2) \qquad \mathfrak{J}_1^1, \quad \mathfrak{J}_1^2, \quad \mathfrak{J}_2^1, \quad \mathfrak{J}_1^3, \quad \mathfrak{J}_2^2, \quad \mathfrak{J}_3^1, \quad \mathfrak{J}_1^4, \quad \mathfrak{J}_2^3, \quad \mathfrak{J}_3^2, \quad \mathfrak{J}_4^1, \quad \ldots$$

and see that

$$s_{(n(n+1))/2} = \sum_{k=1}^{n} (l(\mathfrak{J}_1^k) + l(\mathfrak{J}_2^{k-1}) + \cdots + l(\mathfrak{J}_k^1))$$

represents the sum of the lengths of all these intervals up to and including $\mathfrak{J}_n^1$, which is the $[(n(n+1)/2]$th element in the sequence (48.2). (See exercise 3.)

Since $l(\mathfrak{J}_i^k) \geqslant 0$ for all $i, k \in \mathbf{N}$, we have

$$s_{(n(n+1))/2} = \sum_{k=1}^{n} (l(\mathfrak{J}_1^k) + l(\mathfrak{J}_2^{k-1}) + \cdots + l(\mathfrak{J}_k^1)) \leqslant \sum_{k=1}^{n} \left( \sum_{i=1}^{n} l(\mathfrak{J}_i^k) \right)$$

$$\leqslant \sum_{k=1}^{n} \left( \sum_{i=1}^{\infty} l(\mathfrak{J}_i^k) \right) < \sum_{k=1}^{n} \frac{\varepsilon}{2^{k+1}} < \frac{\varepsilon}{2} < \varepsilon.$$

Since the sequence $\{s_{(n(n+1))/2}\}$ is increasing, and, as we have just seen, bounded above by $\varepsilon$, we obtain, from Theorem 18.1, that the limit exists, and

$$\lim s_{(n(n+1))/2} = s < \varepsilon.$$

Since $\{s_{(n(n+1))/2}\}$ is a subsequence of the sequence

$$\begin{aligned}
s_1 &= l(\mathfrak{J}_1^1), \\
s_2 &= l(\mathfrak{J}_1^1) + l(\mathfrak{J}_1^2), \\
s_3 &= l(\mathfrak{J}_1^1) + l(\mathfrak{J}_1^2) + l(\mathfrak{J}_2^1), \\
s_4 &= l(\mathfrak{J}_1^1) + l(\mathfrak{J}_1^2) + l(\mathfrak{J}_2^1) + l(\mathfrak{J}_1^3), \\
&\vdots
\end{aligned}$$

and since the sequence $\{s_k\}$ is increasing, we obtain from Theorem 18.2 that the limit exists and that

$$\lim_{\infty} s_k = s < \varepsilon.$$

Hence, by definition 19.1, the sum of the lengths of the intervals $\mathfrak{J}_i^k$, taken in the order in which they appear in (48.2), converges to $s$ which, in turn, is less than $\varepsilon$. This, together with (48.1), shows that $\lambda(S) = 0$.

The only sets of Lebesgue measure zero which we have encountered thus far are the finite sets and the denumerable sets. If there were no other sets of Lebesgue measure zero, then Lemma 48.2 would be trivial, because the union of denumerably many denumerable sets is again denumerable (see Theorem 13.2) and has, therefore, Lebesgue measure zero by Lemma 48.1.

However, Lemma 48.2 is *not* trivial because there are sets of Lebesgue measure zero which have cardinality greater than $\aleph_0$ and the Cantor set $C$ (section 1.14) is one of them. This may be seen as follows:

By (14.2), $C \subseteq C_k$ for all $k \in \mathbf{N}$, where $C_k = c(D_k)$ and where $D_k$ is defined in (14.1). $C_k$ may be (comfortably) embedded in $2^k$ open intervals of length $2/3^k$ each. Hence, the sum of the lengths of all these intervals is $2(\frac{2}{3})^k$. Since we may choose, for every $\varepsilon > 0$, a $k \in \mathbf{N}$ so large that $2(\frac{2}{3})^k < \varepsilon$, it follows that *the Cantor set has Lebesgue measure zero* but has, as we have seen in section 1.14, cardinality $\aleph$.

For the record, and for later reference, we state the following lemma, which yields an equivalent definition of the concept of Lebesgue measure zero, utilizing closed intervals instead of open intervals:

***Lemma 48.3***

The point set $S \subset \mathbf{R}$ has Lebesgue measure zero if and only if one can find, for every $\varepsilon > 0$, a sequence of *closed intervals* $\{\mathfrak{J}_k\}$ such that

(a) $$S \subseteq \bigcup_{k=1}^{\infty} \mathfrak{J}_k,$$

(b) $$\sum_{k=1}^{\infty} l(\mathfrak{J}_k) < \varepsilon.$$

For a proof, see exercises 4 and 5.

We are now ready to give precise meaning to the phrases "almost everywhere" and "almost nowhere."

***Definition 48.2***

A function $f : [a, b] \to \mathbf{R}$ is continuous *almost everywhere* (a.e.) on $[a, b]$, if and only if the set $D$ of points where $f$ has discontinuities is a set of Lebesgue measure zero: $\lambda(D) = 0$. ($f$ is continuous *except* on a set of Lebesgue measure zero, that is, $f$ has discontinuities almost nowhere.)

---

▲ *Example 2*

Let $f : [0, 1] \to \mathbf{R}$ be defined by

$$f(x) = \begin{cases} 0 & \text{if } x = 0, \\ 0 & \text{if } \dfrac{1}{2k+1} < x \leqslant \dfrac{1}{2k}, \quad k \in \mathbf{N}, \\ 1 & \text{if } \dfrac{1}{2k} < x \leqslant \dfrac{1}{2k-1}, \quad k \in \mathbf{N}. \end{cases}$$

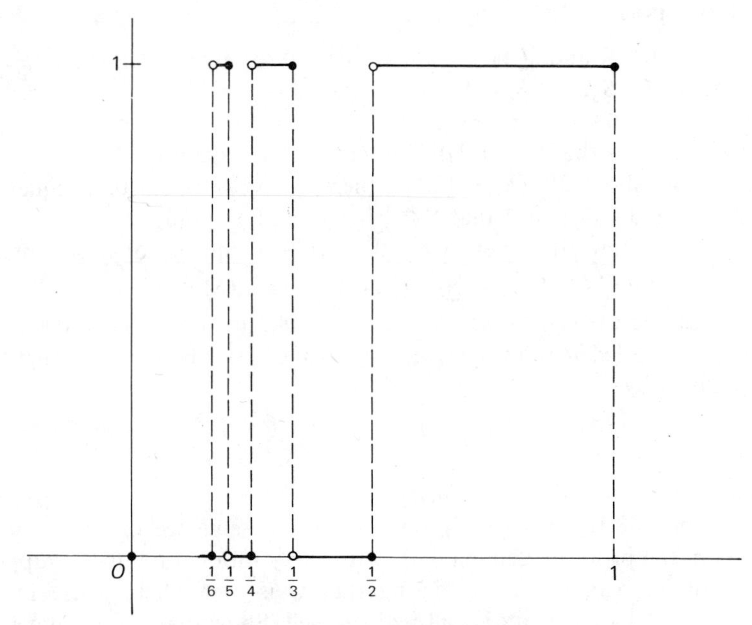


*Fig. 48.1*

(See Fig. 48.1.) This function is continuous on every open interval $(1/(k+1), 1/k)$, where $k \in \mathbf{N}$ and has discontinuities at $x \in \{0, \frac{1}{2}, \frac{1}{3}, \frac{1}{4}, \ldots\}$. This is, by Lemma 48.1, a set of Lebesgue measure zero. Hence, by definition 48.2, $f$ is continuous a.e. on $[0, 1]$.

▲ *Example 3*

Let $g : [0, 1] \to \mathbf{R}$ be defined by

$$g(x) = \begin{cases} 0 & \text{if } x \in C, \\ x^2 & \text{if } x \in [0, 1]\backslash C, \end{cases}$$

where $C$ denotes the Cantor set. Since $C$ is closed (see section 2.29, exercise 2), $[0, 1]\backslash C$ is open, and $g$ is continuous for all $x \in [0, 1]\backslash C$. (See exercise 7.) The only points where $g$ could possibly be discontinuous are in $C$, and $\lambda(C) = 0$. Hence, $g$ is continuous a.e. on $[0, 1]$.

▲ *Example 4*

Let $f : \mathbf{R} \to \mathbf{R}$ be defined by

$$f(x) = \begin{cases} 0 & \text{for all } x \in \mathbf{I}, \\ \dfrac{1}{n} & \text{for all } x = \dfrac{m}{n}, \quad \text{where} \quad m \in \mathbf{Z}, n \in \mathbf{N}, (m, n) = 1. \end{cases}$$

We have seen in example 8, section 2.26, that $f$ is continuous on $\mathbf{I}$ and not continuous on $\mathbf{Q}$. Hence, $f$ is continuous a.e. on $\mathbf{R}$.

## 4.48 Exercises

1. Prove: If $A \subset \mathbf{R}$ and $\lambda(A) = 0$, then $\lambda(B) = 0$ for every set $B \subseteq A$.
2. Prove: If $S_1, S_2, \ldots, S_n \subset \mathbf{R}$ and $\lambda(S_k) = 0$ for $k = 1, 2, 3, \ldots, n$, then $\lambda(\bigcup_{k=1}^{n} S_k) = 0$.
3. Show that $\mathfrak{J}_n^1$ is the $[(n(n+1))/2]$th term in the sequence (48.2).
4. Let $S \subset \mathbf{R}$. Show: If $\lambda(S) = 0$, then there is, for every $\varepsilon > 0$, a sequence of closed intervals $\{\mathfrak{J}_k\}$ such that $S \subseteq \bigcup_{k \in \mathbf{N}} \mathfrak{J}_k$ and $\sum_{k=1}^{\infty} l(\mathfrak{J}_k) < \varepsilon$.
5. Let $S \subset \mathbf{R}$. Show: If, for every $\varepsilon > 0$, there is a sequence of closed intervals $\{\mathfrak{J}_k\}$ such that $S \subseteq \bigcup_{k \in \mathbf{N}} \mathfrak{J}_k$, $\sum_{k=1}^{\infty} l(\mathfrak{J}_k) < \varepsilon$, then $\lambda(S) = 0$.
6. Let $C$ denote the Cantor set. Use Lemma 48.3 to show $\lambda(C) = 0$.
7. Show: If $\mathcal{O} \subseteq [a, b]$ is an open set and $\lambda([a, b] \backslash \mathcal{O}) = 0$, and if $g : [a, b] \to \mathbf{R}$ is defined by
$$g(x) = \begin{cases} x^2 & \text{for } x \in \mathcal{O}, \\ 0 & \text{for } x \notin \mathcal{O}, \end{cases}$$
then $g$ is continuous a.e. on $[a, b]$.
8. By Lemma 48.1, $\lambda(\mathbf{Q}) = 0$. If we construct a sequence of open intervals $\mathfrak{J}_k$, as in the proof of Lemma 48.1, with every rational number wrapped in such an interval, then, considering that $\mathbf{Q}$ is everywhere dense in $\mathbf{R}$, it looks as if $\bigcup_{k \in \mathbf{N}} \mathfrak{J}_k$ contains all reals as well. Show that this cannot be the case.

## 4.49 Characterization of Integrable Functions

The contents of this section represent the culmination of all our efforts. We shall establish a complete characterization of the Riemann-integrable functions as the bounded functions that are continuous almost everywhere. This will be done in two stages:

***Lemma 49.1***

If $f : [a, b] \to \mathbf{R}$ is bounded on $[a, b]$ and continuous almost everywhere on $[a, b]$, then $f$ is Riemann-integrable on $[a, b]$.

*Proof*

We shall prove this lemma by demonstrating that one can find, for every $\varepsilon > 0$, a partition $P_\varepsilon$ of $[a, b]$ such that the Riemann condition is satisfied. Then, by Theorem 46.1, $f$ is integrable. Let

$$D = \{x \in [a, b] \mid f \text{ is } \textit{not} \text{ continuous at } x\}.$$

By hypothesis, $\lambda(D) = 0$, i.e., for every $\varepsilon > 0$, there is a sequence of *open* intervals $\{\mathfrak{J}_k\}$ such that $D \subseteq \bigcup_{k=1}^{\infty} \mathfrak{J}_k$ and

$$\sum_{k=1}^{\infty} l(\mathfrak{J}_k) < \frac{\varepsilon}{2(M - m)} \tag{49.1}$$

where $M = \sup_{[a,b]} f(x)$ and $m = \inf_{[a,b]} f(x)$.

If $\xi \in [a, b]$, then either $\xi \in \bigcup_{k=1}^{\infty} \mathfrak{J}_k$ or $\xi \in c(\bigcup_{k=1}^{\infty} \mathfrak{J}_k)$. In the latter case, $f$ is continuous at $\xi$. Since

$$[a, b] \cap c\left(\bigcup_{k \in \mathbf{N}} \mathfrak{J}_k\right)$$

is compact, $f$ is uniformly continuous, and we have, for every $\varepsilon > 0$, some $\delta(\varepsilon) > 0$ such that

$$(49.2) \quad |f(\xi) - f(x)| < \frac{\varepsilon}{8(b-a)} \qquad \text{for all} \quad x \in N_{\delta(\varepsilon)}(\xi) \cap [a, b] \cap c\left(\bigcup_{k \in \mathbf{N}} \mathfrak{J}_k\right).$$

Next, we note that

$$\Omega = \left\{\mathfrak{J}_k, N_{\delta(\varepsilon)/2}(\xi) | k \in \mathbf{N}, \quad \xi \in c\left(\bigcup_{k=1}^{\infty} \mathfrak{J}_k\right)\right\}$$

is an open cover of $[a, b]$. Since $[a, b]$ is compact, a finite subcover

$$\Omega' = \{\mathfrak{J}_{k_1}, \ldots, \mathfrak{J}_{k_r}, N_{\delta/2}(\xi_1), \ldots, N_{\delta/2}(\xi_s)\}$$

exists. We have written $\delta$ for $\delta(\varepsilon)$. We consider the set of all endpoints of the $(r + s)$ intervals in $\Omega'$. We pick out all those that are distinct and lie in $(a, b)$ and arrange them in their natural order:

$$x_1 < x_2 < \cdots < x_{n-1}.$$

Then

$$P_\varepsilon = \{x_0, x_1, \ldots, x_{n-1}, x_n\}, \qquad x_0 = a, \quad x_n = b,$$

is a partition of $[a, b]$. We shall demonstrate that this partition is the one for which the Riemann condition is satisfied.

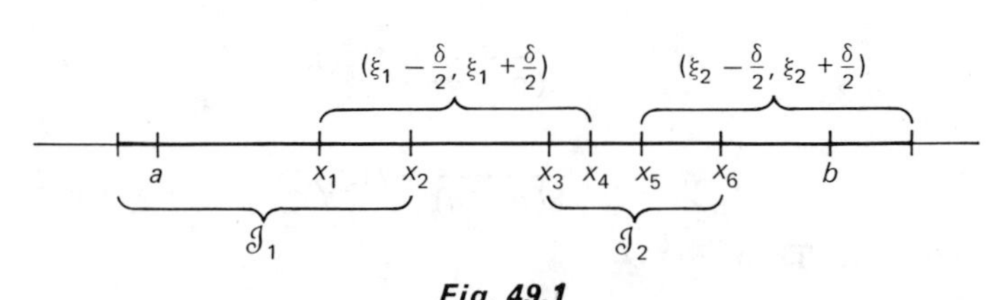


***Fig. 49.1***

By construction (see also Fig. 49.1), either

$$(x_{i-1}, x_i) \subseteq \mathfrak{J}_{k_l} \qquad \text{for some} \quad l \in \{1, 2, \ldots, r\}$$

or

$$[x_{i-1}, x_i] \cap \mathfrak{J}_{k_l} = \varnothing \qquad \text{for all} \quad l \in \{1, 2, \ldots, r\}.$$

If $[x_{i-1}, x_i] \cap \mathfrak{J}_{k_l} = \varnothing$ for all $l \in \{1, 2, \ldots, r\}$, then $(x_{i-1}, x_i) \subseteq N_{\delta/2}(\xi_j)$ for some $j \in \{1, 2, \ldots, s\}$, and hence,

$$[x_{i-1}, x_i] \subset N_\delta(\xi_j) \qquad \text{for some} \quad j \in \{1, 2, \ldots, s\}.$$

Let

(49.3) $$A_1 = \{k \in \{1, 2, \dots, n\} \mid (x_{k-1}, x_k) \subseteq \mathfrak{J}_{k_l} \text{ for some } l \in \{1, 2, \dots, r\}\},$$

and $$A_2 = \{1, 2, \dots, n\} \setminus A_1.$$

If $k \in A_1$, then $(x_{k-1}, x_k) \subseteq \mathfrak{J}_{k_l}$ for some $l \in \{1, 2, \dots, r\}$, and if $k \in A_2$, then $[x_{k-1}, x_k] \subset N_\delta(\xi_j)$ for some $j \in \{1, 2, \dots, s\}$. We have

(49.4)
$$\begin{aligned}\bar{S}(f; P_\varepsilon) - \underline{S}(f; P_\varepsilon) &= \sum_{k=1}^{n} (M_k(f) - m_k(f))\, \Delta x_k \\ &= \sum_{k \in A_1} (M_k(f) - m_k(f))\, \Delta x_k + \sum_{k \in A_2} (M_k(f) - m_k(f))\, \Delta x_k.\end{aligned}$$

By (49.3) and from (49.1), we have

(49.5)
$$\begin{aligned}\sum_{k \in A_1} (M_k(f) - m_k(f))\, \Delta x_k &\leqslant (M - m) \sum_{k \in A_1} \Delta x_k \\ &\leqslant (M - m) \sum_{k=1}^{r} l(\mathfrak{J}_k) \leqslant (M - m) \sum_{k=1}^{\infty} l(\mathfrak{J}_k) \\ &< (M - m) \frac{\varepsilon}{2(M - m)} = \frac{\varepsilon}{2}.\end{aligned}$$

If $k \in A_2$, then $[x_{k-1}, x_k] \subset N_\delta(\xi_j)$ for some $j \in \{1, 2, \dots, s\}$. Hence, if $\xi_k'$, $\xi_k'' \in [x_{k-1}, x_k]$, then $|\xi_k' - \xi_j| < \delta$, $|\xi_k'' - \xi_j| < \delta$, and we obtain, from (49.2), that

(49.6) $$|f(\xi_k') - f(\xi_k'')| \leqslant |f(\xi_k') - f(\xi_j)| + |f(\xi_k'') - f(\xi_j)| < \frac{\varepsilon}{4(b - a)}.$$

We determine $\xi_k'$, $\xi_k'' \in [x_{k-1}, x_k]$ such that

$$M_k(f) - \frac{\varepsilon}{8(b - a)} < f(\xi_k'),$$

$$m_k(f) + \frac{\varepsilon}{8(b - a)} > f(\xi_k'').$$

(See Theorem 9.1.) Then,

(49.7) $$M_k(f) - m_k(f) < f(\xi_k') - f(\xi_k'') + \frac{\varepsilon}{4(b - a)}.$$

From (49.6) and (49.7) we obtain

$$\begin{aligned}\sum_{k \in A_2} (M_k(f) - m_k(f))\, \Delta x_k &< \sum_{k \in A_2} |f(\xi_k') - f(\xi_k'')|\, \Delta x_k + \frac{\varepsilon}{4(b - a)} \sum_{k \in A_2} \Delta x_k \\ &< \frac{\varepsilon}{2(b - a)} \sum_{k \in A_2} \Delta x_k \leqslant \frac{\varepsilon}{2(b - a)} \sum_{k=1}^{n} \Delta x_k \\ &= \frac{\varepsilon}{2(b - a)} (b - a) = \frac{\varepsilon}{2}.\end{aligned}$$

This result, together with (49.5), yields, in view of (49.4),

$$\bar{S}(f;P_\varepsilon) - \underline{S}(f;P_\varepsilon) < \varepsilon,$$

and we see that $f$ is Riemann-integrable on $[a, b]$.

In order to prove the converse of Lemma 49.1, we need a simple characterization of the discontinuities of $f$ in $[a, b]$. If $f$ is *not* continuous at $\xi \in [a, b]$, then there exists an $\varepsilon > 0$ such that every $\delta$-neighborhood of $\xi$ contains some $x$ such that

$$|f(\xi) - f(x)| \geqslant \varepsilon,$$

and vice versa. Surely, if there is such an $\varepsilon > 0$, then there is an $m \in \mathbf{N}$, $(1/m) \leqslant \varepsilon$, such that $|f(\xi) - f(x)| \geqslant 1/m$, and if there is an $m \in \mathbf{N}$ for which this inequality holds, then there is an $\varepsilon = 1/m$ for which it holds. Hence, we can say: $f$ is *not* continuous at $\xi \in [a, b]$ if and only if there is an $m \in \mathbf{N}$ such that, in every $N_\delta(\xi) \cap [a, b]$, we can find an $x$ for which

$$|f(\xi) - f(x)| \geqslant \frac{1}{m}.$$

Let

(49.8)

$$E_m = \{\xi \in [a, b] \mid |f(\xi) - f(x)| \geqslant 1/m \text{ for at least one } x \text{ in every } N_\delta(\xi) \cap [a, b]\}.$$

Then the set $D$ of discontinuities of $f$ on $[a, b]$ is given by

$$D = \bigcup_{k=1}^{\infty} E_k.$$

This may be seen as follows: If $\xi \in D$, then there is an $m \in \mathbf{N}$ such that $|f(\xi) - f(x)| \geqslant 1/m$ for at least one $x$ in every $N_\delta(\xi) \cap [a, b]$. Hence, $\xi \in E_m$, and therefore, $\xi \in \bigcup_{k=1}^{\infty} E_k$. Conversely, if $\xi \in \bigcup_{k=1}^{\infty} E_k$, then $\xi \in E_m$ for some $m \in \mathbf{N}$ and hence, $|f(\xi) - f(x)| \geqslant 1/m$ for at least one $x$ in every $N_\delta(\xi) \cap [a, b]$, i.e., $f$ is *not* continuous at $\xi$, and $\xi \in D$.

For any partition $P = \{x_0, x_1, \ldots, x_n\}$ of $[a, b]$ and any given $m \in \mathbf{N}$, the subintervals $[x_{k-1}, x_k]$ generated by $P$ fall into one of two categories: Either $(x_{k-1}, x_k) \cap E_m \neq \varnothing$ or $(x_{k-1}, x_k) \cap E_m = \varnothing$. Accordingly, we define

$$A_1 = \{k \in \{1, 2, \ldots, n\} \mid (x_{k-1}, x_k) \cap E_m \neq \varnothing\},$$
$$A_2 = \{k \in \{1, 2, \ldots, n\} \mid (x_{k-1}, x_k) \cap E_m = \varnothing\}.$$

By construction,

(49.9) $$A_1 \cap A_2 = \varnothing, \qquad A_1 \cup A_2 = \{1, 2, \ldots, n\}.$$

With these preparations out of the way, we are ready to prove the converse of Lemma 49.1:

***Lemma 49.2***

If the bounded function $f: [a, b] \to \mathbf{R}$ is Riemann-integrable on $[a, b]$, then $f$ is continuous almost everywhere on $[a, b]$.

***Proof***

We have to prove that $\lambda(D) = 0$. Since $D = \bigcup_{k=1}^{\infty} E_k$, it suffices to show that $\lambda(E_m) = 0$ for all $m \in \mathbf{N}$. It follows then, from Lemma 48.2, that $\lambda(D) = 0$.

By hypothesis, $f$ is Riemann-integrable on $[a, b]$. Hence, by the Riemann criterion (Theorem 46.1), there is, for every $\varepsilon > 0$, a partition $P_\varepsilon$ such that

$$\bar{S}(f;P_\varepsilon) - \underline{S}(f;P_\varepsilon) = \sum_{k=1}^{n} (M_k(f) - m_k(f))\, \Delta x_k < \frac{\varepsilon}{2m}, \tag{49.10}$$

where $m \in \mathbf{N}$ is fixed.

By (49.9),

$$\sum_{k=1}^{\infty} (M_k(f) - m_k(f))\, \Delta x_k = \sum_{k \in A_1} (M_k(f) - m_k(f))\, \Delta x_k + \sum_{k \in A_2} (M_k(f) - m_k(f))\Delta x_k.$$

If $k \in A_1$, then, by the definition of $A_1$, there exists a $\xi_k \in E_m$ such that $\xi_k \in (x_{k-1}, x_k)$ and, by the definition of $E_m$, there is a $\xi_k' \in (x_{k-1}, x_k)$ such that

$$|f(\xi_k) - f(\xi_k')| \geqslant \frac{1}{m}, \qquad k \in A_1. \tag{49.11}$$

Since we have, for any $\xi_k, \xi_k' \in (x_{k-1}, x_k)$, that

$$M_k(f) - m_k(f) \geqslant |f(\xi_k) - f(\xi_k')|, \qquad k = 1, 2, \ldots, n,$$

we obtain from (49.11) that

$$\begin{aligned} \sum_{k \in A_1} (M_k(f) - m_k(f))\, \Delta x_k &\geqslant \sum_{k \in A_1} |f(\xi_k) - f(\xi_k')|\, \Delta x_k \\ &\geqslant \sum_{k \in A_1} \frac{1}{m}\, \Delta x_k = \frac{1}{m} \sum_{k \in A_1} \Delta x_k. \end{aligned} \tag{49.12}$$

Since $\sum_{k \in A_2}(M_k(f) - m_k(f))\Delta x_k \geqslant 0$, we obtain, from (49.10) and (49.12), that

$$\frac{\varepsilon}{2m} > \bar{S}(f; P_\varepsilon) - \underline{S}(f; P_\varepsilon) \geqslant \frac{1}{m} \sum_{k \in A_1} \Delta x_k,$$

that is,

$$\sum_{k \in A_1} \Delta x_k < \frac{\varepsilon}{2}. \tag{49.13}$$

By construction,

$$E_m = \left( E_m \cap \bigcup_{k \in A_1} (x_{k-1}, x_k) \right) \cup (E_m \cap P_\varepsilon).$$

By (49.13), $E_m \cap \bigcup_{k \in A_1}(x_{k-1}, x_k)$ has Lebesgue measure zero. Since $E_m \cap P_\varepsilon \subseteq P_\varepsilon$ consists of finitely many points only, it also has Lebesgue measure zero. Hence, $E_m$, being the union of two sets of Lebesgue measure zero, has itself Lebesgue measure zero. Therefore, $\lambda(D) = \lambda\left(\bigcup_{k \in \mathbf{N}} E_k\right) = 0$.

Lemma 49.1 and 49.2 together yield *Lebesgue's criterion for the Riemann-integrability* of bounded functions.

***Theorem 49.1***

The bounded function $f: [a, b] \to \mathbf{R}$ is Riemann-integrable if and only if $f$ is continuous almost everywhere on $[a, b]$.

The reader who encountered integrals of the type

$$\int_1^\infty \frac{dx}{x^2} \quad \text{or} \quad \int_0^1 \frac{dx}{\sqrt{x}}$$

in calculus may have some misgivings about the theory of Riemann integration which we have developed in this chapter, since neither of these integrals fits into our theory. Nevertheless, the theory of Riemann integration which we have developed is the *most general* theory of the Riemann integral. The Riemann integral is not defined in point sets other than bounded ones (see also section 11.107) and is not defined for unbounded functions. The above-mentioned integrals are not Riemann integrals but improper integrals, which are defined in terms of the Riemann integral by a limit process. We shall discuss improper integrals in section 4.53.

## 4.49 Exercises

1. Derive Theorem 47.1 from Lemma 49.1.
2. Prove: If $f: [a, b] \to \mathbf{R}$ is monotonically increasing (or decreasing), then $f$ is continuous almost everywhere on $[a, b]$.
3. $f: [0, 3] \to \mathbf{R}$ is defined by

$$f(x) = \begin{cases} 0 & \text{for } 0 \leqslant x < 1, \\ 1 & \text{for } x = 1, \\ 0 & \text{for } 1 < x < 2, \\ \frac{1}{2} & \text{for } x = 2, \\ 0 & \text{for } 2 < x < 3, \\ \frac{1}{3} & \text{for } x = 3. \end{cases}$$

   Find $E_1$, $E_2$, $E_3$, $E_m$, for $m > 3$. ($E_m$ is defined in (49.8).)
4. Show that $E_m \subseteq E_{m+1}$ for all $m \in \mathbf{N}$, where $E_m$ is defined in (49.8).

## 4.50 The Linearity of the Integral

The integral is a linear function from the set of all Riemann-integrable functions on $[a, b]$ into the reals. (See also definition 39.1.) This means that the integral of a constant times a function is equal to the constant times the integral of the function (the integral is *homogeneous*), and that the integral of a sum of functions is equal to the sum of the integrals of the individual functions (the integral is *additive*).

Whenever $f, g : [a, b] \to \mathbf{R}$ are continuous, then $(f + g)$ and $\lambda f$ for all $\lambda \in R$, are continuous. Hence, the set of discontinuities of $(f + g)$ is contained in the union of the sets of discontinuities of $f$ and $g$, and the set of discontinuities of $\lambda f$ is contained in the set of discontinuities of $f$. Therefore, if $f$, $g$ are integrable on $[a, b]$, then $(f + g)$, $\lambda f$ are continuous a.e. on $[a, b]$, and hence, $\int_a^b (f + g)$ and $\int_a^b \lambda f$ exist. Since $\inf_{(P)} \bar{S}(-f; P) = -\sup_{(P)} \underline{S}(f; P)$ (see exercise 10, section 1.9), we have, for every integrable function $f : [a, b] \to \mathbf{R}$, that

$$\int_a^b (-f) = -\int_a^b f. \tag{50.1}$$

From Lemma 9.1, we obtain, for every $\lambda > 0$, that

$$\sup_{[x_{k-1}, x_k]} \lambda f(x) = \lambda \sup_{[x_{k-1}, x_k]} f(x) \quad \text{and} \quad \inf_{[x_{k-1}, x_k]} \lambda f(x) = \lambda \inf_{[x_{k-1}, x_k]} f(x), \quad k = 1, 2, \ldots, n.$$

Hence,

$$\bar{S}(\lambda f; P) = \lambda \bar{S}(f; P) \quad \text{and} \quad \underline{S}(\lambda f; P) = \lambda \underline{S}(f; P), \qquad \lambda > 0. \tag{50.2}$$

From (50.1), (50.2), and Lemma 9.1, we obtain readily that

$$\int_a^b \lambda f = \lambda \int_a^b f \qquad \text{for all } \lambda \in \mathbf{R}. \tag{50.3}$$

Let $f, g : [a, b] \to \mathbf{R}$ denote bounded functions and let $P = \{x_0, x_1, \ldots, x_n\}$ denote a partition of $[a, b]$. By Lemma 9.1,

$$\sup_{[x_{k-1}, x_k]} (f(x) + g(x)) \leqslant \sup_{[x_{k-1}, x_k]} f(x) + \sup_{[x_{k-1}, x_k]} g(x), \qquad k = 1, 2, \ldots, n,$$

and

$$\inf_{[x_{k-1}, x_k]} (f(x) + g(x)) \geqslant \inf_{[x_{k-1}, x_k]} f(x) + \inf_{[x_{k-1}, x_k]} g(x), \qquad k = 1, 2, \ldots, n.$$

Hence,

$$\bar{S}(f + g; P) \leqslant \bar{S}(f; P) + \bar{S}(g; P), \tag{50.4}$$

$$\underline{S}(f + g; P) \geqslant \underline{S}(f; P) + \underline{S}(g; P). \tag{50.5}$$

To find $\int_a^b (f + g)$ we proceed as follows: For every $\varepsilon > 0$, there are partitions $P_1$ and $P_2$ such that

$$\bar{S}(f; P_1) - \frac{\varepsilon}{2} < \int_a^b f \quad \text{and} \quad \bar{S}(g; P_2) - \frac{\varepsilon}{2} < \int_a^b g.$$

With $Q = P_1 \cup P_2$ we obtain, from (50.4), that

$$\int_a^b (f+g) - \varepsilon \leqslant \bar{S}(f+g;Q) - \varepsilon \leqslant \bar{S}(f;Q) + \bar{S}(g;Q) - \varepsilon < \int_a^b f + \int_a^b g.$$

Hence,

$$\int_a^b (f+g) < \int_a^b f + \int_a^b g + \varepsilon. \tag{50.6}$$

From (50.5) we obtain, by a similar argument, that

$$\int_a^b (f+g) > \int_a^b f + \int_a^b g - \varepsilon. \tag{50.7}$$

Since (50.6) and (50.7) have to hold for all $\varepsilon > 0$, we obtain $\int_a^b (f+g) = \int_a^b f + \int_a^b g$. This result, together with (50.3) yields:

***Theorem 50.1***

The Riemann integral is a linear function from the set of all Riemann-integrable functions on $[a,b]$ into $\mathbf{R}$, i.e., for any Riemann-integrable functions $f$, $g$: $[a,b] \to \mathbf{R}$ and any $\lambda \in \mathbf{R}$,

$$\int_a^b (f+g) = \int_a^b f + \int_a^b g, \tag{50.8}$$

$$\int_a^b \lambda f = \lambda \int_a^b f. \tag{50.9}$$

(50.8) and (50.9) are equivalent to the statement that

$$\int_a^b (\lambda f + \mu g) = \lambda \int_a^b f + \mu \int_a^b g \tag{50.10}$$

for any Riemann-integrable functions $f$, $g : [a,b] \to \mathbf{R}$ and any $\lambda$, $\mu \in \mathbf{R}$. (See exercise 2.) (Note the connection between (50.8) and ($\mu_3$) of definition 41.1.)

## 4.50 Exercises

1. Supply the details in the derivation of (50.3).
2. Show that (50.10) is satisfied if and only if (50.8) and (50.9) are satisfied.
3. Find two functions $f$, $g$ and two constants $\lambda \neq 0$, $\mu \neq 0$ such that $\lambda f + \mu g$ is integrable on $[0,1]$, but neither $f$ nor $g$ is integrable. Is this consistent with Theorem 50.1?
4. Use the results of exercises 6 and 7, section 4.46 to prove Theorem 50.1 without reference to upper and lower sums.
5. Let $f_1, f_2, g : [a,b] \to \mathbf{R}$. Prove: If $f_1, f_2$ are integrable with respect to $g$ and if $\lambda$, $\mu \in \mathbf{R}$, then

   $$\int_a^b (\lambda f_1 + \mu f_2)\, dg = \lambda \int_a^b f_1\, dg + \mu \int_a^b f_2\, dg.$$

   (See also exercise 8, section 4.46.)

6. Let $f, g_1, g_2 : [a,b] \to \mathbf{R}$. Prove: If $f$ is integrable with respect to $g_1$ and with respect to $g_2$, and $\lambda, \mu \in \mathbf{R}$, then

$$\int_a^b f(\lambda\, dg_1 + \mu\, dg_2) = \lambda \int_a^b f\, dg_1 + \mu \int_a^b f\, dg_2.$$

## 4.51 Properties of the Riemann Integral

The properties of the Riemann integral which we shall establish in this section are not only of theoretical interest but of considerable practical importance, as we shall see in the next section.

***Theorem 51.1***

If $h : [a,b] \to \mathbf{R}$ is Riemann-integrable on $[a,b]$ and if $h(x) \geqslant 0$ for all $x \in [a,b]$, then $\int_a^b h \geqslant 0$.

*Proof*

$\int_a^b h = \sup_{(P)} \underline{S}(h;P) \geqslant 0$, since $\underline{S}(h;P) \geqslant 0$ for all $P$.

***Corollary 1 to Theorem 51.1***

If $f, g : [a,b] \to \mathbf{R}$ are Riemann-integrable on $[a,b]$, and if $f(x) \leqslant g(x)$ for all $x \in [a,b]$, then

$$\int_a^b f \leqslant \int_a^b g.$$

(Compare this statement with Theorem 41.1.)

*Proof*

Let $h = g - f$ and apply Theorems 51.1 and 50.1.

***Corollary 2 to Theorem 51.1***

If $f : [a,b] \to \mathbf{R}$ is Riemann-integrable on $[a,b]$, and if $m \leqslant f(x) \leqslant M$ for all $x \in [a,b]$, then

$$m(b-a) \leqslant \int_a^b f \leqslant M(b-a).$$

*Proof*

Apply Corollary 1 to $m, f$ and to $f, M$, where $m, M : [a,b] \to \mathbf{R}$ are defined by $m(x) = m$, $M(x) = M$ for all $x \in [a,b]$.

***Corollary 3 to Theorem 51.1*** *(Triangle Inequality for Integrals)*

If $f : [a,b] \to \mathbf{R}$ is Riemann-integrable on $[a,b]$, then $|f| : [a,b] \to \mathbf{R}$ is Riemann-integrable and

$$\left| \int_a^b f \right| \leqslant \int_a^b |f|,$$

where $|f|$ is defined by $|f|(x) = |f(x)|$ for all $x \in [a,b]$.

*Proof*

If $f$ is continuous at $c \in [a,b]$, then, because of

$$||f(c)| - |f(x)|| \leqslant |f(c) - f(x)|,$$

$|f|$ is continuous at $c$. Hence, the set of points where $f$ is continuous is contained in the set of points where $|f|$ is continuous. Therefore, the set of points where $|f|$ is not continuous is contained in the set of points where $f$ is not continuous. Since $f$ is integrable, $f$ is continuous a.e. on $[a,b]$; i.e., the set $D$ where $f$ is not continuous has Lebesgue measure zero. Hence, the set where $|f|$ is not continuous, being a subset of $D$, also has Lebesgue measure zero; that is, $|f|$ is integrable on $[a,b]$. (Note that the converse is not necessarily true. See exercise 3.)

Since $\pm f(x) \leqslant |f(x)|$ for all $x \in [a,b]$, Corollary 3 follows now from Corollary 1.

***Theorem 51.2*** *(Mean-Value Theorem of the Integral Calculus)*

If $f : [a.b] \to \mathbf{R}$ is *continuous* on $[a,b]$, then there is a $c \in [a,b]$ such that

$$\int_a^b f = f(c)(b-a).$$

*Proof*

$f \in C[a,b]$, $f$ is bounded on $[a,b]$—(see Theorem 34.3)—and hence, $m \leqslant f(x) \leqslant M$ for all $x \in [a,b]$, where $m = \min_{[a,b]} f(x)$, $M = \max_{[a,b]} f(x)$. By Corollary 2,

$$m \leqslant \frac{1}{b-a} \int_a^b f \leqslant M.$$

By the intermediate-value theorem (Theorem 27.1), $f$ assumes every value between $m$ and $M$ (observe that $f$ assumes the values $m$, $M$ on the closed interval $[a,b]$). Hence, there is a $c \in [a,b]$ such that $f(c) = 1/(b-a) \int_a^b f$ and Theorem 51.2 follows readily.

By similar reasoning, we obtain the following, more general, mean-value theorem:

***Theorem 51.3*** *(Second Mean-Value Theorem of the Integral Calculus)*

If $f, g : [a,b] \to \mathbf{R}$ are continuous on $[a,b]$, and if $g(x) \geqslant 0$ for all $x \in [a,b]$, then there exists a $c \in [a,b]$ such that

$$\int_a^b fg = f(c) \int_a^b g.$$

*Proof*

As in the proof of Theorem 51.2, we have $m \leqslant f(x) \leqslant M$ for all $x \in [a,b]$. Since $g(x) \geqslant 0$ for all $x \in [a,b]$, we also have $mg(x) \leqslant f(x)g(x) \leqslant Mg(x)$, and hence, by Corollary 1 to Theorem 51.1,

$$m \int_a^b g \leqslant \int_a^b fg \leqslant M \int_a^b g.$$

If $\int_a^b g = 0$, then $g(x) = 0$ for all $x \in [a,b]$ (see exercise 8), and consequently, $\int_a^b fg = 0$. In this case, the theorem is trivially true. If $\int_a^b g \neq 0$, we divide the above inequality by $\int_a^b g$, and see that there has to be a $c \in [a,b]$ such that $f(c) = \int_a^b fg / \int_a^b g$. The theorem follows readily.

***Theorem 51.4***

If $c \in (a,b)$ and if $f : [a,b] \to \mathbf{R}$ is Riemann-integrable on $[a,b]$, then $f$ is Riemann-integrable on $[a,c]$ and on $[c,b]$. If $f$ is Riemann-integrable on $[a,c]$ and on $[c,b]$, then $f$ is Riemann-integrable on $[a,b]$. In either case,

$$\int_a^b f = \int_a^c f + \int_c^b f. \tag{51.1}$$

(See also $(\mu_3)$ of definition 41.1.)

*Proof*

If $f$ is Riemann-integrable on $[a,b]$, then $f$ is continuous a.e. on $[a,b]$ and hence, continuous a.e. on $[a,c]$ and on $[c,b]$. Hence, $f$ is Riemann-integrable on $[a,c]$ and on $[c,b]$. The converse follows by a similar argument.

In order to establish (51.1), we may therefore assume that $f$ is Riemann-integrable on $[a,b]$, $[a,c]$, and on $[c,b]$. Let

$$f_1(x) = \begin{Bmatrix} f(x) & \text{for } x \in [a,c) \\ 0 & \text{for } x \in [c,b] \end{Bmatrix}, \quad f_2(x) = \begin{Bmatrix} 0 & \text{for } x \in [a,c) \\ f(x) & \text{for } x \in [c,b] \end{Bmatrix}.$$

For any partition of $[a,b]$, there is a refinement that contains $c$ as a partition point. It follows readily that

$$\int_a^b f_1 = \int_a^c f, \qquad \int_a^b f_2 = \int_c^b f.$$

Since $f(x) = f_1(x) + f_2(x)$ for all $x \in [a,b]$, we have from (50.8) that

$$\int_a^b f = \int_a^b f_1 + \int_a^b f_2 = \int_a^c f + \int_c^b f.$$

Thus far we have considered only integrals from $a$ to $b$, where $a < b$. We free ourselves from this restriction by extending the definition of the integral as follows:

(51.2) $$\int_a^b f = -\int_b^a f,$$

(51.3) $$\int_a^a f = 0.$$

In view of this definition we have the following:

***Corollary to Theorem 51.4***

If $f : \mathbf{R} \to \mathbf{R}$ is integrable on every closed interval, and if $a, b, c \in \mathbf{R}$, then, regardless of the order of $a, b, c$,

$$\int_a^b f = \int_a^c f + \int_c^b f.$$

(The proof is left to exercise 10.)

## 4.51 Exercises

1. Show that $\pi/6 \leqslant \int_{\pi/6}^{\pi/2} \sin x \, dx \leqslant \pi/3$ without evaluating the integral.
2. Show that Theorem 51.2 need not be true if $f$ is not continuous in $[a, b]$.
3. Show: If $|f| : [a, b] \to \mathbf{R}$ is Riemann-integrable, then $f$ is not necessarily Riemann-integrable.
4. Show: If $c \in (a, b)$ and $f : [a, b] \to \mathbf{R}$ is Riemann-integrable on $[a, c]$ and on $[c, b]$, then $f$ is Riemann-integrable on $[a, b]$.
5. Show: If $f : [a, b] \to \mathbf{R}$ is Riemann-integrable, and if the values of $f$ are changed at a finite number of points, then the new function $f_1$ is also Riemann-integrable and $\int_a^b f = \int_a^b f_1$.
6. Let $f_1, f_2$ be defined as in the proof of Theorem 51.4. Show that $\int_a^b f_1 = \int_a^c f, \int_a^b f_2 = \int_c^b f$.
7. Let $f, g : [a, b] \to \mathbf{R}$. Prove: If $f$ is integrable with respect to $g$ on $[a, c]$ and on $[c, b]$, then $f$ is integrable with respect to $g$ on $[a, b]$, and
$$\int_a^b f dg = \int_a^c f dg + \int_c^b f dg.$$
8. Let $g : [a, b] \to \mathbf{R}$. Prove: If $g(x) \geqslant 0$ for all $x \in [a, b]$, if $g \in C[a, b]$, and if $\int_a^b g = 0$, then $g(x) = 0$ for all $x \in [a, b]$.
9. Show that Theorem 51.3 need not be true if $g$ assumes positive and negative values in $[a, b]$.
10. Prove the Corollary to Theorem 51.4.

## 4.52 The Riemann Integral with a Variable Upper Limit

If $f : [a,b] \to \mathbf{R}$ is Riemann-integrable on $[a,b]$, then

$$F(x) = \int_a^x f \tag{52.1}$$

defines a function $F : [a,b] \to \mathbf{R}$. Note that, by (51.3),

$$F(a) = 0.$$

***Theorem 52.1***

If $f : [a,b] \to \mathbf{R}$ is Riemann-integrable on $[a,b]$, then the function $F : [a,b] \to \mathbf{R}$, as defined in (52.1), is continuous on $[a,b]$.

*Proof*

Let $c$, $x \in [a,b]$. Then, by the Corollary to Theorem 51.4,

$$F(c) - F(x) = \int_a^c f - \int_a^x f = \int_a^c f + \int_x^a f = \int_x^c f. \tag{52.2}$$

We obtain from Corollary 3 to Theorem 51.1 that

$$\left| \int_x^c f \right| \leqslant \begin{cases} \displaystyle\int_x^c |f| & \text{if } c \geqslant x, \\ \displaystyle\int_c^x |f| & \text{if } c < x. \end{cases}$$

Since $f$ is bounded, there is an $M > 0$ such that

$$|F(c) - F(x)| \leqslant \left| \int_c^x |f| \right| \leqslant M\,|c - x|,$$

and we see that, for every $\varepsilon > 0$, there is a $\delta < (\varepsilon/M)$ such that

$$|F(c) - F(x)| < \varepsilon \qquad \text{for all } x \in N_\delta(c) \cap [a,b];$$

that is, $F$ is continuous at $c$. This argument applies to every point $c \in [a,b]$. Hence, $F$ is continuous on $[a,b]$.

---

▲ *Example 1*

Let $f : [0,2] \to \mathbf{R}$ be given by

$$f(x) = \begin{cases} 0 & \text{for } 0 \leqslant x \leqslant 1, \\ 1 & \text{for } 1 < x \leqslant 2. \end{cases}$$

$f$ is integrable and we obtain from Theorem 52.1 that

$$F(x) = \int_0^x f = \begin{cases} 0 & \text{for } 0 \leqslant x \leqslant 1, \\ x - 1 & \text{for } 1 < x \leqslant 2, \end{cases}$$

is continuous on $[0,2]$ which, indeed, it is. Note that $F \notin C'[0,2]$ since $F'(1)$ does not exist. (See Fig. 52.1.)

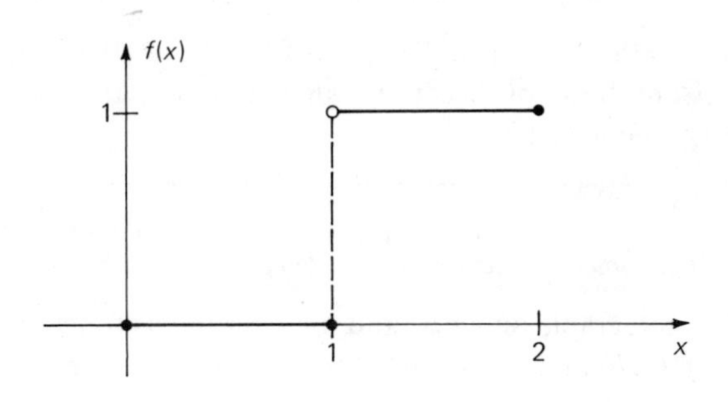


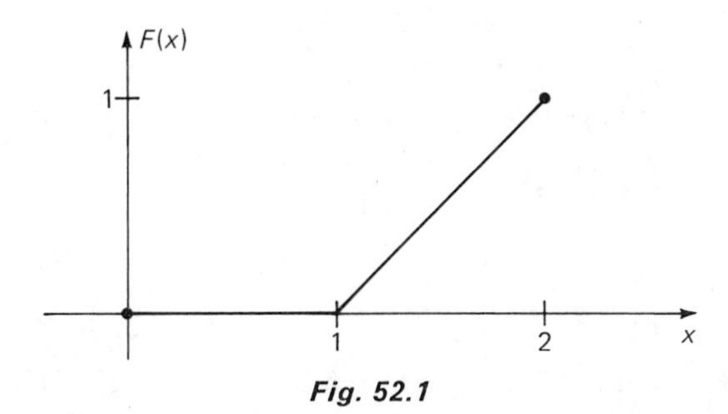


*Fig. 52.1*

---

***Theorem 52.2***

If $f : [a,b] \to \mathbf{R}$ is continuous on $[a,b]$ and if $F : [a,b] \to \mathbf{R}$ is defined as in (52.1), then $F'$ exists and $F'(x) = f(x)$ for all $x \in [a,b]$.

*Proof*

From (52.2) and Theorem 51.2, observing (51.2),

$$F(c) - F(x) = \int_x^c f = f(\xi)(c - x)$$

for some $\xi \in [c,x]$ or $\xi \in [x,c]$. Since $\lim_{x \to c} \xi = c$, and since $f$ is continuous at $c$, we obtain

$$F'(c) = \lim_{x \to c} \frac{F(x) - F(c)}{x - c} = \lim_{x \to c} f(\xi) = f(c)$$

for every $c \in [a,b]$.

▲ *Example 2*

If $f$ is not continuous, $F'$ may still exist for all $x \in [a, b]$ but Theorem 52.2 need not necessarily be true. Let $f : [0, 1] \to \mathbf{R}$ be given by

$$f(x) = \begin{cases} 0 & \text{if } x \neq \frac{1}{2}, \\ 1 & \text{if } x = \frac{1}{2}; \end{cases}$$

then, $F(x) = \int_0^x f = 0$, and hence $F'(x) = 0$ for all $x \in [0, 1]$, but $f$ cannot be the derivative of any function because it does not have the intermediate-value property. (See Theorem 38.1.)

***Theorem 52.3*** *(Fundamental Theorem of the Integral Calculus)*

If $f : [a, b] \to \mathbf{R}$ is integrable on $[a, b]$ and if a continuous function $F : [a, b] \to \mathbf{R}$ is differentiable in $(a, b)$ and $F'(x) = f(x)$ for all $x \in (a, b)$, then

$$\int_a^b f = F(b) - F(a). \tag{52.3}$$

(52.3) is usually written in the symbolic form

$$\int_a^b f = F(x)\Big|_a^b. \tag{52.4}$$

*Proof*

We take any partition $P$ of $[a, b]$ and represent $F(b) - F(a)$ by the following telescoping sum:

$$F(b) - F(a) = F(x_1) - F(a) + F(x_2) - F(x_1) + \cdots + F(b) - F(x_{n-1}).$$

We obtain, from the mean-value theorem of the differential calculus (Theorem 37.3),

$$F(x_k) - F(x_{k-1}) = f(\xi_k)(x_k - x_{k-1}), \qquad x_{k-1} \leqslant \xi_k \leqslant x_k, \quad k = 1, 2, \ldots, n.$$

Hence,

$$F(b) - F(a) = \sum_{k=1}^{n} f(\xi_k)\, \Delta x_k.$$

Since $m_k(f) \leqslant f(\xi_k) \leqslant M_k(f)$ for all $k = 1, 2, \ldots, n$, we have

$$\underline{S}(f; P) \leqslant F(b) - F(a) \leqslant \bar{S}(f; P).$$

Since $f$ is integrable on $[a, b]$, we have

$$\int_a^b f = \sup_{(P)} \underline{S}(f; P) = \inf_{(P)} \bar{S}(f; P) = F(b) - F(a).$$

▲ *Example 3*

The function $F : [0,1] \to \mathbf{R}$, given by

$$F(x) = \begin{cases} 0 & \text{if } x = 0 \\ x^2 \sin \dfrac{1}{x^2} & \text{if } x \neq 0 \end{cases}$$

is differentiable on $[0,1]$ and

$$F'(x) = \begin{cases} 0 & \text{if } x = 0, \\ 2x \sin \dfrac{1}{x^2} - \dfrac{2}{x} \cos \dfrac{1}{x^2} & \text{if } x \neq 0. \end{cases}$$

$F'$ is *not* bounded on $[0,1]$ and hence, not integrable on $[0,1]$. Therefore Theorem 52.3 does not apply.

▲ *Example 4*

Let $f$, $F : [0,1] \to \mathbf{R}$ be defined by

$$f(x) = 1, \qquad F(x) = \begin{cases} x & \text{for } 0 \leqslant x < 1, \\ 0 & \text{for } x = 1. \end{cases}$$

(See Fig. 52.2.) While $\int_0^1 f = 1$, we have $F(1) - F(0) = 0$. Although $F'(x) = f(x)$ for all $x \in (0, 1)$, $F$ is *not* continuous on $[0,1]$ and Theorem 52.3 does not apply.

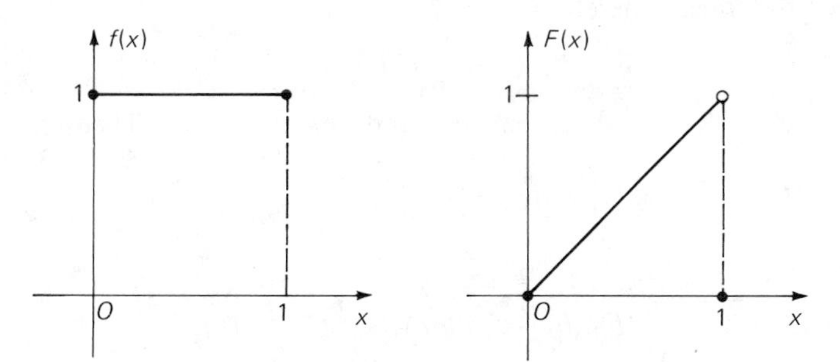


**Fig. 52.2**

Theorems 52.2 and 52.3 establish, to some extent, a connective link between the operations of differentiation and integration.

Whenever two functions $f$, $F : D \to \mathbf{R}$ are related to each other by $F'(x) = f(x)$ for all $x \in D$, one calls $F$ the *antiderivative* of $f$ on $D$.

We have seen in example 1 that, while an integrable function $f$ possesses an integral with a variable upper limit $F$, this $F$ need *not* be the antiderivative of $f$, because it may not be differentiable.

We have seen in example 2 that the integral $F$ of a function $f$ with a variable upper limit may exist and $F$ may be differentiable, but still, it may not be true that $F'(x) = f(x)$ for all $x \in [a, b]$.

Finally, we have seen in example 3 that a function $f$ may have an antiderivative $F$ but $F(x) \neq \int_a^x f$ because $\int_a^x f$ may not even exist.

All we can really say with any certainty is that, by Theorem 52.2, an antiderivative $F$ of $f$ exists on $[a, b]$ when $f$ is continuous on $[a, b]$ and that there are functions that are not continuous on $[a, b]$ but still have an antiderivative on $[a, b]$ (example 3).

***Corollary 1 to Theorem 52.3*** *(Integration by Parts)*

Let $f, g \in C'[a, b]$. If $f', g'$ are integrable on $[a, b]$, then

$$\int_a^b fg' = f(b)g(b) - f(a)g(a) - \int_a^b f'g. \tag{52.5}$$

In the notation which was introduced in (52.4), we may write this as

$$\int_a^b fg' = f(x)g(x)\Big|_a^b - \int_a^b f'g.$$

*Proof*

Since $f, g \in C'[a, b]$, it follows that $f, g \in C[a, b]$ (see Corollary to Theorem 35.2). Hence, $fg'$ and $f'g$ are integrable. (The set of points where $fg'$ and $f'g$ are not continuous is contained in the set of points where $g'$ and $f'$, respectively, are not continuous.) Therefore,

$$(fg)' = f'g + fg'$$

is integrable on $[a, b]$ by Theorem 50.1, and we obtain, from Theorem 52.3, that

$$\int_a^b (fg)' = f(b)g(b) - f(a)g(a).$$

Hence,

$$f(b)g(b) - f(a)g(a) = \int_a^b f'g + \int_a^b fg',$$

and (52.5) follows readily. (For a generalization of (52.5), see exercise 4.)

The following corollary involves the concept of a composite function and the attendant terminology that was introduced in section 3.36.

***Corollary 2 to Theorem 52.3*** *(Change of Integration Variable)*

Let $g : [\alpha, \beta] \to \mathbf{R}$, $g' \in C[\alpha, \beta]$, and let $f : \mathcal{R}(g) \to \mathbf{R}$ where $f \in C(\mathcal{R}(g))$. Then,

$$\int_{g(\alpha)}^{g(\beta)} f = \int_\alpha^\beta (f \circ g)g'. \tag{52.6}$$

*Proof*

Note that, by the Corollary to Theorem 34.3, $\mathfrak{R}(g)$ is an interval or a point. If $\mathfrak{R}(g)$ is just a point, then $g(t) = g(\alpha)$ for all $t \in [\alpha, \beta]$. Hence, $g'(t) = 0$ for all $t \in [\alpha, \beta]$ and (52.6) is trivially true because $\int_\alpha^\beta (f \circ g)g' = \int_\alpha^\beta 0\, dt = 0$ and since $g(\alpha) = g(\beta)$, $\int_{g(\alpha)}^{g(\beta)} f = 0$. Hence, we may assume that $\mathfrak{R}(g)$ is an interval.

We define $F : \mathfrak{R}(g) \to \mathbf{R}$ by

$$F(x) = \int_{g(\alpha)}^{x} f,$$

and $H : [\alpha, \beta] \to \mathbf{R}$ by

$$H(t) = F(g(t)).$$

By Theorem 36.1, $H'(t) = F'(g(t))g'(t)$ for all $t \in [\alpha, \beta]$, and, by Theorem 52.2, $F'(x) = f(x)$ for all $x \in \mathfrak{R}(g)$. Hence, $H'(t) = f(g(t))g'(t)$. Since $H(\alpha) = F(g(\alpha)) = 0$, we have, from Theorem 52.3, that

$$\int_{g(\alpha)}^{g(\beta)} f = F(g(\beta)) = H(\beta) = H(\beta) - H(\alpha) = \int_\alpha^\beta f(g(t))g'(t)\, dt = \int_\alpha^\beta (f \circ g)g'.$$

(Note that, by (51.2) and (51.3), it does not matter whether $g(\beta) \geqslant g(\alpha)$ or $g(\beta) < g(\alpha)$. Note also that $\mathfrak{R}(g)$ need not be the interval $[g(\alpha), g(\beta)]$ or $[g(\beta), g(\alpha)]$, but that this interval is always contained in $\mathfrak{R}(g)$. See also exercise 6.)

---

▲ *Example 5* (*The Natural Logarithm*)

The function $\log : (0, \infty) \to \mathbf{R}$, which is defined by

(52.7) $$\log x = \int_1^x \frac{dt}{t}, \qquad x > 0,$$

is called the *natural logarithm*. Let $x, y > 0$ and define $g : [y, xy] \to (0, \infty)$ (or $[xy, y] \to (0, \infty)$) by $g(u) = u/y$. With $f(t) = 1/t$, we obtain, from the above corollary that

$$\int_1^x \frac{dt}{t} = \int_y^{xy} \frac{du}{u},$$

that is, $\log x = \log(xy) - \log y$. Hence,

(52.8) $$\log(xy) = \log x + \log y \qquad \text{for all} \quad x, y > 0.$$

From (52.7), $\log 1 = 0$. From (52.8),

$$0 = \log 1 = \log\left(x \frac{1}{x}\right) = \log x + \log \frac{1}{x},$$

and hence

(52.9) $$\log \frac{1}{x} = -\log x \qquad \text{for all} \quad x > 0.$$

From Theorem 52.2,

$$(\log x)' = \frac{1}{x} \quad \text{for all} \quad x > 0. \tag{52.10}$$

By Lemma 37.1, Theorem 36.2, and the Corollary to Lemma 37.1, we see that the natural logarithm has an inverse function, that the inverse function is differentiable, and that

$$(\log^{-1})'(y) = \frac{1}{\log'(\log^{-1} y)} = \log^{-1}(y).$$

The inverse function to the natural logarithm is called the *exponential function* $E(y) = \log^{-1}(y)$, and we have just shown that

$$E'(y) = E(y) \qquad \text{for all } y \in \mathfrak{R}(\log^{-1}) = \mathbf{R}.$$

(See also example 2, section 1.5.)

---

## 4.52 Exercises

1. Prove: If $f : [a, b] \to \mathbf{R}$ is continuous, then

$$F(x) - F(a) = \int_a^x f \qquad \text{for all } x \in [a, b]$$

if and only if $F'(x) = f(x)$ for all $x \in [a, b]$.

2. Prove: If $f \in C[a, b]$, then $f$ has an antiderivative in $[a, b]$. If $F$, $G$ are two antiderivatives of $f$, then $F(x) = G(x) + C$ for all $x \in [a, b]$ and some constant $C$.

3. Evaluate, by the Fundamental Theorem of the Integral Calculus:

(a) $\displaystyle\int_a^b x^n \, dx, \quad n \in \mathbf{N}$ (b) $\displaystyle\int_1^2 \frac{dx}{x}$ (see example 5)

(c) $\displaystyle\int_0^{2\pi} \sin x \, dx$ (d) $\displaystyle\int_0^{2\pi} \cos x \, dx$

(e) $\displaystyle\int_0^1 E(x) \, dx$ (see example 5)

4. Let $f, g : [a, b] \to \mathbf{R}$. Prove: $f$ is integrable with respect to $g$ on $[a, b]$ if and only if $g$ is integrable with respect to $f$ on $[a, b]$, and

$$\int_a^b f \, dg = f(b)g(b) - f(a)g(a) - \int_a^b g \, df.$$

5. Let $f, g \in C[a, b]$. Suppose there is a partition $P$ of $[a, b]$ such that

$$f, g \in C'[x_{k-1}, x_k]$$

and that $f'$, $g'$ are Riemann-integrable on $[x_{k-1}, x_k]$ for all $k = 1, 2, \ldots, n$. Show that (52.4) holds. Where is the continuity of $f$, $g$ on $[a, b]$ used?

6. Let $f:[-1,1]\to\mathbf{R}$ denote a continuous function. Show that

$$\int_0^1 f(x)\,dx = \int_0^{\pi/2} f(\sin x)\cos x\,dx = \int_0^{5\pi/2} f(\sin x)\cos x\,dx$$
$$= \int_0^{9\pi/2} f(\sin x)\cos x\,dx = \cdots.$$

7. Use integration by parts twice in succession to show that

$$\int_0^x e^t \sin t\,dt = \tfrac{1}{2}e^x(\sin x - \cos x) + \tfrac{1}{2}.$$

8. Use integration by parts twice in succession to show that

$$\int_0^\pi \sin^k x\,dx = \frac{k-1}{k}\int_0^\pi \sin^{k-2} x\,dx \qquad \text{for all integers } k \geqslant 2.$$

9. Use the result in exercise 8 to show that

$$\int_0^\pi \sin^k x\,dx = \begin{cases} \dfrac{(k-1)(k-3)\cdots 4\cdot 2}{k(k-2)\cdots 5\cdot 3}\cdot 2 & \text{if } k \text{ is odd,}\\[2ex] \dfrac{(k-1)(k-3)\cdots 5\cdot 3}{k(k-2)\cdots 4\cdot 2}\cdot \pi & \text{if } k \text{ is even.}\end{cases}$$

10. Use integration by parts to evaluate $\int_1^x t\log t\,dt,\quad x>0$.
11. Use Corollary 2 to Theorem 52.3 (change of integration variable) to evaluate the following integrals:

(a) $\displaystyle\int_0^x \tan t\,dt,\quad 0<x<\frac{\pi}{2}$ (b) $\displaystyle\int_1^x \frac{1}{t}\log t\,dt,\quad x>0$

(c) $\displaystyle\int_1^x \frac{t^2-2t+1}{t^3-3t^2+3t}\,dt,\quad x>0.$

12. Let $f:[0,1]\to\mathbf{R}$ be defined by $f(x)=1$, and let $F:[0,1]\to\mathbf{R}$ be defined by:

$$F(x) = \begin{cases} 1 & \text{for } x=0,\\ x & \text{for } 0<x<1,\\ 0 & \text{for } x=1.\end{cases}$$

Then $F'(x)=f(x)$ for all $x\in(0,1)$. Still, $F(1)-F(0)=-1\neq\int_0^1 f$. Explain.
13. Prove the following stronger version of Theorem 52.3: If $f:[a,b]\to\mathbf{R}$ is integrable and if a function $F:[a,b]\to\mathbf{R}$ is differentiable in the interval $(a,b)$, $\lim_a F(x)$, $\lim_b F(x)$ exist, and $F'(x)=f(x)$ for all $x\in(a,b)$, then

$$\int_a^b f = \lim_b F(x) - \lim_a F(x).$$

14. Apply the theorem in exercise 13 to the example in exercise 12.

## 4.53 Improper Integrals*

As we have indicated in section 4.49, we shall now generalize the concept of the Riemann integral so that it also applies to integration intervals that are not bounded and to integrands that are not bounded.

For the purpose of studying integrals on unbounded integration intervals, we need the concept of the limit of a function at infinity:

***Definition 53.1***

Let $f : [a, \infty) \to \mathbf{R}$. Then $f$ is said to approach the limit $L$ at $\infty$ (at $-\infty$),

$$\lim_{\infty} f(x) = L, \left( \lim_{-\infty} f(x) = L \right),$$

if and only if, for every $\varepsilon > 0$, there is an $A(\varepsilon) > 0$ such that

$$|f(x) - L| < \varepsilon \qquad \text{for all} \quad x > A(\varepsilon) \qquad (\text{for all } x < -A(\varepsilon)).$$

---

▲ *Example 1*

Let $f : (0, \infty) \to \mathbf{R}$ be defined by $f(x) = 1/x$. Clearly, $\lim_{\infty} f(x) = 0$ because $|f(x) - 0| = 1/|x| < \varepsilon$ for all $x > 1/\varepsilon = A(\varepsilon)$.

▲ *Example 2*

Let $f, g : \mathbf{R} \to \mathbf{R}$ be defined by $f(x) = x^n$, $n \in \mathbf{N}$, and $g(x) = e^{ax}$, $a > 0$, respectively. We consider $\lim_{\infty} (f(x)/g(x))$. By an obvious modification of l'Hôpital's rule (see exercise 25), we obtain that $\lim_{\infty}(x^n/e^{ax}) = \lim_{\infty} (f(x)/g(x)) = \lim_{\infty}(f'(x)/g'(x)) = \lim_{\infty} (nx^{n-1}/ae^{ax})$. Repeated application of l'Hôpital's rule ultimately yields:

$$\lim_{\infty} \frac{x^n}{e^{ax}} = \lim_{\infty} \frac{n!}{a^n e^{ax}} = 0. \tag{53.1}$$

---

Suppose that $f : [a, \infty) \to \mathbf{R}$ is Riemann-integrable on $[a, \omega]$ for all $\omega > a$. Then, $F(\omega) = \int_a^\omega f$, for fixed $a$, defines a function of $\omega$ on $[a, \infty)$. This function may or may not have a limit at infinity:

***Definition 53.2***

If $f$: $[a, \infty) \to \mathbf{R}$ is Riemann-integrable on $[a, \omega]$ for all $\omega > a$, and if $\lim_{\omega \to \infty} \int_a^\omega f$ exists, then this limit is called an *improper integral of the first kind* of $f$ on $[a, \infty)$, and is denoted by

$$\int_a^\infty f = \lim_{\omega \to \infty} \int_a^\omega f.$$

▲ *Example 3*

Let $f : [1, \infty) \to \mathbf{R}$ be defined by $f(x) = x^\alpha$, $\alpha \in \mathbf{R}$. Then,

$$\int_1^\omega x^\alpha \, dx = \begin{cases} \log \omega & \text{for } \alpha = -1 \\ \dfrac{\omega^{\alpha+1}}{\alpha+1} - \dfrac{1}{\alpha+1} & \text{for } \alpha \neq -1 \end{cases}$$

exists for all $\omega > 1$. $\lim_{\omega \to \infty} \int_1^\omega x^\alpha \, dx$ exists for $\alpha < -1$ and does not exist for $\alpha \geqslant -1$. We obtain

$$\int_1^\infty x^\alpha \, dx = \lim_{\omega \to \infty} \left( \frac{\omega^{\alpha+1}}{\alpha+1} - \frac{1}{\alpha+1} \right) = -\frac{1}{\alpha+1} \quad \text{for } \alpha < -1.$$

▲ *Example 4*

Let $f : [0, \infty) \to \mathbf{R}$ be defined by $f(x) = e^{-x}$. Then $\int_0^\omega e^{-x} \, dx = 1 - e^{-\omega}$ exists for all $\omega > 0$, and

$$\int_0^\infty e^{-x} \, dx = \lim_{\omega \to \infty} (1 - e^{-\omega}) = 1.$$

▲ *Example 5*

Let $f : [0, \infty) \to \mathbf{R}$ be defined by $f(x) = x^n e^{-x}$, with $n = 0, 1, 2, 3, \ldots$. For every $\omega > 0$, we obtain by integration by parts (Corollary 1 to Theorem 52.3), for $n \geqslant 1$,

$$\int_0^\omega x^n e^{-x} \, dx = -\omega^n e^{-\omega} \Big|_0^\omega + n \int_0^\omega x^{n-1} e^{-x} \, dx.$$

Since $\lim_{\omega \to \infty} \omega^n / e^\omega = 0$ for all $n = 0, 1, 2, \ldots$ (see (53.1)), we obtain

$$\int_0^\infty x^n e^{-x} \, dx = n \int_0^\infty x^{n-1} e^{-x} \, dx.$$

For $n = 0$ we obtain directly $\int_0^\infty e^{-x} \, dx = 1$ (see example 4). Hence, $\int_0^\infty x e^{-x} \, dx = 1$, $\int_0^\infty x^2 e^{-x} \, dx = 2 \cdot 1$, etc... and, in general,

(53.2) $$\int_0^\infty x^n e^{-x} dx = n! \qquad \text{for} \qquad n = 0, 1, 2, \ldots.$$

If $f : (a, b] \to \mathbf{R}$ is Riemann-integrable on $[\varepsilon, b]$ for all $\varepsilon \in (a, b]$, then,

$$F(\varepsilon) = \int_\varepsilon^b f,$$

for fixed $b$, defines a function of $\varepsilon$ on $(a, b]$. The limit of this function as $\varepsilon$ approaches $a$ (from the right: $\varepsilon \to a + 0$) may or may not exist:

***Definition 53.3***

If $f:(a,b] \to \mathbf{R}$ is Riemann-integrable on $[\varepsilon, b]$ for all $\varepsilon \in (a,b]$, and if $\lim_{\varepsilon \to a+0} \int_\varepsilon^b f$ exists, then this limit is called an *improper integral of the second kind* of $f$ on $[a,b]$, and is denoted by

$$\int_a^b (f) = \lim_{\varepsilon \to a+0} \int_\varepsilon^b f.$$

Note that we put the integrand in parentheses to indicate that the integral is improper. Such a distinction was not necessary in the case of improper integrals of the first kind because the upper integration limit $\infty$ already signifies that the integral is improper.

Note also that the notation $\varepsilon \to a+0$ is quite unnecessary in view of our definition of a limit (definition 24.1) of a function, because the domain of $\int_\varepsilon^b f$ is $(a,b]$. It would suffice to write $\varepsilon \to a$. Still, we have chosen to use the more elaborate notation $\varepsilon \to a+0$ for emphasis.

More general versions of improper integrals of the first and the second kind are discussed in exercises 14 through 17.

---

▲ *Example 6*

Let $f:(0,1] \to \mathbf{R}$ be defined by $f(x) = x^\beta$, $\beta \in \mathbf{R}$. Then,

$$\int_\varepsilon^1 x^\beta dx = \begin{cases} -\log \varepsilon & \text{for } \beta = -1, \\ \dfrac{1}{\beta+1}(1-\varepsilon^{\beta+1}) & \text{for } \beta \neq -1, \end{cases}$$

exists for all $\varepsilon \in (0,1]$. $\lim_{\varepsilon \to 0+0} \int_\varepsilon^1 x^\beta dx$ exists for $\beta > -1$ and does not exist for $\beta \leqslant -1$. We obtain

$$\int_0^1 (x^\beta dx) = \lim_{\varepsilon \to 0+0} \frac{1}{\beta+1}(1-\varepsilon^{\beta+1}) = \frac{1}{\beta+1} \qquad \text{for } \beta > -1.$$

Note that for $\beta \geqslant 0$, the integral is a Riemann integral.

---

In the following theorem we state a crude, but practical, sufficient condition for the existence of improper integrals.

***Theorem 53.1***

(a) If $f, g:[a,\infty) \to \mathbf{R}$ are Riemann-integrable on $[a,\omega]$ for all $\omega > a$, if $0 \leqslant f(x) \leqslant g(x)$ for all $x \geqslant a$, and if $\int_a^\infty g$ exists, then $\int_a^\infty f$ exists also.

(b) If $f, g:(a,b] \to \mathbf{R}$ are Riemann-integrable on $[\varepsilon, b]$ for all $\varepsilon \in (a,b]$, if $0 \leqslant f(x) \leqslant g(x)$ for all $x \in (a,b]$, and if $\int_a^b (g)$ exists, then $\int_a^b (f)$ exists also.

*Proof*

We shall prove only part (a) and leave the proof of part (b) to the reader.

Since $0 \leqslant \int_a^\omega f \leqslant \int_a^\omega g \leqslant \int_a^\infty g$, we see that $\{\int_a^\omega f \mid \omega > a\}$ is bounded above, and hence has a least upper bound. Since $F(\omega) = \int_a^\omega f$ increases, the statement of the theorem follows readily. (See also exercise 7.)

---

▲ *Example 7* (*Gamma Function*)

Let $f:(0, \infty) \to \mathbf{R}$ be defined by $f(x) = x^{\alpha-1}e^{-x}$, $\alpha > 0$. We consider

$$\int_\varepsilon^\omega x^{\alpha-1}e^{-x}\,dx,$$

where $0 < \varepsilon < 1 < \omega$, and note that

$$\int_\varepsilon^\omega x^{\alpha-1}e^{-x}\,dx = \int_\varepsilon^1 x^{\alpha-1}e^{-x}\,dx + \int_1^\omega x^{\alpha-1}e^{-x}\,dx. \tag{53.3}$$

Since $0 \leqslant x^{\alpha-1}e^{-x} \leqslant x^{\alpha-1}$ for all $x \geqslant 0$, we have, from Theorem 53.1, part (b), that the first of these two integrals in (53.3) exists as an improper integral of the second kind as $\varepsilon$ approaches zero, when $0 < \alpha < 1$. For $\alpha \geqslant 1$, the integral is a Riemann integral. We shall next investigate the second integral in (53.3) as $\omega$ approaches $\infty$. We let $g(x) = 1/x^2$ and see, from $\lim_{x\to\infty}(f(x)/g(x)) = \lim_{x\to\infty} x^{\alpha+1}/e^x = 0$ (see example 2 and exercise 3), that $0 \leqslant x^{\alpha-1}e^{-x} \leqslant \varepsilon x^{-2}$ for all $x > A(\varepsilon)$ for some $A(\varepsilon) > 0$. Hence, $\int_1^\infty x^{\alpha-1}e^{-x}dx$ exists as an improper integral of the first kind. Then,

$$\int_0^\infty (x^{\alpha-1}e^{-x}\,dx) = \lim_{\varepsilon\to 0+0} \int_\varepsilon^1 x^{\alpha-1}e^{-x}\,dx + \lim_{\omega\to\infty} \int_1^\omega x^{\alpha-1}e^{-x}\,dx$$

defines, for $\alpha > 0$, the so-called *Gamma function*:

$$\Gamma(\alpha) = \int_0^\infty (x^{\alpha-1}e^{-x}\,dx), \qquad \alpha > 0. \tag{53.4}$$

From (53.2),

$$\Gamma(n+1) = n!, \qquad n = 0, 1, 2, 3, \ldots . \tag{53.5}$$

For $\alpha > 1$, we may integrate (53.4) by parts and obtain the recursion formula for the Gamma function:

$$\Gamma(\alpha) = (\alpha - 1)\Gamma(\alpha - 1), \qquad \alpha > 1. \tag{53.6}$$

This formula is used to extend the definition of the Gamma function for all nonintegral negative numbers:

$$\Gamma(\alpha - 1) = \frac{1}{\alpha - 1}\Gamma(\alpha). \tag{53.7}$$

We obtain, for example,

$$\Gamma\left(-\frac{7}{2}\right) = \frac{2^4}{1\cdot 3\cdot 5\cdot 7}\Gamma\left(\frac{1}{2}\right)$$

where $\Gamma(\frac{1}{2})$ is obtained from (53.4) by a process that is discussed in section 12.115, example 1.

---

## 4.53 Exercises

1. Show that $\lim_{x\to\infty} x^\alpha = 0, \quad \alpha < 0$.
2. Show that $e^x \geqslant 1 + x$ for all $x \geqslant 0$.
3. Let $\alpha, a \in \mathbf{R}$, $\alpha \geqslant 0$ and $a > 0$. Show that $\lim_{x\to\infty} x^\alpha/e^{ax} = 0$.
4. Show that $\lim_{x\to 0+0} x^\beta = 0, \quad \beta > 0$.
5. Show that $\int_0^\infty dx/(1+x^2)$ exists.
6. Show that $\int_{-1}^1 (dx/\sqrt{1+x})$ exists, and evaluate it.
7. Let $F: [a,\infty) \to \mathbf{R}$ denote an increasing function. Show: If $|F(x)| \leqslant M$ for some $M > 0$ and all $x \in [a,\infty)$, then $\lim_{x\to\infty} F(x)$ exists.
8. Prove part (b) of Theorem 53.1.
9. Prove: If $f: [a,\infty) \to \mathbf{R}$ is Riemann-integrable on $[a,\omega]$ for all $\omega > a$, and if $|f(x)| \leqslant M$ for some $M > 0$ and all $x \in [a,\infty)$, then $\int_a^\infty f(x)e^{-x}\,dx$ exists.
10. Let $s > 0$. Show that $\int_0^\infty x^n e^{-sx}\,dx$, $n \in \mathbf{N}$, exists.
11. Let $f, g: [a,\infty) \to \mathbf{R}$ and let $\lim_\infty f(x) = L_f$, $\lim_\infty g(x) = L_g$. Show that $\lim_\infty (f(x) \pm g(x)) = L_f + L_g$, $\lim_\infty f(x)g(x) = L_f L_g$.
12. Show: If $f: [a,\infty) \to \mathbf{R}$ is Riemann-integrable on $[a,\omega]$ for all $\omega > a$, and if $|f(x)| \leqslant 1/x^\alpha$ for some $\alpha > 1$ and all $x > A$ for some $A > 0$, then $\int_a^\infty f$ exists.
13. Show: If $f: (a,b] \to \mathbf{R}$ is Riemann-integrable on $[\varepsilon,b]$ for all $\varepsilon \in (a,b]$ and if $|f(x)| \leqslant 1/(x-a)^\beta$ for some $\beta < 1$ and all $x \in (a,\delta)$ for some $\delta \in (a,b)$, then $\int_b^a(f)$ exists.
14. If $f: \mathbf{R} \to \mathbf{R}$ is Riemann-integrable on every interval $[a,b]$, then

$$\int_{-\infty}^{\infty} f = \lim_{A\to\infty}\int_0^A f + \lim_{B\to-\infty}\int_B^0 f$$

is also called an improper integral of the first kind, provided that the two limits exist. Show that $\int_{-\infty}^{\infty} dx/(1+x^2)$ exists and has the value $\pi$.

15. If $f: \mathbf{R} \to \mathbf{R}$ is Riemann-integrable on every interval $[a,b]$, then

$$(CPV)\int_{-\infty}^{\infty} f = \lim_{A\to\infty}\int_{-A}^{A} f$$

is called the *Cauchy Principal Value* of the integral of $f$ on $(-\infty,\infty)$, provided that the limit exists.

(a) Show that $(CPV)\int_{-\infty}^{\infty}\sin x$ exists but that $\int_{-\infty}^{\infty}\sin x$ does not exist.

(b) Let $f: \mathbf{R} \to \mathbf{R}$ be Riemann-integrable on every interval $[a, b]$, and $f(x) = -f(-x)$ for all $x \in \mathbf{R}$. Show that $(CPV) \int_{-\infty}^{\infty} f$ exists.

(c) Show that neither $\int_{-\infty}^{\infty} \cos x$ nor $(CPV) \int_{-\infty}^{\infty} \cos x$ exists.

16. If $f: [a, c) \cup (c, b] \to \mathbf{R}$, $a < c < b$, is Riemann-integrable on $[a, c - \delta] \cup [c + \delta, b)$ for all $0 < \delta < \min(c - a, b - c)$, then

$$\int_a^b (f) = \lim_{\varepsilon_1 \to 0+0} \int_a^{c-\varepsilon_1} f + \lim_{\varepsilon_2 \to 0+0} \int_{c+\varepsilon_2}^b f$$

is also called an improper integral of the second kind, provided the two limits exist. Show that $\int_{-1}^{1} (dx/\sqrt[3]{x})$ exists but that $\int_{-1}^{1} (dx/x)$ does not exist.

17. If $f: [a, c) \cup (c, b] \to \mathbf{R}$, $a < c < b$, is Riemann-integrable on $[a, c - \delta] \cup [c + \delta, b]$ for all $0 < \delta < \min(c - a, b - c)$, then

$$(CPV) \int_a^b f = \lim_{\varepsilon \to 0+0} \left( \int_a^{c-\varepsilon} f + \int_{c+\varepsilon}^b f \right)$$

is called the *Cauchy Principal Value* of the integral of $f$ on $[a, b]$, provided that the limit exists. Show that $(CPV) \int_{-1}^{1} (dx/x)$ exists, and evaluate.

18. Prove: If $f: \mathbf{R} \to \mathbf{R}$ is Riemann-integrable on every interval $[a, b]$ and if $\int_{-\infty}^{\infty} f$ exists, then $(CPV) \int_{-\infty}^{\infty} f$ exists. The converse is not true. Why?

19. Prove: If $f: [a, c) \cup (c, b] \to \mathbf{R}$ is Riemann-integrable on $[a, c - \delta] \cup [c + \delta, b]$ for all $0 < \delta < \min(c - a, b - c)$, and if $\int_a^b (f)$ exists, then $(CPV) \int_a^b (f)$ exists. The converse is not true. Why?

20. Show that

$$\int_0^\infty \frac{\sin^2 x}{x^2} dx$$

exists and is equal to

$$\int_0^\infty \frac{\sin x}{x} dx.$$

21. Show that $\int_0^1 (x \log x \, dx)$ exists and evaluate.

22. For what values of $\alpha \in \mathbf{R}$ does

$$\int_2^\infty \frac{dx}{x (\log x)^\alpha}$$

exist?

23. Let $\alpha > 1$. Show that

$$\int_1^\infty \frac{\sin^2 x}{x^\alpha} dx$$

exists.

24. Show that

$$\int_0^2 \left(\frac{dx}{\sqrt{|x-1|}}\right)$$

exists, and evaluate.

25. Let $\lim_\infty f(x) = \infty$, $\lim_\infty g(x) = \infty$. Impose suitable conditions, and show that $\lim_\infty (f(x)/g(x)) = \lim_\infty (f'(x)/g'(x))$.

26. Prove : If $f$: $[a, \infty) \to \mathbf{R}$ and $\int_a^\infty |f|$ exists, then $\int_a^\infty f$ also exists.

27. $\mathfrak{L}(f(t)) = \int_0^\infty f(t)e^{-st}\,dt$ is called the *Laplace Transform*† of $f: [0, \infty) \to \mathbf{R}$ provided that the integral exists. Find:

(a) $\mathfrak{L}(e^{at})$ (b) $\mathfrak{L}(t^a)$ for $\alpha > 1$, $\alpha \in \mathbf{R}$

(c) $\mathfrak{L}(\sin \omega t)$ (d) $\mathfrak{L}(\cos \omega t)$

(e) $\mathfrak{L}(e^{at}f(t))$ if $\mathfrak{L}(f(t)) = \phi(s)$.

† *Pierne-Simon Laplace*, 1749–1827.

# The Euclidean $n$-Space 5

## 5.54 Introduction

The motivation for a study of vector-valued functions of a vector variable came, as so often in mathematics, from the physical sciences. The location of a point mass moving in space, given by the Cartesian coordinates $(x, y, z)$, depends on the time $t$ that has elapsed since the point started moving. We may represent this relationship in the form

$$\begin{aligned} x &= \phi(t), \\ y &= \psi(t), \\ z &= \chi(t). \end{aligned} \tag{54.1}$$

Since a masspoint cannot be at two different locations at the same time (or so the physicists tell us), (54.1) defines a function from the nonnegative reals into the set of all ordered triples $(x, y, z)$ of real numbers.

The instantaneous velocity with components $(u, v, w)$ of such a moving point may depend on the time $t$ elapsed and the location $(x, y, z)$:

$$\begin{aligned} u &= \phi(t, x, y, z), \\ v &= \psi(t, x, y, z), \\ w &= \chi(t, x, y, z). \end{aligned} \tag{54.2}$$

Since a moving point mass cannot have two different velocities at the same time and place, (54.2) defines a function from a subset of the set of all ordered quadruples $(t, x, y, z)$ of real numbers into the set of all ordered triples $(u, v, w)$ of real numbers.

Current $I$ (amps) and voltage $V$ (volts) in an electric circuit depend on the resistance $R$ (ohms), the inductance $L$ (henries), and the capacitance $C$ (farads):

$$\begin{aligned} I &= \phi(R, L, C), \\ V &= \psi(R, L, C). \end{aligned}$$

This defines a function from a subset of the set of all ordered triples $(R, L, C)$ of real numbers into the set of all ordered pairs $(I, V)$ of real numbers.

In this chapter we shall deal in full generality with functions from a subset of the set of all ordered $n$-tuples of real numbers into the set of all ordered $m$-tuples of real numbers. The theory of real-valued functions of a real variable, as we have developed it in the preceding chapters, will emerge as a special case of the more general theory that is to be developed in this and the following chapters. Many aspects of the new theory will be easily recognizable as straightforward generalizations from the theory that was established in Chapters 1 to 4.

Not all aspects of the earlier theory, however, can be generalized. For example, it would not make sense to ask for the maximum of a vector-valued function. But there are also aspects of the new, more general theory that do not have a meaningful counterpart in the theory of real-valued functions of a real variable. Some are trivial (e.g., "a vector-valued function is continuous if and only if every one of its (real-valued) components is continuous" would reduce to the trivial statement "a function is continuous if and only if it is continuous"), and still others do not even make sense in the earlier theory. (For example, it makes no sense to solve an equation in one unknown for some but not all of the unknowns in terms of the others).

## 5.55 The Cartesian $n$-Space

In section 1.4 we defined the Cartesian product $X \times Y$ of two nonempty sets $X$, $Y$, as:

$$X \times Y = \{(\xi, \eta) \mid \xi \in X, \eta \in Y\},$$

the collection of ordered pairs $(\xi, \eta)$ where the first element hails from $X$ and the second element from $Y$. This concept has an obvious generalization: We define the Cartesian product of $n$ nonempty sets $X_1, X_2, \ldots, X_n$ as:

$$X_1 \times X_2 \times \cdots \times X_n = \{(\xi_1, \xi_2, \ldots, \xi_n) \mid \xi_1 \in X_1, \quad \xi_2 \in X_2, \quad \ldots, \quad \xi_n \in X_n\},$$

the set of all ordered $n$-tuples of elements where the element in the $j$th place is an element of $X_j$. (Note that $(\xi_1, \xi_2, \ldots, \xi_n)$ is an ordered $n$-tuple when $(\xi_1, \xi_2, \ldots, \xi_n) = (\eta_1, \eta_2, \ldots, \eta_n)$ if and only if $\xi_1 = \eta_1$, $\xi_2 = \eta_2$, ..., $\xi_n = \eta_n$.)

If $X_j = \mathbf{R}$ for all $j = 1, 2, \ldots, n$, then, the Cartesian product

$$\mathbf{R} \times \mathbf{R} \times \cdots \times \mathbf{R} = \{(\xi_1, \xi_2, \ldots, \xi_n) \mid \xi_j \in \mathbf{R}, \quad j = 1, 2, \ldots, n\}$$

is the collection of all ordered $n$-tuples of real numbers.

***Definition 55.1***

The collection of all ordered $n$-tuples of real numbers

$$\mathbf{R} \times \mathbf{R} \times \cdots \times \mathbf{R} = \{(\xi_1, \xi_2, \ldots, \xi_n) \mid \xi_j \in \mathbf{R}, \quad j = 1, 2, \ldots, n\}$$

is called the *Cartesian n-space* and is denoted by $\mathbf{R}^n$.

$\mathbf{R}^2$ is called the *real plane* because the elements of $\mathbf{R}^2$ may be interpreted as the Cartesian coordinates of points in the plane. $\mathbf{R}^3$ is called the *real space* because the elements of $\mathbf{R}^3$ may be interpreted as the Cartesian coordinates of points in the three-dimensional space.

We shall denote the elements of $\mathbf{R}^n$ by lower-case roman letters: $x \in \mathbf{R}^n$ shall mean $x = (\xi_1, \xi_2, \ldots, \xi_n)$ for some $\xi_j \in \mathbf{R}$, $j = 1, 2, \ldots, n$. We call the $\xi_j$ the Cartesian components of $x$, and we shall always denote the Cartesian components of any $x \in \mathbf{R}^n$ by lower-case Greek letters, such as $x = (\xi_1, \xi_2, \ldots, \xi_n)$, $y = (\eta_1, \eta_2, \ldots, \eta_n)$, $z = (\zeta_1, \zeta_2, \ldots, \zeta_n)$, etc. We say

$$(55.1) \qquad x = y \qquad \text{if and only if } \xi_j = \eta_j \qquad \text{for all } j = 1, 2, \ldots, n.$$

(Two elements of $\mathbf{R}^n$ are equal if and only if all corresponding components are equal.)

We lend $\mathbf{R}^n$ algebraic structure by introducing the operations of *addition* and *scalar multiplication* according to:

$$(55.2) \qquad x + y = (\xi_1 + \eta_1, \xi_2 + \eta_2, \ldots, \xi_n + \eta_n),$$

$$(55.3) \qquad \lambda x = (\lambda\xi_1, \lambda\xi_2, \ldots, \lambda\xi_n),\ \lambda \in \mathbf{R}.$$

If $n = 2, 3$, then the elements of $\mathbf{R}$ are known as *position vectors* (vectors with head at $(\xi_1, \xi_2)$ or $(\xi_1, \xi_2, \xi_3)$ and tail at $(0,0)$ or $(0, 0, 0)$).

The operations which we defined in (55.2) and (55.3) are the well-known operations of vector addition and multiplication of a vector by a scalar. (See Fig. 55.1.)

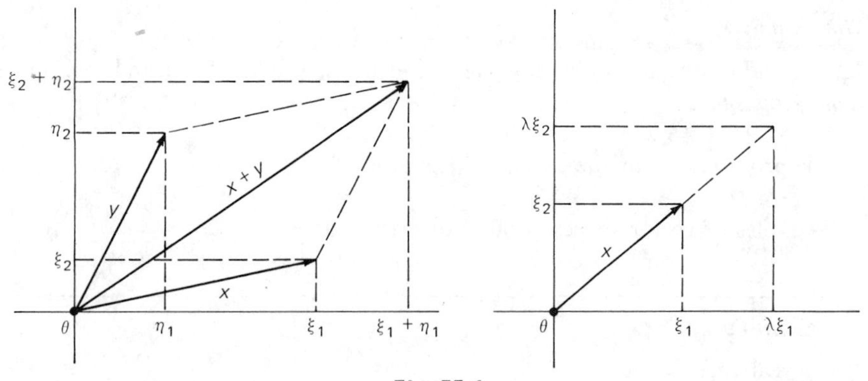


*Fig. 55.1*

We obtain from (55.2) and (55.3) that whenever $x$, $y \in \mathbf{R}^n$ and $\lambda$, $\mu \in \mathbf{R}$, then

$$(55.4) \qquad \lambda x + \mu y = (\lambda\xi_1 + \mu\eta_1,\ \ \lambda\xi_2 + \mu\eta_2,\ \ \ldots,\ \ \lambda\xi_n + \mu\eta_n) \in \mathbf{R}^n.$$

A system with this property is called a *vector space* (or *linear space*). More precisely:

***Definition 55.2***

A collection $V$ of elements $x, y, z, \ldots$, for which addition and multiplication by real numbers is defined so that the following nine conditions hold, is called a *real vector space* (or *real linear space*, or *vector space over the reals*):

1. If $x, y \in V$, then $x + y \in V$.
2. $x + y = y + x$ for all $x, y \in V$. (Addition is *commutative*.)
3. $x + (y + z) = (x + y) + z$ for all $x, y, z \in V$. (Addition is *associative*.)
4. There is a unique element $\theta \in V$ such that $x + \theta = x$ for all $x \in V$. ($\theta$ is called the *additive identity* in $V$.)
5. If $x \in V$, $\lambda \in \mathbf{R}$, then $\lambda x \in V$.
6. If $x \in V$, $\lambda, \mu \in \mathbf{R}$, then $(\lambda\mu)x = \lambda(\mu x) = \mu(\lambda x)$.
7. If $x \in V$, $\lambda, \mu \in \mathbf{R}$, then $(\lambda + \mu)x = \lambda x + \mu x$.
8. If $x, y \in V$, $\lambda \in \mathbf{R}$, then $\lambda(x + y) = \lambda x + \lambda y$.
9. If $x \in V$, then $1(x) = x$ and $0(x) = \theta$.

The elements of $V$ are called *vectors*.

Observe that (4), (7), and (9) imply the existence of a *unique additive inverse* in $V$:

$$x + (-x) = (1 - 1)x = 0(x) = \theta,$$

that is, $(-x) = -x$ is an additive inverse. Suppose $y \in V$ is also an additive inverse of $x$; that is, $x + y = \theta$. Then $(-x) + x + y = (-x)$ and hence, $y = -x$.

Also observe that $\lambda x + \mu y \in V$ for $x, y \in V$ and $\lambda, \mu \in \mathbf{R}$ follows from (1) and (5).

***Theorem 55.1***

$\mathbf{R}^n$ with addition and scalar multiplication defined as in (55.2) and (55.3) is a *real vector space.*

The proof is left for the exercises (exercise 1).

Examples of vector spaces abound in mathematics.

---

▲ *Example 1*

$\mathbf{R}$ is a real vector space.

▲ *Example 2*

The space of continuous functions on $[a, b]$, with addition and scalar multiplication defined by

$$(f + g)(x) = f(x) + g(x), \qquad (\lambda f)(x) = \lambda f(x),$$

is a real vector space. (See Theorem 27.2.)

▲ *Example 3*

The space of Riemann-integrable functions on $[a, b]$, with addition and scalar multiplication defined as in example 2, is a real vector space. (See Theorem 50.1.)

▲ *Example 4*

The space of bounded functions on $[a, b]$, with addition and scalar multiplication defined as in example 2, is a vector space. Note, that if $|f(x)| \leqslant M$, $|g(x)| \leqslant N$ for all $x \in [a, b]$, then $|f(x) + g(x)| \leqslant |f(x)| + |g(x)| \leqslant M + N$, $|\lambda f(x)| \leqslant |\lambda| M$ for all $x \in [a, b]$.

▲ *Example 5*

The space of differentiable functions on $[a, b]$, with addition and multiplication defined as in example 2, is a real vector space. (See exercise 3, section 3.35.)

---

The vectors

(55.5) $\quad e_1 = (1, 0, \ldots, 0), \quad e_2 = (0, 1, 0, \ldots, 0), \quad \ldots, \quad e_n = (0, 0, \ldots, 0, 1) \in \mathbf{R}^n$

are called *fundamental vectors*. They form a *basis* for $\mathbf{R}^n$ in the following sense: In view of (55.2) and (55.3), every vector $x = (\xi_1, \xi_2, \ldots, \xi_n) \in \mathbf{R}^n$ may be written uniquely as a linear combination of the fundamental vectors as follows:

$$\begin{aligned} x = (\xi_1, \xi_2, \ldots, \xi_n) &= \xi_1(1, 0, \ldots, 0) + \xi_2(0, 1, 0, \ldots, 0) + \cdots + \xi_n(0, 0, \ldots, 0, 1) \\ &= \xi_1 e_1 + \xi_2 e_2 + \cdots + \xi_n e_n. \end{aligned}$$

## 5.55 Exercises

1. Prove Theorem 55.1.
2. Prove: $x + y \in V$ and $\lambda x \in V$ for all $x, y \in V$ and all $\lambda \in \mathbf{R}$ if and only if $\lambda x + \mu y \in V$ for all $x, y \in V$ and all $\lambda, \mu \in \mathbf{R}$.
3. Given $x = (1, 0, -3, \pi, \sqrt{7})$, $y = (0, 0, 5, \frac{1}{8}, -27) \in \mathbf{R}^5$. Find $2x - 3y$.
4. Prove that the space in example 2 is a vector space.
5. Prove that the space in example 3 is a vector space.
6. Prove that the space in example 5 is a vector space.
7. Given $u_1 = (1, 0, 1)$, $\quad u_2 = (1, 1, 1)$, $\quad u_3 = (0, 1, 1)$. Show: Every $x \in \mathbf{R}^3$ can be written as a linear combination of $u_1, u_2, u_3$.

## 5.56 Dot Product

Scalar multiplication, as defined in (55.3) is a function from $\mathbf{R} \times \mathbf{R}^n$ into $\mathbf{R}^n$ (or $\mathbf{R} \times V$ into $V$), as it associates with each element $(\lambda, x) \in \mathbf{R} \times \mathbf{R}^n$ a unique element $\lambda x \in \mathbf{R}^n$. We shall now define a different type of multiplication of elements of $\mathbf{R}^n$ as a function from $\mathbf{R}^n \times \mathbf{R}^n$ to $\mathbf{R}$ by:

$$x \cdot y = \xi_1 \eta_1 + \xi_2 \eta_2 + \cdots + \xi_n \eta_n,$$

where $x = (\xi_1, \xi_2, \dots, \xi_n)$, $y = (\eta_1, \eta_2, \dots, \eta_n)$, and call it *dot multiplication.* $x \cdot y$ is called the *dot product* (or *inner product*) of $x$ and $y$.

***Definition 56.1***

If $V$ is a real vector space, then a function from $V \times V$ into $\mathbf{R}$ with values $x \cdot y \in \mathbf{R}$ for $x, y \in V$, is called a *dot product* (or *inner product*) if and only if:

1. $x \cdot x > 0$ if $x \neq \theta$,
2. $\lambda(x \cdot y) = (\lambda x) \cdot y = x \cdot (\lambda y)$ for all $x, y \in V$, $\lambda \in \mathbf{R}$,
3. $x \cdot y = y \cdot x$. (Dot multiplication is *commutative.*)
4. $(x + y) \cdot z = x \cdot z + y \cdot z$. (Dot multiplication is *distributive.*)

Note that (1) and (2) imply that $x \cdot x = 0$ if and only if $x = \theta$.

***Theorem 56.1***

The function from $\mathbf{R}^n \times \mathbf{R}^n$ into $\mathbf{R}$ with values

$$x \cdot y = \xi_1\eta_1 + \xi_2\eta_2 + \cdots + \xi_n\eta_n$$

is a dot product, by definition 56.1.

*Proof*

1. $x \cdot x = \xi_1^2 + \xi_2^2 + \cdots + \xi_n^2 > 0$ if $x \neq \theta$; i.e. at least one $\xi_j \neq 0$, for some $j \in \{1, 2, \dots, n\}$.
2. $\lambda(x \cdot y) = \lambda(\xi_1\eta_1 + \xi_2\eta_2 + \cdots + \xi_n\eta_n) = (\lambda\xi_1)\eta_1 + (\lambda\xi_2)\eta_2 + \cdots + (\lambda\xi_n)\eta_n$
   $= \xi_1(\lambda\eta_1) + \xi_2(\lambda\eta_2) + \cdots + \xi_n(\lambda\eta_n)$.
3. $x \cdot y = \xi_1\eta_1 + \xi_2\eta_2 + \cdots + \xi_n\eta_n = \eta_1\xi_1 + \eta_2\xi_2 + \cdots + \eta_n\xi_n = y \cdot x$.
4. $(x + y) \cdot z = (\xi_1 + \eta_1, \xi_2 + \eta_2, \dots, \xi_n + \eta_n) \cdot (\zeta_1, \zeta_2, \dots, \zeta_n)$
   $= \xi_1\zeta_1 + \xi_2\zeta_2 + \cdots + \xi_n\zeta_n + \eta_1\zeta_1 + \eta_2\zeta_2 + \cdots + \eta_n\zeta_n$
   $= x \cdot z + y \cdot z$.

The following fundamental inequality is indispensable for our future investigations.

***Theorem 56.2***

If $V$ is a real vector space with dot product, then

$$|x \cdot y| \leqslant \sqrt{(x \cdot x)(y \cdot y)}. \tag{56.1}$$

The equal sign holds if and only if $\alpha x + \beta y = \theta$ for some $\alpha, \beta \in \mathbf{R}$, $(\alpha, \beta) \neq (0, 0)$. (56.1) is called the *Cauchy–Buniakovskii–Schwarz*† *inequality, CBS inequality* for short.

†*Victor Buniakovskii*, 1804–1889; *Hermann Amandus Schwarz*, 1843–1921.

*Proof*

We obtain from (1) and (2) of definition 56.1, that

$$(56.2) \qquad (\lambda x - \mu y) \cdot (\lambda x - \mu y) \geqslant 0$$

for any $x, y \in V$ and $\lambda, \mu \in \mathbf{R}$. By (2), (3), and (4) of definition 56.1,

$$(56.3) \qquad (\lambda x - \mu y) \cdot (\lambda x - \mu y) = \lambda^2 (x \cdot x) - 2\lambda\mu(x \cdot y) + \mu^2 (y \cdot y).$$

Let $\lambda = \sqrt{y \cdot y}$, $\mu = \sqrt{x \cdot x} \in \mathbf{R}$. Then, we obtain, from (56.2) and (56.3), that

$$(y \cdot y)(x \cdot x) - 2\sqrt{(x \cdot x)(y \cdot y)}\,(x \cdot y) + (x \cdot x)(y \cdot y) \geqslant 0.$$

If $x = \theta$ and/or $y = \theta$, then (56.1) is trivially satisfied. Hence, we may assume that $x \neq \theta$ and $y \neq \theta$ and divide by $\sqrt{(x \cdot x)(y \cdot y)}$ to obtain

$$x \cdot y \leqslant \sqrt{(x \cdot x)(y \cdot y)}.$$

If we let $\mu = -\sqrt{x \cdot x}$, we obtain

$$-x \cdot y \leqslant \sqrt{(x \cdot x)(y \cdot y)}.$$

(56.1) follows readily.

Since we have equality in (56.2) if and only if $\lambda x - \mu y = \theta$, the same is true for (56.1).

---

▲ *Example 1*

Ordinary multiplication in $\mathbf{R}$ is a dot multiplication. Since any two real numbers are scalar multiples of each other, the CBS inequality reduces to the trivial statement $|xy| = |x|\,|y|$.

▲ *Example 2*

We define a dot product on the space of continuous functions of example 2, section 5.55 by

$$(56.4) \qquad f \cdot g = \int_a^b fg.$$

We note that the dot product exists because the product of continuous functions is continuous (see section 2.27) and hence, integrable. Let us check the four conditions of definition 56.1:

1. $f \cdot f = \int_a^b f^2 > 0$ if $f$ is not the zero function.
2. $\lambda(f \cdot g) = \lambda \int_a^b fg = \int_a^b (\lambda f)g = \int_a^b f(\lambda g)$.
3. $f \cdot g = \int_a^b fg = \int_a^b gf = g \cdot f$.
4. $(f + g) \cdot h = \int_a^b (f + g)h = \int_a^b (fh + gh) = \int_a^b fh + \int_a^b gh = f \cdot h + g \cdot h$.

Since (56.4) is a dot product, the CBS inequality holds:

$$\left|\int_a^b fg\right| \leqslant \sqrt{\int_a^b f^2}\sqrt{\int_a^b g^2}. \tag{56.5}$$

▲ *Example 3*

We may define the same dot product as in (56.4) in the vector space of Riemann-integrable functions (see example 3, section 5.55), with the understanding that all Riemann-integrable functions that are zero a.e. are to be considered as the zero function. Consequently, two Riemann-integrable functions that are equal a.e. have to be considered as the same function. The CBS inequality (56.5) holds for Riemann-integrable functions as well.

---

## 5.56 Exercises

1. Show by using definition 56.1 only that $x \cdot x = 0$ if and only if $x = \theta$.
2. Show that (56.4) defines a dot product in the space of Riemann-integrable functions on $[a, b]$.
3. Show: If $f$ is Riemann-integrable on $[a, b]$, then

$$\left|\int_a^b f\right| \leqslant \sqrt{\int_a^b f^2}\sqrt{(b-a)}.$$

4. Use the CBS inequality to show that

$$|\xi_j| \leqslant \sqrt{x \cdot x}, \qquad \text{where } x = (\xi_1, \xi_2, \ldots, \xi_n),$$

for all $j = 1, 2, 3, \ldots, n$.

5. Show: $(x + y) \cdot (x + y) \leqslant (\sqrt{x \cdot x} + \sqrt{y \cdot y})^2$.
6. Prove: If $a_1, a_2, \ldots, a_n \in \mathbf{R}$, then

$$\frac{\left(\sum_{i=1}^n a_i\right)^2}{\sum_{i=1}^n a_i^2} \leqslant n.$$

## 5.57 Norm

The length $\|x\|$ of the vector $x = (\xi_1, \xi_2, \xi_3) \in \mathbf{R}^3$ is given by

$$\|x\| = \sqrt{\xi_1^2 + \xi_2^2 + \xi_3^3}$$

(see Fig. 57.1).

In an obvious generalization to $\mathbf{R}^n$, we call

$$\|x\| = \sqrt{\xi_1^2 + \xi_2^2 + \cdots + \xi_n^2} = \sqrt{x \cdot x} \tag{57.1}$$

the length or *norm* of $x \in \mathbf{R}^n$.

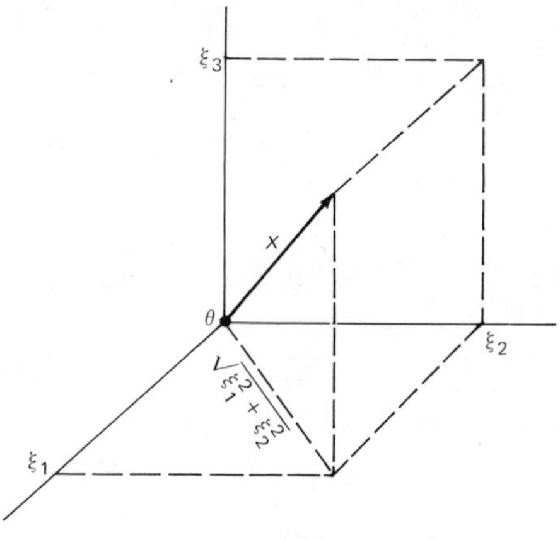


*Fig. 57.1*

***Definition 57.1***

If $V$ is a real vector space, then a function from $V$ into $\mathbf{R}$, with values $\|x\|$ for $x \in V$, is called a *norm* in $V$ if and only if:

1. $\|x\| > 0$ if $x \neq \theta$.
2. $\|\lambda x\| = |\lambda| \, \|x\|$ for all $x \in V$, $\lambda \in \mathbf{R}$.
3. $\|x + y\| \leqslant \|x\| + \|y\|$. (*Triangle inequality.*)

A vector space in which a norm is defined is called a *normed vector space.*

Note that (1) and (2) imply that $\|x\| = 0$ if and only if $x = \theta$. (See exercise 1.)

***Lemma 57.1***

If $\| x \|$ denotes the norm of $x \in V$, a real vector space, then

$$| \|x\| - \|y\| | \leqslant \|x \pm y\|. \tag{57.2}$$

*Proof*

Let x = a + b, y = −b. Then, by (3) of definition 57.1,

$$\|a + b\| \leqslant \|a\| + \|b\|.$$

By (2) of definition 57.1, $\| -y\| = \|y\|$, and hence,

$$\|x\| \leqslant \|x + y\| + \|y\|.$$

(57.2) follows readily. (For the case $x - y$, see exercise 2.)

That a function such as the one defined in (57.1) is a norm follows from the following general theorem.

***Theorem 57.1***

If $V$ is a real vector space with dot product, then the function from $V$ into $\mathbf{R}$ defined by $\|x\| = \sqrt{x \cdot x}$ is a norm.

*Proof*

That (1) of definition 57.1 is satisfied follows immediately from (1) of definition 56.1.

2. $\|\lambda x\| = \sqrt{(\lambda x) \cdot (\lambda x)} = \sqrt{\lambda^2 (x \cdot x)} = |\lambda| \sqrt{x \cdot x} = |\lambda| \, \|x\|.$
3. We obtain, by means of the CBS inequality,

$$\begin{aligned} \|x + y\|^2 &= (x + y) \cdot (x + y) = x \cdot x + 2(x \cdot y) + y \cdot y \\ &\leqslant x \cdot x + 2\sqrt{(x \cdot x)(y \cdot y)} + y \cdot y \\ &= \|x\|^2 + 2\|x\| \, \|y\| + \|y\|^2 = (\|x\| + \|y\|)^2, \end{aligned}$$

and (3) follows readily.

***Corollary to Theorem 57.1***

The function from $\mathbf{R}^n$ to $\mathbf{R}$, defined by

$$\|x\| = \sqrt{\xi_1^2 + \xi_2^2 + \cdots + \xi_n^2},$$

is a norm in $\mathbf{R}^n$ and is called the *Euclidean*† *norm.*

---

▲ *Example 1*

Since ordinary multiplication in $\mathbf{R}$ is a dot multiplication in $\mathbf{R}$ (example 1, section 5.56), the absolute-value function is a norm in $\mathbf{R}$;

$$\|x\| = \sqrt{xx} = \sqrt{x^2} = |x|, \qquad x \in \mathbf{R}.$$

▲ *Example 2*

In the space of continuous functions on $[a, b]$, we may define a norm by:

$$\|f\| = \sqrt{f \cdot f} = \sqrt{\int_a^b f^2}$$

(see example 2, section 5.56). The same represents a norm in the space of Riemann-integrable functions on $[a, b]$ (see example 3, section 5.56).

---

We may now state the *CBS inequality* (56.1) in terms of the norm as follows:

(57.3) $$|x \cdot y| \leqslant \|x\| \, \|y\|.$$

† Εὐκλείδης, ca. 333–257 B.C.

For $n = 2, 3$, the CBS inequality simply states that $|\cos \alpha| \leqslant 1$ for all $\alpha \in \mathbf{R}$: If $\alpha$ is the angle spanned by two vectors $x, y \in \mathbf{R}^2$, then:

$$\cos \alpha = \frac{\xi_1 \eta_1 + \xi_2 \eta_2}{\sqrt{\xi_1^2 + \xi_2^2}\sqrt{\eta_1^2 + \eta_2^2}} = \frac{x \cdot y}{\|x\| \, \|y\|}.$$

(An analogous formula holds for $n = 3$).

A norm need not be defined in terms of a dot product, as the following examples show:

---

▲ *Example 3*

For $x \in \mathbf{R}^n$, let $\|x\|_1 = |\xi_1| + |\xi_2| + \cdots + |\xi_n|$. It is quite obvious that conditions (1) and (2) of definition 57.1 are satisfied. As for (3), we note that

$$\begin{aligned} \|x + y\|_1 &= |\xi_1 + \eta_1| + |\xi_2 + \eta_2| + \cdots + |\xi_n + \eta_n| \\ &\leqslant |\xi_1| + |\eta_1| + |\xi_2| + |\eta_2| + \cdots + |\xi_n| + |\eta_n| \\ &= \|x\|_1 + \|y\|_1. \end{aligned}$$

▲ *Example 4*

For $x \in \mathbf{R}^n$ let

$$\|x\|_\infty = \max_{j \in \{1, 2, \ldots, n\}} |\xi_j|.$$

Again, it is easy to see that (1), (2) of definition 57.1 are satisfied. In regard to (3), we note that

$$\begin{aligned} |\xi_j + \eta_j| \leqslant |\xi_j| + |\eta_j| &\leqslant \max_{j \in \{1, 2, \ldots, n\}} |\xi_j| + \max_{j \in \{1, 2, \ldots, n\}} |\eta_j| \\ &= \|x\|_\infty + \|y\|_\infty. \end{aligned}$$

Hence,

$$\|x + y\|_\infty = \max_{j \in \{1, 2, \ldots, n\}} |\xi_j + \eta_j| \leqslant \|x\|_\infty + \|y\|_\infty.$$

▲ *Example 5*

For all continuous functions $f$ on $[a, b]$ let $\|f\| = \max_{[a,b]} |f(x)|$. The method of example 4 can be modified to show that this is a norm, the so-called *maximum norm* or *Chebyshev*† *norm*.

---

## 5.57 Exercises

1. Show, using definition 57.1 only that $\|x\| = 0$ if and only if $x = \theta$.
2. Show: $\|x\| - \|y\| \leqslant \|x - y\|$, $x, y \in V$, where $V$ is a real vector space.

†*Pafnuti L. Chebyshev*, 1821–1894.

3. Show without reference to Theorem 57.1, that $\|x\| = \sqrt{\xi_1^2 + \xi_2^2 + \xi_3^2}$ is a norm in $\mathbf{R}^3$.
4. Show:

$$\sqrt{\int_a^b (f+g)^2} \leqslant \sqrt{\int_a^b f^2} + \sqrt{\int_a^b g^2}$$

for all Riemann-integrable functions $f$, $g$.
5. Check conditions (1) and (2) of definition 57.1 for the norm that is defined in example 3.
6. Same as in exercise 5 for the norm in example 4.
7. Show that $\|f\| = \max_{[a,b]}|f(x)|$ is a norm in the real vector space of continuous functions on $[a, b]$.
8. Show that $\|f\| = \sup_{[a,b]}|f(x)|$ is a norm in the real vector space of bounded functions on $[a, b]$.
9. Show: for all $x, y \in \mathbf{R}^n$, $\|x+y\|^2 + \|x-y\|^2 = 2(\|x\|^2 + \|y\|^2)$.

## 5.58 The Euclidean $n$-Space

One may define the distance between two points $x, y \in \mathbf{R}^n$ as follows:

(58.1) $d(x, y) = \|x - y\| = \sqrt{(\xi_1 - \eta_1)^2 + (\xi_2 - \eta_2)^2 + \cdots + (\xi_n - \eta_n)^2}$.

A respectable distance satisfies certain fundamental requirements, which we shall list in the following definition:

***Definition 58.1***

If $X$ is a nonempty set, then a function $d\colon X \times X \to \mathbf{R}$ with values $d(x, y)$ for $x, y \in X$, is called a *distance in* $X$, if and only if:

1. $d(x, y) > 0$ if $x \neq y$.
2. $d(x, x) = 0$.
3. $d(x, y) = d(y, x)$. (The distance is *symmetric.*)
4. $d(x, y) \leqslant d(x, z) + d(z, y)$. (*Triangle inequality.*)

A space $X$ in which a distance is defined is called a *metric space.*

Every normed vector space can be turned into a metric space, as we shall demonstrate in the following theorem.

***Theorem 58.1***

If $V$ is a normed vector space, then $d(x, y) = \|x - y\|$, $x, y \in V$, is a distance in $V$.

*Proof*

1. If $x \neq y$, then $x - y \neq \theta$, and hence, by (1) of definition 57.1, $d(x, y) = \|x - y\| > 0$.

2. $d(x, x) = \|x - x\| = \|\theta\| = 0$.
3. $d(x, y) = \|x - y\| = |-1| \, \|x - y\| = \|y - x\| = d(y, x)$, by (2) of definition 57.1.
4. We obtain, from (3) of definition 57.1, that:

$$\begin{aligned} d(x, y) &= \|x - y\| = \|(x - z) + (z - y)\| \leqslant \|x - z\| + \|z - y\| \\ &= d(x, z) + d(z, y). \end{aligned}$$

Since $\|x\| = \sqrt{\xi_1^2 + \xi_2^2 + \cdots + \xi_n^2}$ is a norm in $\mathbf{R}^n$, we have, from Theorem 58.1, that $d(x, y)$ as defined in (58.1) is a distance in $\mathbf{R}^n$. We call (58.1) the *Euclidean distance.*

***Definition 58.2***

The metric space that is obtained from $\mathbf{R}^n$ by introduction of the Euclidean distance (58.1) is called the *Euclidean n-space* and denoted by $\mathbf{E}^n$.

At several later occasions, we shall consider Cartesian products of Euclidean spaces such as

$$\mathbf{E}^n \times \mathbf{E}^m = \{(x, y) \mid x \in \mathbf{E}^n, y \in \mathbf{E}^m\}.$$

We introduce a Euclidean distance in $\mathbf{E}^n \times \mathbf{E}^m$ by means of

$$d((x_1, y_1), (x_2, y_2)) = \sqrt{\|x_1 - x_2\|^2 + \|y_1 - y_2\|^2}, \tag{58.2}$$

and identify $\mathbf{E}^n \times \mathbf{E}^m$ with the Euclidean space $\mathbf{E}^{n+m}$. Note that $d((x_1, \theta), (x_2, \theta)) = \|x_1 - x_2\|$ and $d((\theta, y_1), (\theta, y_2)) = \|y_1 - y_2\|$ are the norms of $x_1 - x_2$ in $\mathbf{E}^n$ and of $y_1 - y_2$ in $\mathbf{E}^m$, respectively.

A distance need not be defined in terms of a norm and, as a matter of fact, may be defined on spaces other than vector spaces, as the following examples show.

---

▲ *Example 1*

Let A denote the set of all apples in the state of Washington, and define a distance in $A$ as follows:

$$d(x, y) = \begin{cases} 0 & \text{if } x = y, \\ 1 & \text{if } x \neq y. \end{cases}$$

This is indeed a distance:

1. If $x \neq y$, then $d(x, y) = 1 > 0$.
2. If $x = y$, then $d(x, y) = 0$.
3. $d(x, y) = d(y, x)$, because if $x = y$, both are 0, and if $x \neq y$, then both are 1.

4. Given $x, y, z \in A$. Since $z$ cannot be equal to $x$ and $y$ at the same time, unless $x = y$, we have

$$d(x, y) \leqslant d(x, z) + d(z, y).$$

$A$ is certainly *not* a vector space because the sum of two apples is not an apple.

▲ *Example 2*

Let P denote the set of all people on earth with the exception of lost tribes. Let $x_i \in P$. We call $\{x_1, x_2, \dots, x_n\}$ a chain in $P$ if $x_1$ is acquainted with $x_2$, $x_2$ is acquainted with $x_3, \dots, x_{n-1}$ is acquainted with $x_n$. If we define

$$d(x, y) = -1 + \text{number of elements in the shortest chain from } x \text{ to } y,$$

then $d(x, y)$ is a distance in $P$:

1. If $x \neq y$, then the shortest possible chain from $x$ to $y$ is $\{x, y\}$, that is, $d(x, y) \geqslant 1 > 0$.
2. If $x = y$, then the shortest chain from $x$ to $x$ is $\{x\}$ (under the presumption that $x$ knows himself), and we have $d(x, x) = 0$.
3. Being acquainted is a symmetric relationship. Hence, if $\{x_1, x_2, \dots, x_n\}$ is a chain from $x_1$ to $x_n$, then $\{x_n, \dots, x_2, x_1\}$ is a chain from $x_n$ to $x_1$. Hence, $d(x, y) = d(y, x)$.
4. If $d(x, y) > d(x, z) + d(z, y)$, then $d(x, y)$ could not be the distance from $x$ to $y$ because a shorter chain via $z$ exists. Hence, $d(x, y) \leqslant d(x, z) + d(z, y)$.

Again, we have succeeded in defining a distance on a space that is not a vector space. Leo Moser† conjectured that the maximum distance between any two people is 6 and, somewhat facetiously, that the maximum distance between any two mathematicians is two—or else at least one of them is not really a mathematician.

---

Summarizing the results of this and previous sections we may state:

A vector space with dot product can always be normed by $\|x\| = \sqrt{x \cdot x}$, but a norm need not be defined in terms of a dot product. (See examples 3 and 4, section 5.57.)

A normed vector space is a metric space with $d(x, y) = \|x - y\|$, but a metric space need not be a normed vector space, or, for that matter, a vector space. (See examples 1 and 2.)

A vector space where the distance is defined in terms of a norm, and the norm, in turn, is defined in terms of a dot product, is called a *pre-Hilbert‡ space*.

The Euclidean $n$-space $\mathbf{E}^n$ is a pre-Hilbert space.

†*Leo Moser*, 1922–1970.
‡*David Hilbert*, 1862–1943.

## 5.58 Exercises

1. Show that

$$d(x, y) = \max_{j \in \{1, 2, \ldots, n\}} |\xi_j - \eta_j|, \qquad x, y \in \mathbf{R}^n,$$

is a distance.

2. Same as in exercise 1, for $d(f, g) = \max_{[a,b]} |f(x) - g(x)|$ where $f$, $g$ belong to the space of continuous functions on $[a, b]$.
3. Show: The apples in the space $A$ (example 1) cannot really get very close to each other.
4. Interpret geometrically: $D = [x \in \mathbf{E}^2 \mid d(\theta, x) < 1]$.
5. Same as in exercise 4, for $S = [x \in \mathbf{E}^3 \mid d(\theta, x) < 1]$.
6. Same as in exercise 4, for $B = [x \in A \mid d(a, x) < 1]$, where $A$ is the apple set of example 1, and where $a$ is a given apple in $A$.
7. Show: $d(x, y) = d(x, z) + d(z, y)$, $x, y \in \mathbf{E}^n$ if and only if $z = (1 - t)x + ty$, $t \in [0, 1]$.
8. Show that $d(f, g) = \int_0^1 |f - g|$ defines a metric in the space of continuous functions on $[0, 1]$.

## 5.59 Open and Closed Sets in $\mathbf{E}^n$

In section 2.23, we have introduced the concepts of $\delta$-neighborhood and deleted $\delta$-neighborhood. These concepts have a straightforward generalization in $\mathbf{E}^n$.

***Definition 59.1***

If $a \in \mathbf{E}^n$ and $\delta > 0$, then $N_\delta(a) = \{x \in \mathbf{E}^n \mid \|x - a\| < \delta\}$ is called a *$\delta$-neighborhood of $a$* (or an *open ball* with center $a$ and radius $\delta$), and $N'_\delta(a) = \{x \in \mathbf{E}^n \mid 0 < \|x - a\| < \delta\}$ is called a *deleted $\delta$-neighborhood of $a$*.

Clearly, $N'_\delta(a) = N_\delta(a) \backslash \{a\}$.

As in sections 2.23 and 2.28, we define *accumulation point*, *interior point*, and *isolated point*, as follows:

***Definition 59.2***

$a \in \mathbf{E}^n$ is an *accumulation point* of the set $A \subseteq \mathbf{E}^n$ if and only if, for every $\delta > 0$, $N'_\delta(a)$ contains at least one point of $A$. The set of all accumulation points of $A$ is called the *derived set* of $A$ and is denoted by $A'$.

$a \in \mathbf{E}^n$ is an *interior point* of the set $A \subseteq \mathbf{E}^n$ if and only if there exists a $\delta > 0$ such that $N_\delta(a) \subseteq A$.

$a \in A \subseteq \mathbf{E}^n$ is an *isolated point* of $A$ if and only if there is a $\delta > 0$ such that $N'_\delta(a) \cap A = \varnothing$.

Every point of $A$ is either an accumulation point of $A$ or an isolated point of $A$.

***Definition 59.3***

The set $\mathcal{O} \subseteq \mathbf{E}^n$ is *open* if and only if all $x \in \mathcal{O}$ are *interior points.*

The set $\mathcal{C} \subseteq \mathbf{E}^n$ is *closed*, if and only if $c(\mathcal{C}) = \mathbf{E}^n \backslash \mathcal{C}$ is open.

***Lemma 59.1***

$N_\delta(a)$ is an open set and $\overline{N_\delta(a)} = \{x \in \mathbf{E}^n \mid \|x - a\| \leqslant \delta\}$ is a closed set.

*Proof*

We have to prove that every $x \in N_\delta(a)$ is an interior point. If $x \in N_\delta(a)$, then $\|x - a\| = \rho < \delta$. Choose $\delta_1 = (\delta - \rho)/2 > 0$. If $y \in N_{\delta_1}(x)$, then $y \in N_\delta(a)$, because

$$\|y - a\| \leqslant \|y - x\| + \|x - a\| < \delta_1 + \rho = \frac{\delta - \rho}{2} + \rho = \frac{\delta + \rho}{2} < \delta.$$

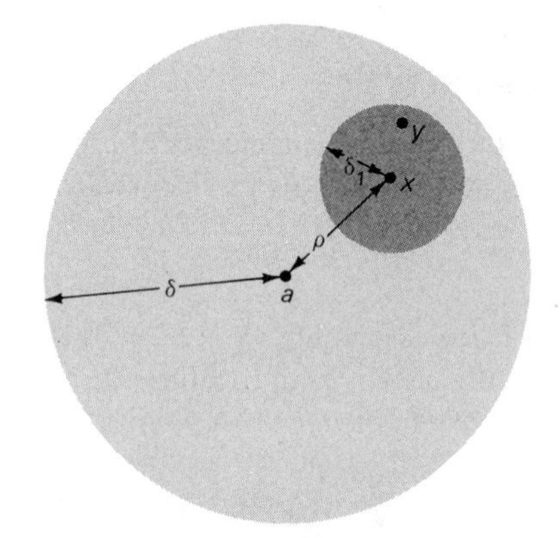


*Fig. 59.1*

(See Fig. 59.1.) Hence, $N_{\delta_1}(x) \subseteq N_\delta(a)$, that is, $x$ is an interior point of $N_\delta(a)$. Since this is true for all $x \in N_\delta(a)$, it follows that $N_\delta(a)$ is open.

In order to show that $\overline{N_\delta(a)}$ is closed, one has to show that $c(\overline{N_\delta(a)})$ is open. The details of the proof are left to the reader. (See Fig. 59.2.)

As in section 2.28, one can prove the following theorem merely by reinterpreting the symbols in the proof of Theorem 28.2.

***Theorem 59.1***

The set $A \subseteq \mathbf{E}^n$ is closed if and only if $A' \subseteq A$; that is, $A$ contains all its accumulation points.

As in section 2.29 we obtain

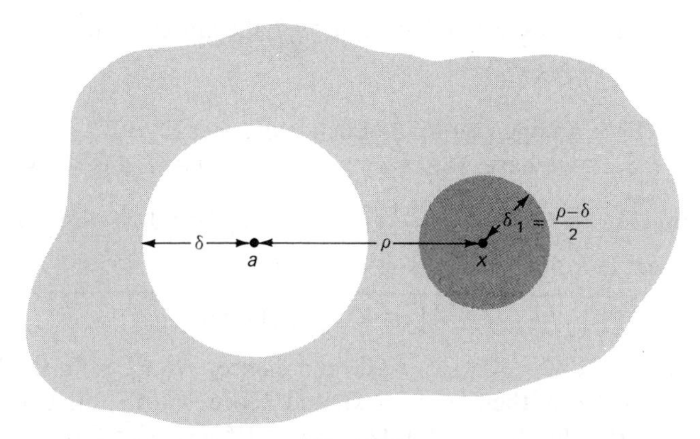


*Fig. 59.2*

***Theorem 59.2***

If $A$, $B$ are given index sets and $\{\Omega_\alpha \mid \alpha \in A\}$ is a collection of open subsets of $\mathbf{E}^n$ and $\{C_\beta \mid \beta \in B\}$ is a collection of closed subsets of $\mathbf{E}^n$, then:

$$\bigcup_{\alpha \in A} \Omega_\alpha \qquad \text{is open}$$

and

$$\bigcap_{\beta \in B} C_\beta \qquad \text{is closed.}$$

If $\Omega_1, \Omega_2, \ldots, \Omega_k \subseteq \mathbf{E}^n$ are open and $C_1, C_2, \ldots, C_l \subseteq \mathbf{E}^n$ are closed, then

$$\bigcap_{j=1}^{k} \Omega_j \qquad \text{is open}$$

and

$$\bigcup_{j=1}^{l} C_j \qquad \text{is closed.}$$

The following theorem on the representation of open sets is not only of theoretical interest, but also of practical importance, as we shall see in later developments.

***Theorem 59.3***

$\Omega \subseteq \mathbf{E}^n$ is open if and only if it may be represented as the union of $\delta$-neighborhoods; i.e., there is a collection $\{N_{\delta_\alpha}(\alpha) \mid \alpha \in A\}$ for some index set $A$ and some $\delta_\alpha > 0$ such that

$$\Omega = \bigcup_{\alpha \in A} N_{\delta_\alpha}(\alpha).$$

*Proof*

Since $\Omega$ is open, we have, for every $x \in \Omega$, a $\delta_x > 0$ such that $N_{\delta_\alpha}(x) \subseteq \Omega$. Hence,

$$\Omega \subseteq \bigcup_{x \in \Omega} N_{\delta_x}(x) \subseteq \Omega.$$

The converse is trivial since, by Theorem 59.2, the union of open sets, and hence of $\delta$-neighborhoods, is open.

If $\Omega \subseteq \mathbf{E}^n \times \mathbf{E}^m$ is an open set in the Euclidean space $\mathbf{E}^n \times \mathbf{E}^m$, with the distance defined as in (58.2), we have, for every point $(x_0, y_0) \in \Omega$, a $\delta > 0$ such that

$$N_\delta(x_0, y_0) = \{(x, y) \in \mathbf{E}^n \times \mathbf{E}^m \mid \|x - x_0\|^2 + \|y - y_0\|^2 < \delta^2\} \subseteq \Omega.$$

Let $y_1 \in \mathbf{E}^m$ denote a fixed vector and consider

$$\Omega_{y_1} = \{x \in \mathbf{E}^n \mid (x, y_1) \in \Omega\}.$$

Then, for any $x_0 \in \Omega_{y_1}$, $(x_0, y_1) \in \Omega$; and hence, $N_\delta(x_0, y_1) \in \Omega$. Therefore, whenever $\|x - x_0\| < \delta$, then $(x, y_1) \in \Omega$, and hence, $x \in \Omega_{y_1}$; that is, $\Omega_{y_1}$ is an open set. The same holds true for $\Omega_{x_1} = \{y \in \mathbf{E}^m \mid (x_1, y) \in \Omega\}$. (See also Fig.

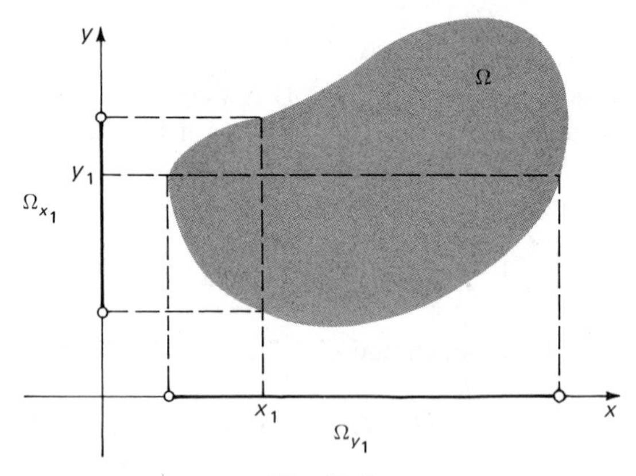


*Fig. 59.3*

59.3.) If there is no $x$ such that $(x, y_1) \in \Omega$, then $\Omega_{y_1} = \varnothing$ and is trivially open. The same is true for $\Omega_{x_1}$. Hence:

***Lemma 59.2***

If $\Omega \subseteq \mathbf{E}^n \times \mathbf{E}^m$ is open, then the sets $\Omega_{y_1} = \{x \in \mathbf{E}^n \mid (x, y_1) \in \Omega\}$, for fixed $y_1 \in \mathbf{E}^m$, and $\Omega_{x_1} = \{y \in \mathbf{E}^m \mid (x_1, y) \in \Omega\}$, for fixed $x_1 \in \mathbf{E}^n$, are open sets in $\mathbf{E}^n$ and $\mathbf{E}^m$, respectively.

If $\Omega_1 \subseteq \mathbf{E}^n$, $\Omega_2 \subseteq \mathbf{E}^m$ are open sets, then for $x_0 \in \Omega_1$, $y_0 \in \Omega_2$, there exist $\delta_1 > 0$, $\delta_2 > 0$ such that $N_{\delta_1}(x_0) \subseteq \Omega_1$, $N_{\delta_2}(y_0) \subseteq \Omega_2$. Hence, if $\delta = \min(\delta_1, \delta_2)$ and $(x, y) \in N_\delta(x_0, y_0)$, then $\|x - x_0\|^2 + \|y - y_0\|^2 < \delta^2$; that is, $x \in N_{\delta_1}(x_0)$, $y \in N_{\delta_2}(y_0)$; and, consequently, $(x, y) \in \Omega_1 \times \Omega_2$. Therefore, $\Omega_1 \times \Omega_2$ is open. (See also Fig. 59.4.)

***Lemma 59.3***

If $\Omega_1 \subseteq \mathbf{E}^n$, $\Omega_2 \subseteq \mathbf{E}^m$ are open, then $\Omega_1 \times \Omega_2$ is open in $\mathbf{E}^n \times \mathbf{E}^m$.

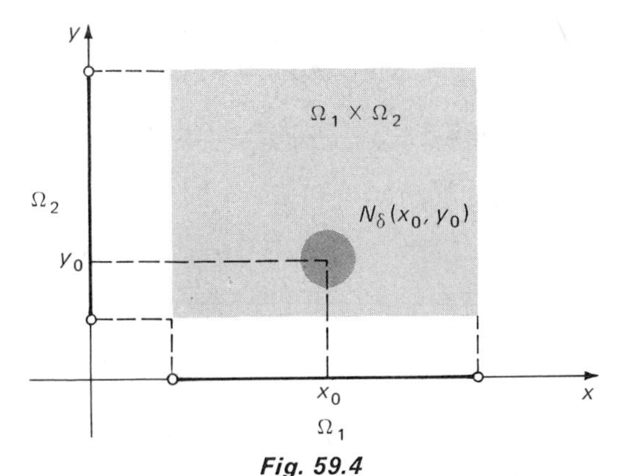


*Fig. 59.4*

## 5.59 Exercises

1. Show that an accumulation point need not be an interior point.
2. Prove: $a$ is an accumulation point of $S \subseteq \mathbf{E}^n$ if and only if every deleted $\delta$-neighborhood of $a$ contains infinitely many points from $S$.
3. Prove Theorem 59.1.
4. Prove Theorem 59.2.
5. Show: Every open set in $\mathbf{E}^2$ may be represented as the union of open rectangles, i.e., sets of the type $\{(\xi_1, \xi_2) \in \mathbf{E}^2 \mid a < \xi_1 < b, c < \xi_2 < d\}$.
6. Prove: A set $A \subseteq \mathbf{E}^2$ is open if and only if it can be represented as the union of open rectangles.
7. Prove that $\overline{N_\delta(a)}$ is closed.
8. Prove that $N'_\delta(a)$ is open.
9. Prove: Every open set $\Omega$ in $\mathbf{E}^n$ may be represented as the union of denumerably many $\delta$-neighborhoods.
10. Show: If $N_\delta(a) = \{x \in \mathbf{E}^n \mid \|x - a\| < \delta\}$, then the $n$-dimensional cube $C^n = \{x \in \mathbf{E}^n \mid \alpha_j - \delta/2\sqrt{n} < \xi_j < \alpha_j + \delta/2\sqrt{n}, \quad j = 1, 2, \ldots, n\}$, where $a = (\alpha_1, \alpha_2, \ldots, \alpha_n)$, is contained in $N_\delta(a)$. Show: If $C^n = \{x \in \mathbf{E}^n \mid \alpha_j - \delta < \xi_j < \alpha_j + \delta, j = 1, 2, \ldots, n\}$, where $a = (\alpha_1, \alpha_2, \ldots, \alpha_n)$, then $N_\delta(a) \subseteq C^n$.

## 5.60 The Nested-Interval Property in $\mathbf{E}^n$ and the Bolzano-Weierstrass Theorem

We have seen in section 2.21 that an ordered Archimedean field has the least-upper-bound property if and only if it has the nested-interval property. Either property characterizes a complete ordered field.

$\mathbf{E}^n$ is not an ordered field. It is not even a field, and hence, it is meaningless to talk about a least-upper-bound property of $\mathbf{E}^n$. However, it does make sense to talk about nested intervals in $\mathbf{E}^n$, and it turns out that $\mathbf{E}^n$ inherits the nested-interval property from $\mathbf{R}$, as we shall demonstrate in this section.

***Definition 60.1***

If $\alpha_j < \beta_j$ for all $j = 1, 2, \ldots, n$, where $\alpha_j, \beta_j \in \mathbf{R}$, then

(60.1) $$I = \{x \in \mathbf{E}^n \mid \alpha_j \leqslant \xi_j \leqslant \beta_j, \quad j = 1, 2, \ldots, n\}$$

is called a *closed interval* in $\mathbf{E}^n$, and

(60.2) $$J = \{x \in \mathbf{E}^n \mid \alpha_j < \xi_j < \beta_j, \quad j = 1, 2, \ldots, n\}$$

is called an *open interval* in $\mathbf{E}^n$.

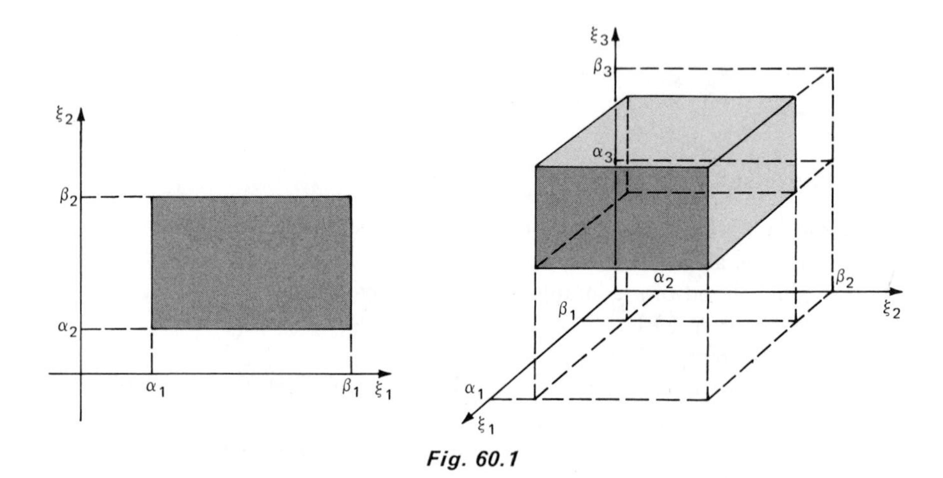


***Fig. 60.1***

An interval in $\mathbf{E}^2$ is a rectangle, and in $\mathbf{E}^3$ a right parallelepiped. (See Fig. 60.1.)

Clearly, a closed interval is a closed set and an open interval is an open set. (See exercises 3 and 4.)

(If, for at least one $j \in \{1, 2, \ldots, n\}$, $\alpha_j = \beta_j$ in (60.1), then we obtain an interval in $\mathbf{E}^m$ for $m < n$.)

***Theorem 60.1*** *(Nested-Interval Property of* $\mathbf{E}^n$*)*

If $\{I_k\}$ is a nested sequence of closed nonempty intervals in $\mathbf{E}^n$, that is, if

(60.3) $$I_1 \supseteq I_2 \supseteq I_3 \supseteq \cdots \supseteq I_k \supseteq \ldots,$$

$$\text{where } \mathfrak{J}_k = \{x \in \mathbf{E}^n \mid \alpha_j^{(k)} \leqslant \xi_j \leqslant \beta_j^{(k)}, j = 1, 2, \ldots, n\},$$

then there exists at least one point $x \in \mathbf{E}^n$ such that $x \in \mathfrak{J}_k$ for all $k \in \mathbf{N}$. If

$$\lim_{\infty} | \beta_j^{(k)} - \alpha_j^{(k)} | = 0$$

for all $j = 1, 2, \ldots, n$, then there is exactly one such point.

*Proof*

We note that (60.3) defines $n$ sequences of nonempty closed nested intervals in $\mathbf{R}$, namely,

$$[\alpha_j^{(1)}, \beta_j^{(1)}] \supseteq [\alpha_j^{(2)}, \beta_j^{(2)}] \supseteq \cdots \supseteq [\alpha_j^{(k)}, \beta_j^{(k)}] \supseteq \cdots$$

for $j = 1, 2, \ldots, n$. By Theorem 21.1, there are $n$ real numbers $\xi_j \in [\alpha_j^{(k)}, \beta_j^{(k)}]$ for all $k \in \mathbf{N}$, $j = 1, 2, \ldots, n$. Hence $x = (\xi_1, \xi_2, \ldots, \xi_n) \in \mathfrak{J}_k$ for all $k \in \mathbf{N}$. If

$$\lim_{\infty} (\beta_j^{(k)} - \alpha_j^{(k)}) = 0, \tag{60.4}$$

then, by the Corollary to Theorem 21.1, there is exactly one $\xi_j \in \mathbf{R}$ such that $\xi_j \in [\alpha_j^{(k)}, \beta_j^{(k)}]$ for all $k \in \mathbf{N}$. If (60.4) holds for all $j = 1, 2, \ldots, n$, then there is exactly one $x = (\xi_1, \xi_2, \ldots, \xi_n) \in \mathfrak{J}_k$ for all $k \in \mathbf{N}$.

In section 1.8 we called an ordered field complete if and only if it has the least-upper-bound property. In view of Theorem 21.1 we might just as well call an ordered Archimedean field complete if and only if it has the nested-interval property. In an obvious generalization of this concept, we call $\mathbf{E}^n$ *complete* because it has the nested-interval property.

As in section 2.30, we can now prove a generalized version of the Bolzano–Weierstrass theorem. For this purpose we need a precise definition of bounded sets in $\mathbf{E}^n$.

***Definition 60.2***

The set $X \subseteq \mathbf{E}^n$ is bounded if and only if there is an $M > 0$ such that $\| x \| \leqslant M$ for all $x \in X$.

***Theorem 60.2*** (*Bolzano–Weierstrass Theorem*)

If $A \subseteq \mathbf{E}^n$ contains infinitely many elements and is bounded, then $A$ has at least one accumulation point in $\mathbf{E}^n$ (that need not lie in $A$).

The proof follows the same pattern as the proof of Theorem 30.1. The only difference is that instead of splitting the interval at each step into two intervals, we have to split it now into $2^n$ intervals (see Fig. 60.2) due to the fact that the intervals are in $\mathbf{E}^n$ rather than in $\mathbf{R}$.

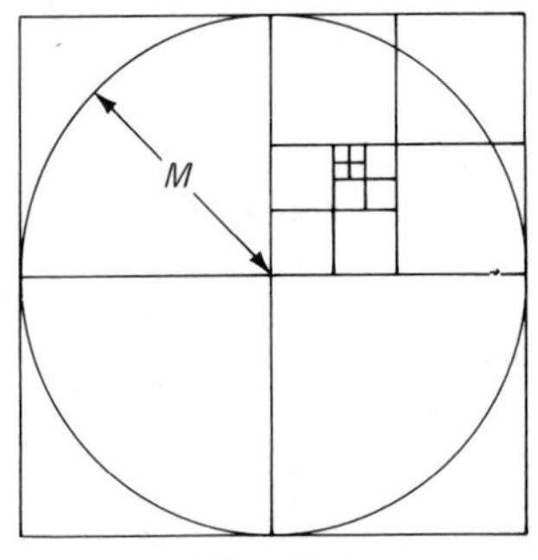


*Fig. 60.2*

## 5.60 Exercises

1. $A \subseteq \mathbf{E}^3$ is given by $\{(1, \frac{1}{2}, \frac{1}{3}), (\frac{1}{4}, \frac{1}{5}, \frac{1}{6}), (\frac{1}{7}, \frac{1}{8}, \frac{1}{9}), \ldots\}$,
   (a) Show that $A$ is bounded.
   (b) Find an accumulation point $a$ of $A$. Is $a \in A$?
2. Carry out the proof of Theorem 60.2 in detail.
3. Prove that a closed interval in $\mathbf{E}^n$ is a closed set.
4. Prove that an open interval in $\mathbf{E}^n$ is an open set.

## 5.61 Sequences in $\mathbf{E}^n$

In section 2.15 we defined $a$ to be the limit of a sequence $\{a_k\}$ of real numbers if and only if the distance from $a_k$ to $a$, namely $|a_k - a|$, can be made as small as one pleases by choosing $k \in \mathbf{N}$ sufficiently large. When stated in these words, the generalization to sequences in $\mathbf{E}^n$ is quite obvious:

***Definition 61.1***

The sequence $\{a_k\}$ of elements $a_k \in \mathbf{E}^n$ converges to the limit $a \in \mathbf{E}^n$,

$$\lim_{\infty} a_k = a,$$

if and only if for every $\varepsilon > 0$ there is an $N(\varepsilon) \in \mathbf{N}$ such that

$$d(a_k, a) = \|a_k - a\| < \varepsilon$$

for all $k > N(\varepsilon)$ or, equivalently, if $a_k \in N_\varepsilon(a)$ for all $k > N(\varepsilon)$. (See Fig. 61.1.)

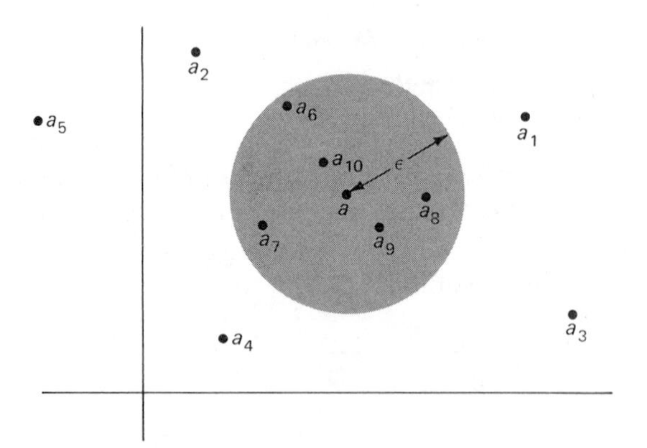


*Fig. 61.1*

Note that if, for every $\varepsilon > 0$, there is an $N(\varepsilon) \in \mathbf{N}$ such that $a_k \in N'_\varepsilon(a)$ for all $k > N(\varepsilon)$, then $\{a_k\}$ cannot be a constant sequence, i.e., a sequence where $a_k = a_{k+1} = a_{k+2} = \cdots$ for some $k$.

As in section 2.23, we can now prove the following generalized version of Theorem 23.1.

***Theorem 61.1***

$a \in \mathbf{E}^n$ is an accumulation point of $S \subseteq \mathbf{E}^n$ if and only if there exists a sequence $\{x_k\}$, $x_k \in S \backslash \{a\}$, such that $\lim_\infty x_k = a$.

In the next section we shall make use of the following stronger result:

***Theorem 61.2***

If $a \in \mathbf{E}^n$ is an accumulation point of the nonconstant sequence $\{a_k\}$, $a_k \in \mathbf{E}^n$, then there exists a subsequence of $\{a_k\}$ that converges to $a$.

*Proof*

Choose $a_{n_1} \in \{a_k\}$ such that $a_{n_1} \in N_1'(a)$. Next, choose $a_{n_2} \in \{a_{n_1+1}, a_{n_1+2}, \ldots\}$ such that $a_{n_2} \in N_{1/2}'(a)$, choose $a_{n_3} \in \{a_{n_2+1}, a_{n_2+2}, \ldots\}$ such that $a_{n_3} \in N_{1/3}'(a)$, etc. In general, choose $a_{n_k} \in \{a_{n_{k-1}+1}, a_{n_{k-1}+2}, \ldots\}$ such that $a_{n_k} \in N_{1/k}'(a)$. By Theorem 61.1, such a choice is always possible. By construction, $n_1 < n_2 < n_3 < \cdots$. Hence, $\{a_{n_k}\}$ is a subsequence of $\{a_k\}$. Since $a_{n_k} \in N_{1/k}'(a), \lim_{k\to\infty} a_{n_k} = a$.

Since an element $a_k \in \mathbf{E}^n$ is actually an $n$-tuple of real numbers $(\alpha_1^{(k)}, \alpha_2^{(k)}, \ldots, \alpha_n^{(k)})$, we see that a sequence in $\mathbf{E}^n$ is really a sequence of $n$-tuples of real numbers. It is reasonable to expect that such a sequence converges if and only if the $n$ sequences of components converge. This is indeed true, as we shall prove by means of the following lemma:

***Lemma 61.1***

If $x = (\xi_1, \xi_2, \ldots, \xi_n) \in \mathbf{E}^n$, then

$$|\xi_j| \leqslant \|x\| \leqslant \sqrt{n} \max (|\xi_1|, |\xi_2|, \ldots, |\xi_n|) \tag{61.1}$$

for all $j = 1, 2, \ldots, n$. (We shall henceforth refer to (61.1) as the *fundamental inequality* or *FIE*.)

*Proof*

We have, from the CBS inequality, that

$$|x \cdot e_j| \leqslant \|x\| \, \|e_j\|$$

for each fundamental vector $e_j$, $j = 1, 2, \ldots, n$. Since $\|e_j\| = 1$ and $x \cdot e_j = \xi_j$, $|\xi_j| \leqslant \|x\|$ follows. Since $|\xi_j| \leqslant \max (|\xi_1|, |\xi_2|, \ldots, |\xi_n|)$, we obtain, after squaring and summing from $j = 1$ to $j = n$:

$$\|x\|^2 = \sum_{j=1}^{n} |\xi_j|^2 \leqslant n \, [\max (|\xi_1|, |\xi_2|, \ldots, |\xi_n|)]^2,$$

from which the second part of the FIE follows readily.

***Theorem 61.3***

If $a_k = (\alpha_1^{(k)}, \alpha_2^{(k)}, \ldots, \alpha_n^{(k)}) \in \mathbf{E}^n$ and $a = (\alpha_1, \alpha_2, \ldots, \alpha_n) \in \mathbf{E}^n$, then $\lim_\infty a_k = a$ if and only if $\lim_\infty \alpha_j^{(k)} = \alpha_j$ for all $j = 1, 2, \ldots, n$.

*Proof*

(a) If $\lim_\infty a_k = a$, we have from definition 61.1 and from the FIE (61.1) that

$$|\alpha_j^{(k)} - \alpha_j| \leqslant \|a_k - a\| < \varepsilon \qquad \text{for } k > N(\varepsilon).$$

Hence, $\lim_\infty \alpha_j^{(k)} = \alpha_j$ for all $j = 1, 2, \ldots, n$.

(b) If $\lim_\infty \alpha_j^{(k)} = \alpha_j$ for all $j = 1, 2, \ldots, n$, we have, for every $\varepsilon > 0$, an $N_j(\varepsilon) \in \mathbf{N}$ such that $|\alpha_j^{(k)} - \alpha_j| < \varepsilon/\sqrt{n}$ for all $k > N_j(\varepsilon)$.

Let $N = \max(N_1(\varepsilon), N_2(\varepsilon), \ldots, N_n(\varepsilon))$. Then

$$|\alpha_j^{(k)} - \alpha_j| < \frac{\varepsilon}{\sqrt{n}} \qquad \text{for all } k > N(\varepsilon) \quad \text{and all } j = 1, 2, \ldots, n.$$

Hence, by the FIE,

$$\|a_k - a\| \leqslant \sqrt{n}\max\left(|\alpha_1^{(k)} - \alpha_1|, |\alpha_2^{(k)} - \alpha_2|, \ldots, |\alpha_n^{(k)} - \alpha_n|\right) < \varepsilon$$

for all $k > N(\varepsilon)$, and we see that $\lim_\infty a_k = a$.

Theorem 61.3 makes it abundantly clear that nothing can be learned from a study of sequences in $\mathbf{E}^n$ that cannot be learned from a study of sequences of real numbers. Theorem 61.3 implies, in particular, that the *limit of a sequence is unique*, that a *convergent sequence is bounded* (see definition 60.2), and that *any subsequence of a convergent sequence converges to the same limit*. (See section 2.16.)

---

▲ *Example 1*

The sequence $\{a_k\}$, where $a_k = (1/k, k/(k+1), (-1)^k/k^2) \in \mathbf{E}^3$ converges to the limit $a = (0, 1, 0)$ because $\lim_\infty 1/k = 0$, $\lim_\infty k/(k+1) = 1$, and $\lim_\infty (-1)^k/k^2 = 0$.

▲ *Example 2*

The sequence $\{a_k\}$, where $a_k = ((k+1)/k, k^2/(2k^2+1), (-1)^k, 1/(k+1)) \in \mathbf{E}^4$ does not converge because $\{(-1)^k\}$ does not converge.

---

In section 2.15 we promised to develop a criterion for the convergence of sequences that does not require the knowledge of the limit of the sequences. This will be accomplished in the next section.

## 5.61 Exercises

1. Prove: If $\{a_k\}$, $a_k \in \mathbf{E}^n$, converges, then the limit is unique.
2. Prove: If $\{a_k\}$, $a_k \in \mathbf{E}^n$, converges, then there is an $M > 0$ such that $\|a_k\| \leqslant M$ for all $k \in \mathbf{N}$.
3. Prove: If $\lim_\infty a_k = a$, $a_k, a \in \mathbf{E}^n$, then $\lim_\infty a_{k_j} = a$ where $k_1 < k_2 < k_3 < \cdots$ is a sequence of integers.
4. Show: If $\lim_\infty a_k = a$, then, for every $\varepsilon > 0$, there is an $N(\varepsilon) \in \mathbf{N}$ such that $\|a_k - a_l\| < \varepsilon$ for all $k, l > N(\varepsilon)$.
5. Prove Theorem 61.1.
6. Let $\| \ \|_1$, $\| \ \|_\infty$ denote the norms that were defined in examples 3 and 4, section 5.57. Show that $(1/n)\|x\|_1 \leqslant \|x\| \leqslant \sqrt{n}\,\|x\|_\infty \leqslant \sqrt{n}\,\|x\|$.
7. Show: $S \subseteq \mathbf{E}^n$ is open if and only if, for every $a \in S$, there is a $\delta > 0$ such that $\{x \in \mathbf{E}^n \mid \|x - a\|_1 < \delta\} \subseteq S$. Same for $\| \ \|_\infty$ instead of $\| \ \|_1$. (See exercise 6.)

## 5.62 Cauchy Sequences

It is intuitively clear that the elements of a convergent sequence crowd closer together as $k$ becomes large. This is indeed so. Let $\{a_k\}$ represent a sequence in $\mathbf{E}^n$ that converges to $a$. Then, by definition 61.1, we have, for every $\varepsilon > 0$, an $N(\varepsilon) \in \mathbf{N}$ such that $\|a_k - a\| < \varepsilon/2$ for all $k > N(\varepsilon)$. Hence, by the triangle inequality,

$$\|a_k - a_l\| \leqslant \|a_k - a\| + \|a - a_l\| < \varepsilon \qquad \text{for all } k, l > N(\varepsilon).$$

A sequence with this property is called a *Cauchy sequence*:

***Definition 62.1***

The sequence $\{a_k\}$, where $a_k \in \mathbf{E}^n$, is called a *Cauchy sequence* if and only if, for every $\varepsilon > 0$, there is an $N(\varepsilon) \in \mathbf{N}$ such that

$$\|a_k - a_l\| < \varepsilon \qquad \text{for all } k, l > N(\varepsilon), \tag{62.1}$$

or, equivalently,

$$\|a_{k+p} - a_k\| < \varepsilon \qquad \text{for all } k > N(\varepsilon) \text{ and all } p \in \mathbf{N}. \tag{62.2}$$

Clearly, every constant sequence is a Cauchy sequence and, as we have seen above, every convergent sequence is a Cauchy sequence:

***Lemma 62.1***

If $a_k, a \in \mathbf{E}^n$ and $\lim_\infty a_k = a$, then $\{a_k\}$ is a Cauchy sequence.

The question arises: is every Cauchy sequence in $\mathbf{E}^n$ a convergent sequence in the sense of definition 61.1? The answer is *yes*. We shall establish this fact in three stages. Since a constant sequence converges trivially, we may exclude constant sequences from our considerations without loss of generality.

***Lemma 62.2***

If $\{a_k\}$ is a Cauchy sequence in $\mathbf{E}^n$, then $\{a_k\}$ is bounded, i.e., there is an $M > 0$ such that $\|a_k\| \leqslant M$ for all $k \in \mathbf{N}$.

*Proof*

By definition 62.1, we have, for every $\varepsilon > 0$, and, in particular, for $\varepsilon = 1$, an $N(1)$ such that

$$\|a_{k+p} - a_k\| < 1 \qquad \text{for all } k > N(1) \text{ and all } p \in \mathbf{N}.$$

Hence, by Lemma 57.1,

$$\|a_{N(1)+1+p}\| - \|a_{N(1)+1}\| < 1,$$

or,

$$\|a_{N(1)+1+p}\| < 1 + \|a_{N(1)+1}\| \qquad \text{for all } p \in \mathbf{N}.$$

Let $M = \max\,(\|a_1\|, \|a_2\|, \ldots, \|a_{N(1)}\|, 1 + \|a_{N(1)+1}\|)$. Then $\|a_k\| \leqslant M$ for all $k \in \mathbf{N}$.

***Lemma 62.3***

If $\{a_k\}$ is a Cauchy sequence in $\mathbf{E}^n$ and is not a constant sequence, then $\{a_k\}$ has exactly one accumulation point.

*Proof*

By Lemma 62.2, $\{a_k\}$ is bounded. Hence, by the Bolzano–Weierstrass Theorem (Theorem 60.2), $\{a_k\}$ has at least one accumulation point $a \in \mathbf{E}^n$. Suppose $\{a_k\}$ has at least two accumulation points $a, b \in \mathbf{E}^n$, where $a \neq b$. Since $a, b$ are accumulation points, we have, by Theorem 61.1 for $\delta = \|a - b\| > 0$, two subsequences $\{a_{k_j}\}$ and $\{a_{l_j}\}$ such that $\lim_\infty a_{k_j} = a$, $\lim_\infty a_{l_j} = b$, and hence, $a_{k_j} \in N'_{\delta/3}(a)$ and $a_{l_j} \in N'_{\delta/3}(b)$ for $j > N_1(\delta)$ for some $N_1(\delta) \in \mathbf{N}$.

Since $\{a_k\}$ is a Cauchy sequence, there is an $N_2(\delta) \in \mathbf{N}$ such that

$$\|a_k - a_l\| < \frac{\delta}{3} \qquad \text{for all } k, l > N_2(\delta).$$

Hence,

$$\delta = \|a - b\| \leqslant \|a - a_{k_j}\| + \|a_{k_j} - a_{l_i}\| + \|a_{l_i} - b\| < \delta$$

for all $j$, $i > \max(N_1(\delta), N_2(\delta))$, that is, $\delta < \delta$, which is not possible. Hence, $a = b$.

***Lemma 62.4***

If $\{a_k\}$ is a Cauchy sequence in $\mathbf{E}^n$, then $\lim_\infty a_k = a$ for some $a \in \mathbf{E}^n$.

*Proof*

If $\{a_k\}$ is a constant sequence, then the lemma is trivial. We may therefore assume w.l.o.g. that $\{a_k\}$ is not a constant sequence. By Theorem 61.2 and Lemma 62.3, there is a subsequence $\{a_{n_k}\}$ of $\{a_k\}$ such that $\lim_{k\to\infty} a_{n_k} = a$. Hence, for every $\varepsilon > 0$, there is some $N_1(\varepsilon)$ such that $\|a_{n_k} - a\| < \varepsilon/2$ for all $k > N_1(\varepsilon)$. Since $\{a_k\}$ is a Cauchy sequence, there is some $N_2(\varepsilon)$ such that $\|a_k - a_{n_k}\| < \varepsilon/2$ for all $k > N_2(\varepsilon)$. (Note that $n_k \geqslant k$ for all $k \in \mathbf{N}$.) Hence,

$$\|a_k - a\| \leqslant \|a_k - a_{n_k}\| + \|a_{n_k} - a\| < \varepsilon$$

for all $k > N(\varepsilon) = \max(N_1(\varepsilon), N_2(\varepsilon))$; that is, $\lim_\infty a_k = a$.

Lemmas 62.1 and 62.4 together yield:

***Theorem 62.1*** *(Cauchy Criterion for Convergence)*

The sequence $\{a_k\}$ in $\mathbf{E}^n$ converges to a limit $a \in \mathbf{E}^n$ if and only if it is a Cauchy sequence.

This criterion is of the utmost importance, as it enables us to establish the convergence of a sequence without *a priori* knowledge of its limit. This will eventually make it possible to speak of functions that are defined as limits of sequences, and to differentiate and integrate such functions without knowing them explicitly.

Note that we could have established Theorem 62.1 by proving that a sequence of real numbers converges if and only if it is a Cauchy sequence, and then invoking Theorem 61.3. The proof for sequences of real numbers is not any easier or any more complicated than for sequences in $\mathbf{E}^n$, because the norm can be handled like an absolute value. One may obtain the one proof from the other by replacing $|$ by $\|$, or vice versa.

One calls a metric space *complete* if and only if every Cauchy sequence in the space converges to an element of that space. Theorem 62.1 is based on the Bolzano–Weierstrass Theorem which, in turn, is based on the nested-interval property. Conversely, the nested-interval property can be established on the basis of Theoren 62.1. (See exercise 4.) Hence, the above-mentioned concept of completeness is, at least as far as the $\mathbf{E}^n$ is concerned, equivalent to the concept of completeness which we mentioned in section 5.60.

That there are metric spaces that are not complete can be seen from the following example:

---

▲ *Example*

$\mathbf{Q}$, the space of all rational numbers with

$$d(r, s) = |r - s|, \qquad r, s \in \mathbf{Q},$$

is obviously a metric space. The sequence

$$\{a_k\} = \{1, 1.01, 1.01001, 1.010010001, 1.01001000100001, \ldots\}$$

is a Cauchy sequence in $\mathbf{Q}$ since

$$|a_k - a_l| < \frac{1}{10^n} \qquad \text{for all } k, l > n.$$

However, $\lim_\infty a_k = 1.01001000100001\ldots \notin \mathbf{Q}$. (See exercise 7, section 1.8.)

---

Before closing this section, we wish to establish a lemma that will prove useful in our later investigations:

***Lemma 62.5***

If $c \in \mathbf{E}^n$ and $a_k \in \mathbf{E}^n$ such that

$$\|a_k - c\| \leqslant r$$

for some $r > 0$ and all $k > N(r)$ for some $N(r) \in \mathbf{N}$, then, for all accumulation points $a$ of $\{a_k\}$,

$$\|a - c\| \leqslant r.$$

In particular, if $\lim_\infty a_k = a$, then $\|a - c\| \leqslant r$.

*Proof*

$A = \{x \in \mathbf{E}^n \mid \|x - c\| \leqslant r\}$ is closed and $a_k \in A$ for all $k > N(r)$. By Theorem 59.1, $a \in A$.

## 5.62 Exercises

1. Show that $\{1/k\}$ is a Cauchy sequence without appealing to Theorem 62.1.
2. Same as in exercise 1 for $\{q^k\}$ where $|q| < 1$.
3. State and prove Lemma 62.4 for Cauchy sequences of real numbers.
4. Prove: If every Cauchy sequence in $\mathbf{E}^n$ converges to an element in $\mathbf{E}^n$, then $\mathbf{E}^n$ has the nested-interval property.
5. Let $f_k(x) = 0$ for $0 \leqslant x \leqslant \frac{1}{2} - 1/k$, $f_k(x) = kx + 1 - k/2$ for $\frac{1}{2} - 1/k < x < \frac{1}{2}$, $f_k(x) = 1$ for $\frac{1}{2} \leqslant x \leqslant 1$. Find $\lim_\infty f_k(x)$.
6. Use the sequence in exercise 5 to demonstrate that the metric space of continuous functions on $[0, 1]$, with the distance defined by $d(f, g) = \int_0^1 |f - g|$, is not complete. (See also exercise 8, section 5.58.)
7. Show that the statements (62.1) and (62.2) are equivalent.
8. Show: $\{a_k\}$, where $a_k = (\alpha_1^{(k)}, \alpha_2^{(k)}, \ldots, \alpha_n^{(k)})$, is a Cauchy sequence, if and only if each $\{\alpha_j^{(k)}\}$, $j = 1, 2, \ldots, n$, is a Cauchy sequence.
9. Show: $\{a_k\}$, where $a_k = (\alpha_1^{(k)}, \alpha_2^{(k)}, \ldots, \alpha_n^{(k)})$ converges if and only if each $\{\alpha_j^{(k)}\}$, $j = 1, 2, \ldots, n$, is a Cauchy sequence.

10. The sequence $\{a_k\}$ is recursively defined by $a_1 = 1,\ a_2 = 2, \ldots, a_k = (a_{k-1} + a_{k-2})/2$.

    (a) Show that $1 \leqslant a_k \leqslant 2$ for all $k \in \mathbf{N}$.
    (b) Show that $\{a_k\}$ converges.
    (c) Find $\lim_\infty a_k$.

11. Let $a_k = 1 + \frac{1}{2} + \cdots + 1/k$. Show that $\{a_k\}$ diverges.
12. Prove: $\{a_k\}$ is not a Cauchy sequence if and only if there is an $\varepsilon > 0$ such that, for every $N \in \mathbf{N}$, there are integers $l$, $k > N$ for which $\|a_k - a_l\| \geqslant \varepsilon$.

# Vector-Valued Functions of a Vector Variable 6

## 6.63 Notation

In this and the following four chapters, we shall study functions with domain in $\mathbf{E}^n$ and range in $\mathbf{E}^m$:

$$f : \mathfrak{D} \to \mathbf{E}^m, \qquad \text{where } \mathfrak{D} \subseteq \mathbf{E}^n.$$

If $x \in \mathfrak{D}$, then $y = f(x) \in \mathbf{E}^m$ is the value of the function $f$ at the point $x$.

For $n > 1$, $m > 1$, the supply set and target set are *sets of vectors*. We speak, therefore, of *vector-valued functions of a vector variable*. If $n = 1$, $m > 1$, we speak of vector-valued functions of a real variable, and if $n > 1$, $m = 1$, we speak of real-valued functions of a vector variable.

Vector-valued functions of a vector variable may be represented in various ways, as the following example shows:

---

▲ *Example 1*

We consider a function $f : \mathbf{E}^2 \to \mathbf{E}^3$, and represent $x \in \mathbf{E}^2$ by its Cartesian coordinates $x = (\xi_1, \xi_2)$ and the function value $y = f(x) \in \mathbf{E}^3$ by its Cartesian coordinates $y = (\eta_1, \eta_2, \eta_3)$. The function shall be defined by

$$\eta_1 = \xi_1,$$
$$\eta_2 = \xi_2, \tag{63.1}$$
$$\eta_3 = \begin{cases} 1 & \text{for } \xi_1^2 + \xi_2^2 \leqslant 1, \\ 0 & \text{for } \xi_1^2 + \xi_2^2 > 1. \end{cases}$$

We call $(\phi_1, \phi_2, \phi_3)$, where $\phi_1(\xi_1, \xi_2) = \xi_1$, $\phi_2(\xi_1, \xi_2) = \xi_2$, and

$$\phi_3(\xi_1, \xi_2) = \left\{\begin{matrix} 1 & \text{for } \xi_1^2 + \xi_2^2 \leqslant 1 \\ 0 & \text{for } \xi_1^2 + \xi_2^2 > 1 \end{matrix}\right\},$$

the Cartesian components of $f : f = (\phi_1, \phi_2, \phi_3)$.

The same function may be represented in an entirely different manner if we represent $x \in \mathbf{E}^2$ by its polar coordinates $(\theta, r)$, and $y = f(x) \in \mathbf{E}^3$ by its spherical coordinates $(\alpha, \beta, \rho)$. Then,

$$\alpha = \theta,$$

$$\beta = \begin{cases} \mathrm{Sin}^{-1} \dfrac{r}{\sqrt{r^2+1}} & \text{for } |r| \leqslant 1, \\ \dfrac{\pi}{2} & \text{for } |r| > 1, \end{cases} \tag{63.2}$$

$$\rho = \begin{cases} \sqrt{r^2+1} & \text{for } |r| \leqslant 1, \\ r & \text{for } |r| > 1, \end{cases}$$

where $\mathrm{Sin}^{-1}$ denotes the principal value of $\sin^{-1}$. (See Fig. 63.1.)

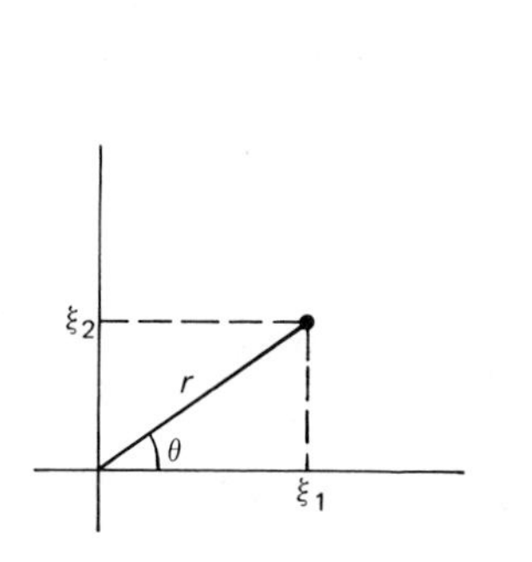


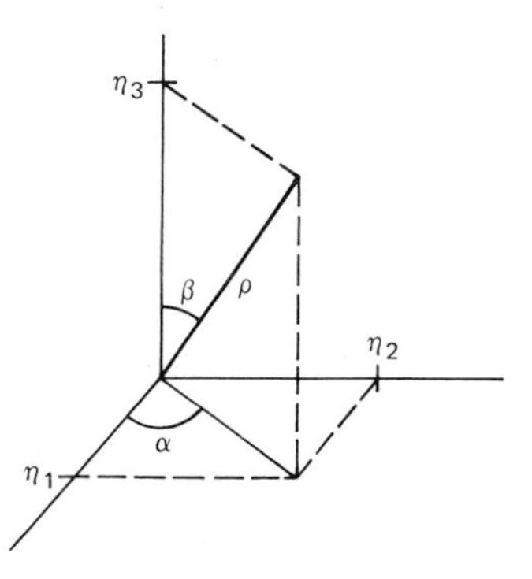


***Fig. 63.1***

While the representation (63.1) and (63.2) have no resemblance to each other, both represent the same function. However, if we interpret $(\theta, r)$ and $(\alpha, \beta, \rho)$ as Cartesian coordinates, then (63.1) and (63.2) represent entirely different functions.

It is, of course, most convenient to represent the independent variable $x \in \mathfrak{D} \subseteq \mathbf{E}^n$ by its Cartesian coordinates $(\xi_1, \xi_2, \ldots, \xi_n)$, and the function value $y \in \mathbf{E}^m$ by its Cartesian coordinates $(\eta_1, \eta_2, \ldots, n_m)$. In this representation, $y = f(x)$ will appear as

$$\begin{aligned} \eta_1 &= \phi_1(\xi_1, \xi_2, \ldots, \xi_n), \\ \eta_2 &= \phi_2(\xi_1, \xi_2, \ldots, \xi_n), \\ &\vdots \\ \eta_m &= \phi_m(\xi_1, \xi_2, \ldots, \xi_n). \end{aligned}$$

Here, $(\phi_1, \phi_2, \ldots, \phi_m)$ are called the Cartesian components of $f$. The Cartesian components of $f$ are functions in their own right, namely,

$$\phi_j : \mathfrak{D} \to \mathbf{E},$$

$j = 1, 2, \ldots, m$, where $\mathbf{E}$ denotes the Cartesian space $\mathbf{R}$ with Euclidean metric $d(y_1, y_2) = |y_1 - y_2|$.

In future we shall drop the adjective "Cartesian" and simply speak of the components of $x, y, f, \ldots$.

---

▲ *Example 2*

If the function $g : \mathbf{E}^3 \to \mathbf{E}^2$ is given by its components $\psi_1(\xi_1, \xi_2, \xi_3) = \xi_1 + \xi_2 + \xi_3$, $\psi_2(\xi_1, \xi_2, \xi_3) = \xi_1^2 - \xi_2\xi_3$, then $y = g(x)$, in component representation, reads

$$\eta_1 = \xi_1 + \xi_2 + \xi_3,$$
$$\eta_2 = \xi_1^2 - \xi_2\xi_3,$$

or, equivalently,

$$\begin{pmatrix} \eta_1 \\ \eta_2 \end{pmatrix} = \begin{pmatrix} \xi_1 + \xi_2 + \xi_3 \\ \xi_1^2 - \xi_2\xi_3 \end{pmatrix}.$$

---

Since $x \in \mathbf{E}^n$, we may represent $x$ by $(\xi_1, \xi_2, \ldots, \xi_n)$, and we have

$$\|x\| = \sqrt{\xi_1^2 + \xi_2^2 + \cdots + \xi_n^2},$$

and since $y = f(x) \in \mathbf{E}^m$, we may represent $f$ by $(\phi_1, \phi_2, \ldots, \phi_m)$ and we have

(63.3) $$\|f(x)\| = \sqrt{[\phi_1(\xi_1, \xi_2, \ldots, \xi_n)]^2 + [\phi_2(\xi_1, \xi_2, \ldots, \xi_n)]^2 + \cdots + [\phi_m(\xi_1, \xi_2, \ldots, \xi_n)]^2}.$$

---

▲ *Example 3*

We have, for the function $f : \mathbf{E}^2 \to \mathbf{E}^3$ defined by (63.1),

$$\|f(x)\| = \begin{cases} \sqrt{\xi_1^2 + \xi_2^2 + 1} & \text{for } \|x\| \leqslant 1, \\ \sqrt{\xi_1^2 + \xi_2^2} & \text{for } \|x\| > 1; \end{cases}$$

and for the function $g : \mathbf{E}^3 \to \mathbf{E}^2$, as defined in example 2,

$$\|g(x)\| = \sqrt{(\xi_1 + \xi_2 + \xi_3)^2 + (\xi_1^2 - \xi_2\xi_3)^2}.$$

---

## 6.63 Exercises

1. Given the function $f: \mathbf{E}^2 \to \mathbf{E}^4$ by $\phi_1(\xi_1, \xi_2) = \xi_1 - \xi_2$, $\phi_2(\xi_1, \xi_2) = \xi_1 + \xi_2$, $\phi_3(\xi_1, \xi_2) = \xi_1$, $\phi_4(\xi_1, \xi_2) = \xi_2$. Find $\|f(x)\|$.
2. Show that, for the function $f$ defined in exercise 1,

$$f(x_1 + x_2) = f(x_1) + f(x_2) \qquad \text{for all } x_1, x_2 \in \mathbf{E}^2$$

and

$$f(\lambda x) = \lambda f(x) \qquad \text{for every } \lambda \in \mathbf{R} \text{ and all } x \in \mathbf{E}^2.$$

3. Given the function $g: \mathbf{E}^3 \to \mathbf{E}^2$ by $\psi_1(\xi_1, \xi_2, \xi_3) = \xi_2^2$, $\psi_2(\xi_1, \xi_2, \xi_3) = \xi_1 - \xi_3^2$. Find $\|g(x)\|$, and show that, for some $x_1 \neq \theta$, $x_2 \neq \theta$, $g(x_1 + x_2) \neq g(x_1) + g(x_2)$.

## 6.64 Limit of a Function

In defining the limit of a function $f: \mathfrak{D} \to \mathbf{E}^m$, $\mathfrak{D} \subseteq \mathbf{E}^n$, we shall generalize the criterion that was established in Theorem 25.1, rather than generalize definition 24.1. We do this merely to provide for some variety.

***Definition 64.1***

Let $f: \mathfrak{D} \to \mathbf{E}^m$, $\mathfrak{D} \subseteq \mathbf{E}^n$, and let $a \in \mathbf{E}^n$ represent an accumulation point of $\mathfrak{D}$. Then $\lim_a f(x) = b$, where $b \in \mathbf{E}^m$, if and only if, for every $\varepsilon > 0$, there is a $\delta(\varepsilon) > 0$ such that:

(64.1) $$f(x) \in N_\varepsilon(b) \qquad \text{for all } x \in N'_{\delta(\varepsilon)}(a) \cap \mathfrak{D},$$

or, equivalently,

(64.2) $$\|f(x) - b\| < \varepsilon \qquad \text{for all } x \in \mathfrak{D} \text{ for which } 0 < \|x - a\| < \delta(\varepsilon).$$

(See also Theorem 25.1a.)

Because of this new approach, we obtain the appropriate generalization of definition 24.1 as a theorem.

***Theorem 64.1***

If $f: \mathfrak{D} \to \mathbf{E}^m$, $\mathfrak{D} \subseteq \mathbf{E}^n$, and if $a$ is an accumulation point of $\mathfrak{D}$, then $\lim_a f(x) = b$ if and only if

$$\lim_\infty f(x_k) = b$$

for all sequences $\{x_k\}$, where $x_k \in \mathfrak{D} \backslash \{a\}$ for all $k \in \mathbf{N}$, for which $\lim_\infty x_k = a$.

The proof of this theorem consists of a verbatim restatement of the proof of Theorem 25.1 with an appropriate interpretation of the symbols to fit the more general setting of Theorem 64.1.

If $f = (\phi_1, \phi_2, \dots, \phi_m)$ and $b = (\beta_1, \beta_2, \dots, \beta_m)$, we obtain, from the FIE (61.1):

***Lemma 64.1***

$\lim_a f(x) = b$ if and only if $\lim_a \phi_j(x) = \beta_j$ for all $j = 1, 2, \dots, m$.
The proof is left for the exercises (exercise 3).

Lemma 64.1, in effect, reduces the problem of finding the limit of a vector-valued function to the problem of finding the limit of a real-valued function. In some instances, this can be broken down even further by computing the limits by iteration, i.e., by taking the limits of $f$ as each individual component of $x$ approaches the corresponding component of $a$. (See exercise 4.)

## 6.64 Exercises

1. Show that the statements (64.1) and (64.2) are equivalent.
2. Prove Theorem 64.1.
3. Prove Lemma 64.1.
4. If $f : \mathfrak{D} \to \mathbf{E}^2$, where $\mathfrak{D} \subseteq \mathbf{E}^2$, and if $(\alpha_1, \alpha_2) \in \mathbf{E}^2$ is an accumulation point of $\mathfrak{D}$, we say that $\lim_{\xi_1 \to \alpha_1} f(\xi_1, \xi_2) = g(\xi_2)$ *uniformly* in $\xi_2$ for all $\xi_2 \in N_\delta(\alpha_2)$ if and only if, for all $\xi_2 \in N_\delta(\alpha_2)$ and every $\varepsilon > 0$, there is a $\delta_1(\varepsilon) > 0$ independent of $\xi_2$ such that $|f(\xi_1, \xi_2) - g(\xi_2)| < \varepsilon$ for all $\xi_1 \in N'_{\delta_1(\varepsilon)}(\alpha_1)$. Prove: If $\lim_a f(x) = b$ exists and if, for some $\delta > 0$, $\lim_{\xi_1 \to \alpha_1} f(\xi_1, \xi_2) = g(\xi_2)$ uniformly in $\xi_2$ for all $\xi_2 \in N_\delta(\alpha_2)$, then $\lim_a f(x) = \lim_{\xi_2 \to \alpha_2} (\lim_{\xi_1 \to \alpha_1} f(\xi_1, \xi_2))$. (The limit on the right is called an *iterated limit*.)
5. Let $f : \mathbf{E}^2 \backslash \{\theta\} \to \mathbf{E}$ be defined by $f(\xi_1, \xi_2) = (\xi_1^2 - \xi_2^2)/(\xi_1^2 + \xi_2^2)$. Does $\lim_\theta f(x)$ exist? Do the iterated limits $\lim_{\xi_1 \to 0} (\lim_{\xi_2 \to 0} f(x))$ and $\lim_{\xi_2 \to 0} (\lim_{\xi_1 \to 0} f(x))$ exist?
6. Let $f : \{x \in \mathbf{E}^2 \mid \xi_1 \neq 0\} \to \mathbf{E}$ be defined by $f(\xi_1, \xi_2) = \xi_1^2 + \xi_2^2 \sin(1/\xi_1)$. Show that $\lim_\theta f(x)$ exists but that $\lim_{\xi_1 \to 0} f(x)$ does not exist for $\xi_2 \neq 0$.
7. Let $f(x) = (\xi_1 |\xi_2|)/\xi_2$ for $\xi_2 \neq 0$. Show:

   (a) $\lim_{\xi_1 \to 0} f(x) = 0$ uniformly in $\xi_2$.
   (b) $\lim_\theta f(x) = \lim_{\xi_2 \to 0} (\lim_{\xi_1 \to 0} f(x))$.

8. Prove: If $f, g : \mathfrak{D} \to \mathbf{E}^m$, $\mathfrak{D} \subseteq \mathbf{E}^n$, if $a \in \mathbf{E}^n$ is an accumulation point of $\mathfrak{D}$, and if $\lim_a f(x) = b$, $\quad \lim_a g(x) = c$,
then
$$\lim_a (f \pm g)(x) = b \pm c,$$
$$\lim_a \lambda f(x) = \lambda b,\ \lambda \in \mathbf{R},$$
$$\lim_a f(x) \cdot g(x) = b \cdot c.$$

9. Let $h : \mathfrak{D} \to \mathbf{E}$, $\mathfrak{D} \subseteq \mathbf{E}^n$ and let $a \in \mathbf{E}^n$ denote an accumulation point of $\mathfrak{D}$. *Prove*: If $\lim_a h(x) = d > 0$, then there is a $\delta > 0$ such that $h(x) > 0$ for all $x \in N'_\delta(a) \cap \mathfrak{D}$.

10. Let $f$ be defined as in exercise 8 and $h$ as in exercise 9. *Prove*: If $\lim_a h(x) = d > 0$, then

$$\lim_a \frac{f(x)}{h(x)} = \frac{b}{d}.$$

11. $f: \mathbf{E}^2 \backslash \{\theta\} \to \mathbf{E}^3$ is given by

$$\phi_1(\xi_1, \xi_2) = \frac{\xi_1 \xi_2}{\sqrt{\xi_1^2 + \xi_2^2}},$$

$$\phi_2(\xi_1, \xi_2) = \begin{cases} \xi_1^2 + \xi_2^2 & \text{if } \xi_1, \xi_2 \in \mathbf{Q}, \\ 0 & \text{otherwise}, \end{cases}$$

$$\phi_3(\xi_1, \xi_2) = (\xi_1^2 + \xi_2^2) \sin \frac{1}{\xi_1^2 + \xi_2^2}.$$

(a) Does $\lim_\theta f(x)$ exist?
(b) If the answer to (a) is yes, find $\lim_\theta f(x)$.

## 6.65 Continuity

Definition 26.1 of continuity of a function at a point has the following straightforward generalization.

#### *Definition 65.1*

If $f: \mathfrak{D} \to \mathbf{E}^m$, where $\mathfrak{D} \subseteq \mathbf{E}^n$ and $a \in \mathfrak{D}$, then $f$ is continuous at $a$ if and only if, for every $\varepsilon > 0$, there is a $\delta(\varepsilon) > 0$ such that

(65.1) $$f(x) \in N_\varepsilon(f(a)) \qquad \text{for all } x \in N_{\delta(\varepsilon)}(a) \cap \mathfrak{D},$$

or, equivalently,

(65.2) $$\|f(x) - f(a)\| < \varepsilon \qquad \text{for all } x \in \mathfrak{D} \quad \text{for which } \|x - a\| < \delta(\varepsilon).$$

$f$ is continuous on $\mathcal{A} \subseteq \mathfrak{D}$ if and only if it is continuous for all $x \in \mathcal{A}$.

Theorems 26.1 and 26.2 retain their validity in this more general setting:

#### *Theorem 65.1*

If $f: \mathfrak{D} \to \mathbf{E}^m$, $\mathfrak{D} \subseteq \mathbf{E}^n$, and if $a \in \mathfrak{D}$ is an accumulation point of $\mathfrak{D}$, then $f$ is continuous at $a$ if and only if $\lim_a f(x) = f(a)$. If $a$ is an isolated point of $\mathfrak{D}$, then $f$ is continuous at $a$.

From the FIE (61.1) we obtain the following counterpart to Lemma 64.1:

#### *Lemma 65.1*

$f: \mathfrak{D} \to \mathbf{E}^m$, $\mathfrak{D} \subseteq \mathbf{E}^n$, is continuous at $a \in \mathfrak{D}$ if and only if all components $\phi_1, \phi_2, \ldots, \phi_m$ of $f$ are continuous at $a$.

▲ *Example 1*

Let $f: \mathbf{E}^2 \to \mathbf{E}^2$ be given by $\phi_1(\xi_1, \xi_2) = \xi_1 + \xi_2$, $\phi_2(\xi_1, \xi_2) = 2\xi_1 \xi_2$. In order to show that $f$ is continuous at $\theta$, we need only show, according to Lemma 65.1, that $\phi_1$ and $\phi_2$ are continuous at $\theta$. We have:

$$|\phi_1(\xi_1, \xi_2) - \phi_1(0, 0)| = |\xi_1 + \xi_2| \leqslant |\xi_1| + |\xi_2| < \varepsilon$$

whenever $\|x - \theta\| = \sqrt{\xi_1^2 + \xi_2^2} < \varepsilon/2 = \delta_1(\varepsilon)$;

$$|\phi_2(\xi_1, \xi_2) - \phi_2(0, 0)| = |2\xi_1 \xi_2| < \xi_1^2 + \xi_2^2 < \varepsilon$$

whenever $\|x - \theta\| = \sqrt{\xi_1^2 + \xi_2^2} < \sqrt{\varepsilon} = \delta_2(\varepsilon)$. Hence, $f$ is continuous at $\theta$.

If we wish to apply definition 65.1 directly, then we have to proceed as follows:

$$\begin{aligned}\|f(x) - f(\theta)\| &= \sqrt{(\phi_1(\xi_1, \xi_2) - \phi_1(0, 0))^2 + (\phi_2(\xi_1, \xi_2) - \phi_2(0, 0))^2} \\ &= \sqrt{(\xi_1 + \xi_2)^2 + 4\xi_1^2\xi_2^2} < \varepsilon,\end{aligned}$$

whenever $\|x - \theta\| = \sqrt{\xi_1^2 + \xi_2^2} < \delta(\varepsilon) = \min\left(\sqrt{2\varepsilon}/2, \varepsilon/4\right)$. This second approach appears simple only because of all the preparatory work which we have already done in the first approach. In general, it is easier to establish the continuity of the individual components than to establish the continuity of the function directly, because it is quite awkward to work with the norm of the function in $\mathbf{E}^m$ for $m \geqslant 2$.

Lemma 65.1 may suggest to some that the investigation of the continuity of a function may be further broken down into an investigation of the *continuity of the components* as a function of each component of $x$, taken one at a time. This is *not* possible, as the following example shows:

▲ *Example 2*

Let $f: \mathbf{E}^2 \to \mathbf{E}$ by given by

$$f(\xi_1, \xi_2) = \begin{cases} \dfrac{\xi_1 \xi_2}{\xi_1^2 + \xi_2^2} & \text{for } (\xi_1, \xi_2) \neq (0, 0), \\ 0 & \text{for } (\xi_1, \xi_2) = (0, 0); \end{cases}$$

this function is *not* continuous at $\theta$, as we shall demonstrate by invoking Theorem 65.1: The sequence $\{x_k\} = \{(1, 0), (\frac{1}{2}, 0), (\frac{1}{3}, 0), \ldots, (1/k, 0), \ldots\}$ has all the properties prescribed by Theorem 64.1, and we obtain

$$\lim_{\infty} f(x_k) = \lim_{\infty} 0 = 0.$$

On the other hand, also, $\{x_k'\} = \{(1, 1), (\frac{1}{2}, \frac{1}{2}), (\frac{1}{3}, \frac{1}{3}), \ldots, (1/k, 1/k), \ldots\}$ has the required properties, and we obtain

$$\lim_{\infty} f(x_k') = \tfrac{1}{2} \neq \lim_{\infty} f(x_k).$$

Hence, $f$ is *not* continuous at $\theta$.

Still, for fixed $\xi_1 = \alpha_1$, $f$ is a continuous function of $\xi_2$, and for fixed $\xi_2 = \alpha_2$, $f$ is a continuous function of $\xi_1$, as we can see by invoking Theorem 65.1:

We have $f(\alpha_1, 0) = 0$ for all $\alpha_1 \in \mathbf{E}$ and

$$\lim_{\xi_2 \to 0} f(\alpha_1, \xi_2) = \begin{cases} \lim\limits_{0} \dfrac{\alpha_1 \xi_2}{\alpha_1^2 + \xi_2^2} = 0 & \text{if } \alpha_1 \neq 0 \\ \lim\limits_{0} 0 = 0 & \text{if } \alpha_1 = 0. \end{cases}$$

Hence, $f$ is a continuous function of $\xi_2$ at $(0, 0)$. By symmetry, it follows that $f$ is also a continuous function of $\xi_1$ at $(0, 0)$.

---

For later reference, we state the following lemma:

***Lemma 65.2***

If $f : \mathfrak{D} \to \mathbf{E}$, $\mathfrak{D} \subseteq \mathbf{E}^n$, is continuous at $a \in \mathfrak{D}$ and if $f(a) > 0$, then there is a $\delta > 0$ such that $f(x) > 0$ for all $x \in N_\delta(a) \cap \mathfrak{D}$.

The proof is left for the exercises.

## 6.65 Exercises

1. Show that the statements (65.1) and (65.2) are equivalent.
2. Prove Theorem 65.1.
3. Prove Lemma 65.1.
4. Prove Lemma 65.2.
5. $f : \mathbf{E} \to \mathbf{E}$ has the property that $f(x_1 + x_2) = f(x_1)f(x_2)$ for all $x_1, x_2 \in \mathbf{E}$.
   (a) Show: If $f(x_0) = 0$ for some $x_0 \in \mathbf{E}$, then $f(x) = 0$ for all $x \in \mathbf{E}$.
   (b) Show: If $f$ is continuous at $x = 0$, then $f$ is continuous on $\mathbf{E}$.
   (c) Which nonzero function satisfies the above requirements?
6. Show: If $f, g : [0, 1] \to \mathbf{E}$ are continuous on $[0, 1]$, then $f(x) = g(x)$ for all $x \in [0, 1]$ if and only if $f(x) = g(x)$ for all $x \in [0, 1] \cap \mathbf{Q}$.

## 6.66 Direct and Inverse Images

It will give us a practical advantage in our presentations in subsequent sections to be in possession of the concepts of the *direct image* and the *inverse image* of a set under a function, and the behavior of these images under the set operations.

To introduce these concepts and to explain how to manipulate them is the objective of this section.

We consider a function $f: \mathcal{D}(f) \to Y$. Let $X$ denote the supply set of $f$. If $A \subseteq X$, then the set $\{f(x) \in \mathcal{R}(f) \mid x \in A \cap \mathcal{D}(f)\}$ is called the *direct image of $A$ under $f$*. It is the collection of all function values $f(x)$ for which $x \in A \cap \mathcal{D}(f)$. Since there is no $f(x)$ when $x \in A \backslash \mathcal{D}(f)$ and since $f(x)$, by definition, lies in $\mathcal{R}(f)$, we may just as well write $\{f(x) \mid x \in A\}$ instead. Similarly, if $B \subseteq Y$, where $Y$ is the target set of $f$, then $\{x \in \mathcal{D}(f) \mid f(x) \in B \cap \mathcal{R}(f)\}$ is called the *inverse image of $B$ under $f$*. Since there is no $x$ for which $f(x) \in B \backslash \mathcal{R}(f)$ and since $x$ cannot lie anywhere but in $\mathcal{D}(f)$, we may write $\{x \mid f(x) \in B\}$ instead. (Note the symmetry in the definitions of direct and inverse image.)

In the following formal definition, we introduce a notation which is not very commonly used but which we find very important for conceptual as well as practical reasons:

***Definition 66.1***

Let $f: \mathcal{D}(f) \to Y$ and let $X$ denote the supply set of $f$. Then,

$$(66.1)\quad f_*(A) = \{f(x) \in \mathcal{R}(f) \mid x \in A \cap \mathcal{D}(f)\} = \{f(x) \mid x \in A\} \qquad \text{for } A \subseteq X$$

is called the *direct image of $A$* under $f$; and

$$(66.2)\quad f^*(B) = \{x \in \mathcal{D}(f) \mid f(x) \in B \cap \mathcal{R}(f)\} = \{x \mid f(x) \in B\} \qquad \text{for } B \subseteq Y$$

is called the *inverse image of $B$* under $f$.

Note that (66.1) and (66.2) define set-valued functions $f_*$ and $f^*$ of a set variable from the set of all subsets of the supply set into the set of all subsets of the target set, and from the set of all subsets of the target set into the set of all subsets of the supply set, respectively. One often finds the notation $f(A)$ instead of $f_*(A)$ and $f^{-1}(B)$ instead of $f^*(B)$. This is not quite legitimate in view of the above remark, and it is also confusing at times, because the notation $f^{-1}$ is generally used to denote the inverse function of $f$. Note that $f$ need not be invertible, but still $f^*$ exists as a set-valued function of a set variable.

Since

$$(66.3)\qquad f_*(A) = f_*(A \cap \mathcal{D}(f)), \qquad f^*(B) = f^*(B \cap \mathcal{R}(f)),$$

there seems to be no point in allowing $A$ and $B$ to be anything but subsets of $\mathcal{D}(f)$ and $\mathcal{R}(f)$, respectively. However, our more general formulation results in a simplification of notation. For example, definition 64.1 of the limit of a function and definition 65.1 of continuity may now be formulated as follows:

Let $f: \mathcal{D}(f) \to \mathbf{E}^m$, $\mathcal{D}(f) \subseteq \mathbf{E}^n$, and let $a \in \mathbf{E}^n$ denote an accumulation point of $\mathcal{D}(f)$. $\lim_a f(x) = b$ if and only if, for every $\varepsilon > 0$, there is a $\delta(\varepsilon) > 0$ such that

$$(66.4)\qquad f_*(N'_{\delta(\varepsilon)}(a)) \subseteq N_\varepsilon(b).$$

Let $a \in \mathcal{D}(f)$. $f$ is continuous at $a$ if and only if, for every $\varepsilon > 0$, there is a $\delta(\varepsilon) > 0$ such that

(66.5) $$f_*(N_{\delta(\varepsilon)}(a)) \subseteq N_\varepsilon(f(a)).$$

Since we allow $A$ and $B$ in definition 66.1 to be sets that are not necessarily subsets of the domain of $f$ and the range of $f$, respectively, it is important at this point to be in possession of the concept of supply set and target set. Otherwise, $f_*$ and $f^*$ would emerge as set-valued functions with the set of all sets as their domain. Since functions are defined only on sets, and since the set of all sets is not a set (see section 1.3), this would put us into an untenable position.

Since a function can assume only one value $f(x)$ for each $x \in \mathfrak{D}(f)$, we have, from (66.2), that $x \in f^*(B)$ if and only if $f(x) \in B$. However, a function (unless it is one-to-one) may assume the same value $f(x)$ for different values of $x$ and hence, $f(x) \in f_*(A)$ is possible without $x$ being a member of $A \cap \mathfrak{D}(f)$. (There could be a $y \in A \cap \mathfrak{D}(f)$ such that $f(x) = f(y)$.) These facts are reflected in the asymmetry between the following two relations:

(66.6) $$f^*(f_*(A)) \supseteq A \cap \mathfrak{D}(f), \qquad f_*(f^*(B)) = B \cap \mathfrak{R}(f).$$

The converse of the first relationship need not hold, as the following example shows:

---

▲ *Example 1*

Let $f: \mathbf{E} \to \mathbf{E}$ be defined by $f(x) = x^2$, and let $A = [0, 1]$. Then, $f_*(A) = [0, 1]$ and $f^*(f_*(A)) = f^*([0, 1]) = [-1, 1] \supset [0, 1] = A \cap \mathfrak{D}(f)$. (See Fig. 66.1.)

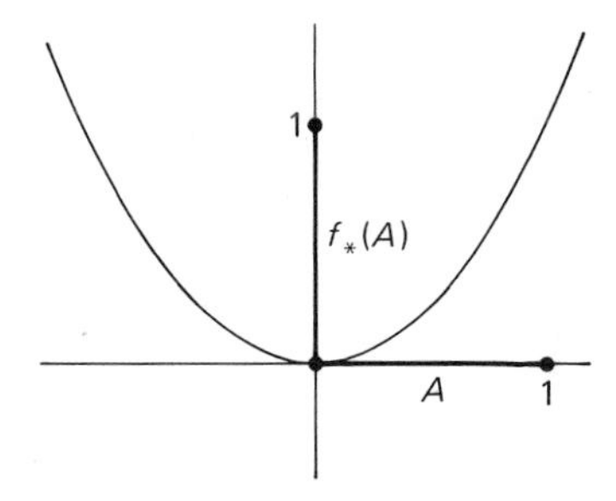


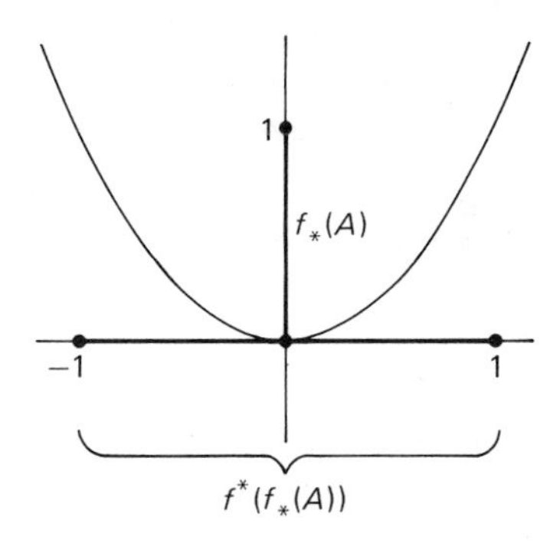


***Fig. 66.1***

---

To prove the first formula in (66.6), we note that, by (66.1), $x \in A \cap \mathfrak{D}(f)$ implies that $f(x) \in f_*(A)$. From (66.2), $x \in f^*(f_*(A))$.

To prove the second formula, we note that, in view of (66.1) and (66.2),

$$f_*(f^*(B)) = \{f(x) \mid x \in f^*(B)\} = \{f(x) \mid f(x) \in B\} = B \cap \mathfrak{R}(f),$$

since $x \in f^*(B)$ if and only if $f(x) \in B$.

If $f$ is one-to-one, then $x \in A \cap \mathfrak{D}(f)$ if and only if $f(x) \in f_*(A)$ and hence,

$$f^*(f_*(A)) = \{x \mid f(x) \in f_*(A)\} = \{x \mid x \in A \cap \mathfrak{D}(f)\} = A \cap \mathfrak{D}(f).$$

We list this result for later reference:

$$(66.7) \qquad f^*(f_*(A)) = A \cap \mathfrak{D}(f) \qquad \text{if } f \text{ is one-to-one.}$$

***Theorem 66.1***

Let $f : \mathfrak{D}(f) \to Y$ and let $X$ denote the supply set of $f$. Then,

$$(66.8) \qquad f_*(\varnothing) = \varnothing, \qquad f^*(\varnothing) = \varnothing.$$

$$(66.9) \qquad \begin{array}{lll} \text{If } f_*(A) = \varnothing, & \text{then} & A \cap \mathfrak{D}(f) = \varnothing, \\ \text{If } f^*(B) = \varnothing, & \text{then} & B \cap \mathfrak{R}(f) = \varnothing. \end{array}$$

If $A_1 \subseteq A_2 \subseteq X$, $B_1 \subseteq B_2 \subseteq Y$, then

$$(66.10) \qquad f_*(A_1) \subseteq f_*(A_2), \qquad f^*(B_1) \subseteq f^*(B_2).$$

If $\Lambda$, $M$ are given index sets and $A_\lambda \subseteq X$, $B_\mu \subseteq Y$ for all $\lambda \in \Lambda$, $\mu \in M$, then

$$(66.11) \qquad f_*\left(\bigcup_{\lambda \in \Lambda} A_\lambda\right) = \bigcup_{\lambda \in \Lambda} f_*(A_\lambda), \qquad f^*\left(\bigcup_{\mu \in M} B_\mu\right) = \bigcup_{\mu \in M} f^*(B_\mu).$$

*Proof*

(66.8) and (66.9) follow directly from definition 66.1.

If $y \in f_*(A_1)$, then there is an $x \in A_1 \subseteq A_2$ such that $f(x) = y$. Hence, $f(x) \in f_*(A_2)$. Therefore, $f_*(A_1) \subseteq f_*(A_2)$. If $x \in f^*(B_1)$, then $f(x) \in B_1 \subseteq B_2$. Hence, $x \in f^*(B_2)$, and we have $f^*(B_1) \subseteq f^*(B_2)$. This establishes (66.10).

Since $A_\lambda \subseteq \bigcup_{\lambda \in \Lambda} A_\lambda$, we have, from (66.10), that $f_*(A_\lambda) \subseteq f_*(\bigcup_{\lambda \in \Lambda} A_\lambda)$ for all $\lambda \in \Lambda$. Hence,

$$(66.12) \qquad \bigcup_{\lambda \in \Lambda} f_*(A_\lambda) \subseteq f_*\left(\bigcup_{\lambda \in \Lambda} A_\lambda\right).$$

Conversely, if $y \in f_*(\bigcup_{\lambda \in \Lambda} A_\lambda)$, then there is an $x \in \bigcup_{\lambda \in \Lambda} A_\lambda$ such that $f(x) = y$ and hence, $x \in A_\lambda$ for some $\lambda \in \Lambda$. Therefore, $f(x) \in f_*(A_\lambda) \subseteq \bigcup_{\lambda \in \Lambda} f_*(A_\lambda)$ and we have

$$(66.13) \qquad f_*\left(\bigcup_{\lambda \in \Lambda} A_\lambda\right) \subseteq \bigcup_{\lambda \in \Lambda} f_*(A_\lambda).$$

(66.12) and (66.13) together yield that part of (66.11) that pertains to direct images. The other part is proved by similar reasoning and is left to the exercises (exercise 6).

When it comes to intersections and complementations, then direct and inverse image part company, as we shall see from the following examples:

▲ *Example 2*

Let $f: \mathbf{E} \to \mathbf{E}$ be defined by $f(x) = x^2$. Let $A_1 = (-1, 0)$, $A_2 = (0, 1)$. Then, $A_1 \cap A_2 = \varnothing$ and hence, $f_*(A_1 \cap A_2) = \varnothing$. However, $f_*(A_1) = (0, 1)$, $f_*(A_2) = (0, 1)$, and hence, $f_*(A_1) \cap f_*(A_2) = (0, 1) \cap (0, 1) = (0, 1)$. We see that, in this case,

$$f_*(A_1 \cap A_2) \neq f_*(A_1) \cap f_*(A_2).$$

Let $B_1 = [0, 1]$, $B_2 = [1, 4]$. Then, $B_1 \cap B_2 = \{1\}$ and $f^*(B_1 \cap B_2) = \{-1, 1\}$. Since $f^*(B_1) = [-1, 1]$, $f^*(B_2) = [-2, -1] \cup [1, 2]$, we have $f^*(B_1) \cap f^*(B_2) = \{-1, 1\}$, and we see that, in this case,

$$f^*(B_1 \cap B_2) = f^*(B_1) \cap f^*(B_2).$$

▲ *Example 3*

Let $f: \mathbf{E} \to \mathbf{E}$ be defined as in example 2. Let $B_1 = [-1, 4]$, $B_2 = [1, 4]$. Then, $B_1 \backslash B_2 = [-1, 1)$ and $f^*(B_1 \backslash B_2) = (-1, 1)$. Since $f^*(B_1) = [-2, 2]$, $f^*(B_2) = [-2, -1] \cup [1, 2]$, we have $f^*(B_1) \backslash f^*(B_2) = (-1, 1)$, and we see that, in this case,

$$f^*(B_1 \backslash B_2) = f^*(B_1) \backslash f^*(B_2).$$

A similar relation does not hold, however, for direct images. Let $A_1 = \mathbf{E}$, $A_2 = [0, 1]$. Then, $A_1 \backslash A_2 = (-\infty, 0) \cup (1, \infty)$ and $f_*(A_1 \backslash A_2) = (0, \infty)$. Since $f_*(A_1) = [0, \infty)$, $f_*(A_2) = [0, 1]$, we have $f_*(A_1) \backslash f_*(A_2) = (1, \infty)$, and we see that in this case,

$$f_*(A_1 \backslash A_2) \neq f_*(A_1) \backslash f_*(A_2).$$

What we have seen in examples 2 and 3 is typical for direct and inverse images of intersections and complements. (We note that the inverse image is better behaved than the direct image.) In general, we have:

***Theorem 66.2***

Let $f: \mathfrak{D}(f) \to Y$ and let $X$ denote the supply set of $f$. If $\Lambda$ is some given index set and if $A_\lambda \subseteq X$ for all $\lambda \in \Lambda$, then

$$f_*\left(\bigcap_{\lambda \in \Lambda} A_\lambda\right) \subseteq \bigcap_{\lambda \in \Lambda} f_*(A_\lambda). \tag{66.14}$$

If $A_1, A_2 \subseteq X$, then

$$f_*(A_1 \backslash A_2) \subseteq f_*(A_1). \tag{66.15}$$

If $M$ is some given index set and if $B_\mu \subseteq Y$ for all $\mu \in M$, then

$$f^*\left(\bigcap_{\mu \in M} B_\mu\right) = \bigcap_{\mu \in M} f^*(B_\mu). \tag{66.16}$$

If $B_1, B_2 \subseteq Y$, then

$$f^*(B_1 \backslash B_2) = f^*(B_1) \backslash f^*(B_2). \tag{66.17}$$

In particular,

$$f^*(c(B)) = \mathfrak{D}(f) \backslash f^*(B). \tag{66.18}$$

*Proof*

Since $\bigcap_{\lambda \in \Lambda} A_\lambda \subseteq A_\lambda$ for all $\lambda \in \Lambda$, we have, from (66.10), that $f_*(\bigcap_{\lambda \in \Lambda} A_\lambda) \subseteq f_*(A_\lambda)$ for all $\lambda \in \Lambda$ and (66.14) follows readily.

Since $A_1 \backslash A_2 \subseteq A_1$, (66.15) follows from (66.10).

Since $\bigcap_{\mu \in M} B_\mu \subseteq B_\mu$ for all $\mu \in M$, we have, from (66.10), that $f^*(\bigcap_{\mu \in M} B_\mu) \subseteq f^*(B_\mu)$ for all $\mu \in M$, and hence,

$$f^*\left(\bigcap_{\mu \in M} B_\mu\right) \subseteq \bigcap_{\mu \in M} f^*(B_\mu). \tag{66.19}$$

Conversely, if $x \in \bigcap_{\mu \in M} f^*(B_\mu)$, then $x \in f^*(B_\mu)$ for all $\mu \in M$ and $f(x) \in B_\mu$ for all $\mu \in M$. Hence, $f(x) \in \bigcap_{\mu \in M} B_\mu$ and therefore, $x \in f^*(\bigcap_{\mu \in M} B_\mu)$. Consequently,

$$\bigcap_{\mu \in M} f^*(B_\mu) \subseteq f^*\left(\bigcap_{\mu \in M} B_\mu\right). \tag{66.20}$$

(66.16) follows from (66.19) and (66.20).

If $x \in f^*(B_1 \backslash B_2)$, then $f(x) \in B_1 \backslash B_2$; that is, $f(x) \in B_1$ but $f(x) \notin B_2$. Hence, $x \in f^*(B_1)$ but $x \notin f^*(B_2)$:

$$f^*(B_1 \backslash B_2) \subseteq f^*(B_1) \backslash f^*(B_2). \tag{66.21}$$

Conversely, if $x \in f^*(B_1) \backslash f^*(B_2)$, then $x \in f^*(B_1)$; that is, $f(x) \in B_1$, but $x \notin f^*(B_2)$, that is, $f(x) \notin B_2$. Hence, $f(x) \in B_1 \backslash B_2$ and consequently, $x \in f^*(B_1 \backslash B_2)$. Therefore,

$$f^*(B_1) \backslash f^*(B_2) \subseteq f^*(B_1 \backslash B_2). \tag{66.22}$$

(66.17) follows from (66.21) and (66.22).

If $U$ denotes the universal set, then $c(B) = U \backslash B$ and $f^*(U) = \mathfrak{D}(f)$. (66.18) follows now from (66.17) with $B_1 = U$, $B_2 = B$.

We note that the statements of Theorem 66.2 concerning the direct image are weaker than the corresponding statements concerning the inverse image. However, if $f$ is invertible, then the direct image is the inverse image under the inverse function $f^{-1}$ and, as such, satisfies the relation for intersection and complementation in the stronger form. Indeed, if $f: \mathfrak{D}(f) \xrightarrow{1\text{-}1} Y$, then $f^{-1}: \mathfrak{R}(f) \to \mathfrak{D}(f)$ exists and we obtain from (66.16) and (66.17) that

$$(f^{-1})^*\left(\bigcap_{\lambda \in \Lambda} A_\lambda\right) = \bigcap_{\lambda \in \Lambda} (f^{-1})^*(A_\lambda)$$

and

$$(f^{-1})^*(A_1\backslash A_2) = (f^{-1})^*(A_1)\backslash(f^{-1})^*(A_2).$$

Since $y = f(x)$ if and only if $x = f^{-1}(y)$, we have, for any set $A \subseteq X$,

(66.23) $$(f^{-1})^*(A) = \{y \mid f^{-1}(y) \in A\} = \{f(x) \mid x \in A\} = f_*(A).$$

Hence:

***Corollary to Theorem 66.2***

If $f: \mathfrak{D}(f) \xrightarrow{1\text{-}1} Y$, and if $X$ denotes the supply set of $f$, then

(66.24) $$f_*\left(\bigcap_{\lambda \in \Lambda} A_\lambda\right) = \bigcap_{\lambda \in \Lambda} f_*(A_\lambda), \qquad \text{where } A_\lambda \subseteq X \text{ for all } \lambda \in \Lambda;$$

(66.25) $$f_*(A_1\backslash A_2) = f_*(A_1)\backslash f_*(A_2) \qquad \text{for any sets } A_1, A_2 \subseteq X.$$

## 6.66 Exercises

1. Let $f: \mathbf{E} \to \mathbf{E}$ be defined by $f(x) = x^2$. Let $A_1 = (-1, 1)$, $A_2 = [0, 1)$, $A_3 = [-1, 0]$. Find $f_*(A_1), f_*(A_2), f_*(A_3)$.
2. Let $f: \mathbf{E} \to \mathbf{E}$ be defined by $f(x) = x^2$. Let $B_1 = (-1, 0)$, $B_2 = (-\frac{1}{4}, \frac{1}{4})$, $B_3 = \{-1, 0, 1\}$. Find $f^*(B_1), f^*(B_2), f^*(B_3)$.
3. Let $f: \mathbf{E}^2 \to \mathbf{E}^2$ be defined by $\phi_1(\xi_1, \xi_2) = 4\xi_1, \phi_2(\xi_1, \xi_2) = 4\xi_2$. Let $A = \{(\xi_1, \xi_2) \in \mathbf{E}^2 \mid \xi_1^2 + \xi_2^2 < 1\}$. Find $f_*(A)$.
4. Let $f: \mathbf{E}^2 \to \mathbf{E}^2$ be defined by $\phi_1(\xi_1, \xi_2) = 4\xi_1$, $\phi_2(\xi_1, \xi_2) = 4\xi_2$. Let $B = \{(\eta_1, \eta_2) \in \mathbf{E}^2 \mid \eta_1^2 + \eta_2^2 < 1\}$. Find $f^*(B)$.
5. Let $f: \mathbf{E} \to \mathbf{E}$ be defined by $f(x) = x^2$.

   (a) Let $A_1 = (0, 1)$, $A_2 = [-1, 1]$. Show by direct computation that $f_*(A_1) \subseteq f_*(A_2)$.
   (b) Let $B_1 = \{0\} \cup (1, 4)$, $B_2 = [0, 4]$. Show by direct computation that $f^*(B_1) \subseteq f^*(B_2)$.
   (c) Let $\Lambda = \{2, 3, 4, \ldots\}$ and $A_\lambda = (1/\lambda, \lambda)$. Show by direct computation that $f_*(\bigcup_{\lambda \in \Lambda} A_\lambda) = \bigcup_{\lambda \in \Lambda} f_*(A_\lambda)$.
   (d) Let $M = \mathbf{R}$ and $B_\mu = (-\mu^2, \mu^2)$. Show by direct computation that

   $$f^*\left(\bigcup_{\mu \in \mathbf{R}} B_\mu\right) = \bigcup_{\mu \in \mathbf{R}} f^*(B_\mu) \qquad \text{and} \qquad f^*\left(\bigcap_{\mu \in \mathbf{R}} B_\mu\right) = \bigcap_{\mu \in \mathbf{R}} f^*(B_\mu).$$

6. Prove that

   $$f^*\left(\bigcup_{\mu \in M} B_\mu\right) = \bigcup_{\mu \in M} f^*(B_\mu).$$

7. Let $f: \mathbf{E}\backslash\{0\} \to \mathbf{E}$ be defined by $f(x) = 1/|x|$, and let $B = [-7, 1)$. Find $f^*(B)$.
8. Show that $A\backslash B = \varnothing$ if and only if $A \subseteq B$.
9. Show that (66.17) is trivially true if $B_1 = \varnothing$.

10. Let $f:\mathbf{E}^2\to\mathbf{E}^2$ be defined by $\phi_1(\xi_1,\xi_2)=\xi_1\cos\xi_2, \phi_2(\xi_1,\xi_2)=\xi_1\sin\xi_2$. Let

$$A=\{(\xi_1,0)\in\mathbf{E}^2\,|\,0\leqslant\xi_1\leqslant 1\}\cup\{(1,\xi_2)\in\mathbf{E}^2\,|\,0\leqslant\xi_2\leqslant 2\pi\}$$
$$\cup\{(\xi_1,2\pi)\in\mathbf{E}^2\,|\,0\leqslant\xi_1\leqslant 1\}\cup\{(0,\xi_2)\in\mathbf{E}^2\,|\,0\leqslant\xi_2\leqslant 2\pi\}.$$

Find $f_*(A)$ and interpret geometrically.
11. Same as in exercise 10, for the function $f:\mathbf{E}^2\to\mathbf{E}^2$ which is defined by $\phi_1(\xi_1,\xi_2)=\cos\xi_1, \phi_2(\xi_1,\xi_2)=\sin\xi_1\cos\xi_2$ and $\xi_1,\xi_2\in[0,\pi]$.
12. Given $f:\mathbf{E}\to\mathbf{E}$ by $f(x)=\sin x$ and $B_1=[0,2]$, $B_2=(-1,4)$. Show by direct computation, that

$$f^*(B_1)\subseteq f^*(B_2),\qquad f^*(B_1\cup B_2)=f^*(B_1)\cup f^*(B_2),$$
$$f^*(B_1\cap B_2)=f^*(B_1)\cap f^*(B_2),\qquad f^*(B_1\backslash B_2)=f^*(B_1)\backslash f^*(B_2).$$

13. Let $f:\mathbf{E}\to\mathbf{E}$ and suppose that $f$ is such that for any open set $B\subseteq\mathbf{E}, f^*(B)$ is open. Show that for any closed set $C\subseteq\mathbf{E}, f^*(C)$ is closed.
14. Show that $f^*(f_*(\mathfrak{D}(f)))=\mathfrak{D}(f)$.

## 6.67 Composite Functions

If $f:\mathbf{E}\to\mathbf{E}$ is given by $f(x)=x^2$ and $g:\mathbf{E}\to\mathbf{E}$ by $g(y)=\sin y$, then $g(f(x))=\sin(x^2)$ defines a function from $\mathbf{E}$ into $\mathbf{E}$, the so-called composition of $g$ and $f$ or the *composite function* of $g$ and $f$.

In general, if $f:\mathfrak{D}(f)\to\mathbf{E}^p$ where $\mathfrak{D}(f)\subseteq\mathbf{E}^n$, if $g:\mathfrak{D}(g)\to\mathbf{E}^m$ where $\mathfrak{D}(g)\subseteq\mathbf{E}^p$, and if $\mathfrak{R}(f)\cap\mathfrak{D}(g)\neq\varnothing$, then, by (66.9), $f^*(\mathfrak{D}(g))\neq\varnothing$, and $g(f(x))$ is defined for all $x\in f^*(\mathfrak{D}(g))$.

Since $f$, $g$ are functions, we have, from $x_1=x_2$, where $x_1,x_2\in f^*(\mathfrak{D}(g))$, that $f(x_1)=f(x_2)$ and hence, $g(f(x_1))=g(f(x_2))$. Also $g(f(x))\in\mathbf{E}^m$ for all $x\in f^*(\mathfrak{D}(g))$. Hence, $g(f(x))$ defines a function from $f^*(\mathfrak{D}(g))$ into $\mathbf{E}^m$, the composite function $g\circ f$. (See also Fig. 67.1.)

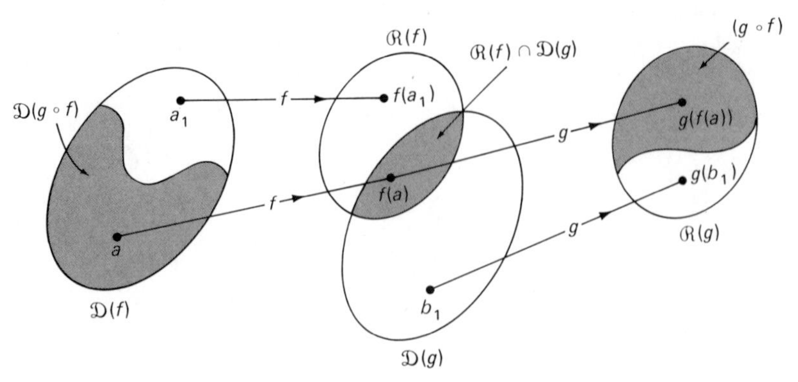


*Fig. 67.1*

***Definition 67.1***

If $f : \mathfrak{D}(f) \to \mathbf{E}^p$, where $\mathfrak{D}(f) \subseteq \mathbf{E}^n$, and $g : \mathfrak{D}(g) \to \mathbf{E}^m$, where $\mathfrak{D}(g) \subseteq \mathbf{E}^p$, and if $f^*(\mathfrak{D}(g)) \neq \varnothing$, then the function

$$g \circ f : \mathfrak{D}(g \circ f) \to \mathbf{E}^m$$

where $\mathfrak{D}(g \circ f) = f^*(\mathfrak{D}(g)) \subseteq \mathbf{E}^n$, which is defined by $(g \circ f)(x) = g(f(x))$, is called the *composite function* of $g$ and $f$ (or the *composition* of $g$ and $f$).

---

▲ *Example 1*

Let $f : \mathbf{E}^2 \to \mathbf{E}^3$ be given by $\phi_1(\xi_1, \xi_2) = \xi_1 - \xi_2, \phi_2(\xi_1, \xi_2) = \xi_2^2, \phi_3(\xi_1, \xi_2) = -\xi_1$ and let $g : \mathbf{E}^3 \to \mathbf{E}$ be given by $g(\eta_1, \eta_2, \eta_3) = \eta_1 + \eta_2 + \eta_3^2$. Then

$$\mathfrak{R}(f) \cap \mathfrak{D}(g) = \mathfrak{R}(f) \cap \mathbf{E}^3 = \mathfrak{R}(f) \neq \varnothing.$$

$\mathfrak{D}(g \circ f) = f^*(\mathfrak{D}(g)) = f^*(\mathbf{E}^3) = \mathbf{E}^2$. Hence, $g \circ f$ is defined on $\mathbf{E}^2$ by

$$(g \circ f)(x) = g(f(x)) = g(\xi_1 - \xi_2, \xi_2^2, -\xi_1) = \xi_1 - \xi_2 + \xi_2^2 + \xi_1^2.$$

---

If $A \subseteq \mathbf{E}^n$ is any set, then

$$\begin{aligned} g_*(f_*(A)) &= \{g(y) \mid y \in f_*(A)\} = \{g(y) \mid y \in \{f(x) \mid x \in A\}\} \\ &= \{g(f(x)) \mid x \in A\} = (g \circ f)_*(A). \end{aligned}$$

Similarly, for any set $B \subseteq \mathbf{E}^m$

$$\begin{aligned} f^*(g^*(B)) &= \{x \mid f(x) \in g^*(B)\} = \{x \mid f(x) \in \{y \mid g(y) \in B\}\} \\ &= \{x \mid g(f(x)) \in B\} = (g \circ f)^*(B). \end{aligned}$$

Hence,

$$(g \circ f)_*(A) = g_*(f_*(A)), \qquad (g \circ f)^*(B) = f^*(g^*(B)). \tag{67.1}$$

Loosely speaking, the composition of continuous functions is a continuous function. More precisely:

***Theorem 67.1***

If $f : \mathfrak{D}(f) \to \mathbf{E}^p$, $\mathfrak{D}(f) \subseteq \mathbf{E}^n$, is continuous at $a \in \mathfrak{D}(f)$ and if $g : \mathfrak{D}(g) \to \mathbf{E}^m$, $\mathfrak{D}(g) \subseteq \mathbf{E}^p$, is continuous at $b = f(a) \in \mathfrak{D}(g)$, then $g \circ f : \mathfrak{D}(g \circ f) \to \mathbf{E}^m$ is continuous at $a$.

*Proof*

By hypothesis, we have, for every $\varepsilon > 0$, a $\delta_1 > 0$ such that $g_*(N_{\delta_1}(b)) \subseteq N_\varepsilon(g(b))$ and a $\delta > 0$ such that $f_*(N_\delta(a)) \subseteq N_{\delta_1}(f(a)) = N_{\delta_1}(b)$. Hence, $(g \circ f)_*(N_\delta(a)) = g_*(f_*(N_\delta(a))) \subseteq g_*(N_{\delta_1}(b)) \subseteq N_\varepsilon(g(b))$; that is, $g \circ f$ is continuous at $a$. (See also Fig. 67.2.)

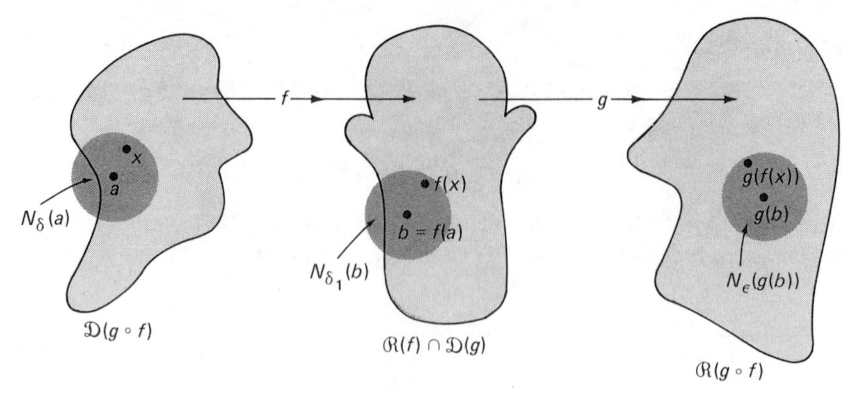


***Fig. 67.2***

That the condition in Theorem 67.2 is not necessary may be seen from the following example:

---

▲ *Example 2*

Let $f : \mathbf{E} \to \mathbf{E}$ be given by

$$f(x) = \begin{cases} 1 & \text{if } x \in \mathbf{Q}, \\ -1 & \text{if } x \in \mathbf{I}, \end{cases}$$

and $g$: $\mathbf{E} \to \mathbf{E}$ by $g(y) = y^2$. Then $(g \circ f)(x) = 1$ for all $x \in \mathbf{E}$ and is continuous everywhere, although $f$ is not continuous anywhere.

---

## 6.67 Exercises

1. Given $f : \mathbf{E}^3 \to \mathbf{E}$ by $f(\xi_1, \xi_2, \xi_3) = \sqrt{\xi_1^2 + \xi_2^2 + \xi_3^2}$ and $g : [0, 1] \to \mathbf{E}^2$ by $\psi_1(y) = \sqrt{1 - y^2}$, $\psi_2(y) = \sqrt{y}$. Find $\mathfrak{D}(g \circ f)$, $g \circ f : \mathbf{E}^3 \to \mathbf{E}^2$, $\lim_\theta (g \circ f)(x)$, and check whether or not $g \circ f$ is continuous at $\theta$.
2. Let $f : [0, 4\pi] \to \mathbf{E}^2$ be defined by

$$\phi_1(x) = \begin{cases} x & \text{for } 0 \leqslant x < \pi, \\ \pi & \text{for } \pi \leqslant x < 2\pi, \\ -x + 3\pi & \text{for } 2\pi \leqslant x < 3\pi, \\ 0 & \text{for } 3\pi \leqslant x \leqslant 4\pi; \end{cases}$$

$$\phi_2(x) = \begin{cases} 0 & \text{for } 0 \leqslant x < \pi, \\ x - \pi & \text{for } \pi \leqslant x < 2\pi, \\ \pi & \text{for } 2\pi \leqslant x < 3\pi, \\ -x + 4\pi & \text{for } 3\pi \leqslant x \leqslant 4\pi; \end{cases}$$

and $g : \mathbf{E}^2 \to \mathbf{E}^2$ by

$$\psi_1(\eta_1, \eta_2) = \cos \eta_1,$$
$$\psi_2(\eta_1, \eta_2) = \sin \eta_1 \cos \eta_2.$$

Sketch $(g \circ f)_*([0, 4\pi])$.

3. Let $f : \mathbf{E}^2 \to \mathbf{E}^2$ be defined by

$$\phi_1(\xi_1, \xi_2) = \xi_1 \cos \xi_2,$$
$$\phi_2(\xi_1, \xi_2) = \xi_1 \sin \xi_2,$$

and $g : \mathbf{E}^2 \to \mathbf{E}^2$ by

$$\psi_1(\eta_1, \eta_2) = \eta_1^2 + \eta_2^2,$$
$$\psi_2(\eta_1, \eta_2) = \eta_1^2 - \eta_2^2.$$

Show that $g \circ f : \mathbf{E}^2 \to \mathbf{E}^2$ is continuous on $\mathbf{E}^2$.

4. Let $f : \mathfrak{D}(f) \to \mathbf{E}^p$, $\mathfrak{D}(f) \subseteq \mathbf{E}^n$, $g : \mathfrak{D}(g) \to \mathbf{E}^m$, $\mathfrak{D}(g) \subseteq \mathbf{E}^p$. Let $a$ denote an accumulation point of $\mathfrak{D}(g \circ f) = f^*(\mathfrak{D}(g)) \neq \varnothing$. *Prove*:
   (a) If $\lim_a f(x) = b$ exists, if $b \in \mathfrak{D}(g)$, and if $g$ is continuous at $b$, then $\lim_a (g \circ f)(x) = g(b)$.
   (b) If $\lim_a f(x) = b$ exists and if $f(x) \neq b$ for all $x \in N'_\delta(a) \cap \mathfrak{D}(g \circ f)$ for some $\delta > 0$, then $b$ is an accumulation point of $\mathfrak{D}(g)$.
   (c) If the hypotheses in (b) hold and if $\lim_b g(y) = c$, then $\lim_a (g \circ f)(x) = \lim_b g(y) = c$.

5. Let $f, g : \mathbf{E} \to \mathbf{E}$ be defined by $f(z) = g(z) = 0$ if $z \neq 0$ and $= 1$ if $z = 0$. Find $\lim_0 (g \circ f)(x)$ and show that the results in exercise 4 do not apply.

6. Let $f : \mathbf{E} \backslash \{0\} \to \mathbf{E}$ be given by $f(x) = x \sin(1/x)$ and let $g : \mathbf{E} \to \mathbf{E}$ be given by

$$g(y) = \sqrt{1 + y^2}.$$

Use the result of exercise 4 to find $\lim_0 (g \circ f)(x)$. Check your result by evaluating $(g \circ f)(x)$ and computing the limit directly.

7. Let $f(x) = x$ if $x \in \mathbf{Q}$, $-x$ if $x \in \mathbf{I}$, and $g(y) = -y$ if $y \in \mathbf{Q}$ and $y$ if $y \in \mathbf{I}$. Use parts (b) and (c) of exercise 4 to find $\lim_0 (g \circ f)(x)$.

8. Let $f : \mathbf{E} \to \mathbf{E}$ be defined by $f(x) = x$ for $x \in \mathbf{Q}$, $1 - x$ for $x \in \mathbf{I}$, and let $g : \mathbf{E} \to \mathbf{E}$ be defined by $g(y) = y^2$ for $y \neq \frac{1}{2}$, 1 for $y = \frac{1}{2}$. Use exercise 4 to find $\lim_{1/2} (g \circ f)(x)$.

9. Let $f : (0, \infty) \times (0, \pi/2) \to \mathbf{E}^2$ be defined by

$$\phi_1(x) = \xi_1 \cos \xi_2, \qquad \phi_2(x) = \xi_1 \sin \xi_2,$$

and let $g : (0, \infty) \times (0, \infty) \to \mathbf{E}^2$ be defined by $\psi_1(y) = \sqrt{\eta_1^2 + \eta_2^2}$, $\psi_2(y) = \operatorname{Tan}^{-1}(\eta_2/\eta_1)$ where $\operatorname{Tan}^{-1}$ is the principal value of the inverse tangent function. Find $g \circ f$.

## 6.68 Global Characterization of Continuous Functions

There is a widespread belief among students of mathematics that continuous functions map open sets onto open sets. The genealogy of this rumor is easily traced. According to an often encountered and sloppy interpretation of continuity, points that are near each other are mapped by a continuous function onto points that are again near each other. "Near" becomes "neighborhood" and "neighborhood" becomes "open set," and there it is. As we have pointed out on a previous occasion (see section 2.31), continuous functions need not map open sets onto open sets, nor do they necessarily map closed sets onto closed sets, as we can see from the following examples:

---

▲ *Example 1*

The continuous function $f : \mathbf{E} \to \mathbf{E}$, defined by

$$f(x) = \begin{cases} x & \text{for } x \in (-\infty, 1), \\ 1 & \text{for } x \in [1, 2], \\ x - 1 & \text{for } x \in (2, \infty), \end{cases}$$

maps the open set $(1, 2)$ onto the closed set $\{1\}$. (See Fig. 68.1.)

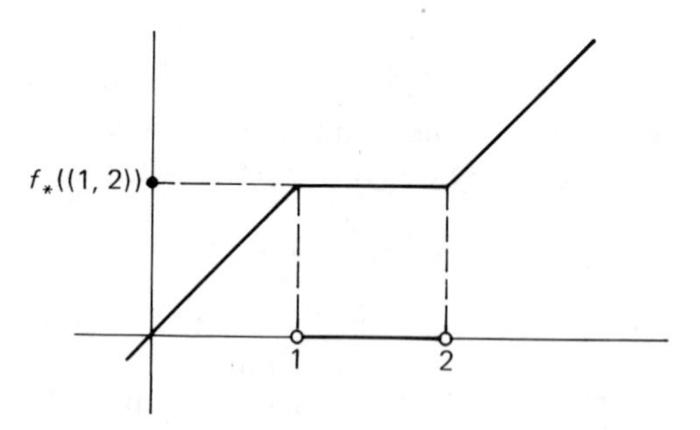


*Fig. 68.1*

▲ *Example 2*

The continuous function $f : \mathbf{E} \to \mathbf{E}$, defined by $f(x) = x/(1 + x^2)$, maps the closed set $[1, \infty)$ onto $(0, \frac{1}{2}]$, which is neither open nor closed. (See Fig. 68.2.)

---

What is not true for the direct images of continuous functions is, however, true, with a slight modification, for the inverse images of continuous functions. We examine this situation in the following examples:

---

▲ *Example 3*

Let $f : \mathbf{E} \to \mathbf{E}$ be defined by $f(x) = x^2$. $G = (-1, 1)$ is open and

$$f^*(G) = \{x \in \mathbf{E} \mid x^2 \in (-1, 1)\} = (-1, 1)$$

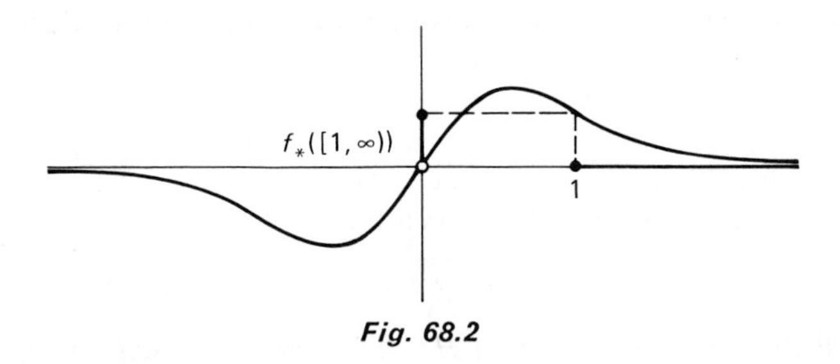


*Fig. 68.2*

is open. $H = [-1, 1]$ is closed and

$$f^*(H) = \{x \in \mathbf{E} \mid x^2 \in [-1, 1]\} = [-1, 1]$$

is closed.

▲ *Example 4*

When the domain of $f$ is not all of $\mathbf{E}$, then the situation is not so simple. Suppose we restrict the domain of $f$, defined as above, to $\mathfrak{D}(f) = [0, 1]$. $G = (\frac{1}{4}, 2)$ is open and $f^*(G) = \{x \in \mathfrak{D}(f) \mid x^2 \in (\frac{1}{4}, 2)\} = (\frac{1}{2}, 1]$ is *not* open (as a subset of $\mathbf{E}$). Still, $(\frac{1}{2}, 1]$ may be viewed as an open set if the domain of $f$, namely $[0, 1]$, is viewed as the universal set. Then $(\frac{1}{2}, 1]$ is an open set in $[0, 1]$ because now, the point 1 which causes the difficulties appears as an interior point of $(\frac{1}{2}, 1]$. The reasoning for this goes as follows: Since $\mathfrak{D}(f)$ is the universal set, any $\delta$-neighborhood of 1 is defined as

$$N_\delta(1) = \{x \in [0, 1] \mid |x - 1| < \delta\}.$$

Hence, there exists a $\delta > 0$ such that $N_\delta(1) \subseteq (\frac{1}{2}, 1]$ and hence, 1 is an interior point of $(\frac{1}{2}, 1]$. One says, in this case, that $(\frac{1}{2}, 1]$ is *relatively open* in $[0, 1]$.

We shall take a different approach in this treatment. Although $(\frac{1}{2}, 1]$ is not open in $\mathbf{E}$, there is, nevertheless, an open set, say $G_1 = (\frac{1}{2}, \frac{3}{2}) \subseteq \mathbf{E}$ such that $G_1 \cap \mathfrak{D}(f) = (\frac{1}{2}, \frac{3}{2}) \cap [0, 1] = (\frac{1}{2}, 1] = f^*(G)$.

Similarly, if we consider $f$ on the domain $\mathfrak{D}(f) = (0, 1)$, then the inverse image of $H = [\frac{1}{4}, 2]$, namely

$$f^*(H) = \{x \in (0, 1) \mid x^2 \in [\tfrac{1}{4}, 2]\} = [\tfrac{1}{2}, 1),$$

may be viewed either as *relatively closed* in $(0, 1)$, or as the intersection of $\mathfrak{D}(f)$ with some closed set, say $H_1 = [\frac{1}{2}, 27]$:

$$f^*(H) = [\tfrac{1}{2}, 1) = \mathfrak{D}(f) \cap H_1 = (0, 1) \cap [\tfrac{1}{2}, 27] = [\tfrac{1}{2}, 1).$$

---

We formalize the ideas which we developed in example 4 in the following theorem which characterizes functions that are continuous on their domain.

***Theorem 68.1***

$f : \mathfrak{D} \to \mathbf{E}^m$, where $\mathfrak{D} \subseteq \mathbf{E}^n$, is continuous on $\mathfrak{D}$ if and only if for *every open set* $G \subseteq \mathbf{E}^m$ *there is an open set* $G_1 \subseteq \mathbf{E}^n$ such that

$$f^*(G) = G_1 \cap \mathfrak{D}, \tag{68.1}$$

or, equivalently, for *every closed set* $H \subseteq \mathbf{E}^m$ *there is a closed set* $H_1 \subseteq \mathbf{E}^n$ such that

$$f^*(H) = H_1 \cap \mathfrak{D}. \tag{68.2}$$

(*Note*: If one substitutes for $\mathfrak{D}$ a subset $\mathcal{A}$ of $\mathfrak{D}$ in the above theorem, one obtains a criterion for the continuity of $f$ on $\mathcal{A}$.)

*Proof*

(a) Suppose $f$ is continuous on $\mathfrak{D}$ and $G \subseteq \mathbf{E}^m$ is an open set. If $f^*(G) = \varnothing$, we need not go further because $\varnothing$ is open. Hence, we may assume without loss of generality, that $f^*(G) \neq \varnothing$. Then, for every $a \in f^*(G) \subseteq \mathfrak{D}$, we have $f(a) \in G$. Since $G$ is open, there is an $\varepsilon > 0$ such that $N_\varepsilon(f(a)) \subseteq G$. Since $f$ is continuous at $a$, there is a $\delta(\varepsilon, a) > 0$ such that

$$f_*(N_{\delta(\varepsilon, a)}(a)) \subseteq N_\varepsilon(f(a)) \subseteq G,$$

that is,

$$N_{\delta(\varepsilon, a)}(a) \cap \mathfrak{D} \subseteq f^*(G). \tag{68.3}$$

Let

$$G_1 = \bigcup_{a \in f^*(G)} N_{\delta(\varepsilon, a)}(a).$$

$G_1$ is open, being the union of open sets. If $x \in f^*(G)$, then $x \in G_1$. Therefore, $f^*(G) \subseteq G_1$. Since $f^*(G) \subseteq \mathfrak{D}$,

$$f^*(G) \subseteq G_1 \cap \mathfrak{D}. \tag{68.4}$$

On the other hand, if $x \in G_1 \cap \mathfrak{D}$, then $x \in N_{\delta(\varepsilon, a)}(a)$ for some $a \in f^*(G)$ and, by (68.3), $x \in f^*(G)$. Therefore,

$$G_1 \cap \mathfrak{D} \subseteq f^*(G). \tag{68.5}$$

(68.4) and (68.5) together yield (68.1).

(b) Suppose that, for every open $G \subseteq \mathbf{E}^m$, there is an open set $G_1 \subseteq \mathbf{E}^n$ such that (68.1) is true.

Let $a \in \mathfrak{D}$. For every $\varepsilon > 0$, $N_\varepsilon(f(a)) = G$ is an open set in $\mathbf{E}^m$ and $a \in f^*(N_\varepsilon(f(a)))$. By hypothesis, there is an open set $G_1 \subseteq \mathbf{E}^n$ such that $f^*(N_\varepsilon(f(a))) = G_1 \cap \mathfrak{D}$. Since $a \in G_1$ and since $G_1$ is open, there is a $\delta(\varepsilon) > 0$ such that $N_{\delta(\varepsilon)}(a) \subseteq G_1$.

Since $N_{\delta(\varepsilon)}(a) \cap \mathfrak{D} \subseteq G_1 \cap \mathfrak{D} = f^*(N_\varepsilon(f(a)))$, we have

$$f_*(N_{\delta(\varepsilon)}(a)) \subseteq N_\varepsilon(f(a)),$$

which means, according to (66.5), that $f$ is continuous at $a$. Since $a$ is any point in $\mathfrak{D}$, $f$ is continuous on $\mathfrak{D}$.

(c) We shall now demonstrate that (68.1) implies (68.2). Let $H \subseteq \mathbf{E}^m$ denote a closed set. Then $c(H) = G$ is open and, by (68.1), there is an open set $G_1 \subseteq \mathbf{E}^n$ such that

$$f^*(c(H)) = G_1 \cap \mathfrak{D}. \tag{68.6}$$

By (66.18), $f^*(c(H)) = \mathfrak{D} \backslash f^*(H) = \mathfrak{D} \cap c(f^*(H))$. Hence,

$$\mathfrak{D} \cap c(f^*(H)) = G_1 \cap \mathfrak{D}.$$

We take the complement on both sides and obtain, by DeMorgan's law, that $c(\mathfrak{D}) \cup f^*(H) = c(G_1) \cup c(\mathfrak{D})$. If we intersect both sides by $\mathfrak{D}$, we obtain $f^*(H) \cap \mathfrak{D} = c(G_1) \cap \mathfrak{D}$. (See also exercise 3.) Since $f^*(H) \subseteq \mathfrak{D}$, and since $H_1 = c(G_1)$ is closed, we have $f^*(H) = H_1 \cap \mathfrak{D}$, which is (68.2).

In order to show that (68.2) also implies (68.1), one only needs to reverse the roles of $H$, $G$, $H_1$, $G_2$, closed, open in the above argument (exercise 4).

This global characterization of continuous functions as presented in Theorem 68.1 will be used in section 6.69 to establish the continuity of the inverse function of a one-to-one open map, in section 6.71 to demonstrate that continuous functions map compact sets onto compact sets, and in section 6.73 to show that continuous functions map connected sets onto connected sets. What we mean by open maps, compact sets, and connected sets will be explained in sections 6.69, 6.70, and 6.72.

## 6.68 Exercises

1. Given the function $f : \mathbf{E} \to \mathbf{E}$ by

$$f(x) = \begin{cases} \dfrac{1}{x} & \text{for } x \neq 0, \\ 0 & \text{for } x = 0. \end{cases}$$

Use Theorem 68.1 to show that $f$ is not continuous on $\mathbf{E}$.

2. Given the function $f : \mathbf{E}^2 \to \mathbf{E}$ by

$$f(\xi_1, \xi_2) = \begin{cases} 1 & \text{for } \xi_1^2 + \xi_2^2 < 1, \\ 0 & \text{for } \xi_1^2 + \xi_2^2 \geqslant 1. \end{cases}$$

Use Theorem 68.1 to show that $f$ is not continuous on $\mathbf{E}^2$.

3. Show that $\mathfrak{D} \cap (c(\mathfrak{D}) \cup f^*(H)) = \mathfrak{D} \cap f^*(H)$ and $\mathfrak{D} \cap (c(G_1) \cup c(\mathfrak{D})) = \mathfrak{D} \cap c(G_1)$, where $\mathfrak{D}$, $H$, $G_1$ are defined in the proof of Theorem 68.1.
4. Show that (68.2) implies (68.1).
5. Let $\mathfrak{D}(f) = \mathbf{E}^n$. Show that $f : \mathfrak{D}(f) \to \mathbf{E}^m$ is continuous if and only if, for every open set $G \subseteq \mathbf{E}^m$, there is an open set $G_1 \subseteq \mathbf{E}^n$ such that $f^*(G) = G_1$.
6. Let $\mathfrak{D}(f) = \mathbf{E}^n$. Show that $f : \mathfrak{D}(f) \to \mathbf{E}^m$ is continuous if and only if, for every closed set $H \subseteq \mathbf{E}^m$, there is a closed set $H_1 \subseteq \mathbf{E}^n$ such that $f^*(H) = H_1$.
7. Prove: $f : \mathfrak{D} \to \mathbf{E}^m$, $\mathfrak{D} \subseteq \mathbf{E}^n$, is continuous on $\mathfrak{D}$ if and only if, for every $\varepsilon > 0$ and every $y \in \mathbf{E}^m$, there is an open set $G_1 \subseteq \mathbf{E}^n$ such that $f^*(N_\varepsilon(y)) = G_1 \cap \mathfrak{D}$.

## 6.69 Open Maps

### *Definition 69.1*

The function $f : \mathfrak{D} \to \mathbf{E}^m$, $\mathfrak{D} \subseteq \mathbf{E}^n$, is an open map if and only if for every open set $\Omega \subseteq \mathbf{E}^n$ there is an open set $\Omega_1 \subseteq \mathbf{E}^m$ such that

$$f_*(\Omega) = \Omega_1 \cap \mathfrak{R}(f).$$

Clearly, if $\mathfrak{D} = \mathbf{E}^n$ and $\mathfrak{R}(f) = \mathbf{E}^m$, then $f$ is an open map if and only if it maps all open subsets of $\mathbf{E}^n$ onto open subsets of $\mathbf{E}^m$ because, in this case, $\Omega \cap \mathfrak{D} = \Omega \cap \mathbf{E}^n = \Omega$ and $\Omega_1 \cap \mathfrak{R}(f) = \Omega_1 \cap \mathbf{E}^m = \Omega_1$.

If $\mathfrak{D}$ is open and if $f$ maps all open subsets of $\mathfrak{D}$ onto open sets in $\mathbf{E}^m$ then $f$ is also an open map.

---

▲ *Example 1*

The function defined in example 1, section 6.68, is not an open map because $f_*((1, 2)) = \{1\}$ and there is no open set $\Omega_1 \subseteq \mathbf{E}$ such that $\Omega_1 \cap \mathbf{E} = \{1\}$.

▲ *Example 2*

$f$: $\mathbf{E} \to \mathbf{E}$, defined by $f(x) = x^3$, is an open map. In view of Theorem 59.3, we need only show that $f_*(N_\delta(x))$ is open for every $\delta > 0$ and every $x \in \mathbf{E}$. We have $N_\delta(x) = (x - \delta, x + \delta)$ and hence, $f_*(N_\delta(x)) = ((x - \delta)^3, (x + \delta)^3$, which is open. By Theorem 59.3, every open set $\Omega \subset \mathbf{E}$ may be represented as $\Omega = \bigcup_{x \in \Omega} N_{\delta_x}(x)$, where $N_{\delta_x}(x) \subseteq \Omega$. Hence, by Theorem 66.1, $f_*(\Omega) = f_* \bigcup_{x \in \Omega} N_{\delta_x}(x)) = \bigcup_{x \in \Omega} f_*(N_{\delta_x}(x))$, which, as the union of open sets, is open.

---

### *Theorem 69.1*

If $\mathfrak{D} \subseteq \mathbf{E}^n$ and $f : \mathfrak{D} \to \mathbf{E}^m$ is a one-to-one open map, then the inverse function $f^{-1} : \mathfrak{R}(f) \to \mathfrak{D}$ exists and is continuous.

*Proof*

Since $f : \mathfrak{D} \xrightarrow[1\text{-}1]{} \mathbf{E}^m$, $f^{-1} : \mathfrak{R}(f) \rightarrow \mathbf{E}^n$ exists. We have, from (66.23), that

$$(f^{-1})^*(A) = f_*(A). \tag{69.1}$$

If $G \subseteq \mathbf{E}^n$ is open, there exists an open set $G_1 \subseteq \mathbf{E}^m$ such that $f_*(G) = G_1 \cap \mathfrak{R}(f) = G_1 \cap \mathfrak{D}(f^{-1})$. Hence, by (69.1), $(f^{-1})^*(G) = G_1 \cap \mathfrak{D}(f^{-1})$, that is, $f^{-1}$ is continuous on $\mathfrak{R}(f)$ by Theorem 68.1.

Note that $f$ itself need not be continuous. If $f$ is open and one-to-one, then $f^{-1}$ is continuous even if $f$ is not. (See example 2, section 2.27.)

## 6.69 Exercises

1. Given the function $f : \mathbf{E} \rightarrow \mathbf{E}$ by $f(x) = x^2$. Is $f$ an open map?
2. Give an example of a function that is an open map but is not one-to-one.
3. Is $f : \mathbf{E} \rightarrow \mathbf{E}$, given by $f(x) = \sin x$, an open map?
4. Show that the function $f : \mathbf{E}^2 \rightarrow \mathbf{E}^3$ which is given by $\phi_1(\xi_1, \xi_2) = \xi_1 + \xi_2$, $\phi_2(\xi_1, \xi_2) = \xi_1 - \xi_2$, $\phi_3(\xi_1, \xi_2) = \xi_1^2$, is one-to-one and open but does not map open sets onto open sets in $\mathbf{E}^3$.
5. Show: If $f : \mathfrak{D} \rightarrow \mathbf{E}^m$ is an open map and if $\mathfrak{R}(f)$ is open in $\mathbf{E}^m$, then $f$ maps all open subsets of $\mathfrak{D}$ onto open subsets of $\mathbf{E}^m$.
6. Let $f \in C'(\mathbf{R})$ and $f'(x) > 0$ for all $x \in \mathbf{R}$. Show that $f : \mathbf{R} \rightarrow \mathbf{R}$ is an open map.
7. Show: If $\mathfrak{D} \subseteq \mathbf{E}^n$ is open and $f : \mathfrak{D} \rightarrow \mathbf{E}^m$ maps all open subsets of $\mathfrak{D}$ onto open sets in $\mathbf{E}^m$, then $f$ is an open map.

## 6.70 Compact Sets and the Heine–Borel Theorem

In a straightforward generalization of definition 31.1, we call a collection $\Omega$ of open sets $\Omega_\alpha \subseteq \mathbf{E}^n$,

$$\Omega = \{\Omega_\alpha \mid \alpha \in A\},$$

where $A$ is a given index set, an *open cover* of the set $S \subseteq \mathbf{E}^n$ if and only if

$$S \subseteq \bigcup_{\alpha \in A} \Omega_\alpha.$$

We say that $S \subseteq \mathbf{E}^n$ has the *Heine–Borel property* if and only if every open cover $\Omega$ of $S$ contains finitely many sets

$$\Omega' = \{\Omega_{\alpha_j} \mid j = 1, 2, \ldots, m\}, \qquad \alpha_j \in A,$$

such that $\Omega'$ is also an open cover of $S$; that is,

$$S \subseteq \bigcup_{j=1}^{m} \Omega_{\alpha_j}.$$

($\Omega'$ is called a *finite subcover*.)

***Definition 70.1***

$K \subseteq \mathbf{E}^n$ is called *compact* if and only if it has the *Heine–Borel property.*

We obtain, as in section 2.32:

***Theorem 70.1*** (*Heine–Borel Theorem*)

$K \subseteq \mathbf{E}^n$ is compact if and only if it is *bounded* and *closed.*

*Proof*

(a) One can show that a compact set is bounded and closed exactly as in part (a) of the proof of Theorem 32.1. No words need to be changed. Only the absolute-value symbols have to be replaced by the norm symbols. (See also Fig. 70.1 and exercise 1.)

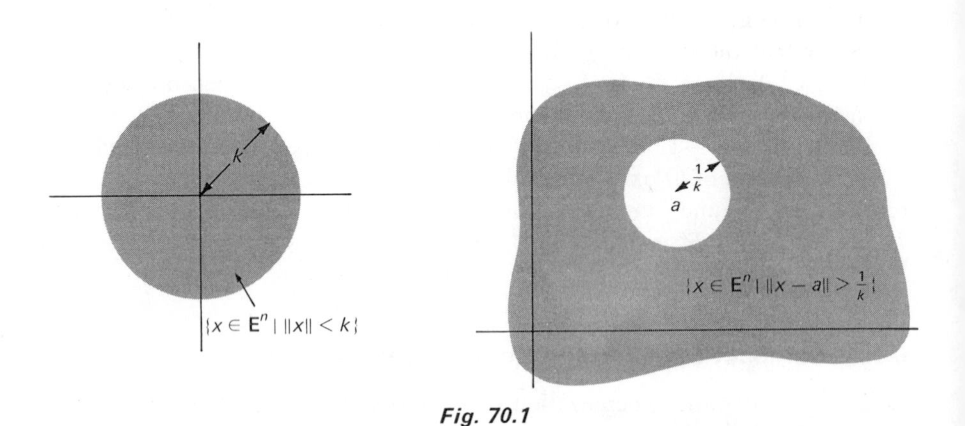


*Fig. 70.1*

(b) If $K$ is bounded, then there is an $M > 0$ such that $\|x\| \leqslant M$ for all $x \in K$. Hence, whenever $x \in K$, then $x \in [-M, M] \times [-M, M] \times \cdots \times [-M, M] = C^n$, which is an $n$-dimensional cube. (See Fig. 70.2.)

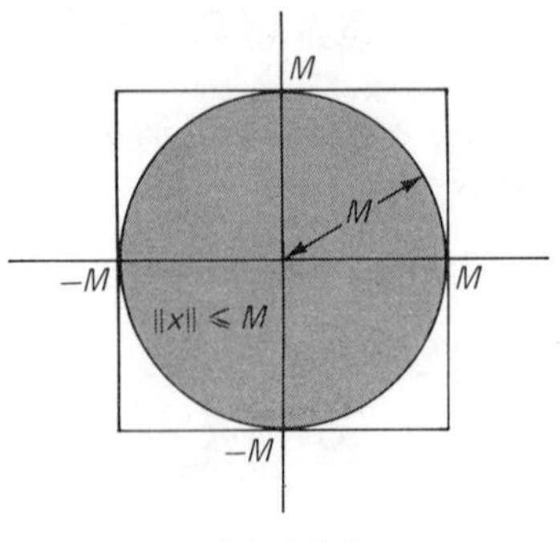


*Fig. 70.2*

We observe, as in the proof of Theorem 32.1, that if $\Omega$ is an open cover of $K$, then $\{\Omega, c(K)\}$ is an open cover of $C^n$, because $c(K)$ is open and $(\bigcup_{\alpha \in A} \Omega_\alpha) \cup c(K) = \mathbf{E}^n$. If we can show that $C^n$ has the Heine–Borel property, then we can select from $\{\Omega, c(K))$ a finite subcover $\{\Omega_{\alpha_1}, \Omega_{\alpha_2}, \ldots, \Omega_{\alpha_m}, c(K)\}$ of $C^n$. Hence, $\{\Omega_{\alpha_1}, \Omega_{\alpha_2}, \ldots, \Omega_{\alpha_m}\}$ is a finite subcover of $K$, and the theorem is proved.

Therefore, we need only show that $C^n$ has the Heine–Borel property. We proceed as in part (b) of the proof of Theorem 32.1 by partitioning $C^n$ into $2^n$ congruent cubes with nonoverlapping interiors and repeating this process *ad infinitum*. This straightforward generalization of the earlier proof for the one-dimensional case yields a valid proof of the sufficiency part of Theorem 70.1, and we may consider the theorem as proved.

As in Theorem 32.2, we obtain:

***Theorem 70.2***

$K \subseteq \mathbf{E}^n$ is compact if and only if one can select from every infinite subset $A$ of $K$ a nonconstant sequence $\{a_k\}$ which converges to a point in $K$.

The proof is left for the exercises (exercise 3).

## 6.70 Exercises

1. Prove that, if $K \subseteq \mathbf{E}^n$ is compact, then it is bounded and closed.
2. Prove that a bounded and closed subset of $\mathbf{E}^n$ has the Heine–Borel property, by modifying the proof of Theorem 32.1 to fit this more general setting.
3. Prove Theorem 70.2.
4. Given $A \subseteq \mathbf{E}^n \times \mathbf{E}^m$. By the projection $p_{n.}(A)$ into $\mathbf{E}^n$ we mean the set

$$p_{n.}(A) = \{(\xi_1, \xi_2, \ldots, \xi_n) \in \mathbf{E}^n \mid (\xi_1, \xi_2, \ldots, \xi_n, \xi_{n+1}, \ldots, \xi_{n+m}) \in A \quad \text{for some } (\xi_{n+1}, \ldots, \xi_{n+m}) \in \mathbf{E}^m\}.$$

   Prove: If $K \subseteq \mathbf{E}^n \times \mathbf{E}^m$ is compact, then $p_{n.}(A)$ and $p_{.m}(A)$ are compact subsets of $\mathbf{E}^n$ and $\mathbf{E}^m$, respectively.
5. Given $K \subseteq \mathbf{E}^n$, $H \subseteq \mathbf{E}^m$. Prove: if $K$, $H$ are compact, then $K \times H$ is compact in $\mathbf{E}^n \times \mathbf{E}^m$.
6. Prove the statements in exercises 4 and 5 using the Heine–Borel Theorem.
7. Prove: $K \subseteq \mathbf{E}^n$ is compact if and only if every cover of $K$ by $\delta$-neighborhoods $\{N_{\delta_\alpha}(\alpha) \mid \alpha \in A\}$, where $A$ is a given index set, contains finitely many $\delta$-neighborhoods that cover $K$.
8. Given an index set $A$. The collection $\{C_\alpha \mid \alpha \in A\}$ of closed subsets of $\mathbf{E}^n$ is said to have the *finite intersection property* with respect to some given set $S \subseteq \mathbf{E}^n$ if and only if, for every finite subset $B \subset A$, $S \cap (\bigcap_{\alpha \in B} C_\alpha) \neq \varnothing$. Prove: $S \subseteq \mathbf{E}^n$ is compact if and only if, for every collection $\{C_\alpha \mid \alpha \in A\}$ of closed subsets of $\mathbf{E}^n$ with the finite intersection property with respect to $S$, $S \cap (\bigcap_{\alpha \in A} C_\alpha) \neq \varnothing$ holds true.

## 6.71 The Continuous Image of a Compact Set

That the continuous image of a compact set is compact can be established by adapting the proof of Theorem 34.1 to the more general setting of this chapter. We shall present, instead, a much simpler and more elegant proof that is based on the global characterization of continuous functions.

***Theorem 71.1***

If $K \subseteq \mathbf{E}^n$ is compact and if $f: K \to \mathbf{E}^m$ is continuous on $K$, then $f_*(K)$ is compact.

*Proof*

Let $\Omega = \{\Omega_\alpha \mid \alpha \in A\}$ denote an open cover of $f_*(K)$. Since $f$ is continuous on $K$, we have, from Theorem 68.1, that for each $\Omega_\alpha \in \Omega$, there is an open set $G_\alpha \in \mathbf{E}^n$ such that

$$f^*(\Omega_\alpha) = G_\alpha \cap K.$$

If $x \in K$, then $f(x) \in f_*(K)$ and hence, $f(x) \in \Omega_\alpha$ for some $\alpha \in A$. Hence, $x \in G_\alpha$, and it follows that $G = \{G_\alpha \mid \alpha \in A\}$ is an open cover of $K$. Since $K$ is compact, finitely many sets from $G$, say

$$G' = \{G_{\alpha_1}, G_{\alpha_2}, \ldots, G_{\alpha_l}\},$$

cover $K$. For every $f(x) \in f_*(K)$, we have $x \in G_{\alpha_j}$ for some $j \in \{1, 2, \ldots, l\}$, and hence, $f(x) \in \Omega_{\alpha_j}$. Therefore, $\Omega' = \{\Omega_{\alpha_1}, \Omega_{\alpha_2}, \ldots, \Omega_{\alpha_l}\}$ covers $f_*(K)$ and we see that $f_*(K)$ is compact.

Since $\mathbf{E}^m$, $m > 1$, is not ordered, it makes no sense to speak of a maximum or minimum of $f: \mathfrak{D} \to \mathbf{E}^m$. However, if $f$ is a real-valued function on some $\mathfrak{D} \subseteq \mathbf{E}^n$, then it does make sense to talk about a maximum or a minimum of $f$. As in section 2.34, we have:

***Theorem 71.2***

If $f: K \to \mathbf{E}$ is continuous on the compact set $K \subseteq \mathbf{E}^n$, then $f$ assumes its maximum and its minimum on $K$, that is, there are points $x_M, x_m \in K$ such that $f(x_M) \geqslant f(x)$ and $f(x_m) \leqslant f(x)$ for all $x \in K$.

*Proof*

Since, by Theorem 71.1, $f_*(K)$ is a compact set and since $f_*(K) \subseteq \mathbf{E}$, we have from Theorem 34.2, that $\sup f_*(K)$, $\inf f_*(K) \in f_*(K)$.

***Corollary to Theorem 71.2***

If $f: K \to \mathbf{E}^m$ is continuous on the compact set $K \subseteq \mathbf{E}^n$, then the function $\|f\|$: $K \to \mathbf{E}$, defined by $\|f\|(x) = \|f(x)\|$, assumes its maximum and its minimum on $K$.

*Proof*

We need only show that $\|f\| : K \to \mathbf{E}$ is continuous on $K$. Let $a \in K$. By Lemma 57.1,

$$\big| \|f\|(x) - \|f\|(a) \big| \leqslant \|f(x) - f(a)\|.$$

Hence, $\|f\|$ is continuous wherever $f$ is continuous. The conclusion of the corollary follows now from Theorem 71.2.

## 6.71 Exercises

1. Prove Theorem 71.1 by adapting the proof of Theorem 34.1.
2. Given the function $f : \mathfrak{D} \to \mathbf{E}$, where

$$\mathfrak{D} = \{x \in \mathbf{E}^n \mid \|x\| \leqslant 1\},$$

   and $f(x) = \xi_1 \xi_2 \ldots \xi_n$. Show that $f$ assumes a maximum and a minimum on $\mathfrak{D}$.
3. Given $f : \mathfrak{D} \to \mathbf{E}$ where

$$\mathfrak{D} = \{x \in \mathbf{E}^n \mid \xi_1 \xi_2 \ldots \xi_n = 1\}, \qquad n > 1,$$

   and $f(x) = (1/n) \sum_{k=1}^{n} \xi_k$. Show that $f$ does *not* assume a maximum on $\mathfrak{D}$. What conclusions can you draw about $\mathfrak{D}$?
4. Let $f : \mathfrak{D} \to \mathbf{E}^m$, $\mathfrak{D} \subseteq \mathbf{E}^n$, where $\mathfrak{D}$ and $\mathfrak{R}(f)$ are bounded sets. *Prove*: $f$ is continuous on $\mathfrak{D}$ if and only if, for every compact set $K \subseteq \mathbf{E}^m$, there is a compact set $C \subseteq \mathbf{E}^n$ such that $f^*(K) = C \cap \mathfrak{D}$.

## 6.72 Connected Sets

The open interval $(a, a)$ is empty and the closed interval $[a, a]$ contains one point only. We shall disregard such degenerate intervals in our discussion and consider only intervals that contain at least two distinct points. Such intervals $\mathfrak{J} \subseteq \mathbf{R}$ (open, or closed, or neither) may be characterized by the property that, with any two points $x, y \in \mathfrak{J}$, with $x < y$, all points between $x$ and $y$, namely $\{z \in \mathbf{R} \mid z = (1 - t)x + ty,\ t \in [0, 1]\}$ also lie in $\mathfrak{J}$. It is clear that every interval has this property. Conversely, suppose that a set $S \subseteq \mathbf{R}$ has this property. If $a = \inf S$, $b = \sup S$ exist, then, depending on whether $a, b$ lie in $S$, or one does and the other one does not, or neither lies in $S$, we obtain $S = [a, b]$ or $S = [a, b)$ or $S = (a, b]$ or $S = (a, b)$. If $\inf S$ does not exist or $\sup S$ does not exist, or neither exists, we obtain

$$S = (-\infty, b] \text{ or } S = (-\infty, b) \text{ or } S = [a, \infty) \text{ or } S = (a, \infty) \text{ or } S = (-\infty, \infty).$$

Intervals are *connected subsets* of $\mathbf{E}$. In order to arrive at a characterization of connected sets in $\mathbf{E}^n$, let us first establish an alternate characterization of intervals.

***Lemma 72.1***

$\mathfrak{J} \subseteq \mathbf{R}$ is an interval if and only if it is *not* possible to find two open sets $A, B \subseteq \mathbf{R}$ such that

$$A \cap \mathfrak{J} \neq \varnothing, \qquad B \cap \mathfrak{J} \neq \varnothing, \tag{72.1}$$

$$(A \cap \mathfrak{J}) \cup (B \cap \mathfrak{J}) = \mathfrak{J}, \tag{72.2}$$

$$(A \cap \mathfrak{J}) \cap (B \cap \mathfrak{J}) = \varnothing. \tag{72.3}$$

*Note*: Two such sets exist, for example, for $S = \{x \in \mathbf{R} \mid -1 \leqslant x < 0 \text{ or } 0 < x \leqslant 1\}$, as we have depicted in Fig. 72.1.

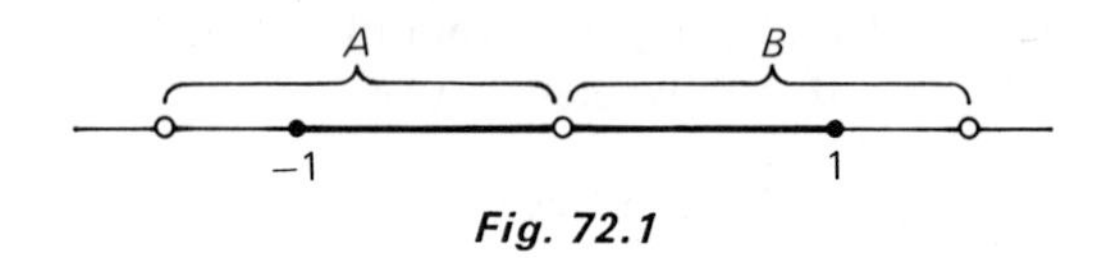


***Fig. 72.1***

*Proof* (by contradiction)

(a) Let $\mathfrak{J}$ denote an interval and suppose that there are open sets $A$, $B$ such that (72.1), (72.2), (72.3) are satisfied. Let $a \in A \cap \mathfrak{J}$, $b \in B \cap \mathfrak{J}$, where we assume without loss of generality that $a < b$, and consider the set

$$S = \{z \in [a, b] \mid z \in A \cap \mathfrak{J}\}.$$

$S$ is not empty because $a \in S$. $S$ is bounded above and hence,

$$s_0 = \sup S \in [a, b] \tag{72.4}$$

exists.

Suppose that $s_0 \in B \cap \mathfrak{J}$. Since $B$ is open, there is a $\delta > 0$ such that $(s_0 - \delta, s_0] \subseteq B \cap \mathfrak{J}$. ($\mathfrak{J}$ contains all points between $a$ and $b$!) Since $s_0 = \sup S$, there is a $z \in A \cap \mathfrak{J}$, $a \leqslant z \leqslant b$, such that $s_0 - \delta < z \leqslant s_0$. This is not possible because $(A \cap \mathfrak{J}) \cap (B \cap \mathfrak{J}) = \varnothing$. Hence $s_0 \notin B \cap \mathfrak{J}$ and therefore, $s_0 < b$ because $b \in B \cap \mathfrak{J}$. (See also Fig. 72.2.)

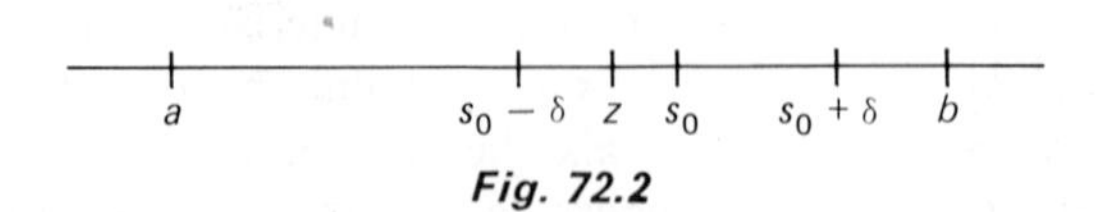


***Fig. 72.2***

Suppose that $s_0 \in A \cap \mathfrak{J}$. Since $A$ is open and $s_0 < b$, there is a $\delta > 0$ such that $[s_0, s_0 + \delta) \subseteq A \cap \mathfrak{J}$, in contradiction to (72.4).

Thus, $a \in \mathfrak{J}$, $b \in \mathfrak{J}$, $a < s_0 < b$ where $s_0 \notin (A \cap \mathfrak{J}) \cup (B \cap \mathfrak{J}) = \mathfrak{J}$. Hence, $\mathfrak{J}$ cannot be an interval.

(b) Suppose that $\mathfrak{J}$ is not an interval but contains at least two points. Then there are two points $a, b \in \mathfrak{J}$ and a point $z \notin \mathfrak{J}$ satisfying the condition that $a < z < b$. Let $A = \{x \in \mathbf{R} \mid x < z\}$, $B = \{x \in \mathbf{R} \mid x > z\}$. $A$, $B$ are open and satisfy conditions (72.1), (72.2), and (73.3), as the reader can easily verify.

Lemma 72.1 provides the motivation for the following definition:

***Definition 72.1***

The set $S \subseteq \mathbf{E}^n$ is *disconnected* if and only if there are two open sets $A$, $B \subseteq \mathbf{E}^n$ such that

$$A \cap S \neq \emptyset, \qquad B \cap S \neq \emptyset,$$
$$(A \cap S) \cup (B \cap S) = S,$$
$$(A \cap S) \cap (B \cap S) = \emptyset.$$

Such sets $A$, $B$ are said to establish a *disconnection* of $S$.

The set $S \subseteq \mathbf{E}^n$ is called *connected* if and only if it is not disconnected.

---

▲ *Example 1*

The set $S \subseteq \mathbf{E}^2$ which is darkly shaded in Fig. 72.3 is disconnected.
The sets $A$, $B$ which are lightly shaded in Fig. 72.3 establish a disconnection of $S$.

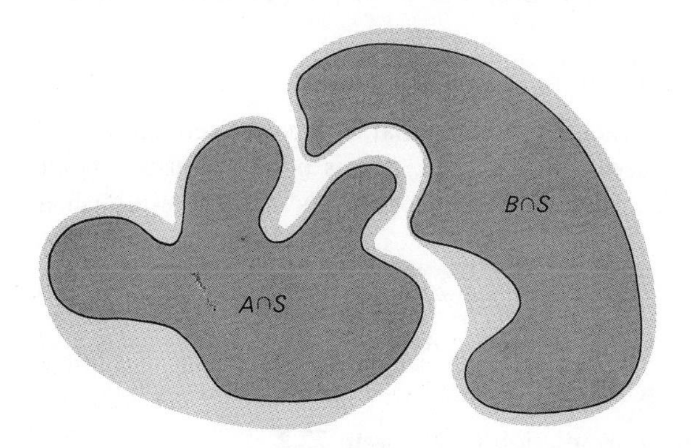


***Fig. 72.3***

---

We obtain from Lemma 72.1 and definition 72.1 immediately,

***Theorem 72.1***

A set $S \subseteq \mathbf{R}$ is connected if and only if it is an interval. In particular, $\mathbf{R}$, being an interval, is connected.

More generally, we have:

***Theorem 72.2***

$\mathbf{E}^n$ is connected.

*Proof* (by contradiction)

Suppose that the open sets $A, B \subseteq \mathbf{E}^n$ establish a disconnection of $\mathbf{E}^n$. Let $a \in A \cap \mathbf{E}^n = A$, $b \in B \cap \mathbf{E}^n = B$, and consider the sets

$$A_1 = \{t \in \mathbf{R} \mid (1-t)a + tb \in A\},$$
$$B_1 = \{t \in \mathbf{R} \mid (1-t)a + tb \in B\}.$$

Now, $A_1, B_1$ are open subsets of $\mathbf{R}$ because $f: \mathbf{E} \to \mathbf{E}^n$, defined by $f(t) = (1-t)a + tb$, is continuous and $A_1 = f^*(A)$, $B_1 = f^*(B)$, where $0 \in A_1$, $1 \in B_1$. Hence $A_1 \cap [0,1] \neq \emptyset$, $B_1 \cap [0,1] \neq \emptyset$. Since $A \cup B = \mathbf{E}^n$, we have $(A_1 \cap [0,1]) \cup (B_1 \cap [0,1]) = [0,1]$. Since $A \cap B = \emptyset$, we have $(A_1 \cap [0,1]) \cap (B_1 \cap [0,1]) = \emptyset$. Hence, $A_1$, $B_1$ establish a disconnection for $[0,1]$, in contradiction to Theorem 72.1.

We mentioned in section 2.28 that $\emptyset$ and $\mathbf{R}$ are the only subsets of $\mathbf{R}$ that are both open and closed, and indicated that a proof would be forthcoming at a later occasion. This *is* the later occasion.

***Theorem 72.3***

$\mathbf{E}^n$ and $\emptyset$ are the only subsets of $\mathbf{E}^n$ that are both open and closed.

*Proof* (by contradiction)

Suppose $A \neq \emptyset$ and $A \subset \mathbf{E}^n$, and that $A$ is both open and closed. Then, $c(A) = \mathbf{E}^n \setminus A$ is closed because $A$ is open, and is open because $A$ is closed. $c(A) \neq \emptyset$ because $A \subset E^n$. Hence, the two open sets $A$, $c(A)$ establish a disconnection for $\mathbf{E}^n$:

$$A \neq \emptyset \qquad c(A) \neq \emptyset,$$
$$A \cup c(A) = \mathbf{E}^n,$$
$$A \cap c(A) = \emptyset.$$

This stands in contradiction to Theorem 72.2.

## 6.72 Exercises

1. Show that the Cantor set $C$ is not connected. (See section 1.14.)
2. In example 4, section 6.68, we explained the terms *relatively open* and *relatively closed*. Show that definition 72.1 is equivalent to the following definition: $S \subseteq \mathbf{E}^n$ is disconnected if and only if there are two sets $A$, $B \subseteq S$ which are *relatively open* in $S$ such that $A \neq \emptyset$, $B \neq \emptyset$, $A \cup B = S$, $A \cap B = \emptyset$.
3. Show that definition 72.1 is equivalent to the following definition: $S \subseteq \mathbf{E}^n$ is disconnected if and only if there are two sets $A$, $B \subseteq \mathbf{E}^n$ such that $A \cap S \neq \emptyset$, $B \cap S \neq \emptyset$, $S \subseteq A \cup B$ and $(A \cup A') \cap B = \emptyset$, as well as $A \cap (B \cup B') = \emptyset$.

4. $S \subseteq \mathbf{E}^n$ is called *convex* if and only if, whenever $x, y \in S$, then $z = (1-t)x + ty \in S$ for all $t \in [0, 1]$. Prove: $\mathbf{E}^n$ is convex.
5. Prove: A convex set is connected.
6. Prove: the $n$-dimensional interval $\mathfrak{J} = (\alpha_1, \beta_1) \times (\alpha_2, \beta_2) \times \cdots \times (\alpha_n, \beta_n)$, with $\alpha_j < \beta_j$ for $j = 1, 2, \ldots, n$, is connected.
7. Prove: $N_\delta(a)$, $\overline{N_\delta(a)} \subseteq \mathbf{E}^n$ are connected sets.
8. Which of the following sets are disconnected?

   (a) $S = \{x \in \mathbf{E}^2 \mid \xi_1 > 0\} \cup \{\theta\}$
   (b) $S = \{x \in \mathbf{E}^3 \mid \xi_3 \leqslant 0\} \cup \{x \in \mathbf{E}^3 \mid \xi_3 > \xi_1^2 + \xi_2^2\}$
   (c) $S = \{x \in \mathbf{E}^3 \mid \xi_3 > \xi_1^2 + \xi_2^2\} \cup \{x \in \mathbf{E}^3 \mid \xi_3 < \xi_1^2 + \xi_2^2\}$

## 6.73 The Continuous Image of a Connected Set

We have shown, in section 6.71, that a continuous function preserves compactness. We shall now demonstrate that a continuous function also preserves connectedness.

### *Theorem 73.1*

If $f: \mathfrak{D} \to \mathbf{E}^m$ is continuous on the connected set $\mathfrak{D} \subseteq \mathbf{E}^n$, then $f_*(\mathfrak{D})$ is connected.

*Proof* (by contradiction)

Suppose that $f_*(\mathfrak{D})$ is not connected. Then, there are two open sets $A, B \subseteq \mathbf{E}^m$ such that $A \cap f_*(\mathfrak{D}) \neq \varnothing, B \cap f_*(\mathfrak{D}) \neq \varnothing, (A \cap f_*(\mathfrak{D})) \cup (B \cap f_*(\mathfrak{D})) = f_*(\mathfrak{D})$, and $(A \cap f_*(\mathfrak{D})) \cap (B \cap f_*(\mathfrak{D})) = \varnothing$. Since $f$ is continuous, we obtain, from Theorem 68.1, that there are two open sets $A_1, B_1 \subseteq E^n$ such that

$$f^*(A) = A_1 \cap \mathfrak{D}, \qquad f^*(B) = B_1 \cap \mathfrak{D}. \tag{73.1}$$

Since $A \cap f_*(\mathfrak{D}) \neq \varnothing$, $B \cap f_*(\mathfrak{D}) \neq \varnothing$, we have, from (66.9), that

$$A_1 \cap \mathfrak{D} \neq \varnothing, \ B_1 \cap \mathfrak{D} \neq \varnothing. \tag{73.2}$$

Since $f_*(\mathfrak{D}) = (A \cap f_*(\mathfrak{D})) \cup (B \cap f_*(\mathfrak{D}))$ we have,

$$\begin{aligned} \mathfrak{D} = f^*(A \cap f_*(\mathfrak{D})) \cup f^*(B \cap f_*(\mathfrak{D})) &= f^*(A) \cup f^*(B) \\ &= (A_1 \cap \mathfrak{D}) \cup (B_1 \cap \mathfrak{D}). \end{aligned} \tag{73.3}$$

By (73.1), and since $A \cap B \cap f_*(\mathfrak{D}) = \varnothing$,

$$(A_1 \cap \mathfrak{D}) \cap (B_1 \cap \mathfrak{D}) = f^*(A) \cap f^*(B) = f^*(A \cap B) = \varnothing. \tag{73.4}$$

We see from (73.2), (73.3), and (73.4), that $\mathfrak{D}$ is disconnected, contrary to our hypothesis.

### *Corollary 1 to Theorem 73.1*

If $f: \mathfrak{D} \to \mathbf{E}^m$, $\mathfrak{D} \subseteq \mathbf{E}^n$, is continuous on $\mathfrak{D}$ and if $A \subseteq \mathfrak{D}$ is connected, then $f_*(A)$ is connected.

*Proof*

Apply Theorem 73.1 to $f : A \to \mathbf{E}^m$.

***Corollary 2 to Theorem 73.1***

If $f : [a, b] \to \mathbf{R}$ is continuous on $[a, b]$, then $f_*([a, b])$ is an interval. (The same is true if $[a, b]$ is replaced by $(a, b)$, or $(a, b]$, or $[a, b)$, or $(-\infty, b)$, or $(a, \infty)$, or $(-\infty, b]$, or $[a, \infty)$, or $(-\infty, \infty)$.)

*Proof*

By Theorem 73.1, $f_*([a, b])$ is connected. By Theorem 72.1, $f_*([a, b])$ is an interval. (See also the Corollary to Theorem 34.2.)

We can see now that the intermediate-value theorem for continuous functions (Theorem 27.1) is an immediate consequence of the above corollary: $f(a)$, $f(b)$ lie in the interval $f_*([a, b])$, and hence, $f$ has to assume every value between $f(a)$ and $f(b)$. For this reason, Theorem 73.1 is often referred to as the generalized intermediate-value theorem.

The reader may have come across some loose talk about a set being connected whenever any two points $a$, $b$ in the set can be joined by a polygonal path (broken line with finitely many vertices) that lies entirely in the set. (See Fig. 73.1.)

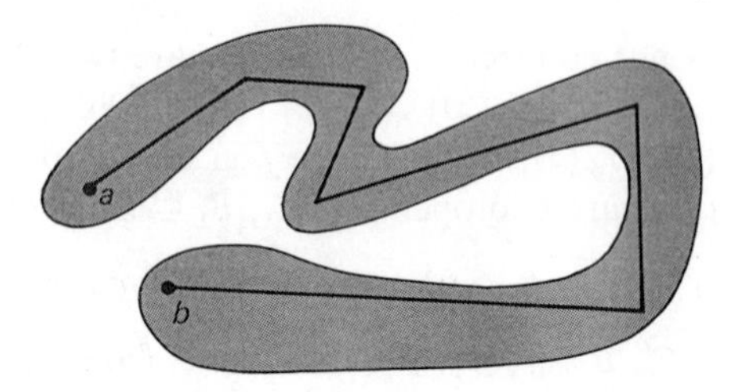


*Fig. 73.1*

It is true that such a set is connected (see exercise 2). However, the converse is not always true; i.e., a connected set need not have this property, as we shall show in the following example.

---

▲ *Example 1*

The set $\mathfrak{D} = \{x \in \mathbf{E}^2 \mid \xi_2 = \xi_1^2\}$, being a continuous image of the set $\mathbf{E}$, is connected. (See Fig. 73.2.) Clearly, no two points in $\mathfrak{D}$ can be joined by a polygonal path that lies in $\mathfrak{D}$.

The set $\mathfrak{D}$ in the above example is a closed set. In connected sets that are not open, it is not always possible to join two points by a polygonal path that lies in the set. However, an open set is connected if and only if any two points in the set can be joined by a polygonal path that lies in the set. (See exercises 2 and 3.)

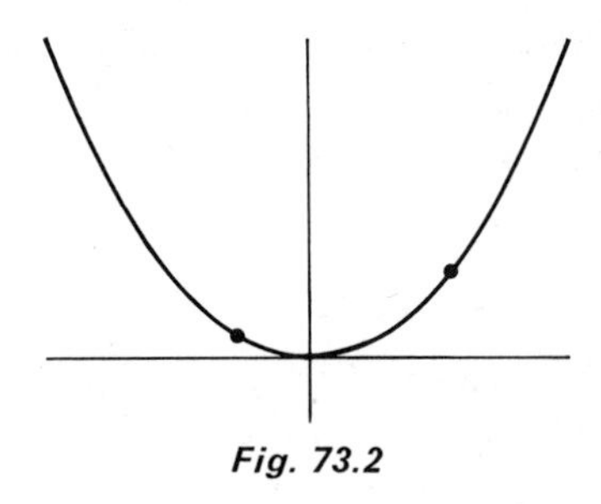

*Fig. 73.2*

▲ *Example 2* (*Brouwer's Fixed-Point Theorem*)

If one fills a glass with water (or any other liquid of one's choice) and moves it gently so that surface particles stay on the surface and submerged particles remain submerged, then one may view the configuration of particles at any given instant as the continuous image of the configuration at a previous instant. It turns out that there is at least one particle that occupies the same place at both instants. This, in essence, is the content of Brouwer's fixed-point theorem.† In general,

> *If* $f: S \to S$, *where* $S = \{x \in \mathbf{E}^n \mid \|x\| \leqslant R\}$ *for some* $R > 0$, *is continuous on* $S$, *then there is at least one* $x_0 \in S$ *such that* $f(x_0) = x_0$. $x_0$ is called the *fixed point* of $f: S \to S$.

A proof of this theorem in its full generality is beyond the scope of this treatment. Therefore, we shall present a proof only for the case $n = 1$, as an interesting application of Theorem 73.1.

Note that we may choose $S = [0, 1]$ without loss of generality. The theorem states that the graph of $f$ has to intersect the graph of $l : [0, 1] \to [0, 1]$ that is given by $l(x) = x$. (See Fig. 73.3.)

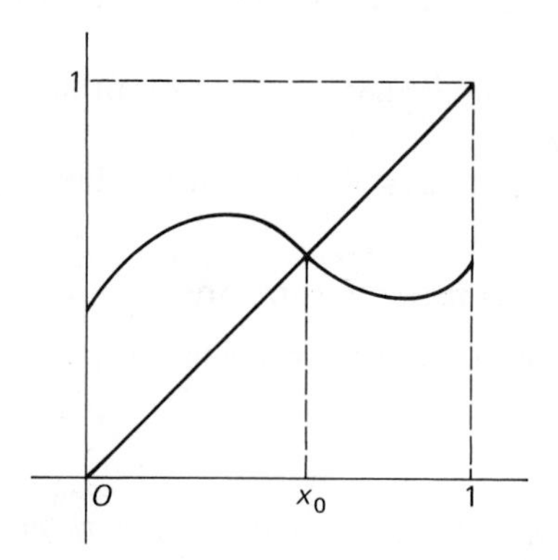


*Fig. 73.3*

† *L. E. J. Brouwer*, 1881–1967.

If $f(0) = 0$ or $f(1) = 1$, then the theorem is trivially true. Hence, we may assume without loss of generality that $f(0) > 0$ and $f(1) < 1$.

The function $g : [0, 1] \to \mathbf{E}^2$ which is defined by

$$\psi_1(x) = x,$$
$$\psi_2(x) = f(x),$$

is continuous on $[0, 1]$. By Theorem 73.1, $g_*([0, 1])$ is connected. Suppose that $f(x) \neq x$ for all $x \in [0, 1]$; that is, $(x, x) \notin g_*([0, 1])$ for all $x \in [0, 1]$. Then, the sets $A = \{y \in \mathbf{E}^2 \mid \eta_1 > \eta_2\}$, $B = \{y \in \mathbf{E}^2 \mid \eta_1 < \eta_2\}$ establish a disconnection of $g_*([0, 1])$ contrary to the fact that $g_*([0, 1])$ is connected. Hence, $(x_0, x_0) \in g_*([0, 1])$ for some $x_0 \in [0, 1]$.

A simpler, but less interesting proof is obtained from an application of the intermediate-value theorem to the continuous function $h : [0, 1] \to \mathbf{E}$ that is defined by $h(x) = f(x) - x$. The details are left to the reader.

---

## 6.73 Exercises

1. Show by Theorem 73.1 that $f : \mathbf{E} \to \mathbf{E}$, defined by

$$f(x) = \begin{cases} \dfrac{1}{x} & \text{for } x \neq 0 \\ 0 & \text{for } x = 0 \end{cases}$$

   is not continuous on $\mathbf{E}$.
2. *Prove*: If $S \subseteq \mathbf{E}^n$ and if any two points in $S$ can be joined by a polygonal path that lies entirely in $S$, than $S$ is connected.
3. *Prove*: If $S \subseteq \mathbf{E}^n$ is open and connected, then any two points in $S$ can be joined by a polygonal path that lies entirely in $S$.
4. Show that the set $\mathfrak{D} = \{x \in \mathbf{E}^2 \mid 0 < \xi_2 \leqslant 1\}$ is connected.
5. Show that the set $\mathfrak{D} = \{x \in \mathbf{E}^2 \mid 0 < \xi_2 \leqslant \xi_1^2\} \cup \{(0, 0)\}$ is connected but contains pairs of points that cannot be joined by a polygonal path that lies in the set.
6. Use the intermediate-value theorem to prove Brouwer's fixed-point theorem in example 2 for $n = 1$.
7. Show: If $n > 1$, then $N'_\delta(a) \subseteq \mathbf{E}^n$ is connected but not convex.

## 6.74 Uniformly Continuous Functions

We generalize definition 33.1 of uniform continuity as follows:

***Definition 74.1***

The function $f : \mathfrak{D} \to \mathbf{E}^m$, $\mathfrak{D} \subseteq \mathbf{E}^n$, is *uniformly continuous* on $\mathfrak{D}$ if and only if, for every $\varepsilon > 0$, there is a $\delta(\varepsilon) > 0$ such that

$$(74.1) \quad \|f(x_1) - f(x_2)\| < \varepsilon \qquad \text{for all } x_1, x_2 \in \mathfrak{D} \text{ for which } \|x_1 - x_2\| < \delta(\varepsilon).$$

If we negate (74.1), we obtain that $f$ is not uniformly continuous on $\mathfrak{D}$ if and only if there is an $\varepsilon > 0$ such that for every $\delta > 0$ there are two elements $x_1, x_2 \in \mathfrak{D}$ such that $\|x_1 - x_2\| < \delta$ but $\|f(x_1) - f(x_2)\| \geqslant \varepsilon$. Choosing the sequence $\{\delta_k\} = \{1/k\}$ and noting that for each such $\delta_k$, there have to be elements $x_k^{(1)}, x_k^{(2)} \in \mathfrak{D}$ such that $\|x_k^{(1)} - x_k^{(2)}\| < \delta_k$ but $\|f(x_k^{(1)}) - f(x_k^{(2)})\| \geqslant \varepsilon$, we obtain:

***Theorem 74.1***

The function $f: \mathfrak{D} \to \mathbf{E}^m$, $\mathfrak{D} \subseteq \mathbf{E}^n$, is not uniformly continuous on $\mathfrak{D}$ if and only if there exists an $\varepsilon > 0$ and two sequences $\{x_k^{(1)}\}$, $\{x_k^{(2)}\}$, with $x_k^{(1)}, x_k^{(2)} \in \mathfrak{D}$ for all $k \in \mathbf{N}$, such that $\|x_k^{(1)} - x_k^{(2)}\| < 1/k$ and $\|f(x_k^{(1)}) - f(x_k^{(2)})\| \geqslant \varepsilon$.
(See also Theorem 33.2.)

---

▲ *Example 1*

The function $f: (0,1) \times (0,1) \to \mathbf{E}$ which is given by $f(\xi_1, \xi_2) = 1/\xi_1 + \xi_2$ is not uniformly continuous on $(0,1) \times (0,1)$ because we have, for $x_k^{(1)} = (1/k, 0)$, $x_k^{(2)} = (1/(k+1), 0)$ that

$$\|x_k^{(1)} - x_k^{(2)}\| = \sqrt{\left(\frac{1}{k} - \frac{1}{k+1}\right)^2} = \frac{1}{k(k+1)} < \frac{1}{k}$$

and $\|f(x_k^{(1)}) - f(x_k^{(2)})\| = \|k - (k+1)\| = 1 \geqslant 1 = \varepsilon$.

---

We have proved in section 2.33 that if a real-valued function of a real variable is continuous on a compact subset of $\mathbf{R}$, then it is uniformly continuous on that compact subset (Theorem 33.1). It is easy to generalize the proof of Theorem 33.1 to fit the present situation. To provide for some variety, we shall proceed differently. We suggested in section 2.33 that such a theorem could be proved by contradiction, using the negation of uniform continuity. This is what we shall do.

***Theorem 74.2***

If $K \subseteq \mathbf{E}^n$ is compact and if $f: K \to \mathbf{E}^m$ is continuous on $K$, then $f$ is uniformly continuous on $K$.

*Proof* (by contradiction)

If $K$ contains finitely many elements only, then the theorem is trivial. We may therefore assume without loss of generality that $K$ contains infinitely many elements.

Suppose that $f$ is not uniformly continuous on $K$. Then, by Theorem 74.1, there is an $\varepsilon > 0$ and two sequences $\{x_k^{(1)}\}$, $\{x_k^{(2)}\}$, with $x_k^{(1)}, x_k^{(2)} \in K$, such that $\|x_k^{(1)} - x_k^{(2)}\| < 1/k$ but

$$\|f(x_k^{(1)}) - f(x_k^{(2)})\| \geqslant \varepsilon \qquad \text{for all } k \in \mathbf{N}. \tag{74.2}$$

By Theorem 70.2, $\{x_k^{(1)}\}$ has a subsequence $\{x_{n_k}^{(1)}\}$ such that $\lim_{k\to\infty} x_{n_k}^{(1)} = x_0^{(1)} \in K$ and $\{x_{n_k}^{(2)}\}$ contains a subsequence $\{x_{n_{kj}}^{(2)}\}$ such that

$$\lim_{j\to\infty} x_{n_{kj}}^{(2)} = x_0^{(2)} \in K.$$

Since a subsequence of a convergent sequence converges to the same limit, we have

$$\lim_{j\to\infty} x_{n_{kj}}^{(1)} = x_0^{(1)}.$$

Hence, we have, for some $N(\varepsilon) \in \mathbf{N}$, that

$$\|x_0^{(1)} - x_{n_{kj}}^{(1)}\| < \frac{\varepsilon}{3}, \quad \|x_0^{(2)} - x_{n_{kj}}^{(2)}\| < \frac{\varepsilon}{3} \qquad \text{for all } j > N(\varepsilon).$$

We also have, by construction,

$$\|x_{n_{kj}}^{(1)} - x_{n_{kj}}^{(2)}\| < \frac{1}{n_{k_j}} < \frac{\varepsilon}{3} \qquad \text{for } n_{k_j} > M$$

for some $M \in \mathbf{N}$. With $N_1 = \max(N(\varepsilon), M)$, we obtain

$$\|x_0^{(1)} - x_0^{(2)}\| \leqslant \|x_0^{(1)} - x_{n_{kj}}^{(1)}\| + \|x_{n_{kj}}^{(1)} - x_{n_{kj}}^{(2)}\| + \|x_{n_{kj}}^{(2)} - x_0^{(2)}\| < \varepsilon$$

for all $j > N_1$. Hence, $x_0^{(1)} = x_0^{(2)}$.

Since $f$ is continuous at $x_0^{(1)} \in K$, we have a $\delta(\varepsilon) > 0$ such that

$$(74.3) \quad f(x_{n_{kj}}^{(1)}) - f(x_{n_{kj}}^{(2)})\| \leqslant \|f(x_{n_{kj}}^{(1)}) - f(x_0^{(1)})\| + \|f(x_0^{(1)}) - f(x_{n_{kj}}^{(2)})\| < \varepsilon$$

provided that $\|x_{n_{kj}}^{(1)} - x_0^{(1)}\| < \delta(\varepsilon)$, $\|x_{n_{kj}}^{(2)} - x_0^{(1)}\| < \delta(\varepsilon)$, which can be assured for sufficiently large $j$. (74.3) stands in contradiction to (7.42). Hence, $f$ is uniformly continuous on $K$.

## 6.74 Exercises

1. *Prove*: $f: \mathfrak{D} \to \mathbf{E}^m$, $\mathfrak{D} \subseteq \mathbf{E}^n$ is uniformly continuous on $\mathfrak{D}$ if and only if all components of $f$ are uniformly continuous on $\mathfrak{D}$.
2. Generalize the proof of Theorem 33.1 to prove Theorem 74.2.
3. Prove that the function $f: \mathbf{E}^4 \to \mathbf{E}^3$ that is given by $\phi_1(\xi_1, \xi_2, \xi_3, \xi_4) = \xi_1 - 2\xi_2 + \xi_3 + 2\xi_4$, $\phi_2(\xi_1, \xi_2, \xi_3, \xi_4) = \xi_2 - \xi_4$, $\phi_3(\xi_1, \xi_2, \xi_3, \xi_4) = \xi_1 + \xi_2 + 6\xi_4$ is uniformly continuous on $\mathbf{E}^4$.
4. Generalize the result of exercise 3 to functions $f: \mathbf{E}^n \to \mathbf{E}^m$ that are given by $\phi_j(\xi_1, \xi_2, \ldots, \xi_n) = a_{j1}\xi_1 + a_{j2}\xi_2 + \cdots + a_{jn}\xi_n$, $j = 1, 2, \ldots, m$, $a_{jk} \in \mathbf{R}$.

## 6.75 Contraction Mappings

In this section we shall prove a fixed-point theorem that will be instrumental in our development of the inverse-function theorem in Chapter 10. The theorem we are about to discuss is less general than Brouwer's theorem, which we stated in example 2, section 6.73, because it applies only to functions that represent *contraction mappings*. To explain this term is the object of the following definition.

***Definition 75.1***

The function $f: \mathfrak{D} \to \mathbf{E}^m$, $\mathfrak{D} \subseteq \mathbf{E}^n$, satisfies a *global Lipschitz*† condition on $\mathfrak{D}$ if and only if there exists a constant $L > 0$ such that

$$\|f(x_1) - f(x_2)\| \leqslant L\|x_1 - x_2\| \qquad \text{for all } x_1, x_2 \in \mathfrak{D}. \tag{75.1}$$

If $0 < L < 1$, then $f$ is called a *contraction mapping* on $\mathfrak{D}$.

The name contraction mapping derives from the fact that the images of two points under such a mapping are closer together than the two points themselves. Indeed, take any $x_1, x_2 \in \mathfrak{D}$. Then

$$d(f(x_1), f(x_2)) = \|f(x_1) - f(x_2)\| \leqslant L\,\|x_1 - x_2\| = Ld(x_1, x_2) < d(x_1, x_2).$$

***Lemma 75.1***

If $f: \mathfrak{D} \to \mathbf{E}^m$, $\mathfrak{D} \subseteq \mathbf{E}^n$, satisfies a global Lipschitz condition on $\mathfrak{D}$, then $f$ is uniformly continuous on $\mathfrak{D}$.

*Proof*

$$\|f(x_1) - f(x_2)\| \leqslant L\|x_1 - x_2\| < \varepsilon \qquad \text{whenever } \|x_1 - x_2\| < \frac{\varepsilon}{L} = \delta(\varepsilon).$$

***Theorem 75.1*** *(Contraction Mapping Theorem)*

Let $C \subseteq \mathbf{E}^n$ denote a closed set. If $f: C \to C$ is a contraction mapping on $C$, then there is a *unique* point $x_0 \in C$ such that $f(x_0) = x_0$. (A contraction mapping has a *unique fixed point*.)

(Note that this theorem applies in particular to contraction mappings from $\mathbf{E}^n$ into $\mathbf{E}^n$ because $\mathbf{E}^n$ is a closed subset of itself.)

*Proof*

Choose any $x_1 \in C$ and let

$$x_2 = f(x_1), \quad x_3 = f(x_2), \quad \ldots, \quad x_{k+1} = f(x_k), \quad \ldots. \tag{75.2}$$

Clearly, $x_k \in C$ for all $k \in \mathbf{N}$.

Since $f$ is a contraction mapping, there is an $L \in (0, 1)$ such that $\|f(x') - f(x'')\| \leq L\|x' - x''\|$ for all $x', x'' \in C$. Hence,

$$\begin{aligned}
\|x_3 - x_2\| &= \|f(x_2) - f(x_1)\| \leqslant L\|x_2 - x_1\|, \\
\|x_4 - x_3\| &= \|f(x_3) - f(x_2)\| \leqslant L\|x_3 - x_2\| \leqslant L^2\|x_2 - x_1\|, \\
&\vdots \\
\|x_{k+1} - x_k\| &\leqslant L^{k-1}\|x_2 - x_1\|.
\end{aligned}$$

†*Rudolph Lipschitz*, 1832–1903.

By means of the triangle inequality

$$\begin{aligned}\|x_{k+p} - x_k\| &\leqslant \|x_{k+p} - x_{k+p-1}\| + \|x_{k+p-1} - x_{k+p-2}\| + \cdots + \|x_{k+1} - x_k\| \\ &\leqslant (L^{k+p-2} + L^{k+p-3} + \cdots + L^{k-1})\|x_2 - x_1\| \\ &= \left(\frac{1 - L^{k+p-1}}{1 - L} - \frac{1 - L^{k-1}}{1 - L}\right)\|x_1 - x_2\| \leqslant \frac{L^{k-1}}{1 - L}\|x_1 - x_2\|.\end{aligned}$$

Since $0 < L < 1$, we can find, for every $\varepsilon > 0$, an $N(\varepsilon) \in \mathbf{N}$ such that

$$\frac{L^{k-1}}{1 - L}\|x_1 - x_2\| < \varepsilon \qquad \text{for all } k > N(\varepsilon).$$

Hence,

$$\|x_{k+p} - x_k\| < \varepsilon \qquad \text{for all } k > N(\varepsilon);$$

that is, $\{x_k\}$ is a Cauchy sequence in $C$. Since $C$ is closed, $\lim_\infty x_k = x_0$ lies in $C$. By construction,

$$f(x_k) = x_{k+1}.$$

We pass to the limit and obtain in view of the continuity of $f$ (Lemma 75.1) that

$$f(x_0) = x_0,$$

i.e., there is at least one fixed point. Suppose, for the moment, that there is another point $x_0' \neq x_0, f(x_0') = x_0'$. Then, by (75.1)

$$\|f(x_0) - f(x_0')\| \leqslant L\|x_0 - x_0'\| < \|x_0 - x_0'\| = \|f(x_0) - f(x_0')\|$$

which is *not* possible. Hence, the fixed point is unique.

Note that for any $x_1 \in C$, the sequence defined in (75.2) converges to the fixed point $x_0$.

## 6.75 Exercises

1. Given $f: \mathbf{E}^n \to \mathbf{E}^n$ by $\phi_j(\xi_1, \xi_2, \ldots, \xi_n) = a_{j1}\xi_1 + a_{j2}\xi_2 + \ldots + a_{jn}\xi_n$, where $j = 1, 2, 3, \ldots, n$, $a_{jk} \in \mathbf{R}$. Impose suitable conditions on the coefficients $a_{jk}$ to ensure that $f$ is a contraction mapping on $\mathbf{E}^n$.
2. Given $f \in C^1(\mathbf{R})$ with $f(0) = 0, f'(0) = 0$. Show that there is a $\delta > 0$ such that $f$ is a contraction mapping from $[-\delta, \delta]$ into $[-\delta, \delta]$.
3. Show that $x^3 + x^2 - x + \frac{1}{8} = 0$ has a solution in $[-\frac{1}{4}, \frac{1}{4}]$.
4. Let $f: \mathbf{E}^2 \to \mathbf{E}^2$ be defined by $\phi_1(\xi_1, \xi_2) = \frac{1}{2}\xi_1 + \frac{1}{3}\xi_2$, $\phi_2(\xi_1, \xi_2) = \frac{1}{2}\xi_1 - \frac{1}{3}\xi_2$. Let $x_1 = (4,17)$, $x_k = f(x_{k-1})$ for $k > 1$. Show that $\lim_\infty x_k = \theta$.

# Sequences of Functions 7

## 7.76 Pointwise Convergence

Let $\mathfrak{D} \subseteq \mathbf{E}^n$ and let the functions $f_k : \mathfrak{D} \to \mathbf{E}^m$ be defined for all $k \in \mathbf{N}$. Then, $\{f_k\} = \{f_1, f_2, f_3, \ldots\}$ is called a *sequence of functions* on $\mathfrak{D}$. For every given $a \in \mathfrak{D}$, $\{f_k(a)\}$ is a sequence in $\mathbf{E}^m$ which may or not converge. If the sequences $\{f_k(x)\}$ converge for all $x \in A \subseteq \mathfrak{D}$, we say that $\{f_k\}$ converges on $A$. We have seen in Theorem 61.3 that a sequence in $\mathbf{E}^m$ converges if and only if the sequences of components converge and, consequently, that not much can be learned from a study of sequences in $\mathbf{E}^m$ that cannot be learned from a study of sequences of real numbers. For this reason we restrict our considerations in this chapter to sequences of real-valued functions. We shall also assume, without much loss of generality, that the domain is a subset of the reals. Various generalizations of our theory to vector-valued functions of a real variable and vector-valued functions of a vector variable will be dealt with in the exercises.

***Definition 76.1***

If $f_k : \mathfrak{D} \to \mathbf{E}$, $\mathfrak{D} \subseteq \mathbf{E}$, for all $k \in \mathbf{N}$, then the sequence of functions $\{f_k\}$ *converges pointwise* on $A \subseteq \mathfrak{D}$ to a function $f : A \to \mathbf{E}$ if and only if $\lim_\infty f_k(x) = f(x)$ for all $x \in A$. $f$ is called the *limit function* of $\{f_k\}$ on $A$. We express the fact that $\{f_k\}$ converges to $f$ on $A$ by writing

$$\lim_\infty f_k = f \text{ on } A \quad \text{or} \quad \{f_k\} \to f \text{ on } A \quad \text{or} \quad \lim_\infty f_k(x) = f(x) \quad \text{for all } x \in A.$$

In view of definition 61.1 we can say:

***Lemma 76.1***

The sequence $\{f_k\}$ of functions $f_k : \mathfrak{D} \to \mathbf{E}$, $\mathfrak{D} \subseteq \mathbf{E}$, converges pointwise on $A \subseteq \mathfrak{D}$ to the limit function $f : A \to \mathbf{E}$ if and only if, for all $x \in A$ and for every $\varepsilon > 0$, there is an $N(\varepsilon, x) \in \mathbf{N}$ such that

$$|f_k(x) - f(x)| < \varepsilon \qquad \text{for all } k > N(\varepsilon, x).$$

▲ *Example 1*

Let $f_k : [0, \pi] \to \mathbf{E}$ be defined by $f_k(x) = k \sin x$. Then $\{f_k\} = \{k \sin x\}$ is a sequence of functions on $\mathfrak{D} = [0, \pi]$. We see that

$$\lim_{\infty} f_k(0) = \lim_{\infty} 0 = 0,$$

$$\lim_{\infty} f_k(\pi) = \lim_{\infty} 0 = 0,$$

and that for $a \in (0, \pi)$, $\{k \sin a\}$ diverges. Hence, with $A = \{0, \pi\}$, we have

$$\lim_{\infty} f_k = f \text{ on } A,$$

where $f(x) = 0$ for all $x \in A$.

▲ *Example 2*

Let $\phi_k : [0, 1] \to \mathbf{E}$ be defined by $\phi_k(x) = x^k$. Then $\{x^k\}$ is a sequence on $\mathfrak{D} = [0, 1]$. We shall show that $\{x^k\}$ converges pointwise on $A = \mathfrak{D} = [0, 1]$.

(a) Let $x = 1 \in [0, 1]$. Then $\phi_k(1) = 1^k = 1$, and we have

$$\lim_{\infty} \phi_k(1) = 1.$$

(b) Let $0 < a < 1$. Then, $\phi_k(a) = a^k$. We have seen in section 2.15 (exercise 3) that $\lim_{\infty} a^k = 0$ for all $a \in (0, 1)$. Hence,

$$\lim_{\infty} \phi_k(a) = 0, \qquad a \in (0, 1).$$

(c) Let $x = 0 \in [0, 1]$. Then $\phi_k(0) = 0$, and we have

$$\lim_{\infty} \phi_k(0) = 0.$$

In view of these results, we see that

$$\lim_{\infty} \phi_k = \phi \qquad \text{on } A = [0, 1],$$

where

$$\phi(x) = \begin{cases} 0 & \text{for } 0 \leqslant x < 1, \\ 1 & \text{for } x = 1, \end{cases}$$

or, equivalently,

$$\lim_{\infty} x^k = \begin{cases} 0 & \text{for } 0 \leqslant x < 1 \\ 1 & \text{for } x = 1. \end{cases}$$

We observe that $\phi_k$ is continuous on $[0, 1]$ for all $k \in \mathbf{N}$ but that the limit function $\phi$ is *not* continuous on $[0, 1]$.

▲ *Example 3*

Let $\psi_k : [0, 1] \to \mathbf{E}$ be defined by

$$\psi_k(x) = \begin{cases} 1 - kx & \text{for } 0 \leqslant x \leqslant \dfrac{1}{k}, \\ 0 & \text{for } \dfrac{1}{k} < x \leqslant 1. \end{cases}$$

(See Fig. 76.1.)

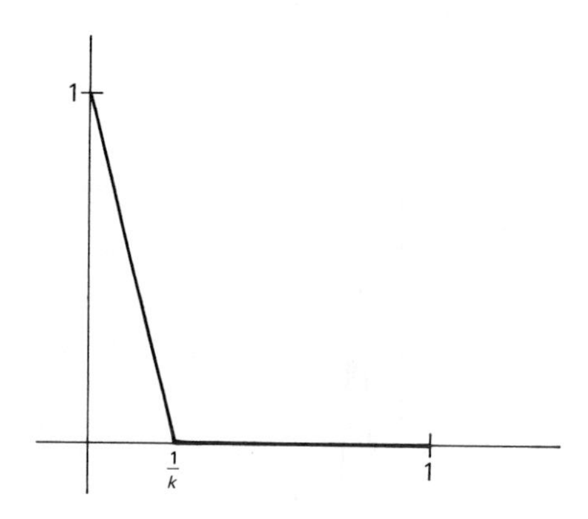


*Fig. 76.1*

We shall demonstrate that $\{\psi_k\}$ converges pointwise on $[0, 1]$.

(a) Let $x = 0$. Then $\psi_k(0) = 1$ for all $k \in \mathbf{N}$, and we have

$$\lim_{\infty} \psi_k(0) = 1.$$

(b) Let $0 < a \leqslant 1$. Then, for every $\varepsilon > 0$, there is an $N(\varepsilon) \in \mathbf{N}$ such that $1/k < a$ for all $k > N(\varepsilon)$ and hence,

$$|\psi_k(a) - 0| = |0 - 0| = 0 < \varepsilon \qquad \text{for all } k > N(\varepsilon).$$

Hence,

$$\lim_{\infty} \psi_k(a) = 0, \qquad a \in (0, 1].$$

Therefore,

$$\lim_{\infty} \psi_k = \psi \qquad \text{on } A = [0, 1],$$

where $\psi$ is given by

$$\psi(x) = \begin{cases} 1 & \text{for } x = 0, \\ 0 & \text{for } x \in (0, 1]. \end{cases}$$

Again we note that the functions $\psi_k$ are continuous on $[0, 1]$ for all $k \in \mathbf{N}$ but that the limit function $\psi$ is *not* continuous on $[0, 1]$.

▲ *Example 4*

Let $\chi_k : [0, 1] \to \mathbf{E}$ be defined by

$$\chi_k(x) = \begin{cases} 4k^2x & \text{for } 0 \leqslant x \leqslant \dfrac{1}{2k}, \\ -4k^2x + 4k & \text{for } \dfrac{1}{2k} < x \leqslant \dfrac{1}{k}, \\ 0 & \text{for } \dfrac{1}{k} < x \leqslant 1. \end{cases}$$

(See Fig. 76.2.)

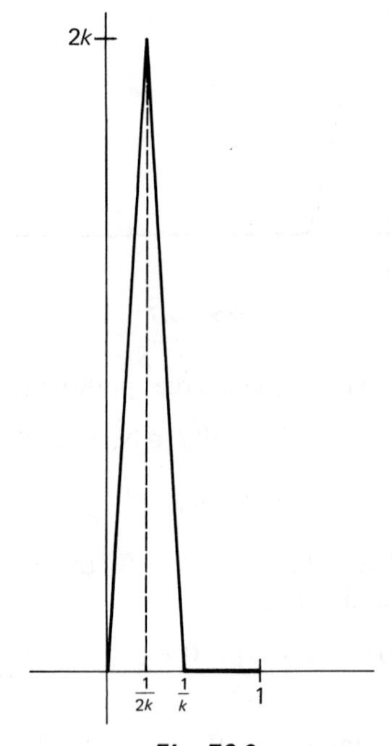


*Fig. 76.2*

Clearly, $\lim_\infty \chi_k(0) = 0$. As in example 3, we can see also that $\lim_\infty \chi_k(a) = 0$ for all $a \in (0, 1]$. Hence, with $\chi : [0, 1] \to \mathbf{E}$ defined by $\chi(x) = 0$, we see that

$$\lim_\infty \chi_k = \chi \qquad \text{on } A.$$

▲ *Example 5*

Let $\beta_k : (0, 1] \to \mathbf{E}$ be defined by

$$\beta_k(x) = \begin{cases} k & \text{for } 0 < k \leqslant \dfrac{1}{k}, \\ \dfrac{1}{x} & \text{for } \dfrac{1}{k} < x \leqslant 1. \end{cases}$$

(See Fig. 76.3.)

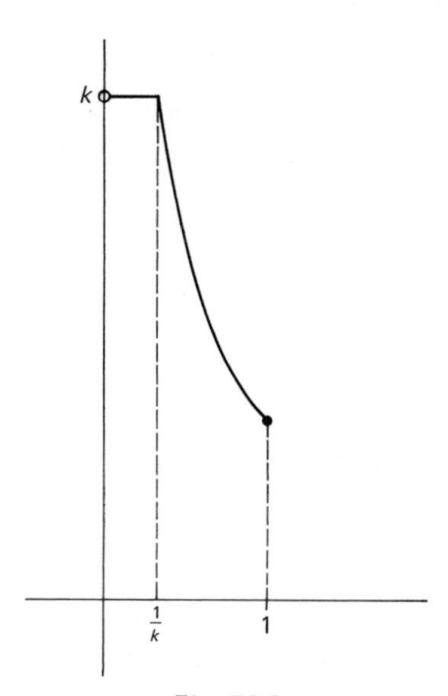


*Fig. 76.3*

Let $a \in (0, 1]$. Then, there is an $N(\varepsilon) \in \mathbf{N}$ such that $1/k < a$ for all $k > N(\varepsilon)$, and hence,

$$\left| \beta_k(a) - \frac{1}{a} \right| = \left| \frac{1}{a} - \frac{1}{a} \right| = 0 \qquad \text{for all } k > N(\varepsilon).$$

Hence,

$$\lim_{\infty} \beta_k(x) = \frac{1}{x}, \qquad x \in (0, 1];$$

or, equivalently,

$$\lim_{\infty} \beta_k = \beta \qquad \text{on } (0, 1],$$

where $\beta$ is given by $\beta(x) = 1/x$. We observe that for each $k \in \mathbf{N}$, $\beta_k$ is bounded on $(0, 1]$ because

$$|\beta_k(x)| \leqslant k \qquad \text{for all } x \in (0, 1].$$

However, the limit function $\beta$, given by $\beta(x) = 1/x$, is *not* bounded on $(0, 1]$.

▲ *Example 6*

Let $\sigma_k : \mathbf{E} \to \mathbf{E}$ be defined by $\sigma_k(x) = (1/k) \sin kx$. For every $a \in \mathbf{E}$,

$$\left| \frac{1}{k} \sin ka - 0 \right| \leqslant \frac{1}{k} < \varepsilon \qquad \text{for all } k > N(\varepsilon)$$

and hence,

$$\lim_{\infty} \frac{1}{k} \sin kx = 0, \qquad x \in \mathbf{E}.$$

---

By Theorem 62.1, $\{f_k(a)\}$ converges to a limit if and only if it is a Cauchy sequence. Hence:

***Theorem 76.1***

The sequence of functions $\{f_k\}$, where $f_k : \mathfrak{D} \to \mathbf{E}$, $\mathfrak{D} \subseteq \mathbf{E}$ for all $k \in \mathbf{N}$, converges pointwise on $A \subseteq \mathfrak{D}$ to a limit function $f : A \to \mathbf{E}$ if and only if $\{f_k(x)\}$ is a Cauchy sequence for each $x \in A$; that is, if and only if, for each $x \in A$ and every $\varepsilon > 0$, there is an $N(\varepsilon, x)$ such that

$$| f_k(x) - f_l(x) | < \varepsilon \qquad \text{for all } k, l > N(\varepsilon, x).$$

## 7.76 Exercises

1. Given the functions $f_k : \mathbf{E}\backslash\{1\} \to \mathbf{E}$ by $f_k(x) = (1 - x^k)/(1 - x)$. Find the largest set $A \subseteq \mathbf{E}\backslash\{1\}$ on which the sequence $\{f_k\}$ converges pointwise, and find the limit function.
2. The sequence $\{f_k\}$ of functions $f_k : \mathfrak{D} \to \mathbf{E}^m$, $\mathfrak{D} \subseteq \mathbf{E}^n$, converges pointwise on $A \subseteq \mathfrak{D}$ if and only if there exists a function $f : A \to \mathbf{E}^m$ such that $\lim_\infty f_k(x) = f(x)$ for all $x \in A$. *Prove*: $\lim_\infty f_k(x) = f(x)$ for all $x \in A$ if and only if $\lim_\infty \phi_j^{(k)}(x) = \phi_j(x)$ for all $x \in A$ and all $j \in \{1, 2, \ldots, m\}$, where $f_k = (\phi_1^{(k)}, \phi_2^{(k)}, \ldots, \phi_m^{(k)})$ and $f = (\phi_1, \phi_2, \ldots, \phi_m)$.
3. *Prove*: The sequence $\{f_k\}$ of functions $f_k : \mathfrak{D} \to \mathbf{E}^m$, $\mathfrak{D} \subseteq \mathbf{E}^n$, converges to $f : A \to \mathbf{E}^m$ on $A \subseteq \mathfrak{D}$ if and only if each sequence formed from the components of $f_k$ is a Cauchy sequence for every $x \in A$. (See exercise 2.)
4. Given $f_k : [0, 1] \to \mathbf{E}$ by

$$f_k(x) = \begin{cases} 0 & \text{for } x = 0, \\ k - k^2 x & \text{for } 0 < x \leqslant \dfrac{1}{k}, \\ 0 & \text{for } \dfrac{1}{k} < x \leqslant 1. \end{cases}$$

   Where does $\{f_k\}$ converge? What is the limit function?
5. *Prove*: If $\{f_k\} \to f$ on $A$, then the limit function is unique.
6. *Prove*: If $\{f_k\} \to f$ and $\{g_k\} \to g$ on $A$, then $\{f_k \pm g_k\} \to f \pm g$ on $A$, where

$f_k, g_k : \mathfrak{D} \to \mathbf{E}$, $\mathfrak{D} \subseteq \mathbf{E}$ and $A \subseteq \mathfrak{D}$. Also prove that $\{f_k g_k\} \to fg$ on $A$. Generalize this theorem to vector-valued functions of a vector variable, replacing $f_k g_k$ and $fg$ by $f_k \cdot g_k$ and $f \cdot g$, respectively.

## 7.77 Uniform Convergence

We have seen in the preceding section that a pointwise-convergent sequence of continuous functions need not converge to a continuous limit function (example 2), and that a pointwise-convergent sequence of bounded functions need not converge to a bounded limit function (example 5). In order to guarantee that the limit function of a pointwise-convergent sequence of continuous functions is continuous, and of bounded functions is bounded, we need a stronger concept of convergence than merely pointwise convergence. To introduce such a stronger concept is the goal of this section.

***Definition 77.1***

Let $f_k : \mathfrak{D} \to \mathbf{E}$, $\mathfrak{D} \subseteq \mathbf{E}$, for all $k \in \mathbf{N}$. The sequence $\{f_k\}$ *converges uniformly* on $A \subseteq \mathfrak{D}$ to a limit function $f : A \to \mathbf{E}$ if and only if, for every $\varepsilon > 0$, there is an $N(\varepsilon) \in \mathbf{N}$ such that

$$|f_k(x) - f(x)| < \varepsilon \qquad \text{for all } k > N(\varepsilon) \quad \text{and all } x \in A.$$

We express the fact that $\{f_k\}$ converges uniformly on $A$ to $f$ by writing

$$\lim_{\infty} f_k = f \quad \text{uniformly on } A, \qquad \text{or} \qquad \{f_k\} \xrightarrow{\text{unif}} f \text{ on } A.$$

(See also Fig. 77.1.)

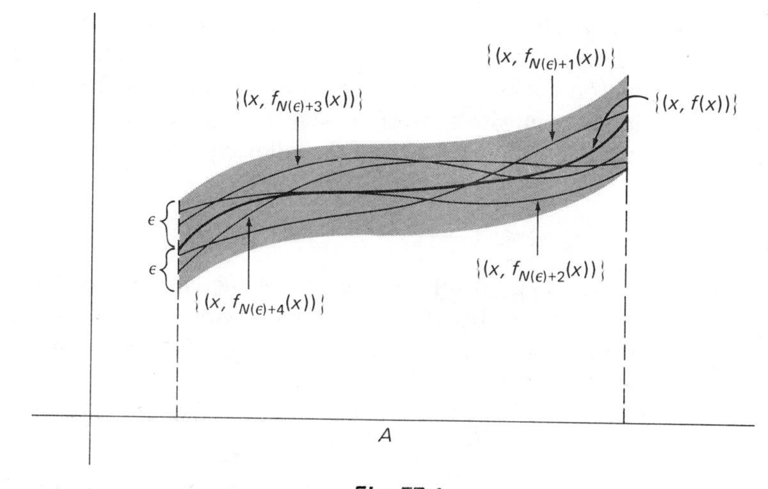


*Fig. 77.1*

▲ *Example 1*

Let $f_k : (0, 1] \to \mathbf{E}$ be given by

$$f_k(x) = \frac{1}{x} - \frac{x}{k}.$$

For $f : (0, 1] \to \mathbf{E}$, given by $f(x) = 1/x$, we have

$$|f_k(x) - f(x)| = \left|\frac{x}{k}\right| \leqslant \frac{1}{k} < \varepsilon \qquad \text{for all } k > N(\varepsilon) \quad \text{and all } x \in (0, 1].$$

Hence, $\{f_k\} \xrightarrow{\text{unif}} f$ on $(0, 1]$.

What distinguishes uniform convergence from pointwise convergence is that, for a given $\varepsilon > 0$, there is an $N(\varepsilon) \in \mathbf{N}$ that serves for *all* $x \in A$ while, in the case of mere pointwise convergence which is not uniform, $N(\varepsilon)$ depends on $x$ and there is no $N(\varepsilon)$ that serves for all $x \in A$.

From definitions 76.1 and 77.1, we obtain immediately:

***Theorem 77.1***

If the sequence of functions $\{f_k\}$, where $f_k : \mathfrak{D} \to \mathbf{E}$, $\mathfrak{D} \subseteq \mathbf{E}$, for all $k \in \mathbf{N}$, converges uniformly on $A \subseteq \mathfrak{D}$ to a limit function $f : A \to \mathbf{E}$, then $\{f_k\}$ converges pointwise on $A$ to the limit function $f$:

$$\{f_k\} \xrightarrow{\text{unif}} f \text{ on } A \qquad \text{implies} \quad \{f_k\} \to f \text{ on } A.$$

In order to decide whether or not a given sequence converges uniformly, we need only investigate pointwise-convergent sequences because, by Theorem 77.1, sequences that do not even converge pointwise cannot possibly converge uniformly. Among the pointwise-convergent sequences, we can eliminate those that do not converge uniformly by means of the following theorem.

***Theorem 77.2***

The sequence $\{f_k\}$, where $f_k : \mathfrak{D} \to \mathbf{E}$, $\mathfrak{D} \subseteq \mathbf{E}$ for all $k \in \mathbf{N}$, which is presumed to converge pointwise on $A \subseteq \mathfrak{D}$ to $f : A \to \mathbf{E}$, does *not* converge uniformly on $A$ if and only if there exists an $\varepsilon > 0$ such that, for every $n \in \mathbf{N}$, there is an integer $k > n$ such that for, at least one $x_k \in A$,

$$|f_k(x_k) - f(x_k)| \geqslant \varepsilon.$$

The proof of this theorem consists simply in the negation of definition 77.1.

▲ *Example 2*

The sequence $\{\phi_k\}$ of example 2, section 7.76, does not converge uniformly on $[0, 1]$. Let $x_k = (\frac{1}{2})^{1/k} \in [0, 1]$. Then, $\phi_k(x_k) = x_k^k = \frac{1}{2}$, and we have

$$|\phi_k(x_k) - \phi(x_k)| = |\tfrac{1}{2} - 0| = \tfrac{1}{2} \geqslant \varepsilon = \tfrac{1}{2} \qquad \text{for all } k \in \mathbf{N}.$$

(Note that $x_k \in (0, 1)$ and hence, $\phi(x_k) = 0$.)

From the above argument, it is clear that $\{\phi_k\}$ does not converge uniformly on $(0, 1)$ either.

▲ *Example 3*

The sequence $\{\psi_k\}$ of example 3, section 7.76, does not converge uniformly on $[0, 1]$. Let $x_k = (1/2k) \in [0, 1]$. Then,

$$|\psi_k(x_k) - \psi(x_k)| = |\tfrac{1}{2} - 0| = \tfrac{1}{2} \geqslant \varepsilon = \tfrac{1}{2} \qquad \text{for all } k \in \mathbf{N}.$$

(Note that $x_k > 0$ and hence, $\psi(x_k) = 0$.)

This sequence does not converge uniformly on $(0, 1]$ either.

▲ *Example 4*

The sequence $\{\chi_k\}$ of example 4, section 7.76 does not converge uniformly on $[0, 1]$. Let $x_k = (1/2k) \in [0, 1]$. Then,

$$|\chi_k(x_k) - \chi(x_k)| = 2k \geqslant 2 = \varepsilon \qquad \text{for all } k \in \mathbf{N}.$$

▲ *Example 5*

The sequence $\{\beta_k\}$ of example 5, section 7.76, does not converge uniformly on $(0, 1]$. Let $x_k = 1/2k \in (0, 1]$. Then,

$$|\beta_k(x_k) - \beta(x_k)| = |k - 2k| = k \geqslant 1 = \varepsilon \qquad \text{for all } k \in \mathbf{N}.$$

▲ *Example 6*

The sequence $\{\phi_k\}$ of example 2 converges uniformly on $[0, a]$ for $a \in (0, 1)$:

$$|\phi_k(x) - \phi(x)| = |x^k - 0| \leqslant a^k < \varepsilon \quad \text{for } k > N(\varepsilon),$$

because $\lim_\infty a^k = 0$ for all $a \in (0, 1)$.

***Definition 77.2***

The sequence $\{f_k\}$, where $f_k : \mathfrak{D} \to \mathbf{E}$, $\mathfrak{D} \subseteq \mathbf{E}$ for all $k \in \mathbf{N}$, is called a *uniformly convergent Cauchy sequence* on $A \subseteq \mathfrak{D}$ if and only if, for every $\varepsilon > 0$, there is an $N(\varepsilon) \in \mathbf{N}$ such that

$$|f_k(x) - f_l(x)| < \varepsilon \qquad \text{for all } l, k > N(\varepsilon) \quad \text{and all } x \in A.$$

***Theorem 77.3***

The sequence $\{f_k\}$, where $f_k : \mathfrak{D} \to \mathbf{E}$, $\mathfrak{D} \subseteq \mathbf{E}$ for all $k \in \mathbf{N}$, converges uniformly on $A \subseteq \mathfrak{D}$ to a limit function $f: A \to \mathbf{E}$ if and only if $\{f_k\}$ is a uniformly convergent Cauchy sequence on $A$.

*Proof*

(a) If $\{f_k\}$ converges uniformly on $A$ to $f$, then we obtain, in view of

$$|f_k(x) - f_l(x)| \leqslant |f_k(x) - f(x)| + |f(x) - f_l(x)|,$$

that $\{f_k\}$ is a uniformly convergent Cauchy sequence on $A$.

(b) Suppose that $\{f_k\}$ is a uniformly convergent Cauchy sequence on $A$. Then, for every $\varepsilon > 0$, there is an $N(\varepsilon) \in \mathbf{N}$ such that

$$(77.1) \qquad |f_k(x) - f_l(x)| < \frac{\varepsilon}{2} \qquad \text{for all } k, l > N(\varepsilon) \quad \text{and all } x \in A.$$

Since $\{f_k\}$ is a uniformly convergent Cauchy sequence, it follows that $\{f_k(x)\}$ is a Cauchy sequence for each $x \in A$. By Theorem 76.1, $\{f_k(x)\}$ converges pointwise on $A$ to some function $f: A \to \mathbf{E}$. We choose in (77.1) a fixed $l_0 > N(\varepsilon)$ and, for each $x \in A$, let $k$ tend to infinity. Then we obtain from Lemma 62.5 that, for all $x \in A$,

$$|f(x) - f_{l_0}(x)| \leqslant \frac{\varepsilon}{2} < \varepsilon.$$

Since $l_0$ is any integer that is greater than $N(\varepsilon)$, we have

$$|f(x) - f_l(x)| < \varepsilon \qquad \text{for all } l > N(\varepsilon).$$

Since this is true for all $x \in A$, it follows that $\{f_k\} \xrightarrow{\text{unif}} f$ on $A$.

In the next section we shall investigate uniformly convergent sequences of continuous functions and of bounded functions.

## 7.77 Exercises

1. Does the sequence $\{kx/(1 + kx)\}$ converge uniformly on $\mathbf{E}$? on $[1, 2]$?
2. Does the sequence $\{f_k\}$, where

$$f_1(x) = 1, f_k(x) = \begin{cases} kx & \text{for } 0 \leqslant x \leqslant \dfrac{1}{k} \\ \dfrac{1}{k-1}(1 - x) & \text{for } \dfrac{1}{k} < x \leqslant 1 \end{cases} \qquad \text{for } k > 1$$

converge uniformly on $[0, 1]$? on $(0, 1)$?

3. Show: If $\{f_k\}$ is a uniformly convergent Cauchy sequence on $A$, then $\{f_k(x)\}$ is a Cauchy sequence for each $x \in A$.

4. Let $f_k, g_k : \mathfrak{D} \to \mathbf{E}$ and

$$\{f_k\} \xrightarrow{\text{unif}} f, \qquad \{g_k\} \xrightarrow{\text{unif}} g \text{ on } \mathfrak{D}.$$

Prove that $\{f_k \pm g_k\} \xrightarrow{\text{unif}} f \pm g$.

5. Generalize definition 77.2 to sequences of vector-valued functions of a vector variable, and show that Theorem 77.3 is still valid in such a general setting. (See also exercise 2, section 7.76.)
6. Show that the sequence in example 6, section 7.76, converges uniformly on **R**.

## 7.78 Uniformly Convergent Sequences of Continuous Functions and of Bounded Functions

We are now ready to establish the continuity of the limit function of a uniformly convergent sequence of continuous functions.

### *Theorem 78.1*

Let $f_k : \mathfrak{D} \to \mathbf{E}$, $\mathfrak{D} \subseteq \mathbf{E}$ for all $k \in \mathbf{N}$. If the functions $f_k$ are continuous on $A \subseteq \mathfrak{D}$ for all $k \in \mathbf{N}$, and if $\{f_k\}$ converges uniformly on $A$ to $f: A \to \mathbf{E}$, then $f$ is continuous on $A$.

If $\{f_k\}$ is a uniformly convergent Cauchy sequence of continuous functions on $A$, then $\{f_k\}$ converges uniformly on $A$ to a continuous function $f: A \to \mathbf{E}$.

### *Proof*

We need to prove only the first part of Theorem 78.1. The second part will then follow in view of Theorem 77.3. We have to show that, whenever $a \in A$, there is, for every $\varepsilon > 0$, a $\delta(\varepsilon, a) > 0$ such that

$$|f(x) - f(a)| < \varepsilon \qquad \text{for all } x \in N_{\delta(\varepsilon,a)}(a) \cap A.$$

Since $\{f_k\} \xrightarrow{\text{unif}} f$ on $A$, we have, for every $\varepsilon > 0$, an $N(\varepsilon) \in \mathbf{N}$ such that

$$|f(x) - f_k(x)| < \frac{\varepsilon}{3} \qquad \text{for all } k > N(\varepsilon) \quad \text{and all } x \in A.$$

Let $k_0 > N(\varepsilon)$. Then, we have

$$|f(x) - f_{k_0}(x)| < \frac{\varepsilon}{3} \qquad \text{for all } x \in A \quad \text{and, in particular,} \quad |f(a) - f_{k_0}(a)| < \frac{\varepsilon}{3}. \tag{78.1}$$

By the triangle inequality,

$$|f(x) - f(a)| \leqslant |f(x) - f_{k_0}(x)| + |f_{k_0}(x) - f_{k_0}(a)| + |f_{k_0}(a) - f(a)|. \tag{78.2}$$

Since $f_{k_0}$ is continuous on $A$, we have a $\delta(\varepsilon, a)$ such that

$$(78.3) \qquad |f_{k_0}(x) - f_{k_0}(a)| < \frac{\varepsilon}{3} \qquad \text{for all } x \in N_{\delta(\varepsilon,a)}(a) \cap A.$$

Collecting the partial results in (78.1), (78.2), and (78.3), we obtain

$$|f(x) - f(a)| < \varepsilon \qquad \text{for all } x \in N_{\delta(\varepsilon,a)}(a) \cap \mathfrak{D};$$

that is, $f$ is continuous at $a$. Since $a$ is any point in $A$, $f$ is continuous on $A$.

We record the following result for later reference:

***Corollary to Theorem 78.1***

If $\{f_k\} \xrightarrow{\text{unif}} f$ on $A \subseteq \mathfrak{D}$ where $f_k : \mathfrak{D} \to \mathbf{E}$, $\mathfrak{D} \subseteq \mathbf{E}$, or if $\{f_k\}$ is a uniformly convergent Cauchy sequence on $A$, and if $f_k$ is continuous at $a \in A$ for all $k \in \mathbf{N}$, then the limit function $f$ is continuous at $a$.

The proof of this corollary is identical with the proof of Theorem 78.1 except for the last sentence, which is to be omitted.

Theorem 78.1 enables us to rule out the possibility of uniform convergence in examples 2 and 3, section 7.76, without having to invoke Theorem 77.2 because the limit functions in these examples are not continuous.

A result that is analogous to Theorem 78.1 holds for uniformly convergent sequences of bounded functions:

***Theorem 78.2***

Let $f_k : \mathfrak{D} \to \mathbf{E}$, $\mathfrak{D} \subseteq \mathbf{E}$, represent bounded functions on $A \subseteq \mathfrak{D}$ for all $k \in \mathbf{N}$. If $\{f_k\}$ converges uniformly on $A$ to $f: A \to \mathbf{E}$, then $f$ is bounded on $A$. There also is an $M > 0$ such that $|f_k(x)| \leqslant M$ for all $x \in A$ and all $k \in \mathbf{N}$.

The same is true if $\{f_k\}$ is a uniformly convergent Cauchy sequence of bounded functions on $A$.

*Proof*

Because of Theorem 77.3, we need only prove the first part. By hypothesis, we have for $\varepsilon = 1$ some $N(1) \in \mathbf{N}$ such that

$$(78.4) \qquad |f_k(x) - f(x)| < 1 \qquad \text{for all } x \in A \quad \text{and all } k > N(1).$$

Let $k_0 > N(1)$. Then,

$$|f(x)| \leqslant |f_{k_0}(x)| + 1 \qquad \text{for all } x \in A.$$

By hypothesis, $|f_k(x)| \leqslant M_k$ for some $M_k > 0$ and all $x \in A$. Hence,

$$(78.5) \qquad |f(x)| \leqslant M_{k_0} + 1 \qquad \text{for all } x \in A;$$

that is, $f$ is bounded on $A$.

That $|f_k(x)| \leqslant M$ for $M = \max\{M_1, M_2, \ldots, M_{N(1)}, M_{k_0} + 2\}$, for all $k \in \mathbf{N}$ and all $x \in A$, follows from (78.4) and (78.5).

Theorem 78.2 rules out the possibility of uniform convergence of the sequence of bounded functions in example 5, section 7.76, because the limit function is not bounded.

## 7.78 Exercises

1. *Prove*: If the functions $f_k : K \to \mathbf{E}$, where $K \subseteq \mathbf{E}$ is compact, are continuous on $K$ and if $\{f_k\} \xrightarrow{\text{unif}} f$ on $K$, then $f$ is uniformly continuous on $K$.
2. Given $f_k : K \to \mathbf{E}$ where $K \subseteq \mathbf{E}$ is compact and where $f_1(x) \leqslant f_2(x) \leqslant f_3(x) \leqslant \cdots$ for all $x \in K$. Suppose that all $f_k$ are continuous on $K$ and that $\{f_k\}$ converges pointwise on $K$ to a continuous function $f$. *Prove*: $\{f_k\}$ converges *uniformly* on $K$ to $f$.
3. Generalize Theorem 78.1 to sequences of vector-valued functions of a vector variable, and adapt the proof of Theorem 78.1 to fit this more general situation.
4. Same as in exercise 3, for Theorem 78.2.
5. Let $f_k : [0, 1] \to \mathbf{E}^3$ be defined by $\phi_1^{(k)}(x) = \phi_k(x)$, $\phi_2^{(k)}(x) = \psi_k(x)$, $\phi_3^{(k)}(x) = \chi_k(x)$, where $\phi_k, \psi_k, \chi_k$ are defined in examples 2, 3, 4, of section 7.76.

   (a) Find $\lim_\infty f_k$ on $[0, 1]$.
   (b) Find the largest subset of $[0, 1]$ on which $\{f_k\}$ converges uniformly.

6. Let $K \subset \mathbf{E}$ denote a compact set and let $A$ denote an index set. The collection of functions $f_\alpha : K \to \mathbf{E}$, $\alpha \in A$ is *equicontinuous* on $K$ if and only if, for every $\varepsilon > 0$, there is a $\delta(\varepsilon) > 0$ such that $|f_\alpha(x) - f_\alpha(y)| < \varepsilon$ for all $x, y \in K$ for which $|x - y| < \delta(\varepsilon)$ and *all* $\alpha \in A$. (Note that $\delta(\varepsilon)$ is independent of $\alpha$. Hence the term *equicontinuity*.) Let $\{f_\alpha | \alpha \in A\}$ be equicontinuous on $K$.

   (a) Show: Every function $f_\alpha$, $\alpha \in A$, is uniformly continuous on $K$.
   (b) Show: If $f_\alpha : K \to \mathbf{E}$ is uniformly continuous and if $A$ is a finite set, then $\{f_\alpha | \alpha \in A\}$ is equicontinuous.
   (c) Let $\{p_\alpha | \alpha \in A\}$ denote a collection of continuous, piecewise-linear functions on $[a, b]$, i.e., functions whose graphs are polygonal lines with finitely many vertices. Show: If none of the lines from which the $p_\alpha$ are pieced together has a slope that is greater than some given $M > 0$ and less than $-M$, then $\{p_\alpha | \alpha \in A\}$ is equicontinuous.

7. Given a sequence $\{f_k\}$ of functions $f_k : [a, b] \to \mathbf{E}$. Show: If there is an $M > 0$ such that $|f_k(x)| \leqslant M$ for all $x \in [a, b]$, that is, $\{f_k\}$ is *uniformly bounded* on $[a, b]$, then there exists a subsequence $\{f_{n_k}\}$ of $\{f_k\}$ which converges on $\mathbf{Q} \cap [a, b]$.

8. *Prove*: If $f_\alpha : [a,b] \to \mathbf{E}$, if $A$ is some infinite index set, and if $\{f_\alpha | \alpha \in A\}$ is *uniformly bounded* and *equicontinuous* (see exercises 6 and 7), then there exists a sequence $\alpha_1, \alpha_2, \alpha_3, \ldots \in A$ such that $\{f_{\alpha_k}\}$ converges uniformly on $[a,b]$. This is known as the Lemma of Ascoli.†
9. Let $B(f, \mathfrak{D})$ denote the space of bounded functions $f : \mathfrak{D} \to \mathbf{E}$. Show that $B(f, \mathfrak{D})$ is a normed vector space if addition and scalar multiplication are defined by $(f+g)(x) = f(x) + g(x)$, $(\lambda f)(x) = \lambda f(x)$, and the norm is defined by $\|f\|_{\mathfrak{D}} = \sup_{x \in \mathfrak{D}} |f(x)|$.
10. $\{f_k\}$ is a Cauchy sequence in $B(f, \mathfrak{D})$ if and only if $f_k \in B(f, \mathfrak{D})$ for all $k \in \mathbf{N}$ and for every $\varepsilon > 0$, there is an $N(\varepsilon) \in \mathbf{N}$ such that $\|f_k - f_l\|_{\mathfrak{D}} < \varepsilon$ for all $k, l > N(\varepsilon)$. *Prove*: If $\{f_k\}$ is a Cauchy sequence in $B(f, \mathfrak{D})$, then $\{f_k\}$ converges uniformly to a bounded function $f : \mathfrak{D} \to \mathbf{E}$ and for every $\varepsilon > 0$, there is an $N(\varepsilon) \in \mathbf{N}$ such that $\|f_k - f\|_{\mathfrak{D}} < \varepsilon$ for all $k > N(\varepsilon)$. (In section 5.62 we called a metric space *complete* if and only if every Cauchy sequence converged to an element of the space. A complete normed vector space is called a *Banach space*.‡ Hence, $B(f, \mathfrak{D})$ is a Banach space.)
11. Let $C(f, \mathfrak{D})$ denote the space of bounded continuous functions $f : \mathfrak{D} \to \mathbf{E}$ with addition, scalar multiplication, and norm defined as in exercise 9. Show that $C(f, \mathfrak{D})$ is a Banach space. (See exercise 10.)

### 7.79 Weierstrass' Approximation Theorem*

We have seen, in the preceding section, that a uniformly convergent sequence of continuous functions converges to a continuous function. Since polynomials are continuous functions, it follows, in particular, that a uniformly convergent sequence of polynomials converges to a continuous function. Weierstrass has shown that, on a closed interval, the converse is also true; namely, that every continuous function on a closed interval is the limit of a uniformly convergent sequence of polynomials. This result is of great significance, because it implies that every continuous function can be uniformly approximated on a closed interval to any degree of accuracy by a polynomial.

We shall prove Weierstrass' theorem constructively by making use of *Bernstein§ polynomials*:

If $f : [0,1] \to \mathbf{E}$, then

$$(79.1) \qquad B_k(x, f) = \sum_{j=0}^{k} f\left(\frac{j}{k}\right)\binom{k}{j} x^j (1-x)^{k-j}, \text{ where } \binom{k}{j} = \frac{k!}{j!(k-j)!}$$

is called the $k$th Bernstein polynomial for $f$.

For example,

$$B_0(x, f) = f(0),\ B_1(x, f) = f(0)(1-x) + f(1)x,$$
$$B_2(x, f) = f(0)(1-x)^2 + 2f(\tfrac{1}{2})\, x\, (1-x) + f(1)x^2.$$

† *Giulio Ascoli*, 1843–1896.
‡ *Stephen Banach*, 1894–1964.
§ *Serge N. Bernstein*, 1880–1968.

One obtains from the binomial formula

$$(a+b)^k = \sum_{j=0}^{k} \binom{k}{j} a^j b^{k-j},$$

by formal manipulations, that

(79.2)
$$\sum_{j=0}^{k} \binom{k}{j} x^j (1-x)^{k-j} = 1,$$

(79.3)
$$\sum_{j=0}^{k} \left(x - \frac{j}{k}\right)^2 \binom{k}{j} x^j (1-x)^{k-j} = \frac{1}{k} x(1-x) \qquad \text{for } k \geqslant 1.$$

(See exercises 1, 2, 3, 4, 5.)

With these preparations out of the way we are ready to state and prove:

***Theorem 79.1*** *(Weierstrass' Approximation Theorem)*

If $f : [a,b] \to \mathbf{E}$ is continuous, then there exists a sequence $\{p_k(x)\}$ of polynomials $p_k : [a,b] \to \mathbf{E}$ which converges uniformly to $f$ on $[a,b]$.

*Proof*

Since the coordinate transformation $q : [a,b] \leftrightarrow [0,1]$, which is defined by $q(x) = (x-a)/(b-a)$, is continuous and has a continuous inverse, we may assume w.l.o.g. that $[a,b] = [0,1]$.

We shall demonstrate that

$$\lim_{\infty} B_k(x,f) = f(x) \quad \text{uniformly on } [0,1];$$

i.e., that, for every $\varepsilon > 0$, there is an $N(\varepsilon) \in \mathbf{N}$ such that

(79.4)
$$|f(x) - B_k(x,f)| < \varepsilon \qquad \text{for all } k > N(\varepsilon) \quad \text{and all } x \in [0,1].$$

From (79.1) and (79.2),

(79.5)
$$\begin{aligned} |f(x) - B_k(x,f)| &= \left| \sum_{j=0}^{k} \left( f(x) - f\left(\frac{j}{k}\right)\right) \binom{k}{j} x^j (1-x)^{k-j} \right| \\ &\leqslant \sum_{j=0}^{k} \left| f(x) - f\left(\frac{j}{k}\right) \right| \binom{k}{j} x^j (1-x)^{k-j}. \end{aligned}$$

Since $f$ is continuous on $[0,1]$, $f$ is also bounded on $[0,1]$, i.e.,

(79.6)
$$|f(x)| \leqslant M \quad \text{for all } x \in [0,1] \quad \text{and some } M > 0,$$

and $f$ is uniformly continuous on $[0,1]$, i.e., there is a $\delta(\varepsilon) > 0$ such that

(79.7)
$$|f(x) - f(y)| < \frac{\varepsilon}{2} \qquad \text{for all } x, y \in [0,1] \quad \text{for which } |x-y| < \delta(\varepsilon).$$

We choose

$$N(\varepsilon) = \left[ \max\left( \frac{M^2}{4\varepsilon^2}, \frac{4}{(\delta(\varepsilon))^4} \right) \right],$$

i.e., the largest integer that is less than or equal to the number in brackets. Then, we have, for $k > N(\varepsilon)$,

$$\delta^2(\varepsilon) > \frac{2}{\sqrt{k}}, \; \varepsilon > \frac{M}{2\sqrt{k}}. \tag{79.8}$$

We let

$$J_1 = \left\{ j \in \{1,2,\dots,k\} \,\middle|\, \left| x - \frac{j}{k} \right| < \delta(\varepsilon) \right\}, \quad J_2 = \left\{ j \in \{1,2,\dots,k\} \,\middle|\, \left| x - \frac{j}{k} \right| \geqslant \delta(\varepsilon) \right\}.$$

Then, $J_1 \cup J_2 = \{1, 2, \dots, k\}$ and $J_1 \cap J_2 = \varnothing$. By (79.2) and (79.7),

$$\sum_{J_1} \left| f(x) - f\left(\frac{j}{k}\right) \right| \binom{k}{j} x^j (1-x)^k < \frac{\varepsilon}{2} \sum_{J_1} \binom{k}{j} x^j (1-x)^{k-j} \leqslant \frac{\varepsilon}{2}; \tag{79.9}$$

and, by (79.3), (79.6), and (79.8), and, since $\max_{[0,1]} x(1-x) = \frac{1}{4}$,

$$\sum_{J_2} \left| f(x) - f\left(\frac{j}{k}\right) \right| \binom{k}{j} x^j (1-x)^{k-j} \leqslant 2M \sum_{J_2} \frac{\left(x - \frac{j}{k}\right)^2}{\left(x - \frac{j}{k}\right)^2} \binom{k}{j} x^j (1-x)^{k-j} \tag{79.10}$$

$$\leqslant 2M \frac{\sqrt{k}}{2} \frac{1}{k} x(1-x) \leqslant \frac{M}{4\sqrt{k}} < \frac{\varepsilon}{2}.$$

(79.4) follows readily from (79.5), (79.9), and (79.10).

If $f: [a, b] \to \mathbf{E}$ is continuous, then $g : [0, 1] \to \mathbf{E}$, defined by

$$g(t) = f((b-a)t + a),$$

is also continuous; and $\{B_k(t, g)\}$ converges uniformly to $g$ on $[0, 1]$:

$$f((b-a)t + a) = \lim_{\infty} B_k(t, g) = \lim_{\infty} \sum_{j=0}^{k} g\left(\frac{j}{k}\right) \binom{k}{j} t^j (1-t)^{k-j}$$

uniformly on $[0, 1]$.

Hence, with $x = (b-a)\,t + a$,

$$f(x) = \lim_{\infty} \sum_{j=0}^{k} f\left((b-a)\frac{j}{k} + a\right) \binom{k}{j} \left(\frac{x-a}{b-a}\right)^j \left(\frac{b-x}{b-a}\right)^{k-j} \quad \text{uniformly on } [a, b].$$

Therefore, with

$$p_k(x) = \sum_{j=0}^{k} f\left((b-a)\frac{j}{k} + a\right) \binom{k}{j} \left(\frac{x-a}{b-a}\right)^j \left(\frac{b-x}{b-a}\right)^{k-j},$$

which is a polynomial of degree $k$, we have

$$f(x) = \lim_{\infty} p_k(x) \qquad \text{uniformly on } [a, b].$$

## 7.79 Exercises

1. Establish the validity of (79.2).
2. Show that $x = \sum_{j=0}^{k}(j/k)\binom{k}{j}x^j(1-x)^{k-j}$ for $k \geqslant 1$.
3. Show that $(k^2-k)x^2 = \sum_{j=0}^{k}(j^2-j)\binom{k}{j}x^j(1-x)^{k-j}$ for $k \geqslant 1$.
4. Show that $(1-1/k)x^2 + (1/k)x = \sum_{j=0}^{k}(j/k)^2\binom{k}{j}x^j(1-x)^{k-j}$ for $k \geqslant 1$.
5. Establish the validity of (79.3).
6. Represent $B_3(x,f)$, $B_4(x,f)$ explicitly.
7. *Prove*: If $f:[a,b] \to \mathbf{R}$ is continuous, then there is, for every $\varepsilon > 0$, a piecewise linear function (function whose graph is a polygon) $p$ such that $|f(x) - p(x)| < \varepsilon$ for all $x \in [a,b]$.

## 7.80 A Continuous Function that is Nowhere Differentiable*

That there could be continuous functions that are nowhere differentiable was inconceivable to 18th- and 19th-century mathematicians until K. Weierstrass produced an example of just such a function in 1872.

We now have the means at our disposal to discuss a function of this type. The function is defined as the limit of a uniformly convergent sequence of continuous functions. The elements of this sequence, in turn, are defined recursively in terms of the periodic function $f_0 : \mathbf{E} \to \mathbf{E}$, which is given by

$$f_0(x) = \begin{cases} x & \text{for } 0 \leqslant x < \frac{1}{2}, \\ 1-x & \text{for } \frac{1}{2} \leqslant x \leqslant 1, \end{cases} \tag{80.1}$$

$$f_0(x+1) = f_0(x). \tag{80.2}$$

(See Fig. 80. 1.)

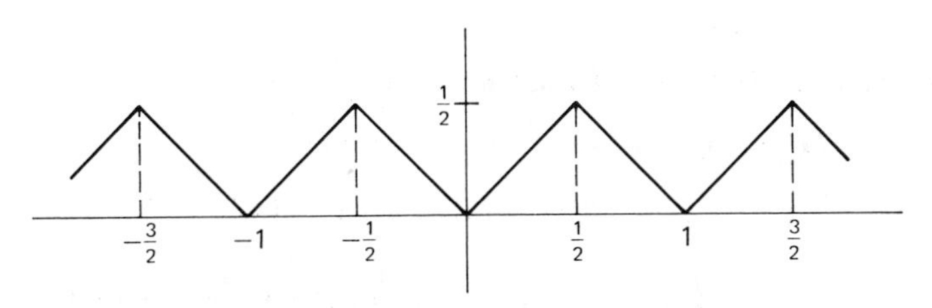


*Fig. 80.1*

Observe that

$$|f_0(x)| \leqslant \tfrac{1}{2} \qquad \text{for all } x \in \mathbf{E}. \tag{80.3}$$

We define functions $f_k : \mathbf{E} \to \mathbf{E}$ for all $k \in \mathbf{N}$ in terms of $f_0$ by

$$f_k(x) = \frac{1}{4^k} f_0(4^k x), \qquad k \in \mathbf{N}, \tag{80.4}$$

and note that, in view of (80.2),

$$f_k(x+1) = \frac{1}{4^k} f_0(4^k x + 4^k) = \frac{1}{4^k} f_0(4^k x) = f_k(x), \tag{80.5}$$

i.e., all the functions $f_k$ have period 1. (See Fig. 80.2.)

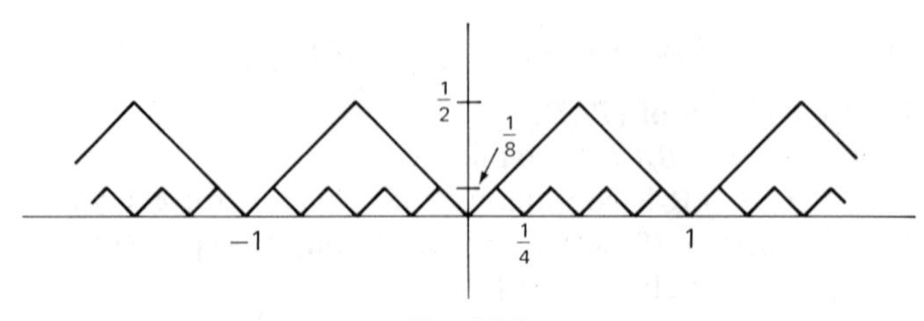


*Fig. 80.2*

Finally, we define a sequence of functions $\{w_k\}$ in terms of $f_k$ by

$$w_k = f_0 + f_1 + \cdots + f_k; \tag{80.6}$$

and note that, in view of (80.5),

$$w_k(x+1) = w_k(x). \tag{80.7}$$

Since $f_0$ is continuous on **E**, it follows that $f_k$ is continuous on **E** for all $k \in \mathbf{N}$ and hence, $w_k$ is continuous on **E** for all $k \in \mathbf{N}$.

***Lemma 80.1***

The sequence $\{w_k\}$ of continuous functions converges uniformly on **E** and the limit function $w$, defined by $w(x) = \lim_\infty w_k(x)$ for all $x \in \mathbf{E}$, is continuous on **E**.

*Proof*

We have, from (80.3), for $k > l$ and for all $x \in \mathbf{E}$, that

$$\begin{aligned} |w_k(x) - w_l(x)| &= |f_{l+1}(x) + f_{l+2}(x) + \cdots + f_k(x)| \\ &\leqslant |f_{l+1}(x)| + |f_{l+2}(x)| + \cdots + |f_k(x)| \\ &\leqslant \frac{1}{2}\left(\frac{1}{4^{l+1}} + \frac{1}{4^{l+2}} + \cdots + \frac{1}{4^k}\right) = \frac{2}{3}\left(\frac{1}{4^{l+1}} - \frac{1}{4^{k+1}}\right) < \frac{1}{3}\cdot\frac{1}{4^l}. \end{aligned}$$

Hence, $\{w_k\}$ is a uniformly convergent Cauchy sequence on **E** and, by Theorem 78.1, converges uniformly to a continuous limit function $w$ on **E**.

We shall now demonstrate that the continuous limit function $w: \mathbf{E} \to \mathbf{E}$ is nowhere differentiable.

Towards this end, we note first that the function $f_k$, defined in (80.4), is pieced together from line segments with slope 1 alternating with line segments with slope $-1$ (see Fig. 80.2). The points where the slope switches from 1 to $-1$, or vice versa, have the abscissas

$$0, \quad \pm\frac{1}{2}\cdot\frac{1}{4^k}, \quad \pm\frac{1}{4^k}, \quad \pm\frac{3}{2}\cdot\frac{1}{4^k}, \quad \ldots .$$

We shall call these points *switching points.* If $a, b \in \mathbf{E}$ both lie between two consecutive switching points of $f_k$, then they also lie between two consecutive switching points of $f_j$ for all $j \leqslant k$, and we have

$$\frac{f_j(a) - f_j(b)}{a - b} = \pm 1, \quad a, b \in \left[\frac{n}{2}\frac{1}{4^k}, \frac{n+1}{2}\frac{1}{4^k}\right], \quad n \in \mathbf{Z}, \quad \text{for all } j \leqslant k. \tag{80.8}$$

In order to show that $w$ is nowhere differentiable, it suffices to show that for any $c \in \mathbf{E}$,

$$\lim_{\infty} \frac{w(x_k) - w(c)}{x_k - c}$$

does *not* exist for *one* sequence $\{x_k\}$ which converges to $c$. It then follows, by definitions 24.1 and 35.1, that

$$\lim_{c} \frac{w(x) - w(c)}{x - c}$$

does *not* exist; i.e., $w$ is not differentiable at $c$.

We choose the sequence $\{x_k\}$ as follows:

$$x_k = c \pm \frac{1}{4^{k+1}},$$

where we choose the sign in such a manner that $x_k, c$ both lie between the same consecutive switching points of $f_k$ (and, in view of (80.8), of all $f_j$ for $j \leqslant k$). This is always possible because consecutive switching points of $f_k$ are $\frac{1}{2}(1/4^k)$ units apart.

Clearly,

$$\lim_{\infty} x_k = c.$$

From (80.8),

$$\frac{f_j(x_k) - f_j(c)}{x_k - c} = \pm 1 \qquad \text{for all } j \leqslant k. \tag{80.9}$$

If $j > k$, we have

$$\begin{aligned} f_j(x_k) = f_j\left(c \pm \frac{1}{4^{k+1}}\right) &= \frac{1}{4^j} f_0\left(4^j\left(c \pm \frac{1}{4^{k+1}}\right)\right) \\ &= \frac{1}{4^j} f_0(4^j c \pm 4^{j-k-1}) = \frac{1}{4^j} f_0(4^j c) \\ &= f_j(c) \end{aligned}$$

because of (80.5) and $j - k - 1 \geqslant 0$.

Hence,

$$\frac{f_j(x_k) - f_j(c)}{x_k - c} = 0 \qquad \text{for all } j > k. \tag{80.10}$$

Since

$$\begin{aligned}\frac{w(x_k) - w(c)}{x_k - c} &= \lim_{l \to \infty} \frac{w_l(x_k) - w_l(c)}{x_k - c} \\ &= \lim_{l \to \infty} \frac{f_0(x_k) - f_0(c) + f_1(x_k) - f_1(c) + \cdots + f_l(x_k) - f_l(c)}{x_k - c},\end{aligned}$$

we obtain, from (80.9) and (80.10), that

$$\begin{aligned}\frac{w(x_k) - w(c)}{x_k - c} &= \frac{f_0(x_k) - f_0(c)}{x_k - c} + \frac{f_1(x_k) - f_1(c)}{x_k - c} + \cdots + \frac{f_k(x_k) - f_k(c)}{x_k - c} \\ &= \pm 1 \pm 1 \cdots \pm 1;\end{aligned}$$

and we see that

$$\Delta_k = \frac{w(x_k) - w(c)}{x_k - c} = \begin{cases} \text{even} & \text{if } k \text{ is odd,} \\ \text{odd} & \text{if } k \text{ is even.} \end{cases}$$

Therefore, no matter how large $k$ is, $|\Delta_k - \Delta_{k+1}| \geqslant 1$, that is, $\{\Delta_k\}$ is *not* a Cauchy sequence and hence, does not converge. Therefore,

$$\lim_{c} \frac{w(x) - w(c)}{x - c}$$

does not exist; and we have

***Theorem 80.1***

The function $w : \mathbf{E} \to \mathbf{E}$, which is defined by $w(x) = \lim_{\infty} w_k(x)$ for all $x \in \mathbf{E}$ where $w_k$ is defined in (80.6), is continuous on $\mathbf{E}$ but nowhere differentiable.

In Fig. 80.3, we have made an attempt to exhibit the graph of $w_3$. This should give some indication of what is going to happen when $k$ becomes large. In view of this, the result of Theorem 80.1 is not really surprising.

## 7.80 Exercises

1. Show: If $|\Delta_k - \Delta_{k+1}| \geqslant 1$ for all $k \in \mathbf{N}$, then $\{\Delta_k\}$ is not a Cauchy sequence.
2. Consider the function $f_0$ as defined in (80.1) and let $f_k(x) = (1/10^k) f_0(10^k x)$ for all $k \in \mathbf{N}$. Show that the function $w$, defined by $w = \lim_{\infty} (f_0 + f_1 + \cdots + f_k)$ is continuous on $\mathbf{E}$ but nowhere differentiable.
3. Prove that the function $w$, defined in Lemma 80.1, is uniformly continuous on $\mathbf{E}$.

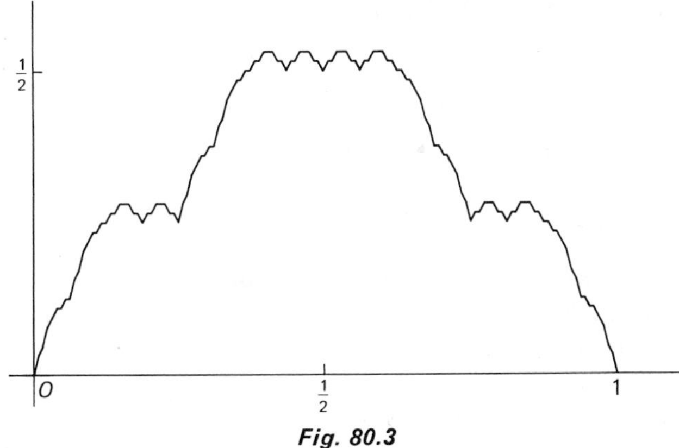


*Fig. 80.3*

## 7.81 Termwise Integration of Sequences

A pointwise convergent sequence of Riemann-intergrable functions need not converge to a Riemann-integrable function.

---

▲ *Example 1*

Let $\{r_1, r_2, r_3, \ldots\}$ represent an enumeration of all rationals in $[0, 1]$ and let $f_k : [0, 1] \to \mathbf{E}$ be defined by

$$f_k(x) = \begin{cases} 0 & \text{if } x \in \{r_1, r_2, r_3, \ldots, r_k\}, \\ 1 & \text{if } x \in [0, 1] \backslash \{r_1, r_2, r_3, \ldots, r_k\}; \end{cases}$$

$f_k$ is Riemann-integrable on $[0, 1]$ because it is bounded and continuous almost everywhere. (Actually, $f_k$ is continuous except at finitely many points.) The sequence $\{f_k\}$ converges pointwise on $[0, 1]$:

If $x$ is rational, then there is a smallest $l \in \mathbf{N}$ such that $x = r_l$ and hence, $\{f_k(x)\} = \{1, 1, \ldots, 1, 0, 0, 0, \ldots\}$ where 1 appears in the first $(l - 1)$ places. Hence,

$$\lim_{\infty} f_k(x) = 0 \qquad \text{if } x \in \mathbf{Q} \cap [0, 1].$$

If $x$ is irrational, then $x \neq r_l$ for all $l \in \mathbf{N}$ and hence, $\{f_k(x)\} = \{1, 1, 1, \ldots\}$. Therefore,

$$\lim_{\infty} f_k(x) = 1 \qquad \text{if } x \in [0, 1] \backslash \mathbf{Q}.$$

Hence,

$$\lim_{\infty} f_k = \delta \text{ on } [0, 1],$$

where $\delta\colon [0, 1] \to \mathbf{E}$ is the Dirichlet function which is defined by

$$\delta(x) = \begin{cases} 0 & \text{if } x \in \mathbf{Q} \cap [0, 1], \\ 1 & \text{if } x \in \mathbf{I} \cap [0, 1]. \end{cases}$$

We have seen in section 4.45, example 2, that the Dirichlet function is *not* Riemann-integrable.

---

Even if a sequence of Riemann-integrable functions converges pointwise to a Riemann-integrable function, the integral of the limit function need not be equal to the limit of the sequence of integrals of the individual elements of the sequence of functions:

---

▲ *Example 2*

We consider the sequence $\{\chi_k\}$ of example 4, section 7.76, where

$$\chi_k(x) = \begin{cases} 4k^2 x & \text{for } 0 \leqslant x \leqslant \dfrac{1}{2k}, \\[2ex] -4k^2 x + 4k & \text{for } \dfrac{1}{2k} < x \leqslant \dfrac{1}{k}, \\[2ex] 0 & \text{for } \dfrac{1}{k} < x \leqslant 1. \end{cases}$$

(See Fig. 76.2.) We have seen that

$$\lim_{\infty} \chi_k = \chi \text{ on } [0, 1],$$

where $\chi : [0, 1] \to \mathbf{E}$ is defined by

$$\chi(x) = 0.$$

Each $\chi_k$, as well as the limit function, are Riemann-integrable. (They are bounded and continuous.) We see, from Fig. 76.2, that

$$\int_0^1 \chi_k = 1 \qquad \text{for all } k \in \mathbf{N};$$

and hence,

$$\lim_{\infty} \int_0^1 \chi_k = 1.$$

On the other hand,

$$\int_0^1 \lim_{\infty} \chi_k = \int_0^1 \chi = 0.$$

Hence,

$$\lim_{\infty} \int_0^1 \chi_k \neq \int_0^1 \lim_{\infty} \chi_k.$$

---

If we impose the condition of uniform convergence on a sequence of functions, then it develops that a uniformly convergent sequence of Riemann-integrable functions converges to a Riemann-integrable function, and that limit and integral are interchangeable; i.e., the integral of the limit function is equal to the limit of the sequence of integrals of the elements of the sequence of functions. This latter property is often expressed by stating that the sequence is *termwise-integrable.*

The next two theorems will establish the above mentioned facts.

### *Theorem 81.1*

If the functions $f_k : [a, b] \to \mathbf{E}$ are Riemann-integrable and if the sequence $\{f_k\}$ converges uniformly on $[a, b]$ to a limit function $f$, then $f$ is Riemann-integrable.

### *Proof*

A Riemann-integrable function is bounded, by definition. Hence, by Theorem 78.2, the limit function $f$ is also bounded.

It remains to be shown that $f$ is continuous a.e. on $[a, b]$. Suppose that $f$ is *not* continuous at $x_0 \in [a, b]$, then at least one of the $f_k$ is not continuous at $x_0$. Otherwise, $f$ would be continuous at $x_0$, by the Corollary to Theorem 78.1. Hence, if $D(f)$ denotes the set of discontinuities of $f$ in $[a, b]$ and $D(f_k)$ denotes the set of discontinuities of $f_k$ in $[a, b]$, then

$$D(f) \subseteq \bigcup_{k=1}^{\infty} D(f_k).$$

Since, by hypothesis, each $f_k$ is Riemann-integrable, we have $\lambda(D(f_k)) = 0$ for all $k \in \mathbf{N}$, and, from Lemma 48.2. $\lambda(\bigcup_{k=1}^{\infty} D(f_k)) = 0$.

So we see that $D(f)$ is contained in a set of Lebesgue measure zero and, therefore, $\lambda(D(f)) = 0$. Since $f$ is bounded on $[a, b]$ and continuous a.e. on $[a, b]$, it follows, by Lemma 49.1, that $f$ is Riemann-integrable on $[a,b]$.

### *Corollary to Theorem 81.1*

If the functions $f_k : [a, b] \to \mathbf{E}$ are Riemann-integrable and if the sequence $\{f_k\}$ is a uniformly convergent Cauchy sequence on $[a, b]$, then $\{f_k\} \to f$, where $f$ is Riemann-integrable on $[a, b]$.

The proof consists in an application of Theorems 77.3 and 81.1.

Theorem 81.1 has a trivial generalization to sequences $\{f_k\}$, where $f_k: [a,b] \to \mathbf{E}^m$, if one defines the integral of a function $f: [a,b] \to \mathbf{E}^m$ as

$$\int_a^b f = \left(\int_a^b \phi_1, \int_a^b \phi_2, \ldots, \int_a^b \phi_m\right),$$

the $\phi_j$ representing the components of $f$.

***Theorem 81.2***

If the functions $f_k: [a,b] \to \mathbf{E}$ are Riemann-integrable and if the sequence $\{f_k\}$ converges uniformly to $f$ on $[a,b]$ (or if $\{f_k\}$ is a uniformly convergent Cauchy sequence on $[a,b]$), then

$$\int_a^b \lim_{\infty} f_k = \lim_{\infty} \int_a^b f_k.$$

*Proof*

We have, from Theorem 81.1, that $f = \lim_\infty f_k$ is Riemann-integrable.
From Corollary 3 to Theorem 51.1, we have

$$(81.1) \qquad \left|\int_a^b (f - f_k)\right| \leqslant \int_a^b |f - f_k|.$$

Since $\{f_k\}$ converges uniformly to $f$ on $[a,b]$, we have, for every $\varepsilon > 0$, an $N(\varepsilon) \in \mathbf{N}$ such that

$$|f(x) - f_k(x)| < \frac{\varepsilon}{b-a} \qquad \text{for all } k > N(\varepsilon).$$

Hence, we obtain, from Corollary 2 to Theorem 51.1, that

$$(81.2) \qquad \int_a^b |f - f_k| < \frac{\varepsilon}{b-a}(b-a) = \varepsilon \qquad \text{for all } k > N(\varepsilon).$$

Collection of the partial results in (81.1) and (81.2) yields

$$\left|\int_a^b f - \int_a^b f_k\right| = \left|\int_a^b (f - f_k)\right| < \varepsilon \qquad \text{for all } k > N(\varepsilon);$$

that is,

$$\int_a^b f = \lim_{\infty} \int_a^b f_k,$$

where $\int_a^b f = \int_a^b \lim_\infty f_k$.

Theorem 81.2 has a trivial generalization in the same sense as we have indicated in the paragraph following the Corollary to Theorem 81.1.

Theorem 81.2 enables us to integrate the limit function of a uniformly convergent sequence of functions without explicit knowledge of the limit function.

▲ *Example 3*

We consider the function $w: \mathbf{E} \to \mathbf{E}$ that was defined in Lemma 80.1. This function is defined as the limit function of a uniformly convergent sequence of continuous functions $w_k$. Hence, $w$ is Riemann-integrable on $[0, 1]$. From (80.6),

$$\int_0^1 w_k = \int_0^1 f_0 + \int_0^1 f_1 + \cdots + \int_0^1 f_k.$$

We see, from Fig. 80.2, that

$$\int_0^1 f_j = \frac{1}{4^{j+1}}.$$

Hence,

$$\int_0^1 w_k = \frac{1}{4} + \frac{1}{4^2} + \cdots + \frac{1}{4^{k+1}} = \frac{1}{4}\left(1 + \frac{1}{4} + \cdots + \frac{1}{4^k}\right)$$

$$= \frac{1}{4} \cdot \frac{4}{3}\left(1 - \frac{1}{4^{k+1}}\right) = \frac{1}{3}\left(1 - \frac{1}{4^{k+1}}\right).$$

Therefore, we obtain, from Theorem 81.2, that

$$\int_0^1 w = \lim_{\infty} \int_0^1 w_k = \lim_{\infty} \frac{1}{3}\left(1 - \frac{1}{4^{k+1}}\right) = \frac{1}{3}.$$

Thus we succeeded in integrating a function of which we have no explicit representation and the graph of which is beyond human perception.

## 7.81 Exercises

1. Let $\{\phi_k\}$ represent the sequence of example 2, section 7.76. Find $\lim_{\infty} \int_0^1 \phi_k$ and $\int_0^1 \lim_{\infty} \phi_k$. Does the result conflict with Theorem 81.2?
2. Same as in exercise 1, for the sequence $\{\psi_k\}$ of example 3, section 7.76.
3. Let $R([a, b], \mathbf{E})$ denote the collection of Riemann-integrable functions from $[a, b]$ into $\mathbf{E}$. Define addition and scalar multiplication as in example 2, section 5.55, and show that $R([a, b], \mathbf{E})$ is a vector space.
4. Show that
$$\|f\|_{[a,b]} = \sup_{[a,b]} |f(x)|$$
is a norm in $R([a, b], \mathbf{E})$.
5. Show that $R([a, b], \mathbf{E})$ with the sup-norm is a complete metric space, i.e., each Cauchy sequence in $R([a, b], \mathbf{E})$ converges to an element of $R([a, b], \mathbf{E})$.
6. Let $g: [a, b] \to \mathbf{E}^m$, where the components of $g$ are given by $(\psi_1, \psi_2, \ldots, \psi_m)$. Let $\int_a^b g = (\int_a^b \psi_1, \int_a^b \psi_2, \ldots, \int_a^b \psi_m)$. State and prove Theorem 81.1 for a sequence of functions $f_k: [a, b] \to \mathbf{E}^m$.
7. State and prove Theorem 81.2 for a sequence of functions $f_k: [a, b] \to \mathbf{E}^m$. (See also exercise 6.)
8. Find $\int_0^1 w$ where $w: \mathbf{E} \to \mathbf{E}$ is defined in exercise 2, section 7.80.

## 7.82 Termwise Differentiation of Sequences

A pointwise-convergent sequence of differentiable functions need not converge to a differentiable function.

---

▲ *Example 1*

The sequence $\{f_k\}$, where $f_k(x) = x^k$ converges on $[0, 1]$ to the function

$$f(x) = \begin{cases} 0 & \text{for } 0 \leqslant x < 1, \\ 1 & \text{for } x = 1. \end{cases}$$

While each $f_k$ is differentiable on $[0, 1]$, the limit function is not differentiable on $[0, 1]$.

---

In contrast to our experience in section 7.81, a *uniformly* convergent sequence of differentiable functions need not converge to a differentiable function either.

---

▲ *Example 2*

The functions $w_k$ which are defined in (80.6) are differentiable for all irrational $x$ (and most rational $x$). The sequence $\{w_k\}$ converges uniformly to $w$, but $w$ is nowhere differentiable.

▲ *Example 3*

Consider the sequence $\{f_k\}$, where $f_k : \mathbf{E} \to \mathbf{E}$ is given by

$$f_k(x) = \cos x + \frac{1}{2}\cos 3x + \frac{1}{4}\cos 9x + \cdots + \frac{1}{2^k}\cos 3^k x.$$

$f_k$ is differentiable on $\mathbf{E}$ for all $k \in \mathbf{N}$. $\{f_k\}$ converges uniformly on $\mathbf{E}$ because for $k > l$,

$$|f_k(x) - f_l(x)| = \left|\frac{1}{2^{l+1}}\cos 3^{l+1}x + \cdots + \frac{1}{2^k}\cos 3^k x\right|$$

$$\leqslant \frac{1}{2^l} + \cdots + \frac{1}{2^k} = \frac{1}{2^{l-1}} - \frac{1}{2^k} < \frac{1}{2^{l-1}}.$$

Still, one can show that $f = \lim_\infty f_k$ is *nowhere* differentiable on $\mathbf{E}$.†

Let it suffice here to show that $f$ is not differentiable at 0 and, by implication at $2k\pi$, $k \in \mathbf{N}$ (because $f$ is $2\pi$-periodic). We consider the sequence

$$\{x_k\} = \left\{\frac{\pi}{3^k}\right\}.$$

† G. H. Hardy, *Proc. London Math. Soc.* **2**, 9, (1909), pp. 126–144.

Clearly, $\lim_\infty x_k = 0$. Since

$$\cos 3^j x_k = \cos 3^{j-k}\pi = -1 \qquad \text{for all } j \geqslant k,$$

we have

$$\frac{f_j(x_k) - f_j(0)}{x_k - 0}$$

$$= \frac{\cos\dfrac{\pi}{3^k} + \dfrac{1}{2}\cos\dfrac{\pi}{3^{k-1}} + \cdots + \dfrac{1}{2^{k-1}}\cos\dfrac{\pi}{3} - \dfrac{1}{2^k} - \cdots - \dfrac{1}{2^j} - 1 - \dfrac{1}{2} - \cdots - \dfrac{1}{2^j}}{\dfrac{\pi}{3^k}}.$$

Since

$$\frac{f(x_k) - f(0)}{x_k - 0} = \lim_{j\to\infty} \frac{f_j(x_k) - f_j(0)}{x_k - 0},$$

we have

$$\frac{f(x_k) - f(0)}{x_k - 0} = \frac{3^k}{\pi}\left(\cos\frac{\pi}{3^k} + \frac{1}{2}\cos\frac{\pi}{3^{k-1}} + \cdots + \frac{1}{2^{k-1}}\cos\frac{\pi}{3} - \frac{1}{2^{k-1}} - 2\right)$$

$$\leqslant \frac{3^k}{\pi}\left(1 + \frac{1}{2} + \cdots + \frac{1}{2^{k-1}} - \frac{1}{2^{k-1}} - 2\right) = -\frac{3^k}{\pi 2^{k-2}} = -\frac{9}{\pi}\left(\frac{3}{2}\right)^{k-2};$$

and we see that $(f(x_k) - f(0))/(x_k - 0)$ diverges to $-\infty$. (See definition 15.2.) Hence, $\lim_0 \{(f(x) - f(0))/(x - 0)\}$ does not exist; i.e., $f$ is *not* differentiable at 0.

---

In order to guarantee the differentiability of the limit function of a sequence of functions, we have to impose a much stronger condition, namely uniform convergence of the sequence of the derivatives of the functions. In addition, we have to require convergence of the sequence of functions at at least one point, because the uniform convergence of a sequence of functions does *not* imply convergence of the sequence of antiderivatives.

---

▲ *Example 4*

Let $f_k : \mathbf{E} \to \mathbf{E}$ be given by $f_k(x) = k$. Then $f_k'(x) = 0$ and $\{f_k'\}$ converges uniformly on $\mathbf{E}$. However, $\{f_k\}$ does not converge anywhere.

---

***Theorem 82.1***

Let $\mathfrak{J}$ denote a bounded interval (open, or closed, or neither) and let $f_k : \mathfrak{J} \to \mathbf{E}$ for all $k \in \mathbf{N}$.

If $\{f_k(x_0)\}$ converges for some $x_0 \in \mathfrak{J}$, if $f_k'(x)$ exists for all $x \in \mathfrak{J}$, and if $\{f_k'\}$ converges uniformly on $\mathfrak{J}$ to a limit function $g : \mathfrak{J} \to \mathbf{E}$, then

$$\begin{aligned} &\{f\} \xrightarrow{\text{unif}} f \text{ on } \mathfrak{J}, \\ &f \quad \text{is differentiable on } \mathfrak{J}, \\ &f'(x) = g(x) \qquad \text{for all } x \in \mathfrak{J}. \end{aligned}$$

(The last statement expresses the fact that $(\lim_\infty f_k)' = \lim_\infty f_k'$, i.e., limit and differentiation are interchangeable, or, as it is often expressed, $\{f_k\}$ is *termwise differentiable*.)

*Proof*

First we shall demonstrate that $\{f_k\}$ converges uniformly on $\mathfrak{J}$. We apply the mean-value theorem of the differential calculus, Theorem 37.3, to the function $f_k - f_l$ between $x, x_0 \in \mathfrak{J}$:

$$(82.1) \qquad f_k(x) - f_l(x) - f_k(x_0) + f_l(x_0) = (f_k - f_l)'(\xi)(x - x_0),$$

where $\xi$ is between $x$ and $x_0$ and depends on $k$ and $l$. Hence,

$$|f_k(x) - f_l(x)| \leqslant |f_k(x_0) - f_l(x_0)| + |f_k'(\xi) - f_l'(\xi)|\, l(\mathfrak{J}),$$

where $l(\mathfrak{J})$ denotes the length of $\mathfrak{J}$.

Since $\{f_k(x_0)\}$ converges and since $\{f_k'\}$ converges uniformly on $\mathfrak{J}$, we have, for every $\varepsilon > 0$, an $N(\varepsilon) \in \mathbf{N}$ such that

$$|f_k(x_0) - f_l(x_0)| < \frac{\varepsilon}{2} \qquad \text{for all } k, l > N(\varepsilon);$$

and

$$|f_k'(\xi) - f_l'(\xi)| < \frac{\varepsilon}{2l(\mathfrak{J})} \qquad \text{for all } k, l > N(\varepsilon) \quad \text{and all } \xi \in \mathfrak{J}.$$

Hence,

$$|f_k(x) - f_l(x)| < \varepsilon \qquad \text{for all } k, l > N(\varepsilon) \text{ and all } x \in \mathfrak{J};$$

i.e., $\{f_k\}$ is a uniformly convergent Cauchy sequence on $\mathfrak{J}$ and converges uniformly to a function $f$ by Theorem 77.3.

Since $f_k$ is differentiable on $\mathfrak{J}$ for all $k$, it follows that $f_k$ is continuous on $\mathfrak{J}$ and hence, by Theorem 78.1, $f$ is continuous on $\mathfrak{J}$.

Next, and finally, we shall prove that $f$ is differentiable on $\mathfrak{J}$ and that $(\lim_\infty f_k)' = \lim_\infty f_k'$.

By hypothesis, $\{f_k'\} \xrightarrow{\text{unif}} g$ on $\mathfrak{J}$. In order to show that, for every $c \in \mathfrak{J}$, $f'(c) = g(c)$ and, by implication, exists, i.e., that for every $\varepsilon > 0$ there is a $\delta(\varepsilon) > 0$ such that

$$\left| \frac{f(x) - f(c)}{x - c} - g(c) \right| < \varepsilon \qquad \text{for all } x \in N_{\delta(\varepsilon)}(c) \cap \mathfrak{J},$$

we split the left side of this inequality as follows:

$$\left|\frac{f(x)-f(c)}{x-c}-g(c)\right| \leqslant \left|\frac{f(x)-f(c)}{x-c}-\frac{f_k(x)-f_k(c)}{x-c}\right| + \left|\frac{f_k(x)-f_k(c)}{x-c}-f_k'(c)\right| + |f_k'(c)-g(c)|. \tag{82.2}$$

If we replace in (82.1) $x_0$ by $c$ and $\xi$ by $\eta$, we obtain

$$f_k(x)-f_l(x)-f_k(c)+f_l(c)=(f_k-f_l)'(\eta)(x-c);$$

and hence,

$$\left|\frac{f_k(x)-f_k(c)}{x-c}-\frac{f_l(x)-f_l(c)}{x-c}\right| \leqslant |f_k'(\eta)-f_l'(\eta)|.$$

Since $\{f_k'\}$ converges uniformly on $\mathfrak{J}$, we have, for every $\varepsilon > 0$, an $N_1(\varepsilon)$ such that

$$|f_k'(\eta)-f_l'(\eta)| < \frac{\varepsilon}{3} \qquad \text{for all } k, l > N_1(\varepsilon) \quad \text{and all } \eta \in \mathfrak{J}.$$

Hence,

$$\left|\frac{f_k(x)-f_k(c)}{x-c}-\frac{f_l(x)-f_l(c)}{x-c}\right| < \frac{\varepsilon}{3}$$

for all $k, l > N_1(\varepsilon)$ and all $x \in \mathfrak{J}$. For each $x \in \mathfrak{J}$, we keep $k$ fixed and let $l$ tend to infinity. Then we obtain, in view of Lemma 62.5,

$$\left|\frac{f(x)-f(c)}{x-c}-\frac{f_k(x)-f_k(c)}{x-c}\right| \leqslant \frac{\varepsilon}{3} \qquad \text{for all } k > N_1(\varepsilon) \quad \text{and all } x \in \mathfrak{J}. \tag{82.3}$$

Since $f_k$ is differentiable on $\mathfrak{J}$, we have a $\delta(\varepsilon) > 0$ such that

$$\left|\frac{f_k(x)-f_k(c)}{x-c}-f_k'(c)\right| < \frac{\varepsilon}{3} \qquad \text{for all } x \in N_{\delta(\varepsilon)} \cap \mathfrak{J}. \tag{82.4}$$

Since $\lim_\infty f_k' = g$ on $\mathfrak{J}$, we have some $N_2(\varepsilon) \in \mathbf{N}$ such that

$$|f_k'(c)-g(c)| < \frac{\varepsilon}{3} \qquad \text{for all } k > N_2(\varepsilon). \tag{82.5}$$

Let $k_0 > \max(N_1(\varepsilon), N_2(\varepsilon))$. Then we choose $\delta(\varepsilon)$ in (82.4) for this particular $k_0$ and we obtain for (82.2) in view of (82.3), (82.4), and (82.5) that

$$\left|\frac{f(x)-f(c)}{x-c}-g(c)\right| < \varepsilon \qquad \text{for all } x \in N_{\delta(\varepsilon)}(c) \cap \mathfrak{J},$$

that is, $f'(c) = g(c)$. Since this argument is valid for all $c \in \mathfrak{J}$, we have $f'(x) = g(x)$ for all $x \in \mathfrak{J}$.

A trivial generalization of Theorem 82.1 may be obtained for sequences of functions $f_k : \mathfrak{J} \to \mathbf{E}^m$ if one defines $f'(c) = (\phi_1'(c), \phi_2'(c), \dots, \phi_m'(c))$ where $\phi_1, \phi_2, \dots, \phi_m$ are the components of $f$.

Theorem 82.1 enables us to differentiate functions of which we possess no explicit representation.

---

▲ *Example 5*

Let $f_k : [0, a] \to \mathbf{E}$, $a \in (0, 1)$, be given by

$$f_k(x) = 1 + x + \frac{x^2}{2} + \cdots + \frac{x^k}{k}. \tag{82.6}$$

Clearly, $\{f_k(0)\} = \{1\}$ converges. We have

$$f_k'(x) = 1 + x + \cdots + x^{k-1} = \frac{1 - x^k}{1 - x} \qquad \text{for all } x \in [0, a].$$

For $k > l$,

$$|f_k'(x) - f_l'(x)| = \frac{1}{1 - x}|x^l - x^k| \leqslant \frac{1}{1 - a}|x^l - x^k| < \varepsilon$$

for all $k, l > N(\varepsilon)$ and all $x \in [0, a]$, because the sequence $\{x^k\}$ converges uniformly to 0 on $[0, a]$ (see example 6, section 7.77). Hence $\{f_k'\}$ converges uniformly on $[0, a]$. Hence, by Theorem 82.1,

$$\left(\lim_{\infty} f_k\right)' = \lim_{\infty} f_k' = f',$$

where $f'(x) = \lim_{\infty}(1 - x^k)/(1 - x) = 1/(1 - x)$. Thus we found $f'$ without being in possession of an explicit representation of $f$. We may now, after the fact, recover $f$ from $f'$ by integration. We have from (82.6) that $f(0) = \lim_{\infty} f_k(0) = 1$, and hence, we obtain from Theorem 52.3, that

$$f(x) = \int_0^x f' + 1.$$

Since $\int_0^x dt/(1 - t) = -\log(1 - x)$ for $x \in [0, a]$, where $0 < a < 1$, we see that

$$f(x) = \lim_{\infty}\left(1 + x + \frac{x^2}{2} + \cdots + \frac{x^k}{k}\right) = 1 - \log(1 - x),$$

and hence,

$$\log(1 - x) = -\lim_{\infty}(x + \tfrac{1}{2}x^2 + \tfrac{1}{3}x^3 + \cdots + (1/k)x^k).$$

---

The limit function of a sequence may be differentiable while none of the elements of the sequence are differentiable.

---

▲ *Example 6*

Let $f_k : [0, 1] \to \mathbf{E}$ be defined by

$$f_k(x) = \begin{cases} 0 & \text{for all } x \in \mathbf{Q} \cap [0, 1], \\ \dfrac{1}{k} & \text{for all } x \in \mathbf{I} \cap [0, 1]. \end{cases}$$

Since $f_k$ is nowhere continuous on $[0, 1]$, $f_k$ is nowhere differentiable on $[0, 1]$. Still, the limit function

$$f = \lim_{\infty} f_k \text{ on } [0, 1]$$

is everywhere differentiable on $[0, 1]$ because $f(x) = 0$ for all $x \in [0, 1]$.

---

The reader may now proceed directly to Chapter 14.

## 7.82 Exercises

1. Given the sequence $\{f_k\}$ on $[0, a]$, where $0 < a < 1$ and

$$f_k(x) = 1 + x + \frac{x^2}{2} + \frac{x^3}{2 \cdot 3} + \cdots + \frac{x^k}{(k-1)k}.$$

   Find an explicit representation of the limit function $f$.
2. Given the sequence $\{f_k\}$ on $[0, 1]$, where $f_k(x) = 1 + x + x^2/2! + \cdots + x^k/k!$. Show that $f'(x) = f(x)$ for all $x \in [0, 1]$ where $f$ denotes the limit function of $\{f_k\}$.
3. Given the sequence $\{f_k\}$ on $[1 - a, 1 + a]$, $0 < a < 1$, where

$$f_k(x) = (x - 1) - \frac{1}{2}(x - 1)^2 + \frac{1}{3}(x - 1)^3 - \cdots + \frac{(-1)^{k+1}}{k}(x - 1)^k.$$

   Find an explicit representation of the limit function $f$.
4. Let $g : \mathfrak{J} \to \mathbf{E}^m$ where $\mathfrak{J}$ is a bounded interval and define $g' = (\psi'_1, \psi'_2, \ldots, \psi'_m)$, with $\psi_j$ being the components of $g$. State and prove Theorem 82.1 for a sequence $\{f_k\}$ where $f_k : \mathfrak{J} \to \mathbf{E}^m$.
5. Given the sequence $\{f_k\}$, where $f'_k \in C[a, b]$ and where $\{f_k(x_0)\}$ converges for some $x_0 \in [a, b]$, and where $\{f'_k\}$ converges uniformly on $[a, b]$. Use Theorem 52.3 to prove that $\lim_\infty f_k = f$ is differentiable on $[a, b]$ and $f' = \lim_\infty f'_k$. (Note that this is a weaker version of Theorem 82.1, because of the new and stronger hypothesis $f'_k \in C[a, b]$.)

# 8 Linear Functions

## 8.83 Definition and Representation

We consider the function $f: \mathbf{E} \to \mathbf{E}$ defined by

$$f(x) = x^2,$$

in some neighborhood $N_\delta(0)$ of 0. (See Fig. 83.1.)

We see that, for every $y \in f_*(N_\delta(0)) \setminus \{0\}$, there are two values $x$, $-x \in N_\delta(0)$ such that $x^2 = y$. Hence, $f: N_\delta(0) \to \mathbf{E}$ is *not* one-to-one, or, as we may put it, the equation $x^2 = y$ does not have a unique solution in $N_\delta(0)$ for any $y \in f_*(N_\delta(0)) \setminus \{0\}$.

On the other hand, if we consider $f$ in some sufficiently small neighborhood $N_\delta(1)$ of 1 (see Fig. 83.2), then, for every $y \in f_*(N_\delta(1))$, there is precisely one $x \in N_\delta(1)$ such that $x^2 = y$. Hence, $f: N_\delta(1) \to \mathbf{E}$ is one-to-one, or, as we may put it, the equation $x^2 = y$ has a unique solution in $N_\delta(1)$ for every $y \in f_*(N_\delta(1))$.

The most striking difference between these two cases is, that the derivative of $f$ at $x = 0$ is 0 and at $x = 1$ is not zero. In section 3.39 we characterized the derivative $f'(a)$ of $f$ at $a$ as a linear function $l_a: \mathbf{E} \to \mathbf{E}$ which is defined by $l_a(h) = f'(a)h$. In case the derivative is different from 0, then $l_a$ is one-to-one; and in case the derivative is zero, then $l_a: \mathbf{E} \to \{0\}$ is definitely *not* one-to-one. It appears from our discussion that a function inherits the property of being one-to-one in a neighborhood of a point from its derivative at that point. We observe also that $l_a$ is onto $\mathbf{E}$ if $f'(a) \neq 0$, a property which $f$ inherits to the extent that it maps open sets in some neighborhood of $a$ onto open sets.

In order to pursue the ideas which were expressed above, in the more general setting of vector-valued functions of a vector variable, we have to deal first with linear functions from $\mathbf{E}^n$ into $\mathbf{E}^m$ for any $n, m \in \mathbf{N}$. In an obvious and straightforward generalization of definition 39.1, we define linear functions as follows.

***Definition 83.1***

$l: \mathbf{E}^n \to \mathbf{E}^m$ is a *linear function* if and only if $l$ is additive:

(83.1) $$l(x + y) = l(x) + l(y) \qquad \text{for all } x, y \in \mathbf{E}^n$$

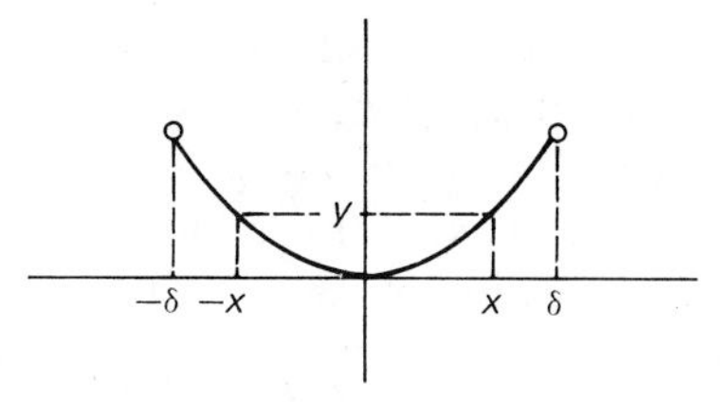


*Fig. 83.1*

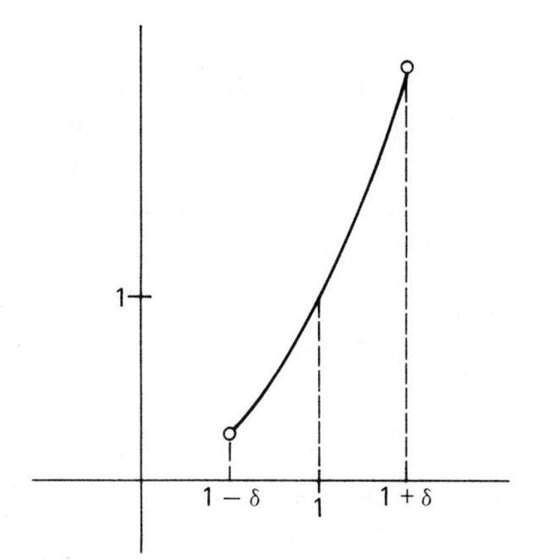


*Fig. 83.2*

and *homogeneous*:

(83.2) $$l(\lambda x) = \lambda l(x) \qquad \text{for all } x \in \mathbf{E}^n \text{ and every } \lambda \in \mathbf{R}.$$

Conditions (83.1) and (83.2) are equivalent to the condition that

(83.3) $$l(\lambda x + \mu y) = \lambda l(x) + \mu l(y) \qquad \text{for all } x, y \in \mathbf{E}^n \text{ and every } \lambda, \mu \in \mathbf{R}.$$

The reader is presumed to be familiar with the concepts of *linear dependence of vectors, linear subspace, rank of a linear function, matrix multiplication, rank of a matrix*, and the *representation of a linear function by a matrix*. For the convenience of the reader, and in order to synchronize the terminology, we present here a very brief review of these concepts.

The vectors $x_1, x_2, \ldots, x_p \in \mathbf{E}^n$ are *linearly dependent*, if and only if there exists a vector $c = (\gamma_1, \gamma_2, \ldots, \gamma_p) \neq (0, 0, \ldots, 0) \in \mathbf{E}^p$ such that $\gamma_1 x_1 + \gamma_2 x_2 + \cdots + \gamma_p x_p = \theta$.

The vectors $x_1, x_2, \ldots, x_p \in \mathbf{E}^n$ are *linearly independent* if and only if they are not linearly dependent; i.e., if and only if $\gamma_1 x_1 + \gamma_2 x_2 + \cdots + \gamma_p x_p = \theta$ implies $(\gamma_1, \gamma_2, \ldots, \gamma_p) = (0, 0, \ldots, 0)$.

A *linear subspace* $V$ of $\mathbf{E}^n$ is a subset of $\mathbf{E}^n$ which is itself a vector space (see definition 55.2) under the same operations.

The *dimension* of a vector space $V$ is the maximum number of linearly independent vectors in $V$.

The *range* of a linear function $l : \mathbf{E}^n \to \mathbf{E}^m$ is a linear subspace of $\mathbf{E}^m$. Indeed, if $e_1 = (1, 0, \ldots, 0)$, $e_2 = (0, 1, 0, \ldots, 0), \ldots, e_n = (0, 0, \ldots, 0, 1)$ denote the fundamental vectors in $\mathbf{E}^n$, and if $l(e_1) = y_1$, $l(e_2) = y_2$, $\ldots, l(e_n) = y_n$, then every vector $y$ in the range of $l$, that is, $l(x) = y$, $x = (\xi_1, \xi_2, \ldots, \xi_n) \in \mathbf{E}^n$, may be represented by a linear combination of $y_1, y_2, \ldots, y_n$:

$$(83.4) \quad y = l(x) = l(\xi_1 e_1 + \xi_2 e_2 + \cdots + \xi_n e_n) = \xi_1 l(e_1) + \xi_2 l(e_2) + \cdots + \xi_n l(e_n)$$
$$= \xi_1 y_1 + \xi_2 y_2 + \cdots + \xi_n y_n,$$

and every linear combination of such vectors is again a linear combination of $y_1, y_2, \ldots, y_n$. Since $y_j \in \mathbf{E}^m$ for all $j = 1, 2, \ldots, n$, we see that $\xi_1 y_1 + \xi_2 y_2 + \cdots + \xi_n y_n \in \mathbf{E}^m$ and hence, the range of $l$ is a linear subspace of $\mathbf{E}^m$.

The *dimension* of the range of $l$, the number of linearly independent vectors among $y_1, y_2, \ldots, y_n$, is called the *rank of l*. Since $m + p$, $p \geqslant 1$, vectors in $\mathbf{E}^m$ are always linearly dependent, the rank of $l$ is always less than or equal to $m$.

If we represent the vectors $y_1 = l(e_1), y_2 = l(e_2), \ldots, y_n = l(e_n)$ in terms of the fundamental vectors $i_1 = (1, 0, \ldots, 0)$, $i_2 = (0, 1, 0, \ldots, 0), \ldots, i_m = (0, 0, \ldots, 0, 1) \in \mathbf{E}^m$, we obtain

$$y_k = a_{1k} i_1 + a_{2k} i_2 + \cdots + a_{mk} i_m, \qquad k = 1, 2, \ldots, n,$$

for some coefficients $a_{jk} \in \mathbf{R}$. Hence, we have, from (83.4),

$$(83.5) \quad l(x) = \xi_1(a_{11} i_1 + \cdots + a_{m1} i_m) + \cdots + \xi_n(a_{1n} i_1 + \cdots + a_{mn} i_m)$$
$$= \sum_{j=1}^{m} \left[ \sum_{k=1}^{n} a_{jk} \xi_k \right] i_j.$$

The *matrix product* of the $p \times q$ matrix $A$ with the $q \times r$ matrix $B$ is defined by

$$(83.6) \quad \begin{pmatrix} a_{11} a_{12} \ldots a_{1q} \\ a_{21} a_{22} \ldots a_{2q} \\ \vdots \\ a_{p1} a_{p2} \ldots a_{pq} \end{pmatrix} \begin{pmatrix} b_{11} b_{12} \ldots b_{1r} \\ b_{21} b_{22} \ldots b_{2r} \\ \vdots \\ b_{q1} b_{q2} \ldots b_{qr} \end{pmatrix}$$
$$= \begin{pmatrix} \sum_1^q a_{1k} b_{k1} & \sum_1^q a_{1k} b_{k2} & \cdots & \sum_1^q a_{1k} b_{kr} \\ \sum_1^q a_{2k} b_{k1} & \sum_1^q a_{2k} b_{k2} & \cdots & \sum_1^q a_{2k} b_{kr} \\ \vdots & & & \\ \sum_1^q a_{pk} b_{k1} & \sum_1^q a_{pk} b_{k2} & \cdots & \sum_1^q a_{pk} b_{kr} \end{pmatrix}.$$

(The element in the $j$th row and $k$th column of the matrix product is the dot product of the $j$th row vector of $A$ with the $k$th column vector of $B$.)

With this understanding, we may write (83.5) as follows:

$$l(x) = \begin{pmatrix} a_{11} a_{12} \dots a_{1n} \\ a_{21} a_{22} \dots a_{2n} \\ \vdots \\ a_{m1} a_{m2} \dots a_{mn} \end{pmatrix} \begin{pmatrix} \xi_1 \\ \xi_2 \\ \vdots \\ \xi_n \end{pmatrix}. \tag{83.7}$$

The matrix

$$L = \begin{pmatrix} a_{11} a_{12} \dots a_{1n} \\ a_{21} a_{22} \dots a_{2n} \\ \vdots \\ a_{m1} a_{m2} \dots a_{mn} \end{pmatrix} \tag{83.8}$$

represents the linear function $l: \mathbf{E}^n \to \mathbf{E}^m$ that maps the fundamental vectors $e_1, e_2, \dots, e_n$ into the vectors $(a_{11}, a_{21}, \dots, a_{m1})$, $(a_{12}, a_{22}, \dots, a_{m2}), \dots,$ $(a_{1n}, a_{2n}, \dots, a_{mn})$ (the columns of $L$) with the understanding that the elements in $\mathbf{E}^n$ are represented in terms of the fundamental vectors $e_1, e_2, \dots, e_n$ of $\mathbf{E}^n$ and the images in $\mathbf{E}^m$ are represented in terms of the fundamental vectors $i_1, i_2, \dots, i_m$ of $\mathbf{E}^m$.

Conversely, if a function $l: \mathbf{E}^n \to \mathbf{E}^m$ is defined by $l(x) = Lx$, where $L$ is an $m \times n$ matrix as in (83.8), then $l$ is a linear function from $\mathbf{E}^n$ into $\mathbf{E}^m$ because, by the rules for the basic matrix manipulations,

$$l(\lambda x + \mu y) = L(\lambda x + \mu y) = \lambda Lx + \mu Ly = \lambda l(x) + \mu l(y)$$

for all $x, y \in \mathbf{E}^n$ and any $\lambda, \mu \in \mathbf{R}$.

If $L$ is the matrix representing the linear function $l: \mathbf{E}^n \to \mathbf{E}^m$, then the dimension of the range of $l$ (rank of $l$) is equal to the number of linearly independent column vectors in $L$, which in turn is equal to the number of linearly independent row vectors in $L$, which in turn is equal to the order of the largest nonvanishing determinant that can be formed from a square submatrix of $L$. We call this number the *rank of the matrix L*. Hence, the rank of $l$ (dimension of the range of $l$) and the rank of its matrix representation $L$ are the same.

Do *not* identify $L$ with $l$. $l$ is a linear function that may be represented in a variety of manners, while $L$ is merely one of its manifestations. In this and the following chapters, we shall concern ourselves to a considerable extent with linear functions. We shall always use lower-case Roman letters to denote the linear function, and the corresponding capital letters for is matrix representation in terms of the fundamental vectors of $\mathbf{E}^n$ and $\mathbf{E}^m$.

## 8.83 Exercises

1. The linear function $l: \mathbf{E}^3 \to \mathbf{E}^4$ maps the vectors $(1, 1, 0)$, $(1, 1, 2)$, and $(1, 0, 2) \in \mathbf{E}^3$ into $(1, 1, 2, 1)$, $(3, 0, -1, 2)$, and $(4, 1, 1, 3)$, respectively.
   (a) Find $l(e_1)$, $l(e_2)$, $l(e_3)$.
   (b) Find the matrix representation of $l$ in terms of $e_1, e_2, e_3, i_1, i_2, i_3, i_4$.
   (c) Find $l(x)$ for $x = (1, -1, 3)$.
   (d) Find rank $(l)$.

2. *Prove*: If $x_1, x_2, \dots, x_p \in \mathbf{E}^n$ are linearly dependent, then $x_1, x_2, \dots, x_p, x_{p+1}, \dots, x_q \in \mathbf{E}^n$ are linearly dependent.
3. *Prove*: $x_1, x_2, \dots, x_n, x_{n+1}, \dots, x_{n+p} \in \mathbf{E}^n$, $p \geqslant 1$, are linearly dependent.
4. *Prove*: If $x_1, x_2, \dots, x_p \in \mathbf{E}^n$ are linearly dependent and none is the zero vector $\theta = (0, 0, \dots, 0)$, then, for some $j \in \{1, 2, \dots, p\}$, $x_j$ may be expressed as a linear combination of all the other vectors.
5. *Prove*: $x_1, x_2, \dots, x_p, x_p + x_1 \in \mathbf{E}^n$ are linearly dependent.
6. Show: If $l: \mathbf{E}^n \to \mathbf{E}^m$, then $l(\theta) = \theta$, where the first $\theta$ is the additive identity in $\mathbf{E}^n$ and the second one is the additive identity in $\mathbf{E}^m$. Give an example of a linear function of rank greater than zero where $l(x) = \theta$ for some $x \neq \theta$.
7. Given

$$A = \begin{pmatrix} 1 & 1 & 0 & 3 \\ -1 & 2 & 0 & 1 \\ 1 & 1 & 2 & 4 \end{pmatrix}, \qquad B = \begin{pmatrix} 1 & 3 \\ 0 & 1 \\ 1 & -1 \\ 2 & 0 \end{pmatrix}.$$

Find $AB$.

8. Given

$$A = \begin{pmatrix} 1 & 3 & 2 & 1 \\ 0 & 1 & 1 & 2 \end{pmatrix}, \qquad B = \begin{pmatrix} 1 & 1 & -1 & 0 \\ 3 & 3 & 2 & 1 \end{pmatrix}.$$

Find $3A - 2B$.

9. *Prove*: If $x_1, x_2, \dots, x_p, \dots, x_q \in \mathbf{E}^n$ are linearly independent, then $x_1, x_2, \dots, x_p$ are linearly independent.
10. Given the linear function $l: \mathbf{E}^2 \to \mathbf{E}^3$ which has the matrix representation

$$L = \begin{pmatrix} 1 & 1 \\ 0 & -1 \\ 1 & 0 \end{pmatrix}.$$

Show that every vector in the range of $l$ can be represented by a linear combination of the vectors $(1, 0, 1)$ and $(1, -1, 0)$.

11. Given the linear functions $l_1 : \mathbf{E}^n \to \mathbf{E}^p$, $l_2 : \mathbf{E}^p \to \mathbf{E}^m$. Show that $l_2 \circ l_1$: $\mathbf{E}^n \to \mathbf{E}^m$ is a linear function.
12. Let $L_1, L_2$ be the matrix representation of the linear functions $l_1, l_2$ in exercise 11. Show that $L_2 L_1$ is the matrix representation of $l_2 \circ l_1$.

## 8.84 Linear Onto Functions and Linear One-to-One Functions

The linear function $l: \mathbf{E}^n \to \mathbf{E}^m$ is *onto* $\mathbf{E}^m$ if and only if the dimension of the range of $l$ is $m$. Then the range of $l$ and $\mathbf{E}^m$ coincide. Since the dimension of the range of $l$ is the rank of $l$ which, in turn, is equal to the rank of the $m \times n$ matrix $L$ representing $l$, and since the rank of an $m \times n$ matrix cannot exceed min $(m, n)$, we have the following.

***Lemma 84.1***

The linear function $l\colon \mathbf{E}^n \to \mathbf{E}^m$ is onto $\mathbf{E}^m$ if and only if rank $l = m$ where $m \leqslant n$.

If $L$ is the matrix representation of $l$, this is equivalent to the statement that the system of $m$ nonhomogeneous linear equations in $n$ unknowns $Lx = y$ has at least one solution $x \in \mathbf{E}^n$ for every $y \in \mathbf{E}^m$ if and only if rank $L = m \leqslant n$.

---

▲ *Example 1*

Let $l : \mathbf{E}^2 \to \mathbf{E}^3$ denote a linear function which maps $e_1$, $e_2$, into

$$l(e_1) = \begin{pmatrix} 1 \\ 1 \\ 1 \end{pmatrix}, \qquad l(e_2) = \begin{pmatrix} 1 \\ -1 \\ 0 \end{pmatrix}.$$

Then, $l$ may be represented by the matrix

$$L = \begin{pmatrix} 1 & 1 \\ 1 & -1 \\ 1 & 0 \end{pmatrix}.$$

$l$ is *not* onto $\mathbf{E}^3$ because rank $L = 2 < m = 3$ and $n = 2 < 3 = m$.

▲ *Example 2*

Let $l : \mathbf{E}^3 \to \mathbf{E}^2$ denote the linear function which maps $e_1$, $e_2$, $e_3$ into

$$l(e_1) = \begin{pmatrix} 1 \\ 0 \end{pmatrix}, \qquad l(e_2) = \begin{pmatrix} -1 \\ 0 \end{pmatrix}, \qquad l(e_3) = \begin{pmatrix} 0 \\ 0 \end{pmatrix}.$$

Then, $l$ may be represented by the matrix

$$L = \begin{pmatrix} 1 & -1 & 0 \\ 0 & 0 & 0 \end{pmatrix}.$$

$l$ is *not* onto $\mathbf{E}^2$ because rank $L = 1 < 2 = m$.

▲ *Example 3*

Let $l : \mathbf{E}^3 \to \mathbf{E}^2$ denote the linear function which maps $e_1$, $e_2$, $e_3$ into

$$l(e_1) = \begin{pmatrix} 1 \\ 0 \end{pmatrix}, \qquad l(e_2) = \begin{pmatrix} 1 \\ 1 \end{pmatrix}, \qquad l(e_3) = \begin{pmatrix} 2 \\ 1 \end{pmatrix}.$$

Then, $l$ may be represented by the matrix

$$L = \begin{pmatrix} 1 & 1 & 2 \\ 0 & 1 & 1 \end{pmatrix},$$

and we see that $l$ is onto $\mathbf{E}^2$ because rank $L = 2 = m < n = 3$.

---

If $l: \mathbf{E}^n \to \mathbf{E}^m$ is a linear function and if there exists a linear function $l^R$: $\mathbf{E}^m \to \mathbf{E}^n$ such that $l \circ l^R(y) = y$ for all $y \in \mathbf{E}^m$, then $l^R$ is called a *right inverse* of $l$. Linear onto functions possess right inverses:

***Theorem 84.1***

If the linear function $l: \mathbf{E}^n \to \mathbf{E}^m$ is onto $\mathbf{E}^m$, then it has at least one right inverse $l^R: \mathbf{E}^m \to \mathbf{E}^n$ such that $l \circ l^R(y) = y$ for all $y \in \mathbf{E}^m$.

*Proof*

Let $i_1, i_2, \ldots, i_m$ denote the fundamental vectors in $\mathbf{E}^m$. Since $l$ is onto $\mathbf{E}^m$, there exist (not necessarily unique) vectors $u_1, u_2, \ldots, u_m$ such that $l(u_j) = i_j$ for $j = 1, 2, \ldots, m$. We define the function $l^R: \mathbf{E}^m \to \mathbf{E}^n$ by

$$l^R(y) = \sum_{j=1}^{m} \eta_j u_j \qquad \text{for all } y = (\eta_1, \eta_2, \ldots, \eta_m) \in \mathbf{E}^m.$$

This function is obviously linear and, moreover,

$$l \circ l^R(y) = \sum_{j=1}^{m} \eta_j l(u_j) = \sum_{j=1}^{m} \eta_j i_j = y.$$

Hence, $l^R$ is the right inverse of $l$.

Unless $l$ is one-to-one, there is a choice of vectors $u_1, u_2, \ldots, u_m$ with the required property in the above proof. Hence, $l$ possesses in such a case more than one inverse. We shall illustrate such a situation in the following example.

---

▲ *Example 4*

Let $l: \mathbf{E}^3 \to \mathbf{E}^2$ be defined by the matrix

$$L = \begin{pmatrix} 1 & 0 & 1 \\ 0 & -1 & 0 \end{pmatrix}.$$

Since rank $L = 2$, $l$ is onto $E^2$. Let

$$A = \begin{pmatrix} 1 & 0 \\ 0 & -1 \end{pmatrix};$$

then,

$$A^{-1} = \begin{pmatrix} 1 & 0 \\ 0 & -1 \end{pmatrix}.$$

Hence, if we let

$$L^R = \begin{pmatrix} & A^{-1} & \\ \cdots & \cdots & \cdots \\ 0 & & 0 \end{pmatrix} = \begin{pmatrix} 1 & 0 \\ 0 & -1 \\ 0 & 0 \end{pmatrix},$$

then

$$LL^R = \begin{pmatrix} 1 & 0 & 1 \\ 0 & -1 & 0 \end{pmatrix} \begin{pmatrix} 1 & 0 \\ 0 & -1 \\ 0 & 0 \end{pmatrix} = \begin{pmatrix} 1 & 0 \\ 0 & 1 \end{pmatrix};$$

that is, $L^R$ represents a right inverse of $l$. We obtain other right inverses of $l$ as follows: The matrix

$$L^N = \begin{pmatrix} 1 & 1 \\ 0 & 0 \\ -1 & -1 \end{pmatrix}$$

maps $\mathbf{E}^2$ into the null space of $l$, i.e., the linear subspace of all vectors $x \in \mathbf{E}^3$ for which $lx = \theta$. Indeed, if $y = (\eta_1, \eta_2) \in \mathbf{E}^2$, then

$$LL^N y = \begin{pmatrix} 1 & 0 & 1 \\ 0 & -1 & 0 \end{pmatrix} \begin{pmatrix} 1 & 1 \\ 0 & 0 \\ -1 & -1 \end{pmatrix} \begin{pmatrix} \eta_1 \\ \eta_2 \end{pmatrix} = \begin{pmatrix} 0 & 0 \\ 0 & 0 \end{pmatrix} \begin{pmatrix} \eta_1 \\ \eta_2 \end{pmatrix} = \theta.$$

Hence, for every $\lambda \in \mathbf{R}$,

$$L(L^R + \lambda L^N)y = LL^R y + \lambda LL^N y = y;$$

i.e., if $L^R$ represents a right inverse of $l$, then $L^R + \lambda L^N$ also represents a right inverse of $l$ for all $\lambda \in \mathbf{R}$.

---

If a linear function is to be one-to-one it is reasonable to expect that the dimensions of domain and range have to be the same. Otherwise, some vectors would appear to be left out either in the range or in the domain. This is indeed so.

***Lemma 84.2***

The linear function $l: \mathbf{E}^n \to \mathbf{E}^m$ is one-to-one on $\mathbf{E}^n$ if and only if the dimension of the range of $l$ is equal to $n$: rank $l = n \leqslant m$.

If $L$ is the matrix representation of $l$, this is equivalent to the statement that the system of $m$ nonhomogeneous linear equations in $n$ unknowns $Lx = y$ has a unique solution $x \in \mathbf{E}^n$ for all $y \in \mathfrak{R}(l)$ if and only if rank $L = n \leqslant m$. (If $y \in \mathbf{E}^m \backslash \mathfrak{R}(l)$, then $Lx = y$ does not have any solution at all.)

*Proof*

(a) Suppose that the dimension of the range of $l$ is $n$ (which is, of course, only possible if $n \leqslant m$). Let $x_1, x_2 \in \mathbf{E}^n$ and assume that $l(x_1) = l(x_2)$; that is, $l(x_1) - l(x_2) = l(x_1 - x_2) = \theta$. Suppose that $x_1 - x_2 = \xi_1 e_1 + \xi_2 e_2 + \cdots + \xi_n e_n \neq \theta$. Then, by the linearity of $l$, $\xi_1 l(e_1) + \xi_2 l(e_2) + \cdots + \xi_n l(e_n) = \theta$; that is, $l(e_1), l(e_2), \ldots, l(e_n)$ are linearly dependent. Since every vector in $\mathfrak{R}(l)$ can be written as a linear combination of $l(e_1)$, $l(e_2), \ldots, l(e_n)$, and since $l(e_1)$, $l(e_2), \ldots,$ $l(e_n)$ are linearly dependent, it follows that the dimension of the range of $l$ is less than $n$, contrary to our hypothesis. Therefore, $x_1 - x_2 = \theta$, i.e., $x_1 = x_2$. Since $l(x_1) = l(x_2)$ implies $x_1 = x_2$, we have, from definition 4.4, that $l$ is one-to-one.

(b) Suppose that $l$ is one-to-one. Every element in the range of $l$ is a linear combination of $l(e_1), l(e_2), \ldots, l(e_n)$. Suppose that these vectors are linearly dependent: Then, there exists a vector $(\gamma_1, \gamma_2, \ldots, \gamma_n) \neq (0, 0, \ldots, 0)$ such that

$$\gamma_1 l(e_1) + \gamma_2 l(e_2) + \cdots + \gamma_n l(e_n) = \theta$$

or, equivalently,

$$l(\gamma_1 e_1 + \gamma_2 e_2 + \cdots + \gamma_n e_n) = \theta.$$

Since $l$ is, by hypothesis, one-to-one, $l(x) = \theta$ implies $x = \theta$ because $l(\theta) = \theta$ and $\theta$ cannot have another pre-image. Hence,

$$\gamma_1 e_1 + \gamma_2 e_2 + \cdots + \gamma_n e_n = \theta;$$

that is, $e_1, e_2, \ldots, e_n$ are linearly dependent, which is absurd. Hence, $l(e_1), l(e_2),$ $\ldots, l(e_n)$ have to be linearly independent, which means that the dimension of the range of $l$ is $n$.

---

▲ *Example 5*

Let $l : \mathbf{E}^2 \to \mathbf{E}^3$ represent the linear function of example 1. Since

$$\text{rank } l = \text{rank} \begin{pmatrix} 1 & 1 \\ 1 & -1 \\ 1 & 0 \end{pmatrix} = 2 = n \leqslant m,$$

we see that $l$ is one-to-one on $\mathbf{E}^2$. The range of $l$ consists of all vectors

$$y = Lx = \begin{pmatrix} 1 & 1 \\ 1 & -1 \\ 1 & 0 \end{pmatrix} \begin{pmatrix} \xi_1 \\ \xi_2 \end{pmatrix} = \begin{pmatrix} \xi_1 + \xi_2 \\ \xi_1 - \xi_2 \\ \xi_1 \end{pmatrix}.$$

---

If $l : \mathbf{E}^n \to \mathbf{E}^m$ is a linear function and if there exists a linear function $l^L$: $\mathbf{E}^m \to \mathbf{E}^n$ such that $l^L \circ l(x) = x$ for all $x \in \mathbf{E}^n$, then $l^L$ is called a *left inverse* of $l$. Linear one-to-one functions possess at least one left inverse. This is to be proved in exercise 5.

The above-mentioned left inverse is a function on $\mathbf{E}^m$ and not to be confused with the *inverse* of $l$ on the range of $l$ which, for $n < m$, is a subspace of $\mathbf{E}^m$. The left inverse on $\mathbf{E}^m$ is, for $n < m$, not unique. The inverse on $\mathfrak{R}(l)$, however, is unique. In the following theorem we show that the inverse of a linear one-to-one function is a linear function.

*Theorem 84.2*

If the linear function $l: \mathbf{E}^n \to \mathbf{E}^m$ is one-to-one on $\mathbf{E}^n$, then it possesses a unique inverse $l^{-1}: \mathcal{R}(l) \to \mathbf{E}^n$, and $l^{-1}$ is a linear function.

*Proof*

By Theorem 5.1, there exists a unique inverse function $l^{-1}: \mathcal{R}(l) \to \mathbf{E}^n$. If $y_1$, $y_2 \in \mathcal{R}(l)$, and if $\lambda_1, \lambda_2 \in \mathbf{R}$, then

$$l^{-1}(\lambda_1 y_1 + \lambda_2 y_2) = x_1$$

for some $x_1 \in \mathbf{E}^n$. This is equivalent to

$$\lambda_1 y_1 + \lambda_2 y_2 = l(x_1).$$

Since $\lambda_1 l^{-1}(y_1) + \lambda_2 l^{-1}(y_2) = x_2$ for some $x_2 \in \mathbf{E}^n$ is equivalent to $l(\lambda_1 l^{-1}(y_1) + \lambda_2 l^{-1}(y_2)) = \lambda_1 y_1 + \lambda_2 y_2 = l(x_2)$, we have $l(x_1) = l(x_2)$. Since $l$ is one-to-one, this implies $x_1 = x_2$; that is,

$$l^{-1}(\lambda_1 y_1 + \lambda_2 y_2) = \lambda_1 l^{-1}(y_1) + \lambda_2 l^{-1}(y_2).$$

Hence, $l^{-1}$ is a linear function.

A combination of Lemmas 84.1 and 84.2 yields:

*Theorem 84.3*

$l: \mathbf{E}^n \leftrightarrow \mathbf{E}^m$ if and only if rank $l = m = n$.

If $L$ is the matrix representation of $l$, this statement is equivalent to $Lx = y$ has a unique solution $x \in \mathbf{E}^n$ for all $y \in \mathbf{E}^m$ if and only if rank $L = m = n$; that is, $L$ is a nonsingular square matrix.

---

▲ *Example 6*

Let $l: \mathbf{E}^3 \to \mathbf{E}^3$ denote the linear function that maps $e_1, e_2, e_3$ into

$$l(e_1) = \begin{pmatrix} 1 \\ 0 \\ 1 \end{pmatrix}, \qquad l(e_2) = \begin{pmatrix} 0 \\ 1 \\ 1 \end{pmatrix}, \qquad l(e_3) = \begin{pmatrix} 1 \\ 1 \\ 0 \end{pmatrix}.$$

Since

$$\det(l(e_1), l(e_2), l(e_3)) = \begin{vmatrix} 1 & 0 & 1 \\ 0 & 1 & 1 \\ 1 & 1 & 0 \end{vmatrix} = -2 \neq 0,$$

we have rank $l =$ rank $L = 3 = m = n$, and we see that $l: \mathbf{E}^3 \leftrightarrow \mathbf{E}^3$.

---

## 8.84 Exercises

1. Given

$$L = \begin{pmatrix} 1 & 1 & 3 \\ 4 & 0 & 1 \\ 3 & -1 & -2 \\ 1 & 0 & 0 \end{pmatrix}.$$

Find all $y \in \mathbf{E}^4$ for which $Lx = y$ has a unique solution.

2. The linear function $l : \mathbf{E}^n \to \mathbf{E}^m$ is represented by

$$L = \begin{pmatrix} a_{11} & a_{12} \cdots a_{1n} \\ a_{21} & a_{22} \cdots a_{2n} \\ \vdots & \\ a_{m1} & a_{m2} \cdots a_{mn} \end{pmatrix}.$$

Find $\|l(x)\|^2$.

3. Given

$$L = \begin{pmatrix} 1 & 1 & 3 \\ 4 & 0 & 1 \\ 1 & 0 & 0 \end{pmatrix}.$$

Find rank $L$ and if rank $L = 3$, find $L^{-1}$; that is, a $3 \times 3$ matrix such that $LL^{-1} = L^{-1}L = I$ where $I$ is the identity matrix

$$I = \begin{pmatrix} 1 & 0 & 0 \\ 0 & 1 & 0 \\ 0 & 0 & 1 \end{pmatrix}.$$

4. Let $l : \mathbf{E}^4 \to \mathbf{E}^2$ be represented by

$$L = \begin{pmatrix} 1 & 1 & 1 & 0 \\ 1 & 0 & 0 & 1 \end{pmatrix}.$$

(a) Show that $l$ is onto $\mathbf{E}^2$.
(b) Find the matrix representation of a right inverse of $l$.
(c) Find the matrix representation of a linear function that maps $\mathbf{E}^2$ into the null space of $l$.
(d) Find the matrix representation of infinitely many right inverses of $l$.

5. *Prove*: If $l : \mathbf{E}^n \to \mathbf{E}^m$ is one-to-one on $\mathbf{E}^n$, then $l$ possesses a left inverse $l^L : \mathbf{E}^m \to \mathbf{E}^n$; that is, there exists a linear function $l^L : \mathbf{E}^m \to \mathbf{E}^n$ such that $l^L \circ l(x) = x$ for all $x \in \mathbf{E}^n$.
6. *Prove*: If $l : \mathbf{E}^n \leftrightarrow \mathbf{E}^m$, then $l^R = l^L = l^{-1}$.

## 8.85 Properties of Linear Functions

Let $l : \mathbf{E}^n \to \mathbf{E}^m$ represent a linear function. For every $x = \xi_1 e_1 + \xi_2 e_2 + \cdots + \xi_n e_n$, where $e_j, j = 1, 2, \ldots, n$, are the fundamental vectors in $\mathbf{E}^n$, we have from the FIE (61.1) that

$$\begin{aligned}\|l(x)\| &= \|\xi_1 l(e_1) + \xi_2 l(e_2) + \cdots + \xi_n l(e_n)\| \leqslant |\xi_1|\,\|l(e_1)\| + |\xi_2|\,\|l(e_2)\| + \cdots \\ &\quad + |\xi_n|\,\|l(e_n)\| \\ &\leqslant \max_{(j)} |\xi_j|(\|l(e_1)\| + \|l(e_2)\| + \cdots + \|l(e_n)\|) \\ &= M \max_{(j)} |\xi_j| \leqslant M\|x\|,\end{aligned}$$

where

$$M = \|l(e_1)\| + \|l(e_2)\| + \cdots + \|l(e_n)\| \tag{85.1}$$

is a fixed number that is fully determined by $l$. Hence:

***Lemma 85.1***

If $l: \mathbf{E}^n \to \mathbf{E}^m$ is a linear function, then there exists an $M > 0$ such that

$$\|l(x)\| \leqslant M\,\|x\| \tag{85.2}$$

for all $x \in \mathbf{E}^n$. $M$ is given by (85.1).

We obtain from Lemma 85.1:

***Theorem 85.1***

If $l: \mathbf{E}^n \to \mathbf{E}^m$ is a linear function, then $l$ is uniformly continuous on $\mathbf{E}^n$.

*Proof*

Since $l(x_1) - l(x_2) = l(x_1 - x_2)$ for all $x_1, x_2 \in \mathbf{E}^n$, we have, from Lemma 85.1, that

$$\|l(x_1) - l(x_2)\| = \|l(x_1 - x_2)\| \leqslant M\,\|x_1 - x_2\| \qquad \text{for all} \quad x_1, x_2 \in \mathbf{E}^n$$

for some $M > 0$. Hence, if $\varepsilon > 0$ and $\delta(\varepsilon) = \varepsilon/M$, then

$$\|l(x_1) - l(x_2)\| < \varepsilon \qquad \text{for all } x_1,\ x_2 \in \mathbf{E}^n \text{ with } \|x_1 - x_2\| < \delta(\varepsilon);$$

that is, $l$ is uniformly continuous on $\mathbf{E}^n$.

The following characterization of linear one-to-one functions will turn out to be of considerable importance in later developments.

***Theorem 85.2***

The linear function $l: \mathbf{E}^n \to \mathbf{E}^m$ is *one-to-one on* $\mathbf{E}^n$ and hence, invertible, if and only if there exists a $\mu > 0$ such that

$$\|l(x)\| \geqslant \mu\,\|x\| \qquad \text{for all } x \in \mathbf{E}^n. \tag{85.3}$$

*Proof*

(a) By Theorem 84.2, there exists a unique inverse $l^{-1}: \mathfrak{R}(l) \to \mathbf{E}^n$ of $l$ which is a linear function. Hence, by Lemma 85.1, there exists an $M > 0$ such that

$$\|l^{-1}(y)\| \leqslant M\,\|y\| \qquad \text{for all } y \in \mathfrak{R}(l). \tag{85.4}$$

For every $x \in \mathbf{E}^n$, $y = l(x) \in \mathbf{E}^m$ and, consequently, $x = l^{-1}(y)$. Hence we obtain from (84.4) that

$$\|x\| \leqslant M \|l(x)\| \qquad \text{for all } x \in \mathbf{E}^n$$

from which (85.3) follows readily for $\mu = 1/M$.

(b) Suppose (85.3) holds. If $l(x_1) = l(x_2)$, then $l(x_1 - x_2) = \theta$ and, by (85.3), $0 = \|l(x_1 - x_2)\| \geqslant \mu \|x_1 - x_2\|$ for some $\mu > 0$. Hence $x_1 = x_2$; that is, $l$ is one-to-one.

---

▲ *Example 1*

The function $l: \mathbf{E}^2 \to \mathbf{E}^3$ that is represented by the matrix

$$L = \begin{pmatrix} 1 & 1 \\ 1 & -1 \\ 1 & 0 \end{pmatrix}$$

is one-to-one because

$$\begin{aligned} \|l(x)\|^2 = \|Lx\|^2 &= \|(\xi_1 + \xi_2), (\xi_1 - \xi_2), \xi_1\|^2 \\ &= (\xi_1 + \xi_2)^2 + (\xi_1 - \xi_2)^2 + \xi_1^2 \\ &= 3\xi_1^2 + 2\xi_2^2 \geqslant \xi_1^2 + \xi_2^2 = \|x\|^2; \end{aligned}$$

that is,

$$\|l(x)\| \geqslant \|x\| \qquad \text{for all } x \in \mathbf{E}^2 \qquad (\mu = 1).$$

Hence, $l^{-1}$ exists on the range of $l$. Since

$$l(e_1) = \begin{pmatrix} 1 \\ 1 \\ 1 \end{pmatrix}, \qquad l(e_2) = \begin{pmatrix} 1 \\ -1 \\ 0 \end{pmatrix},$$

we have

$$l(\lambda e_1 + \lambda e_2) = \lambda \begin{pmatrix} 1 \\ 1 \\ 1 \end{pmatrix} + \mu \begin{pmatrix} 1 \\ -1 \\ 0 \end{pmatrix}$$

for all $\lambda, \mu \in \mathbf{R}$, and hence, $l^{-1}: \mathcal{R}(l) \to \mathbf{E}^2$ is given by

$$l^{-1} \left[ \lambda \begin{pmatrix} 1 \\ 1 \\ 1 \end{pmatrix} + \mu \begin{pmatrix} 1 \\ -1 \\ 0 \end{pmatrix} \right] = \lambda e_1 + \mu e_2$$

for all $\lambda, \mu \in \mathbf{R}$.

---

Theorem 85.2 has, in a certain sense, a counterpart pertaining to linear onto functions:

***Theorem 85.3***

If the linear function $l : \mathbf{E}^n \to \mathbf{E}^m$ is *onto* $\mathbf{E}^m$, then, for every $y \in \mathbf{E}^m$, there is some $x \in \mathbf{E}^n$ such that $y = l(x)$ and

$$\|x\| \leqslant \nu \|y\| \tag{85.5}$$

for some $\nu > 0$ which is independent of $y$. (Note that there could be more than one $x \in \mathbf{E}^n$ such that $y = l(x)$. Also note that the converse of this theorem is trivial.)

*Proof*

If $l$ is onto $\mathbf{E}^n$, then, by Theorem 84.1, a right inverse $l^R : \mathbf{E}^m \to \mathbf{E}^n$ exists. Since $l^R$ is a linear function, we have, from Lemma 85.1, a $\nu > 0$ such that

$$\|l^R(y)\| \leqslant \nu \|y\| \qquad \text{for all } y \in \mathbf{E}^m.$$

Since $l^R(y) = x \in \mathbf{E}^n$, we obtain, for all $y \in \mathbf{E}^m$, an $x \in \mathbf{E}^n$ such that

$$\|x\| \leqslant \nu \|y\|.$$

---

▲ *Example 2*

Let $l : \mathbf{E}^3 \to \mathbf{E}^2$ be defined by

$$L = \begin{pmatrix} 1 & 0 & 1 \\ 0 & 1 & 1 \end{pmatrix}.$$

$l$ is onto because rank $l = 2 = m < n$. For every $y = (\eta_1, \eta_2) \in \mathbf{E}^2$, there is an $x \in \mathbf{E}^3$, namely $x = (\eta_1, \eta_2, 0)$, such that $y = l(x)$ and $\|x\| = \sqrt{n_1^2 + \eta_2^2 + 0} = \|y\|$; that is, $\|x\| \leqslant \nu \|y\|$ for $\nu = 1$.

---

***Corollary to Theorem 85.3***

If $l : \mathbf{E}^n \to \mathbf{E}^m$ is *onto* $\mathbf{E}^m$ then $l$ maps all open subsets of $\mathbf{E}^n$ onto open subsets of $\mathbf{E}^m$.

*Proof*

Let $\Omega \subseteq \mathbf{E}^n$ denote an open set and let $a \in \Omega$. We need only show that $l(a) = b$ is an interior point of $l_*(\Omega)$. Since $a \in \Omega$, there is a $\delta > 0$ such that $N_\delta(a) \subseteq \Omega$. By Theorem 85.3, there is, for every $(y - b) \in \mathbf{E}^m$, an $(x - a) \in \mathbf{E}^n$ such that $l(x - a) = l(x) - l(a) = y - b$ and $\|x - a\| \leqslant \nu \|y - b\|$. Hence, for every $y \in N_{\delta/\nu}(b)$, there is an $x \in N_\delta(a)$ such that $y = f(x)$; that is, $b$ is an interior point of $l_*(\Omega)$. (See also Fig. 85.1.)

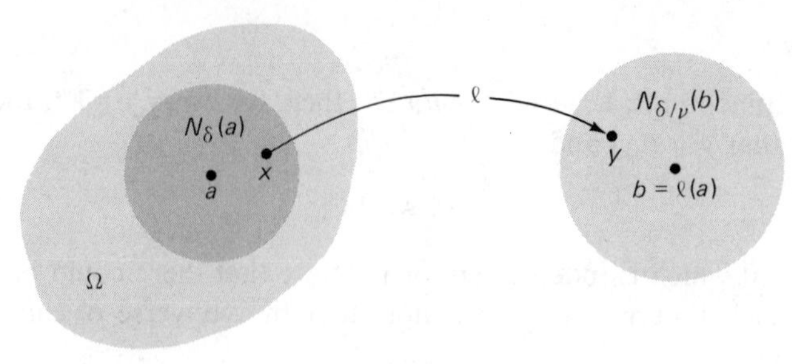


*Fig. 85.1*

## 8.85 Exercises

1. Given $l : \mathbf{E}^4 \to \mathbf{E}^4$ by

$$L = \begin{pmatrix} 1 & 1 & 0 & 0 \\ 0 & 1 & 1 & 0 \\ 1 & 1 & 1 & 1 \\ 1 & 0 & -1 & 1 \end{pmatrix}.$$

   (a) Find an $M > 0$ such that $\|l(x)\| \leqslant M\|x\|$ for all $x \in \mathbf{E}^4$.
   (b) Show that $l$ is one-to-one, and find a $\mu > 0$ such that $\|l(x)\| \geqslant \mu\|x\|$ for all $x \in \mathbf{E}^4$.
   (c) Show that $l$ is onto $\mathbf{E}^4$, and find a $\nu > 0$ such that, for each $y \in \mathbf{E}^4$, there is an $x \in \mathbf{E}^4$ for which $l(x) = y$ and

$$\|x\| \leqslant \nu\|y\|.$$

2. Let

$$L = \begin{pmatrix} a_{11} & a_{12} \dots a_{1n} \\ a_{21} & a_{22} \dots a_{2n} \\ \vdots & \\ a_{m1} & a_{m2} \dots a_{mn} \end{pmatrix}$$

represent $l : \mathbf{E}^n \to \mathbf{E}^m$, and let

$$K = \max\{|a_{ik}|, \quad i = 1, 2, \dots, m; \quad k = 1, 2, \dots, n\}.$$

Show that

$$\|l(x)\| \leqslant n\sqrt{m}K\|x\| \qquad \text{for all } x \in \mathbf{E}^n.$$

3. Let $l : \mathbf{E}^n \to \mathbf{E}^m$ be represented as in exercise 2. Show that

$$\|l(x)\| \leqslant \sqrt{\sum_{j=1}^{n} \sum_{i=1}^{m} a_{ij}^2}\ \|x\|.$$

4. Let $l : \mathbf{E}^n \leftrightarrow \mathbf{E}^n$ be represented by the matrix

$$L = \begin{pmatrix} a_{11} \dots a_{1n} \\ a_{n1} \dots a_{nn} \end{pmatrix}.$$

Let $\Delta = \det L^{-1}$ and let $A_{ij}$ denote the cofactors of $a_{ij}$. Show that

$$\left(\sum_{i=1}^{n}\sum_{j=1}^{n} a_{ij}^2\right)^{-1/2} \|l(x)\| \leqslant \|x\| \leqslant \frac{1}{\Delta}\left(\sum_{i=1}^{n}\sum_{j=1}^{n} A_{ij}^2\right)^{1/2} \|l(x)\|.$$

5. Let $l: \mathbf{E}^n \to \mathbf{E}^m$ be represented as in exercise 2 and let rank $(l) = m \leqslant n$. Show: For each $y \in \mathbf{E}^n$ there is some $x \in \mathbf{E}^n$ such that

$$\|x\| \leqslant (\|u_1\| + \|u_2\| + \cdots + \|u_m\|)\|y\|,$$

where $l(u_j) = i_j$ and where $i_j, j = 1, 2, \ldots, m$ are the fundamental vectors in $\mathbf{E}^m$.
6. Let $l: \mathbf{E}^2 \to \mathbf{E}^3$ be defined as in example 1. Find the matrix representation of the left inverse of $l$ on $\mathbf{E}^3$ and the inverse of $l$ on $\mathcal{R}(l)$.
7. Let $l: \mathbf{E}^3 \to \mathbf{E}^2$ be defined as in example 2. Find the matrix representation of a right inverse of $l$.

# 9 The Derivative of a Vector-Valued Function of a Vector Variable

## 9.86 Definition of the Derivative

We have seen in Chapter 3 that the derivative $f'(a)$ of a real-valued function $f: \mathfrak{D} \to \mathbf{E}$, $\mathfrak{D} \subseteq \mathbf{E}$ of a real variable at an accumulation point $a \in \mathfrak{D}$ of $\mathfrak{D}$ may be characterized as follows: $f$ is differentiable at $a$ and $f'(a)$ is the derivative if and only if there exists a linear function $l_a: \mathbf{E} \to \mathbf{E}$ with the property that for every $\varepsilon > 0$, there is a $\delta(\varepsilon) > 0$ such that

$$|f(x) - f(a) - l_a(x - a)| \leqslant \varepsilon\,|x - a|$$

for all $x \in N_\delta(a) \cap \mathfrak{D}$ and $f'(a)h = l_a(h)$ for all $h \in \mathbf{E}$. (See Theorem 39.2.)

In generalizing the concept of a derivative to vector-valued functions of a vector variable from $\mathfrak{D} \subseteq \mathbf{E}^n$ into $\mathbf{E}^m$, we proceed in a straightforward manner, replacing the absolute value by the norm and $l_a: \mathbf{E} \to \mathbf{E}$ by a linear function $l: \mathbf{E}^n \to \mathbf{E}^m$. However, we shall restrict our discussion to interior points of $\mathfrak{D}$ rather than consider the more general case where $a \in \mathfrak{D}$ is merely an accumulation point of $\mathfrak{D}$. Otherwise we would not obtain a unique derivative when $n > 1$.

***Definition 86.1***

If $f: \mathfrak{D} \to \mathbf{E}^m$, $\mathfrak{D} \subseteq \mathbf{E}^n$, and if $a$ is an interior of $\mathfrak{D}$, then $f$ is differentiable at $a$ if and only if there exists a linear function $l: \mathbf{E}^n \to \mathbf{E}^m$ with the property that for every $\varepsilon > 0$ there is a $\delta(\varepsilon) > 0$ where $N_{\delta(\varepsilon)}(a) \subseteq \mathfrak{D}$ such that

$$\|f(x) - f(a) - l(x - a)\| \leqslant \varepsilon\|x - a\| \tag{86.1}$$

for all $x \in N_{\delta(\varepsilon)}(a)$. The function $l$ is called the derivative of $f$ at $a$ and is denoted by $f'(a)$.

Note that $f'(a)$ is, for given $a$, a linear function from $\mathbf{E}^n$ into $\mathbf{E}^m$, that is $f'(a)(h + k) = f'(a)(h) + f'(a)(k)$ for all $h, k \in \mathbf{E}^n$, and $f'(a)(\lambda h) = \lambda f'(a)(h)$ for all $h \in \mathbf{E}^n$ and any $\lambda \in \mathbf{R}$.

▲ *Example 1*

Let $f: \mathbf{E}^n \to \mathbf{E}^m$ be defined by

$$f(x) = c + l(x)$$

where $c \in \mathbf{E}^m$ and where $l: \mathbf{E}^n \to \mathbf{E}^m$ is a linear function. Then, for any $a \in \mathbf{E}^n$,

$$f'(a) = l$$

because

$$\|f(x) - f(a) - l(x - a)\| = \|l(x) - l(a) - l(x - a)\| = 0$$

for all $x \in \mathbf{E}^n$. (This is a generalization of the well-known result that the derivative of $f: \mathbf{E} \to \mathbf{E}$, given by $f(x) = a + bx$, is $b$.)

▲ *Example 2*

Let $f: \mathbf{E}^2 \to \mathbf{E}$ be defined by $f(\xi_1, \xi_2) = \xi_1^2 + \xi_2$ and let $a = (\alpha_1, \alpha_2)$. We propose to show that $f'(a): \mathbf{E}^2 \to \mathbf{E}$ is the linear function that is defined by

(86.2) $$f'(a)(u) = 2\alpha_1\omega_1 + \omega_2$$

for all $u = (\omega_1, \omega_2) \in \mathbf{E}^2$. We have

$$\begin{aligned} |f(x) - f(a) - f'(a)(x - a)| &= |\xi_1^2 + \xi_2 - \alpha_1^2 - \alpha_2 - 2\alpha_1(\xi_1 - \alpha_1) - \xi_2 - \alpha_2| \\ &= |\xi_1^2 + \alpha_1^2 - 2\alpha_1\xi_1| = (\xi_1 - \alpha_1)^2 \\ &= |\xi_1 - \alpha_1|\,|\xi_1 - \alpha_1| \leqslant \varepsilon\|x - a\| \end{aligned}$$

for all $x \in N_\delta(a)$ for $\delta = \varepsilon$. Hence (86.1) is satisfied and the linear function in (86.2) is the derivative of $f$ at $a$.

---

As in the case of a real-valued function of a real variable, the derivative, when it exists, is unique, and a function that is differentiable at a point is continuous at that point. This will be shown in the next two theorems.

***Theorem 86.1***

If $f: \mathfrak{D} \to \mathbf{E}^m$, $\mathfrak{D} \subseteq \mathbf{E}^n$, is differentiable at the interior point $a \in \mathfrak{D}$, then the derivative $f'(a)$ is unique.

*Proof* (by contradiction)

Suppose that $f_1'(a)$ as well as $f_2'(a)$ represent the derivative of $f$ at $a$. Then, by definition 86.1, we have, for every $\varepsilon > 0$, a $\delta > 0$ such that

$$\|f(x) - f(a) - f_1'(a)(x - a)\| \leqslant (\varepsilon/2)\|x - a\|,$$
$$\|f(x) - f(a) - f_2'(a)(x - a)\| \leqslant (\varepsilon/2)\|x - a\|$$

for all $x \in N_\delta(a)$.

If $f_1'(a) \neq f_2'(a)$, then there is at least one $u \in \mathbf{E}^n$ such that $f_1'(a)(u) \neq f_2'(a)(u)$; that is,

$$(86.3) \qquad \|f_1'(a)(u) - f_2'(a)(u)\| > 0.$$

Since the norm as well as $f_1'(a), f_2'(a)$ are homogeneous, we have, from (86.3), for $z = u/\|u\|$ that

$$(86.4) \qquad \|f_1'(a)(z) - f_2'(a)(z)\| > 0, \qquad \|z\| = 1.$$

We choose $\alpha > 0$ such that $x = a + \alpha z \in N_\delta(a)$; that is, $|\alpha| < \delta$. Then, we have from (86.4) that

$$\begin{aligned} 0 < |\alpha|\,\|f_1'(a)(z) - f_2'(a)(z)\| &= \|f_1'(a)(\alpha z) - f_2'(a)(\alpha z)\| \\ &= \|f_1'(a)(x-a) - f_2'(a)(x-a)\| \\ &\leqslant \|f(x) - f(a) - f_1'(a)(x-a)\| \\ &\quad + \|f(x) - f(a) - f_2'(a)(x-a)\| \\ &\leqslant \varepsilon\|x - a\| = \varepsilon|\alpha|. \end{aligned}$$

Cancellation by $|\alpha|$ yields

$$0 < \|f_1'(a)(z) - f_2'(a)(z)\| \leqslant \varepsilon;$$

that is, $f_1'(a)(z) = f_2'(a)(z)$, and hence, $f_1'(a)(u) = f_2'(a)(u)$, contrary to our assumption.

If $a$ is not an interior point of $\mathfrak{D}$, then the above argument collapses, because then the existence of an $\alpha > 0$ such that $a + \alpha z \in N_\delta(a) \cap \mathfrak{D}$ is not assured. The following example will show that the derivative, defined as in definition 86.1, need not be unique if $a$ is not an interior point:

---

▲ *Example 3*

Let $f: \mathfrak{D} \to \mathbf{E}$, $\mathfrak{D} = \{x \in \mathbf{E}^2 \mid \xi_1 = \xi_2\}$, be defined by $f(x) = \xi_1$ and let $a = (\alpha_1, \alpha_2) \in \mathfrak{D}$; that is, $\alpha_1 = \alpha_2$. Clearly, $a$ is not an interior point because $\mathfrak{D}$ has no interior points. Then, the following two linear functions $l_1, l_2 : \mathbf{E}^2 \to \mathbf{E}$, defined by

$$l_1(u) = (1, 0)\begin{pmatrix}\omega_1\\ \omega_2\end{pmatrix}, \qquad l_2(u) = (0, 1)\begin{pmatrix}\omega_1\\ \omega_2\end{pmatrix}, \qquad \text{where } u = \begin{pmatrix}\omega_1\\ \omega_2\end{pmatrix},$$

satisfy (86.1), because

$$|f(x) - f(a) - l_1(x-a)| = \left|\xi_1 - \alpha_1 - (1, 0)\begin{pmatrix}\xi_1 - \alpha_1\\ \xi_1 - \alpha_1\end{pmatrix}\right| = 0$$

and

$$|f(x) - f(a) - l_2(x-a)| = \left|\xi_1 - \alpha_1 - (0, 1)\begin{pmatrix}\xi_1 - \alpha_1\\ \xi_1 - \alpha_1\end{pmatrix}\right| = 0.$$

Hence, both $l_1$ and $l_2$ would have to be considered to be the derivative of $f$ at $a$.

---

***Theorem 86.2***

If $f: \mathfrak{D} \to \mathbf{E}^m$, $\mathfrak{D} \subseteq \mathbf{E}^n$, is differentiable at the interior point $a \in \mathfrak{D}$, then $f$ is continuous at $a$.

*Proof*

Since $f$ is differentiable at $a$, we have from definition 86.1 that for $\varepsilon_1 = 1$ there is a $\delta_1 > 0$ such that

$$\|f(x) - f(a) - f'(a)(x - a)\| \leqslant \|x - a\|$$

for all $x \in N_{\delta_1}(a)$. Hence,

$$\|f(x) - f(a)\| \leqslant \|f'(a)(x - a)\| + \|x - a\|.$$

Since $f'(a)$ is a linear function from $\mathbf{E}^n$ into $\mathbf{E}^m$, we obtain, from Lemma 85.1, that

$$\|f'(a)(x - a)\| \leqslant M \|x - a\|$$

for some $M > 0$. Hence,

$$\|f(x) - f(a)\| \leqslant (M + 1) \|x - a\| < \varepsilon \tag{86.5}$$

for all $x \in N_\delta(a)$ where $\delta = \min(\delta_1, \varepsilon/(M + 1))$. Hence, $f$ is continuous at $a$.

If $f$ is differentiable for all $x \in A$, where $A \subseteq \mathfrak{D}$ is an open set, then $f$ is said to be differentiable in $A$.

***Definition 86.2***

If $f: \mathfrak{D} \to \mathbf{E}^m$, $\mathfrak{D} \subseteq \mathbf{E}^n$, is differentiable in the open set $A \subseteq \mathfrak{D}$, then the function $df: A \times \mathbf{E}^n \to \mathbf{E}^m$ which is defined by

$$df(x, u) = f'(x)(u), \qquad x \in A, u \in \mathbf{E}^n,$$

is called the *differential of f*. Specifically, for $a \in A$, $u_0 \in \mathbf{E}^n$, $df(a, u_0)$ is called the differential of $f$ at $a$ with increment $u_0$.

In the following three sections we shall seek, and eventually find, representations of the derivative and the differential.

## 9.86 Exercises

1. Given the function $f: \mathbf{E}^n \to \mathbf{E}^m$ by $\phi_j(\xi_1, \xi_2, \ldots, \xi_n) = a_{j1}\xi_1 + a_{j2}\xi_2 + \cdots + a_{jn}\xi_n$, where $j = 1, 2, \ldots, m$; $a_{jk} \in \mathbf{R}$. Represent $f'(a)$, $a \in \mathbf{E}^n$, by an $m \times n$ matrix.
2. Given the function $f: \mathbf{E}^2 \to \mathbf{E}$ by

$$f(\xi_1, \xi_2) = \xi_1^2 + \xi_2^2.$$

Show, by invoking definition 86.1 that $f'(0,0)$ is given by

$$f'(0,0)(h) = 0$$

for all $h \in \mathbf{E}^2$.

3. Given the function of example 3. Show that the linear function $l : \mathbf{E}^2 \to \mathbf{E}$ which is defined by $l(u) = (\frac{1}{2}, \frac{1}{2})\binom{\omega_1}{\omega_2}$ for all $u \in \mathbf{E}^2$ also satisfies definition 86.1.
4. Given $f : \mathbf{E}^2 \to \mathbf{E}^2$ by

$$\phi_1(\xi_1, \xi_2) = \xi_1^2 \xi_2,$$
$$\phi_2(\xi_1, \xi_2) = \xi_1 - \xi_2.$$

Show that the linear function $l : \mathbf{E}^2 \to \mathbf{E}^2$ which is given by

$$l(u) = \begin{pmatrix} 0 & 0 \\ 1 & -1 \end{pmatrix} \begin{pmatrix} \omega_1 \\ \omega_2 \end{pmatrix}$$

is the derivative of $f$ at $\theta$.

5. Let $f : \mathfrak{D} \to \mathbf{E}^n$, $\mathfrak{D} \subseteq \mathbf{E}^n$, represent the identity function. Show that $f'(a)$ is the identity function for all $a \in \mathfrak{D}$.
6. *Prove*: For every fixed $a \in A$, the differential $df(a, u)$ is a linear function of $u \in \mathbf{E}^n$.
7. Let $f : \mathfrak{D} \to \mathbf{E}$, $\mathfrak{D} \subseteq \mathbf{E}$, and let $a \in \mathfrak{D}$ denote an interior point. Show that definitions 86.1 and 35.1 are equivalent.

## 9.87 The Directional Derivative

If $f$ is a vector-valued function of a real variable, i.e., if $f : \mathfrak{D} \to \mathbf{E}^m$ where $\mathfrak{D} \subseteq \mathbf{E}$, then we obtain, from definition (86.1), that

$$(87.1) \qquad \|f(x) - f(a) - f'(a)(x - a)\| \leqslant \varepsilon |x - a|,$$

and, after division by $|x - a|$, provided that $x \neq a$,

$$\left\| \frac{f(x) - f(a)}{x - a} - f'(a) \right\| \leqslant \varepsilon$$

for all $x \in N'_\delta(a)$. Hence, by definition 64.1, (87.1) is equivalent to:

$$(87.2) \qquad f'(a) = \lim_{a} \frac{f(x) - f(a)}{x - a}.$$

If we represent $f$ by the column vector

$$f = \begin{pmatrix} \phi_1 \\ \phi_2 \\ \vdots \\ \phi_m \end{pmatrix},$$

then we have, from Lemma 64.1, that (87.2) holds if and only if

$$[f'(a)]_j = \lim_a \frac{\phi_j(x) - \phi_j(a)}{x - a}$$

for all $j = 1, 2, \ldots, m$, where $[\ ]_j$ denotes the $j$th component of $[\ ]$. Since

$$\lim_a \frac{\phi_j(x) - \phi_j(a)}{x - a} = \phi_j'(a),$$

if the limit exists, we arrive at the conclusion that a vector-valued function of a real variable is differentiable if and only if each of its components is differentiable and we obtain the following representation of $f'(a)$:

$$F'(a) = \begin{pmatrix} \phi_1'(a) \\ \phi_2'(a) \\ \vdots \\ \phi_m'(a) \end{pmatrix}. \tag{87.3}$$

(For the notation, we refer to the remark at the end of section 8.83.) Recall that, by definition 86.1, $f'(a)$ is supposed to be a linear function from $\mathbf{E}$ into $\mathbf{E}^m$. Linear functions from $\mathbf{E}$ into $\mathbf{E}^m$ can be represented by an $m \times 1$ matrix, and we see that (87.3) is indeed an $m \times 1$ matrix.

If $f$ is not a function of a real variable but of a vector variable, then one cannot divide by $x - a$ in (86.1) because division by a vector is not defined and an explicit representation of the derivative is not immediately accessible.

Rather than consider (86.1) for arbitrary points $x$ in $N_\delta(a)$, we shall first investigate the simple case where $x$ deviates from $a$ in a given direction represented by a *fixed* vector $u \in \mathbf{E}^n$:

$$x = a + \tau \frac{u}{\|u\|}, \qquad u \neq \theta.$$

We choose $\tau \in \mathbf{R}$ such that $x = a + \tau u/\|u\| \in N_\delta(a)$; that is, $|\tau| < \delta$, and denote the unit vector $u/\|u\|$ by $z$:

$$z = \frac{u}{\|u\|}, \qquad \|z\| = 1.$$

If $f$ is differentiable at $a$, then we obtain from (86.1) that

$$\|f(a + \tau z) - f(a) - f'(a)(\tau z)\| \leqslant \varepsilon \|\tau z\|.$$

Utilizing the homogeneity of the norm and the homogeneity of the linear function $f'(a)$, we obtain the equivalent statement

$$\left\| \frac{f(a + \tau z) - f(a)}{\tau} - f'(a)(z) \right\| \leqslant \varepsilon$$

for all $0 < |\tau| < \delta$. By definition 64.1, this is equivalent to

$$f'(a)(z) = \lim_0 \frac{f(a + \tau z) - f(a)}{\tau}.$$

We replace $z$ by $u/\|u\|$, let $t = \tau/\|u\|$, note that $\lim_{\tau\to 0} t = 0$ and $\lim_{t\to 0} \tau = 0$. and obtain

$$f'(a)\left(\frac{u}{\|u\|}\right) = \lim_{t\to 0} \frac{f(a + tu) - f(a)}{t\|u\|}.$$

Hence,

$$f'(a)(u) = \lim_0 \frac{f(a + tu) - f(a)}{t}. \tag{87.4}$$

***Definition 87.1***

If $f : \mathfrak{D} \to \mathbf{E}^m$, $\mathfrak{D} \subseteq \mathbf{E}^n$, if $a \in \mathfrak{D}$ is an interior point, and if $u \in \mathbf{E}^n, u \neq \theta$, is a fixed vector, then

$$D_u f(a) = \lim_0 \frac{f(a + tu) - f(a)}{t} \tag{87.5}$$

is called the *directional derivative* of $f$ at $a$ in the direction of $u$, provided that the limit in (87.5) exists.

Observing that our conclusion is valid for every $u \in \mathbf{E}^n$, $u \neq \theta$, we have from (87.4), provided that $f'(a)$ exists, the following result:

***Theorem 87.1***

If $f : \mathfrak{D} \to \mathbf{E}^m$, $\mathfrak{D} \subseteq \mathbf{E}^n$, is differentiable at the interior point $a \in \mathfrak{D}$, then the directional derivative $D_u f(a)$ exists for all $u \in \mathbf{E}^n$, $u \neq \theta$, and

$$f'(a)(u) = D_u f(a) = \lim_0 \frac{f(a + tu) - f(a)}{t}.$$

Note that Theorem 87.1 is strictly a one-way street. The existence of the derivative implies the existence of the directional derivative, for all $u \in \mathbf{E}^n$. However, the directional derivative may exist for all $u \in \mathbf{E}^n$ and the function may still not be differentiable, as the following example shows.

---

▲ *Example 1*

Let $f : \mathbf{E}^2 \to \mathbf{E}$ be defined by

$$f(\xi_1, \xi_2) = \begin{cases} \dfrac{\xi_1 \xi_2^2}{\xi_1^2 + \xi_2^2} & \text{for } (\xi_1, \xi_2) \neq (0,0), \\ 0 & \text{for } (\xi_1, \xi_2) = (0,0). \end{cases}$$

We obtain, for $u = (\omega_1, \omega_2) \neq (0, 0)$,

$$D_u f(0,0) = \lim_{0} \frac{f(t\omega_1, t\omega_2) - f(0,0)}{t}$$

$$= \lim_{0} \frac{t^3 \omega_1 \omega_2^2}{t^3(\omega_1^2 + \omega_2^2)} = \frac{\omega_1 \omega_2^2}{\omega_1^2 + \omega_2^2};$$

that is, $D_u f(0,0)$ exists for all $u \in \mathbf{E}^2$, $u \neq \theta$. However, $f$ is *not* differentiable at $(0,0)$ because if $f'(0,0)$ existed, then

$$f'(0,0)(u) = D_u f(0,0) = \frac{\omega_1 \omega_2^2}{\omega_1^2 + \omega_2^2}, \tag{87.6}$$

and $f'(0,0)$ would be a linear function from $\mathbf{E}^2$ into $\mathbf{E}$; that is,

$$f'(0,0)(u + v) = f'(0,0)(u) + f'(0,0)(v),$$
$$f'(0,0)(\lambda u) = \lambda f'(0,0)(u)$$

for all $u$, $v \in \mathbf{E}^2$ and every $\lambda \in \mathbf{R}$. Let $u = (1,0)$, $v = (1,1)$. Then, we obtain, from (87.6),

$$f'(0,0)(u) = D_u f(0,0) = 0, \qquad f'(0,0)(v) = D_v f(0,0) = \tfrac{1}{2},$$
$$f'(0,0)(u + v) = D_{u+v} f(0,0) = \tfrac{2}{5}.$$

Since $0 + \frac{1}{2} \neq \frac{2}{5}$, we see that $f'(0,0)$ cannot exist.

---

The directional derivative may exist for some directions and not for others, as the following example shows.

---

▲ *Example 2*

Let $f \colon \mathbf{E}^2 \to \mathbf{E}$ be given by

$$f(\xi_1, \xi_2) = \begin{cases} 1 & \text{for } \xi_1 \xi_2 = 0, \\ 0 & \text{for } \xi_1 \xi_2 \neq 0. \end{cases}$$

(See Fig. 87.1.)

We obtain for $u_1 = (\omega_1, 0)$, $\omega_1 \neq 0$,

$$D_u f(0,0) = \lim_{0} \frac{f(t\omega_1, 0) - f(0,0)}{t} = \lim_{0} \frac{1 - 1}{t} = 0 \tag{87.7}$$

and, by symmetry, for $u_2 = (0, \omega_2)$, $\omega_2 \neq 0$,

$$D_u f(0,0) = 0. \tag{87.8}$$

However, if $u = (\omega_1, \omega_2)$, where $\omega_1 \neq 0$, $\omega_2 \neq 0$, we have

$$D_u f(0,0) = \lim_{0} \frac{f(t\omega_1, t\omega_2) - f(0,0)}{t} = \lim_{0} \frac{0 - 1}{t}$$

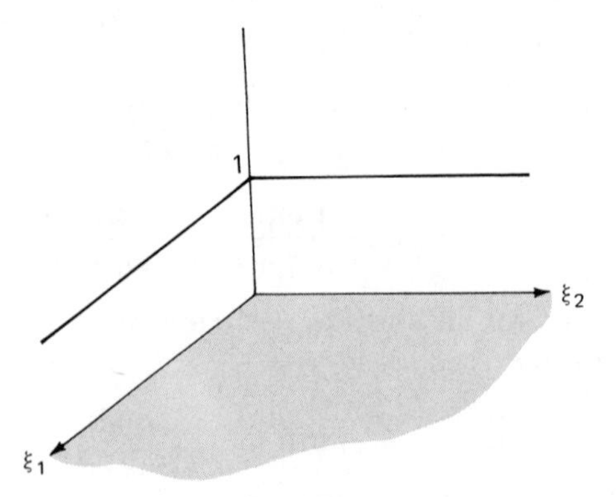


*Fig. 87.1*

which does *not* exist. So we see that the directional derivatives of $f$ exist only in the directions of the coordinate axes and do not exist in any other direction. Clearly, $f$ is *not* differentiable at $(0,0)$, in view of Theorem 87.1. The same result is obtained from Theorem 86.2 because $f$ is not continuous at $\theta$.

---

## 9.87 Exercises

1. Given $f: \mathbf{E}^3 \to \mathbf{E}^2$ by

$$\begin{aligned}\phi_1(\xi_1, \xi_2, \xi_3) &= \xi_1 - \xi_2\xi_3,\\ \phi_2(\xi_1, \xi_2, \xi_3) &= \xi_1^2 + \xi_2^2 + \xi_3^2.\end{aligned}$$

   Find $D_u f(0,0,0)$ for all $u \in \mathbf{E}^3$.

2. Given $f: \mathbf{E}^2 \to \mathbf{E}^3$ by

$$\begin{aligned}\phi_1(\xi_1, \xi_2) &= \xi_2,\\ \phi_2(\xi_1, \xi_2) &= \xi_1^2,\\ \phi_3(\xi_1, \xi_2) &= 2\xi_1.\end{aligned}$$

   Find $D_u f(0,0)$ for all $u \in \mathbf{E}^2$.

3. Given $f: \mathbf{E}^n \to \mathbf{E}$. Express $D_{e_j} f(\theta)$, $j = 1, 2, \ldots, n$, in terms of

$$\frac{\partial f}{\partial \xi_1}(\theta), \qquad \frac{\partial f}{\partial \xi_2}(\theta), \qquad \ldots, \qquad \frac{\partial f}{\partial \xi_n}(\theta)$$

   where $e_j$, $j = 1, 2, \ldots, n$, are the fundamental vectors in $\mathbf{E}^n$.

4. Given $f: \mathbf{E}^2 \to \mathbf{E}$ by

$$f(\xi_1, \xi_2) = \begin{cases} (\xi_1^2 + \xi_2^2)\sin\dfrac{1}{\xi_1^2 + \xi_2^2} & \text{for } (\xi_1, \xi_2) \neq (0,0),\\[2ex] 0 & \text{for } (\xi_1, \xi_2) = (0,0).\end{cases}$$

   Find $D_{e_1} f(0,0)$, $D_{e_2} f(0,0)$.

5. Let $f: \mathfrak{D} \to \mathbf{E}^m$, $\mathfrak{D} \subseteq \mathbf{E}^n$ and let $a = (\alpha_1, \alpha_2, \dots, \alpha_n) \in \mathfrak{D}$ denote an interior point. Show: If $D_u f(a)$ exists for all $u = (\omega_1, \omega_2, \dots, \omega_n) \in \mathbf{E}^n$, then

$$D_u f(a) = \lim_{0} \frac{f(\alpha_1 + t\omega_1, \alpha_2 + t\omega_2, \dots, \alpha_n + t\omega_n) - f(\alpha_1, \alpha_2, \dots, \alpha_n)}{t}.$$

## 9.88 Partial Derivatives

The partial derivative of a real-valued function of a vector variable $\phi : \mathfrak{D} \to \mathbf{E}$, $\mathfrak{D} \subseteq \mathbf{E}^n$, with respect to $\xi_j$, at an interior point $a = (\alpha_1, \alpha_2, \dots, \alpha_n) \in \mathfrak{D}$, is defined by

(88.1) $$\frac{\partial \phi}{\partial \xi_j}(a)$$

$$= \lim_{0} \frac{\phi(\alpha_1, \alpha_2, \dots, \alpha_{j-1}, \alpha_j + t, \alpha_{j+1}, \dots, \alpha_n) - \phi(\alpha_1, \alpha_2, \dots, \alpha_{j-1}, \alpha_j, \alpha_{j+1}, \dots, \alpha_n)}{t}.$$

If we consider the directional derivatives $D_u f(a)$ of $f: \mathfrak{D} \to \mathbf{E}^m$, $\mathfrak{D} \subseteq \mathbf{E}^n$ at an interior point $a \in \mathfrak{D}$ in the directions of the fundamental vectors $e_1, e_2, \dots, e_n \in \mathbf{E}^n$, we obtain from definition 87.1

(88.2) $$D_{e_j} f(a)$$

$$= \lim_{0} \frac{f(\alpha_1, \alpha_2, \dots, \alpha_{j-1}, \alpha_j + t, \alpha_{j+1}, \dots, \alpha_n) - f(\alpha_1, \alpha_2, \dots, \alpha_{j-1}, \alpha_j, \alpha_{j+1}, \dots, \alpha_n)}{t},$$

where $j = 1, 2, \dots, n$. We have, from Lemma 64.1, that the limit in (88.2) exists if and only if the corresponding limits for the components $\phi_1, \phi_2, \dots, \phi_m$ of $f$ exist, which are, in view of (88.1):

$$\frac{\partial \phi_k}{\partial \xi_j}(a)$$

$$= \lim_{0} \frac{\phi_k(\alpha_1, \alpha_2, \dots, \alpha_{j-1}, \alpha_j + t, \alpha_{j+1}, \dots, \alpha_n) - \phi_k(\alpha_1, \alpha_2, \dots, \alpha_{j-1}, \alpha_j, \alpha_{j+1}, \dots, \alpha_n)}{t}$$

where $k = 1, 2, \dots, m$, $j = 1, 2, \dots, n$.

Therefore we can say that the directional derivatives in the direction of the fundamental vectors exist, if and only if all partial derivatives of all components of $f$ exist.

***Definition 88.1***

The directional derivatives of $f: \mathfrak{D} \to \mathbf{E}^m$, $\mathfrak{D} \subseteq \mathbf{E}^n$ at an interior point $a \in \mathfrak{D}$ in the direction of the fundamental vectors $e_1, e_2, \dots, e_n \in \mathbf{E}^n$, namely $D_{e_j} f(a)$, are called the partial derivatives of $f$ with respect to $\xi_1, \xi_2, \dots, \xi_n$ at $a$, and are denoted by

$$D_j f(a) = D_{e_j} f(a), \qquad j = 1, 2, \dots, n.$$

We have, from the argument that preceded the above definition:

***Lemma 88.1***

The partial derivatives $D_j f(a)$, $j = 1, 2, \ldots, n$, of $f: \mathfrak{D} \to \mathbf{E}^m$, $\mathfrak{D} \subseteq \mathbf{E}^n$ at the interior point $a \in \mathfrak{D}$ exist if and only if all partial derivatives of the components of $f$, $D_j \phi_k(a)$, $j = 1, 2, \ldots, n$, and $k = 1, 2, \ldots, m$, exist.

***Theorem 88.1***

If $f: \mathfrak{D} \to \mathbf{E}^m$, $\mathfrak{D} \subseteq \mathbf{E}^n$, is differentiable at the interior point $a \in \mathfrak{D}$, then all partial derivatives $D_j f(a)$, $j = 1, 2, \ldots, n$, exist.

*Proof*

By Theorem 87.1, $D_u f(a)$ exists for all $u \in \mathbf{E}^n$. Since $e_j \in \mathbf{E}^n$ for all $j = 1, 2, \ldots, n$, the theorem follows.

The existence of all partial derivatives does *not* imply the existence of the directional derivatives for all $u \in \mathbf{E}^n$, as the following example shows:

---

▲ *Example 1*

In example 2, section 9.87, we considered the function $f: \mathbf{E}^2 \to \mathbf{E}$ given by

$$f(\xi_1, \xi_2) = \begin{cases} 1 & \text{for } \xi_1 \xi_2 = 0, \\ 0 & \text{for } \xi_1 \xi_2 \neq 0. \end{cases}$$

We have, from (87.7) and (87.8), that

$$D_1 f(0,0) = D_{e_1} f(0,0) = 0, \qquad D_2 f(0,0) = D_{e_2} f(0,0) = 0.$$

Hence, the partial derivatives exist. However, $D_u f(0,0)$ does *not* exist for $u = (\omega_1, \omega_2)$, where $\omega_1 \neq 0$ and $\omega_2 \neq 0$, as we have seen in section 9.87. Clearly, $f$ is *not* differentiable at $(0,0)$.

---

Summarizing, we can state: *The existence of the derivative implies the existence of the directional derivative for all directions, and this, in turn, implies the existence of all partial derivatives. However, the existence of all partial derivatives does not imply the existence of the directional derivative for all directions, and the existence of the directional derivative for all directions does not imply the existence of the derivative.*

## 9.88 Exercises

1. Let $f: \mathbf{E}^2 \to \mathbf{E}$ be given by

$$f(\xi_1, \xi_2) = \begin{cases} (\xi_1^2 + \xi_2^2) \sin \dfrac{1}{\xi_1^2 + \xi_2^2} & \text{for } (\xi_1, \xi_2) \neq (0,0), \\ 0 & \text{for } (\xi_1, \xi_2) = (0,0). \end{cases}$$

Show that $D_j f(x)$ exist for all $x \in \mathbf{E}^2$ and all $j = 1, 2$, but that $D_j f(x)$, as functions of $x$, are *not* continuous at $\theta$.

2. Given $f: \mathbf{E}^2 \to \mathbf{E}$ by

$$f(\xi_1, \xi_2) = \begin{cases} \xi_1^2 + \xi_2^2 & \text{for } (\xi_1, \xi_2) \in \mathbf{Q} \times \mathbf{Q}, \\ 0 & \text{for } (\xi_1, \xi_2) \notin \mathbf{Q} \times \mathbf{Q}. \end{cases}$$

Find $D_1 f(0,0)$, $D_2 f(0,0)$, and show that

$$F'(\theta) = (D_1 f(0,0),\ D_2 f(0,0))$$

represents $f'(\theta)$.

3. Given $f: \mathbf{E}^2 \to \mathbf{E}$ by

$$f(\xi_1, \xi_2) = \begin{cases} \dfrac{\xi_1 \xi_2}{\sqrt{\xi_1^2 + \xi_2^2}} & \text{for } (\xi_1, \xi_2) \neq (0,0), \\ 0 & \text{for } (\xi_1, \xi_2) = (0,0). \end{cases}$$

Find $D_1 f(\theta)$, $D_2 f(\theta)$.

4. Given $f: \mathbf{E}^2 \to \mathbf{E}$ by

$$f(\xi_1, \xi_2) = \begin{cases} \dfrac{\xi_1 \xi_2^2}{\xi_1^2 + \xi_2^2} & \text{for } \xi_1 \neq 0, \\ 0 & \text{for } \xi_1 = 0. \end{cases}$$

Find $D_1 f(0,0)$, $D_2 f(0,0)$.

5. Show: If $f: \mathfrak{D} \to \mathbf{E}^m$, $\mathfrak{D} \subseteq \mathbf{E}$, is differentiable at the interior point $a \in \mathfrak{D}$, then $f'(a) = D_1 f(a)$.

## 9.89 Representation of the Derivative

We have seen, in section 9.87, that if $f: \mathfrak{D} \to \mathbf{E}^m$, $\mathfrak{D} \subseteq \mathbf{E}^n$, is differentiable at an interior point $a \in \mathfrak{D}$, then

$$f'(a)(u) = D_u f(a) \tag{89.1}$$

for all $u \in \mathbf{E}^n$. Since $u \in \mathbf{E}^n$ may be represented by

$$u = \omega_1 e_1 + \omega_2 e_2 + \cdots + \omega_n e_n,$$

where $e_1, e_2, \ldots, e_n$ are the fundamental vectors in $\mathbf{E}^n$, and since $f'(a)$ is a linear function from $\mathbf{E}^n$ into $\mathbf{E}^m$, we have

$$\begin{aligned} f'(a)(u) &= f'(a)(\omega_1 e_1 + \omega_2 e_2 + \cdots + \omega_n e_n) \\ &= \omega_1 f'(a)(e_1) + \omega_2 f'(a)(e_2) + \cdots + \omega_n f'(a)(e_n). \end{aligned}$$

From (89.1) and definition 88.1,

$$f'(a)(e_j) = D_{e_j} f(a) = D_j f(a).$$

Hence,

$$f'(a)(u) = \omega_1 D_1 f(a) + \omega_2 D_2 f(a) + \cdots + \omega_n D_n f(a) \tag{89.2}$$

for all $u = (\omega_1, \omega_2, \ldots, \omega_n) \in \mathbf{E}^n$. If

$$f = \begin{pmatrix} \phi_1 \\ \phi_2 \\ \vdots \\ \phi_m \end{pmatrix},$$

we have, from Lemma 88.1, that all $D_j \phi_k(a), j = 1, 2, \ldots, n$, and $k = 1, 2, \ldots, m$, exist, and we obtain the representation

$$D_j f(a) = \begin{pmatrix} D_j\phi_1(a) \\ D_j\phi_2(a) \\ \vdots \\ D_j\phi_m(a) \end{pmatrix}, \quad j = 1, 2, \ldots, n.$$

Hence, we may write (89.2) as

$$(89.3) \quad f'(a)(u) = \omega_1 \begin{pmatrix} D_1\phi_1(a) \\ D_1\phi_2(a) \\ \vdots \\ D_1\phi_m(a) \end{pmatrix} + \omega_2 \begin{pmatrix} D_2\phi_1(a) \\ D_2\phi_2(a) \\ \vdots \\ D_2\phi_m(a) \end{pmatrix} + \cdots + \omega_n \begin{pmatrix} D_n\phi_1(a) \\ D_n\phi_2(a) \\ \vdots \\ D_n\phi_m(a) \end{pmatrix}.$$

In view of the definition of matrix multiplication in (83.6), (89.3) may be written as

$$f'(a)(u) = \begin{pmatrix} D_1\phi_1(a) & D_2\phi_1(a) \cdots D_n\phi_1(a) \\ D_1\phi_2(a) & D_2\phi_2(a) \cdots D_n\phi_2(a) \\ \vdots & \\ D_1\phi_m(a) & D_2\phi_m(a) \cdots D_n\phi_m(a) \end{pmatrix} \begin{pmatrix} \omega_1 \\ \omega_2 \\ \vdots \\ \omega_n \end{pmatrix}$$

for all $u = (\omega_1, \omega_2, \ldots, \omega_n) \in \mathbf{E}^n$, and we see that the derivative $f'(a)$ may be represented by the $m \times n$ matrix

$$F'(a) = \begin{pmatrix} D_1\phi_1(a) & D_2\phi_1(a) \cdots D_n\phi_1(a) \\ D_1\phi_2(a) & D_2\phi_2(a) \cdots D_n\phi_2(a) \\ \vdots & \\ D_1\phi_m(a) & D_2\phi_m(a) \cdots D_n\phi_m(a) \end{pmatrix}.$$

($f'(a)$ is, by definition, a linear function from $\mathbf{E}^n$ into $\mathbf{E}^m$, and its matrix representation is, accordingly, an $m \times n$ matrix.)

***Definition 89.1***

If $f: \mathfrak{D} \to \mathbf{E}^m$, $\mathfrak{D} \subseteq \mathbf{E}^n$, and if $a \in \mathfrak{D}$ is an interior point and if $D_j \phi_k(a)$ exist for all $j = 1, 2, \ldots, n$, and $k = 1, 2, \ldots, m$, then

$$(89.4) \quad \begin{pmatrix} D_1\phi_1(a) & D_2\phi_1(a) \cdots D_n\phi_1(a) \\ D_1\phi_2(a) & D_2\phi_2(a) \cdots D_n\phi_2(a) \\ \vdots & \\ D_1\phi_m(a) & D_2\phi_m(a) \cdots D_n\phi_m(a) \end{pmatrix}$$

is called the *Jacobian matrix*† of $f$ at $a$.

† *Carl Gustav Jacob Jacobi*, 1804–1851.

***Theorem 89.1***

If $f: \mathfrak{D} \to \mathbf{E}^m$, $\mathfrak{D} \subseteq \mathbf{E}^n$, is differentiable at the interior point $a \in \mathfrak{D}$, then $f'(a)$ may be represented by the Jacobian matrix of $f$ at $a$, which is defined in (89.4).

Observe that the Jacobian matrix in (89.4) is *not* the derivative $f'(a)$ but only one representation (of many possible representations) of the derivative. (See also section 8.83.)

All the entries in the Jacobian matrix (89.4) may exist but the function still need not be differentiable. (See example 1, section 9.88.) Only when the function $f$ *is* differentiable does the Jacobian matrix in (89.4) represent the derivative in terms of $e_1, e_2, \ldots, e_n \in \mathbf{E}^n$ and $i_1, i_2, \ldots, i_m \in \mathbf{E}^m$.

---

▲ *Example 1*

If $f: \mathfrak{D} \to \mathbf{E}$, $\mathfrak{D} \subseteq \mathbf{E}$, is differentiable at the interior point $a \in \mathfrak{D}$, then $f'(a)$ is represented by the $1 \times 1$ Jacobian matrix $f'(a)$.

▲ *Example 2*

If $f: \mathfrak{D} \to \mathbf{E}$, $\mathfrak{D} \subseteq \mathbf{E}^n$, is differentiable at the interior point $a \in \mathfrak{D}$, then $f'(a)$ may be represented by the $1 \times n$ Jacobian matrix

$$F'(a) = (D_1 f(a),\ D_2 f(a), \ldots,\ D_n f(a)).$$

▲ *Example 3*

If $f: \mathfrak{D} \to \mathbf{E}^m$, $\mathfrak{D} \subseteq \mathbf{E}$ is differentiable at the interior point $a \in \mathfrak{D}$, then $f'(a)$ may be represented by the $m \times 1$ Jacobian matrix

$$F'(a) = \begin{pmatrix} D_1\phi_1(a) \\ D_1\phi_2(a) \\ \vdots \\ D_1\phi_m(a) \end{pmatrix} = \begin{pmatrix} \phi_1'(a) \\ \phi_2'(a) \\ \vdots \\ \phi_m'(a) \end{pmatrix}.$$

(See also exercise 5, section 9.88.)

---

Since the Jacobian matrix represents the derivative, and since the Jacobian matrix is easily accessible, it may be used to show that a given function either is, or is not, differentiable. This will be illustrated in the following two examples.

---

▲ *Example 4*

Let $f: \mathbf{E}^2 \to \mathbf{E}$ be defined by

$$f(\xi_1, \xi_2) = \xi_1 + \xi_2^2.$$

Then

$$(D_1 f(\xi_1, \xi_2), D_2 f(\xi_1, \xi_2)) = (1, 2\xi_2)$$

is the Jacobian matrix. Hence, if $f'(a)$ exists for $a = (\alpha_1, \alpha_2)$, then it may be represented by

$$F'(a) = (1, 2\alpha_2). \tag{89.5}$$

We check this out with definition 86.1 and obtain

$$\begin{aligned}|f(x) - f(a) - f'(a)(x - a)| &= |\xi_1 + \xi_2^2 - \alpha_1 - \alpha_2^2 - (\xi_1 - \alpha_1) - 2\alpha_2(\xi_2 - \alpha_2)| \\ &= (\xi_2 - \alpha_2)^2 \leqslant \varepsilon \|x - a\| \\ &\quad \text{for } x \in N_\delta(a) \text{ where } \delta = \varepsilon.\end{aligned}$$

Hence, $f'(a)$ exists and is represented by (89.5).

▲ *Example 5*

Let $f: \mathbf{E}^2 \to \mathbf{E}$ be defined by

$$f(\xi_1, \xi_2) = \begin{cases} 0 & \text{for } \xi_1 \xi_2 \neq 0, \\ \xi_2 + 1 & \text{for } \xi_1 = 0, \\ 1 & \text{for } \xi_2 = 0. \end{cases}$$

Then,

$$D_1 f(0,0) = \lim_{t \to 0} \frac{f(t,0) - f(0,0)}{t} = \lim_{t \to 0} \frac{1 - 1}{t} = 0$$

and

$$D_2 f(0,0) = \lim_{t \to 0} \frac{f(0,t) - f(0,0)}{t} = \lim_{t \to 0} \frac{t + 1 - 1}{t} = 1.$$

Hence,

$$(D_1 f(0,0), D_2 f(0,0)) = (0, 1)$$

represents the derivative $f'(\theta)$ *if* the derivative exists. For $x = (\xi_1, \xi_2)$ with $\xi_2 > 0$, we have

$$\left| f(x) - f(\theta) - (0, 1)\begin{pmatrix} \xi_1 \\ \xi_2 \end{pmatrix} \right| = |-1 - \xi_2| = 1 + \xi_2 \geqslant 1.$$

Hence, by (86.1), $f$ is not differentiable at $\theta$.

---

If $f: \mathfrak{D} \to \mathbf{E}^m$, $\mathfrak{D} \subseteq \mathbf{E}^n$, is differentiable in the open set $A \subseteq \mathfrak{D}$, then the value of the differential $df: A \times \mathbf{E}^n \to \mathbf{E}^m$ for any $x \in A$, $u = (\omega_1, \omega_2, \ldots, \omega_n) \in \mathbf{E}^n$ is given by

$$df(x, u) = F'(x)u = \begin{pmatrix} D_1 \phi_1(x) & D_2 \phi_1(x) \cdots D_n \phi_1(x) \\ D_1 \phi_2(x) & D_2 \phi_2(x) \cdots D_n \phi_2(x) \\ \vdots & \\ D_1 \phi_m(x) & D_2 \phi_m(x) \cdots D_n \phi_m(x) \end{pmatrix} \begin{pmatrix} \omega_1 \\ \omega_2 \\ \vdots \\ \omega_n \end{pmatrix}.$$

This follows directly from definition 86.2 and Theorem 89.1.

## 9.89 Exercises

1. Given $f: \mathbf{E}^2 \to \mathbf{E}^2$ by
$$\phi_1(\xi_1, \xi_2) = \xi_1^2 - \xi_2,$$
$$\phi_2(\xi_1, \xi_2) = \xi_1 + 2\xi_2.$$
Show that $f'(a)$, $a = (\alpha_1, \alpha_2) \in \mathbf{E}^2$, exists and give the matrix representation of $f'(a)$.
2. Given $f: \mathbf{E}^2 \to \mathbf{E}^3$ by
$$\phi_1(\xi_1, \xi_2) = \xi_1^2 + \xi_1 \cos \xi_2,$$
$$\phi_2(\xi_1, \xi_2) = \xi_1\xi_2 + \sin \xi_2,$$
$$\phi_3(\xi_1, \xi_2) = \xi_1^2 + \xi_2^2 + 2\xi_1.$$
Find the Jacobian matrix of $f$ at $(0, 0)$.
3. Same as in exercise 2, for $f: \mathbf{E}^2 \to \mathbf{E}^2$, given by
$$\phi_1(\xi_1, \xi_2) = \begin{cases} (\xi_1^2 + \xi_2^2) \sin \dfrac{1}{\xi_1^2 + \xi_2^2} & \text{for } (\xi_1, \xi_2) \neq (0, 0), \\ 0 & \text{for } (\xi_1, \xi_2) = 0; \end{cases}$$
$$\phi_2(\xi_1, \xi_2) = \begin{cases} \xi_2 & \text{for } \xi_1 = 0, \\ \dfrac{\xi_1\xi_2^2}{\xi_1^2 + \xi_2^2} & \text{for } \xi_1 \neq 0. \end{cases}$$
4. Given: $f: \mathbf{E}^n \to \mathbf{E}^m$ by $\phi_j(\xi_1, \xi_2, \ldots, \xi_n) = a_{j1}\xi_1 + a_{j2}\xi_2 + \cdots + a_{jn}\xi_n + b_j$, where $j = 1, 2, \ldots, m$, and $a_{jk}$, $b_j \in \mathbf{R}$. Find the Jacobian matrix of $f$ at $x \in \mathbf{E}^n$, and show that $f$ is differentiable for all $x \in \mathbf{E}^n$. (Compare your result with example 1, section 9.86, and draw the proper conclusions.)

## 9.90 Existence of the Derivative

It is our goal in this section to establish a practical test for the existence of the derivative. Among derivative, directional derivative, and partial derivatives, only the latter are readily accessible because they can be computed for a given function by well-known rules. It is, therefore, fortunate that a simple sufficient condition for the existence of the derivative, based on the behavior of the partial derivatives, can be established.

First, let us show that an investigation into the existence of the derivative of a vector-valued function can be reduced to an investigation into the existence of the derivative of its components. We have, in analogy to Lemmas 64.1 and 65.1:

***Lemma 90.1***

If $f: \mathfrak{D} \to \mathbf{E}^m$, $\mathfrak{D} \subseteq \mathbf{E}^n$, and $a \in \mathfrak{D}$ is an interior point, then $f'(a)$ exists if and only if $\phi_j'(a)$ exists for all $j = 1, 2, \ldots, m$, where $\phi_1, \phi_2, \ldots, \phi_m$ are the components of $f$.

*Proof*

(a) Suppose that $f'(a)$ exists. Then, for every $\varepsilon > 0$, there is a $\delta > 0$ such that

(90.1) $$\|f(x) - f(a) - f'(a)(x - a)\| \leqslant \varepsilon \|x - a\| \qquad \text{for all } x \in N_\delta(a),$$

where $f'(a) : \mathbf{E}^n \to \mathbf{E}^m$ is a linear function.

The functions $l_j : \mathbf{E}^n \to \mathbf{E}$, $j = 1, 2, \ldots, m$, which are defined by

(90.2) $$l_j(u) = j\text{th component of } f'(a)(u),\ u \in \mathbf{E}^n$$

are also linear functions because

$$\begin{aligned} l_j(u + v) &= j\text{th component of } f'(a)(u + v) \\ &= j\text{th component of } (f'(a)(u) + f'(a)(v)) \\ &= j\text{th component of } f'(a)(u) + j\text{th component of } f'(a)(v) \\ &= l_j(u) + l_j(v); \\ l_j(\lambda u) &= j\text{th component of } f'(a)(\lambda u) \\ &= j\text{th component of } \lambda f'(a)(u) \\ &= \lambda \text{ times the } j\text{th component of } f'(a)(u) = \lambda l_j(u). \end{aligned}$$

In view of (90.2), we may write the $j$th component of

$$f(x) - f(a) - f'(a)(x - a)$$

as

$$\phi_j(x) - \phi_j(a) - l_j(x - a),$$

and we obtain from (90.1) and the FIE (61.1) that

$$|\phi_j(x) - \phi_j(a) - l_j(x - a)| \leqslant \varepsilon \|x - a\| \qquad \text{for all } x \in N_\delta(a),$$

where $l_j : \mathbf{E}^n \to \mathbf{E}$ is a linear function. Hence, by definition 86.1, $\phi_j : \mathfrak{D} \to \mathbf{E}$ is differentiable at $a$ and $l_j = \phi_j'(a)$ is its derivative.

(b) Suppose that $\phi_j : \mathfrak{D} \to \mathbf{E}$, is differentiable at $a$ for all $j = 1, 2, \ldots, m$. Then, by definition 86.1, we have, for every $\varepsilon > 0$, a $\delta_j > 0$ such that

(90.3) $$|\phi_j(x) - \phi_j(a) - \phi_j'(a)(x - a)| \leqslant (\varepsilon/\sqrt{m}) \|x - a\| \qquad \text{for all } x \in N_{\delta_j}(a),$$

where $\phi_j'(a) : \mathbf{E}^n \to \mathbf{E}$ is a linear function.

The function $l : \mathbf{E}^n \to \mathbf{E}^m$, defined by

(90.4) $$l(u) = \begin{pmatrix} \phi_1'(a)(u) \\ \phi_2'(a)(u) \\ \vdots \\ \phi_m'(a)(u) \end{pmatrix} \text{ for all } u \in \mathbf{E}^n$$

is a linear function, as the reader can easily verify. We obtain, from (90.3) and the FIE (61.1), that

$$\begin{aligned} \|f(x) - f(a) - l(x - a)\| \leqslant \sqrt{m} \max_{j \in \{1, 2, \ldots, m\}} (|\phi_j(x) - \phi_j(a) - \phi_j'(a)(x - a)|) \\ \leqslant \varepsilon \|x - a\| \end{aligned}$$

for all $x \in N_\delta(a)$, where $\delta = \min(\delta_1, \delta_2, \ldots, \delta_m)$ and where $l : \mathbf{E}^n \to \mathbf{E}^m$ is a linear function. Hence, $f$ is differentiable at $a$ and $l = f'(a)$ is its derivative.

***Definition 90.1***

Given $f\colon \mathfrak{D} \to \mathbf{E}^m$ and $\mathcal{A} \subseteq \mathfrak{D} \subseteq \mathbf{E}^n$.

$f \in C'(\mathcal{A})$ if and only if all partial derivatives of $f$ exist for all $x \in \mathcal{A}$.

$f \in C^1(\mathcal{A})$ if and only if all partial derivatives of $f$ exist for all $x \in \mathcal{A}$, and are continuous in $\mathcal{A}$. (See also definition 35.3.)

***Theorem 90.1***

If $f\colon \mathfrak{D} \to \mathbf{E}^m$, $\mathfrak{D} \subseteq \mathbf{E}^n$, if $a \in \mathfrak{D}$ is an interior point, and if, for some $\delta > 0$, with $N_\delta(a) \subseteq \mathfrak{D}$,

$$f \in C'(N_\delta(a)), \tag{90.5}$$

$$f \in C^1(\{a\}), \tag{90.6}$$

then $f'(a)$ exists.

*Proof*

In view of Lemma 90.1, we may assume w.l.o.g. that $f$ is a real-valued function. We shall demonstrate that the linear function $l\colon \mathbf{E}^n \to \mathbf{E}$ that is defined by

$$l(u) = \sum_{j=1}^{n} D_j f(a)\omega_j \qquad \text{for all } u = (\omega_1, \omega_2, \ldots, \omega_n) \in \mathbf{E}^n$$

is the derivative of $f$ at $a$.

Because of (90.5) and (90.6), we have, for every $\varepsilon > 0$, a $\delta_1 > 0$ such that

$$|D_j f(x) - D_j f(a)| < \frac{\varepsilon}{\sqrt{n}} \tag{90.7}$$

for all $x \in N_{\delta_1}(a) \subseteq \mathfrak{D}$.

Let $x \in N_{\delta_1}(a)$, and consider the following finite sequence of points:

$$\begin{aligned} x_0 = x &= (\xi_1, \xi_2, \ldots, \xi_n), \\ x_1 \quad &= (\alpha_1, \xi_2, \ldots, \xi_n), \\ &\vdots \\ x_{n-1} \quad &= (\alpha_1, \alpha_2, \ldots, \alpha_{n-1}, \xi_n), \\ x_n = a &= (\alpha_1, \alpha_2, \ldots, \alpha_n). \end{aligned}$$

(See Fig. 90.1.) Since $x_j - a = (0, 0, \ldots, 0, \xi_{j+1} - \alpha_{j+1}, \ldots, \xi_n - \alpha_n)$, we have

$$\|x_j - a\| = \sqrt{\sum_{i=j+1}^{n} (\xi_i - \alpha_i)^2} \leqslant \|x - a\| < \delta_1;$$

that is, $x_j \in N_{\delta_1}(a)$ for all $j = 1, 2, \ldots, n$.

We represent $f(x) - f(a)$ by the following telescoping sum

$$f(x) - f(a) = \sum_{j=1}^{n} (f(x_{j-1}) - f(x_j)), \tag{90.8}$$

and note that $x_{j-1}$ and $x_j$ differ only in their $j$th component:

$$x_{j-1} = (\alpha_1, \ldots, \alpha_{j-1}, \xi_j, \xi_{j+1}, \ldots, \xi_n), \qquad x_j = (\alpha_1, \ldots, \alpha_j, \xi_{j+1}, \ldots, \xi_n).$$

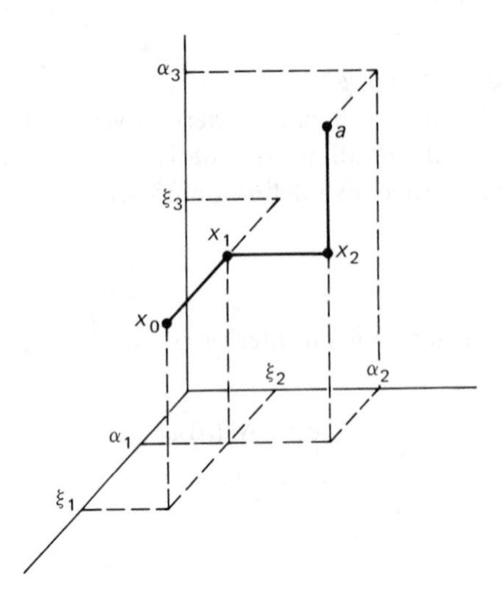


*Fig. 90.1*

Hence, we may view $f$ in $f(x_{j-1}) - f(x_j)$ as a (real-valued) function of one real variable only, namely $\xi_j$. Since $x_{j-1}, x_j \in N_{\delta_1}(a)$, the line segment from $x_{j-1}$ to $x_j$ lies in $N_\delta(a)$; and since $f \in C'(N_\delta(a))$, we may apply the mean-value theorem of the differential calculus (Theorem 37.3) to $f(x_{j-1}) - f(x_j)$:

(90.9) $$f(x_{j-1}) - f(x_j) = D_j f(c_j)(\xi_j - \alpha_j),$$

where $c_j = (1-t)x_{j-1} + tx_j, 0 \leqslant t \leqslant 1$. (Note that the derivative of $f$ with respect to $\xi_j$ is $D_j f$.)

From (90.8) and (90.9):

$$f(x) - f(a) = \sum_{j=1}^{n} D_j f(c_j)(\xi_j - \alpha_j),$$

and consequently,

$$f(x) - f(a) - \sum_{j=1}^{n} D_j f(a)(\xi_j - \alpha_j) = \sum_{j=1}^{n} (D_j f(c_j) - D_j f(a_j))(\xi_j - \alpha_j).$$

Since $c_j \in N_{\delta_1}(a)$ for all $j = 1, 2, \ldots, m$, we have, from (90.7), that

$$|D_j f(c_j) - D_j f(a)| < \frac{\varepsilon}{\sqrt{n}},$$

and hence, using the CBS inequality,

$$\begin{aligned} \left| f(x) - f(a) - \sum_{j=1}^{n} D_j f(a)(\xi_j - \alpha_j) \right| &= \left| \sum_{j=1}^{n} (D_j f(c_j) - D_j f(a))(\xi_j - \alpha_j) \right| \\ &\leqslant \sqrt{\sum_{j=1}^{n} (D_j f(c_j) - D_j f(a))^2} \sqrt{\sum_{j=1}^{n} (\xi_j - \alpha_j)^2} \\ &< \varepsilon \|x - a\|. \end{aligned}$$

The function $l: \mathbf{E}^n \to \mathbf{E}$ that is defined by

$$(90.10) \qquad l(u) = \sum_{j=1}^{n} D_j f(a)\omega_j \qquad \text{for all } u = (\omega_1, \omega_2, \ldots, \omega_n) \in \mathbf{E}^n$$

is obviously a linear function. Hence, $f$ is differentiable at $a$ and $l: \mathbf{E}^n \to \mathbf{E}$, defined in (90.10), is its derivative.

---

▲ *Example 1*

In example 2, section 9.87, we considered the function $f: \mathbf{E}^2 \to \mathbf{E}$ defined by

$$f(\xi_1, \xi_2) = \begin{cases} 1 & \text{for } \xi_1\xi_2 = 0, \\ 0 & \text{for } \xi_1\xi_2 \neq 0. \end{cases}$$

Let us consider the point $(0, \xi_2)$, where $\xi_2 \neq 0$ may be chosen as close to 0 as one pleases. We have that

$$D_1 f(0, \xi_2) = \lim_{t \to 0} \frac{f(t, \xi_2) - f(0, \xi_2)}{t} = \lim_{t \to 0} \frac{1 - 0}{t}$$

does *not* exist. Theorem 90.1 does not guarantee the existence of $f'(0,0)$ in that case, and we have seen, in section 9.87, that, as a matter of fact, $f'(0,0)$ does not exist.

---

The derivative may exist even though not all the conditions in Theorem 90.1 are satisfied, as the following example shows:

---

▲ *Example 2*

(This example is a two-dimensional version of example 3, section 3.35. See also exercise 1, section 9.88.) Let $f: \mathbf{E}^2 \to \mathbf{E}$ be defined by

$$f(\xi_1, \xi_2) = \begin{cases} (\xi_1^2 + \xi_2^2) \sin \dfrac{1}{\xi_1^2 + \xi_2^2} & \text{for } (\xi_1, \xi_2) \neq (0,0), \\ 0 & \text{for } (\xi_1, \xi_2) = (0,0). \end{cases}$$

We have

$$D_1 f(0,0) = \lim_{0} \frac{f(t,0) - f(0,0)}{t} = \lim_{0} \frac{t^2 \sin \dfrac{1}{t^2}}{t} = 0,$$

$$D_2 f(0,0) = 0 \qquad \text{(by reasons of symmetry).}$$

If $(\xi_1, \xi_2) \neq (0,0)$, we obtain, for $D_1 f(\xi_1, \xi_2)$, $D_2 f(\xi_1, \xi_2)$, by the well-known differentiation rules,

$$D_1 f(\xi_1, \xi_2) = 2\xi_1 \sin \frac{1}{\xi_1^2 + \xi_2^2} - \frac{2\xi_1}{\xi_1^2 + \xi_2^2} \cos \frac{1}{\xi_1^2 + \xi_2^2},$$

$$D_2 f(\xi_1, \xi_2) = 2\xi_2 \sin \frac{1}{\xi_1^2 + \xi_2^2} - \frac{2\xi_2}{\xi_1^2 + \xi_2^2} \cos \frac{1}{\xi_1^2 + \xi_2^2}.$$

Hence, all partial derivatives exist for all $x \in \mathbf{E}^2$. However, neither $D_1 f$ nor $D_2 f$ are continuous at $(0,0)$, as we can see by considering the sequence $\{x_k\} = \{((1/k), 0)\}$. Then

$$\lim_{\infty} D_1 f\left(\frac{1}{k}, 0\right) = \lim_{\infty}\left(\frac{2}{k} \sin k^2 - \frac{2}{k(1/k)^2} \cos k^2\right)$$

does *not* exist. (If $D_1 f$ were continuous at $(0,0)$, then this limit would have to be 0.) Similarly, one can see that $D_2 f$ is not continuous at $(0,0)$, by taking the sequence $\{x_k'\} = \{(0, (1/k))\}$. In view of this, Theorem 90.1 does not apply. Still, $f'(0,0)$ exists, and is given by

$$f'(0,0) : \mathbf{E}^2 \to \{0\} \qquad \text{or, equivalently,} \qquad F'(0,0) = (0,0)$$

as one can see from a direct appeal to definition 86.1.

---

While Theorem 90.1 is quite practical and relatively easy to prove, it is not the best sufficient condition one can come up with. The hypothesis of Theorem 90.1 may be replaced by the following weaker hypothesis:

For $n$ linearly independent vectors $u_1, u_2, \dots, u_n$ in $\mathbf{E}^n$, all directional derivatives in the directions $u_1, u_2, \dots, u_n$ shall exist at $a$ and all but one of these directional derivatives shall exist in some neighborhood of $a$ and be continuous at $a$.†

That these conditions are still not necessary may be seen from example 2, where neither directional derivative is continuous at $\theta$ but where the derivative at $\theta$ still exists. (See exercise 1.)

The above-mentioned weaker sufficient condition applies to the following example, to which Theorem 90.1 is not applicable.

---

▲ *Example 3*

Let $f: \mathbf{E}^2 \to \mathbf{E}^2$ be defined by

$$\phi_1(\xi_1, \xi_2) = \begin{cases} \xi_1 + \xi_1^2 \sin \dfrac{1}{\xi_1} & \text{for } \xi_1 \neq 0, \\ 0 & \text{for } \xi_1 = 0, \end{cases}$$

$$\phi_2(\xi_1, \xi_2) = \xi_2.$$

† A. Devinatz, *Advanced Calculus*. (New York: Holt, Reinhart and Winston, 1968), p. 312.

We have

$$D_1\phi_1(0,0) = \lim_{0} \frac{t + t^2 \sin \frac{1}{t}}{t} = 1, \qquad D_2\,\phi_1(0,0) = 0,$$

$$D_1\phi_2(0,0) = 0, \qquad D_2\,\phi_2(0,0) = 1.$$

Obviously,

$$D_2\,\phi_1(\xi_1, \xi_2), \qquad D_1\phi_2(\xi_1, \xi_2), \qquad D_2\,\phi_2(\xi_1, \xi_2)$$

exist in a neighborhood of $\theta$ and are continuous at $\theta$. However, for $\xi_1 \neq 0$,

$$D_1\phi_1(\xi_1, \xi_2) = 1 + 2\xi_1 \sin \frac{1}{\xi_1} - \cos \frac{1}{\xi_1},$$

and hence, $D_1\phi_1$ is *not* continuous at $\theta$. The above-mentioned conditions are satisfied and the existence of $f'(\theta)$ is assured but Theorem 90.1 does *not* apply.

That $f'(\theta)$ exists may also be seen by a direct appeal to definition 86.1. By Lemma 90.1, we need only check the existence of $\phi_1'(\theta)$, $\phi_2'(\theta)$. There is no question about the existence of $\phi_2'(\theta)$ which may be represented by its Jacobian matrix $\Phi_2'(\theta) = (0, 1)$. That $\phi_1'(\theta)$ exists and may be represented by

$$\Phi_1'(\theta) = (1, 0)$$

follows from a direct appeal to definition 86.1:

$$|\phi_1(\xi_1, \xi_2) - \phi_1(0,0) - \phi_1'(0,0)(\xi_1 - 0, \xi_2 - 0)|$$

$$= \left|\xi_1 + \xi_1^2 \sin \frac{1}{\xi_1} - \xi_1\right| \leqslant \xi_1^2 < \varepsilon|\xi_1| \leqslant \varepsilon\|x\|$$

for all $x \in N_\varepsilon(\theta)$. Hence, $\phi_1'(\theta)$ exists and therefore, $f'(\theta)$ exists.

---

## 9.90 Exercises

1. Consider the function $f: \mathbf{E}^2 \to \mathbf{E}$ of example 2 and find $D_u f(0,0)$, as well as $D_u f(\xi_1, \xi_2)$, for $(\xi_1, \xi_2) \neq (0,0)$. Show that $D_u f$ is not continuous at $\theta$ for all $u \in \mathbf{E}^n$.
2. Consider the function of exercise 2, section 9.88. Does Theorem 90.1 apply at $\theta$? Does the weaker version of Theorem 90.1, which is quoted preceding example 3, apply?
3. Consider the function $f: \mathbf{E}^2 \to \mathbf{E}$ of exercise 3, section 9.88. Do $D_1 f$, $D_2 f$, $D_u f$ exist in a neighborhood of $\theta$? Are they continuous at $\theta$? Does $f'(\theta)$ exist?
4. Same as in exercise 3, for the function in exercise 4, section 9.88.
5. Apply the results of exercises 1 and 4 to the function in exercise 3, section 9.89.

## 9.91 Differentiation Rules

In this obligatory section we shall, for the sake of completeness, dutifully list the customary differentiation rules for the differentiation of sums, products, and quotients.

First, let us note that if $f, g : \mathfrak{D} \to \mathbf{E}^m$, $\mathfrak{D} \subseteq \mathbf{E}^n$, then $f \pm g : \mathfrak{D} \to \mathbf{E}^m$ is defined by $(f \pm g)(x) = f(x) \pm g(x)$. If $(f \pm g)'(a)$ exists, then it is a linear function from $\mathbf{E}^n$ into $\mathbf{E}^m$. If $\lambda : \mathfrak{D} \to \mathbf{E}$, $f$ as before, then $\lambda f : \mathfrak{D} \to \mathbf{E}^m$ is defined by $(\lambda f)(x) = \lambda(x) f(x)$. If $(\lambda f)'(a)$ exists, then it is a linear function from $\mathbf{E}^n$ into $\mathbf{E}^m$. If $f, g$ are as before, then $f \cdot g : \mathfrak{D} \to \mathbf{E}$ is defined by $(f \cdot g)(x) = f(x) \cdot g(x)$ (dot product). If $(f \cdot g)'(a)$ exists, then it is a linear function from $\mathbf{E}^n$ into $\mathbf{E}$. Finally, if $\lambda$ is as before and $\lambda(x) \neq 0$ for all $x \in \mathfrak{D}$, then $(1/\lambda) : \mathfrak{D} \to \mathbf{E}$ is defined by $(1/\lambda)(x) = 1/\lambda(x)$. If $(1/\lambda)'(a)$ exists, then it is a linear function from $\mathbf{E}^n$ into $\mathbf{E}$.

***Theorem 91.1***

If $f, g : \mathfrak{D} \to \mathbf{E}^m$, $\lambda, \mu : \mathfrak{D} \to \mathbf{E}$, $\mathfrak{D} \subseteq \mathbf{E}^n$, if $\mu(x) \neq 0$ for all $x \in \mathfrak{D}$, and if $f'(a)$ $g'(a), \lambda'(a), \mu'(a)$ exist at an interior point $a \in \mathfrak{D}$, then $(f \pm g)'(a)$, $(\lambda f)'(a)$, $(f \cdot g)'(a)$, and $(1/\mu)'(a)$ exist and

$$(91.1) \qquad (f \pm g)'(a) = f'(a) \pm g'(a),$$

$$(91.2) \qquad (\lambda f)'(a)(u) = \lambda(a) f'(a)(u) + \lambda'(a)(u) f(a) \qquad \text{for all } u \in \mathbf{E}^n,$$

(Observe that $\lambda(a)$, $\lambda'(a)(u)$ are scalars. See also exercise 1.)

$$(91.3) \qquad (f \cdot g)'(a)(u) = f(a) \cdot g'(a)(u) + g(a) \cdot f'(a)(u) \qquad \text{for all } u \in \mathbf{E}^n,$$

(Observe that $f(a)$, $g(a) \in \mathbf{E}^m$ and $f'(a), g'(a) : \mathbf{E}^n \to \mathbf{E}^m$. See also exercise 2.)

$$(91.4) \qquad \left(\frac{1}{\mu}\right)'(a) = -\frac{\mu'(a)}{\mu^2(a)}.$$

Before we present a proof, we discuss a number of examples to illustrate the four differentiation rules in Theorem 91.1. The existence of the derivatives in all the following examples is assured by Theorem 90.1 because all the functions that are being considered possess continuous partial derivatives everywhere.

---

▲ *Example 1*

Let $f, g : \mathbf{E}^2 \to \mathbf{E}^2$ be given by

$$\phi_1(\xi_1, \xi_2) = \xi_1^2 + \xi_2^2, \qquad \phi_2(\xi_1, \xi_2) = \xi_1^2,$$
$$\psi_1(\xi_1, \xi_2) = 2\xi_1\xi_2, \qquad \psi_2(\xi_1, \xi_2) = -\xi_1^2.$$

With $a = (\alpha_1, \alpha_2)$, $f'(a), g'(a)$ are represented by their Jacobian matrices

$$F'(a) = \begin{pmatrix} 2\alpha_1 & 2\alpha_2 \\ 2\alpha_1 & 0 \end{pmatrix} \quad \text{and} \quad G'(a) = \begin{pmatrix} 2\alpha_2 & 2\alpha_1 \\ -2\alpha_1 & 0 \end{pmatrix}.$$

$f + g$ is given by

$$(\phi_1 + \psi_1)(x) = (\xi_1 + \xi_2)^2, \qquad (\phi_2 + \psi_2)(x) = 0.$$

Hence, $(f + g)'(a)$ is represented by the Jacobian matrix

$$(F + G)'(a) = \begin{pmatrix} 2(\alpha_1 + \alpha_2) & 2(\alpha_1 + \alpha_2) \\ 0 & 0 \end{pmatrix},$$

and we see that, indeed,

$$F'(a) + G'(a) = (F + G)'(a),$$

which also follows from (91.1) directly.

▲ *Example 2*

Let $\lambda : \mathbf{E}^3 \to \mathbf{E}$ be given by $\lambda(\xi_1, \xi_2, \xi_3) = \xi_1 + \xi_2 + \xi_3$ and let $f : \mathbf{E}^3 \to \mathbf{E}^2$ be given by $\phi_1(\xi_1, \xi_2, \xi_3) = \xi_1 - \xi_3$, $\phi_2(\xi_1, \xi_2, \xi_3) = \xi_2$. Then, $\lambda'(a)$ is represented by its Jacobian matrix

$$\Lambda'(a) = (1, 1, 1),$$

and $f'(a)$ is represented by its Jacobian matrix

$$F'(a) = \begin{pmatrix} 1 & 0 & -1 \\ 0 & 1 & 0 \end{pmatrix}.$$

$\lambda f$ is given by

$$(\lambda\phi_1)(\xi_1, \xi_2, \xi_3) = \xi_1^2 + \xi_1\xi_2 - \xi_2\xi_3 - \xi_3^2,$$
$$(\lambda\phi_2)(\xi_1, \xi_2, \xi_3) = \xi_1\xi_2 + \xi_2^2 + \xi_2\xi_3.$$

Hence, $(\lambda f)'(a)$ is represented by its Jacobian matrix

$$(\Lambda F)'(a) = \begin{pmatrix} 2\alpha_1 + \alpha_2 & \alpha_1 - \alpha_3 & -\alpha_2 - 2\alpha_3 \\ \alpha_2 & \alpha_1 + 2\alpha_2 + \alpha_3 & \alpha_2 \end{pmatrix}.$$

Therefore, with $u = (\omega_1, \omega_2, \omega_3) \in \mathbf{E}^3$ we have

$$(91.5) \quad (\Lambda F)'(a)(u) = \begin{pmatrix} (2\alpha_1 + \alpha_2)\omega_1 + (\alpha_1 - \alpha_3)\omega_2 - (\alpha_2 + 2\alpha_3)\omega_3 \\ \alpha_2\omega_1 + (\alpha_1 + 2\alpha_2 + \alpha_3)\omega_2 + \alpha_2\omega_3 \end{pmatrix}.$$

On the other hand,

$$\lambda(a)F'(a)(u) + \Lambda'(a)(u)f(a)$$

$$= (\alpha_1 + \alpha_2 + \alpha_3)\begin{pmatrix} 1 & 0 & -1 \\ 0 & 1 & 0 \end{pmatrix}\begin{pmatrix} \omega_1 \\ \omega_2 \\ \omega_3 \end{pmatrix} + (1, 1, 1)\begin{pmatrix} \omega_1 \\ \omega_2 \\ \omega_3 \end{pmatrix}\begin{pmatrix} \alpha_1 - \alpha_3 \\ \alpha_2 \end{pmatrix}$$

$$= \begin{pmatrix} (2\alpha_1 + \alpha_2)\omega_1 + (\alpha_1 - \alpha_3)\omega_2 - (\alpha_2 + 2\alpha_3)\omega_3 \\ \alpha_2\omega_1 + (\alpha_1 + 2\alpha_2 + \alpha_3)\omega_2 + \alpha_2\omega_3 \end{pmatrix},$$

hich agrees with (91.5), as it is supposed to, by (91.2).

▲ *Example 3*

Let $f, g : \mathbf{E}^2 \to \mathbf{E}^2$ by given by

$$\phi_1(\xi_1, \xi_2) = \xi_1, \qquad \phi_2(\xi_1, \xi_2) = \xi_2,$$
$$\psi_1(\xi_1, \xi_2) = \xi_2, \qquad \psi_2(\xi_1, \xi_2) = -\xi_1.$$

Then, $f'(a), g'(a)$ are represented by

$$F'(a) = \begin{pmatrix} 1 & 0 \\ 0 & 1 \end{pmatrix}, \qquad G'(a) = \begin{pmatrix} 0 & 1 \\ -1 & 0 \end{pmatrix},$$

and we obtain with $u = (\omega_1, \omega_2) \in \mathbf{E}^2$,

$$(91.6) \qquad f(a) \cdot G'(a)(u) + g(a) \cdot F'(a)(u)$$
$$= (\alpha_1, \alpha_2)\begin{pmatrix} 0 & 1 \\ -1 & 0 \end{pmatrix}\begin{pmatrix} \omega_1 \\ \omega_2 \end{pmatrix} + (\alpha_2, -\alpha_1)\begin{pmatrix} 1 & 0 \\ 0 & 1 \end{pmatrix}\begin{pmatrix} \omega_1 \\ \omega_2 \end{pmatrix} = 0.$$

On the other hand, $f \cdot g$ is given by

$$(f \cdot g)(x) = f(x) \cdot g(x) = \xi_1\xi_2 - \xi_2\xi_1 = 0;$$

and hence, $(f \cdot g)'(a) = 0$, in agreement with (91.6), and as predicted by (91.3).

▲ *Example 4*

Let $\mu : \mathbf{E}^2 \to \mathbf{E}$ by given by $\mu(x) = 1/(1 + \xi_1^2 + \xi_2^2)$. Then, $\mu(x) \neq 0$ for all $x \in \mathbf{E}^2$. Since $\mu'(a)$ is represented by

$$M'(a) = \left(\frac{-2\alpha_1}{(1 + \alpha_1^2 + \alpha_2^2)^2}, \frac{-2\alpha_2}{(1 + \alpha_1^2 + \alpha_2^2)^2}\right),$$

we obtain, from (91.4), that $(1/\mu)'(a)$ is represented by

$$(91.7) \qquad \left(\frac{1}{M}\right)'(a) = (1 + \alpha_1^2 + \alpha_2^2)^2\left(\frac{2\alpha_1}{(1 + \alpha_1^2 + \alpha_2^2)^2}, \frac{2\alpha_2}{(1 + \alpha_1^2 + \alpha_2^2)^2}\right)$$
$$= (2\alpha_1, 2\alpha_2).$$

Since $(1/\mu)(x) = 1 + \xi_1^2 + \xi_2^2$, we obtain directly,

$$\left(\frac{1}{M}\right)'(a) = (2\alpha_1, 2\alpha_2),$$

in agreement with (91.7).

---

*Proof of Theorem 91.1*

*Proof of* (91.1): See exercise 3.

*Proof of* (91.2): According to definition 86.1, we have to show that, for every $\varepsilon > 0$, there is a $\delta > 0$ such that

$$\|\lambda(x)f(x) - \lambda(a)f(a) - \lambda(a)f'(a)(x - a) - \lambda'(a)(x - a)f(a)\| \leqslant \varepsilon \|x - a\|$$

for all $x \in N_\delta(a)$. By the triangle inequality,

$$(91.8)\qquad \begin{aligned} \|\lambda(x)f(x) - \lambda(a)f(a) - \lambda(a)f'(a)(x-a) - \lambda'(a)(x-a)f(a)\| \\ \leqslant |\lambda(x)|\,\|f(x)-f(a)-f'(a)(x-a)\| \\ + \|f(a)\|\,|\lambda(x)-\lambda(a)-\lambda'(a)(x-a)| \\ + |\lambda(x)-\lambda(a)|\,\|f'(a)(x-a)\|. \end{aligned}$$

We have, from Lemma 85.1, that

$$(91.9)\qquad \|f'(a)(x-a)\| \leqslant M_1\|x-a\| \qquad \text{for some } M_1 > 0 \text{ and all } x \in \mathbf{E}^n.$$

Since $\lambda$ is differentiable at $a$, then $\lambda$ is also continuous at $a$ by Theorem 86.2. Hence, for some $\delta_1 > 0$,

$$(91.10)\qquad |\lambda(x)-\lambda(a)| < \frac{\varepsilon}{3M_1} \qquad \text{for all } x \in N_{\delta_1}(a).$$

Therefore,

$$(91.11)\qquad |\lambda(x)| < \frac{\varepsilon}{3M_1} + |\lambda(a)| = M_2 \qquad \text{for all } x \in N_{\delta_1}(a).$$

Since $f$ is differentiable at $a$,

$$(91.12)\qquad \|f(x)-f(a)-f'(a)(x-a)\| \leqslant \frac{\varepsilon}{3M_2}\|x-a\| \qquad \text{for all } x \in N_{\delta_2}(a)$$

for some $\delta_2 > 0$. If $f(a) \neq \theta$, let $M_3 = \|f(a)\|$. Since $\lambda$ is differentiable at $a$, we have, for some $\delta_3 > 0$,

$$(91.13)\qquad |\lambda(x)-\lambda(a)-\lambda'(a)(x-a)| \leqslant \frac{\varepsilon}{3M_3}\|x-a\| \qquad \text{for all } x \in N_{\delta_3}(a).$$

Hence, we obtain, for (91.8), in view of (91.9) through (91.13), that

$$\begin{aligned} &\|\lambda(x)f(x) - \lambda(a)f(a) - \lambda(a)f'(a)(x-a) - \lambda'(a)(x-a)f(a)\| \\ &\qquad \leqslant M_2\frac{\varepsilon}{3M_2}\|x-a\| + M_3\frac{\varepsilon}{3M_3}\|x-a\| + \frac{\varepsilon}{3M_1}M_1\|x-a\| \\ &\qquad = \varepsilon\|x-a\| \end{aligned}$$

for all $x \in N_\delta(a)$, where $\delta = \min(\delta_1, \delta_2, \delta_3)$, and (91.2) is established.

*Proof of* (91.3): Adding and subtracting the terms $f(a)\cdot g(x)$, $f'(a)(x-a)\cdot g(x)$, we obtain, from the triangle inequality,

$$\begin{aligned} &\|f(x)\cdot g(x) - f(a)\cdot g(a) - f(a)\cdot g'(a)(x-a) - g(a)\cdot f'(a)(x-a)\| \\ &\quad \leqslant \|(f(x)-f(a)-f'(a)(x-a))\cdot g(x)\| + \|f'(a)(x-a)\cdot(g(x)-g(a))\| \\ &\qquad + \|f(a)\cdot(g(x)-g(a)-g'(a)(x-a))\| \\ &\quad \leqslant \|f(x)-f(a)-f'(a)(x-a)\|\,\|g(x)\| + \|f'(a)(x-a)\|\,\|g(x)-g(a)\| \\ &\qquad + \|f(a)\|\,\|g(x)-g(a)-g'(a)(x-a)\|. \end{aligned}$$

If we observe that $f, g$ are differentiable at $a$, that $g$ is continuous at $a$, and that $\|f'(a)(x-a)\| \leqslant M\|x-a\|$ for some $M > 0$, by Lemma 85.1, then (91.3) follows readily.

*Proof of* (91.4): See exercise 4.

## 9.91 Exercises

1. Let $\lambda, f$ be as in Theorem 91.1. Show that the function $l : \mathbf{E}^n \to \mathbf{E}^m$, which is defined by

$$l(u) = \lambda(a)f'(a)(u) + \lambda'(a)(u)f(a) \qquad \text{for all } u \in \mathbf{E}^n,$$

is a linear function.

2. Let $f, g$ be as in Theorem 91.1. Show that the function $l : \mathbf{E}^n \to \mathbf{E}$, which is defined by

$$l(u) = f(a) \cdot g'(a)(u) + g(a) \cdot f'(a)(u) \qquad \text{for all } u \in \mathbf{E}^n,$$

is a linear function.

3. Prove (91.1).
4. Prove (91.4).
5. Given $f, g : \mathbf{E}^2 \to \mathbf{E}^2$ by $\phi_1(\xi_1, \xi_2) = \xi_1 - \xi_2$, $\phi_2(\xi_1, \xi_2) = 1$, $\psi_1(\xi_1, \xi_2) = \xi_1^2$, $\psi_2(\xi_1, \xi_2) = \xi_1 + \xi_2$. Let $a = (\alpha_1, \alpha_2)$. Show that $f'(a), g'(a)$ exist, and find $(f+g)'(a)$, $(f \cdot g)'(a)$, by Theorem 91.1.

## 9.92 Differentiation of Composite Functions

In section 3.36, we derived the chain rule for the differentiation of the *composition* of real-valued functions of a real variable. In order to generalize this rule to the composition of vector-valued functions of a vector variable, we need the following lemma:

### *Lemma 92.1*

Let $f : \mathfrak{D}(f) \to \mathbf{E}^p$, $\mathfrak{D}(f) \subseteq \mathbf{E}^n$, and let $a$ denote an interior point of $\mathfrak{D}(f)$. If $f$ is continuous at $a$ and if $f(a) = b$ is an interior point of $A \subseteq \mathbf{E}^p$, then $a$ is also an interior point of $f^*(A)$.

### *Proof*

Since $b = f(a)$ is an interior point of $A$, there is an $\varepsilon > 0$ such that

(92.1) $$N_\varepsilon(f(a)) \subseteq A.$$

Since $f$ is continuous at $a$ and since $a$ is an interior point of $\mathfrak{D}(f)$, there is a $\delta > 0$ such that

(92.2) $$f_*(N_\delta(a)) \subseteq N_\varepsilon(f(a)) \qquad \text{where } N_\delta(a) \subseteq \mathfrak{D}(f).$$

From (66.6), (66.10), (92.1), and (92.2),

$$N_\delta(a) = N_\delta(a) \cap \mathfrak{D}(f) \subseteq f^*(f_*(N_\delta(a))) \subseteq f^*(N_\varepsilon(f(a))) \subseteq f^*(A),$$

and the lemma is proved.

In view of this lemma, we may state the chain rule for the differentiation of the composition of vector-valued functions as follows:

***Theorem 92.1***

Let $f: \mathfrak{D}(f) \to \mathbf{E}^p$, $\mathfrak{D}(f) \subseteq \mathbf{E}^n$, and $g: \mathfrak{D}(g) \to \mathbf{E}^m$, $\mathfrak{D}(g) \subseteq \mathbf{E}^p$. Let $a$ be an interior point of $\mathfrak{D}(f)$, and let $f(a) = b$ be an interior point of $\mathfrak{D}(g)$. If $f'(a)$ and $g'(b)$ exist, then the composite function $g \circ f$ is differentiable at $a$ and

$$(g \circ f)'(a) = g'(f(a)) \circ f'(a).$$

*Proof*

If we replace $A$ in Lemma 92.1 by $\mathfrak{D}(g)$, we obtain that $a$ is an interior point of $f^*(\mathfrak{D}(g)) = \mathfrak{D}(g \circ f)$. Hence it is meaningful to talk about the derivative of $g \circ f$ at $a$.

We want to show that, for every $\varepsilon > 0$, there is a $\delta > 0$ such that $N_\delta(a) \subseteq \mathfrak{D}(g \circ f)$, and

$$\|(g \circ f)(x) - (g \circ f)(a) - g'(f(a)) \circ f'(a)(x - a)\| \leqslant \varepsilon \|x - a\| \tag{92.3}$$

for all $x \in N_\delta(a)$, and that $g'(f(a)) \circ f'(a)$ is a linear function from $\mathbf{E}^n$ into $\mathbf{E}^m$. The latter is trivial, since $f'(a)$ is a linear function from $\mathbf{E}^n$ into $\mathbf{E}^p$ and $g'(f(a))$ is a linear function from $\mathbf{E}^p$ into $\mathbf{E}^m$. (See also exercise 11, section 8.83.)

In order to show that (92.3) holds, we transform the expression under the norm, using the notation $f(a) = b$, $f(x) = y$, as follows:

$$\begin{aligned} &(g \circ f)(x) - (g \circ f)(a) - g'(f(a)) \circ f'(a)(x - a) \\ &\quad = g(y) - g(b) - g'(b) \circ f'(a)(x - a) \\ &\quad = g(y) - g(b) - g'(b)(y - b) + g'(b)(y - b) - g'(b) \circ f'(a)(x - a) \\ &\quad = g(y) - g(b) - g'(b)(y - b) + g'(b)(f(x) - f(a) - f'(a)(x - a)). \end{aligned} \tag{92.4}$$

(Note that $g'(b) \circ f'(a)(x - a) = g'(b)(f'(a)(x - a))$.) Since $f$ is differentiable at $a$, we have, from (86.5), that

$$\|f(x) - f(a)\| \leqslant M \|x - a\| \tag{92.5}$$

for some $M > 0$ and all $x \in N_{\delta_1}(a)$ for some $\delta_1 > 0$. Since $a$ is an interior point of $\mathfrak{D}(g \circ f)$, we may choose $\delta_1$ such that $N_{\delta_1}(a) \subseteq \mathfrak{D}(g \circ f)$.

Since $g$ is differentiable at $b$, we have a $\delta_2 > 0$ such that

$$\|g(y) - g(b) - g'(b)(y - b)\| \leqslant \frac{\varepsilon}{2M} \|y - b\| \tag{92.6}$$

for all $y \in N_{\delta_2}(b)$ where we may choose $\delta_2$ such that $N_{\delta_2}(b) \subseteq \mathfrak{D}(g)$. If we choose $\delta_3 = \min(\delta_1, \delta_2/M)$, then we have, from (92.5), that

$$\|y - b\| = \|f(x) - f(a)\| \leqslant M\|x - a\| < M\frac{\delta_2}{M} = \delta_2 \qquad \text{for all } x \in N_{\delta_3}(a),$$

and we obtain, from (92.5) and (92.6), that

$$\|g(y) - g(b) - g'(b)(y - b)\| \leqslant \frac{\varepsilon}{2}\|x - a\| \tag{92.7}$$

for all $x \in N_{\delta_3}(a)$. Since $g'(b)$ is a linear function, we have, from Lemma 85.1, that

$$\|g'(b)(u)\| \leqslant K\|u\| \qquad \text{for all } u \in \mathbf{E}^p \tag{92.8}$$

and some $K > 0$. Since $f$ is differentiable at $a$, we have a $\delta_4 > 0$ such that $N_{\delta_4}(a) \subseteq \mathfrak{D}(g \circ f)$ and

$$\|f(x) - f(a) - f'(a)(x - a)\| \leqslant \frac{\varepsilon}{2K}\|x - a\| \tag{92.9}$$

for all $x \in N_{\delta_4}(a)$.

From (92.8) and (92.9),

$$\begin{aligned} &\|g'(b)(f(x) - f(a) - f'(a)(x - a))\| \\ &\qquad \leqslant K\|f(x) - f(a) - f'(a)(x - a)\| \leqslant \frac{\varepsilon}{2}\|x - a\| \end{aligned} \tag{92.10}$$

for all $x \in N_{\delta_4}(a)$.

Finally, we obtain for (92.3), via (92.4), utilizing (92.7) and (92.10), that

$$\|(g \circ f)(x) - (g \circ f)(a) - g'(f(a)) \circ f'(a)(x - a)\| \leqslant \varepsilon\|x - a\|$$

for all $x \in N_\delta(a)$, where $\delta = \min(\delta_3, \delta_4)$, and our theorem is proved.

---

▲ *Example 1*

Let $f: \mathbf{E}^2 \to \mathbf{E}^3$ be given by

$$\begin{aligned} \phi_1(\xi_1, \xi_2) &= \xi_1^2, \\ \phi_2(\xi_1, \xi_2) &= \xi_2, \\ \phi_3(\xi_1, \xi_2) &= \xi_1 + \xi_2, \end{aligned}$$

and let $g : \mathbf{E}^3 \to \mathbf{E}$ by given by

$$g(\eta_1, \eta_2, \eta_3) = \eta_1 + \eta_2^2 + \eta_3.$$

Since $\mathfrak{D}(g \circ f) = f^*(\mathfrak{D}(g)) = f^*(\mathbf{E}^3) = \mathbf{E}^2$, we see that $g \circ f: \mathbf{E}^2 \to \mathbf{E}$.

$f$ is differentiable on $\mathbf{E}^2$ and $g$ is differentiable on $\mathbf{E}^3$, since all partial derivatives exist and are continuous.

$f'(a)$, $a = (\alpha_1, \alpha_2)$, is represented by its Jacobian matrix

$$F'(a) = \begin{pmatrix} 2\alpha_1 & 0 \\ 0 & 1 \\ 1 & 1 \end{pmatrix},$$

and $g'(b)$, $b = (\beta_1, \beta_2)$, is represented by its Jacobian matrix

$$G'(b) = (1, 2\beta_2, 1).$$

From the chain rule in Theorem 92.1, we obtain the following Jacobian matrix representing $(g \circ f)'(a)$:

$$(G \circ F)'(a) = G'(f(a))F'(a) = (1, 2\alpha_2, 1)\begin{pmatrix} 2\alpha_1 & 0 \\ 0 & 1 \\ 1 & 1 \end{pmatrix}$$

$$= (2\alpha_1 + 1, 2\alpha_2 + 1).$$

Since $g \circ f$ is given by

$$(g \circ f)(x) = \xi_1^2 + \xi_2^2 + \xi_1 + \xi_2,$$

we may also obtain this result directly:

$$(G \circ F)'(a) = (2\alpha_1 + 1, 2\alpha_2 + 1).$$

---

Under certain circumstances, it is easier to compute the derivative of a composite function than it is to compute the derivative of one of the functions it is composed of, as, for example, when the composite function is the identity function and the functions it is composed of are inverse to each other and only one of them is known explicitly. (See also Theorem 36.2.) When and how the derivative of the one function can be recovered from the derivatives of the other function and the composite function, is explained in the following theorem:

***Theorem 92.2***

Let $\mathfrak{D}(f) \subseteq \mathbf{E}^n$, $\mathfrak{D}(g) \subseteq \mathbf{E}^m$. If $f: \mathfrak{D}(f) \to \mathbf{E}^m$, $g: \mathfrak{D}(g) \to \mathbf{E}^m$, if $a$ is an interior point of $\mathfrak{D}(f)$ and $f(a) = b$ an interior point of $\mathfrak{D}(g)$, if $g'(b)$, $(g \circ f)'(a)$ exist, and if $g'(b): \mathbf{E}^m \leftrightarrow \mathbf{E}^m$, then $[g'(b)]^{-1}$ and $f'(a)$ exist, and

$$f'(a) = [g'(f(a))]^{-1} \circ (g \circ f)'(a). \tag{92.11}$$

*Proof*

Since $g'(b)$ is one-to-one from $\mathbf{E}^m$ onto $\mathbf{E}^m$, $[g'(b)]^{-1}$ exists on $\mathbf{E}^m$ and is a linear function. (See Theorem 84.2.) $f'(a)$, as defined in (92.11), being the composition of linear functions, is again a linear function.

It remains to be shown that, for every $\varepsilon > 0$, there is a $\delta > 0$ such that $N_\delta(a) \subseteq \mathfrak{D}(f)$ and

$$\|f(x) - f(a) - [g'(f(a))]^{-1} \circ (g \circ f)'(a)(x - a)\| \leqslant \varepsilon \|x - a\| \tag{92.12}$$

for all $x \in N_\delta(a)$.

Since $g'(f(a))$ is one-to-one, there exists a $\mu > 0$ (Theorem 85.2) such that

(92.13) $$\|g'(f(a))(u)\| \geqslant \mu \|u\| \qquad \text{for all } u \in \mathbf{E}^m.$$

Since $[g'(f(a)]^{-1} \circ (g \circ f)'(a)$ is a linear function, there is an $M > 0$ (Lemma 85.1) such that

(92.14) $$\|[g'(f(a))]^{-1} \circ (g \circ f)'(a)(v)\| \leqslant M \|v\| \qquad \text{for all } v \in \mathbf{E}^n.$$

Since $g'(f(a)), (g \circ f)'(a)$ exist and since $f$ is continuous at $a$, we have, for every $\varepsilon_1 > 0$, which we choose such that

(92.15) $$1 - \frac{\varepsilon_1}{\mu} > 0, \qquad \frac{\varepsilon_1}{\mu}\left[\frac{M + \varepsilon_1/\mu}{1 - \varepsilon_1/\mu} + 1\right] < \varepsilon,$$

a $\delta > 0$ such that

(92.16) $$\begin{aligned} \mu \|f(x) - f(a) - [g'(f(a))]^{-1} \circ (g \circ f)'(a)(x - a)\| \\ \leqslant \|g'(f(a))(f(x) - f(a)) - (g \circ f)'(a)(x - a)\| \\ \leqslant \|g(f(x)) - g(f(a)) - g'(f(a))(f(x) - f(a))\| \\ + \|g(f(x)) - g(f(a)) - (g \circ f)'(a)(x - a)\| \\ \leqslant \varepsilon_1(\|f(x) - f(a)\| + \|x - a\|) \end{aligned}$$

for all $x \in N_\delta(a)$.

Considering the first and last member of this inequality only, we obtain, in view of (92.14), that

$$\mu \|f(x) - f(a)\| \leqslant M\mu \|x - a\| + \varepsilon_1(\|f(x) - f(a)\| + \|x - a\|),$$

and hence

$$\|f(x) - f(a)\| \leqslant \frac{M + \varepsilon_1/\mu}{1 - \varepsilon_1/\mu} \|x - a\|.$$

Therefore, we have, from (92.16) and in view of (92.15), that (92.12) holds. Hence, $f'(a)$ exists and is given by (92.11).

The following theorem on the derivative of the inverse function, which will play an important role in the proof of the inverse-function theorem in Chapter 10, is a simple consequence of Theorem 92.2.

***Theorem 92.3***

Let $\mathfrak{D} \subseteq \mathbf{E}^n$ denote an open set and let $f: \mathfrak{D} \xrightarrow[1-1]{} \mathbf{E}^n$ map all open subsets of $\mathfrak{D}$ onto open subsets of $\mathbf{E}^n$. If $a \in \mathfrak{D}$ and if $f'(a) : \mathbf{E}^n \leftrightarrow \mathbf{E}^n$, then $(f^{-1})'(f(a))$ exists, and is given by

(92.17) $$(f^{-1})'(f(a)) = [f'(a)]^{-1},$$

or, equivalently, by

$$(f^{-1})'(b) = [f'(f^{-1}(b))]^{-1} \qquad \text{where } b = f(a).$$

*Proof*

Since $f$ maps all open sets of $\mathfrak{D}$ (including $\mathfrak{D}$ itself) onto open subsets of $\mathbf{E}^n$, $f$ is an open map (exercise 7, section 6.69.) Since $f$ is also one-to-one, we have, from Theorem 69.1, that $f^{-1}: \mathfrak{R}(f) \to \mathbf{E}^n$ exists and is continuous. Since $a$ is an interior point of $\mathfrak{D}$, $b = f(a)$ is an interior point of $\mathfrak{R}(f)$. Since $f'(a): \mathbf{E}^n \leftrightarrow \mathbf{E}^n$, $[f'(a)]^{-1}$ exists on $\mathbf{E}^n$ and is a linear function. Since $f \circ f^{-1}$ is the identity function on $\mathfrak{R}(f)$, $(f \circ f^{-1})'(b)$ is the identity function on $\mathbf{E}^n$ (exercise 5, section 9.86).

If we substitute, in Theorem 92.2, $f^{-1}$ for $f$, $f$ for $g$, $b$ for $a$, $a$ for $b$, and let $n = m$, we see that all hypotheses of Theorem 92.2 are met and (92.17) follows readily from (92.11).

Meaningful and nontrivial applications of this theorem for $n > 1$ cannot be discussed at this time because we have not yet developed any tests that indicate when a function is one-to-one and maps open sets onto open sets. (See example 2, section 10.97.)

For $n = 1$ we obtain the following, somewhat more general, version of Theorem 36.2:

***Corollary to Theorem 92.3***

Let $\mathfrak{J} \subseteq \mathbf{E}$ denote an open interval and let $f: \mathfrak{J} \xrightarrow[1-1]{} \mathbf{E}$ map all open subsets of $\mathfrak{J}$ onto open subsets of $\mathbf{E}$. If $a \in \mathfrak{J}$ and if $f'(a) \neq 0$, then $(f^{-1})'(f(a)) = 1/f'(a)$. (See also exercise 6.)

## 9.92 Exercises

1. Given $\mu > 0$, $M > 0$, $\varepsilon > 0$. Show that one may choose $\varepsilon_1 > 0$ such that

$$1 - \frac{\varepsilon_1}{\mu} > 0 \qquad \text{and} \qquad \frac{\varepsilon_1}{\mu}\left[\frac{M + \varepsilon_1/\mu}{1 - \varepsilon_1/\mu} + 1\right] < \varepsilon.$$

2. Given $f: \mathbf{E}^2 \to \mathbf{E}^3$ by $\phi_1(\xi_1, \xi_2) = \xi_1 - \xi_2$, $\phi_2(\xi_1, \xi_2) = \xi_2$, $\phi_3(\xi_1, \xi_2) = \xi_1^2 + \xi_2^2$ and $g: \mathbf{E}^3 \to \mathbf{E}^3$ by $\psi_1(\eta_1, \eta_2, \eta_3) = \eta_2$, $\psi_2(\eta_1, \eta_2, \eta_3) = \eta_1 - \eta_3$, $\psi_3(\eta_1, \eta_2, \eta_3) = \eta_2^2 + \eta_3$. Use Theorem 92.1 to find Jacobian matrix of $(g \circ f)'(\theta)$.
3. $f = \{(x, y) \in \mathbf{E}^2 \mid y = \cos x, x \in [0, \pi]\}$ is one-to-one, and hence $f^{-1}$ exists on $[-1, 1]$. We denote $f^{-1}(x) = \text{Cos}^{-1} x$, $x \in [-1, 1]$. Use the Corollary to Theorem 92.3 to find $(\text{Cos}^{-1})'(b)$ for $b \in (-1, 1)$.
4. Show: If one replaces, in Lemma 92.1, "interior point" by "accumulation point" wherever it occurs, then the lemma is false.
5. Given the function $f: \mathbf{E}^2 \to \mathbf{E}^2$ by

$$\phi_1(\xi_1, \xi_2) = \xi_1^2 + 2\xi_2,$$
$$\phi_2(\xi_1, \xi_2) = \xi_1 - \xi_1\xi_2.$$

We shall see, in section 10.96, that $f$ is one-to-one in some $\delta$-neighborhood of $\theta$ and, in section 10.97, that it maps all open subsets of that neighborhood onto open subsets of $\mathbf{E}^2$. Taking this for granted at the moment, find $(f^{-1})'(0,0)$. (See also exercise 5, section 10.96 and exercise 6, section 10.97.)

6. Show that the Corollary to Theorem 92.3 implies Theorem 36.2.

## 9.93 Mean-Value Theorems

If $f$ is a vector-valued function, a direct generalization of the mean-value theorem of the differential calculus, Theorem 37.3, is not possible, as the following example shows:

---

▲ *Example 1*

Let $f: \mathbf{E} \to \mathbf{E}^2$ be given by $\phi_1(x) = x - x^2$, $\phi_2(x) = x - x^3$. Then, $f'(a)$ is represented by the Jacobian matrix

$$F'(a) = \begin{pmatrix} 1 - 2a \\ 1 - 3a^2 \end{pmatrix},$$

and

$$f(1) - f(0) = \begin{pmatrix} 0 \\ 0 \end{pmatrix}.$$

If a theorem of the type of Theorem 37.3 were to hold, then there would exist a point $\xi \in (0,1)$ such that

$$f(1) - f(0) = \begin{pmatrix} 0 \\ 0 \end{pmatrix} = f'(\xi)(1 - 0) = \begin{pmatrix} 1 - 2\xi \\ 1 - 3\xi^2 \end{pmatrix},$$

Clearly, there is *no* $\xi \in (0, 1)$ such that $1 - 2\xi = 0$ and $1 - 3\xi^2 = 0$.

---

If $f$ is a *real-valued* function of a vector variable, then Theorem 37.3 may be generalized as follows.

***Theorem 93.1***

Let $f: \mathfrak{D} \to \mathbf{E}$, $\mathfrak{D} \subseteq \mathbf{E}^n$, and let $a, b \in \mathfrak{D}$ denote interior points. If $(1 - t)a + tb$ are interior points of $\mathfrak{D}$ for all $t \in (0, 1)$, and if $f$ is differentiable at all points $(1 - t)a + tb$, $t \in (0, 1)$, then there exists a point $c = (1 - t_0)a + t_0 b$ for some $t_0 \in (0, 1)$ such that

$$f(b) - f(a) = f'(c)(b - a). \tag{93.1}$$

(Note that $(1 - t)a + tb$, $t \in [0, 1]$ represents the line segment that joins $a$ to $b$.)

*Proof*

Consider the function $\phi : [0, 1] \to \mathbf{E}$, defined by $\phi(t) = f((1-t)a + tb)$. Then

$$\phi(0) = f(a), \qquad \phi(1) = f(b). \tag{93.2}$$

The function $g : [0, 1] \to \mathbf{E}^n$, defined by $g(t) = (1-t)a + tb$ is differentiable for all $t \in (0, 1)$ and $g'(t) = -a + b$. (See example 1, section 9.86.) By hypothesis, $f$ is differentiable at all points $(1-t)a + tb$, $t \in (0, 1)$. Hence, by the chain rule (Theorem 92.1), the derivative of the composite function $\phi = f \circ g$ exists and is given by

$$\phi'(t) = f'((1-t)a + tb)(b - a). \tag{93.3}$$

By Theorem 37.3, $\phi(1) - \phi(0) = \phi'(t_0)$ for some $t_0 \in (0, 1)$, and (93.1) follows readily from (93.2) and (93.3).

***Corollary 1 to Theorem 93.1***

Let $f : \mathfrak{D} \to \mathbf{E}^m$, $\mathfrak{D} \subseteq \mathbf{E}^n$, and let $a, b \in \mathfrak{D}$ denote interior points of $\mathfrak{D}$. If $(1-t)a + tb$ are interior points of $\mathfrak{D}$ for all $t \in (0, 1)$, and if $f$ is differentiable at all points $(1-t)a + tb$, $t \in (0, 1)$, then, for all $u \in \mathbf{E}^m$, there is a $c_u = (1-t_u)a + t_u b$ for some $t_u \in (0, 1)$, depending on $u$, such that

$$(f(b) - f(a)) \cdot u = f'(c_u)(b - a) \cdot u.$$

*Proof*

Let $g : \mathfrak{D} \to \mathbf{E}$ be defined by $g(x) = f(x) \cdot u$ and apply Theorem 93.1 to the real-valued function $g$. (That $t_u$ and hence, $c_u$, depend on $u$ is quite clear because, for different vectors $u$, one obtains different functions $g$.)

***Corollary 2 to Theorem 93.1***

Under the condition of Corollary 1,

$$f(b) - f(a) = l(b - a), \tag{93.4}$$

where $l : \mathbf{E}^n \to \mathbf{E}^m$ is a linear function which is represented by the matrix

$$L = \begin{pmatrix} D_1\phi_1(c_1) & D_2\phi_1(c_1) & \cdots & D_n\phi_1(c_1) \\ D_1\phi_2(c_2) & D_2\phi_2(c_2) & \cdots & D_n\phi_2(c_2) \\ \vdots & & & \\ D_1\phi_m(c_m) & D_2\phi_m(c_m) & \cdots & D_n\phi_m(c_m) \end{pmatrix}, \tag{93.5}$$

where $c_j = (1-t_j)a + t_j b$, $j = 1, 2, \ldots, m$, for some $t_j \in (0, 1)$.

*Proof*

By Corollary 1, there is a $c_j = (1-t_j)a + t_j b$, with $t_j \in (0, 1)$ such that

$$(f(b) - f(a)) \cdot i_j = f'(c_j)(b - a) \cdot i_j,$$

where $i_1, i_2, \ldots, i_m$ are the fundamental vectors in $\mathbf{E}^m$. Since, for every $v \in \mathbf{E}^m$, $v \cdot i_j$ is the $j$th component of $v$, the statement of the corollary follows readily.

***Corollary 3 to Theorem 93.1***

Let $f : \mathfrak{D} \to \mathbf{E}^m$, $\mathfrak{D} \subseteq \mathbf{E}^n$. If $\mathfrak{D}$ is open and if, whenever $a, b \in \mathfrak{D}$, then $(1 - t)a + tb \in \mathfrak{D}$ for all $t \in (0, 1)$ ($\mathfrak{D}$ is *convex*), and if $f'(x)$ exists and is the zero function for all $x \in \mathfrak{D}$, that is, $f'(x)(z) = \theta$ for all $z \in \mathbf{E}^n$ and each $x \in \mathfrak{D}$, then $f(x) = c$ for all $x \in \mathfrak{D}$ where $c \in \mathbf{E}^m$ is a constant vector.

*Proof*

Clearly, all hypotheses of Corollary 2 are met. Since $f'(x)$ is the zero function for all $x \in \mathfrak{D}$, we have, from Theorem 87.1, that, for each $x \in \mathfrak{D}$ and all $u \in \mathbf{E}^n$,

$$D_u f(x) = f'(x)(u) = \theta,$$

and hence, $D_i \phi_j(x) = D_{e_i} \phi_j(x) = 0$ for all $x \in \mathfrak{D}$. Therefore, the matrix $L$ in (93.5) is the zero matrix, and we obtain, from (93.4), that $f(a) = f(b)$ for any $a, b \in \mathfrak{D}$. That $f(x) = c$ for all $x \in \mathfrak{D}$ follows readily. (For a generalization of this corollary, see exercise 1.)

## 9.93 Exercises

1. *Prove*: If $f : \mathfrak{D} \to \mathbf{E}^m$, $\mathfrak{D} \subseteq \mathbf{E}^n$, if $\mathfrak{D}$ is open and connected, and if $f'(x)$ exists and is the zero function for all $x \in \mathfrak{D}$, then $f(x) = c$ for all $x \in \mathfrak{D}$, where $c \in \mathbf{E}^m$ is a constant vector.
2. Give an example to show that the conclusion of exercise 1 need not be valid if $\mathfrak{D}$ is not connected.
3. Show: Under the hypotheses of Corollary 1, and if $\|f'(c)(u)\| \leqslant M\|u\|$ for some $M > 0$, all $u \in \mathbf{E}^n$, and all $c = (1 - t)a + tb$, $t \in [0, 1]$, then

$$\|f(b) - f(a)\| \leqslant M\|b - a\|.$$

4. Show: If $\mathfrak{D} \subseteq \mathbf{E}^n$ is connected and if $\|f'(x)(u)\| \leqslant M\|u\|$ for all $x \in \mathfrak{D}$ and all $u \in \mathbf{E}^n$, then, for any two points $a, b \in \mathfrak{D}$, there is a constant $K > 0$ such that $\|f(a) - f(b)\| \leqslant K\|b - a\|$. (See also exercise 3.)

## 9.94 Partial Derivatives of Higher Order

Let $f : \mathfrak{D} \to \mathbf{E}^m$, $\mathfrak{D} \subseteq \mathbf{E}^n$ and let $a \in \mathfrak{D}$ represent an interior point. Suppose that $D_1 f$ exists for all $x \in N_\delta(a)$ for some $\delta > 0$. The function $g : N_\delta(a) \to \mathbf{E}^m$, which is defined by

$$g(x) = D_1 f(x),$$

may or may not possess partial derivatives at $a$. If $D_2 g(a)$ exists, we call

$$D_2 g(a) = D_2 D_1 f(a)$$

the second partial derivative of $f$ at $a$ with respect to the first and the second variable. (The reader is probably more familiar with the notation $\partial^2 f/(\partial\xi_2\,\partial\xi_1)$.)

***Definition 94.1***

If $f\colon \mathfrak{D} \to \mathbf{E}^m$, $\mathfrak{D} \subseteq \mathbf{E}^n$, if $a \in \mathfrak{D}$ is an interior point, if $D_j f$ exists for all $x \in N_\delta(a)$ for some $\delta > 0$, and if $D_k g_j(a)$ exists where $g_j\colon N_\delta(a) \to \mathbf{E}^m$ is defined by $g_j(x) = D_j f(x)$, then

$$D_k\, D_j\, f(a) = D_k\, g_j(a)$$

is called the *second partial derivative* of $f$ at $a$ with respect to the $j$th variable and the $k$th variable.

If $D_{j_l} D_{j_{l-1}} \cdots D_{j_1} f$ exists for all $x \in N_\delta(a)$ for some $\delta > 0$ and if $D_{j_{l+1}} g_{j_1 j_2 \cdots j_l}(a)$ exists, where $g_{j_1 j_2 \cdots j_l}\colon N_\delta(a) \to \mathbf{E}^m$ is defined by $g_{j_1 j_2 \cdots j_l}(x) = D_{j_l} D_{j_{l-1}} \cdots D_{j_1} f(x)$, then

$$D_{j_{l+1}} D_{j_l} \cdots D_{j_1} f(a) = D_{j_{l+1}} g_{j_1 j_2 \cdots j_l}(a)$$

is called the $(l + 1)$st partial derivative of $f$ at $a$ with respect to the $j_1$th, $j_2$th, ..., $j_{l+1}$th variable, where $j_1, j_2, \ldots, j_{l+1} \in \{1, 2, \ldots, n\}$. (Note that some or all of the $j_k$ may be equal.)

---

▲ *Example 1*

Let $f\colon \mathbf{E}^2 \to \mathbf{E}^2$ be defined by

$$\phi_1(\xi_1, \xi_2) = \xi_1^2 - \xi_1\xi_2,$$
$$\phi_2(\xi_1, \xi_2) = \xi_1 \cos \xi_2.$$

Since

$$D_1 f(\xi_1, \xi_2) = \begin{pmatrix} 2\xi_1 - \xi_2 \\ \cos \xi_2 \end{pmatrix}$$

exists for all $x \in \mathbf{E}^2$, it makes sense to pose the question as to the existence of a second partial derivative. We have

$$D_1 D_1 f(x) = \begin{pmatrix} 2 \\ 0 \end{pmatrix}, \qquad D_2 D_1 f(x) = \begin{pmatrix} -1 \\ -\sin \xi_2 \end{pmatrix}.$$

Since

$$D_2 f(x) = \begin{pmatrix} -\xi_1 \\ -\xi_1 \sin \xi_2 \end{pmatrix},$$

we obtain also

$$D_1 D_2 f(x) = \begin{pmatrix} -1 \\ -\sin \xi_2 \end{pmatrix}, \qquad D_2 D_2 f(x) = \begin{pmatrix} 0 \\ -\xi_1 \cos \xi_2 \end{pmatrix}.$$

We observe that $D_1 D_2 f(x) = D_2 D_1 f(x)$.

▲ *Example 2*

Let $f: \mathbf{E}^2 \to \mathbf{E}$ be given by

$$f(x) = \xi_1 \xi_2 + \sqrt{1 + \xi_1^2} \sin \frac{\xi_1}{\sqrt{1 + \xi_1^2}}.$$

We obtain, after cumbersome manipulations, that

$$D_1 f(x) = \xi_2 + \frac{\xi_1}{\sqrt{1 + \xi_1^2}} \sin \frac{\xi_1}{\sqrt{1 + \xi_1^2}} + \frac{1}{1 + \xi_1^2} \cos \frac{\xi_1}{1 + \xi_1^2}$$

and

$$D_2 D_1 f(x) = 1.$$

Since $D_2 f(x) = \xi_1$, we obtain immediately that $D_1 D_2 f(x) = 1$. If we had some assurance that $D_1 D_2 f(x) = D_2 D_1 f(x)$, we could have found $D_2 D_1 f(x)$ with considerably less effort. That this is not always possible will be seen from the next example.

▲ *Example 3*

Let $f: \mathbf{E}^2 \to \mathbf{E}$ be defined by

$$f(x) = \begin{cases} \dfrac{\xi_1 \xi_2 (\xi_1^2 - \xi_2^2)}{\xi_1^2 + \xi_2^2} & \text{if } (\xi_1, \xi_2) \neq (0, 0), \\ 0 & \text{if } (\xi_1, \xi_2) = (0, 0). \end{cases}$$

We have

$$D_1 f(\theta) = \lim_0 \frac{0}{t^2} = 0,$$

and for $x \neq \theta$,

$$D_1 f(x) = \xi_2 \frac{(\xi_1^2 - \xi_2^2)}{\xi_1^2 + \xi_2^2} + 4 \frac{\xi_1^2 \xi_2^3}{(\xi_1^2 + \xi_2^2)^2}.$$

Hence,

$$D_2 D_1 f(\theta) = \lim_0 \frac{D_1 f(0, t) - D_1 f(0, 0)}{t} = \lim_0 \left( -\frac{t^3}{t^3} \right) = -1.$$

Since

$$D_2 f(\theta) = \lim_0 \frac{0}{t^2} = 0$$

and for $x \neq \theta$,

$$D_2 f(x) = \xi_1 \frac{(\xi_1^2 - \xi_2^2)}{\xi_1^2 + \xi_2^2} - 4 \frac{\xi_1^3 \xi_2^2}{(\xi_1^2 + \xi_2^2)^2},$$

we obtain

$$D_1 D_2 f(\theta) = \lim_0 \frac{D_2 f(t,0) - D_2 f(0,0)}{t} = \lim_0 \frac{t^3}{t^3} = 1,$$

and see that

$$D_2 D_1 f(\theta) \neq D_1 D_2 f(\theta).$$

What is responsible for this blatant misbehavior of the second partial derivatives is that neither $D_1 D_2 f$ nor $D_2 D_1 f$ are continuous at $\theta$. We have

$$D_2 D_1 f(x) = \frac{\xi_1^2 - \xi_2^2}{\xi_1^2 + \xi_2^2} - 4\frac{\xi_1^2 \xi_2^2}{(\xi_1^2 + \xi_2^2)^2} + 4\frac{\xi_1^2 \xi_2^2 (3\xi_1^4 + 2\xi_1^2 \xi_2^2 - 4\xi_2^4)}{(\xi_1^2 + \xi_2^2)^4}$$

and $\lim_0 D_2 D_1 f(t,0) = 1$. Hence, if $\lim_\theta D_2 D_1 f(x)$ exists at all, it is certainly different from $D_2 D_1 f(\theta) = -1$; that is, $D_2 D_1 f$ is *not* continuous at $\theta$. Similarly, one can see that $D_1 D_2 f$ is not continuous at $\theta$. (See exercise 1.)

---

In the next theorem we shall establish the continuity of $D_1 D_2 f$ (or, equivalently, of $D_2 D_1 f$) as a sufficient condition for the interchangeability of the order of partial differentiations.

***Theorem 94.1***

Let $f: \mathfrak{D} \to \mathbf{E}^m$, $\mathfrak{D} \subseteq \mathbf{E}^n$, and let $a$ denote an interior point of $\mathfrak{D}$. If for some $\delta > 0$, $D_i f(x)$, $D_k f(x)$, $D_i D_k f(x)$ exist in $N_\delta(a)$ and if $D_i D_k f$ is continuous at $a$, then $D_k D_i f(a)$ exists and

$$D_k D_i f(a) = D_i D_k f(a).$$

*Proof*

We observe that the hypotheses as well as the conclusion hold if and only if they hold for each component of $f$. Hence, we may assume without loss of generality that $f$ is a real-valued function. Since only partial derivatives with respect to two distinct variables occur in this theorem, we may also assume without loss of generality that $\mathfrak{D} \subseteq \mathbf{E}^2$. Then, we may reformulate our hypotheses to read that $D_1 f(x)$, $D_2 f(x)$, $D_1 D_2 f(x)$ exist in $N_\delta(a)$ and that $D_1 D_2 f$ is continuous at $a$. We want to prove that $D_2 D_1 f(a)$ exists and that $D_2 D_1 f(a) = D_1 D_2 f(a)$.

With the abbreviation

$$\Delta f(x) = f(x + te_1) - f(x), \tag{94.1}$$

we may write

$$D_1 f(x) = \lim_{t \to 0} \frac{\Delta f(x)}{t}.$$

We want to show that

$$D_2 D_1 f(a) = \lim_{\tau \to 0} \frac{D_1 f(a + \tau e_2) - D_1 f(a)}{\tau}$$

$$= \lim_{\tau \to 0} \frac{1}{\tau} \left[ \lim_{t \to 0} \frac{1}{t} (\Delta f(a + \tau e_2) - \Delta f(a)) \right]$$

exists and is equal to $D_1 D_2 f(a)$.

If we choose $|t|$, $|\tau|$ sufficiently small such that $a + te_1$, $a + \tau e_2$, $a + te_1 + \tau e_2 \in N_\delta(a)$ (see Fig. 94.1), then $D_2(\Delta f(x))$ exists for all $x$ on the line segment

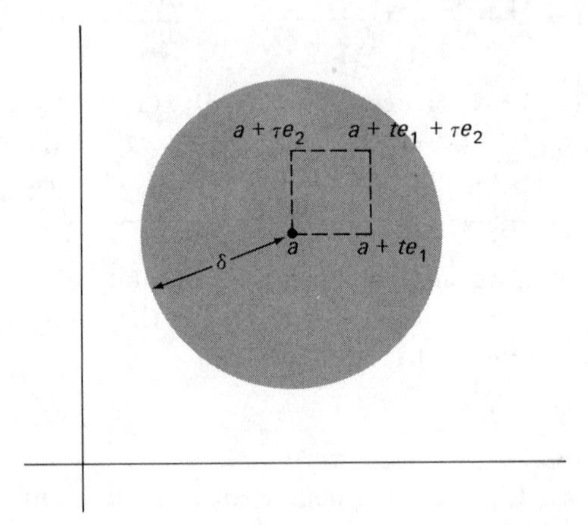


*Fig. 94.1*

joining $a$ to $(a + \tau e_2)$, endpoints included, and we obtain, from the mean-value theorem of the differential calculus (Theorem 37.3), that

$$\Delta f(a + \tau e_2) - \Delta f(a) = D_2 \Delta f(a + \theta(\tau)\tau e_2)\tau \qquad \text{for some } \theta(\tau) \in [0, 1],$$

and we have

$$D_2 D_1 f(a) = \lim_{\tau \to 0} \lim_{t \to 0} \frac{1}{t} D_2 \Delta f(a + \theta(\tau)\tau e_2). \tag{94.2}$$

By (94.1),

$$D_2 \Delta f(a + \theta(\tau)\tau e_2) = D_2 f(a + \theta(\tau)\tau e_2 + te_1) - D_2 f(a + \theta(\tau)\tau e_2).$$

Since $a + \theta(\tau)\tau e_2 \in N_\delta(a)$ and since $D_1 D_2 f(x)$ exists in $N_\delta(a)$, we have

$$\begin{aligned} \lim_{t \to 0} \frac{1}{t} D_2 \Delta f(a + \theta(\tau)\tau e_2) &= \lim_{t \to 0} \frac{1}{t} [D_2 f(a + \theta(\tau)\tau e_2 + te_1) \\ &\qquad - D_2 f(a + \theta(\tau)\tau e_2)] \\ &= D_1 D_2 f(a + \theta(\tau)\tau e_2). \end{aligned} \tag{94.3}$$

Since $D_1 D_2 f$ is continuous at $a$, we have

$$\lim_{\tau \to 0} D_1 D_2 f(a + \theta(\tau)\tau e_2) = D_1 D_2 f(a). \tag{94.4}$$

(Observe that $\|\theta(\tau)\tau e_2\| = |\theta(\tau)\tau| \leqslant |\tau|$.) From (94.2), (94.3), and (94.4), we obtain the desired result that

$$D_2 D_1 f(a) = D_1 D_2 f(a).$$

---

▲ *Example 4*

Suppose we wish to interchange the order of differentiation with respect to $\xi_4$ and $\xi_5$ in $D_5 D_4 D_1 D_1 D_2 D_3 D_1 f(a)$. In order to apply Theorem 94.1, we have to check that the function $g$, defined by $g(x) = D_1 D_1 D_2 D_3 D_1 f(x)$, exists in some $N_\delta(a)$, that $D_4 g(x)$, $D_5 g(x)$, and $D_5 D_4 g(x)$ exist in $N_\delta(a)$, and that $D_5 D_4 g$ is continuous at $a$. If all this is true, then

$$D_4 D_5 D_1 D_1 D_2 D_3 D_1 f(a) = D_5 D_4 D_1 D_1 D_2 D_3 D_1 f(a).$$

---

## 9.94 Exercises

1. Given the function $f$ in example 3. Show that $D_1 D_2 f$ is not continuous at $\theta$.
2. Given $f: \mathbf{E}^2 \to \mathbf{E}$ by

$$f(x) = \begin{cases} \xi_1^2 \operatorname{Tan}^{-1} \dfrac{\xi_2}{\xi_1} - \xi_2^2 \operatorname{Tan}^{-1} \dfrac{\xi_1}{\xi_2} & \text{for } \xi_1 \xi_2 \neq 0, \\ 0 & \text{for } \xi_1 \xi_2 = 0. \end{cases}$$

   Show that $D_1 D_2 f(\theta) \neq D_2 D_1 f(\theta)$. ($\operatorname{Tan}^{-1}$ denotes the principal value of the inverse-tangent function.)
3. Let $u = (\omega_1, \omega_2, \ldots, \omega_n) \in \mathbf{E}^n$ and define

$$D^1 f(a)(u) = \sum_{i=1}^{n} D_i f(a) \omega_i$$

$$D^2 f(a)(u)^2 = \sum_{k=1}^{n} \sum_{i=1}^{n} D_i D_k f(a) \omega_i \omega_k$$

$$\vdots$$

$$D^k f(a)(u)^k = \sum_{j_k=1}^{n} \sum_{j_{k-1}=1}^{n} \cdots \sum_{j_1=1}^{n} D_{j_1} D_{j_2} \cdots D_{j_k} f(a) \omega_{j_1} \omega_{j_2} \cdots \omega_{j_k}.$$

   Assume that $f: \mathfrak{D} \to \mathbf{E}$, $\mathfrak{D} \subseteq \mathbf{E}^n$, that $a, b \in \mathfrak{D}$ are interior points of $\mathfrak{D}$, and that all partial derivatives of $f$ up to and including those of $p$th order are continuous in an open set $\Omega \subseteq \mathfrak{D}$ which contains all points $(1 - t)a + tb$,

$t \in [0, 1]$. Show that there is a point $c = (1 - t_0)a + t_0 b$ for some $t_0 \in [0, 1]$ such that

$$f(a) = f(b) + D^1 f(b)(a - b) + \frac{1}{2!} D^2 f(b)(a - b)^2 + \cdots$$
$$+ \frac{1}{(p-1)!} D^{p-1} f(b)(a - b)^{p-1} + \frac{1}{p!} D^p f(c)(a - b)^p.$$

This is *Taylor's formula* for real-valued functions of a vector variable.

4. Let $\mathfrak{D} \subseteq \mathbf{E}^2$ in exercise 3 and write out Taylor's formula in detail for $p = 3$.

## 9.95 The Inverse-Function Theorem and the Implicit-Function Theorem*

We have seen, in section 8.84 (Theorem 84.3) that the system of linear equations

$$\begin{aligned} a_{11}\xi_1 + a_{12}\xi_2 + \cdots + a_{1n}\xi_n &= \eta_1, \\ a_{21}\xi_1 + a_{22}\xi_2 + \cdots + a_{2n}\xi_n &= \eta_2, \\ &\vdots \\ a_{m1}\xi_1 + a_{m2}\xi_2 + \cdots + a_{mn}\xi_n &= \eta_m, \end{aligned}$$

has a unique solution $x = (\xi_1, \xi_2, \ldots, \xi_n) \in \mathbf{E}^n$ for every $y = (\eta_1, \eta_2, \ldots, \eta_m) \in \mathbf{E}^m$ if and only if rank $A = n = m$, where

$$A = \begin{pmatrix} a_{11} a_{12} & \cdots & a_{1n} \\ a_{21} a_{22} & \cdots & a_{2n} \\ \vdots & & \\ a_{m1} a_{m2} & \cdots & a_{mn} \end{pmatrix}.$$

If we let $\lambda_j(x) = a_{j1}\xi_1 + a_{j2}\xi_2 + \cdots + a_{jn}\xi_n$ and $l = (\lambda_1, \lambda_2, \ldots, \lambda_m)$, then $A$ is the Jacobian matrix of $l: \mathbf{E}^n \to \mathbf{E}^m$,

$$L'(x) = A;$$

and we may rephrase the above statement as follows: $l(x) = y$ has a unique solution $x \in \mathbf{E}^n$ for every $y \in \mathbf{E}^m$ if and only if rank $L'(x) = n = m$.

We consider a function $f: \mathfrak{D} \to \mathbf{E}^m$, $\mathfrak{D} \subseteq \mathbf{E}^n$, which is not necessarily linear, and pose the more general problem: When and for what vectors $y \in \mathbf{E}^m$ does the system of equations

$$f(x) = y$$

have a unique solution $x \in \mathbf{E}^n$? An equivalent formulation of this problem is the following: When does $f$ possess a unique inverse $f^{-1}$, and what are the domain and range of $f^{-1}$? As in the case of real-valued functions of a real variable, it develops that the invertibility of the derivative of a well-behaved function $f$ at some point $a \in \mathfrak{D}$ guarantees the existence of a unique inverse in some neigh-

* This section need not be taken up unless Chapter 10 is omitted.

borhood of the point $f(a) \in \mathbf{E}^m$. This reduces the problem of establishing the invertibility of a function to the problem of establishing the invertibility of its derivative. Since the derivative of a function is a linear function, the latter problem had already been solved in Chapter 8.

In section 10.98 we shall establish a sufficient condition for the invertibility of a function in the so-called *inverse-function theorem.*

***Theorem 95.1*** (*Inverse-Function Theorem*)

If $f: \mathfrak{D} \to \mathbf{E}^n$, where $\mathfrak{D} \subseteq \mathbf{E}^n$ is an open set, if $f \in C^1(\mathfrak{D})$, and if, for some point $a \in \mathfrak{D}$, rank $f'(a) = n$, then there exist open sets $U, V \subseteq \mathbf{E}^n$ such that $a \in U$, $f(a) \in V$, and

$$f: U \leftrightarrow V,$$
$$f^{-1}: V \leftrightarrow U \text{ exists},$$
$$f^{-1} \in C^1(V).$$

(See Fig. 95.1.) Hence, if $f$ satisfies the conditions of this theorem, then the system of equations $f(x) = y$ has a unique solution $x \in U$ for all $y \in V$. (Note also that $f^{-1}$ inherits its differentiability properties from $f$.)

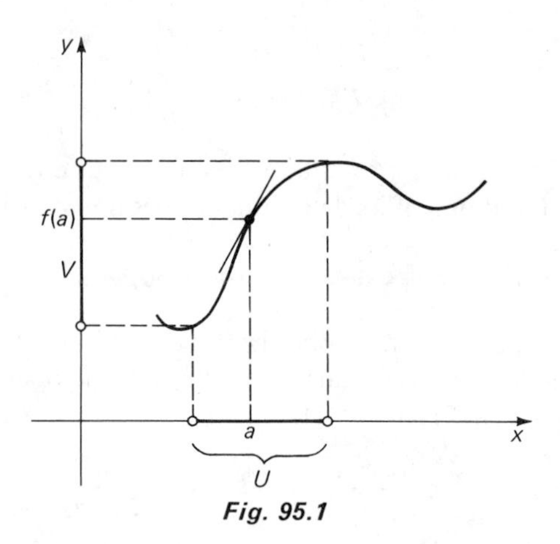

***Fig. 95.1***

For $n = 1$, the inverse-function theorem follows from Theorem 36.2 and the Corollary to Lemma 37.1. For $n > 1$, the proof will be presented in section 10.98. Here, we merely sketch an argument for $n = 2$ (which may be generalized to any $n$ without difficulty) to show that $f$ is one-to-one on some $\delta$-neighborhood of $a$. If

$$G(x_1, x_2) = \begin{vmatrix} D_1\phi_1(x_1) & D_2\phi_1(x_1) \\ D_1\phi_2(x_2) & D_2\phi_2(x_2) \end{vmatrix},$$

then

$$G(a, a) = \det F'(a).$$

Since rank $f'(a) = 2$, we have det $F'(a) \neq 0$. Hence, there is a $\delta > 0$ such that $G(x_1, x_2) \neq 0$ for all $x_1, x_2 \in N_\delta(a)$ (why?). For any $x_1, x_2 \in N_\delta(a)$, we have, from Corollary 2 to Theorem 93.1, that

$$f(x_1) - f(x_2) = \begin{pmatrix} D_1\phi_1(c_1) & D_2\phi_1(c_1) \\ D_1\phi_2(c_2) & D_2\phi_2(c_2) \end{pmatrix}(x_1 - x_2)$$

for some $c_1, c_2$ on the line segment joining $x_1$ and $x_2$. Since $G(c_1, c_2) \neq 0$, the above matrix represents a one-to-one linear function, and we obtain, from $f(x_1) = f(x_2)$, that $x_1 = x_2$. This is true for any $x_1, x_2 \in N_\delta(a)$. Therefore, $f$ is one-to-one on $N_\delta(a)$.

Clearly, the sufficiency part of Theorem 84.3 is a special case of the inverse function theorem with $U = V = \mathbf{E}^n$.

---

▲ *Example 1*

To be solved is the system of equations

$$\begin{aligned} \xi_1^2 + \xi_2^2 &= \eta_1, \\ \xi_1^2 - \xi_2^2 &= \eta_2, \end{aligned} \tag{95.1}$$

for $(\xi_1, \xi_2)$. We have, with $f = (\xi_1^2 + \xi_2^2, \xi_1^2 - \xi_2^2)$,

$$\operatorname{rank} f'(x) = \operatorname{rank} F'(x) = \operatorname{rank}\begin{pmatrix} 2\xi_1 & 2\xi_2 \\ 2\xi_1 & -2\xi_2 \end{pmatrix} = 2$$

for all $(\xi_1, \xi_2) \in \mathbf{E}^2$ for which $\xi_1 \neq 0$ and $\xi_2 \neq 0$, because the rank of a square matrix is maximal if and only if its determinant does not vanish and

$$\begin{vmatrix} 2\xi_1 & 2\xi_2 \\ 2\xi_1 & -2\xi_2 \end{vmatrix} = -8\xi_1\xi_2.$$

We choose a point $a = (\alpha_1, \alpha_2)$ in the first quadrant $U_1 = \{x \in \mathbf{E}^2 \mid \xi_1 > 0, \xi_2 > 0\}$. Then rank $f'(a) = 2$ and $f^{-1}$ exists in some open set $V$ containing $f(a) = (\alpha_1^2 + \alpha_2^2, \alpha_1^2 - \alpha_2^2)$. From (95.1):

$$\xi_1^2 = \frac{\eta_1 + \eta_2}{2}, \qquad \xi_2^2 = \frac{\eta_1 - \eta_2}{2}.$$

Hence, $f^{-1}(y) = (\sqrt{(\eta_1 + \eta_2)/2}, \sqrt{(\eta_1 - \eta_2)/2})$ is uniquely determined in $V$. (The signs of the square roots have been chosen such that

$$f^{-1}(\alpha_1^2 + \alpha_2^2, \alpha_1^2 - \alpha_2^2) = (\alpha_1, \alpha_2) \in U_1.)$$

We see easily that $U = U_1$ and $V = \{y \in \mathbf{E}^2 \mid \eta_1 > \eta_2 > -\eta_1\}$. (See Fig. 95.2.)

We obtain, for any $(\alpha_1, \alpha_2)$ in the second quadrant, that $f^{-1}(y) = (-\sqrt{(\eta_1 + \eta_2)/2}, \sqrt{(\eta_1 - \eta_2)/2})$; in the third quadrant $f^{-1}(y) = (-\sqrt{(\eta_1 + \eta_2)/2}, -\sqrt{(\eta_1 - \eta_2)/2})$; and in the fourth quadrant, $f^{-1}(y) = (\sqrt{(\eta_1 + \eta_2)/2}, -\sqrt{(\eta_1 - \eta_2)/2})$. $V$ is in all instances as in Fig. 95.2.

Clearly, $f^{-1} \in C^1(V)$.

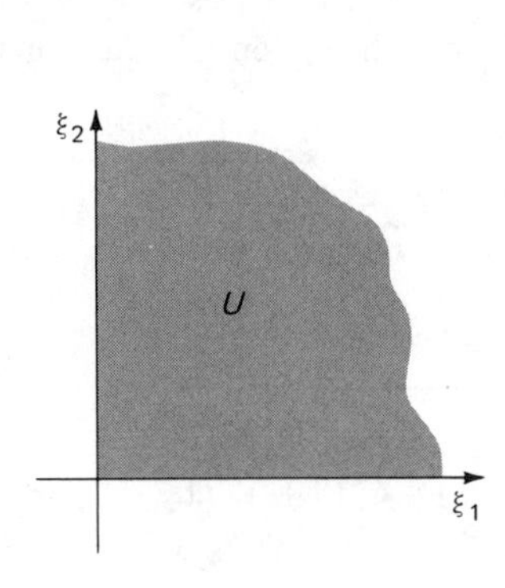


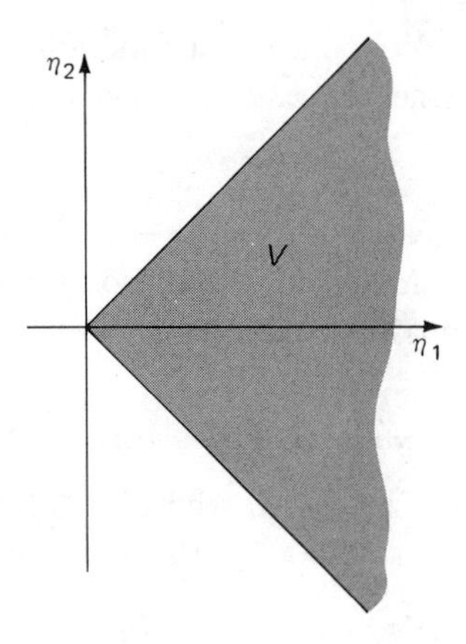


*Fig. 95.2*

---

We consider next a system of linear equations where the number $n + m$ of unknowns exceeds the number $n$ of equations. If the coefficient matrix has rank $n$, then one may choose certain $m$ of the unknowns freely and then solve for the remaining $n$ unknowns. The value of the remaining unknowns is uniquely determined by the choice of the $m$ unknowns. This may also be put as follows: One may solve for certain $n$ unknowns in terms of the remaining $m$ unknowns. To express this with more precision, we denote the unknowns for which we want to solve, by $x = (\xi_1, \xi_2, \ldots, \xi_n)$, and the remaining ones by $y = (\eta_1, \eta_2, \ldots, \eta_m)$. Then, the system of equations, which we assume w.l.o.g. to be homogeneous, looks as follows:

$$\begin{aligned}
&a_{11}\xi_1 + a_{12}\xi_2 + \cdots + a_{1n}\xi_n + b_{11}\eta_1 + b_{12}\eta_2 + \cdots + b_{1m}\eta_m = 0,\\
&a_{21}\xi_1 + a_{22}\xi_2 + \cdots + a_{2n}\xi_n + b_{21}\eta_1 + b_{22}\eta_2 + \cdots + b_{2m}\eta_m = 0,\\
&\vdots\\
&a_{n1}\xi_1 + a_{n2}\xi_2 + \cdots + a_{nn}\xi_n + b_{m1}\eta_1 + b_{m2}\eta_2 + \cdots + b_{mm}\eta_m = 0.
\end{aligned}$$

We have, from Theorem 84.3, that this system has a unique solution $x \in \mathbf{E}^n$ for every $y \in \mathbf{E}^m$ if and only if

$$\operatorname{rank}\begin{pmatrix} a_{11}a_{12} & \cdots & a_{1n} \\ a_{21}a_{22} & \cdots & a_{2n} \\ \vdots & & \\ a_{n1}a_{n2} & \cdots & a_{nn} \end{pmatrix} = n.$$

If we let $\lambda_j(x, y) = a_{j1}\xi_1 + a_{j2}\xi_2 + \cdots + a_{jn}\xi_n + b_{j1}\eta_1 + b_{j2}\eta_2 + \cdots + b_{jm}\eta_m$ and $l = (\lambda_1, \lambda_2, \ldots, \lambda_n)$, we may express this result as follows: $l(x, y) = \theta$ has a unique solution $x \in \mathbf{E}^n$ for every $y \in \mathbf{E}^m$ if and only if

$$\operatorname{rank}\begin{pmatrix} D_1\lambda_1(x) & D_2\lambda_1(x) & \cdots & D_n\lambda_1(x) \\ D_1\lambda_2(x) & D_2\lambda_2(x) & \cdots & D_n\lambda_2(x) \\ \vdots & & & \\ D_1\lambda_n(x) & D_2\lambda_n(x) & \cdots & D_n\lambda_n(x) \end{pmatrix} = n.$$

We consider a function $f\colon \mathfrak{D} \to \mathbf{E}^n$, where $\mathfrak{D} \subseteq \mathbf{E}^n \times \mathbf{E}^m$, and pose the more general question: When and for what vectors $y \in \mathbf{E}^m$ does the system of equations

$$f(x, y) = \theta$$

have a solution $x = h(y) \in \mathbf{E}^n$ such that $f(h(y), y) = \theta$?

A sufficient condition is stated in the *implicit-function theorem*, which we shall prove in section 10.99.

***Theorem 95.2*** (*Implicit-Function Theorem*)

If $f\colon \mathfrak{D} \to \mathbf{E}^n$, where $\mathfrak{D} \subseteq \mathbf{E}^n \times \mathbf{E}^m$ is an open set, if $f \in C^1(\mathfrak{D})$, if $f(x_0, y_0) = \theta$ for some $(x_0, y_0) \in \mathfrak{D}$, and if

$$\operatorname{rank}\begin{pmatrix} D_1\phi_1(x_0, y_0) & D_2\phi_1(x_0, y_0) & \cdots & D_n\phi_1(x_0, y_0) \\ D_1\phi_2(x_0, y_0) & D_2\phi_2(x_0, y_0) & \cdots & D_n\phi_2(x_0, y_0) \\ \vdots & & & \\ D_1\phi_n(x_0, y_0) & D_2\phi_n(x_0, y_0) & \cdots & D_n\phi_n(x_0, y_0) \end{pmatrix} = n,$$

then there exists an open set $U \subseteq \mathbf{E}^n \times \mathbf{E}^m$ containing $(x_0, y_0)$, an open set $V \subseteq \mathbf{E}^m$ containing $y_0$, and a function $h\colon V \to \mathbf{E}^n$ such that $h(y_0) = x_0$ and $f(h(y), y) = \theta$ for all $y \in V$. Furthermore, $h \in C^1(V)$ and $h$ is uniquely determined by $(h(y), y) \in U$ for all $y \in V$. (See Fig. 95.3.)

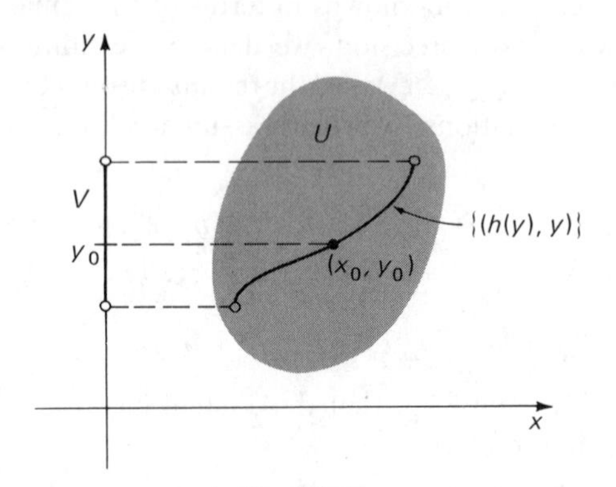


***Fig. 95.3***

Noting that $l(x, y) = \theta$ always has the trivial solution $x = \theta$, $y = \theta$, we see that the sufficiency part of the theorem pertaining to linear functions is a special case of the implicit-function theorem with $U = \mathbf{E}^n \times \mathbf{E}^m$ and $V = \mathbf{E}^m$.

A proof of the implicit-function theorem will be presented in section 10.99. Here, we shall sketch a proof for $n = 1$, $m = 1$ that is based on the inverse-function theorem:

We consider the system of equations

$$\begin{aligned} f(x, y) &= z_1, \\ y &= z_2, \end{aligned} \tag{95.2}$$

which, by hypothesis, is satisfied for $x = x_0$, $y = y_0$, $z_1 = 0$, $z_2 = y_0$. Let $g(x, y) = (f(x, y), y)$. Then,

$$\operatorname{rank} g'(x_0, y_0) = \operatorname{rank}\begin{pmatrix} D_1 f(x_0, y_0) & D_2 f(x_0, y_0) \\ 0 & 1 \end{pmatrix} = 2,$$

since, by hypothesis, rank $D_1 f(x_0, y_0) = 1$; that is, $D_1 f(x_0, y_0) \neq 0$. From Theorem 95.1, there are open sets $U, V_1 \subseteq \mathbf{E}^2$ with $(x_0, y_0) \in U$, $(0, z_2) \in V_1$, such that $g : U \leftrightarrow V_1$; that is, (95.2) has a unique solution $h_1 : V_1 \leftrightarrow U$. Let $h_1(0, z_2) = \mathrm{h}(z_2)$. Since $z_2 = y$, we obtain readily that $h(y_0) = x_0$ and $f(h(y), y) = 0$ for all $y \in V = \{y \in \mathbf{E} \mid (0, y) \in V_1\}$.

---

▲ *Example 2*

$$\begin{aligned} \xi_3 \sin \xi_1 \cos \xi_2 - \eta_1 &= 0, \\ \xi_3 \sin \xi_1 \sin \xi_2 - \eta_2 &= 0, \\ \xi_3 \cos \xi_1 - \sqrt{1 - \eta_1^2 - \eta_2^2} &= 0, \end{aligned} \tag{95.3}$$

is to be solved for $x = (\xi_1, \xi_2, \xi_3)$ in terms of $y = (\eta_1, \eta_2)$. We observe that

(95.4) $F'(x, y) =$

$$\begin{pmatrix} \xi_3 \cos \xi_1 \cos \xi_2 & -\xi_3 \sin \xi_1 \sin \xi_2 & \sin \xi_1 \cos \xi_2 & -1 & 0 \\ \xi_3 \cos \xi_1 \sin \xi_2 & \xi_3 \sin \xi_1 \cos \xi_2 & \sin \xi_1 \sin \xi_2 & 0 & -1 \\ \xi_3 \sin \xi_1 & 0 & \cos \xi_1 & \dfrac{\eta_1}{\sqrt{1 - \eta_1^2 - \eta_2^2}} & \dfrac{\eta_2}{\sqrt{1 - \eta_1^2 - \eta_2^2}} \end{pmatrix}$$

and hence, $f \in C^1(\mathfrak{D})$, where $\mathfrak{D} = \{(x, y) \in \mathbf{E}^3 \times \mathbf{E}^2 \mid \eta_1^2 + \eta_2^2 < 1\}$; and that the determinant of the matrix with the first three columns in (95.4) has the value $\xi_3^2 \sin \xi_1$. Hence,

$$\operatorname{rank}\begin{pmatrix} D_1\phi_1(x) & D_2\phi_1(x) & D_3\phi_1(x) \\ D_1\phi_2(x) & D_2\phi_2(x) & D_3\phi_2(x) \\ D_1\phi_3(x) & D_2\phi_3(x) & D_3\phi_3(x) \end{pmatrix} = 3$$

for all $x \in \{x \in \mathbf{E}^3 \mid 0 < \xi_1 < \pi, \xi_3 > 0\} = \mathcal{E}$.

Since $(x_0, y_0) = (\xi_1^\circ, \xi_2^\circ, \xi_3^\circ, \eta_1^\circ, \eta_2^\circ) = (\pi/4, 0, 1, \sqrt{2}/2, 0) \in \mathfrak{D} \cap (\mathcal{E} \times \mathbf{E}^2)$ satisfies 95.3, we obtain, from the implicit-function theorem, two open sets $U \subseteq \mathbf{E}^5$, $V \subseteq \mathbf{E}^2$ such that $(x_0, y_0) \in U$, $y_0 \in V$, and a solution $h : V \to \mathbf{E}^3$ of (95.3). By elementary computations,

$$\begin{aligned} \xi_1 &= \cos^{-1} \sqrt{1 - \eta_1^2 - \eta_2^2}, \\ \xi_2 &= \sin^{-1} \frac{\eta_2}{\sqrt{\eta_1^2 + \eta_2^2}}, \\ \xi_3 &= 1; \end{aligned}$$

and we see that $V = \{y \in \mathbf{E}^2 \mid \eta_1^2 + \eta_2^2 < 1 \text{ and } \eta_1 > 0\}$, and

$$U = \{(x, y) \in \mathbf{E}^3 \times \mathbf{E}^2 \mid 0 < \xi_1 < \pi, \quad -\frac{\pi}{2} < \xi_2 < \frac{\pi}{2}, \quad \xi_3 > 0, \quad \eta_1^2 + \eta_2^2 < 1, \quad \eta_1 > 0\}.$$

$(x_0, y_0) = (\pi/2, 0, 1, 1, 0)$ also satisfies (95.3). The implicit-function theorem, however, does not apply, because not all partial derivatives of $f$ are continuous at this point. One may still solve (95.3) for $(\xi_1, \xi_2, \xi_3)$, but then $V$ is no longer an open set and not all the partial derivatives of $h$ are continuous. (See exercise 3.)

---

## 9.95 Exercises

1. Given: $x = r\cos\theta$, $y = r\sin\theta$. Solve for $(r, \theta)$. Invoke the inverse-function theorem, find $U$, $V$, and show that $f^{-1} \in C^1(V)$.
2. Same as in exercise 1, for $x = r\sin\phi\cos\theta$, $y = r\sin\phi\sin\theta$, $z = r\cos\phi$, which is to be solved for $(r, \phi, \theta)$.
3. Solve (95.3) for $x$ at and near $y = (1, 0)$, and discuss the breakdown of the implicit-function theorem in detail.
4. Solve $(\eta_1^2 + \eta_2^2)\xi - \xi^3 = 0$ for $\xi$ in terms of $(\eta_1, \eta_2)$. Check the hypotheses and conclusions of the implicit-function theorem.
5. Prove the inverse-function theorem (Theorem 95.1) for $n = 1$ by means of Theorem 36.2 and the Corollary to Lemma 37.1.

# 10 Nonlinear Functions

## 10.96 One-to-One Functions

We have seen, in section 8.84, that a linear function $l: \mathbf{E}^n \to \mathbf{E}^m$ is one-to-one if and only if rank $(l) = n \leqslant m$. As in the case of real-valued functions of a real variable, we obtain some information about the local behavior of a function from the behavior of its derivative. Specifically, we intend to demonstrate in this section that $f: \mathfrak{D}(f) \to \mathbf{E}^m$, $\mathfrak{D}(f) \subseteq \mathbf{E}^n$, is one-to-one in some neighborhood of an interior point $a \in \mathfrak{D}(f)$ if $f'(a)$ is one-to-one and if some additional differentiability conditions are satisfied. It will then follow that the system of equations $f(x) = y$ has a unique solution in some neighborhood of $a$ for every $y$ that is sufficiently close to $f(a)$ and lies in the range of $f$.

---

▲ *Example 1*

Let $f: \mathbf{E}^2 \to \mathbf{E}^3$ be given by

$$\begin{aligned}
\phi_1(\xi_1, \xi_2) &= \xi_1 + \xi_2,\\
\phi_2(\xi_1, \xi_2) &= \xi_1 - \xi_2,\\
\phi_3(\xi_1, \xi_2) &= \xi_1^2.
\end{aligned}$$

The Jacobian matrix of $f$ at $x = \theta$ is given by

$$F'(\theta) = \begin{pmatrix} 1 & 1 \\ 1 & -1 \\ 0 & 0 \end{pmatrix}.$$

Since rank $F'(\theta) = 2 = n \leqslant m$, we see that $f'(\theta): \mathbf{E}^2 \to \mathbf{E}^3$ is one-to-one. (See Lemma 84.2.) The range of $f'(\theta)$ is a two-dimensional subspace of $\mathbf{E}^3$ which may be generated by the vectors $f'(\theta)(e_1) = (1, 1, 0)$ and $f'(\theta)(e_2) = (1, -1, 0)$.

The function $f$ itself is one-to-one and has an inverse on the range of $f$. $y = (\eta_1, \eta_2, \eta_3)$ is in the range of $f$ if

$$\eta_1 = \xi_1 + \xi_2,$$
$$\eta_2 = \xi_1 - \xi_2,$$
$$\eta_3 = \xi_1^2;$$

that is, if $\eta_3 = \frac{1}{4}(\eta_1 + \eta_2)^2$ which represents a parabolic cylinder. (See Fig. 96.1.)

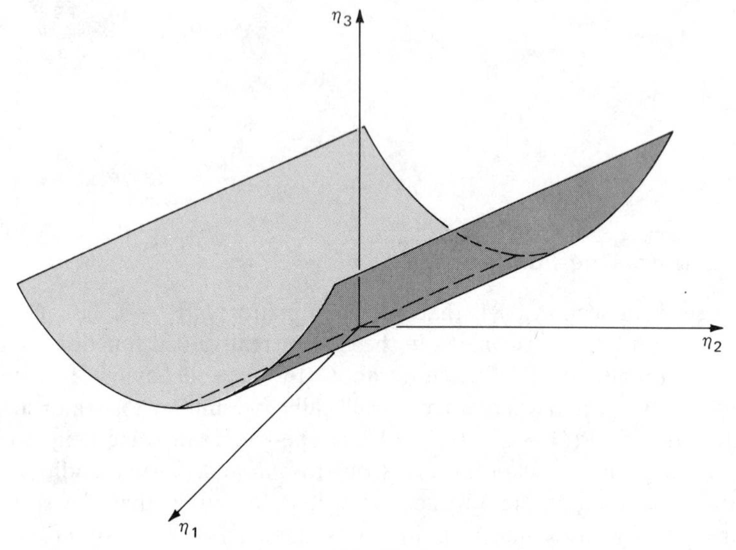


*Fig. 96.1*

We have, for every $y \in \mathcal{R}(f)$, that is, $y = (\eta_1, \eta_2, \frac{1}{4}(\eta_1 + \eta_2)^2)$, a *unique* $x = (\xi_1, \xi_2)$, namely

$$\xi_1 = \frac{\eta_1 + \eta_2}{2}, \qquad \xi_2 = \frac{\eta_1 - \eta_2}{2},$$

such that $f(x) = y$. (Note that if $\xi_1 = (\eta_1 + \eta_2)/2$, then $\eta_3 = \xi_1^2$ is automatically satisfied because $\eta_3 = \frac{1}{4}(\eta_1 + \eta_2)^2$.)

▲ *Example 2*

Let $f: \mathbf{E} \to \mathbf{E}^3$ be given by

$$\phi_1(x) = \cos x,$$
$$\phi_2(x) = \sin x,$$
$$\phi_3(x) = x.$$

The Jacobian matrix of $f$ is given by

$$F'(x) = \begin{pmatrix} -\sin x \\ \cos x \\ 1 \end{pmatrix}.$$

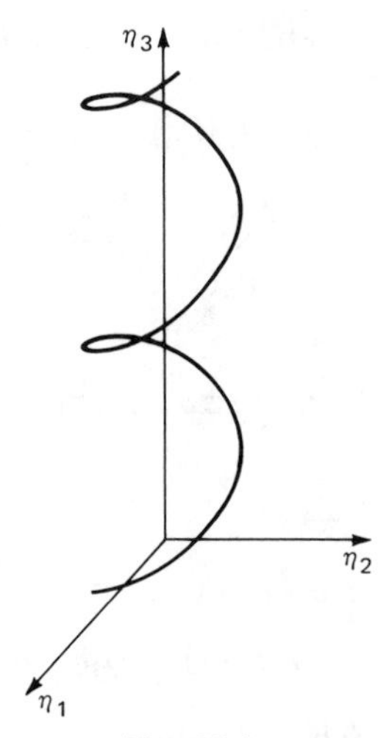


*Fig. 96.2*

Since rank $F'(x) = 1 = n < m$ for all $x \in \mathbf{E}$, we see that $f'(x)$ is one-to-one for every $x \in \mathbf{E}$. The function $f$ is one-to-one on $\mathbf{E}$ and the equation $f(x) = y$ has a unique solution for every $y \in \mathfrak{R}(f)$. The range of $f$ is a helix (see Fig. 96.2).

---

In order to establish a sufficient condition for the existence of a local inverse, we need the following lemma:

***Lemma 96.1***

Let $\mathfrak{D} \subseteq \mathbf{E}^n$ denote an open set and let $f: \mathfrak{D} \to \mathbf{E}^m$. If $f \in C^1(\mathfrak{D})$, then, for every $a \in \mathfrak{D}$ and every $\varepsilon > 0$, there is a $\delta(\varepsilon, a) > 0$ for which $N_\delta(a) \subseteq \mathfrak{D}$, such that

(96.1) $\quad \|f'(a)(u) - f'(x)(u)\| \leqslant \varepsilon \|u\| \qquad$ for all $u \in \mathbf{E}^n$ and all $x \in N_\delta(a)$,

and

(96.2) $$\|f(x) - f(y) - f'(a)(x - y)\| \leqslant \varepsilon \|x - y\|$$

for all $x, y \in N_\delta(a)$.

*Proof*

We let $u = (\omega_1, \omega_2, \ldots, \omega_n)$, note that

$$D_j f(x) = \begin{pmatrix} D_j \phi_1(x) \\ \vdots \\ D_j \phi_m(x) \end{pmatrix},$$

and obtain, from Theorem 89.1 and the CBS inequality, that

$$\begin{aligned}\|(f'(x)-f'(a))u\|^2 &= \left\|\sum_{j=1}^{n}(D_jf(x)-D_jf(a))\omega_j\right\|^2\\ &= \sum_{k=1}^{m}\left(\sum_{j=1}^{n}(D_j\phi_k(x)-D_j\phi_k(a))\omega_j\right)^2\\ &\leqslant \sum_{k=1}^{m}\left[\left(\sum_{j=1}^{n}\omega_j\right)^2\sum_{j=1}^{m}(D_j\phi_k(x)-D_j\phi_k(a))^2\right]\\ &= \sum_{k=1}^{m}\sum_{j=1}^{n}(D_j\phi_k(x)-D_j\phi_k(a))^2\|u\|^2.\end{aligned}$$

Since $f \in C^1(\mathfrak{D})$, we have, for every $\varepsilon > 0$, a $\delta(\varepsilon) > 0$ such that

$$\textstyle\sum_{k=1}^{m}\sum_{j=1}^{n}(D_j\phi_k(x)-D_j\phi_k(a))^2 < \varepsilon^2$$

for all $x \in N_\delta(a)$, and (96.1) follows readily.

To prove (96.2), we use the same $\varepsilon, \delta$ as above. Whenever $x, y \in N_\delta(a)$, then, the line segment joining $x$ to $y$ also lies in $N_\delta(a)$. From Corollary 1 to Theorem 93.1, we have, for every $u \in \mathbf{E}^n$, an $x_u$ on the line segment from $x$ to $y$ (and hence in $N_\delta(a)$) such that

$$(f(x)-f(y))\cdot u = f'(x_u)(x-y)\cdot u.$$

We choose $u = f(x)-f(y)-f'(a)(x-y)$, and obtain from the CBS inequality that

$$\begin{aligned}(96.3)\quad &\|f(x)-f(y)-f'(a)(x-y)\|^2\\ &= (f'(x_u)(x-y)-f'(a)(x-y))\cdot(f(x)-f(y)-f'(a)(x-y))\\ &\leqslant \|f'(x_u)(x-y)-f'(a)(x-y)\|\,\|f(x)-f(y)-f'(a)(x-y)\|.\end{aligned}$$

If $f(x)-f(y)-f'(a)(x-y)=\theta$, then (96.2) is trivially satisfied. If not, we divide (96.3) by $f(x)-f(y)-f'(a)(x-y)$ and obtain (96.2) from (96.3) and (96.1).

The following theorem, the proof of which is based on Lemma 96.1, guarantees the existence of a local inverse:

***Theorem 96.1*** *(Injective Mapping Theorem)*

Let $\mathfrak{D} \subseteq \mathbf{E}^n$ denote an open set and let $f: \mathfrak{D} \to \mathbf{E}^m$. If $f \in C^1(\mathfrak{D})$ and if for some $a \in \mathfrak{D}$, $f'(a)$ is one-to-one on $\mathbf{E}^n$, then there is a $N_\delta(a) \subseteq \mathfrak{D}$ and a $\mu > 0$ such that

$$f'(x) \text{ is one-to-one} \quad \text{for all } x \in N_\delta(a),$$

$$(96.4)\qquad \|f(x)-f(y)\| \geqslant \mu\|x-y\| \qquad \text{for all } x, y \in N_\delta(a).$$

Furthermore, $f$ is one-to-one on $N_\delta(a)$ and the inverse function $f^{-1}$ exists on $f_*(N_\delta(a))$.

(Note that $f_*(N_\delta(a)) \subseteq \mathbf{E}^m$ need not have any interior points. In example 1, $f_*(N_\delta(a))$ is a point set on the surface of a parabolic cylinder and as such, has no interior points because no three-dimensional open ball lies on the surface of a parabolic cylinder. In example 2, $f_*(N_\delta(a))$ is a point set on a helix and, again, has no interior points.)

*Proof*

Since $f'(a)$ is one-to-one, we have, from Theorem 85.2, that there exists a $\mu > 0$ such that

$$\|f'(a)(u)\| \geqslant 2\mu \|u\| \tag{96.5}$$

for all $u \in \mathbf{E}^n$. By (96.1) we have for $\varepsilon = \mu$, a $\delta > 0$ such that

$$\begin{aligned} 2\mu \|u\| - \|f'(x)(u)\| &\leqslant \|f'(a)(u)\| - \|f'(x)(u)\| \\ &\leqslant \|f'(a)(u) - f'(x)(u)\| \leqslant \mu \|u\| \end{aligned}$$

for all $x \in N_\delta(a)$ and hence,

$$\|f'(x)(u)\| \geqslant \mu \|u\| \qquad \text{for all } x \in N_\delta(a). \tag{96.6}$$

Hence, we have, from Theorem 85.2, that $f'(x)$ is one-to-one for all $x \in N_\delta(a)$.

By (96.2),

$$\|f(x) - f(y) - f'(a)(x-y)\| \leqslant \mu \|x - y\| \qquad \text{for all } x \in N_\delta(a).$$

(Note that this is the same $\delta$ as before.) Hence, by the triangle inequality,

$$\|f'(a)(x-y)\| - \|f(x) - f(y)\| \leqslant \mu \|x - y\|,$$

and from (96.5) we obtain

$$\|f(x) - f(y)\| \geqslant \|f'(a)(x-y)\| - \mu \|x-y\| \geqslant \mu \|x - y\|$$

for all $x, y \in N_\delta(a)$, which is (96.4).

If for any $x, y \in N_\delta(a)$, $f(x) = f(y)$ holds, then we have, from (96.4), that $0 \geqslant \mu \|x - y\|$ for some $\mu > 0$. Hence $x = y$; that is, $f$ is one-to-one from $N_\delta(a)$ onto $f_*(N_\delta(a))$. Therefore, $f^{-1}$ exists on $f_*(N_\delta(a))$.

***Corollary to Theorem 96.1***

Under the conditions of Theorem 96.1, there is a $\delta > 0$ such that the system of equations $f(x) = y$ has a unique solution in $N_\delta(a)$ for every $y \in f_*(N_\delta(a))$.

---

▲ *Example 3*

Let $f: \mathbf{E} \to \mathbf{E}^2$ be given by

$$\begin{aligned} \phi_1(x) &= x, \\ \phi_2(x) &= x^2. \end{aligned}$$

Then, the Jacobian matrix of $f$ at 0 is given by

$$F'(0) = \begin{pmatrix} 1 \\ 0 \end{pmatrix},$$

which is one-to-one from $\mathbf{E}$ into $\mathbf{E}^2$. $f$ maps $N_\delta(0) = (-\delta, \delta)$ one-to-one onto $f_*(N_\delta(0)) = \{(\eta_1, \eta_2) \in \mathbf{E}^2 \mid \eta_2 = \eta_1^2, \eta_1 \in (-\delta, \delta)\}$. (See Fig. 96.3.) Clearly no

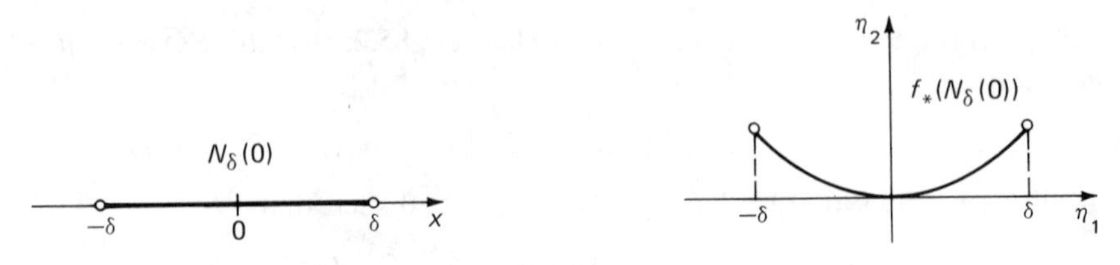


*Fig. 96.3*

point of $f_*(N_\delta(0))$ is an interior point of $f_*(N_\delta(0))$.

---

If the condition $f \in C^1(\mathfrak{D})$ is violated, then it is possible for $f'(a)$ to be one-to-one, but there is *no* neighborhood of $a$ where $f$ is one-to-one, as the following example demonstrates:

---

▲ *Example 4*

Let $f: \mathbf{E} \to \mathbf{E}$ be given by

$$f(x) = \begin{cases} x + x^2 \sin \dfrac{1}{x} & \text{for } x \neq 0, \\ 0 & \text{for } x = 0. \end{cases}$$

We have $f'(0) = 1$ (see also example 3, section 3.35). Hence, $f'(0): \mathbf{E} \to \mathbf{E}$ is one-to-one. Still, no matter how small we may choose $\delta > 0$, $f$ is not one-to-one on $N_\delta(0)$, as we can see as follows:

We have, for $x \neq 0$, $f'(x) = 1 + 2x \sin(1/x) - \cos(1/x)$, $f''(x) = 2 \sin(1/x) - (2/x)\cos(1/x) - (1/x^2)\sin(1/x)$. Hence, $f'(1/2k\pi) = 0$, $f''(1/2k\pi) = -4k\pi < 0$ for all $k \in \mathbf{N}$. Therefore, $f$ assumes a relative maximum for each $x = 1/2k\pi$. (See Theorem 37.1 and exercise 11, section 3.37.) Since $f(1/2k\pi) = 1/2k\pi$, $f$ has to assume a smaller value, say $b$, to the left and to the right of $1/2k\pi$ but somewhere in $(0, 1/(2(k-1)\pi))$. Since we may choose $k$ as large as we please, we see that, no matter how small a neighborhood $N_\delta(0)$ of 0 we select, $f$ will assume some value at least twice in each such neighborhood. Therefore, there is no $N_\delta(0)$ where $f$ is one-to-one. Theorem 96.1 is not applicable because $f'$ is not continuous at 0. (See also exercise 3.)

---

Lemma 96.1, in conjunction with the Heine–Borel Theorem, leads to the following result, which will prove useful in section 12.114.

***Lemma 96.2***

Let $\mathfrak{D} \subseteq \mathbf{E}^n$ denote an open set and let $f: \mathfrak{D} \to \mathbf{E}^m$. If $f \in C^1(\mathfrak{D})$ and if $K \subset \mathfrak{D}$ is compact, then there exists, for every $\varepsilon > 0$, a $\delta > 0$ such that, for all $a \in K$,

$$\|f(x) - f(a) - f'(a)(x - a)\| \leqslant \varepsilon \|x - a\| \tag{96.7}$$

whenever $x \in N_\delta(a) \cap K$.

If, in addition to the above hypotheses, $f'(a)$ is one-to-one for all $a \in K$, then there is a $\mu > 0$ such that

$$\|f'(a)(u)\| \geqslant \mu \|u\| \qquad \text{for all } a \in K \text{ and all } u \in \mathbf{E}^n. \tag{96.8}$$

*Proof*

In order to prove (96.7), one need only observe that, for each $a \in K$, there is a $\delta(\varepsilon, a) > 0$ such that (96.2) is satisfied and that $\{N_{\delta(\varepsilon, a)}(a) \mid a \in K\}$ is an open cover of $K$. (96.8) follows from (96.6) by a similar argument. (Note that (96.6) follows from (96.1) and Theorem 85.2.)

## 10.96 Exercises

1. Find the range of the function $f: \mathbf{E}^2 \to \mathbf{E}^3$ which is given by $\phi_1(\xi_1, \xi_2) = \xi_1 - \xi_2, \phi_2(\xi_1, \xi_2) = \xi_2^2, \phi_3(\xi_1, \xi_2) = \xi_1 + \xi_2$. Is $f$ a one-to-one function? Find $F'(x)$ for every $x \in \mathbf{E}^2$.
2. Same as in exercise 1, for the function $f: \mathbf{E}^2 \to \mathbf{E}^3$ which is given by $\phi_1(\xi_1, \xi_2) = \xi_1^2 + \xi_2^2, \phi_2(\xi_1, \xi_2) = \xi_1, \phi_3(\xi_1, \xi_2) = \xi_2$.
3. Let $f: \mathbf{E}^2 \to \mathbf{E}^2$ be given by

$$\phi_1(\xi_1, \xi_2) = \left\{ \begin{array}{ll} \xi_1 + \xi_1^2 \sin \dfrac{1}{\xi_1} & \text{for } \xi_1 \neq 0 \\ 0 & \text{for } \xi_1 = 0 \end{array} \right\}, \qquad \phi_2(\xi_1, \xi_2) = \xi_2.$$

   Show that $f'(\theta) : \mathbf{E}^2 \xrightarrow[1-1]{} \mathbf{E}^2$ but that there is no $N_\delta(\theta)$ where $f$ is one-to-one. Explain the failure of Theorem 96.1.
4. Given the function $f: \mathbf{E}^3 \to \mathbf{E}^3$ by

$$\begin{aligned} \phi_1(\xi_1, \xi_2, \xi_3) &= \xi_1 \cos \xi_2 \sin \xi_3, \\ \phi_2(\xi_1, \xi_2, \xi_3) &= \xi_1 \sin \xi_2 \sin \xi_3, \\ \phi_3(\xi_1, \xi_2, \xi_3) &= \xi_1 \cos \xi_3. \end{aligned}$$

   Find $F'(\theta)$ and $F'(x)$ for $x \neq 0$. Show that $f$ is *not* one-to-one on $\mathfrak{D} = \{(\xi_1, \xi_2, \xi_3) \in \mathbf{E}^3 \mid 1 \leqslant \xi_1 \leqslant 2, 0 \leqslant \xi_2 \leqslant 4\pi, 0 \leqslant \xi_3 \leqslant 2\pi\}$.
5. Given $f: \mathbf{E}^2 \to \mathbf{E}^2$ by $\phi_1(\xi_1, \xi_2) = \xi_1^2 + 2\xi_2, \phi_2(\xi_1, \xi_2) = \xi_1 - \xi_1 \xi_2$. Show that there is a $\delta > 0$ such that $f: N_\delta(\theta) \xrightarrow[1-1]{} \mathbf{E}^2$. (See also exercise 5, section 9.92.)

## 10.97 Onto Functions

We have seen in section 8.84 that a linear function $l: \mathbf{E}^n \to \mathbf{E}^m$ is *onto* $\mathbf{E}^m$ if and only if rank $(l) = m \leqslant n$. We have seen that such a function maps open sets in $\mathbf{E}^n$ onto open sets in $\mathbf{E}^m$ (Corollary to Theorem 85.3). Again, we shall be able to draw conclusions about the local behavior of a function from the fact that its derivative at a point is onto. In particular, we shall demonstrate that if $f'(a): \mathbf{E}^n \to \mathbf{E}^m$ is onto, and if $f$ satisfies some suitable differentiability conditions, then $f$ maps some $N_\rho(a)$ onto an open set in $\mathbf{E}^m$ and all open subsets of $N_\rho(a)$ onto open subsets of $f_*(N_\rho(a))$.

---

▲ *Example 1*

Given $f: \mathbf{E}^2 \to \mathbf{E}$ by $f(\xi_1, \xi_2) = \xi_1 + \xi_1^2 + \xi_2^2$. Then, $f'(\theta)$ is represented by $F'(\theta) = (1, 0)$. Since rank $F'(\theta) = 1 = m < n$, we see that $f'(\theta)$ is *onto* $\mathbf{E}$. $f$ maps the open disk $N_\delta(0) \subset \mathbf{E}^2$ onto the open interval $(-\delta + \delta^2, \delta + \delta^2) \subset \mathbf{E}$ if $0 < \delta < \frac{1}{2}$. (See Fig. 97.1.)

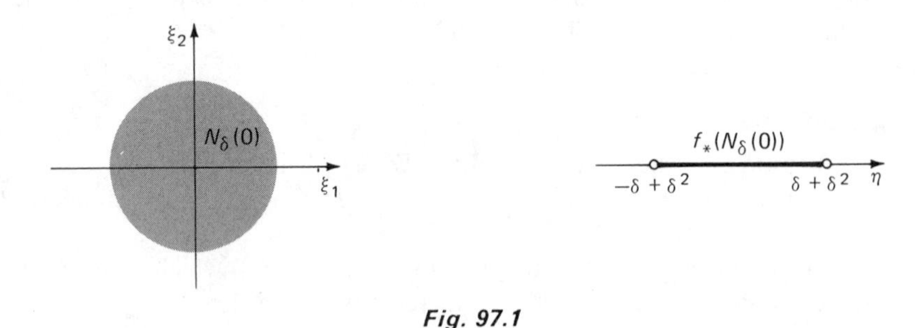


*Fig. 97.1*

---

In order to prove a theorem as indicated above, we need the following lemma.

***Lemma 97.1***

Let $\mathfrak{D} \subseteq \mathbf{E}^n$ denote an open set and let $f: \mathfrak{D} \to \mathbf{E}^m$. If $f \in C^1(\mathfrak{D})$, and if for some $a \in \mathfrak{D}$, $f'(a)$ is onto $\mathbf{E}^m$, then there is a $\rho > 0$ such that $f'(x)$ is onto $\mathbf{E}^m$ for each $x \in N_\rho(a)$.

*Proof*

Since $f'(a)$ maps $\mathbf{E}^n$ onto $\mathbf{E}^m$, we have from Lemma 84.1 that rank $f'(a) = m \leqslant n$. Hence, we can pick an $m \times m$ submatrix $A$ from $F'(a)$ which is nonsingular, i.e., det $A \neq 0$. We may assume without loss of generality that

$$\det A = \begin{vmatrix} D_1\phi_1(a) \ldots D_m\phi_1(a) \\ \vdots \\ D_1\phi_m(a) \ldots D_m\phi_m(a) \end{vmatrix} \neq 0.$$

Since $f \in C^1(\mathfrak{D})$, the functions $D_i \phi_j$ are continuous on $\mathfrak{D}$ and hence, the function det $A$, defined by

$$\det A(x) = \begin{vmatrix} D_1\phi_1(x) \dots D_m \phi_1(x) \\ \vdots \\ D_1\phi_m(x) \dots D_m \phi_m(x) \end{vmatrix}$$

is continuous on $\mathfrak{D}$. Therefore, there is *a* $\rho > 0$ (see Lemma 65.2) such that $\det A(x) \neq 0$ for all $x \in N_\rho(a)$. Hence, $f'(x)$ is onto $\mathbf{E}^m$ for each $x \in N_\rho(a)$.

By Theorem 96.1, $f'(x)$ is one-to-one on $\mathbf{E}^n$ in a neighborhood of $a$ if $f'(a)$ is one-to-one. Lemma 97.1 now assures us that $f'(x)$ is onto $\mathbf{E}^m$ in a neighborhood of $a$ if $f'(a)$ is onto. Hence, if $f'(a)$ is one-to-one *and* onto, then there is a neighborhood of $a$ where $f'(x)$ is also one-to-one and onto.

***Theorem 97.1*** *(Interior Mapping Theorem)*

If $\mathfrak{D} \subseteq \mathbf{E}^n$ represents an open set, if $f: \mathfrak{D} \to \mathbf{E}^m$ where $f \in C^1(\mathfrak{D})$, and if $f'(a)$: $\mathbf{E}^n \xrightarrow{\text{onto}} \mathbf{E}^m$ for some $a \in \mathfrak{D}$, then there is a $\rho > 0$ such that $f$ maps all open subsets of $N_\rho(a)$ onto open sets in $\mathbf{E}^m$ and, in particular, $f_*(N_\rho(a))$ is open in $\mathbf{E}^m$.

*Proof*

We shall demonstrate that $f_*(\Omega)$ is open in $\mathbf{E}^m$ whenever $\Omega$ is an open subset of $N_\rho(a)$ for some $\rho > 0$, by showing first that, for each $c \in N_\rho(a)$, there is a $\delta > 0$ such that $N_\delta(c) \subseteq N_\rho(a)$ and $f(c)$ is an interior point of $f_*(N_\delta(c))$. This will be accomplished by finding, for each $y$ in a sufficiently small neighborhood of $f(c)$, an element $x \in N_\delta(c)$ such that $f(x) = y$.

By Lemma 97.1, there is a $\rho > 0$ such that $f'(x)$ is onto $\mathbf{E}^m$ for each $x \in N_\rho(a)$. Hence, if $c \in N_\rho(a)$, then $f'(c)$ is onto $\mathbf{E}^m$. In order to simplify our proof, we assume without loss of generality that

$$c = \theta, \qquad f(c) = \theta.$$

(This can always be accomplished by a translation of the coordinates in $\mathbf{E}^n$ and in $\mathbf{E}^m$.)

Since $f'(\theta)$ is, hypothesis, onto $\mathbf{E}^m$, there exists, by Theorem 84.1 a right inverse $[f'(\theta)]^R : \mathbf{E}^m \to \mathbf{E}^n$. This right inverse is a linear function. Hence, by Lemma 85.1, there is an $M > 0$ such that

$$\|[f'(\theta)]^R(u)\| \leqslant M\|u\| \qquad \text{for all } u \in \mathbf{E}^m. \tag{97.1}$$

In order to simplify our proof still further, we shall consider the function

$$g = f \circ [f'(\theta)]^R$$

instead of $f$. Note that $g(\theta) = \theta$. We shall explain later how the forthcoming result on $g$ translates into the desired result on $f$. Note that the domain of $g$ lies in $\mathbf{E}^m$ and contains $\theta$ as an interior point. The range of $g$ lies also in $\mathbf{E}^m$. By the chain rule (Theorem 92.1),

$$g'(z) = f'([f'(\theta)]^R(z)) \circ [f'(\theta)]^R,$$

and hence,

$$g'(\theta) = f'([f'(\theta)]^R(\theta)) \circ [f'(\theta)]^R = f'(\theta) \circ [f'(\theta)]^R = i, \tag{97.2}$$

where $i : \mathbf{E}^m \to \mathbf{E}^m$ denotes the identity function. If we let

$$h(z) = z - g(z), \tag{97.3}$$

where we note that $h(\theta) = \theta$, we obtain, in view of (97.2), that $h'(\theta) = \theta$. By Lemma 96.1, we have, for $\varepsilon = \frac{1}{2}$, a $\Delta > 0$ such that

$$\|h'(z)(u)\| \leqslant \tfrac{1}{2}\|u\| \qquad \text{for all } u \in \mathbf{E}^m \text{ and all } z \in N_\Delta(\theta). \tag{97.4}$$

We choose $\Delta > 0$ so small that, in our original notation, and with $M$ defined in (97.1),

$$N_{M\Delta}(c) \subseteq N_\rho(a). \tag{97.5}$$

We obtain, from Corollary 1 to Theorem 93.1, with $u = h(z)$,

$$h(z) \cdot h(z) = \|h(z)\|^2 = |h'(z_1)(z) \cdot h(z)| \leqslant \|h'(z_1)(z)\| \, \|h(z)\|,$$

where $z_1$ depends on $h(z)$ and lies on the line that joins $\theta$ to $z$. If $z \in N_\Delta(\theta)$, then $z_1 \in N_\Delta(\theta)$, and we obtain, from (97.4), that

$$\|h(z)\|^2 \leqslant \tfrac{1}{2}\|z\| \, \|h(z)\|.$$

Either $h(z) = \theta$ or it is not. If it is not, we cancel by $\|h(z)\|$ and obtain in any event

$$\|h(z)\| \leqslant \tfrac{1}{2}\|z\| \qquad \text{for all } z \in N_\Delta(\theta). \tag{97.6}$$

If

$$k(z) = y + h(z), \tag{97.7}$$

then

$$k'(z) = h'(z). \tag{97.8}$$

For all $y \in N_{\Delta/4}(\theta)$, we have, from (97.6) and (97.7), that

$$\|k(z)\| < \frac{\Delta}{2} \qquad \text{for all } \|z\| \leqslant \frac{\Delta}{2};$$

that is, $k$ maps the closed neighborhood $\overline{N_{\Delta/2}(\theta)}$ into the closed neighborhood $\overline{N_{\Delta/2}(\theta)}$.

Application of Corollary 1 to Theorem 93.1 to $k$, with $u = k(z_1) - k(z_2)$, yields

$$\|k(z_1) - k(z_2)\|^2 = |k'(z_3)(z_1 - z_2) \cdot (k(z_1) - k(z_2))|.$$

Hence, we obtain, in view of (97.4) and (97.8), that

$$\|k(z_1) - k(z_2)\| \leqslant \tfrac{1}{2}\|z_1 - z_2\| \qquad \text{for all } z_1, z_2 \in N_\Delta(\theta);$$

that is, $k$ is a contraction mapping from $\overline{N_{\Delta/2}(\theta)}$ into $\overline{N_{\Delta/2}(\theta)}$. By the contraction mapping theorem (Theorem 75.1), there exists a unique fixed point $z_0 \in \overline{N_{\Delta/2}(\theta)} \subset N_{\Delta}(\theta)$ such that $k(z_0) = z_0$. By (97.3) and (97.7), $k(z_0) = z_0 = y + z_0 - g(z_0)$; that is, $g(z_0) = y$. We let $x = [f'(\theta)]^R z_0$ and observe from (97.1) that $\|x\| \leqslant M\Delta$. Hence, we can say that there is a $\delta = M\Delta > 0$ such that $N_\delta(c) \subseteq N_\rho(a)$ (see (97.5)), and that, for every $y \in N_{\delta/4M}(f(c))$, there is an $x \in N_\delta(c)$ such that $f(x) = y$. (Do not let the uniqueness of $z_0$ mislead you into thinking that there is only one $x \in N_\delta(c)$ such that $f(x) = y$. $x$ was obtained from $z_0$ by application of $[f'(\theta)]^R$. If $m < n$, then the right inverse is not unique (see example 4, section 8.84) and different points $x$ may be obtained from different right inverses.)

The above argument reveals that $N_{\delta/4M}(f(c)) \subseteq f_*(N_\delta(c))$; that is, $f(c)$ is an interior point of $f_*(N_\delta(c)) \subseteq f_*(N_\rho(a))$. (See also Fig. 97.2.)

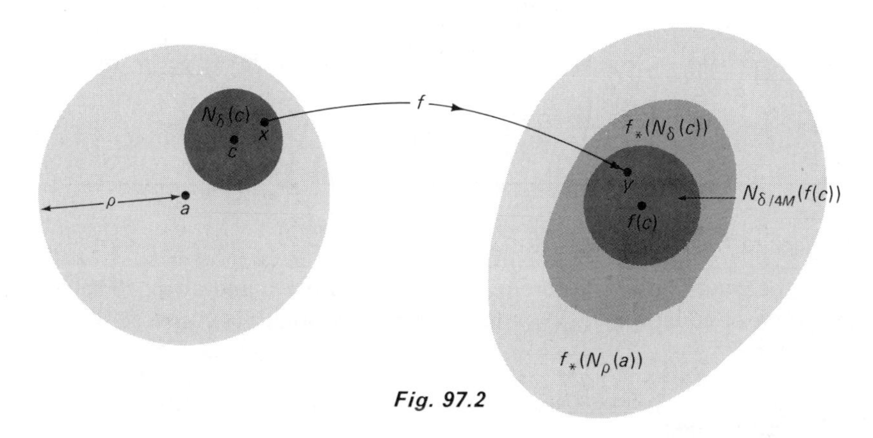


*Fig. 97.2*

Suppose that $\Omega \subseteq N_\rho(a)$ is open in $\mathbf{E}^n$. We want to show that $f_*(\Omega)$ is open in $\mathbf{E}^m$. For every $y \in f_*(\Omega)$, there is an $x \in \Omega$ such that $f(x) = y$. By the foregoing construction, we can choose a $\delta > 0$ such that $N_\delta(x) \subseteq \Omega$ and $N_{\delta/4M}(f(x)) \subseteq f_*(N_\delta(x)) \subseteq f_*(\Omega)$. Hence, $f(x)$ is an interior point of $f_*(\Omega)$; that is, $f_*(\Omega)$ is open and the theorem is proved.

---

▲ *Example 2*

We consider the function $f: \mathbf{E}^2 \to \mathbf{E}^2$ which is defined by

$$\phi_1(\xi_1, \xi_2) = \xi_1 \cos \xi_2,$$
$$\phi_2(\xi_1, \xi_2) = \xi_1 \sin \xi_2.$$

$f$ is differentiable everywhere and $f'(\xi_1, \xi_2)$ is represented by

(97.9)
$$F'(\xi_1, \xi_2) = \begin{pmatrix} \cos \xi_2 & -\xi_1 \sin \xi_2 \\ \sin \xi_2 & \xi_1 \cos \xi_2 \end{pmatrix}.$$

Since det $F'(\xi_1, \xi_2) = \xi_1 \neq 0$ for all $(\xi_1, \xi_2) \in \mathbf{E}^2$ for which $\xi_1 \neq 0$, we see that rank $F'(\xi_1, \xi_2) = 2$ for $\xi \neq 0$, and hence, $f'(\xi_1, \xi_2) : \mathbf{E}^2 \leftrightarrow \mathbf{E}^2$ for $\xi_1 \neq 0$.. Hence, the hypotheses of Theorems 96.1 and 97.1 are satisfied, and there is a $\rho > 0$ such that $f: N_\rho(\xi_1, \xi_2) \xrightarrow[1-1]{} \mathbf{E}^2$ and $f$ maps all open subsets of $N_\rho(x)$ onto open subsets of $\mathbf{E}^2$. Hence, by Theorem 92.3, $(f^{-1})'(f(\xi_1, \xi_2))$ exists whenever $\xi_1 \neq 0$ and is given by $(f^{-1})'(f(\xi_1, \xi_2)) = [f'(\xi_1, \xi_2)]^{-1}$. The inverse of the matrix in (97.9) is given by

$$[F'(\xi_1, \xi_2)]^{-1} = \frac{1}{\xi_1}\begin{pmatrix} \xi_1 \cos \xi_2 & \xi_1 \sin \xi_2 \\ -\sin \xi_2 & \cos \xi_2 \end{pmatrix}. \tag{97.10}$$

Hence, by Theorem 92.3, $(f^{-1})'(\xi_1 \cos \xi_2, \xi_1 \sin \xi_2)$ is given by (97.10) for $\xi_1 \neq 0$. If we express this result in terms of the coordinates $\eta_1 = \phi_1(\xi_1, \xi_2)$, $\eta_2 = \phi_2(\xi_1, \xi_2)$ of the image point, we obtain

$$(F^{-1})'(\eta_1, \eta_2) = \begin{pmatrix} \dfrac{\eta_1}{\sqrt{\eta_1^2 + \eta_2^2}} & \dfrac{\eta_2}{\sqrt{\eta_1^2 + \eta_2^2}} \\[2ex] -\dfrac{\eta_2}{\eta_1^2 + \eta_2^2} & \dfrac{\eta_1}{\eta_1^2 + \eta_2^2} \end{pmatrix}, \qquad (\eta_1, \eta_2) \neq (0, 0).$$

---

***Corollary to Theorem 97.1***

Under the conditions of Theorem 97.1, there is a $\delta > 0$ and an $\varepsilon > 0$ such that the system of equations $f(x) = y$ has at least one solution $x \in N_\delta(a)$ for every $y \in N_\varepsilon(f(a))$.

## 10.97 Exercises

1. Let $f: \mathbf{E} \to \mathbf{E}$ be defined by $f(x) = x^3$. Show that $f'(0) = 0$ but still, $f$ maps all $N_\delta(0)$ onto open subsets of $\mathbf{E}$.
2. Let $f: \mathbf{E}^2 \to \mathbf{E}$ be defined as in example 1. Show that $F'(-\frac{1}{2}, 0) = (0, 0)$ and $f$ does not map any $N_\delta(-\frac{1}{2}, 0)$ onto an open subset of $\mathbf{E}$.
3. Given $f: \mathbf{E}^2 \to \mathbf{E}$ by $f(\xi_1, \xi_2) = \sin(\xi_1^2 + \xi_2^2)$. Find the largest $\delta > 0$ such that $f_*(N_\delta(\theta))$ is open in $\mathbf{E}$.
4. Given $f: \mathbf{E}^2 \to \mathbf{E}^2$ by
$$\phi_1(\xi_1, \xi_2) = \cos \xi_1,$$
$$\phi_2(\xi_1, \xi_2) = \sin \xi_1 \cos \xi_2.$$
Find $(f^{-1})'(f(\xi_1, \xi_2))$ wherever it exists and express the result in terms of the coordinates of the image point. (See also example 2.)
5. Same as in exercise 4, for the function $f: \mathbf{E}^3 \to \mathbf{E}^3$, given by
$$\phi_1(\xi_1, \xi_2, \xi_3) = \cos \xi_1,$$
$$\phi_2(\xi_1, \xi_2, \xi_3) = \sin \xi_1 \cos \xi_2,$$
$$\phi_3(\xi_1, \xi_2, \xi_3) = \sin \xi_1 \sin \xi_2 \cos \xi_3.$$

6. Given $f: \mathbf{E}^2 \to \mathbf{E}^2$ by $\phi_1(\xi_1, \xi_2) = \xi_1^2 + 2\xi_2$, $\phi_2(\xi_1, \xi_2) = \xi_1 - \xi_1 \xi_2$. Show that there is a $\delta > 0$ such that $f$ maps all open subsets of $N_\delta(\theta)$ onto open subsets of $\mathbf{E}^2$. (See also exercise 5, section 9.92, and exercise 5, section 10.96.)

## 10.98 The Inverse-Function Theorem

If we combine the hypotheses of Theorems 96.1 and 97.1, i.e., if we assume that $f'(a) : \mathbf{E}^n \to \mathbf{E}^m$ is both onto and one-to-one, then we obtain that $f$ is one-to-one on some neighborhood of $a$ and that it maps some open neighborhood of $a$ onto an open subset of $\mathbf{E}^m$. Hence, there are open sets $U \subseteq \mathbf{E}^n$, $V \subseteq \mathbf{E}^m$ such that $f: U \leftrightarrow V$, which means, in turn, that $f$ is invertible and the inverse function $f^{-1}$ exists on $V$. For $f'(a)$ to be onto and one-to-one at the same time, however, we have to have rank $f'(a) = m \leqslant n$ and rank $f'(a) = n \leqslant m$ simultaneously. This is only possible if $m = n$. The rank of an $n \times n$ matrix $A$ is $n$ if and only if $\det A \neq 0$. We call

$$\det F'(x) = \begin{vmatrix} D_1\phi_1(x) \ldots D_n\phi_1(x) \\ \vdots \\ D_1\phi_n(x) \ldots D_n\phi_n(x) \end{vmatrix} = J_f(x) \tag{98.1}$$

the *Jacobian* of $f: \mathbf{E}^n \to \mathbf{E}^n$ at $x$. Hence, rank $f'(x) = m = n$ if and only if $J_f(x) \neq 0$.

***Theorem 98.1*** *(Inverse-Function Theorem)*

Let $\mathfrak{D} \subseteq \mathbf{E}^n$ denote an open set and let $f: \mathfrak{D} \to \mathbf{E}^n$. If $f \in C^1(\mathfrak{D})$, and if for some $a \in \mathfrak{D}$, $J_f(a) \neq 0$, then there exist open sets $U$, $V \subseteq \mathbf{E}^n$ such that $a \in U$, $f(a) \in V$, and $f: U \leftrightarrow V$, $f^{-1}: V \leftrightarrow U$ exists, $f^{-1} \in C^1(V)$, and

$$J_f(x) \neq 0 \quad \text{for all } x \in U \qquad \text{and} \qquad J_{f^{-1}}(y) \neq 0 \quad \text{for all } y \in V.$$

*Proof*

Since $J_f(a) \neq 0$, $f'(a)$: $\mathbf{E}^n \to \mathbf{E}^n$ is onto and one-to-one. Hence, by Theorems 96.1 and 97.1, there is a neighborhood $N_\rho(a)$ where $J_f(x) \neq 0$ and where $f$: $N_\rho(a) \to \mathbf{E}^n$ is one-to-one and maps all open subsets $\Omega$ of $N_\rho(a)$, including $N_\rho(a)$, onto open sets in $\mathbf{E}^n$.

Let $N_\rho(a) = U$ and $f_*(N_\rho(a)) = V$, where $V$ is an open set in $\mathbf{E}^n$. Then, $f$: $U \leftrightarrow V$ is an open map. By Theorem 69.1, $f^{-1}$: $V \leftrightarrow U$ exists and is continuous on $V$.

It remains to be shown that $f^{-1} \in C^1(V)$. For every $b \in V$, $f^{-1}(b) = c$ is an interior point of $U$. By Lemma 97.1, $f'(c)$ is onto $\mathbf{E}^n$ and hence, rank $f'(c) = n$. Therefore, $f'(c)$ is also one-to-one. Since $f$: $U \leftrightarrow V$ maps open sets onto open sets, the existence of $(f^{-1})'(b)$ follows from Theorem 92.3 and we obtain

$$(f^{-1})'(y) = [f'(f^{-1}(y))]^{-1} \tag{98.2}$$

for all $y \in V$. If $f^{-1} = (\psi_1, \psi_2, \dots, \psi_n)$, then

$$(F^{-1})'(y) = \begin{pmatrix} D_1\psi_1(y) \dots D_n\psi_1(y) \\ \vdots \\ D_1\psi_n(y) \dots D_n\psi_n(y) \end{pmatrix}. \tag{98.3}$$

Since

$$F'(f^{-1}(y)) = \begin{pmatrix} D_1\phi_1(f^{-1}(y)) \dots D_n\phi_1(f^{-1}(y)) \\ \vdots \\ D_1\phi_n(f^{-1}(y)) \dots D_n\phi_n(f^{-1}(y)) \end{pmatrix},$$

we have

$$[F'(f^{-1}(y))]^{-1} = \frac{1}{J_f(f^{-1}(y))} \begin{pmatrix} \Phi_{11}(f^{-1}(y)) \dots \Phi_{n1}(f^{-1}(y)) \\ \vdots \\ \Phi_{1n}(f^{-1}(y)) \dots \Phi_{nn}(f^{-1}(y)) \end{pmatrix}, \tag{98.4}$$

where $\Phi_{ik}$ is the cofactor of $D_k\phi_i(f^{-1}(y))$ and, as a sum of products of continuous functions on $V$, is continuous on $V$. Since the matrix representation of a linear function is unique, we obtain, in view of (98.2), from (98.3) and (98.4), that

$$D_i\psi_k(y) = \frac{\Phi_{ik}(f^{-1}(y))}{J_f(f^{-1}(y))}. \tag{98.5}$$

By Theorem 96.1, $f'(x)$ is one-to-one for all $x \in U$, and hence, $J_f(f^{-1}(y)) \neq 0$ for all $y \in V$. This Jacobian, being a sum of products of continuous functions of a continuous function, is also continuous on $V$, and hence, $D_i\psi_k(y) \in C(V)$; that is, $f^{-1} \in C^1(V)$.

Finally, we note that $(f^{-1} \circ f) : U \leftrightarrow U$ is the identity function on $U$. By the chain rule $(f^{-1})'(f(x)) \circ f'(x) : \mathbf{E}^n \leftrightarrow \mathbf{E}^n$ is the identity function for each $x \in \mathbf{E}^n$, and therefore, $(F^{-1})'(f(x))F'(x) = I$ is the identity matrix. Therefore, that $J_{f^{-1}}(f(x))J_f(x) = 1$ for all $x \in U$ and $J_{f^{-1}}(y) \neq 0$ for all $y \in V$ follows readily.

Theorem 98.1 tells us, in essence, that under the stated conditions, $f^{-1}$ inherits from $f$ all continuity and differentiability properties. This result can easily be extended to the case where $f \in C^p(\mathfrak{D})$, i.e., has continuous partial derivatives of $p$th order in $\mathfrak{D}$. Then, $f^{-1} \in C^p(V)$, where $V$ is the set defined in the proof of Theorem 98.1. (See (98.5).)

That the conditions which are stated in Theorem 98.1 are not necessary for the existence of an inverse may be seen from the following example.

---

▲ *Example 1*

Let $f : \mathbf{E}^2 \to \mathbf{E}^2$ be given by

$$\phi_1(\xi_1, \xi_2) = \xi_2^3,$$
$$\phi_2(\xi_1, \xi_2) = \xi_1^3.$$

Then, $f'(\theta)$ is represented by its Jacobian matrix

$$F'(\theta) = \begin{pmatrix} 0 & 0 \\ 0 & 0 \end{pmatrix};$$

that is, $f'(\theta)$ is neither one-to-one nor onto. Still, $f^{-1}$ exists everywhere and, in particular, in every neighborhood of $\theta$, and is given by

$$\xi_1 = \sqrt[3]{\eta_2},$$
$$\xi_2 = \sqrt[3]{\eta_1}.$$

(Note that $f^{-1}$ is not differentiable at $\theta$.)

---

It is entirely possible for $f'(x)$ to be onto and one-to-one for each $x \in \mathfrak{D}$, where $\mathfrak{D}$ is some open subset of $\mathbf{E}^n$. One might be tempted to apply the inverse-function theorem to each $x \in \mathfrak{D}$, which is perfectly legitimate, determine in each case a $\rho(x) > 0$ such that $f\colon N_{\rho(x)}(x) \leftrightarrow f_*(N_{\rho(x)}(x))$ where $f_*(N_{\rho(x)}(x))$ is open in $\mathbf{E}^n$ (still valid), and then consider, for some compact $K \subseteq \mathfrak{D}$,

$$\{f_*(N_{\rho(x)}(x)) \mid x \in \mathfrak{D}\}$$

as an open cover of the compact set $f_*(K)$, select a finite subcover, and piece an inverse function together from finitely many inverses that have been obtained by Theorem 98.1, thus concluding that $f^{-1}$ exists on $f_*(K)$. This is *false*, because some of the $f_*(N_{\rho(x)}(x))$, say $f_*(N_{\rho(x_1)}(x_1))$ and $f_*(N_{\rho(x_2)}(x_2))$ with $N_{\rho(x_1)}(x_1) \cap N_{\rho(x_2)}((x_2) = \varnothing$, may coincide or overlap, thus making the points in the overlapping portion appear as images of at least two distinct points in $\mathfrak{D}$. Hence, $f$ need *not* be one-to-one on $K$. This will be demonstrated in the following example.

---

▲ *Example 2*

Let $f\colon \mathbf{E}^2 \to \mathbf{E}^2$ be given by

$$\phi_1(\xi_1, \xi_2) = \xi_1 \cos \xi_2,$$
$$\phi_2(\xi_1, \xi_2) = \xi_1 \sin \xi_2,$$

and let $\mathfrak{D} = \{(\xi_1, \xi_2) \in \mathbf{E}^2 \mid 1 < \xi_1 < 3, 0 < \xi_2 < 4\pi\}$. $f'(x)$ is represented by

$$F'(x) = \begin{pmatrix} \cos \xi_2 & -\xi_1 \sin \xi_2 \\ \sin \xi_2 & \xi_1 \cos \xi_2 \end{pmatrix},$$

and hence,

$$J_f(x) = \xi_1 \neq 0 \qquad \text{for all } x \in \mathfrak{D}.$$

Still, $f$ is *not* one-to-one on $\mathfrak{D}$ because, if $1 < \xi_1 < 3$ and $0 < \xi_2 < 2\pi$, then $(\xi_1, \xi_2)$ as well as $(\xi_1, \xi_2 + 2\pi) \in \mathfrak{D}$ are both mapped into the same image point. If we restrict ourselves, however, to a sufficiently small neighborhood of any given point in $\mathfrak{D}$, then, Theorem 98.1 applies and the existence of $f^{-1}$ and the continuity of its partial derivatives are assured. (In this particular case, $f$ is invertible on every $\mathfrak{D}_a = \{(\xi_1, \xi_2) \mid 1 < \xi_1 < 3,\ a < \xi_2 < a + 2\pi,\ a \in (0, 2\pi)\} \subset \mathfrak{D}$.)

▲ *Example 3*

Let $f: \mathbf{E}^2 \to \mathbf{E}^2$ be given by

$$\phi_1(\xi_1, \xi_2) = \xi_2,$$
$$\phi_2(\xi_1, \xi_2) = \xi_1 + \xi_2^2.$$

$f'(\theta)$ is represented by

$$F'(\theta) = \begin{pmatrix} 0 & 1 \\ 1 & 0 \end{pmatrix}.$$

Since $J_f(\theta) = \det F'(\theta) = -1 \neq 0$ and since all other conditions of Theorem 98.1 are obviously satisfied, we obtain some $N_\rho(\theta)$ where $f$ is invertible. We obtain, from

$$\eta_1 = \xi_2, \qquad \eta_2 = \xi_1 + \xi_2^2,$$

that

$$\xi_1 = \eta_2 - \eta_1^2, \qquad \xi_2 = \eta_1.$$

Hence, the inverse $f^{-1}$ is given by

$$\psi_1(\eta_1, \eta_2) = \eta_2 - \eta_1^2,$$
$$\psi_2(\eta_1, \eta_2) = \eta_1.$$

We may find $(f^{-1})'(y)$ either by direct computation as

$$(F^{-1})'(y) = \begin{pmatrix} -2\eta_1 & 1 \\ 1 & 0 \end{pmatrix},$$

or, from (98.2), and $F'(y) = \begin{pmatrix} 0 & 1 \\ 1 & 2\xi_2 \end{pmatrix}$:

$$(F^{-1})'(y) = [F'(f^{-1}(y))]^{-1} = \begin{pmatrix} 0 & 1 \\ 1 & 2\xi_2 \end{pmatrix}^{-1}_{\xi_2 = \eta_1} = \begin{pmatrix} -2\eta_1 & 1 \\ 1 & 0 \end{pmatrix}.$$

---

***Corollary to Theorem 98.1***

If $f: \mathfrak{D}$, $a$ are as in Theorem 98.1, if $J_f(a) \neq 0$, and if $f(a) = b$, then there are open sets $U, V \in \mathbf{E}^n$, $a \in U$, $b \in V$, such that the system of equations $f(x) = y$ has a unique solution $x \in U$ for all $y \in V$.

For an illustration, see example 1, section 9.95.

## 10.98 Exercises

1. Let $f$, $\mathbf{E}^2 \to \mathbf{E}^2$ be given by $\phi_1(\xi_1, \xi_2) = \xi_1, \phi_2(\xi_1, \xi_2) = \xi_1 \xi_2$. Apply Theorem 98.1 wherever it is applicable; find $f^{-1}$ and $(f^{-1})'$.
2. Let $f: \mathbf{E}^3 \to \mathbf{E}^3$ be given as in exercise 4, section 10.96. Let $\mathfrak{D} = \{(\xi_1, \xi_2, \xi_3) \in \mathbf{E}^3 \mid 1 < \xi_1 < 2, 0 < \xi_2 < 2\pi, 0 < \xi_3 < \pi\}$. Apply Theorem 98.1, find $f^{-1}$ on $f(\mathfrak{D})$ and find $(f^{-1})'$ on $f(\mathfrak{D})$.
3. Let $f: \mathbf{E}^3 \to \mathbf{E}^3$ be given by

$$\phi_1(\xi_1, \xi_2, \xi_3) = 2\xi_1\xi_2 - \xi_3,$$
$$\phi_2(\xi_1, \xi_2, \xi_3) = \xi_1 + \xi_2^2 + \xi_3^2,$$
$$\phi_3(\xi_1, \xi_2, \xi_3) = \xi_1 \xi_2^2 + \xi_2.$$

Apply Theorem 98.1 to some $N_\delta(\theta)$, show that $f$ is invertible, and find $(f^{-1})'(\theta)$.

4. Given $f, g : \mathbf{E}^n \to \mathbf{E}^n$. State and prove a theorem about the unique solution of the equation $f(g(x)) = y$ for suitable $y \in \mathbf{E}^n$.
5. Let $f: \mathbf{E}^2 \to \mathbf{E}^2$ be given by $\phi_1(\xi_1, \xi_2) = \xi_1 - \xi_2, \phi_2(\xi_1, \xi_2) = a\xi_1 + \xi_2$. Show that $f'(x) : \mathbf{E}^2 \leftrightarrow \mathbf{E}^2$ if and only if $a \neq -1$. Find $f_*([0, 1] \times [0, 1])$ for $a = 0, -1, 2$. Sketch. Let $a \neq -1$, apply Theorem 98.1, find $f^{-1}$ and $(f^{-1})'$.
6. Show: If $U, V \subseteq \mathbf{E}^n$ are open sets and $f: U \leftrightarrow V$ such that $f \in C^1(U)$, $f^{-1} \in C^1(V)$, then $f'(x) : \mathbf{E}^n \leftrightarrow \mathbf{E}^n$ for all $x \in U$.
7. Adapt the proof of Theorem 97.1 and prove the inverse-function theorem 98.1.

## 10.99 The Implicit-Function Theorem

The inverse-function theorem may be viewed as a generalization of the theorem on the unique solution of $n$ nonhomogeneous linear equations in $n$ unknowns. Similarly, one may view the implicit-function theorem as a generalization of the theorem on the solution of $n$ homogeneous linear equations in $(n + m)$ unknowns

$$\begin{aligned} a_{11}\xi_1 + \cdots + a_{1n}\xi_n + a_{1n+1}\eta_1 + \cdots + a_{1n+m}\eta_m &= 0, \\ \vdots \qquad\qquad\qquad\qquad & \ \ \vdots \\ a_{n1}\xi_1 + \cdots + a_{nn}\xi_n + a_{nn+1}\eta_1 + \cdots + a_{nn+m}\eta_m &= 0, \end{aligned}$$

for $\xi_1, \xi_2, \ldots, \xi_n$ in terms of $\eta_1, \eta_2, \ldots, \eta_m$. We know that this is possible whenever

$$\det \begin{pmatrix} a_{11} \ldots a_{1n} \\ \vdots \\ a_{n1} \ldots a_{nn} \end{pmatrix} \neq 0.$$

(In general, one may solve such a system for $n$ of the unknowns in terms of the others whenever

$$\operatorname{rank} \begin{pmatrix} a_{11} \ldots a_{1n+m} \\ \vdots \\ a_{n1} \ldots a_{nn+m} \end{pmatrix} = n.)$$

We consider now, more generally, a system of $n$ equations in $(n+m)$ unknowns:

$$\begin{aligned} &\phi_1(\xi_1, \xi_2, \ldots, \xi_n, \eta_1, \eta_2, \ldots, \eta_m) = 0, \\ (99.1) \qquad &\vdots \\ &\phi_n(\xi_1, \xi_2, \ldots, \xi_n, \eta_1, \eta_2, \ldots, \eta_m) = 0, \end{aligned}$$

where we assume, without loss of generality, that

$$\phi_j(0, 0, \ldots, 0, 0, \ldots, 0) = 0, \qquad j = 1, 2, \ldots, n.$$

(If we have, instead, $\phi_j(\overset{\circ}{\xi}_1, \overset{\circ}{\xi}_2, \ldots, \overset{\circ}{\xi}_n, \overset{\circ}{\eta}_1, \overset{\circ}{\eta}_2, \ldots, \overset{\circ}{\eta}_m) = 0$, $j = 1, 2, \ldots, n$, we translate the coordinate system according to $\bar{\xi}_j = \xi_j - \overset{\circ}{\xi}_j$, $\bar{\eta}_k = \eta_k - \overset{\circ}{\eta}_k$.)

We wish to find out under what conditions we can solve (99.1) for $\xi_1, \xi_2, \ldots, \xi_n$ in terms of $\eta_1, \eta_2, \ldots, \eta_m$ or, as we may put it, find unique functions $\chi_j$, $j = 1, 2, \ldots, n$ from a subset of $\mathbf{E}^m$ into $\mathbf{E}$ such that $\chi_j(0, 0, \ldots, 0) = 0$ and $\phi_k(\chi_1(\eta_1, \eta_2, \ldots, \eta_m), \ldots, \chi_n(\eta_1, \eta_2, \ldots, \eta_m), \eta_1, \ldots, \eta_m) = 0$ for all $k = 1, 2, \ldots, n$.

With the notation $f = (\phi_1, \phi_2, \ldots, \phi_n) : \mathfrak{D} \to \mathbf{E}^n$, $\mathfrak{D} \subseteq \mathbf{E}^n \times \mathbf{E}^m$, $x = (\xi_1, \xi_2, \ldots, \xi_n) \in \mathbf{E}^n$, $y = (\eta_1, \eta_2, \ldots, \eta_m) \in \mathbf{E}^m$, we may write (99.1) as

$$(99.2) \qquad f(x, y) = \theta,$$

and formulate our problem as follows: If $f(\theta, \theta) = \theta$, is there a set $V \subseteq \mathbf{E}^m$ with $\theta \in V$ and a function $h = (\chi_1, \chi_2, \ldots, \chi_n) : V \to \mathbf{E}^n$ such that $h(\theta) = \theta$ and $f(h(y), y) = \theta$ for all $y \in V$?

We shall solve this problem by augmenting the function $f$ in such a manner that Theorem 98.1 becomes applicable. We assume that $\mathfrak{D} \subseteq \mathbf{E}^n \times \mathbf{E}^m$ is open (see section 6.59), that $(\theta, \theta) \in \mathfrak{D}$, and that $f \in \mathrm{C}^1(\mathfrak{D})$. Then, the function $g : \mathfrak{D} \to \mathbf{E}^n \times \mathbf{E}^m$, which is defined by

$$(99.3) \qquad g(x, y) = (f(x, y), y),$$

satisfies $g \in C^1(\mathfrak{D})$. Since

$$G'(\theta, \theta) = \begin{pmatrix} D_1\phi_1(\theta,\theta) \ldots D_n\phi_1(\theta,\theta) & D_{n+1}\phi_1(\theta,\theta) \ldots D_{n+m}\phi_1(\theta,\theta) \\ \vdots & \\ D_1\phi_n(\theta,\theta) \ldots D_n\phi_n(\theta,\theta) & D_{n+1}\phi_n(\theta,\theta) \ldots D_{n+m}\phi_n(\theta,\theta) \\ 0 \quad \ldots \quad 0 & 1 \quad \ldots \quad 0 \\ \vdots \qquad\quad \vdots & \vdots \qquad\quad \vdots \\ 0 \quad \ldots \quad 0 & 0 \quad \ldots \quad 1 \end{pmatrix},$$

we have

$$J_g(\theta, \theta) = \begin{pmatrix} D_1\phi_1(\theta,\theta) \ldots D_n\phi_1(\theta,\theta) \\ \vdots \\ D_1\phi_n(\theta,\theta) \ldots D_n\phi_n(\theta,\theta) \end{pmatrix}.$$

Hence, if we assume that $J_g(\theta, \theta) \neq 0$, Theorem 98.1 becomes applicable to $g$. We obtain, from the Corollary to Theorem 98.1, that there are open sets $U, V_1 \subseteq \mathbf{E}^n \times \mathbf{E}^m$ with $(\theta, \theta) \in U$, $(\theta, \theta) \in V_1$ such that

$$(99.4) \qquad g(x, y) = z$$

has a unique solution $(x, y) \in U$ for all $z \in V_1$. Since, by hypothesis, $f(\theta, \theta) = \theta$, we have, from (99.3), that

$$g(\theta, \theta) = \theta. \tag{99.5}$$

With $z = (z_1, z_2)$, where $z_1 = (\zeta_1, \zeta_2, \ldots, \zeta_n)$, $z_2 = (\zeta_{n+1}, \zeta_{n+2}, \ldots, \zeta_{n+m})$, Eq. (99.4) reads

$$f(x, y) = z_1,$$
$$y = z_2.$$

Hence, the solution of (99.4) is of the form

$$\begin{aligned} x &= h_1(z_1, z_2), \\ y &= z_2, \end{aligned} \tag{99.6}$$

and

$$g(h_1(z_1, z_2), z_2) = (z_1, z_2) \tag{99.7}$$

for all $(z_1, z_2) \in V_1$. Since $g^{-1}(z_1, z_2) = (h_1(z_1, z_2), z_2)$ and since, by Theorem 98.1, $g^{-1} \in C^1(V_1)$, we see that $h_1 \in C^1(V_1)$. Since $g^{-1} : V_1 \to U$, we have

$$(h_1(z_1, z_2), z_2) \in U \qquad \text{for all } (z_1, z_2) \in V_1.$$

By Lemma 59.2,

$$V = \{y \in \mathbf{E}^m \mid (\theta, y) \in V_1\}$$

is open in $\mathbf{E}^m$. Since $(\theta, \theta) \in V_1$, we have $\theta \in V$. Let $h : V \to \mathbf{E}^n$ be defined by $h(z_2) = h_1(\theta, z_2)$ for all $z_2 \in V$. By (99.6) and (99.7),

$$g(h(y), y) = (\theta, y) \qquad \text{for all } y \in V,$$

and, by (99.3),

$$f(h(y), y) = \theta \qquad \text{for all } y \in V.$$

From (99.5), $g^{-1}(\theta, \theta) = (\theta, \theta)$ and, consequently $h_1(\theta, \theta) = \theta$ and hence, $h(\theta) = \theta$. Since $h_1 \in C^1(V_1)$, we have $h \in C^1(V)$. Since $(h_1(z_1, z_2), z_2) \in U$ for all $(z_1, z_2) \in V_1$, we have $(h(y), y) \in U$ for all $y \in V$.

This function $h : V \to \mathbf{E}^n$ satisfies all our requirements and we have, writing again $(x_0, y_0)$ instead of $(\theta, \theta)$:

***Theorem 99.1*** *(Implicit-Function Theorem)*

If $f : \mathfrak{D} \to \mathbf{E}^n$, where $\mathfrak{D} \subseteq \mathbf{E}^n \times \mathbf{E}^m$ is an open set, if $f \in C^1(\mathfrak{D})$, if $f(x_0, y_0) = \theta$ for some $(x_0, y_0) \in \mathfrak{D}$, and if

$$\begin{vmatrix} D_1\phi_1(x_0, y_0) \ldots D_n\phi_1(x_0, y_0) \\ \vdots \qquad\qquad \vdots \\ D_1\phi_n(x_0, y_0) \ldots D_n\phi_n(x_0, y_0) \end{vmatrix} \neq 0, \tag{99.8}$$

where $(\phi_1, \phi_2, \ldots, \phi_n)$ are the components of $f$, then there is an open set $U \subseteq \mathbf{E}^n \times \mathbf{E}^m$ with $(x_0, y_0) \in U$, an open set $V \subseteq \mathbf{E}^m$ with $y_0 \in V$, and a function $h: V \to \mathbf{E}^n$ such that $h(y_0) = x_0$ and

$$f(h(y), y) = 0$$

for all $y \in V$. Furthermore, $h \in C^1(V)$, and $h$ is uniquely determined on $V$ by $(h(y), y) \in U$ whenever $y \in V$.

*Proof*

All statements of this theorem, except the last one, have been established in the preceding discussion. In order to prove the uniqueness of $h$, we assume to the contrary, that there *are* two functions $h, k: V \to \mathbf{E}^n$ such that, for at least one $y_1 \in V, h(y_1) \neq k(y_1)$ but still, $(h(y_1), y_1), (k(y_1), y_1) \in U$ and $f(h(y_1), y_1) = \theta, f(k(y_1), y_1) = \theta$. We have, from (99.3), that $g(h(y_1), y_1) = g(k(y_1), y_1)$ which stands in contradiction to the fact that $g: V_1 \leftrightarrow U$.

---

▲ *Example 1*

Let $f(x, y) = x^2 + y^2$. We see that $f(0, 0) = 0$ but that condition (99.8) is *not* satisfied because $D_1 f(0, 0) = 0$. Theorem 99.1 does not apply, and there is indeed no solution to $f(x, y) = 0$.

▲ *Example 2*

Let $f(x, y) = x^2 - y^2$. Again, $f(0, 0) = 0$ but condition (99.8) is not satisfied because $D_1 f(0, 0) = 0$ and Theorem 99.1 does not apply. We obtain immediately two solutions, namely $x = h(y) = y$ and $x = h(y) = -y$. Both solutions are continuous and have a continuous derivative. In addition, there are discontinuous solutions such as

$$x = h(y) = \begin{cases} y & \text{if } y \in \mathbf{Q}, \\ -y & \text{if } y \in \mathbf{I}, \end{cases}$$

and others.

▲ *Example 3*

Let $f: \mathbf{E} \times \mathbf{E} \to \mathbf{E}$ be defined by

$$f(x, y) = \begin{cases} x + x^2 \sin \dfrac{1}{x} & \text{for } x \neq 0, \\ y & \text{for } x = 0. \end{cases}$$

Clearly, $f(0, 0) = 0$. We have, from example 3, section 9.90, that $D_1 f(0,0) = 1$ but that $D_1 f$ is not continuous at $(0,0)$. Note that $x + x^2 \sin(1/x) = 0$ has no solutions for $0 < |x| \leqslant 1$.

▲ *Example 4*

Let $f: \mathbf{E}^3 \to \mathbf{E}^2$ be given by

$$\phi_1(\xi, \eta, \xi) = \xi + \eta + \zeta,$$
$$\phi_2(\xi, \eta, \xi) = \xi - \eta - 2\xi\zeta.$$

We have

$$F'(\theta) = \begin{pmatrix} 1 & 1 & 1 \\ 1 & -1 & 0 \end{pmatrix}.$$

Since every one of the three minors is nonsingular, since all partial derivatives are continuous, and since $f(0, 0, 0) = \theta$, we may solve $f(\xi, \eta, \zeta) = \theta$ for any two of the unknowns in terms of the third one in some neighborhood of 0. We obtain, from

$$\xi + \eta + \zeta = 0,$$
$$\xi - \eta - 2\xi\zeta = 0,$$

that, for $x = (\xi, \eta)$, $y = \zeta$:

(99.9) $$h_1(y) = \left(\frac{\zeta}{2(\zeta - 1)}, \frac{\zeta - 2\zeta^2}{2(\zeta - 1)}\right), \qquad \zeta < 1;$$

that for $x = (\eta, \zeta)$, $y = \xi$:

(99.10) $$h_2(y) = \left(\frac{2\xi^2 + \xi}{1 - 2\xi}, \frac{2\xi}{2\xi - 1}\right), \qquad \xi < \frac{1}{2};$$

and that, for $x = (\xi, \zeta)$, $y = \eta$:

(99.11) $$h_3(y) = \left(\frac{-1 - 2\eta + \sqrt{1 + 12\eta + 4\eta^2}}{2}, \frac{1 - 2\eta - \sqrt{1 + 12\eta + 4\eta^2}}{2}\right),$$
$$\eta > -\frac{3}{2} + \sqrt{2}.$$

(Note that $\xi = \zeta = 0$ for $\eta = 0$ determines the sign of the square root in the above solution.) We note for later reference that

(99.12) $$H_1'(0) = \begin{pmatrix} -\frac{1}{2} \\ -\frac{1}{2} \end{pmatrix}, \qquad H_2'(0) = \begin{pmatrix} 1 \\ -2 \end{pmatrix}, \qquad H_3'(0) = \begin{pmatrix} 1 \\ -2 \end{pmatrix}.$$

## 10.99 Exercises

1. Verify the results in (99.9), (99.10), (99.11), and (99.12).
2. Given $\xi^3 - \xi\eta^2 + \eta\zeta^2 - \zeta^2 - 6 = 0$. This equation is satisfied for $\xi = 2$, $\eta = 1$, $\zeta = 2$. Is it possible to solve for $\xi$, or $\eta$, or $\zeta$ in a neighborhood of this point in terms of the remaining unknowns?

3. Let $f: \mathbf{E}^2 \times \mathbf{E}^2 \to \mathbf{E}^2$ be defined by

$$\phi_1(\xi_1, \xi_2, \eta_1, \eta_2) = \xi_1^3 + \xi_2 \eta_1 + \eta_2,$$
$$\phi_2(\xi_1, \xi_2, \eta_1, \eta_2) = \xi_1 \eta_2 + \xi_2^2 - \eta_1.$$

Can one solve $f(x, y) = \theta$ for $y$ in terms of $x$?

4. Let $f: \mathfrak{D} \to \mathbf{E}$ where $\mathfrak{D} \subseteq \mathbf{E}^3$. Let $f(0, 0, 0) = 0, f \in C^1(\mathfrak{D})$, and $D_1 f(0,0,0) \neq 0$. Prove the implicit-function theorem by applying the inverse-function theorem (Theorem 98.1) to

$$f(\xi_1, \eta_1, \eta_2) = \zeta_1,$$
$$\eta_1 = \zeta_2,$$
$$\eta_2 = \zeta_2.$$

5. In Theorem 99.1, let $U = N_\rho(\theta, \theta)$.

   (a) Show: If $y \in V$, then $\|y\| < \rho$.
   (b) Show: If $y \in V$, then $\|h(y)\| < \sqrt{\rho^2 - \|y\|^2}$.

## 10.100 Implicit Differentiation

Let $f: \mathfrak{D} \to \mathbf{E}$ where $\mathfrak{D} \subseteq \mathbf{E}^2$ is open and $(0, 0) \in \mathfrak{D}$. Suppose that $f \in C^1(\mathfrak{D})$, that $f(0, 0) = 0$, and that $D_1 f(0, 0) = (\partial f/\partial x)(0, 0) \neq 0$. By the implicit-function theorem, there are two open sets $U \subseteq \mathbf{E}^2$, $V \subseteq \mathbf{E}$, such that $(0, 0) \in U$, $0 \in V$, and a uniquely determined function $h : V \to \mathbf{E}$ such that $h(0) = 0$, $h \in C^1(V)$, $(h(y), y) \in U$ whenever $y \in V$, and

$$f(h(y), y) = 0 \tag{100.1}$$

for all $y \in V$.

If we differentiate the identity (100.1) with respect to $y$, we obtain, using the chain rule,

$$D_1 f(h(y), y) h'(y) + D_2 f(h(y), y) = 0$$

or, in the notation the reader is familiar with from calculus,

$$\frac{\partial f}{\partial x}(h(y), y) h'(y) + \frac{\partial f}{\partial y}(h(y), y) = 0.$$

Since $(\partial f/\partial x)(0, 0) \neq 0$, it follows, from Theorem 98.1, that $(\partial f/\partial x)(x, y) \neq 0$ for all $(x, y) \in U$. Hence, we may solve for $h'(y)$ and obtain

$$h'(y) = -\frac{\dfrac{\partial f}{\partial y}(h(y), y)}{\dfrac{\partial f}{\partial x}(h(y), y)} \quad \text{for all } y \in V. \tag{100.2}$$

The reader is familiar with this formula from calculus. The process that yields the result in (100.2) is called *implicit differentiation*. We shall now generalize this process, assuming that $f$, $\mathfrak{D}$ are as in Theorem 99.1.

For any given $(x, y) \in \mathfrak{D}$, we define a linear function $(\partial f/\partial x)(x, y) : \mathbf{E}^n \to \mathbf{E}^n$ by

$$\frac{\partial f}{\partial x}(x, y)(u) = f'(x, y)(u, \theta) \qquad \text{for all } u \in \mathbf{E}^n. \tag{100.3}$$

If we represent $f'(x, y)$ by its Jacobian matrix, we obtain the following representation of $f'(x, y)(u, \theta)$:

$$F'(x, y)(u, \theta) = \begin{pmatrix} D_1\phi_1(x, y) \dots D_n\phi_1(x, y) & D_{n+1}\phi_1(x, y) \dots D_{n+m}\phi_1(x, y) \\ \vdots & \vdots \\ D_1\phi_n(x, y) \dots D_n\phi_n(x, y) & D_{n+1}\phi_n(x, y) \dots D_{n+m}\phi_n(x, y) \end{pmatrix} \begin{pmatrix} \omega_1 \\ \vdots \\ \omega_n \\ 0 \\ \vdots \\ 0 \end{pmatrix}$$

$$= \begin{pmatrix} D_1\phi_1(x, y) \dots D_n\phi_1(x, y) \\ \vdots \\ D_1\phi_n(x, y) \dots D_n\phi_n(x, y) \end{pmatrix} \begin{pmatrix} \omega_1 \\ \vdots \\ \omega_n \end{pmatrix}.$$

We see that, in effect, $(\partial f/\partial x)(x, y)$ is the derivative of $f$, considered as a function of $x$ only. This justifies the notation which suggests a partial derivative. (If $n = 1$, then $(\partial f/\partial x)(x, y) = D_1 f(x, y)$ but conceptually, the two are *not* the same. The existence of $(\partial f/\partial x)(x, y)$ presupposes the existence of $f'(x, y)$ but the existence of $D_1 f(x, y)$ does not.)

We observe that if (99.8) holds, then $J_g(x, y) \neq 0$ for all $(x, y) \in U$. Hence the linear function $(\partial f/\partial x)(x, y)$ is invertible for each $(x, y) \in U$. If $y \in V$, then $(h(y), y) \in U$ and hence,

$$\left[\frac{\partial f}{\partial x}(h(y), y)\right]^{-1} \quad \text{exists} \quad \text{for all } y \in V. \tag{100.4}$$

Similarly, we define a linear function $(\partial f/\partial y)(x, y) : \mathbf{E}^m \to \mathbf{E}^m$, $(x, y) \in \mathfrak{D}$, by

$$\frac{\partial f}{\partial y}(x, y)(v) = f'(x, y)(\theta, v) \qquad \text{for all } v \in \mathbf{E}^m. \tag{100.5}$$

Since $f'(x, y) : \mathbf{E}^n \times \mathbf{E}^m \to \mathbf{E}^n$ is a linear function, we obtain, from (100.3) and (100.5), that

$$\frac{\partial f}{\partial x}(x, y)(u) + \frac{\partial f}{\partial y}(x, y)(v) = f'(x, y)(u, v) \qquad \text{for all } u \in \mathbf{E}^n, v \in \mathbf{E}^m. \tag{100.6}$$

In terms of the function $h : V \to \mathbf{E}^n$, $h \in C^1(V)$, the existence of which is guaranteed by the implicit-function theorem, we define a function $k : V \to \mathbf{E}^n \times \mathbf{E}^m$ by

$$k(y) = (h(y), y), \; y \in V, \tag{100.7}$$

and write $f(h(y), y) = \theta$ as

$$(f \circ k)(y) = \theta.$$

The composite function $f \circ k$, being the zero function, is differentiable, and we obtain

$$(100.8) \qquad (f \circ k)'(y) = \theta \qquad \text{for all } y \in V.$$

Since $h \in C^1(V)$, we have $k \in C^1(V)$ and we obtain, from (100.7),

$$k'(y)(v) = (h'(y)(v), v), \qquad \text{for all } v \in \mathbf{E}^m.$$

Since $f$ is differentiable on $\mathfrak{D}$, since $U \subseteq \mathfrak{D}$, and since $(h(y), y) \in U$ for all $y \in V$, we obtain, differentiating (100.8) by the chain rule,

$$\theta = (f \circ k)'(y)(v) = f'(k(y)) \circ k'(y)(v) = f'(h(y), y) \circ (h'(y)(v), v).$$

(See also exercise 3.) By (100.6),

$$f'(h(y), y)(u, v) = \frac{\partial f}{\partial x}(h(y), y)(u) + \frac{\partial f}{\partial y}(h(y), y)(v).$$

Hence,

$$\frac{\partial f}{\partial x}(h(y), y) \circ (h'(y)(v)) + \frac{\partial f}{\partial y}(h(y), y)(v) = \theta,$$

and we obtain, in view of (100.4), that

$$h'(y)(v) = -\left[\frac{\partial f}{\partial x}(h(y), y)\right]^{-1} \frac{\partial f}{\partial y}(h(y), y)(v) \qquad \text{for all } v \in \mathbf{E}^m,$$

or simply,

$$h'(y) = -\left[\frac{\partial f}{\partial x}(h(y), y)\right]^{-1} \frac{\partial f}{\partial y}(h(y), y), \qquad \text{for all } y \in V.$$

This formula is a direct generalization of (100.2).

We summarize our result in the following theorem:

***Theorem 100.1***

Under the conditions of Theorem 99.1, the derivative of $h : V \to \mathbf{E}^n$ is given by

$$h'(y) = -\left[\frac{\partial f}{\partial x}(h(y), y)\right]^{-1} \frac{\partial f}{\partial y}(h(y), y)$$

for all $y \in V$, where $f(h(y), y) = 0$ for all $y \in V$ and where $(\partial f/\partial x)$ and $(\partial f/\partial y)$ are defined in (100.3) and (100.5), respectively.

---

▲ *Example 1*

Let $f : \mathbf{E}^3 \to \mathbf{E}^2$ be given as in example 4, section 10.99. If $x = (\xi, \eta)$, $y = \zeta$, then

$$\frac{\partial f}{\partial x}(0, 0, 0) = \begin{pmatrix} 1 & 1 \\ 1 & -1 \end{pmatrix}, \qquad \frac{\partial f}{\partial y}(0, 0, 0) = \begin{pmatrix} 1 \\ 0 \end{pmatrix},$$

$$\left[\frac{\partial f}{\partial x}(0, 0, 0)\right]^{-1} = \begin{pmatrix} \frac{1}{2} & \frac{1}{2} \\ \frac{1}{2} & -\frac{1}{2} \end{pmatrix},$$

and hence, by Theorem 100.1,

$$H_1'(0) = -\begin{pmatrix} \frac{1}{2} & \frac{1}{2} \\ \frac{1}{2} & -\frac{1}{2} \end{pmatrix}\begin{pmatrix} 1 \\ 0 \end{pmatrix} = \begin{pmatrix} -\frac{1}{2} \\ -\frac{1}{2} \end{pmatrix}.$$

If $x = (\eta, \zeta)$, $y = \xi$, then

$$\frac{\partial f}{\partial x}(0, 0, 0) = \begin{pmatrix} 1 & 1 \\ -1 & 0 \end{pmatrix}, \quad \frac{\partial f}{\partial y}(0, 0, 0) = \begin{pmatrix} 1 \\ 1 \end{pmatrix},$$

$$\left[\frac{\partial f}{\partial x}(0, 0, 0)\right]^{-1} = \begin{pmatrix} 0 & -1 \\ 1 & 1 \end{pmatrix},$$

and hence,

$$H_2'(0) = -\begin{pmatrix} 0 & -1 \\ 1 & 1 \end{pmatrix}\begin{pmatrix} 1 \\ 1 \end{pmatrix} = \begin{pmatrix} 1 \\ -2 \end{pmatrix}.$$

If $x = (\xi, \zeta)$, $y = \eta$, then

$$\frac{\partial f}{\partial x}(0, 0, 0) = \begin{pmatrix} 1 & 1 \\ 1 & 0 \end{pmatrix}, \quad \frac{\partial f}{\partial y}(0, 0, 0) = \begin{pmatrix} 1 \\ -1 \end{pmatrix},$$

$$\left[\frac{\partial f}{\partial x}(0, 0, 0)\right]^{-1} = \begin{pmatrix} 0 & 1 \\ 1 & -1 \end{pmatrix},$$

and hence,

$$H_3'(0) = -\begin{pmatrix} 0 & 1 \\ 1 & -1 \end{pmatrix}\begin{pmatrix} 1 \\ -1 \end{pmatrix} = \begin{pmatrix} 1 \\ -2 \end{pmatrix}.$$

(Compare these results with (99.12).)

---

## 10.100 Exercises

1. Show: If $(\partial f/\partial x)(\theta, \theta) \neq 0$, then $(\partial f/\partial x)(x, y) \neq 0$ for all $(x, y) \in U$, where $(\partial f/\partial x)$ is defined as in (100.3) and $U$ is defined in the proof of Theorem 99.1.
2. Let $f: \mathbf{E}^3 \to \mathbf{E}^2$ be given by

$$\phi_1(\xi, \eta, \zeta) = \xi\eta + \zeta - \eta^2,$$
$$\phi_2(\xi, \eta, \zeta) = \xi \cos \eta + \zeta^2.$$

   Clearly, $f(0, 0, 0) = (0, 0)$. Show that one can solve $f(\xi, \eta, \zeta) = \theta$ for $\xi, \zeta$ in terms of $\eta$ in a neighborhood of 0 and, without actually solving, find the derivative of the solution at $\eta = 0$.
3. Given $(f \circ k)'(y) = 0$ for all $y \in V \subseteq \mathbf{E}^m$ where $V$ is open. $k: V \to \mathbf{E}^n \times \mathbf{E}^m$ is defined as in (100.7). Show that

$$f'(h(y), y) \circ (h'(y), I): \mathbf{E}^m \to \theta,$$

   where $I: \mathbf{E}^m \to \mathbf{E}^m$ is the identity function.

4. Given

$$\xi_3 \sin \xi_1 \cos \xi_2 - \eta_1 = 0,$$
$$\xi_3 \sin \xi_1 \sin \xi_2 - \eta_2 = 0,$$
$$\xi_3 \cos \xi_1 - \sqrt{1 - \eta_1^2 - \eta_2^2} = 0.$$

Let $x = h(y)$ represent the solution for $\xi_1, \xi_2, \xi_3$ in terms of $\eta_1, \eta_2$ and find $h'(y)$.

5. Given that under the hypotheses of Theorem 99.1, rank $h'(y) = 0$ for all $y \in V$. What can you say about $f$?

## 10.101 Extreme Values*

We generalize definition 37.1 of a relative maximum (minimum) of a real-valued function of a real variable, in a straightforward manner, as follows:

***Definition 101.1***

The function $f: \mathfrak{D} \to \mathbf{E}$, $\mathfrak{D} \subseteq \mathbf{E}^n$, assumes a *relative maximum* (*minimum*) at $a \in \mathfrak{D}$ if there is a $\delta > 0$ such that $f(a) \geqslant f(x)$ ($\leqslant f(x)$) for all $x \in N_\delta(a) \cap \mathfrak{D}$.

---

▲ *Example 1*

Given $f: \mathbf{E}^3 \to \mathbf{E}$ by $f(\xi_1, \xi_2, \xi_3) = \xi_1^2 + \xi_2^2 + \xi_3^2$. Clearly, $f(\theta) = 0 \leqslant f(x)$ for all $x \in \mathbf{E}^3$. ($f$ assumes not only a *relative* minimum at $(0, 0, 0)$ but an *absolute* minimum.)

---

Suppose that $\mathfrak{D}$ is open, that $f \in C^1(\mathfrak{D})$, and that $f$ assumes a relative minimum at $a \in \mathfrak{D}$; then

$$f(a) = \alpha \leqslant f(x) \qquad \text{for all } x \in N_\delta(a) \subseteq \mathfrak{D} \text{ for some } \delta > 0.$$

If rank $f'(a) = 1 = m$, then, according to Theorem 97.1, there is a $\rho > 0$ such that $f: N_\rho(a) \to \mathbf{E}$ maps all open subsets of $N_\rho(a)$ onto open sets in $\mathbf{E}$. Hence, for every $\delta_1 \in (0, \rho)$, $f_*(N_{\delta_1}(a))$ is open in $\mathbf{E}$ and contains, of course, $f(a) = \alpha$. Choose $\beta \neq 0$ such that $\alpha - \beta^2 \in f_*(N_{\delta_1}(a))$. Then there is an $x \in N_{\delta_1}(a)$ such that $f(x) = \alpha - \beta^2 < f(a)$. Since we may choose $\delta_1 > 0$ as small as we please, it follows that $f(a) = \alpha$ cannot be a relative minimum. Hence, for $f$ to assume a relative minimum at $a \in \mathfrak{D}$, it is necessary that rank $f'(a) = 0$; that is, $f'(a)$ is the zero function. (The same condition is obtained for a relative maximum if one chooses, in the above derivation, $\beta \neq 0$ such that $\alpha + \beta^2 \in f_*(N_{\delta_1}(a))$.) We summarize our result as a theorem.

***Theorem 101.1***

Let $\mathfrak{D} \subseteq \mathbf{E}^n$ denote an open set and let $f: \mathfrak{D} \to \mathbf{E}$. If $f \in C^1(\mathfrak{D})$, and if $f$ assumes a relative maximum (minimum) at $a \in \mathfrak{D}$, then it is necessary that $f'(a) = \theta$ which, in turn, implies that $D_j f(a) = 0$ for $j = 1, 2, \ldots, n$.

Theorem 101.1 is somewhat weaker than its counterpart (Theorem 37.1) for a real-valued function of a real variable, where we merely assumed $f$ to be differentiable at $a$. Here we imposed the much stronger hypothesis that $f \in C^1(\mathfrak{D})$. One can obtain a theorem pertaining to real-valued functions of a vector variable that is analogous to Theorem 37.1 (see exercise 1), but we have chosen the approach that led to Theorem 101.1 to serve as a model for our subsequent derivation of a necessary condition for a constrained relative maximum (minimum) of a real-valued function of a vector variable.

---

▲ *Example 2*

To be found is a point with coordinates $(\xi_1, \xi_2, \xi_3)$ on the plane

$$\xi_1 + \xi_2 + \xi_3 = 1, \tag{101.1}$$

which has the shortest distance from the point $(1, 1, 1) \in \mathbf{E}^3$. Since the distance is a minimum if and only if its square is a minimum, our problem amounts to a minimization of the function $d : \mathbf{E}^3 \to \mathbf{E}$, that is given by

$$d(\xi_1, \xi_2, \xi_3) = (\xi_1 - 1)^2 + (\xi_2 - 1)^2 + (\xi_3 - 1)^2 \tag{101.2}$$

under the condition that $(\xi_1, \xi_2, \xi_3)$ satisfy (101.1).

Such a minimum (if it exists) is called a *constrained* minimum, (101.1) being the constraint (*constraining equation*).

A pedestrian approach to the solution of this problem would be to solve (101.1) for one of the unknowns in terms of the others; for example,

$$\xi_1 = 1 - \xi_2 - \xi_3, \tag{101.3}$$

now substitute into (101.2):

$$d(1 - \xi_2 - \xi_3, \xi_2, \xi_3) = (\xi_2 + \xi_3)^2 + (\xi_2 - 1)^2 + (\xi_3 - 1)^2 = \bar{d}(\xi_2, \xi_3);$$

and apply Theorem 101.1:

$$\bar{d}'(\xi_2, \xi_3) = (2(\xi_2 + \xi_3) + 2(\xi_2 - 1), \qquad 2(\xi_2 + \xi_3) + 2(\xi_3 - 1)) = \theta.$$

Hence, if $\bar{d}$ assumes a relative minimum at $(\xi_2, \xi_3)$, then, by necessity,

$$2(\xi_2 + \xi_3) + 2(\xi_2 - 1) = 0, \qquad 2(\xi_2 + \xi_3) + 2(\xi_3 - 1) = 0.$$

These equations have the unique solution $\xi_2 = \frac{1}{3}$, $\xi_3 = \frac{1}{3}$ and, from (101.3), $\xi_1 = \frac{1}{3}$. Whether or not $d$ actually assumes a relative minimum (or maximum) at this point, is another matter. (See exercises 2 through 4.)

---

At times it may not be practical, or it may even be humanly impossible, to solve the constraining equation, or equations, in a closed form, for some of the unknowns in terms of the others (even though the existence of such a solution may be guaranteed by the implicit-function theorem). Therefore we shall now derive a necessary condition for a constrained relative maximum (minimum) that does *not* require the explicit knowledge of a solution of the constraining equations. Before doing so, let us state a precise definition of constrained relative maximum (minimum):

***Definition 101.2***

If $f\colon \mathfrak{D} \to \mathbf{E}$ where $\mathfrak{D} \subseteq \mathbf{E}^n$, and if $\mathfrak{C} \subseteq \mathfrak{D}$ is a *given* subset of $\mathfrak{D}$, then $f$ assumes a *constrained relative maximum* (*minimum*) at $a \in \mathfrak{C}$ if there is a $\delta > 0$ such that $f(a) \geqslant f(x)$ $(\leqslant f(x))$ for all $x \in N_\delta(a) \cap \mathfrak{C}$, the constraint being given by the restriction $x \in \mathfrak{C}$.

Let $g : \mathfrak{D} \to \mathbf{E}^m$ where $\mathfrak{D} \subseteq \mathbf{E}^n$ is open and where $m < n$, $g \in C^1(\mathfrak{D})$, and let

(101.4) $$\mathfrak{C} = \{x \in \mathfrak{D} \mid g(x) = \theta\},$$

where we assume that $\mathfrak{C} \neq \varnothing$; that is, there is at least one $x \in \mathfrak{D}$ such that $g(x) = \theta$.

We assume that $f$ assumes a constrained relative minimum at $a$ on $\mathfrak{C}$, where $\mathfrak{C}$ is given in (101.4); that is, $f(a) \leqslant f(x)$ for all $x \in N_\delta(a) \cap \mathfrak{C}$ for some $\delta > 0$. We consider the function $h : \mathfrak{D} \to \mathbf{E} \times \mathbf{E}^m$, which is defined by

$$h(x) = (f(x), g(x)) \qquad \text{for all } x \in \mathfrak{D}.$$

By hypothesis, $h(a) = (f(a), \theta)$. Suppose that rank $h'(a) = m + 1$. As in the derivation of Theorem 101.1, we invoke Theorem 97.1, according to which there is a $\rho > 0$ such that $h : N_\delta(a) \to \mathbf{E} \times \mathbf{E}^m$ maps all open subsets of $N_\delta(a)$ onto open sets in $\mathbf{E} \times \mathbf{E}^m$. In particular, for every $\delta_1 \in (0, \rho)$, $h_*(N_{\delta_1}(a))$ is open in $\mathbf{E} \times \mathbf{E}^m$ and contains $(f(a), \theta)$. We choose a $\beta \neq 0$, such that $(f(a) - \beta^2, \theta) \in h_*(N_{\delta_1}(a))$. Then, there is an $x \in N_{\delta_1}(a)$ such that $f(x) = f(a) - \beta^2 < f(a)$ and $g(x) = \theta$. Since we may choose $\delta_1 > 0$ as small as we please, $f(a)$ cannot be a constrained relative minimum. Hence, by necessity, rank $h'(a) < m + 1$. (The same condition is obtained for a constrained relative maximum by choosing, in the above argument, $\beta \neq 0$ such that $(f(a) + \beta^2, \theta) \in h_*(N_{\delta_1}(a))$.)

Let $g = (\psi_1, \psi_2, \ldots, \psi_m)$. Since $h'(a) = (f'(a), g'(a))$, we may represent $h'(a)$ by the following Jacobian matrix:

$$H'(a) = \begin{pmatrix} D_1 f(a) \ldots & D_n f(a) \\ D_1\psi_1(a) \ldots & D_n\psi_1(a) \\ \vdots & \\ D_1\psi_m(a) \ldots & D_n\psi_m(a) \end{pmatrix}.$$

rank $h'(a) < m + 1$ is equivalent to rank $H'(a) < m + 1$ which, in turn, means that the rows in $H'(a)$ are linearly dependent. Hence, there are $(m + 1)$ constants $(\lambda_0, \lambda_1, \ldots, \lambda_m) \neq (0, 0, \ldots, 0)$ such that

$$\lambda_0 D_j f(a) + \lambda_1 D_j \psi_1(a) + \cdots + \lambda_m D_j \psi_m(a) = 0, \qquad j = 1, 2, \ldots, n.$$

The result, which we shall summarize in the following theorem, is called the *Lagrange multiplier rule*, and the constants $\lambda_0, \lambda_1, \ldots, \lambda_m$ are called the *Lagrange multipliers*.†

† *Joseph Louis de Lagrange*, 1736–1813.

***Theorem 101.2*** (*Lagrange Multiplier Rule*)

Let $f: \mathfrak{D} \to \mathbf{E}$, $g = (\psi_1, \psi_2, \dots, \psi_m): \mathfrak{D} \to \mathbf{E}^m$ where $\mathfrak{D} \subseteq \mathbf{E}^n$ is an open set and $m < n$, and let $f, g \in C^1(\mathfrak{D})$. If $f$ assumes a constrained relative maximum (minimum) at $a \in \mathfrak{D}$ under the constraint $g(x) = \theta$, then, by necessity, there are $(m + 1)$ constants $(\lambda_0, \lambda_1, \dots, \lambda_m) \neq (0, 0, \dots, 0)$ such that

$$\lambda_0 D_j f(a) + \lambda_1 D_j \psi_1(a) + \cdots + \lambda_m D_j \psi_m(a) = 0 \tag{101.5}$$

for all $j = 1, 2, \dots, n$.

If rank $g'(a) = m$ which, by the implicit-function theorem, guarantees that one can solve $g(x) = \theta$ for $m$ of the $\xi_i$ in terms of the remaining $(n - m)$ of the $\xi_i$, then the condition rank $h'(a) < m + 1$ implies that the first row in $H'(a)$ has to be a linear combination of the remaining $m$ rows. This means, in turn, that one may choose $\lambda_0 = 1$, and that the remaining Lagrange multipliers are uniquely determined.

***Corollary to Theorem 101.2***

If all conditions in Theorem 101.2 are met and if, in addition, rank $g'(a) = m$, then, for $f$ to assume a constrained relative maximum (minimum) at $a \in \mathfrak{D}$ under the constraint $g(x) = \theta$, it is necessary that there be $m$ uniquely determined constants $\lambda_1, \lambda_2, \dots, \lambda_m$ such that

$$D_j f(a) + \lambda_1 D_j \psi_1(a) + \cdots + \lambda_m D_j \psi_m(a) = 0$$

for all $j = 1, 2, \dots, n$.

---

▲ *Example 3*

We consider the same problem as in example 2. We have $g(x) = \xi_1 + \xi_2 + \xi_3 - 1$. Since $g'(x) = (1, 1, 1)$, we have rank $g'(x) = 1 = m$ for all $x \in \mathbf{E}^3$ and the Corollary to Theorem 101.2 applies: There has to be a constant $\lambda_1$ such that

$$2(\xi_1 - 1) + \lambda_1 = 0,$$
$$2(\xi_2 - 1) + \lambda_1 = 0,$$
$$2(\xi_3 - 1) + \lambda_1 = 0.$$

We solve these equations in conjunction with the constraining equation (101.1) for $\lambda_1, \xi_1, \xi_2, \xi_3$, and obtain $\lambda_1 = \frac{4}{3}, \xi_1 = \xi_2 = \xi_3 = \frac{1}{3}$.

▲ *Example 4*

We wish to minimize (or maximize) $(\xi_1 - 1)^2 + \xi_2^2 + \xi_3^2$ under the constraints

$$\xi_1 + \xi_2 + \xi_3 = 0,$$
$$\xi_1^2 \qquad + \xi_3^2 = 0.$$

We have

$$G'(x) = \begin{pmatrix} 1 & 1 & 1 \\ 2\xi_1 & 0 & 2\xi_3 \end{pmatrix}.$$

By Theorem 101.2, there have to be three constants $(\lambda_0, \lambda_1, \lambda_2) \neq (0, 0, 0)$ such that

$$\begin{aligned} 2\lambda_0(\xi_1 - 1) + \lambda_1 + 2\lambda_2 \xi_1 &= 0, \\ 2\lambda_0 \xi_2 \qquad + \lambda_1 \qquad &= 0, \\ 2\lambda_0 \xi_3 \qquad + \lambda_1 + 2\lambda_2 \xi_3 &= 0. \end{aligned}$$

If we solve these equations in conjunction with the constraining equations, we obtain $\xi_1 = \xi_2 = \xi_3 = 0$, $\lambda_0 = 0$, $\lambda_1 = 0$, $\lambda_2$ arbitrary. This is a case where rank $g'(\theta) < 2 = m$.

---

We have taken up the Lagrange multiplier rule because it does not require the explicit knowledge of a solution of the constraining equations in the sense of the implicit-function theorem. It appears now that a solution of the constraining equations cannot be avoided after all if one wants to apply the Lagrange multiplier rule. As a matter of fact, one faces the seemingly more formidable task of having to solve the constraining equations in conjunction with the $n$ equations (101.5). This is not necessarily so. The equations (101.5) may actually simplify the constraining equations. One can also obtain interesting secondary information about the maximum or minimum from the Lagrange multiplier rule without actually solving the equations that result from its application.

---

▲ *Example 5*

We want to find the maximum of the function $f: \mathbf{E}^n \to \mathbf{E}$ which is given by

(101.6) $$f(x) = (\xi_1 \xi_2 \dots \xi_n)^2$$

under the constraint $\|x\| = 1$, which we may write as

(101.7) $$\sum_{k=1}^{n} \xi_k^2 - 1 = 0.$$

With $g(x) = \sum_{k=1}^{n} \xi_k^2 - 1$, we have

$$G'(x) = (2\xi_1, 2\xi_2, \dots, 2\xi_n),$$

and hence, rank $g'(x) = 1$ for all $\|x\| = 1$. Therefore, we may use the Lagrange multiplier rule as stated in the Corollary to Theorem 101.2, and choose $\lambda_0 = 1$. If $f$ assumes a constrained maximum at $x = a = (\alpha_1, \alpha_2, \dots, \alpha_n)$, then, by necessity

(101.8) $$(\alpha_1\alpha_2 \dots \alpha_n)(\alpha_1\alpha_2 \dots \alpha_{k-1}\alpha_{k+1} \dots \alpha_n) + \lambda_1\alpha_k = 0, \qquad k = 1, 2, \dots, n.$$

If $\alpha_j = 0$ for some $j \in \{1, 2, \dots, n\}$, then $f$ clearly assumes a minimum. Hence, we may assume that $\alpha_j \neq 0$ for all $j = 1, 2, \dots, n$, and multiply (101.8) by $\alpha_k$, add the $n$ equations, and obtain, in view of (101.7), that

$$\lambda_1 = -n(\alpha_1 \alpha_2 \dots \alpha_n)^2,$$

and hence,

$$\alpha_k = \pm \frac{1}{\sqrt{n}}. \tag{101.9}$$

We observe that (101.7) defines a compact subset of $\mathbf{E}^n$, namely the surface of the $n$-dimensional unit ball with center at $\theta$. By Theorem 71.2, $f$ assumes a maximum on that set. The only points where a maximum can occur are the ones given in (101.9). Since, for all possible sign combinations, $f(\pm 1/\sqrt{n}, \pm 1/\sqrt{n}, \dots, \pm 1/\sqrt{n}) = 1/n^n$, we see that a maximum is attained at each such point and that $(1/n^n)$ is, in fact, the maximum of $f$ on (101.7).

Hence, for all unit vectors

$$z = \frac{1}{\sqrt{\xi_1^2 + \xi_2^2 + \cdots + \xi_n^2}} (\xi_1, \xi_2, \dots, \xi_n),$$

we have that $f(z) \leqslant (1/n^n)$, which means, in turn, that

$$(\xi_1^2 \xi_2^2 \dots \xi_n^2)^{1/n} \leqslant \frac{1}{n} (\xi_1^2 + \xi_2^2 + \cdots + \xi_n^2).$$

Hence, we have, for any nonnegative numbers $a_1, a_2, \dots, a_n$, that

$$\sqrt[n]{a_1 a_2 \dots a_n} \leqslant \frac{1}{n} (a_1 + a_2 + \cdots + a_n), \qquad a_j \geqslant 0 \qquad \text{for all } j = 1, 2, \dots, n,$$

meaning that the *geometric mean never exceeds the arithmetic mean.* (If some of the $a_j$ are negative, this inequality need not be true: Take $a_1 = a_2 = 1, a_3 = -27$, $n = 3$. Then $\sqrt[3]{-27} = -3$, $\frac{1}{3}(1 + 1 - 27) = -\frac{25}{3}$.)

▲ *Example 6*

We shall derive the CBS inequality by means of the Lagrange multiplier rule. We consider the function $f\colon \mathbf{E}^n \times \mathbf{E}^n \to \mathbf{E}$ which is defined by

$$f(x, y) = \xi_1 \eta_1 + \xi_2 \eta_2 + \cdots + \xi_n \eta_n \tag{101.10}$$

under the constraints $\|x\| = 1$, $\|y\| = 1$, or, as we may put it,

$$\begin{aligned} \xi_1^2 + \xi_2^2 + \cdots + \xi_n^2 &= 1, \\ \eta_1^2 + \eta_2^2 + \cdots + \eta_n^2 &= 1. \end{aligned} \tag{101.11}$$

(101.11) defines a compact subset of $\mathbf{E}^n \times \mathbf{E}^n$ and we see again that $f$ assumes its maximum and minimum on (101.11). The only points where such extreme values may occur are found from the Lagrange multiplier rule. Since $g'(x)$ is represented by

$$G'(x) = \begin{pmatrix} 2\xi_1 \ldots 2\xi_n 0 \ldots 0 \\ 0 \ldots 0 2\eta_1 \ldots 2\eta_n \end{pmatrix},$$

we see that rank $g'(x) = 2$ for all $\|x\| = 1$, $\|y\| = 1$.

Hence, if $f$ assumes a constrained minimum or maximum at $x = a = (\alpha_1, \alpha_2, \ldots, \alpha_n)$, $y = b = (\beta_1, \beta_2, \ldots, \beta_n)$, then we have, from the Corollary to Theorem 101.2, that, by necessity,

$$\beta_k = -2\lambda_1\alpha_k, \quad \alpha_k = -2\lambda_2\beta_k, \qquad k = 1, 2, \ldots, n.$$

We square these equations and sum from 1 to $n$ to obtain $\lambda_1 = \pm\frac{1}{2}$, $\lambda_2 = \pm\frac{1}{2}$, and hence, $\alpha_k = \pm\beta_k$ with all possible sign combinations. If $\alpha_k = \beta_k$ for all $k = 1, 2, \ldots, n$, we obtain, in view of (101.11), that

$$f(\alpha_1, \alpha_2, \ldots, \alpha_n, \alpha_1, \alpha_2, \ldots, \alpha_n) = \alpha_1^2 + \cdots + \alpha_n^2 = 1,$$

and if $\alpha_k = -\beta_k$ for all $k = 1, 2, \ldots, n$, then

$$\begin{aligned} f(-\alpha_1, -\alpha_2, \ldots, -\alpha_n, \alpha_1, \alpha_2, \ldots, \alpha_n) &= f(\alpha_1, \alpha_2, \ldots, \alpha_n, -\alpha_1, -\alpha_2, \ldots, -\alpha_n) \\ &= -(\alpha_1^2 + \cdots + \alpha_n^2) = -1. \end{aligned}$$

Since all other sign combinations yield values of $f$ that lie between $-1$ and $+1$, we have, for any unit vectors $u = (x/\|x\|)$, $v = (y/\|y\|)$, that $-1 \leqslant f(x/\|x\|, y/\|y\|) \leqslant 1$; that is,

$$|\xi_1\eta_1 + \xi_2\eta_2 + \cdots + \xi_n\eta_n| \leqslant \sqrt{\xi_1^2 + \xi_2^2 + \cdots + \xi_n^2}\sqrt{\eta_1^2 + \eta_2^2 + \cdots + \eta_n^2},$$

which is the CBS inequality. (For a generalization of the CBS inequality, see exercise 7.)

---

## 10.101 Exercises

1. Let $f: \mathfrak{D} \to \mathbf{E}$, where $\mathfrak{D} \subseteq \mathbf{E}^n$ is an open set. *Prove*: If $f$ assumes a relative minimum at $a \in \mathfrak{D}$ and if $f'(a)$ exists, then $f'(a) = \theta$.
2. Let $f: \mathfrak{D} \to \mathbf{E}$ where $\mathfrak{D} \subseteq \mathbf{E}^n$ is open, and suppose that $f \in C^2(\mathfrak{D})$; i.e., all partial derivatives of second order are continuous in $\mathfrak{D}$. *Prove*: If $f'(a) = 0$ for $a \in \mathfrak{D}$ and $\sum_{i=1}^n \sum_{j=1}^n D_i D_j f(a)\omega_i\omega_j > 0$ for all $u = (\omega_1, \ldots, \omega_n) \in \mathbf{E}^n$, $u \neq \theta$, then $f$ assumes a relative minimum at $a$.
3. Establish a sufficient condition for a relative maximum. (See exercise 2.)
4. Show: If $n = 2$, then the condition in exercise 2 may be stated in the form $D_1 D_1 f(a) D_2 D_2 f(a) - [D_1 D_2 f(a)]^2 > 0$.
5. Given $f: \mathbf{E} \to \mathbf{E}$ by $f(x) = ax + b$, $a, b \in \mathbf{R}$, and $n$ points $(\xi_i, \eta_i) \in \mathbf{E} \times \mathbf{E}$. Find the relative minimum of $\sum_{i=1}^n (f(\xi_i) - \eta_i)^2$. (*Least square method.*)
6. Find the shortest distance from the point $(3, 3, -1)$ to the surface $\xi_1\xi_2 - \xi_3 = 0$.

7. Derive the *Hölder*† *inequality*

$$a_1 b_1 + \cdots + a_n b_n \leqslant \sqrt[p]{a_1^p + \cdots + a_n^p}\,\sqrt[q]{b_1^q + \cdots + b_n^q},$$

where $a_j \geqslant 0, b_j \geqslant 0$ for all $j \in \{1, 2, \ldots, n\}$ and where $1/p + 1/q = 1$ with $p \geqslant 1$, $q \geqslant 1$.
(Note that the CBS inequality is a special case of the Hölder inequality for $p = q = 2$.)

8. Derive the *Minkowski*‡ *inequality*

$$\sqrt[p]{(\xi_1 + \eta_1)^p + \cdots + (\xi_n + \eta_n)^p} \leqslant \sqrt[p]{\xi_1^p + \cdots + \xi_n^p} + \sqrt[p]{\eta_1^p + \cdots + \eta_n^p}$$

for $\xi_j \geqslant 0$, $\eta_j \geqslant 0$ for all $j \in \{1, 2, \ldots, n\}$ and $p \geqslant 1$. (Note that the triangle inequality is a special case of the Minkowski inequality for $p = 2$.)

† *Otto Hölder*, 1859–1937.
‡ *Hermann Minkowski*, 1864–1909.

# Multiple Integrals 11

## 11.102 The Riemann Integral in $\mathbf{E}^n$

In Chapter 4, we discussed the Riemann integral of a real-valued function of a real variable on a closed interval. Most of the concepts that were introduced at that time can be generalized in a simple and straightforward manner to apply to *real-valued* functions of a vector variable. Once the integral of a real-valued function of a vector variable is established, it is easy and merely a matter of a definition, to extend the notion to *vector-valued* functions of a vector variable. (See also the remark following the Corollary to Theorem 81.1.)

The two-dimensional counterpart to a one-dimensional interval is a rectangle; the three-dimensional counterpart is a right parallelepiped; and the $n$-dimensional counterpart is an $n$-dimensional right parallelepiped. In Chapter 4, we discussed integrals on intervals only. In this chapter, we shall direct a great deal of our attention to integrals on bounded subsets of $\mathbf{E}^n$ other than $n$-dimensional right parallelepipeds. The reader who has already had some experience with double and triple integrals on disks, balls, wedges, and the like, knows that such integrals are not only of theoretical interest but also of considerable practical importance.

Before we enter a discussion of these more complicated matters, we shall define the integral of a real-valued function of a vector variable on $n$-dimensional intervals:

***Definition 102.1***

If $a_i, b_i \in \mathbf{R}, i = 1, 2, \ldots, n$, and $a_i < b_i$ for all $i = 1, 2, \ldots, n$, then

$$(102.1)\quad \mathfrak{J} = \{x \in \mathbf{E}^n \mid a_i < \xi_i < b_i, i = 1, 2, \ldots, n\} = (a_1, b_1) \times (a_2, b_2) \times \cdots \times (a_n, b_n)$$

is called an $n$-dimensional *open interval* and

$$(102.2)\quad \mathfrak{I} = \{x \in \mathbf{E}^n \mid a_i \leqslant \xi_i \leqslant b_i, i = 1, 2, \ldots, n\} = [a_1, b_1] \times [a_2, b_2] \times \cdots \times [a_n, b_n]$$

is called an $n$-dimensional *closed interval.*

If $a_j = b_j$ for some $j \in \{1, 2, \ldots, n\}$, then $\mathfrak{J}$, as defined in (102.2) is called a *degenerate closed interval* in $\mathbf{E}^n$.

By definition 59.3, *a closed n-dimensional interval* (degenerate or not) *is a closed set* in $\mathbf{E}^n$ and *an open n-dimensional interval is an open set* in $\mathbf{E}^n$. (There is no such thing as a degenerate open interval in $\mathbf{E}^n$ other than the empty set.)

In the following definition we generalize the notion of *length* of an interval, *area* measure of a rectangle, and *volume* measure of a right parallelepiped:

### *Definition 102.2*

If $\mathfrak{J}$ is the closed $n$-dimensional interval in (102.2), then

$$|\mathfrak{J}| = (b_1 - a_1)(b_2 - a_2) \ldots (b_n - a_n)$$

is called the *content* of $\mathfrak{J}$ in $\mathbf{E}^n$.

If $\mathfrak{F}$ is the open $n$-dimensional interval in 102.1, then the content $|\mathfrak{F}|$ of $\mathfrak{F}$ is defined by $|\mathfrak{F}| = |\mathfrak{J}|$.

If $\mathfrak{J}$ is degenerate, then its content in $\mathbf{E}^n$ is zero, but, unless $\mathfrak{J}$ is a point, it has positive content in some lower-dimensional Euclidean space.

As in Chapter 4, we shall define the integral in terms of upper and lower sums. For this purpose we need to explain what we mean by a partition of an $n$-dimensional closed interval.

### *Definition 102.3*

If $P_j = \{x_0^{(j)}, x_1^{(j)}, \ldots, x_{k_j}^{(j)}\}$, with $x_0^{(j)} = a_j$, $x_{k_j}^{(j)} = b_j$, $j = 1, 2, \ldots, n$, are partitions of the intervals $[a_j, b_j]$ (see definition 44.1), then

$$P = P_1 \times P_2 \times \cdots \times P_n$$

is called a partition of the $n$-dimensional interval $\mathfrak{J}$, as defined in (102.2).

For $n = 1$, a partition divides a closed interval into $k_1$ subintervals; for $n = 2$, it divides a closed rectangle into $k_1 k_2$ subrectangles, the points of the partition being vertices; for $n = 3$, it divides a right parallelepiped into $k_1 k_2 k_3$ subparallelepipeds, etc. (See Fig. 102.1.)

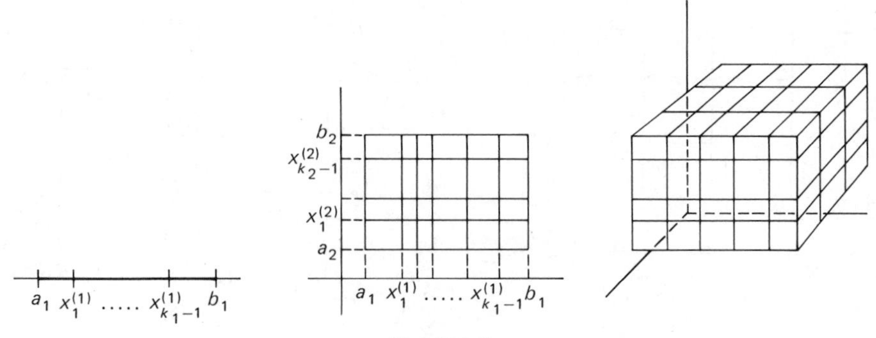


*Fig. 102.1*

Note that a degenerate closed interval cannot be partitioned. (See definition 44.1.)

The concepts of upper and lower sums are the same as in Chapter 4.

***Definition 102.4***

Let $\mathfrak{J}$ denote a closed $n$-dimensional interval and let $f: \mathfrak{J} \to \mathbf{E}$ denote a bounded function. If $P$ is a partition of $\mathfrak{J}$ which divides $\mathfrak{J}$ into $k_1 k_2 \ldots k_n = \mu$ $n$-dimensional closed subintervals $\mathfrak{J}_1, \mathfrak{J}_2, \ldots, \mathfrak{J}_\mu$, and if

$$m_k(f) = \inf_{x \in \mathfrak{J}_k} f(x), \; M_k(f) = \sup_{x \in \mathfrak{J}_k} f(x),$$

then

$$\underline{S}(f;P) = \sum_{k=1}^{\mu} m_k(f)\,|\mathfrak{J}_k|$$

is called a *lower sum* of $f$ on $\mathfrak{J}$, and

$$\bar{S}(f;P) = \sum_{k=1}^{\mu} M_k(f)\,|\mathfrak{J}_k|$$

is called an *upper sum* of $f$ on $\mathfrak{J}$.

Note that for a bounded function, $\underline{S}(f;P)$, $\bar{S}(f;P)$ always exist.

***Definition 102.5***

Let $P$, $Q$ denote partitions of the $n$-dimensional interval $\mathfrak{J}$. $Q$ is a *refinement* of $P$, $Q \supseteq P$, if and only if every point in $P$ is also a point in $Q$.

Note that the configuration in Fig. 102.2b is *not* a refinement of the partition in Fig. 102.2a because Fig. 102.2b does not represent a partition.

Lemmas 44.2, 44.3, 45.1, and 45.2 can easily be generalized to apply to the present, more general, situation. We summarize the results in the following lemma and leave the proof to the reader.

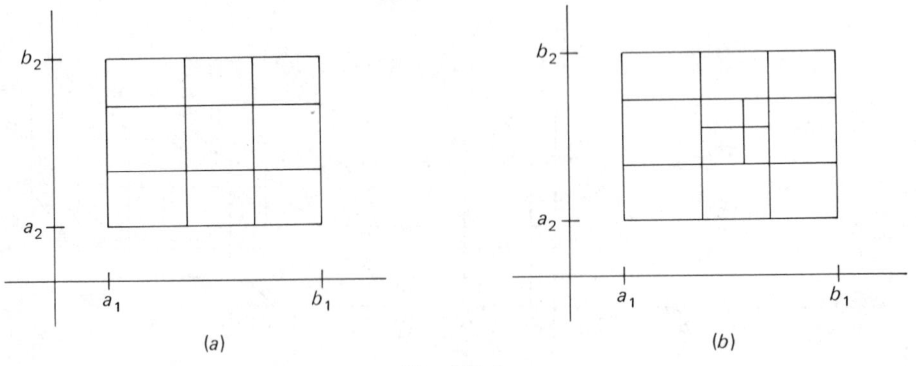


***Fig. 102.2***

***Lemma 102.1***

If $\mathfrak{J}$ is an $n$-dimensional closed interval and if $f: \mathfrak{J} \to \mathbf{E}$ is a bounded function, then

$$\underline{S}(f;P) \leqslant \bar{S}(f;P) \quad \text{for every partition } P \text{ of } \mathfrak{J},$$

$$(102.3) \qquad \underline{S}(f;P) \leqslant \underline{S}(f;Q), \quad \bar{S}(f;P) \geqslant \bar{S}(f;Q) \quad \text{if} \quad Q \supseteq P,$$

$$\underline{S}(f;P) \leqslant \bar{S}(f;Q) \quad \text{for any partitions } P, Q \text{ of } \mathfrak{J};$$

$$(102.4) \qquad \sup_{(P)} \underline{S}(f;P), \inf_{(P)} \bar{S}(f;P) \text{ exist} \quad \text{and} \quad \sup_{(P)} \underline{S}(f;P) \leqslant \inf_{(P)} \bar{S}(f;P).$$

With the statement of this lemma, all preparations for the definition of the Riemann integral in $\mathbf{E}^n$ have been made, and we can now define the integral as in definition 45.1.

***Definition 102.6***

The function $f: \mathfrak{J} \to \mathbf{E}$, which is presumed to be bounded on the $n$-dimensional interval $\mathfrak{J}$ is called *Riemann-integrable* on $\mathfrak{J}$ if and only if

$$\sup_{(P)} \underline{S}(f;P) = \inf_{(P)} \bar{S}(f;P).$$

The number

$$\int_{\mathfrak{J}} f = \sup_{(P)} \underline{S}(f;P) = \inf_{(P)} \bar{S}(f;P)$$

is called the *Riemann integral* of $f$ over $\mathfrak{J}$.

---

▲ *Example 1*

Let $f: \mathfrak{J} \to \mathbf{E}$, where $\mathfrak{J} = [0,1] \times [0,1] \times [0,1]$, be defined by $f(x) = 1$ for all $x \in \mathfrak{J}$. Then, for every partition $P$ of $\mathfrak{J}$, $\underline{S}(f;P) = \bar{S}(f;P) = \sum_{k=1}^{\mu} |\mathfrak{J}_k| = |\mathfrak{J}| = 1$. Hence, $\int_{\mathfrak{J}} f = 1$.

▲ *Example 2*

Let $\mathfrak{J}$ denote the interval in example 1 and let $f: \mathfrak{J} \to \mathbf{E}$ be defined as follows:

$$f(x) = \begin{cases} 0 & \text{for} \quad x \in (\mathbf{Q} \times \mathbf{Q} \times \mathbf{Q}) \cap \mathfrak{J}, \\ 1 & \text{for} \quad x \in c(\mathbf{Q} \times \mathbf{Q} \times \mathbf{Q}) \cap \mathfrak{J}. \end{cases}$$

Then $\underline{S}(f;P) = 0$ and $\bar{S}(f;P) = 1$ for every partition $P$ of $\mathfrak{J}$. Hence, $f$ is *not* Riemann-integrable on $\mathfrak{J}$.

---

We close this section with a generalization of definition 102.6 to *vector-valued* functions of a vector variable:

***Definition 102.7***

The function $f = (\phi_1, \phi_2, \ldots, \phi_m)\colon \mathfrak{J} \to \mathbf{E}^m$ which is presumed bounded on the $n$-dimensional closed interval $\mathfrak{J}$ is called *Riemann-integrable* on $\mathfrak{J}$ if and only if each $\phi_j : \mathfrak{J} \to \mathbf{E}$, $j = 1, 2, \ldots, m$, is Riemann-integrable on $\mathfrak{J}$. The vector

$$\int_{\mathfrak{J}} f = \left( \int_{\mathfrak{J}} \phi_1, \int_{\mathfrak{J}} \phi_2, \ldots, \int_{\mathfrak{J}} \phi_m \right)$$

is called the Riemann integral of $f$ on $\mathfrak{J}$. (Note that $f$ is bounded on $\mathfrak{J}$ if and only if each $\phi_j$ is bounded on $\mathfrak{J}$.)

## 11.102 Exercises

1. Show: If $\mathfrak{J} \subset \mathbf{E}^n$ is a degenerate closed interval, then $|\mathfrak{J}| = 0$ in $\mathbf{E}^n$, and if $a_j < b_j$ for at least one $j \in \{1, 2, \ldots, n\}$, then the projection of $\mathfrak{J}$ has positive content in the space spanned by $e_{j_1}, e_{j_2}, \ldots, e_{j_k}$ where $a_{j_l} < b_{j_l}$ but $a_j = b_j$ for $j \neq j_l$.
2. Show: A closed $n$-dimensional interval is closed in $\mathbf{E}^n$ and an open $n$-dimensional interval is open in $\mathbf{E}^n$.
3. Show: A closed interval (degenerate or not) is a closed set in $\mathbf{E}^n$.
4. Let $\mathfrak{J} = [0, 1] \times [0, 1]$ and let $f\colon \mathfrak{J} \to \mathbf{E}^2$ be defined by
$$\phi_1(x) = 1 \qquad \text{for all } x \in \mathfrak{J},$$
$$\phi_2(x) = \begin{cases} 0 & \text{for } x \in (\mathbf{Q} \times \mathbf{Q}) \cap \mathfrak{J}, \\ 1 & \text{for } x \in c(\mathbf{Q} \times \mathbf{Q}) \cap \mathfrak{J}. \end{cases}$$
Is $f$ integrable on $\mathfrak{J}$?
5. Prove Lemma 102.1.
6. $\underline{\int}_{\mathfrak{J}} f = \sup_{(P)} \underline{S}(f; P)$ and $\overline{\int}_{\mathfrak{J}} f = \inf_{(P)} \overline{S}(f; P)$ are called the *lower* and *upper Darboux integrals*. Let $\mathfrak{J} = \mathfrak{J}_1 \cup \mathfrak{J}_2$ where $\mathfrak{J}, \mathfrak{J}_1, \mathfrak{J}_2$ are closed $n$-dimensional intervals and where $\mathfrak{J}_1, \mathfrak{J}_2$ have no interior points in common. Show:
$$\overline{\int}_{\mathfrak{J}} f = \overline{\int}_{\mathfrak{J}_1} f + \overline{\int}_{\mathfrak{J}_2} f, \qquad \underline{\int}_{\mathfrak{J}} f = \underline{\int}_{\mathfrak{J}_1} f + \underline{\int}_{\mathfrak{J}_2} f.$$

## 11.103 Existence of the Integral

The Riemann condition (definition 46.1) has a straightforward generalization to real-valued functions of a vector variable. The following theorem, establishing the Riemann condition as necessary and sufficient for the existence of the integral, may be proved in the same manner as Theorem 46.1:

***Theorem 103.1***

The function $f\colon \mathfrak{J} \to \mathbf{E}$, which is presumed bounded on the $n$-dimensional closed interval $\mathfrak{J}$, is Riemann-integrable on $\mathfrak{J}$ if and only if, for every $\varepsilon > 0$, there is a partition $P_\varepsilon$ of $\mathfrak{J}$ such that $\overline{S}(f; P) - \underline{S}(f; P) < \varepsilon$ for all $P \supseteq P_\varepsilon$.

As in Chapter 4, we may utilize Theorem 103.1 to establish the existence of the Riemann integral for certain classes of functions.

If $f: \mathfrak{J} \to \mathbf{E}$ is continuous on $\mathfrak{J}$, then it is uniformly continuous on $\mathfrak{J}$ and we can find, for every $\varepsilon > 0$, a $\delta(\varepsilon) > 0$ such that for every $a \in \mathfrak{J}$

$$|f(a) - f(x)| < \frac{\varepsilon}{2|\mathfrak{J}|} \qquad \text{for all } x \in N_{\delta(\varepsilon)}(a) \cap \mathfrak{J}.$$

We inscribe in each $N_{\delta(\varepsilon)/3}(a)$ an open cube $C_a$ with center at $a$ and edges parallel to the coordinates axes. Then, $\{C_a \mid a \in \mathfrak{J}\}$ is an open cover of $\mathfrak{J}$. Since $\mathfrak{J}$ is compact, a finite subcover, say $\{C_{a_j} \mid j = 1, 2, \ldots, l\}$, will do. If $\overline{C}_{a_j}$ is the closed cube obtained from $C_{a_j}$ by adjoining all faces, then $|f(x_1) - f(x_2)| < \varepsilon/|\mathfrak{J}|$ whenever $x_1, x_2 \in \overline{C}_{a_j}$ for some $j \in \{1, 2, \ldots, l\}$.

If we consider a partion $P_\varepsilon$ of $\mathfrak{J}$ that contains all vertices of $\overline{C}_{a_1}, \overline{C}_{a_2}, \ldots, \overline{C}_{a_l}$ (see Fig. 103.1), then we may prove, as in section 4.47, that $f$ satisfies the Riemann condition and hence, is integrable on $\mathfrak{J}$.

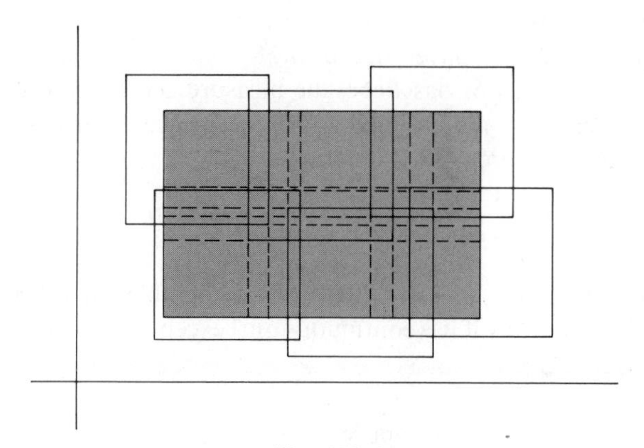

*Fig. 103.1*

***Theorem 103.2***

If $\mathfrak{J}$ is an $n$-dimensional closed interval and if $f: \mathfrak{J} \to \mathbf{E}$ is continuous on $\mathfrak{J}$, then $\int_{\mathfrak{J}} f$ exists. (For a trivial generalization of this theorem see exercise 3.)

In order to characterize the functions that are integrable on an $n$-dimensional interval $\mathfrak{J}$, we generalize the concept of Lebesgue measure zero to $n$-dimensional point sets (see section 4.48) and then proceed as in section 4.49.

***Definition 103.1***

A set $S \subseteq \mathbf{E}^n$ has an $n$-dimensional Lebesgue measure zero, $\lambda(S) = 0$, if and only if one can find, for every $\varepsilon > 0$, a sequence of $n$-dimensional open intervals $\{\mathfrak{J}_1, \mathfrak{J}_2, \mathfrak{J}_3, \ldots\}$ such that

(a)
$$S \subseteq \bigcup_{k \in \mathbf{N}} \mathfrak{J}_k,$$

(b)
$$\sum_{k=1}^{\infty} |\mathfrak{J}_k| < \varepsilon.$$

It is clear that a degenerate closed interval in $\mathbf{E}^n$ has $n$-dimensional Lebesgue measure zero: Let $\mathfrak{I} = \{x \in \mathbf{E}^n \mid a_j \leqslant \xi_j \leqslant b_j,\ j = 1, 2, \ldots, n\}$ and let $a_1 = b_1$. Then $\mathfrak{I} \subseteq \{x \in \mathbf{E}^n \mid a_1 - \Delta < \xi_1 < b_1 + \Delta,\ a_j - \delta < \xi_j < b_j + \delta,\ j = 2, 3, \ldots, n\} = \mathfrak{J}$, where

$$\Delta = \frac{\varepsilon}{3 \prod_{k=2}^{n} (b_k - a_k + 2\delta)}.$$

Then $|\mathfrak{J}| < \varepsilon$.

In view of this we have, as in Lemma 48.3, that the open intervals in definition 103.1 may be replaced by closed intervals, to obtain an equivalent definition of Lebesgue measure zero. (See exercise 4.)

As in Lemma 48.2, we have that if $S_1, S_2, S_3, \ldots \subseteq \mathbf{E}^n$ all have Lebesgue measure zero, then $\bigcup_{k=1}^{\infty} S_k$ has Lebesgue measure zero. (See exercise 5.)

With the concept of $n$-dimensional Lebesgue measure zero established, one can now generalize Theorem 49.1 as follows.

***Theorem 103.3***

The bounded function $f : \mathfrak{I} \to \mathbf{E}$ is Riemann integrable on the $n$-dimensional closed interval $\mathfrak{I}$ if and only if it is continuous on $\mathfrak{I}$ except on a set of $n$-dimensional Lebesgue measure zero.

If a function is continuous except on a set of Lebesgue measure zero, we say, as in the one-dimensional case, that the function is *continuous almost everywhere* (a.e.).

The proof of Theorem 103.3 is similar to the proofs of Lemmas 49.1 and 49.2, and is not repeated here. Note that, in generalizing the proof of Lemma 49.1, it is important to obtain an open cover consisting of $n$-dimensional open intervals. (See also the remark preceding Theorem 103.2.)

As in section 4.50 one can show that the integral is a linear function from the set of all integrable functions on $\mathfrak{I}$ into the reals.

***Theorem 103.4***

If $f, g : \mathfrak{I} \to \mathbf{R}$ are Riemann-integrable on $\mathfrak{I}$ and if $\lambda, \mu \in \mathbf{R}$ are any constants, then $\lambda f + \mu g$ is Riemann-integrable on $\mathfrak{I}$ and

$$\int_{\mathfrak{I}} (\lambda f + \mu g) = \lambda \int_{\mathfrak{I}} f + \mu \int_{\mathfrak{I}} g.$$

## 11.103 Exercises

1. Prove Theorem 103.1.
2. Prove Theorem 103.2.
3. *Prove*: If $f: \mathfrak{J} \to \mathbf{E}^m$ is continuous on $\mathfrak{J}$, then $\int_{\mathfrak{J}} f$ exists. (See definition 102.7.)
4. Show: $S \subseteq \mathbf{E}^n$ has $n$-dimensional Lebesgue measure zero if and only if there is, for every $\varepsilon > 0$, a sequence of closed $n$-dimensional intervals $\{\mathfrak{J}_1, \mathfrak{J}_2, \mathfrak{J}_3, \ldots\}$ such that $S \subseteq \bigcup_{k \in \mathbf{N}} \mathfrak{J}_k$ and $\sum_{k=1}^{\infty} |\mathfrak{J}_k| < \varepsilon$.
5. *Prove*: If $S_1, S_2, S_3, \ldots \subseteq \mathbf{E}^n$ have $n$-dimensional Lebesgue measure zero, then $\bigcup_{k \in \mathbf{N}} S_k$ has $n$-dimensional Lebesgue measure zero.
6. Show that $S = \mathbf{Q} \times \mathbf{Q} \times \cdots \times \mathbf{Q}$ ($n$-tuple Cartesian product) has $n$-dimensional Lebesgue measure zero. (See exercise 5.)
7. Prove Theorem 103.3.
8. Let $S \subseteq \mathfrak{J}$ where $\mathfrak{J}$ is an $n$-dimensional closed interval, and let $f: \mathfrak{J} \to \mathbf{E}$ be defined as follows:

$$f(x) = \begin{cases} 1 & \text{if } x \in S, \\ 0 & \text{if } x \in \mathfrak{J} \backslash S. \end{cases}$$

Let $A$ denote the set of all interior points of $S$, let $B$ denote the set of all interior points of $c(S)$. Prove $\int_{\mathfrak{J}} f$ exists if and only if

$$\lambda(c(A \cup B) \cap \mathfrak{J}) = 0.$$

9. *Prove*: If $\mathfrak{J} = \mathfrak{J}_1 \cup \mathfrak{J}_2$ where $\mathfrak{J}, \mathfrak{J}_1, \mathfrak{J}_2$ are $n$-dimensional closed intervals and $\mathfrak{J}_1, \mathfrak{J}_2$ have no interior points in common, then $f$ is integrable on $\mathfrak{J}$ if and only if $f$ is integrable on $\mathfrak{J}_1$ and on $\mathfrak{J}_2$. In either case, $\int_{\mathfrak{J}} f = \int_{\mathfrak{J}_1} f + \int_{\mathfrak{J}_2} f$.
10. *Prove*: If $f, g : \mathfrak{J} \to \mathbf{R}$ are Riemann-integrable on $\mathfrak{J}$, then $f + g$ is Riemann-integrable on $\mathfrak{J}$ and

$$\int_{\mathfrak{J}} (f + g) = \int_{\mathfrak{J}} f + \int_{\mathfrak{J}} g.$$

11. Show: If $m < n$ and if $S \subset \mathbf{E}^n$ has $m$-dimensional Lebesgue measure zero, then $S$ has $n$-dimensional Lebesgue measure zero. Give an example to demonstrate that the converse need not be true.

## 11.104 The Riemann Integral over Point Sets Other Than Intervals

Let $\mathfrak{D} \subseteq \mathbf{E}^n$ represent a *bounded* point set and let $f: \mathfrak{D} \to \mathbf{E}$ represent a bounded function.

We extend the definition of $f$ to all of $\mathbf{E}^n$ as follows:

$$f_{\mathfrak{D}}(x) = \begin{cases} f(x) & \text{if } x \in \mathfrak{D}, \\ 0 & \text{if } x \in c(\mathfrak{D}). \end{cases} \tag{104.1}$$

Clearly, $f_{\mathfrak{D}} : \mathbf{E}^n \to \mathbf{E}$ is bounded on $\mathbf{E}^n$ and it is meaningful to talk about $\int_{\mathfrak{J}} f_{\mathfrak{D}}$ where $\mathfrak{J}$ is any $n$-dimensional closed interval.

It is our aim to define the integral of $f$ over $\mathfrak{D}$ in terms of an integral of $f_{\mathfrak{D}}$ over an $n$-dimensional closed interval $\mathfrak{J}$ that contains $\mathfrak{D}$. For such a definition to make sense, we have to establish, first, that the choice of $\mathfrak{J}$ is immaterial as long as $\mathfrak{J} \supseteq \mathfrak{D}$.

***Lemma 104.1***

If $f\colon \mathfrak{D} \to \mathbf{E}$ is bounded on $\mathfrak{D}$, where $\mathfrak{D}$ is a bounded subset of $\mathbf{E}^n$ and if $\int_{\mathfrak{J}} f_{\mathfrak{D}}$ exists where $f_{\mathfrak{D}}$ is defined in (104.1) and where $\mathfrak{J}$ is an $n$-dimensional closed interval containing $\mathfrak{D}$, then

$$\int_{\mathfrak{J}} f_{\mathfrak{D}} = \int_{\mathfrak{J}'} f_{\mathfrak{D}}$$

for all closed $n$-dimensional intervals $\mathfrak{J}'$ that contain $\mathfrak{D}$.

*Proof*

Let $\mathfrak{J}''$ denote an $n$-dimensional closed interval such that $\mathfrak{J} \subseteq \mathfrak{J}''$, $\mathfrak{J}' \subseteq \mathfrak{J}''$. We note that any partition of $\mathfrak{J}''$, when restricted to $\mathfrak{J}$, induces a partition of $\mathfrak{J}$, and that any partition of $\mathfrak{J}$ may be extended to a partition of $\mathfrak{J}''$ by projection of the partition points onto the edges of $\mathfrak{J}''$. (See Fig. 104.1.) Let $P_{\mathfrak{J}}$, $P_{\mathfrak{J}''}$ denote

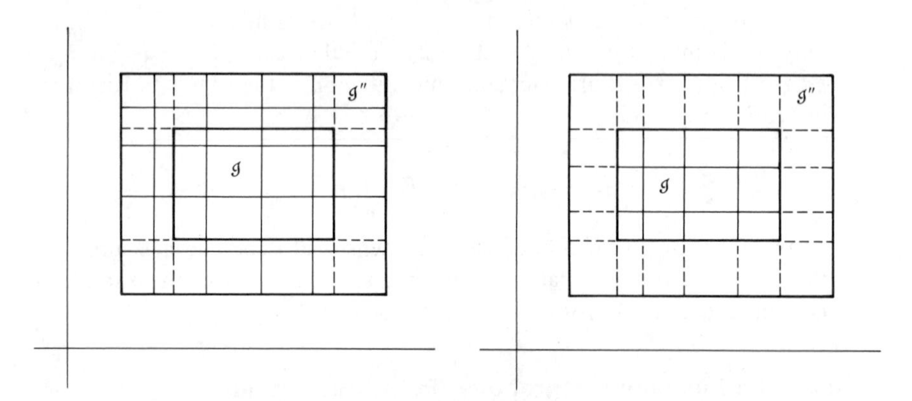


*Fig. 104.1*

corresponding partitions of $\mathfrak{J}$ and $\mathfrak{J}''$, respectively; that is, $P_{\mathfrak{J}}$ is either the restriction of $P_{\mathfrak{J}''}$ to $\mathfrak{J}$ or $P_{\mathfrak{J}''}$ is the extension of $P_{\mathfrak{J}}$ to $\mathfrak{J}''$. Since $f_{\mathfrak{D}}(x) = 0$ for all $x \in \mathfrak{J}''\backslash\mathfrak{J}$, we have $\underline{S}(f_{\mathfrak{D}}; P_{\mathfrak{J}}) = \underline{S}(f_{\mathfrak{D}}; P_{\mathfrak{J}''})$ and $\bar{S}(f_{\mathfrak{D}}; P_{\mathfrak{J}}) = \bar{S}(f_{\mathfrak{D}}; P_{\mathfrak{J}''})$. Hence, $\sup_{(P_{\mathfrak{J}})} \underline{S}(f_{\mathfrak{D}}; P_{\mathfrak{J}}) = \sup_{(P_{\mathfrak{J}''})} \underline{S}(f_{\mathfrak{D}}; P_{\mathfrak{J}''})$ and $\inf_{(P_{\mathfrak{J}})} \bar{S}(f_{\mathfrak{D}}; P_{\mathfrak{J}}) = \inf_{(P_{\mathfrak{J}''})} \bar{S}(f_{\mathfrak{D}}; P_{\mathfrak{J}''})$. Since $\int_{\mathfrak{J}} f_{\mathfrak{D}}$ exists, we have $\int_{\mathfrak{J}} f_{\mathfrak{D}} = \int_{\mathfrak{J}''} f_{\mathfrak{D}}$. In the same manner, one shows that $\int_{\mathfrak{J}'} f_{\mathfrak{D}} = \int_{\mathfrak{J}''} f_{\mathfrak{D}}$, and the statement of the lemma follows readily.

In view of this lemma, we may define the integral over $\mathfrak{D}$ as follows.

***Definition 104.1***

The bounded function $f: \mathfrak{D} \to \mathbf{E}$ on the bounded subset $\mathfrak{D}$ of $\mathbf{E}^n$ is integrable over $\mathfrak{D}$ if and only if $\int_{\mathfrak{J}} f_{\mathfrak{D}}$ exists where $\mathfrak{J} \supseteq \mathfrak{D}$ is any $n$-dimensional closed interval and where $f_{\mathfrak{D}} : \mathbf{E}^n \to \mathbf{E}$ is defined in (104.1). In that case,

$$\int_{\mathfrak{D}} f = \int_{\mathfrak{J}} f_{\mathfrak{D}}.$$

Because of Lemma 104.1, the choice of $\mathfrak{J}$ does not matter as long as $\mathfrak{D} \subseteq \mathfrak{J}$.

In order to establish a sufficient condition for the existence of $\int_{\mathfrak{D}} f$, we examine the structure of the set $\mathfrak{D}$ in some detail. Suppose that $f$ is continuous a.e. on $\mathfrak{D}$. Then, $f_{\mathfrak{D}}$, as defined in (104.1) is continuous at all interior points of $c(\mathfrak{D})$. At interior points of $\mathfrak{D}$ itself, $f_{\mathfrak{D}}$ is continuous if and only if $f$ is continuous. Hence, the only points where $f_{\mathfrak{D}}$ could possibly have discontinuities that are not discontinuities of $f$ are the points which are neither interior points of $\mathfrak{D}$ nor interior points of $c(\mathfrak{D})$. Such points are called boundary points.

***Definition 104.2***

$b \in \mathbf{E}^n$ is a *boundary point* of $\mathfrak{D} \subseteq \mathbf{E}^n$ if and only if every neighborhood of $b$ contains points from $\mathfrak{D}$ as well as points from $c(\mathfrak{D})$; or, as we may put it with more precision: If for every $\delta > 0$, it is true that $N_\delta(b) \cap \mathfrak{D} \neq \varnothing$ and $N_\delta(b) \cap c(\mathfrak{D}) \neq \varnothing$. The set of all boundary points of $\mathfrak{D}$ is called the *boundary of* $\mathfrak{D}$ and is denoted by $\beta\mathfrak{D}$. (Note that a boundary point of $\mathfrak{D}$ need not be in $\mathfrak{D}$. See also Fig. 104.2.)

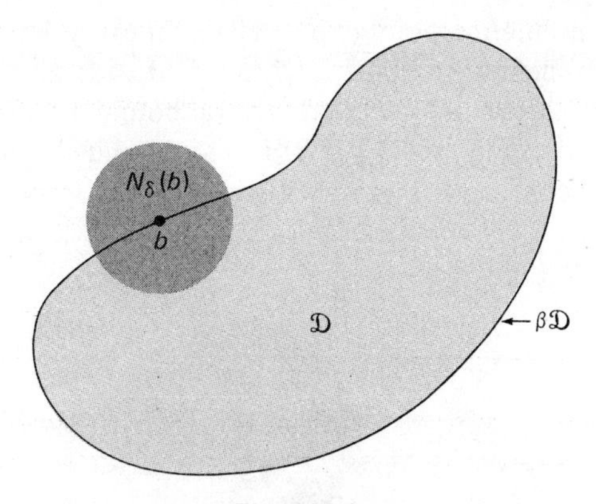


***Fig. 104.2***

Since the definition of boundary point is symmetric in $\mathfrak{D}$ and $c(\mathfrak{D})$, we have

$$\beta\mathfrak{D} = \beta c(\mathfrak{D}). \tag{104.2}$$

By definition 104.2, every point $a \in \mathfrak{D}$ *is either a boundary point of* $\mathfrak{D}$ *or an interior point of* $\mathfrak{D}$.

***Definition 104.3***

The set of all interior points of $\mathfrak{D} \subseteq \mathbf{E}^n$ is called the *interior* of $\mathfrak{D}$ and is denoted by $\mathfrak{D}^\circ$.

Clearly, $\mathfrak{D}^\circ \cap \beta\mathfrak{D} = \varnothing$.

We return now to our problem of the existence of $\int_{\mathfrak{D}} f$. The set of discontinuities of $f_{\mathfrak{D}}$ is a subset of the union of $\beta\mathfrak{D}$ with the set of discontinuities of $f$. If both have Lebesgue measure zero, then $f_{\mathfrak{D}}$ is continuous a.e. on every interval $\mathfrak{J}$ that contains $\mathfrak{D}$ and we have, from Theorem 103.3:

***Theorem 104.1***

If $\mathfrak{D} \subseteq \mathbf{E}^n$ is bounded, if $\lambda(\beta\mathfrak{D}) = 0$, and if the bounded function $f\colon \mathfrak{D} \to \mathbf{E}$ is continuous almost everywhere on $\mathfrak{D}$, then $\int_{\mathfrak{D}} f$ exists.

The boundary of a set will figure prominently in our subsequent investigations. The following result will prove helpful on a number of occasions:

***Lemma 104.2***

$\beta\mathfrak{D}$ is *closed* for every set $\mathfrak{D} \subseteq \mathbf{E}^n$. If $\mathfrak{D}$ is also bounded, then $\beta\mathfrak{D}$ is *compact.*

*Proof*

Let $a$ denote an accumulation point of $\beta\mathfrak{D}$. Then, for every $\delta > 0$, $N_{\delta/2}(a)$ contains a point $a' \in \beta\mathfrak{D}$. By definition 104.2, $N_{\delta/2}(a') \cap \mathfrak{D} \neq \varnothing$ and $N_{\delta/2}(a') \cap c(\mathfrak{D}) \neq \varnothing$. Since $N_{\delta/2}(a') \subseteq N_\delta(a)$, we see that $a$ is a boundary point of $\mathfrak{D}$. Hence, $\beta\mathfrak{D}$ is closed. (See also Fig. 104.3.) If $\mathfrak{D}$ is also bounded, then $\beta\mathfrak{D}$ is bounded and hence compact.

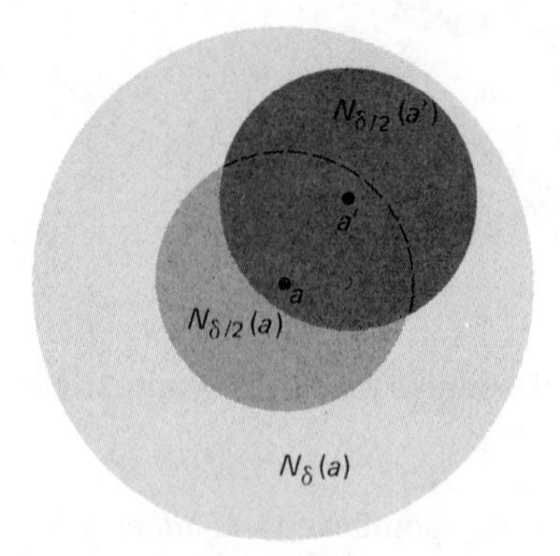


*Fig. 104.3*

---

▲ *Example 1*

Let $\mathfrak{D} = \{x \in \mathbf{E}^3 \mid \|x\| < 1\} \cup (1, 1, 1,)$. Clearly,

$$\beta\mathfrak{D} = \{x \in \mathbf{E}^3 \mid \|x\| = 1\} \cup (1, 1, 1),$$

where $(1, 1, 1)$ is an isolated point of $\mathfrak{D}$ (and of $\beta\mathfrak{D}$) and where all other points of $\beta\mathfrak{D}$ are accumulation points of $\mathfrak{D}$. (See also exercise 2.)

▲ *Example 2*

Let $\mathfrak{D} = \mathbf{Q} \times \mathbf{Q} \subset \mathbf{E}^2$. If $a \in \mathbf{E}^2$, then every $N_\delta(a)$ contains points from $\mathbf{Q} \times \mathbf{Q}$ as well as points from $c(\mathbf{Q} \times \mathbf{Q})$. Hence, every point $a \in \mathbf{E}^2$ is a boundary point of $\mathbf{Q} \times \mathbf{Q}$ and we have $\beta(\mathbf{Q} \times \mathbf{Q}) = \mathbf{E}^2$. Clearly, $\mathbf{E}^2$ is closed. However, $\mathbf{E}^2$ is not compact but this was not to be expected since $\mathbf{Q} \times \mathbf{Q}$ is not bounded.

▲ *Example 3*

Let $\mathfrak{D} = \{x \in \mathbf{E}^2 \mid \|x\| < 1\} \cup \{x \in \mathbf{Q} \times \mathbf{Q} \mid \|x\| = 1\}$. $\mathfrak{D}$ is the open unit disk together with those points on its circumference which have rational coordinates. Clearly $\beta\mathfrak{D} = \{x \in \mathbf{E}^2 \mid \|x\| = 1\}$ and we see that $\beta\mathfrak{D}$ is compact, as predicted by Lemma 104.2.

---

***Lemma 104.3***

For any sets $S_1, S_2 \subseteq \mathbf{E}^n$,

$$\beta(S_1 \cup S_2),\ \beta(S_1 \cap S_2),\ \beta(S_1 \backslash S_2) \subseteq \beta S_1 \cup \beta S_2. \tag{104.3}$$

*Proof*

If $x \in \beta(S_1 \cup S_2)$, then, for every $\delta > 0$, $N_\delta(x) \cap (S_1 \cup S_2) = (N_\delta(x) \cap S_1) \cup (N_\delta(x) \cap S_2) \neq \varnothing$ and $N_\delta(x) \cap c(S_1 \cup S_2) = N_\delta(x) \cap c(S_1) \cap c(S_2) \neq \varnothing$. Hence, $x \in \beta S_1$ or $x \in \beta S_2$.

If $x \in \beta(S_1 \cap S_2)$, then, for every $\delta > 0$, $N_\delta(x) \cap S_1 \cap S_2 \neq \varnothing$ and $N_\delta(x) \cap c(S_1 \cap S_2) = (N_\delta(x) \cap c(S_1)) \cup (N_\delta(x) \cap c(S_2)) \neq \varnothing$. Hence, $x \in \beta S_1$ or $x \in \beta S_2$.

If $x \in \beta(S_1 \backslash S_2)$, then, for every $\delta > 0$, $N_\delta(x) \cap (S_1 \backslash S_2) = N_\delta(x) \cap S_1 \cap c(S_2) \neq \varnothing$ and $N_\delta(x) \cap c(S_1 \backslash S_2) = (N_\delta(x) \cap c(S_1)) \cup (N_\delta(x) \cap S_2) \neq \varnothing$. Hence, $x \in \beta S_1$ or $x \in \beta S_2$.

## 11.104 Exercises

1. Let $\mathfrak{D} = \{x \in \mathbf{E}^3 \mid x = (1/k, 1/k^2, 1/k^3), k \in \mathbf{N}\}$, and let $f\colon \mathbf{E}^3 \to \mathbf{E}$ be given by $f(x) = 1$ for $x \in \mathfrak{D}$, $f(x) = 0$ for $x \notin \mathfrak{D}$. Find $\int_{\mathfrak{D}} f$.
2. Show: Every boundary point of $\mathfrak{D}$ is either an accumulation point of $\mathfrak{D}$ or an isolated point of $\mathfrak{D}$.

3. Show: Every accumulation point of $\mathfrak{D}$ is either an interior point of $\mathfrak{D}$ or a boundary point of $\mathfrak{D}$.
4. Show: If $\mathfrak{D} \subseteq \mathbf{E}^n$, then $\mathfrak{D} \cup \beta\mathfrak{D}$ is closed. If $\mathfrak{D}$ is also bounded, then $\mathfrak{D} \cup \beta\mathfrak{D}$ is compact.
5. Show: $\mathfrak{D} \subseteq \mathbf{E}^n$ is closed if and only if $\beta\mathfrak{D} \subseteq \mathfrak{D}$.
6. Show: If $\mathfrak{D} \subset \beta\mathfrak{D}$, then $\mathfrak{D}$ cannot be closed.
7. Let $\mathfrak{D} \subseteq \mathbf{E}^n$. Show:
   (a) $\beta\mathfrak{D} \cap \mathfrak{D}^\circ = \varnothing$,
   (b) $\beta\mathfrak{D} \cap (c(\mathfrak{D}))^\circ = \varnothing$,
   (c) $\beta\mathfrak{D} = c(\mathfrak{D}^\circ \cup (c(\mathfrak{D}))^\circ)$.
8. Let $\mathfrak{D} \subseteq \mathbf{E}^n$. Show:

$$\beta\mathfrak{D} = (\mathfrak{D} \cup \mathfrak{D}') \cap (c(\mathfrak{D}) \cup (c(\mathfrak{D}))').$$

9. Show: If $S_1^\circ \cap S_2^\circ = \varnothing$ where $S_1, S_2 \subseteq \mathbf{E}^n$, then $S_1 \cap S_2 \subseteq \beta S_1 \cup \beta S_2$.
10. Given the set $\mathbf{Q} \times \mathbf{Q} \times \mathbf{Q} \subset \mathbf{E}^3$. Find $\beta(\mathbf{Q} \times \mathbf{Q} \times \mathbf{Q})$.

## 11.105 Jordan Content

In this section we shall develop a concept of content of point sets in $\mathbf{E}^n$ which contains the concepts of length measure for $n = 1$, area measure for $n = 2$, and volume measure for $n = 3$, as special cases. There are a number of ways of generalizing length measure, area measure, and volume measure. The concept of Jordan† content which we shall develop is based on the concept of the Riemann integral. It is not the most general theory of this kind, but suffices for many practical purposes. The more general theory of Lebesgue measure, which is beyond the scope of this treatment, is applicable to a larger class of point sets in the sense that it assigns content to point sets that do not have content in the Jordan sense. For example, the point set $[0, 1] \cap \mathbf{Q} \subset \mathbf{E}$ has Lebesgue measure zero (see section 4.48) but does not have Jordan content, as we shall see in example 4 of this section.

We have already defined the content of $n$-dimensional closed and open intervals (definition 102.2). We consider now a bounded point set $\mathfrak{D} \subseteq \mathbf{E}^n$ and choose an $n$-dimensional closed interval $\mathfrak{J}$ such that $\mathfrak{J} \supseteq \mathfrak{D}$. The function $\chi_{\mathfrak{D}} : \mathbf{E}^n \to \mathbf{E}$, defined by

$$\chi_{\mathfrak{D}}(x) = \begin{cases} 1 & \text{for } x \in \mathfrak{D}, \\ 0 & \text{for } x \in c(\mathfrak{D}), \end{cases} \tag{105.1}$$

is called the *characteristic function of* $\mathfrak{D}$. Clearly, $\chi_{\mathfrak{D}}$ is bounded on $\mathfrak{J}$.

***Definition 105.1***

$$\underline{J}(\mathfrak{D}) = \sup_{(P)} \underline{S}(\chi_{\mathfrak{D}}, P)$$

is called the *inner* $n$-dimensional *Jordan content* of $\mathfrak{D}$, and

† *Camille Jordan*, 1838–1922.

$$\bar{J}(\mathfrak{D}) = \inf_{(P)} \bar{S}(\chi_{\mathfrak{D}}, P)$$

is called the *outer n*-dimensional *Jordan content* of $\mathfrak{D}$. Here, $P$ denotes a partition of $\mathfrak{J} \supseteq \mathfrak{D}$.

Since $\chi_{\mathfrak{D}}(x) = 0$ for all $x \in c(\mathfrak{D})$, it is clear that the definition of $\underline{J}(\mathfrak{D})$ and $\bar{J}(\mathfrak{D})$ does not depend on the choice of $\mathfrak{J}$ as long as $\mathfrak{D} \subseteq \mathfrak{J}$. (See also Lemma 104.1.)

***Definition 105.2***

The bounded set $\mathfrak{D} \subseteq \mathbf{E}^n$ has $n$-dimensional *Jordan content* $J(\mathfrak{D})$, if and only if

$$\underline{J}(\mathfrak{D}) = \bar{J}(\mathfrak{D}).$$

Then, $J(\mathfrak{D}) = \underline{J}(\mathfrak{D}) = \bar{J}(\mathfrak{D})$. A bounded set which has Jordan content is called *Jordan-measurable*.

By definition 102.6, $\chi_{\mathfrak{D}}$ is Riemann-integrable over $\mathfrak{J} \supseteq \mathfrak{D}$ if and only if $\sup_{(P)} \underline{S}(\chi_{\mathfrak{D}}, P) = \inf_{(P)} \bar{S}(\chi_{\mathfrak{D}}, P)$. Hence,

***Theorem 105.1***

The bounded set $\mathfrak{D} \subseteq \mathbf{E}^n$ is Jordan-measurable if and only if the characteristic function $\chi_{\mathfrak{D}}$ of $\mathfrak{D}$ is integrable on $\mathfrak{J} \supseteq \mathfrak{D}$. Then

$$J(\mathfrak{D}) = \int_{\mathfrak{J}} \chi_{\mathfrak{D}}, \qquad \mathfrak{J} \supseteq \mathfrak{D}.$$

In terms of the Riemann condition, we obtain a more practical criterion by which to judge whether a set is Jordan-measurable:

***Theorem 105.2***

The bounded set $\mathfrak{D} \subseteq \mathbf{E}^n$ is Jordan-measurable if and only if $\beta\mathfrak{D}$ has Jordan content zero.

*Proof*

We have, from Theorem 105.1 and Theorem 103.1, that $\mathfrak{D}$ is Jordan-measurable if and only if, for every $\varepsilon > 0$, there is a partition $P_\varepsilon$ of $\mathfrak{J} \supseteq \mathfrak{D}$ such that

$$\bar{S}(\chi_{\mathfrak{D}}; P) - \underline{S}(\chi_{\mathfrak{D}}; P) < \varepsilon \qquad \text{for all} \quad P \supseteq P_\varepsilon. \tag{105.2}$$

If $\mathfrak{J}_{l_1}, \mathfrak{J}_{l_2}, \ldots, \mathfrak{J}_{l_k}$ are all the intervals generated by $P_\varepsilon$, that contain points from $\beta\mathfrak{D}$, then

$$\bar{S}(\chi_{\mathfrak{D}}; P_\varepsilon) - \underline{S}(\chi_{\mathfrak{D}}; P_\varepsilon) = \sum_{j=1}^{k} |\mathfrak{J}_{l_j}|, \tag{105.3}$$

because all other intervals contain points from $\mathfrak{D}$ only or points from $c(\mathfrak{D}) \cap \mathfrak{J}$ only (see exercise 2) and hence, their contributions to upper and lower sum cancel out. If $\beta : \mathfrak{J} \to \mathbf{E}$ is the characteristic function of $\beta\mathfrak{D}$, then

$$\bar{S}(\beta; P_\varepsilon) = \sum_{j=1}^{k} |\mathfrak{J}_{l_j}|.$$

In view of (105.2) and (105.3) we have

$$\bar{J}(\beta\mathfrak{D}) \leqslant \bar{S}(\beta; P_\varepsilon) < \varepsilon.$$

Since this is true for every $\varepsilon > 0$, we have $\bar{J}(\beta\mathfrak{D}) = 0$. Since $\underline{J}(\beta\mathfrak{D}) \geqslant 0$ and since, in view of (102.4), $\underline{J}(\beta\mathfrak{D}) \leqslant \bar{J}(\beta\mathfrak{D})$, we have $J(\beta\mathfrak{D}) = 0$.

Conversely, if $J(\beta\mathfrak{D}) = 0$, we can find, for every $\varepsilon > 0$, a partition $P_\varepsilon$ such that $\bar{S}(\beta; P_\varepsilon) < \varepsilon$ and hence, (105.2) is satisfied. Consequently, $\chi_{\mathfrak{D}}$ is integrable over $\mathfrak{J}$ and $J(\mathfrak{D})$ exists.

The above argument suggests the following condition for a point set to have Jordan content zero:

***Lemma 105.1***

The bounded point set $S \subseteq \mathbf{E}^n$ has $n$-dimensional Jordan content zero if and only if, for every $\varepsilon > 0$, there are finitely many closed interval $\mathfrak{J}_1, \mathfrak{J}_2, \ldots, \mathfrak{J}_k$ such that

(a) $$S \subseteq \bigcup_{j=1}^{k} \mathfrak{J}_j,$$

(b) $$\sum_{j=1}^{k} |\mathfrak{J}_j| < \varepsilon.$$

***Proof***

Choose an interval $\mathfrak{J} \supseteq S$ and a partition $P$ of $\mathfrak{J}$ that contains all vertices of $\mathfrak{J}_1, \mathfrak{J}_2, \ldots, \mathfrak{J}_k$. Then, $\bar{S}(\chi_S; P) < \varepsilon$, where $\chi_s$ is the characteristic function of $S$. Hence, $J(S) = 0$.

Conversely, if $J(S) = 0$, then, for every $\varepsilon > 0$, there is a partition $P_\varepsilon$ of $\mathfrak{J} \supseteq S$ such that $\bar{S}(\chi_S; P_\varepsilon) < \varepsilon$. (a) and (b) follow readily.

---

▲ *Example 1*

$\mathfrak{J} = \{x \in \mathbf{E}^n \mid a_i \leqslant \xi_i \leqslant b_i;\ i = 1, 2, \ldots, n\}$ has Jordan content

$$J(\mathfrak{J}) = (b_1 - a_1)(b_2 - a_2) \ldots (b_n - a_n) = |\mathfrak{J}|,$$

because, for every partition $P$ of $\mathfrak{J}$ which generates the subintervals $\mathfrak{J}_1, \mathfrak{J}_2, \ldots, \mathfrak{J}_\mu$, we have

$$\bar{S}(\chi_{\mathfrak{J}}; P) = \underline{S}(\chi_{\mathfrak{J}}; P) = \sum_{k=1}^{\mu} |\mathfrak{J}_k| = |\mathfrak{J}|.$$

If $a_j = b_j$ for some $j \in \{1, 2, \ldots, n\}$, then $J(\mathfrak{J}) = 0$. Hence, what we called the content of an $n$-dimensional closed interval in definition 102.2 stands now revealed as the Jordan content of that interval.

▲ *Example 2*

$\mathfrak{Z} = \{x \in \mathbf{E}^n \mid a_i < \xi_i < b_i, i = 1, 2, \ldots, n\}$ has Jordan content

$$J(\mathfrak{Z}) = (b_1 - a_1)(b_2 - a_2) \ldots (b_n - a_n) = |\mathfrak{J}| = |\mathfrak{Z}|,$$

because we have, for every partition $P$ of $\mathfrak{J} \supseteq \mathfrak{Z}$, with $\mathfrak{J}$ as in example 1, that $\bar{S}(\chi_{\mathfrak{Z}}; P) = |\mathfrak{J}|$ and that $\underline{S}(\chi_j; P) = |\mathfrak{J}| - \varepsilon$ where $\varepsilon > 0$ can be made as small as one pleases by suitable choice of $P$. Hence, the content of an open interval is its Jordan content.

▲ *Example 3*

A bounded set $S \subseteq \mathbf{E}^n$ which has Jordan content zero has Lebesgue measure zero. By hypothesis, there are finitely many closed intervals $\{\mathfrak{J}_1, \mathfrak{J}_2, \ldots, \mathfrak{J}_k\}$ such that (a) and (b) in Lemma 105.1 hold. Hence, $S$ has Lebesgue measure zero.

▲ *Example 4*

$\mathbf{Q} \cap [0, 1]$ has (one-dimensional) Lebesgue measure zero (see Lemma 48.1) but does not have one-dimensional Jordan content zero, because $\beta(\mathbf{Q} \cap [0, 1]) = [0, 1]$ and hence, $J(\beta(\mathbf{Q} \cap [0, 1])) = 1$. By Theorem 105.2, $\mathbf{Q} \cap [0, 1]$ is *not* Jordan measurable.

▲ *Example 5*

The set $(\mathbf{Q} \times \mathbf{Q}) \cap ([0, 1] \times [0, 1])$ is not Jordan measurable either, because its boundary is the square $[0, 1] \times [0, 1]$ which has Jordan content 1.

▲ *Example 6*

Let $f: [a, b] \to \mathbf{E}$ denote a continuous function. The set $S = \{(x, f(x)) \mid x \in [a, b]\}$ has two-dimensional Jordan content zero. In order to see this, we need only invoke the fact that $f$ is uniformly continuous on $[a, b]$ and that, for every $\varepsilon > 0$, one can obtain a finite cover of $S$ with closed rectangles of height $\varepsilon/(b - a)$ and nonoverlapping interiors.

We shall see, in section 12.112, however, that the continuous image of a set of $n$-dimensional Jordan content zero need not have $n$-dimensional Jordan content zero.

---

We have, from Theorem 105.2 and Lemma 105.1:

***Theorem 105.3***

The bounded set $\mathfrak{D} \subseteq \mathbf{E}^n$ has $n$-dimensional Jordan content if and only if, for every $\varepsilon > 0$, there are finitely many closed $n$-dimensional intervals $\mathfrak{J}_1, \mathfrak{J}_2, \ldots, \mathfrak{J}_k$ such that $\beta\mathfrak{D} \subseteq \bigcup_{j=1}^{k} \mathfrak{J}_j$ and $\sum_{j=1}^{k} |\mathfrak{J}_j| < \varepsilon$.

Upon superficial examination, there appears to be a discrepancy between the theory that was developed in this section and the one of the preceding section. It follows from Theorem 103.3 that the bounded set $\mathfrak{D} \subseteq \mathbf{E}^n$ has Jordan content if and only if $\lambda(\beta\mathfrak{D}) = 0$. In order to see this, we note that $\chi_{\mathfrak{D}}$ is, in fact, discontinuous at each $b \in \beta\mathfrak{D}$ and continuous everywhere else. Hence, if $\lambda(\beta\mathfrak{D}) = 0$, then $\int_{\mathfrak{J}} \chi_{\mathfrak{D}}$ exists for $\mathfrak{J} \supseteq \mathfrak{D}$.

Conversely, if $\chi_{\mathfrak{D}}$ has discontinuities at all points of the set $\beta\mathfrak{D}$ which does not have Lebesgue measure zero, then $\chi_{\mathfrak{D}}$ is not integrable on $\mathfrak{J} \supseteq \mathfrak{D}$.

On the other hand, we have seen, in Theorem 105.2, that $\mathfrak{D}$ is Jordan-measurable if and only if $\beta\mathfrak{D}$ has Jordan content zero. While it is true that a set of Jordan content zero has Lebesgue measure zero (example 3), the converse is not necessarily true (see example 4). Here is the apparent discrepancy. This is, however, easily resolved since $\beta\mathfrak{D}$ is compact (Lemma 104.2) and since a compact set has Jordan content zero if and only if it has Lebesgue measure zero, as we shall now demonstrate.

***Lemma 105.2***

The compact set $S \subseteq \mathbf{E}^n$ has $n$-dimensional Jordan content zero if and only if it has $n$-dimensional Lebesgue measure zero.

*Proof*

(a) If $S$ has Jordan content zero, then it has Lebesgue measure zero (example 3).

(b) If $S$ has Lebesgue measure zero, then there is, for every $\varepsilon > 0$, a sequence of open intervals $\{\mathfrak{I}_1, \mathfrak{I}_2, \mathfrak{I}_3, \ldots\}$ such that $S \subseteq \bigcup_{k \in \mathbf{N}} \mathfrak{I}_k$ and $\sum_{k=1}^{\infty} |\mathfrak{I}_k| < \varepsilon$. Clearly, $\{\mathfrak{I}_1, \mathfrak{I}_2, \mathfrak{I}_3, \ldots\}$ is an open cover of $S$. Since $S$ is compact, a finite subcover will do, say $\{\mathfrak{I}_{l_1}, \mathfrak{I}_{l_2}, \ldots, \mathfrak{I}_{l_\mu}\}$. If $\mathfrak{I}_{l_j} = \{x \in \mathbf{E}^n \mid a_i^{(l_j)} < \xi_i < b_i^{(l_j)}$, where $i = 1, 2, \ldots, n\}$, let $\mathfrak{J}_{l_j} = \{x \in \mathbf{E}^n \mid a_i^{(l_j)} \leqslant \xi_i \leqslant b_i^{(l_j)}, i = 1, 2, \ldots, n\}$. Then $S \subseteq \bigcup_{k=1}^{\mu} \mathfrak{J}_{l_k}$ and $\sum_{k=1}^{\mu} |\mathfrak{J}_{l_k}| < \sum_{k=1}^{\infty} |\mathfrak{J}_k| = \sum_{k=1}^{\infty} |\mathfrak{I}_k| < \varepsilon$. Hence, $S$ has Jordan content zero.

## 11.105 Exercises

1. Show that definition 105.1 of $\underline{J}(\mathfrak{D})$ and $\bar{J}(\mathfrak{D})$ is independent of the choice of $\mathfrak{J}$ as long as $\mathfrak{J} \supseteq \mathfrak{D}$.
2. Let $\mathfrak{D} \subseteq \mathfrak{J}$ where $\mathfrak{J}$ is an $n$-dimensional closed interval. Let $\mathfrak{J}_1 \subseteq \mathfrak{J}$ be an $n$-dimensional closed interval. Show: If $\mathfrak{J}_1 \cap \beta\mathfrak{D} = \varnothing$, then either $\mathfrak{J}_1 \subseteq \mathfrak{D}$ or $\mathfrak{J}_1 \subseteq c(\mathfrak{D}) \cap \mathfrak{J}$.
3. Show: If $S = \{x_1, x_2, \ldots, x_n\}$, where the $x_j$ are points in $\mathbf{E}^n$, then $J(S) = 0$.

4. Prove the statement in example 6 in detail.
5. *Prove*: If $f: \mathfrak{J} \to \mathbf{E}$ is continuous on the $n$-dimensional interval $\mathfrak{J}$ except on a set $S \subset \mathfrak{J}$ of $n$-dimensional Jordan content zero, then $f$ is integrable on $\mathfrak{J}$.
6. Let $f: \mathfrak{J} \to \mathbf{E}$ where $\mathfrak{J}$ is an $n$-dimensional closed interval and where $f$ is presumed to be bounded on $\mathfrak{J}$. Let

$$E_m = \left\{ x \in \mathfrak{J} \,\middle|\, |f(x) - f(y)| \geqslant \frac{1}{m} \quad \text{for at least one } y \text{ in every } N_\delta(x) \cap \mathfrak{J} \right\}.$$

   *Prove*: If $\int_{\mathfrak{J}} f$ exists, then, for every $m \in \mathbf{N}$, $J(E_m) = 0$.
7. Let $A \subseteq B \subset \mathbf{E}^n$ denote bounded sets. Show that $\bar{J}(A) \leqslant \bar{J}(B)$.

## 11.106 Properties of Jordan-Measurable Sets

As expected, union and intersection of Jordan-measurable sets are again Jordan-measurable.

***Theorem 106.1***

If $S_1, S_2 \subseteq \mathbf{E}^n$ are Jordan-measurable sets, then $S_1 \cup S_2$ and $S_1 \cap S_2$ are Jordan-measurable and

$$J(S_1) + J(S_2) = J(S_1 \cup S_2) + J(S_1 \cap S_2). \tag{106.1}$$

*Proof*

Clearly, $S_1 \cup S_2$, $S_1 \cap S_2$ are bounded. Since $\beta(S_1 \cup S_2), \beta(S_1 \cap S_2) \subseteq \beta(S_1) \cup \beta(S_2)$ (see Lemma 104.3), since $J(\beta S_1) = J(\beta S_2) = 0$ (see Theorem 105.2), since the union of finitely many sets of Jordan content zero has again Jordan content zero (see exercise 1), and since every subset of a set of Jordan content zero has again Jordan content zero (see exercise 7, section 11.105), we see that $S_1 \cup S_2, S_1 \cap S_2$ are Jordan-measurable.

If $\chi_{S_1}, \chi_{S_2}, \chi_{S_1 \cup S_2}, \chi_{S_1 \cap S_2}$ are the characteristic functions of $S_1, S_2, S_1 \cup S_2$, $S_1 \cap S_2$, we have

$$\chi_{S_1} + \chi_{S_2} = \chi_{S_1 \cup S_2} + \chi_{S_1 \cap S_2} \tag{106.2}$$

for the following reason: If $x \in S_1 \backslash S_2$, then $\chi_{S_1}(x) = 1, \chi_{S_2}(x) = 0, \chi_{S_1 \cup S_2}(x) = 1$, $\chi_{S_1 \cap S_2} = 0$. If $x \in S_2 \backslash S_1$, the roles of $S_1, S_2$ in the above argument are reversed. If $x \in S_1 \cap S_2$, then $\chi_{S_1}(x) + \chi_{S_2}(x) = 2$, and $\chi_{S_1 \cup S_2}(x) = 1, \chi_{S_1 \cap S_2}(x) = 1$. Finally, if $x$ is neither in $S_1$ nor in $S_2$, then it is not in $S_1 \cup S_2$ and $S_1 \cap S_2$ either, and we have zero on both sides of (106.2).

Since $\chi_{S_1}, \chi_{S_2}$ both are integrable on every $\mathfrak{J} \supseteq S_1 \cup S_2$, we have that $\chi_{S_1} + \chi_{S_2}$ is also integrable on $\mathfrak{J}$, and we obtain, from Theorem 103.4, that

$$\int_{\mathfrak{J}} (\chi_{S_1} + \chi_{S_2}) = \int_{\mathfrak{J}} \chi_{S_1} + \int_{\mathfrak{J}} \chi_{S_2} = J(S_1) + J(S_2). \tag{106.3}$$

Also

$$\text{(106.4)} \quad \int_{\mathfrak{J}} (\chi_{S_1 \cup S_2} + \chi_{S_1 \cap S_2}) = \int_{\mathfrak{J}} \chi_{S_1 \cup S_2} + \int_{\mathfrak{J}} \chi_{S_1 \cap S_2} = J(S_1 \cup S_2) + J(S_1 \cap S_2).$$

(106.1) follows readily from (106.2), (106.3), and (106.4).

***Corollary 1 to Theorem 106.1***

If $S_1, S_2 \subseteq \mathbf{E}^n$ are Jordan-measurable and if $S_1, S_2$ have no interior points in common, then

$$\text{(106.5)} \qquad J(S_1 \cup S_2) = J(S_1) + J(S_2).$$

*Proof*

If $S_1, S_2$ have no interior points in common, then $S_1 \cap S_2 \subseteq \beta S_1 \cup \beta S_2$. Hence, $J(S_1 \cap S_2) = 0$, and we obtain (106.5) from (106.1).

***Corollary 2 to Theorem 106.1***

If $S_1 \subseteq S_2 \subseteq \mathbf{E}^n$ are Jordan-measurable then $S_2 \backslash S_1$ is Jordan-measurable and

$$\text{(106.6)} \qquad J(S_2 \backslash S_1) = J(S_2) - J(S_1).$$

*Proof*

$S_2 \backslash S_1$ is bounded and $\beta(S_2 \backslash S_1) \subseteq \beta(S_1) \cup \beta(S_2)$ (see Lemma 104.3). Hence, $J(\beta(S_2 \backslash S_1)) = 0$ and therefore, $S_2 \backslash S_1$ is Jordan-measurable.

Since $S_2 = S_1 \cup S_2 = (S_2 \backslash S_1) \cup S_1$, we have, from (106.1), that $J(S_2) = J(S_2 \backslash S_1) + J(S_1) - J((S_2 \backslash S_1) \cap S_1)$. Since $(S_2 \backslash S_1) \cap S_1 = \varnothing$, then (106.6) follows readily.

Corollary 1 and 2 together yield

***Corollary 3 to Theorem 106.1***

If $S_1 \subseteq S_2 \subseteq \mathbf{E}^n$ are Jordan-measurable, then $J(S_1) \leqslant J(S_2)$.

In this and the preceding section, we have established the following properties of the Jordan content:

All $n$-dimensional intervals are Jordan-measurable (examples 1 and 2, section 11.105). The union of Jordan-measurable sets with nonoverlapping interiors has Jordan content, which is the sum of the Jordan contents of the two sets (Corollary 1 to Theorem 106.1). If one set is contained in another set and both have Jordan content, then the set left over after removal of the contained set is again Jordan-measurable (Corollary 2 to Theorem 106.1). With this, we have established that the Jordan-measurable sets satisfy the postulates $(P_1)$,

$(P_3)$, $(P_4)$ of definition 41.1. We shall demonstrate, in section 12.116, that $(P_2)$ is also satisfied.

That $(\mu_1)$, $(\mu_3)$, $(\mu_4)$ of definition 41.1 are also satisfied follows from the fact that $J(S) \geqslant 0$ for every bounded Jordan-measurable set $S \subseteq \mathbf{E}^n$, from Corollary 1 to Theorem 106.1, and from example 1, section 11.105, according to which $J([0,1] \times [0,1] \times \cdots \times [0,1]) = 1$. That $(\mu_2)$ is also satisfied will be shown in section 12.116.

## 11.106 Exercises

1. Show: If $S_1, S_2, \ldots, S_k \subseteq \mathbf{E}^n$ have Jordan content zero, then $\bigcup_{j=1}^k S_j$ has Jordan content zero.
2. Show: If $S_1 \subseteq \mathbf{E}^n$ has Jordan content zero, then $S \subseteq S_1$ has Jordan content zero.
3. Show: If $\mathfrak{D} \subseteq \mathbf{E}^n$ is a Jordan-measurable set, then $\mathfrak{D} \cup (\beta\mathfrak{D})$, and $\mathfrak{D}^\circ$, the set of interior points of $\mathfrak{D}$, are Jordan-measurable, and $J(\mathfrak{D}) = J(\mathfrak{D} \cup \beta\mathfrak{D}) = J(\mathfrak{D}^\circ)$. (Compare this result with Theorem 41.4.)
4. Show: If $S_1$, $S_2 \subset \mathbf{E}^n$ are any two Jordan-measurable sets, then $J(S_1 \backslash S_2) = J(S_1) - J(S_1 \cap S_2)$.

## 11.107 Integrals over Jordan-Measurable Sets

If $\mathfrak{D} \subseteq \mathbf{E}^n$ is bounded and Jordan-measurable, then, by Theorem 105.2, $J(\beta\mathfrak{D}) = 0$ and hence, by Lemma 105.2, $\lambda(\beta\mathfrak{D}) = 0$. Therefore, if $f\colon \mathfrak{D} \to \mathbf{E}$ is bounded and continuous a.e. on $\mathfrak{D}$, then, by Theorem 104.1, $f$ is integrable over $\mathfrak{D}$. We state this result as a theorem.

***Theorem 107.1***

If $f\colon \mathfrak{D} \to \mathbf{E}$ is bounded and continuous a.e. on the Jordan-measurable set $\mathfrak{D} \subseteq \mathbf{E}^n$, then $\int_{\mathfrak{D}} f$ exists.

It is now an easy matter to generalize the theorems of section 4.51 to integrals over Jordan-measurable sets.

***Theorem 107.2***

If $h : \mathfrak{D} \to \mathbf{E}$ is integrable on the Jordan-measurable set $\mathfrak{D} \subseteq \mathbf{E}^n$ and if $h(x) \geqslant 0$ for all $x \in \mathfrak{D}$, then $\int_{\mathfrak{D}} h \geqslant 0$.

*Proof*

Since $\underline{S}(h;P) \geqslant 0$ for all partitions $P$ of any interval $\mathfrak{J} \supseteq \mathfrak{D}$, we have $\int_{\mathfrak{D}} h = \sup_{(P)} \underline{S}(h;P) \geqslant 0$.

***Corollary 1 to Theorem 107.2***

If $f, g : \mathfrak{D} \to \mathbf{E}$ are integrable on the Jordan-measurable set $\mathfrak{D} \subseteq \mathbf{E}^n$ and if $f(x) \leqslant g(x)$ for all $x \in \mathfrak{D}$, then $\int_{\mathfrak{D}} f \leqslant \int_{\mathfrak{D}} g$.

***Corollary 2 to Theorem 107.2***

If $f: \mathfrak{D} \to \mathbf{E}$ is integrable on the Jordan-measurable set $\mathfrak{D}$, and if $m \leqslant f(x) \leqslant M$ for all $x \in \mathfrak{D}$, then

$$mJ(\mathfrak{D}) \leqslant \int_{\mathfrak{D}} f \leqslant MJ(\mathfrak{D}).$$

*Proof*

Note that $\int_{\mathfrak{D}} M = M \int_{\mathfrak{J}} \chi_{\mathfrak{D}} = MJ(\mathfrak{D})$. Same for $m$.

***Corollary 3 to Theorem 107.2***

If $f: \mathfrak{D} \to \mathbf{E}$ is integrable on the Jordan-measurable set $\mathfrak{D} \subseteq \mathbf{E}^n$, then

$$\left| \int_{\mathfrak{D}} f \right| \leqslant \int_{\mathfrak{D}} |f|.$$

***Theorem 107.3*** (*Mean-Value Theorem*)

If $f, g: \mathfrak{D} \to \mathbf{E}$ are bounded and continuous on the Jordan-measurable set $\mathfrak{D} \subset \mathbf{E}^n$, if $\mathfrak{D}$ is connected, and if $g(x) \geqslant 0$ for all $x \in \mathfrak{D}$, then there is a point $c \in \mathfrak{D}$ such that

$$\int_{\mathfrak{D}} fg = f(c) \int_{\mathfrak{D}} g.$$

*Proof*

Let $m = \inf_{x \in \mathfrak{D}} f(x)$, $M = \sup_{x \in \mathfrak{D}} f(x)$. Then $mg(x) \leqslant f(x)g(x) \leqslant Mg(x)$ for all $x \in \mathfrak{D}$. By Corollary 1 to Theorem 107.2, $m \int_{\mathfrak{D}} g \leqslant \int_{\mathfrak{D}} fg \leqslant M \int_{\mathfrak{D}} g$. If $g(x) = 0$ for all $x \in \mathfrak{D}$, the theorem is trivially true. If $g(x_0) > 0$ for some $x_0 \in \mathfrak{D}$, then $\int_{\mathfrak{D}} g \neq 0$, and we can divide by $\int_{\mathfrak{D}} g$. We apply Theorem 73.1 and the result follows readily.

Note that for $g(x) = 1$ for all $x \in \mathfrak{D}$ we obtain

$$\int_{\mathfrak{D}} f = f(c) J(\mathfrak{D}) \qquad \text{for some } c \in \mathfrak{D}.$$

***Theorem 107.4***

Let $\mathfrak{D}, \mathfrak{D}_1, \mathfrak{D}_2 \subseteq \mathbf{E}^n$ be Jordan-measurable sets such that $\mathfrak{D} = \mathfrak{D}_1 \cup \mathfrak{D}_2$ and $J(\mathfrak{D}_1 \cap \mathfrak{D}_2) = 0$. If $f: \mathfrak{D} \to \mathbf{E}$ is integrable over $\mathfrak{D}$, then $f$ is integrable over $\mathfrak{D}_1$ and $\mathfrak{D}_2$ and

$$\int_{\mathfrak{D}} f = \int_{\mathfrak{D}_1} f + \int_{\mathfrak{D}_2} f.$$

(See also exercise 6.)

*Proof*

Let $f_{\mathfrak{D}} : \mathbf{E}^n \to \mathbf{E}$ be defined by

$$f_{\mathfrak{D}}(x) = \begin{cases} f(x) & \text{for } x \in \mathfrak{D}, \\ 0 & \text{for } x \in c(\mathfrak{D}), \end{cases}$$

and let $f_{\mathfrak{D}_1} : \mathbf{E}^n \to \mathbf{E}$ be defined by

$$f_{\mathfrak{D}_1}(x) = \begin{cases} f(x) & \text{for } x \in \mathfrak{D}_1, \\ 0 & \text{for } x \in c(\mathfrak{D}_1). \end{cases}$$

Since $f$ is integrable over $\mathfrak{D}$, $\int_{\mathfrak{D}} f = \int_{\mathfrak{J}} f_{\mathfrak{D}}$ exists for every $\mathfrak{J} \supseteq \mathfrak{D}$. Hence, $f_{\mathfrak{D}}$ is continuous on $\mathfrak{J}$ except on a set of Lebesgue measure zero. Since $\mathfrak{D}_1 \subseteq \mathfrak{D}$, $f_{\mathfrak{D}_1}$ is certainly continuous in the interior of $\mathfrak{D}_1$ except possibly on a set of Lebesgue measure zero and is continuous, by definition, in the interior of $c(\mathfrak{D}_1)$. The only other places it could possibly be discontinuous is on $\beta\mathfrak{D}_1$ but $\lambda(\beta\mathfrak{D}_1) = 0$. Hence, $f_{\mathfrak{D}_1}$ is integrable on $\mathfrak{J}$, and we have

$$\int_{\mathfrak{J}} f_{\mathfrak{D}_1} = \int_{\mathfrak{D}_1} f. \tag{107.1}$$

If $f_{\mathfrak{D}_2}: \mathbf{E}^n \to \mathbf{E}$ is defined by

$$f_{\mathfrak{D}_2}(x) = \begin{cases} f(x) & \text{for } x \in \mathfrak{D}_2, \\ 0 & \text{for } x \in c(\mathfrak{D}_2), \end{cases}$$

we obtain, in the same manner,

$$\int_{\mathfrak{J}} f_{\mathfrak{D}_2} = \int_{\mathfrak{D}_2} f. \tag{107.2}$$

We have $f_{\mathfrak{D}}(x) = f_{\mathfrak{D}_1}(x) + f_{\mathfrak{D}_2}(x)$ for all $x \in \mathfrak{D} \backslash \mathfrak{D}_1 \cap \mathfrak{D}_2$. Hence, if $h : \mathbf{E}^n \to \mathbf{E}$ is defined by

$$h(x) = f_{\mathfrak{D}}(x) - f_{\mathfrak{D}_1}(x) - f_{\mathfrak{D}_2}(x),$$

then $h(x) = 0$ for all $x \in \mathfrak{D} \backslash \mathfrak{D}_1 \cap \mathfrak{D}_2$. Since $h$ is bounded on $\mathfrak{D}$ and $J(\mathfrak{D}_1 \cap \mathfrak{D}_2) = 0$ by hypothesis, it follows that $h$ is integrable over $\mathfrak{D}$ and $\int_{\mathfrak{D}} h = 0$ (see exercise 3). Hence, we obtain, from (107.1) and (107.2), with $h_{\mathfrak{D}} : \mathbf{E}^n \to \mathbf{E}$ defined by

$$h_{\mathfrak{D}}(x) = \begin{cases} h(x) & \text{for } x \in \mathfrak{D}, \\ 0 & \text{for } x \in c(\mathfrak{D}), \end{cases}$$

that, in view of Theorem 103.4,

$$\begin{aligned} \int_{\mathfrak{D}} h = \int_{\mathfrak{J}} h_{\mathfrak{D}} = \int_{\mathfrak{J}} (f_{\mathfrak{D}} - f_{\mathfrak{D}_1} - f_{\mathfrak{D}_2}) &= \int_{\mathfrak{J}} f_{\mathfrak{D}} - \int_{\mathfrak{J}} f_{\mathfrak{D}_1} - \int_{\mathfrak{J}} f_{\mathfrak{D}_2} \\ &= \int_{\mathfrak{D}} f - \int_{\mathfrak{D}_1} f - \int_{\mathfrak{D}_2} f = 0, \end{aligned}$$

and therefore,

$$\int_{\mathfrak{D}} f = \int_{\mathfrak{D}_1} f + \int_{\mathfrak{D}_2} f.$$

***Corollary to Theorem 107.4***

If $\mathfrak{D}, \mathfrak{D}_1 \subseteq \mathbf{E}^n$ are Jordan-measurable, if $\mathfrak{D}_1 \subseteq \mathfrak{D}$, and if $f: \mathfrak{D} \to \mathbf{E}$ is integrable on $\mathfrak{D}$ and $f(x) \geqslant 0$ for all $x \in \mathfrak{D}$, then

$$\int_{\mathfrak{D}_1} f \leqslant \int_{\mathfrak{D}} f.$$

*Proof*

By Corollary 2 to Theorem 106.1, $\mathfrak{D}\backslash\mathfrak{D}_1$ is Jordan-measurable. Since $\mathfrak{D} = \mathfrak{D}_1 \cup \mathfrak{D}\backslash\mathfrak{D}_1$, we have from Theorem 107.4 that

$$\int_{\mathfrak{D}} f = \int_{\mathfrak{D}_1} f + \int_{\mathfrak{D}\backslash\mathfrak{D}_1} f$$

and, from Theorem 107.2, that $\int_{\mathfrak{D}\backslash\mathfrak{D}_1} f \geqslant 0$. The corollary follows readily.

We obtain from Theorem 103.4 and definition 104.1 immediately that,

***Theorem 107.5***

If $f, g : \mathfrak{D} \to \mathbf{E}$ are integrable on the Jordan-measurable set $\mathfrak{D} \subseteq \mathbf{E}^n$, then

$$\int_{\mathfrak{D}} (\lambda f + \mu g) = \lambda \int_{\mathfrak{D}} f + \mu \int_{\mathfrak{D}} g \qquad \text{for all } \lambda, \mu \in \mathbf{R}.$$

## 11.107 Exercises

1. Prove Corollary 1 to Theorem 107.2.
2. Prove Corollary 3 to Theorem 107.2.
3. Let $h : \mathfrak{J} \to \mathbf{E}$ be defined on the closed $n$-dimensional interval $\mathfrak{J}$ by
$$h(x) = \begin{cases} g(x) & \text{for } x \in S, \\ 0 & \text{for } x \in c(S), \end{cases}$$
where $S \subseteq \mathfrak{J}$ has Jordan content zero and where $g : S \to \mathbf{E}$ is bounded on $S$. Prove that $\int_{\mathfrak{J}} h = 0$.
4. Show: If $f: [a, b] \to \mathbf{E}$ is bounded on $[a, b]$ and if $f(x) = 0$ for all $x \in [a, b]$ except at finitely many points, then $\int_a^b f = 0$. (See exercise 3.)
5. Show: If $f, g : [a, b] \to \mathbf{E}$ are integrable on $[a, b]$ and if $f(x) = g(x)$ for all $x \in [a, b]$ except for finitely many points, then $\int_a^b f = \int_a^b g$. (See exercise 4.)
6. Let $\mathfrak{D}, \mathfrak{D}_1, \mathfrak{D}_2 \subseteq \mathbf{E}^n$ be Jordan-measurable sets such that $\mathfrak{D} = \mathfrak{D}_1 \cup \mathfrak{D}_2$ and $J(\mathfrak{D}_1 \cap \mathfrak{D}_2) = 0$. *Prove*: If $f: \mathfrak{D} \to \mathbf{E}$ is integrable over $\mathfrak{D}_1$ and over $\mathfrak{D}_2$, then $f$ is integrable over $\mathfrak{D}$ and
$$\int_{\mathfrak{D}} f = \int_{\mathfrak{D}_1} f + \int_{\mathfrak{D}_2} f.$$

## 11.108 Integration by Iteration—Fubini's Theorem

In section 4.52 we have developed a procedure for the evaluation of single integrals which, in essence, is based on the fact that the integrand is the derivative of the integral with respect to the variable upper integration limit. No such simple relationship exists between a multiple integral and its integrand. Still, for certain domains of integration, one may evaluate a multiple integral by a succession of *single* integrations, each of which may be carried out as in section 4.52.

The reader has already learned in calculus to evaluate double integrals on rectangles by first carrying out the (single) integration with respect to one integration variable, and then by integrating the result (which is a function of the other integration variable) with respect to the other integration variable. This technique may be generalized to multiple integrals where the domain of integration is an interval, as *Guido Fubini*† has shown. In order to establish a theorem that pertains to this matter, we need an intermediate result that is most conveniently expressed in terms of *upper* and *lower Darboux integrals* (see also exercise 6, section 11.102).

***Definition 108.1***

If $g : \mathfrak{J} \to \mathbf{E}$, where $\mathfrak{J} \subseteq \mathbf{E}^p$ is an interval, is bounded on $\mathfrak{J}$, then

$$\sup_{(P)} \underline{S}(g; P) = \underline{\int_{\mathfrak{J}}} g, \qquad \inf_{(P)} \overline{S}(g; P) = \overline{\int_{\mathfrak{J}}} g$$

are called the *lower* and the *upper Darboux integrals*.

(Note that, by Lemma 102.1, the upper and the lower Darboux integrals exist.)

***Lemma 108.1***

Let $\mathcal{A} \subset \mathbf{E}^n$, $\mathcal{B} \subset \mathbf{E}^m$ denote closed intervals and let $f : \mathcal{A} \times \mathcal{B} \to \mathbf{E}$ denote a *bounded* function. For fixed $x \in \mathcal{A}$ we define $\phi_x : \mathcal{B} \to \mathbf{E}$ by

$$\phi_x(y) = f(x, y).$$

Then

$$\phi(x) = \underline{\int_{\mathcal{B}}} \phi_x, \qquad \Phi(x) = \overline{\int_{\mathcal{B}}} \phi_x \tag{108.1}$$

exists for all $x \in \mathcal{A}$. If $P_{\mathcal{A}}$, $Q_{\mathcal{A}}$ are any partitions of $\mathcal{A}$ and $P_{\mathcal{B}}$, $Q_{\mathcal{B}}$ are any partitions of $\mathcal{B}$, then

$$\underline{S}(f; P) \leqslant \underline{S}(\phi; P_{\mathcal{A}}) \leqslant \underline{S}(\Phi; P_{\mathcal{A}}) \leqslant \overline{S}(\Phi; Q_{\mathcal{A}}) \leqslant \overline{S}(f; Q), \tag{108.2}$$

where $P = P_{\mathcal{A}} \times P_{\mathcal{B}}$, $Q = Q_{\mathcal{A}} \times Q_{\mathcal{B}}$.

† *Guido Fubini*, 1879–1943.

If we define, for fixed $y \in \mathcal{B}$, $\psi_y : \mathcal{A} \to \mathbf{E}$ by

$$\psi_y(x) = f(x, y),$$

then

$$\psi(y) = \underline{\int_{\mathcal{A}}} \psi_y, \qquad \Psi(y) = \overline{\int_{\mathcal{A}}} \psi_y \tag{108.3}$$

exist, and

$$\underline{S}(f; P) \leqslant \underline{S}(\psi; P_{\mathcal{B}}) \leqslant \underline{S}(\Psi; P_{\mathcal{B}}) \leqslant \bar{S}(\Psi; Q_{\mathcal{B}}) \leqslant \bar{S}(f; Q). \tag{108.4}$$

*Proof*

Since $f$ is bounded on $\mathcal{A} \times \mathcal{B}$, $\phi_x$ is bounded on $\mathcal{B}$, and $\psi_y$ is bounded on $\mathcal{A}$. Hence, $\phi(x), \Phi(x), \psi(y), \Psi(y)$ exist.

We shall prove only (108.2). (108.4) follows by analogous reasoning.

The partition $P_{\mathcal{A}}$ of $\mathcal{A}$ generates $\mu$ $n$-dimensional closed intervals $I_i^{\mathcal{A}}$, $i = 1, 2, \ldots, \mu$, and the partition $P_{\mathcal{B}}$ of $\mathcal{B}$ generates $\nu$ $m$-dimensional closed intervals $I_j^{\mathcal{B}}$, $j = 1, 2, \ldots, \nu$. Then, $P = P_{\mathcal{A}} \times P_{\mathcal{B}}$ generates $\mu\nu$ closed intervals $I_i^{\mathcal{A}} \times I_j^{\mathcal{B}}$, and we have

$$\underline{S}(f; P) = \sum_{i=1}^{\mu} \sum_{j=1}^{\nu} m_{ij}(f) \, |I_i^{\mathcal{A}} \times I_j^{\mathcal{B}}|$$

where $m_{ij}(f) = \inf \{ f(x, y) \mid (x, y) \in I_i^{\mathcal{A}} \times I_j^{\mathcal{B}} \}$, $x = (\xi_i, \xi_2, \ldots, \xi_n)$, $y = (\eta_1, \eta_2, \ldots, \eta_m)$. We have, from definition 102.2, that $|I_i^{\mathcal{A}} \times I_j^{\mathcal{B}}| = |I_i^{\mathcal{A}}| \, |I_j^{\mathcal{B}}|$, and hence

$$\underline{S}(f; P) = \sum_{i=1}^{\mu} \sum_{j=1}^{\nu} m_{ij}(f) \, |I_i^{\mathcal{A}}| \; I_j^{\mathcal{B}}|.$$

For every fixed $x \in I_i^{\mathcal{A}}$, $m_{ij}(f) \leqslant \inf \{ \phi_x(y) \mid y \in I_j^{\mathcal{B}} \} = m_j(\phi_x)$. Hence,

$$\sum_{j=1}^{\nu} m_{ij}(f) \, |I_j^{\mathcal{B}}| \leqslant \sum_{j=1}^{\nu} m_j(\phi_x) \, |I_j^{\mathcal{B}}| = \underline{S}(\phi_x; P_{\mathcal{B}}) \leqslant \underline{\int_{\mathcal{B}}} \phi_x = \phi(x).$$

Since the left member of this inequality is independent of $x \in I_i^{\mathcal{A}}$, we have

$$\sum_{j=1}^{\nu} m_{ij}(f) \, |I_j^{\mathcal{B}}| \leqslant \inf \{ \phi(x) \mid x \in I_i^{\mathcal{A}} \}.$$

Hence,

$$\underline{S}(f; P) \leqslant \sum_{i=1}^{\mu} \inf \{ \phi(x) \, | x \in I_i^{\mathcal{A}} \} \, |I_i^{\mathcal{A}}| = \underline{S}(\phi; P_{\mathcal{A}}). \tag{108.5}$$

By a similar argument (see exercise 3.),

$$\bar{S}(f; Q) \geqslant \bar{S}(\Phi; Q_{\mathcal{A}}). \tag{108.6}$$

By (102.4), $\phi(x) \leqslant \Phi(x)$ for every $x \in \mathcal{A}$. Hence,

$$\underline{S}(\phi; P_{\mathcal{A}}) \leqslant \underline{S}(\Phi; P_{\mathcal{A}}). \tag{108.7}$$

By Lemma 102.1,

$$\underline{S}(\Phi; P_{\mathcal{A}}) \leqslant \bar{S}(\Phi; Q_{\mathcal{A}}). \tag{108.8}$$

If we string (108.5), (108.7), (108.8), and (108.6) together, we obtain (108.2).

***Theorem 108.1*** (*Fubini's Theorem*)

Let $\mathcal{A} \subset \mathbf{E}^n$, $\mathcal{B} \subset \mathbf{E}^m$ denote closed intervals. If $f : \mathcal{A} \times \mathcal{B} \to \mathbf{E}$ is *integrable* on $\mathcal{A} \times \mathcal{B}$, then

$$\int_{\mathcal{A} \times \mathcal{B}} f = \int_{\mathcal{A}} \phi = \int_{\mathcal{A}} \Phi = \int_{\mathcal{B}} \psi = \int_{\mathcal{B}} \Psi, \tag{108.9}$$

where $\phi$, $\Phi$, $\psi$, $\Psi$ are defined in (108.1) and (108.3).

(Note that, written out in greater detail, (108.9) takes the form

$$\int_{\mathcal{A} \times \mathcal{B}} f(x, y)\, dx\, dy = \int_{\mathcal{A}} \left( \underline{\int_{\mathcal{B}}} f(x, y)\, dy \right) dx = \int_{\mathcal{A}} \left( \overline{\int_{\mathcal{B}}} f(x, y)\, dy \right) dx,$$

$$= \int_{\mathcal{B}} \left( \underline{\int_{\mathcal{A}}} f(x, y)\, dx \right) dy = \int_{\mathcal{B}} \left( \overline{\int_{\mathcal{A}}} f(x, y)\, dx \right) dy.)$$

*Proof*

Since $f$ is integrable on $\mathcal{A} \times \mathcal{B}$, we have $\sup_{(P)} \underline{S}(f; P) = \inf_{(Q)} \bar{S}(f; Q) = \int_{\mathcal{A} \times \mathcal{B}} f$, and the first portion of (108.9) follows from (108.2). The remainder of (108.9) follows from (108.4). (Note that, with any partition $P$ of $\mathcal{A} \times \mathcal{B}$, there correspond partitions $P_{\mathcal{A}}$ of $\mathcal{A}$ and $P_{\mathcal{B}}$ of $\mathcal{B}$ such that $P = P_{\mathcal{A}} \times P_{\mathcal{B}}$ and vice versa. The same holds, of course, for $Q$, $Q_{\mathcal{A}}$, $Q_{\mathcal{B}}$.)

Note that $\int_{\mathcal{A}} \phi = \int_{\mathcal{A}} \Phi$ in (108.9) does *not* imply that $\phi(x) = \Phi(x)$ for all $x \in \mathcal{A}$ because $\phi_x$ need *not* be integrable on $\mathcal{B}$. This will be demonstrated in the following example.

---

▲ *Example 1*

Let $f$: $[0, 1] \times [0, 1] \to \mathbf{E}$ be defined by

$$f(x, y) = \begin{cases} 1 & \text{if } x \text{ is irrational, or if } x \text{ is rational and } y \text{ is irrational,} \\ 1 - \dfrac{1}{q} & \text{if } x = \dfrac{p}{q}, (p, q) = 1, \text{ and } y \text{ is rational.} \end{cases}$$

This function is continuous everywhere except possibly for $(x, y) \in \mathbf{Q} \times \mathbf{Q}$, as we shall now demonstrate:

If $(x_0, y_0) \notin \mathbf{Q} \times \mathbf{Q}$, then $x_0$ is irrational or $y_0$ is irrational and hence, $f(x_0, y_0) = 1$. For every $\varepsilon > 0$, there is a $q_0 \in \mathbf{N}$ such that $(1/q_0) < \varepsilon$. There are only finitely many rational numbers $(p/q) \in [0, 1]$ such that $q < q_0$. Hence, one of them is closest to $x_0$, say at a distance $\delta > 0$. Then, for all rational numbers $p/q$ in $(x_0 - \delta, x_0 + \delta)$ we have $1/q \leqslant 1/q_0 < \varepsilon$. Hence, whenever $(x, y) \in \mathbf{N}_\delta(x_0, y_0)$, then

$$|f(x_0, y_0) - f(x, y)| = \begin{cases} 0 < \varepsilon & \text{if } x \text{ is irrational or } y \text{ is irrational,} \\ \dfrac{1}{q} < \varepsilon & \text{if } y \text{ is rational and } x = \dfrac{p}{q}, \qquad (p, q) = 1. \end{cases}$$

Hence, $f$ is continuous for all $(x_0, y_0) \notin (\mathbf{Q} \times \mathbf{Q}) \cap ([0,1] \times [0,1])$. Since $\mathbf{Q} \times \mathbf{Q}$ is a denumerable set, $\lambda((\mathbf{Q} \times \mathbf{Q}) \cap ([0,1] \times [0,1])) = 0$, and it follows that $f$ is continuous a.e. on $[0,1] \times [0,1]$. Hence, $f$ is integrable on $[0,1] \times [0,1]$.

Since $\bar{S}(f; P) = 1$ for every $P$, we have

$$\int_{[0,1]\times[0,1]} f = 1.$$

Since

$$\phi_x(y) = 1 \qquad \text{for all} \quad y \in [0,1] \text{ if } x \text{ is irrational,}$$

and

$$\phi_x(y) = \left.\begin{cases} 1 & \text{if } \; y \text{ is irrational,} \\ 1 - \dfrac{1}{q} & \text{if } \; y \text{ is rational,} \end{cases}\right\} \text{if } x = \frac{p}{q}, \qquad (p, q) = 1$$

(note that for $x = (p/q)$, $\phi_x$ is *not* integrable (with respect to $y$) on $[0,1]$), we obtain

$$\underline{S}(\phi_x, P_{\mathscr{B}}) = 1 \qquad \text{for all } P_{\mathscr{B}} \text{ if } x \text{ is irrational,}$$

and

$$\underline{S}(\phi_x, P_{\mathscr{B}}) = 1 - \frac{1}{q} \qquad \text{for all } P_{\mathscr{B}} \text{ if } x = \frac{p}{q}.$$

Therefore,

$$\text{(108.10)} \qquad \phi(x) = \underline{\int}_{[0,1]} \phi_x = \sup_{(P_{\mathscr{B}})} \underline{S}(\phi_x, P_{\mathscr{B}}) = \begin{cases} 1 & \text{if } x \text{ is irrational,} \\ 1 - \dfrac{1}{q} & \text{if } x = \dfrac{p}{q}, (p, q) = 1. \end{cases}$$

$\phi$ is continuous a.e. on $[0,1]$, as one can show as in example 8, section 2.26. Hence, $\phi$ is integrable on $[0,1]$ and we obtain

$$\int_{[0,1]} \phi = \inf_{(P_{\mathscr{A}})} \bar{S}(\phi; P_{\mathscr{A}}) = \bar{S}(\phi; P_{\mathscr{A}}) = 1.$$

By contrast,

$$\Phi(x) = \overline{\int}_{[0,1]} \phi_x = \inf_{(P_{\mathscr{B}})} \bar{S}(\phi_x; P_{\mathscr{B}}) = 1 \qquad \text{for all } x \in [0,1]$$

and we see that $\phi(x) \neq \Phi(x)$ for all $x \in \mathbf{Q} \cap [0,1]$. Still,

$$\int_{[0,1]\times[0,1]} f = \int_{[0,1]} \phi = \int_{[0,1]} \Phi = 1.$$

---

Clearly, if $\phi(x) = \Phi(x)$ for all $x \in \mathcal{A}$ and/or $\psi(y) = \Psi(y)$ for all $y \in \mathcal{B}$, then $\phi_x$ is integrable on $\mathcal{B}$ and/or $\psi_y$ is integrable on $\mathcal{A}$ and we have the following, stronger, result.

***Theorem 108.2***

Let $\mathcal{A} \subset \mathbf{E}^n$, $\mathcal{B} \subset \mathbf{E}^m$ denote closed intervals. If $f: \mathcal{A} \times \mathcal{B} \to \mathbf{E}$ is integrable on $\mathcal{A} \times \mathcal{B}$ and if $\phi_x : \mathcal{B} \to \mathbf{E}$, defined by

$$\phi_x(y) = f(x,y) \qquad \text{for each } x \in \mathcal{A},$$

is integrable on $\mathcal{B}$, then

$$\int_{\mathcal{A}\times\mathcal{B}} f = \int_{\mathcal{A}}\left(\int_{\mathcal{B}} \phi_x\right). \tag{108.11}$$

If $\psi_y : \mathcal{A} \to \mathbf{E}$, defined by

$$\psi_y(x) = f(x,y) \qquad \text{for each } y \in \mathcal{B},$$

is integrable on $\mathcal{A}$, then

$$\int_{\mathcal{A}\times\mathcal{B}} f = \int_{\mathcal{B}}\left(\int_{\mathcal{A}} \psi_y\right). \tag{108.12}$$

If both $\phi_x$ and $\psi_y$ are integrable on $\mathcal{B}$ and $\mathcal{A}$, respectively (e.g., if $f$ is continuous on $\mathcal{A} \times \mathcal{B}$), then

$$\int_{\mathcal{A}\times\mathcal{B}} f = \int_{\mathcal{A}}\left(\int_{\mathcal{B}} \phi_x\right) = \int_{\mathcal{B}}\left(\int_{\mathcal{A}} \psi_y\right). \tag{108.13}$$

Written out in greater detail, (108.13) reads

$$\int_{\mathcal{A}\times\mathcal{B}} f(x,y)\,dx\,dy = \int_{\mathcal{A}}\left(\int_{\mathcal{B}} f(x,y)\,dy\right)dx = \int_{\mathcal{B}}\left(\int_{\mathcal{A}} f(x,y)\,dx\right)dy. \tag{108.14}$$

The integrals on the right are called *iterated integrals*.

The formula the reader is familiar with from calculus follows from (108.14) for $\mathcal{A} = [a,b]$, $\mathcal{B} = [c,d]$.

## 11.108 Exercises

1. Prove (108.4).
2. Show that $\phi : [0,1] \to \mathbf{E}$, as defined in (108.10), is continuous a.e. on $[0,1]$.
3. Prove (108.6).

4. With the function $f$ of example 1, let $\psi_y(x) = f(x, y)$ for all $y \in [0, 1]$. Find $\psi(y) = \underline{\int}_{[0,1]} \psi_y, \Psi(y) = \overline{\int}_{[0,1]} \psi_y$ and show that
$$\int_{[0,1]\times[0,1]} f = \int_{[0,1]} \psi = \int_{[0,1]} \Psi.$$
5. Let $\mathfrak{D} = [a_1, b_1] \times [a_2, b_2] \times \cdots \times [a_n, b_n]$, $a_j < b_j$, $j = 1, 2, \ldots, n$. Let $f\colon \mathfrak{D} \to \mathbf{E}$ denote a continuous function and show that
$$\int_{\mathfrak{D}} f = \int_{a_n}^{b_n} \left( \int_{a_{n-1}}^{b_{n-1}} \left( \cdots \left( \int_{a_1}^{b_1} f(\xi_1, \xi_2, \ldots, \xi_n)\, d\xi_1 \right) \cdots \right) d\xi_{n-1} \right) d\xi_n.$$
6. Let $\mathcal{A}, \mathcal{B} \subset \mathbf{E}^3$ denote Jordan-measurable sets. Let $\mathcal{A}_t = \{(x, y) \in \mathbf{E}^2 \mid (x, y, t) \in \mathcal{A}\}$, $\mathcal{B}_t = \{(x, y) \in \mathbf{E}^2 \mid (x, y, t) \in \mathcal{B}\}$. Show: if $\mathcal{A}_t$, $\mathcal{B}_t$ are Jordan-measurable for all $t$ and if $J(\mathcal{A}_t) = J(\mathcal{B}_t)$, then $J(\mathcal{A}) = J(\mathcal{B})$. Give a geometric interpretation.
7. Let $f\colon \mathfrak{D} \to \mathbf{E}$, where $\mathfrak{D} \subseteq \mathbf{E}^2$ denotes an open set. Show without reference to Theorem 94.1: If $D_1 D_2 f$, $D_2 D_1 f$ are continuous in $\mathfrak{D}$, then $D_1 D_2 f(x,y) = D_2 D_1 f(x, y)$ for all $(x, y) \in \mathfrak{D}$.
8. Let $f_i : [a_i, b_i] \to \mathbf{E}$, $i = 1, 2, \ldots, n$, denote continuous functions and let $f\colon \mathfrak{J} = [a_1, b_1] \times [a_2, b_2] \times \cdots \times [a_n, b_n] \to \mathbf{E}$ be defined by $f(x) = f_1(\xi_1) f_2(\xi_2) \ldots f_n(\xi_n)$. Show that
$$\int_{\mathfrak{J}} f = \int_{a_1}^{b_1} f(\xi_1)\, d\xi_1 \int_{a_2}^{b_2} f(\xi_2)\, d\xi_2 \ldots \int_{a_n}^{b_n} f_n(\xi_n)\, d\xi_n.$$

## 11.109 Applications of Fubini's Theorem

We shall now extend the applicability of the stronger version of Fubini's Theorem which was stated in Theorem 108.2 to domains of integration such as the ones depicted in Fig. 109.1.

***Theorem 109.1***

Let $\alpha, \beta : [a, b] \to \mathbf{E}$ denote continuous functions for which $\alpha(x) \leqslant \beta(x)$ for all $x \in [a, b]$ and let
$$\mathfrak{D} = \{(x, y) \in \mathbf{E}^2 \mid a \leqslant x \leqslant b, \alpha(x) \leqslant y \leqslant \beta(x)\}.$$

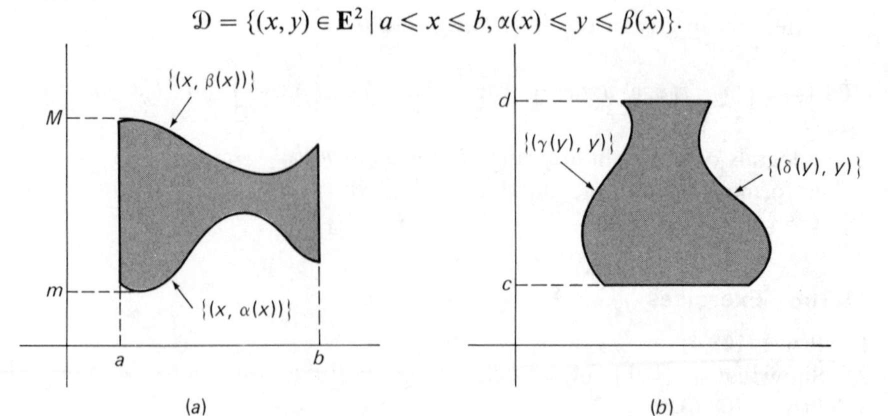


*Fig. 109.1*

If $f : \mathfrak{D} \to \mathbf{E}$ is continuous, then $\int_{\mathfrak{D}} f$ exists and

$$\int_{\mathfrak{D}} f = \int_a^b \left( \int_{\alpha(x)}^{\beta(x)} f(x, y)\, dy \right) dx. \tag{109.1}$$

*Proof*

The boundary of $\mathfrak{D}$ is composed of the intervals $[\alpha(a), \beta(a)]$ on $x = a$, $[\alpha(b), \beta(b)]$ on $x = b$, and the graphs of $\alpha, \beta$ on $[a, b]$. We have seen, in example 6, section 11.105 that these sets have Jordan content zero. Hence, $J(\beta\mathfrak{D}) = 0$. Since $\mathfrak{D}$ is also bounded, we see that $J(\mathfrak{D})$ exists. Hence, $\int_{\mathfrak{D}} f$ exists.

Let

$$m = \min_{[a,b]} \alpha(x), \qquad M = \max_{[a,b]} \beta(x)$$

(see also Fig. 109.1) and let $\mathfrak{I} = [a, b] \times [m, M]$. Then, $f_{\mathfrak{D}} : \mathfrak{I} \to \mathbf{E}$, defined by

$$f_{\mathfrak{D}}(x, y) = \begin{cases} f(x, y) & \text{for } (x, y) \in \mathfrak{D}, \\ 0 & \text{for } (x, y) \notin \mathfrak{D}, \end{cases}$$

is integrable on $\mathfrak{I}$ and $\phi_x : [m, M] \to \mathbf{E}$, defined by $\phi_x(y) = f_{\mathfrak{D}}(x, y)$ for each $x \in [a, b]$ is integrable on $[m, M]$ because it has at most two discontinuities in $[m, M]$. Hence, by Theorem 108.2,

$$\int_{\mathfrak{I}} f_{\mathfrak{D}} = \int_a^b \left( \int_m^M f_{\mathfrak{D}}(x, y)\, dy \right) dx.$$

For fixed $x \in [a, b]$,

$$\int_m^M f_{\mathfrak{D}}(x, y)\, dy = \int_{\alpha(x)}^{\beta(x)} f(x, y)\, dy$$

and (109.1) follows readily.

An analogous theorem holds for $\mathfrak{D} = \{(x, y) \in \mathbf{E}^2 \mid c \leqslant y \leqslant d, \gamma(y) \leqslant x \leqslant \delta(y)\}$. (See Fig. 109.1 and exercise 1.)

Theorem 109.1 can easily be extended to domains of integration which are Cartesian products of domains such as the ones in Fig. 109.1. (See exercise 2.)

---

▲ *Example 1* (*Jordan Content of a Parallelogram*)

Let $P_2 = \{(x, y) \in \mathbf{E}^2 \mid a \leqslant x \leqslant b,\ mx + c \leqslant y \leqslant mx + d\}$ where $c < d$ and $m \neq 0$. (See Fig. 109.2.)

We have, from Theorem 109.1, that

$$J(P_2) = \int_{P_2} \chi_{P_2} = \int_a^b \left( \int_{mx+c}^{mx+d} \chi_{P_2}(x, y)\, dy \right) dx = (b - a)(d - c),$$

$$J(P_2) = (b - a)(d - c). \tag{109.2}$$

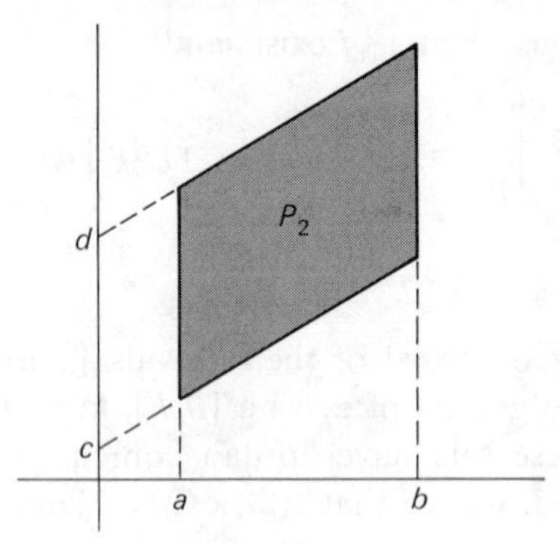


*Fig. 109.2*

▲ *Example 2*

Let $P = \{x \in \mathbf{E}^n \mid a_1 \leqslant \xi_1 \leqslant b_1,\ m\xi_1 + a_2 \leqslant \xi_2 \leqslant m\xi_1 + b_2,\ a_i \leqslant \xi_i \leqslant b_i,\ i = 3, 4, \ldots, n\}$ where $a_j < b_j$ for $j = 1, 2, \ldots, n$ and $m \neq 0$. Since $P = P_2 \times \mathfrak{J}$ where $P_2$ is a parallelogram such as in example 1 and where $\mathfrak{J}$ is the $(n-2)$-dimensional closed interval $\mathfrak{J} = [a_3, b_3] \times [a_4, b_4] \times \cdots \times [a_n, b_n]$, we obtain from Theorem 109.1 and (109.2),

$$J(P) = \int_P \chi_P = \int_{\mathfrak{J}}\left(\int_{P_2} \chi_P \, dx\right) dy = (b_1 - a_1)(b_2 - a_2) \cdots (b_n - a_n),$$

where $x = (\xi_1, \xi_2)$, $y = (\xi_3, \xi_4, \ldots, \xi_n)$. (You should note that $\chi_P(\xi_1, \xi_2, \ldots, \xi_n) = x_{P_2}(\xi_1, \xi_2)\chi_{\mathfrak{J}}(\xi_3, \xi_4, \ldots, \xi_n)$.)

---

One may express a triple integral as an iterated integral in a similar manner if, for example, the domain $\mathfrak{D}$ is encompassed by two surfaces $\{(x, y, \gamma(x, y))\}$, $\{(x, y, \delta(x, y))\}$ and a cylindrical surface whose rulings are orthogonal to the $(x, y)$-plane on the boundary of a region such as in Fig. 109.1a or 109.1b.

Such a domain is represented by

$$(109.3) \quad \mathfrak{D} = \{(x, y, z) \in \mathbf{E}^3 \mid a \leqslant x \leqslant b, \alpha(x) \leqslant y \leqslant \beta(x), \gamma(x, y) \leqslant z \leqslant \delta(x, y)\}.$$

(See Fig. 109.3.)

The reasoning that led to Theorem 109.1 leads, with appropriate modifications, to the following theorem.

***Theorem 109.2***

Let $\alpha, \beta, \gamma, \delta$ represent continuous functions. If $f\colon \mathfrak{D} \to \mathbf{E}$ is continuous, where $\mathfrak{D}$ is defined in (109.3), then $\int_{\mathfrak{D}} f$ exists and

$$\int_{\mathfrak{D}} f = \int_a^b \left(\int_{\alpha(x)}^{\beta(x)} \left(\int_{\gamma(x,y)}^{\delta(x,y)} f(x, y, z)\, dz\right) dy\right) dx.$$

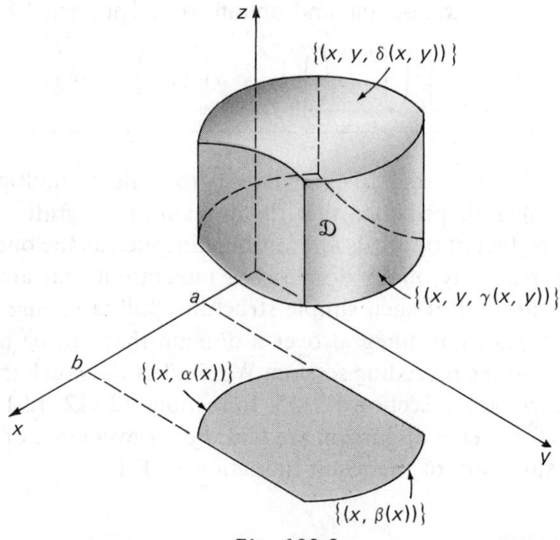


*Fig. 109.3*

Note that the order of integration in the $(x, y)$-plane may be interchanged if the projection of $\mathfrak{D}$ onto the $(x, y)$-plane has a shape as in Fig. 109.1b.

Two analogous versions of this theorem are obtained for the cases where the cylindrical surface that forms part of the boundary of $\mathfrak{D}$ is orthogonal to the $(y, z)$-plane or the $(x, z)$-plane provided that all the other conditions, suitably modified, are met. We shall illustrate one such case in the following example:

---

▲ *Example 3*

We consider the tetrahedron $\mathfrak{T}$ that is formed by the three coordinate planes and the plane $x + y + z = 1$. (See Fig. 109.4.)

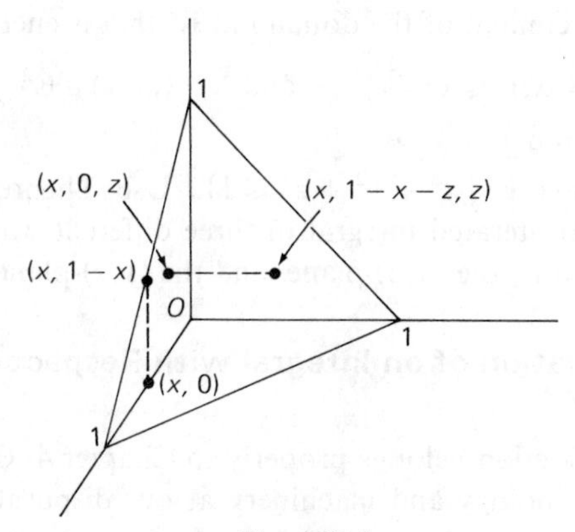


*Fig. 109.4*

We project $\mathcal{T}$ onto the $(x, z)$-plane and obtain from Theorem 109.2 that

$$J(\mathcal{T}) = \int_0^1 \left( \int_0^{1-x} \left( \int_0^{1-x-z} dy \right) dz \right) dx = \frac{1}{6}.$$

---

Theorems 108.2, 109.1, and 109.2 enable us to evaluate multiple integrals in terms of single integrals provided that the domain of integration is an interval or a Cartesian product of intervals and/or domains such as the ones in Fig. 109.1 and 109.3. Unfortunately, many domains of integration that are of interest in the applications are not of such simple structure. Still, a change of integration variables often leads to an integral over a domain that can be handled by the methods of this and the preceding section. We shall discuss such transformations of integration variables in section 12.115. In sections 12.112, 12.113, and 12.114 the foundations for such a discussion are laid. For convenience of the reader, we shall present a summary of the result in section 11.111.

## 11.109 Exercises

1. *Prove*: If $\gamma, \delta : [c, d] \to \mathbf{E}$ are continuous and $\gamma(y) \leqslant \delta(y)$ for all $y \in [c, d]$ and if $f : \mathfrak{D} \to \mathbf{E}$ is continuous, where
$$\mathfrak{D} = \{(x, y) \in \mathbf{E}^2 \mid c \leqslant y \leqslant d, \gamma(y) \leqslant x \leqslant \delta(y)\},$$
then
$$\int_{\mathfrak{D}} f = \int_c^d \left( \int_{\gamma(y)}^{\delta(y)} f(x, y)\, dx \right) dy.$$
2. Given $\mathfrak{D}_1 = \{(x, y) \in \mathbf{E}^2 \mid a \leqslant y \leqslant b, \quad \alpha(x) \leqslant y \leqslant \beta(x)\}, \quad \mathfrak{D}_2 = \{(z, u) \in \mathbf{E}^2 \mid c \leqslant z \leqslant d, \gamma(z) \leqslant u \leqslant \delta(z)\}$ where $\alpha, \beta, \gamma, \delta$ are continuous functions on $[a, b]$ and $[c, d]$, respectively, and where $\alpha(x) \leqslant \beta(x)$ for all $x \in [a, b]$, $\gamma(z) \leqslant \delta(z)$ for all $z \in [c, d]$. Find $J(\mathfrak{D}_1 \times \mathfrak{D}_2)$.
3. Let $\mathfrak{D} = \{(x, y) \in \mathbf{E}^2 \mid x^2 + y^2 \leqslant 1\}$. Write $\int_{\mathfrak{D}} f$ as an iterated integral, assuming that $f$ is continuous.
4. Find the Jordan content of the domain in $\mathbf{E}^2$ that is encompassed by
$$\{(x, y) \in \mathbf{E}^2 \mid y = x, 0 \leqslant x \leqslant 1\} \qquad \text{and} \qquad \{(x, y) \in \mathbf{E}^2 \mid y = x^2, 0 \leqslant x \leqslant 1\}.$$
5. Prove Theorem 109.2.
6. Let $\mathfrak{D} = \{(x, y, z) \in \mathbf{E}^3 \mid x^2 + y^2 + z^2 \leqslant 1\}$. Use Theorem 109.2 to write $\int_{\mathfrak{D}} dx\, dy\, dz$ as an iterated integral in three different ways by projecting $\mathfrak{D}$ onto the $(x, y)$-plane, the $(x, z)$-plane, and the $(y, z)$-plane.

## 11.110 Differentiation of an Integral with Respect to a Parameter*

The content of this section belongs properly to Chapter 4. Only now, however, do we have the terminology and machinery at our disposal to deal with this matter efficiently.

If $g : \mathfrak{D} \to \mathbf{E}$, where

(110.1) $$\mathfrak{D} = \{(x, y) \in \mathbf{E}^2 \mid a \leqslant x \leqslant b, c \leqslant y \leqslant d\},$$

is Riemann-integrable on $[a, b]$ for each $y \in [c, d]$, then

(110.2) $$G(y) = \int_a^b g(x, y)\, dx$$

defines a function $G : [c, d] \to \mathbf{E}$. Under suitable conditions on $g$, this function is differentiable, as we shall show in the following theorem.

***Theorem 110.1***

If the function $g : \mathfrak{D} \to \mathbf{E}$, where $\mathfrak{D}$ is defined in (110.1), is integrable on $[a, b]$ for each $y \in [c, d]$, and if $D_2 g$ is continuous on $\mathfrak{D}$, then the function $G: [c, d] \to \mathbf{E}$, as defined in (110.2), has a continuous derivative $G'$ in $[c, d]$ and

$$G'(y) = \int_a^b D_2 g(x, y)\, dx.$$

*Proof*

For each $x \in [a, b]$, $g$ and $D_2 g$, as functions of $y$, satisfy the hypotheses imposed on $F$, $f$ in Theorem 52.3 (Fundamental Theorem of the Integral Calculus). Hence, for every $y \in [c, d]$ and each $x \in [a, b]$,

$$\int_c^y D_2 g(x, t)\, dt = g(x, y) - g(x, c).$$

By hypothesis, $g(x, y)$, $g(x, c)$ are integrable on $[a, b]$. Hence, $\int_c^y D_2 g(x, t)\, dt$ is integrable on $[a, b]$, and we obtain

$$\int_a^b g(x, y)\, dx = \int_a^b \left( \int_c^y D_2 g(x, t)\, dt \right) dx + \int_a^b g(x, c)\, dx.$$

Since $D_2 g$ is, by hypothesis, continuous on $\mathfrak{D}$, we have, from Theorem 108.2, that

$$\int_a^b \int_c^y D_2 g(x, t)\, dt\, dx = \int_a^b \left( \int_c^y D_2 g(x, t)\, dt \right) dx$$
$$= \int_c^y \left( \int_a^b D_2 g(x, t)\, dx \right) dt.$$

Hence,

(110.3) $$G(y) = \int_c^y \left( \int_a^b D_2 g(x, t)\, dx \right) dt + \int_a^b g(x, c)\, dx.$$

Let

$$f(y) = \int_a^b D_2 g(x, y)\, dx$$

and

$$F(y) = \int_c^y f(y)\, dy.$$

Since $D_2 g$ is continuous on $\mathfrak{D}$, $f$ is continuous on $[c, d]$ and we have, from Theorem 52.2, that

$$F'(y) = f(y) \qquad \text{for all } y \in [c, d].$$

Hence,

$$G'(y) = F'(y) = \int_a^b D_2 g(x, y)\, dx,$$

which was to be proved. (Note that $\int_a^b g(x, c)\, dx$ is independent of $y$ and hence, disappears upon differentiation of (110.3) with respect to $y$.)

That $G'$ is continuous on $[c, d]$ follows readily from the continuity of $D_2 g$ on $\mathfrak{D}$. (See exercise 1.)

***Theorem 110.2***

If $g, \mathfrak{D}$ are as in Theorem 110.1, if $\alpha, \beta : [c, d] \to \mathbf{E}$, $\alpha, \beta \in C^1[c, d]$, $a \leqslant \alpha(y) \leqslant \beta(y) \leqslant b$ for all $y \in [c, d]$, and if for each fixed $y \in [c, d]$, $g$, as a function of $x$, is continuous on $[a, b]$, then the function $G : [c, d] \to \mathbf{E}$, defined by

$$G(y) = \int_{\alpha(y)}^{\beta(y)} g(x, y)\, dx,$$

is differentiable in $[c, d]$ and

$$G'(y) = \int_{\alpha(y)}^{\beta(y)} D_2 g(x, y)\, dx + g(\beta(y), y)\beta'(y) - g(\alpha(y), y)\alpha'(y). \tag{110.4}$$

*Proof*

Let $H : [a, b] \times \mathfrak{D} \to \mathbf{E}$ be defined by

$$H(u, v, y) = \int_u^v g(x, y)\, dx, \qquad u \in [a, b], (v, y) \in \mathfrak{D}.$$

Then

$$G(y) = H(\alpha(y), \beta(y), y).$$

By the chain rule (Theorem 92.1),

$$G'(y) = D_1 H(\alpha(y), \beta(y), y)\alpha'(y) + D_2 H(\alpha(y), \beta(y), y)\beta'(y) + D_3 H(\alpha(y), \beta(y), y). \tag{110.5}$$

By Theorem 52.2 (see also (51.2)),

$$D_1 H(\alpha(y), \beta(y), y) = -g(\alpha(y), y), \qquad D_2 H(\alpha(y), \beta(y), y) = g(\beta(y), y),$$

and by Theorem 110.1,

$$D_3 H(\alpha(y), \beta(y), y) = \int_{\alpha(y)}^{\beta(y)} D_2 g(x, y)\, dx.$$

Substitution of these results into (110.5) yields (110.4).

---

▲ *Example 1*

Let $g : \mathbf{E} \to \mathbf{E}$ represent a continuous function and let

$$g_0(y) = \int_0^y g(x)\, dx, \qquad g_n(y) = \int_0^y \frac{(y-x)^n}{n!} g(x)\, dx.$$

By Theorem 110.2,

$$g_n'(y) = \int_0^y \frac{(y-x)^{n-1}}{(n-1)!} g(x)\, dx = g_{n-1}(y).$$

Hence, by Theorem 52.3

$$g_1(y) = \int_0^y g_0(x)\, dx, \qquad g_2(y) = \int_0^y g_1(x)\, dx, \qquad \ldots, \qquad g_n(y) = \int_0^y g_{n-1}(x)\, dx,$$

or, as we may put it

$$\int_0^y \frac{(y-x)^n}{n!} g(x)\, dx = \int_0^y \left( \int_0^{x_{n-1}} \cdots \left( \int_0^{x_2} \left( \int_0^{x_1} g(x)\, dx \right) dx_1 \right) dx_2 \ldots \right) dx_{n-1}.$$

▲ *Example 2*

Theorem 110.1 may be used in some cases to evaluate an integral. Consider

$$\int_0^{\pi/2} \log(\sin^2 x + a^2 \cos^2 x)\, dx \qquad \text{for } a > 0.$$

We consider

$$G(y) = \int_0^{\pi/2} \log(\sin^2 x + y^2 \cos^2 x)\, dx \qquad \text{for } y > 0 \tag{110.6}$$

and obtain

$$G'(y) = \int_0^{\pi/2} \frac{2y \cos^2 x\, dx}{\sin^2 x + y^2 \cos^2 x}. \tag{110.7}$$

Hence,

$$(y^2-1)G'(y) = 2y\int_0^{\pi/2} \frac{y^2\cos^2 x + \sin^2 x - 1}{\sin^2 x + y^2\cos^2 x}\,dx$$

$$= y\pi - 2y\int_0^{\pi/2} \frac{dx}{\sin^2 x + y^2\cos^2 x}$$

$$= \pi(y-1).$$

(See exercise 2.) Therefore, for $y \neq 1$, $y > 0$,

$$G'(y) = \frac{\pi}{y+1}. \tag{110.8}$$

For $y = 1$, we have, from (110.7), that

$$G'(1) = 2\int_0^{\pi/2} \cos^2 x\,dx = \frac{\pi}{2},$$

which is consistent with (110.8).

We may now recover $G(y)$ by integration:

$$G(y) = \int_1^y \frac{\pi}{t+1}\,dt + G(1) = \pi\log(y+1) - \pi\log 2 = \pi\log\left(\frac{y+1}{2}\right),$$

since, by (110.6), $G(1) = 0$. Hence,

$$\int_0^{\pi/2} \log(\sin^2 x + a^2\cos^2 x)\,dx = \pi\log\left(\frac{a+1}{2}\right) \qquad \text{for } a > 0.$$

---

### 11.110 Exercises

1. Complete the proof of Theorem 110.1 by showing that $G'$ is continuous on $[c, d]$.
2. Integrate

$$\int_0^{\pi/2} \frac{dx}{\sin^2 x + y^2\cos^2 x}.$$

3. Given $F: (0, \infty) \to \mathbf{E}$ by

$$F(x) = \int_x^{x^2} \frac{1}{y}e^{xy}\,dy.$$

   Find $F'(x)$.
4. Given $f, g: \mathbf{E} \to \mathbf{E}$ by

$$f(x) = \left(\int_0^x e^{-y^2}\,dy\right)^2, \qquad g(x) = \int_0^1 \frac{e^{-x^2(y^2+1)}}{y^2+1}\,dy.$$

   Show that $f(x) + g(x) = (\pi/4)$.

5. The *Bessel function* of $n$th order ($n = 0, 1, 2, 3, \ldots$) and of the first kind, $J_n : \mathbf{E} \to \mathbf{E}$, may be defined by

$$J_n(x) = \frac{x^n}{1.3.5 \ldots (2n-1)\pi} \int_{-1}^{1} (\cos xy)(1 - y^2)^{n-(1/2)}\, dy.$$

Show that $J_n$ satisfies the *Bessel*† *equation*

$$J_n'' + \frac{1}{x} J_n' + \left(1 - \frac{n^2}{x^2}\right) J_n = 0,$$

and that $J_1 = -J_0'$, $J_{n+1} = J_{n-1} - 2J_n'$ for $n \in \mathbf{N}$.

6. Given $G : \mathbf{E} \to \mathbf{E}$ by

$$G(x) = \int_{-x}^{x} (x^2 - y^2)^n\, dy.$$

Find $G'(x)$.

7. Given $B$: $[1, \infty) \times [1, \infty) \to \mathbf{E}$ by

$$B(x, y) = \int_0^1 t^{x-1}(1 - t)t^{y-1}\, dt \qquad (\textit{Beta function}).$$

Find $D_1 B$, $D_2 B$.

## 11.111 Transformation of Double Integrals*

We have mentioned in section 11.109 that a transformation cf integration variables sometimes makes an integral more amenable to integration by iteration. This section is devoted to a discussion of such transformations. We have already studied transformations of single integrals (Corollary 2 to Theorem 52.3, change of integration variable), namely

$$\int_{g(\alpha)}^{g(\beta)} f = \int_{\alpha}^{\beta} (f \circ g)g', \tag{111.1}$$

where $g : [\alpha, \beta] \to \mathbf{R}$, $f : \mathfrak{R}(g) \to \mathbf{R}$, where $g$ has a continuous derivative in $[\alpha, \beta]$ and where $f$ is continuous on the range of $g$.

We illustrate this change-of-variable procedure in the following example.

---

▲ *Example 1*

If we let $f(y) = y^2$, $g(x) = \sin x$, $g(\alpha) = 0$, $g(\beta) = 1$, we obtain

$$\int_0^1 y^2\, dy = \int_0^{\pi/2} \sin^2 x \cos x\, dx = \int_0^{5\pi/2} \sin^2 x \cos x\, dx$$
$$= \int_0^{9\pi/2} \sin^2 x \cos x\, dx = \cdots = \frac{1}{3},$$

† *Friedrich Wilhelm Bessel*, 1784–1846.

* This section need not be taken up unless Chapter 12 is omitted.

since $g(0) = 0$, $g(\pi/2) = g(5\pi/2) = g(9\pi/2) = \cdots = 1$. The reader can see, from Fig. 111.1, why the integrals on the right have the same value for the different integration limits $\pi/2$, $5\pi/2$, $9\pi/2$, ... even though $g$ maps, loosely speaking, $[0, (5\pi/2)]$ two and a half times onto $[-1, 1] \supset [0, 1]$, maps $[0,(9\pi/2)]$ four and a half times onto $[-1, 1]$, etc.

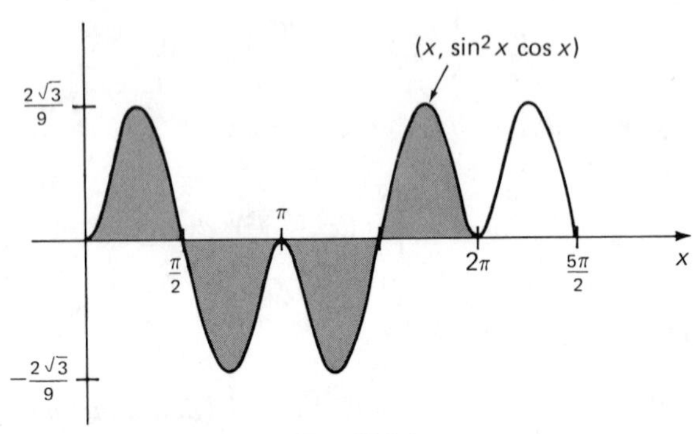


*Fig. 111.1*

---

In calculus, the reader has made the acquaintance of a formula for the change of variables in double and triple integrals which is, in many ways, similar to (111.1), namely

$$\iint_{g_*(\mathfrak{D})} f(\eta_1, \eta_2)\, d\eta_1\, d\eta_2 = \iint_{\mathfrak{D}} f(g(\xi_1, \xi_2))|\,J_g(\xi_1, \xi_2)|\, d\xi_1\, d\xi_2,$$

or, as we may write in our abbreviated notation,

$$\iint_{g_*(\mathfrak{D})} f = \iint_{\mathfrak{D}} (f \circ g)\,|J_g|, \tag{111.2}$$

and an analogous formula for triple integrals. (Observe that $g' = J_g$ when $g$ is a real-valued function of a real variable, but that, in general, $[g(\alpha), g(\beta)]$ or $[g(\beta), g(\alpha)]$ need not be $g_*([\alpha, \beta])$, as we have seen in example 1.)

The next example will show that the hypotheses which one has to impose on $g$ for the two (and more)-dimensional case have to be much stronger than in the one-dimensional case.

---

▲ *Example 2*

Let $\mathfrak{D} = \{(\xi_1, \xi_2) \in \mathbf{E}^2 \mid 0 \leqslant \xi_1 \leqslant 1, 0 \leqslant \xi_2 \leqslant 4\pi\}$, let $g : \mathfrak{D} \to \mathbf{E}$ be given by

$$\eta_1 = \xi_1 \cos \xi_2,$$
$$\eta_2 = \xi_1 \sin \xi_2,$$

and let $f(\eta_1, \eta_2) = 1$ for all $(\eta_1, \eta_2) \in \mathbf{E}^2$. Clearly, $g_*(\mathfrak{D}) = \{(\eta_1, \eta_2) \in \mathbf{E}^2 \mid \eta_1^2 + \eta_2^2 \leq 1\}$ is the unit disk. Further,

$$J_g(\xi_1, \xi_2) = \begin{vmatrix} \cos \xi_2 & \sin \xi_2 \\ -\xi_1 \sin \xi_2 & \xi_1 \cos \xi_2 \end{vmatrix} = \xi_1.$$

Hence, from (111.2),

$$\iint_{g_*(\mathfrak{D})} d\eta_1 \, d\eta_2 = \int_0^{4\pi} \int_0^1 \xi_1 \, d\xi_1 \, d\xi_2 = 2\pi,$$

which is quite absurd if one notes that the integral on the left represents the Jordan content (area) of the unit disk which is $\pi$. While it does not appear to matter in the one-dimensional case how often $g$ maps $[\alpha, \beta]$ onto $[g(\alpha), g(\beta)]$ or $[g(\beta), g(\alpha)]$ or even a larger interval, for that matter, the fact that $g$ maps $\mathfrak{D}$ *twice* onto the unit disk seems clearly responsible for our nonsensical result. For any $\mathfrak{D} = \{(\xi_1, \xi_2) \in \mathbf{E}^2 \mid 0 \leq \xi_1 \leq 1, 0 \leq \xi_2 \leq 2\pi + \alpha^2, \alpha \in \mathbf{R}\}$, we obtain the unit disk with center at $(0, 0)$ for $g_*(\mathfrak{D})$. However, (111.2) only seems applicable when $\alpha = 0$.

For $\mathfrak{D} = \{(\xi_1, \xi_2) \in \mathbf{E}^2 \mid 0 \leq \xi_1 \leq 1, 0 \leq \xi_2 \leq 2\pi\}$, we have depicted $\mathfrak{D}$ and $g_*(\mathfrak{D})$ in Fig. 111.2.

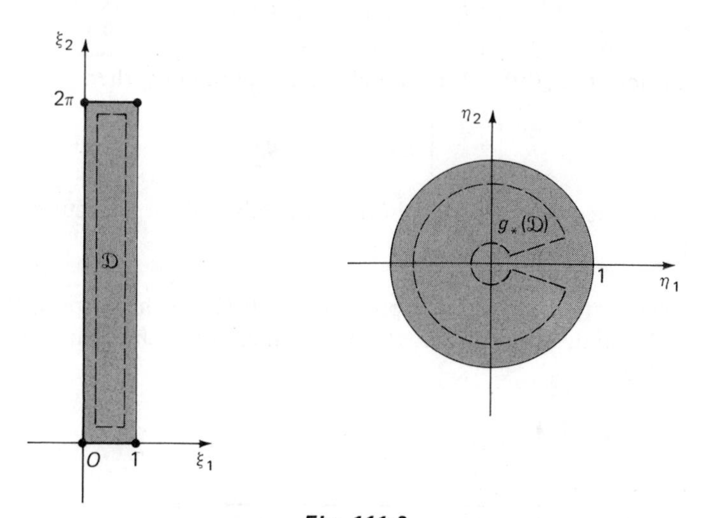


*Fig. 111.2*

Note that the left side of $\mathfrak{D}$ is mapped into the point $(0, 0)$ and that each of the two horizontal sides of $\mathfrak{D}$ is mapped onto the interval $[0, 1]$ but that the mapping is one-to-one on the remaining portion of $\mathfrak{D}$. For example, the right side of $\mathfrak{D}$, exclusive of one endpoint, is mapped one-to-one onto the circumference of the disk and the dotted line in $\mathfrak{D}$ is mapped one-to-one onto the dotted line in $g_*(\mathfrak{D})$. Also observe that $J_g(\xi_1, \xi_2) \neq 0$ for all $\xi_1 > 0$.

A similar situation arises when one introduces spherical coordinates in a triple integral. (See example 4.)

A transformation formula for multiple integrals is only then of practical value when it contains the transformations into polar coordinates and spherical coordinates as special cases because of the importance of these two transformations in the applications. For this purpose, it suffices to require that $g$ is one-to-one in the interior of $\mathfrak{D}$ and that $J_g(x) \neq 0$ in the interior of $\mathfrak{D}$. (Note that neither hypothesis implies the other; see examples 1 and 2, section 10.98.) In section 12.115 we shall prove a theorem (Jacobi's Theorem) on the transformation of $n$-tuple integrals that will satisfy these requirements. Preparatory to this proof we shall discuss the transformation of Jordan-measurable sets in sections 12.112 to 12.114.

At this point we shall merely state this theorem for double integrals and present a heuristic argument to support its plausibility.

***Theorem 111.1*** *(Transformation of Double Integrals)*

Let $\Omega \subseteq \mathbf{E}^2$ denote an open set and let $\mathfrak{D}$ denote a Jordan-measurable set such that $\mathfrak{D} \cup \beta\mathfrak{D} \subset \Omega$. If $g : \Omega \to \mathbf{E}^2$, $g \in C^1(\Omega)$, $g : \mathfrak{D}^\circ \xrightarrow[1-1]{} \mathbf{E}^2$, $J_g(x) \neq 0$ for all $x \in \mathfrak{D}^\circ$, then $g_*(\mathfrak{D})$ is Jordan-measurable and

$$J(g_*(\mathfrak{D})) = \int_{\mathfrak{D}}\int |J_g|.$$

If, in addition, $f : g_*(\mathfrak{D}) \to \mathbf{E}$ is bounded and continuous, then

$$\int_{g_*(\mathfrak{D})}\int f = \int_{\mathfrak{D}}\int (f \circ g)\,|J_g|.$$

(An analogous theorem holds for $n$-tuple integrals and, in particular, for triple integrals.)

The following argument is not a proof and only serves the purpose of making Theorem 111.1 and, in particular, the appearance of the Jacobian in the integrand plausible.

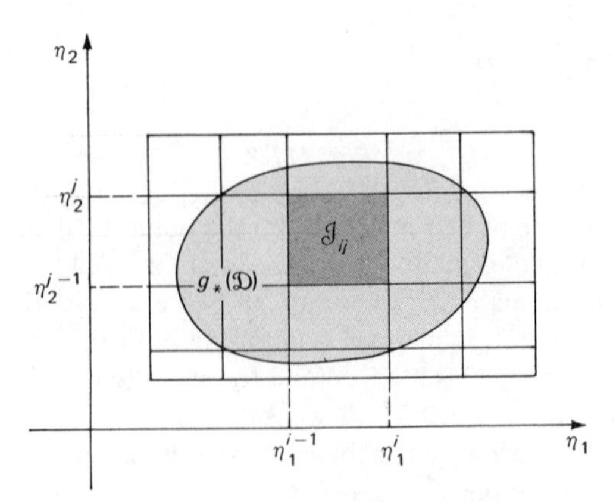


*Fig. 111.3*

In order to compute a lower or upper sum or an intermediate sum [see (46.3)], for that matter, for the integral on the left, we consider a partition $Q$ of an interval $\mathfrak{J}$ containing $g_*(\mathfrak{D})$ in the $\eta_1, \eta_2$-plane (see Fig. 111.3) and obtain, for some $y_{i,j} \in [\eta_1^{i-1}, \eta_1^i] \times [\eta_2^{j-1}, \eta_2^j] = \mathfrak{J}_{ij}$,

(111.3) $$\sum_{i=1}^{n} \sum_{j=1}^{m} f(y_{ij})\,|\mathfrak{J}_{ij}|.$$

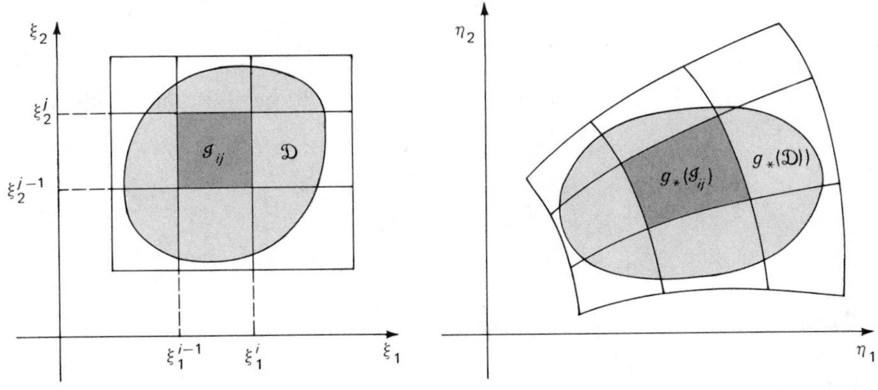


*Fig. 111.4*

On the other hand, we may consider the image under $g$ in the $\eta_1, \eta_2$-plane of a partition of an interval $\mathfrak{I}$ containing $\mathfrak{D}$ in the $\xi_1, \xi_2$-plane (see Fig. 111.4) and argue that the sum in (111.3) is approximately equal to

(111.4) $$\sum_{i=1}^{n} \sum_{j=1}^{m} f(y_{ij})|\mathfrak{J}_{ij}| \cong \sum_{i=1}^{\nu} \sum_{j=1}^{\mu} f(\bar{y}_{ij}) J(g_*(\mathfrak{I}_{ij}))$$

for some $\bar{y}_{ij} \in g_*(\mathfrak{I}_{ij})$.

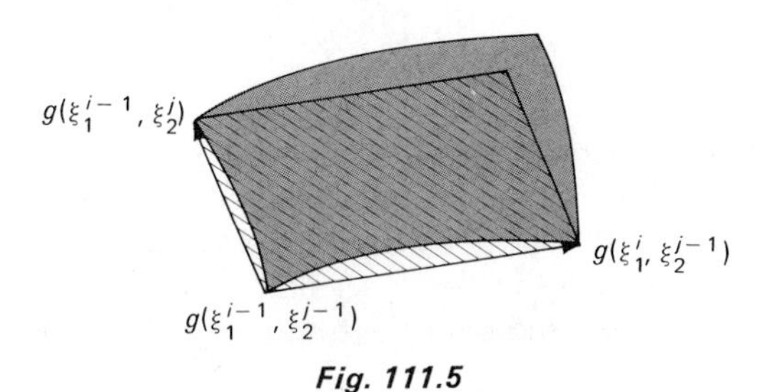


*Fig. 111.5*

We approximate the Jordan content (area measure) of $g_*(\mathfrak{I}_{ij})$ by the area measure of the parallelogram that is spanned by the two vectors in Fig. 111.5:

(111.5)
$$J(g_*(\mathfrak{I}_{ij})) \cong \left| \begin{vmatrix} \psi_1(\xi_1^i, \xi_2^{j-1}) - \psi_1(\xi_1^{i-1}, \xi_2^{j-1}) & \psi_1(\xi_1^{i-1}, \xi_2^j) - \psi_1(\xi_1^{i-1}, \xi_2^{j-1}) \\ \psi_2(\xi_1^i, \xi_2^{j-1}) - \psi_2(\xi_1^{i-1}, \xi_2^{j-1}) & \psi_2(\xi_1^{i-1}, \xi_2^j) - \psi_2(\xi_1^{i-1}, \xi_2^{j-1}) \end{vmatrix} \right|$$

where $\psi_1, \psi_2$ are the components of $g$. By the mean-value theorem of the differential calculus,

$$\psi_1(\xi_1^i, \xi_1^{j-1}) - \psi_1(\xi_1^{i-1}, \xi_2^{j-1}) = D_1\psi_1(\bar{\xi}_1^i, \xi_2^{j-1})(\xi_1^i - \xi_1^{i-1}),$$

$$\psi_1(\xi_1^{i-1}, \xi_2^j) - \psi_1(\xi_1^{i-1}, \xi_1^{j-1}) = D_2\psi_1(\xi_1^{i-1}, \bar{\xi}_2^j)(\xi_2^j - \xi_2^{j-1}),$$

$$\psi_2(\xi_1^i, \xi_2^{j-1}) - \psi_2(\xi_1^{i-1}, \xi_2^{j-1}) = D_1\psi_2(\bar{\xi}_1^i, \xi_2^{j-1})(\xi_1^i - \xi_1^{i-1}),$$

$$\psi_2(\xi_1^{i-1}, \xi_2^j) - \psi_2(\xi_1^{i-1}, \xi_2^{j-1}) = D_2\psi_2(\xi_1^{i-1}, \bar{\xi}_2^j)(\xi_2^j - \xi_2^{j-1}).$$

We approximate these expressions, in turn, by replacing the various arguments of $D_i\psi_k$ by $(\tilde{\xi}_1^i, \tilde{\xi}_2^j) = \tilde{x}_{ij}$ with the understanding that the coordinate differences $(\xi_1^i - \xi_1^{i-1})$, $(\xi_2^j - \xi_2^{j-1})$ are very small. Then

$$(111.6) \qquad J(g_*(\mathfrak{I}_{ij})) \cong \left| \begin{vmatrix} D_1\psi_1(\tilde{\xi}_1^i, \tilde{\xi}_2^j) & D_2\psi_1(\tilde{\xi}_1^i, \tilde{\xi}_2^j) \\ D_1\psi_2(\tilde{\xi}_1^i, \tilde{\xi}_2^j) & D_2\psi_2(\tilde{\xi}_1^i, \tilde{\xi}_2^j) \end{vmatrix} \right| (\xi_1^i - \xi_1^{i-1})(\xi_2^j - \xi_2^{j-1})$$

$$= |J_g(\tilde{x}_{ij})|\,|\mathfrak{I}_{ij}|.$$

Since $g$ is one-to-one in the interior of $\mathfrak{D}$, we have, for the $\bar{y}_{ij} \in g_*(\mathfrak{I}_{ij})$ in (111.4), a unique $\bar{x}_{ij} \in \mathfrak{I}_{ij}$ such that $g(\bar{x}_{ij}) = \bar{y}_{ij}$ and hence, $(f \circ g)(\bar{x}_{ij}) = f(\bar{y}_{ij})$. Therefore, we may approximate (111.4) by

$$(111.7) \qquad \sum_{i=1}^{\nu} \sum_{j=1}^{\mu} f(\bar{y}_{ij}) J(g_*(\mathfrak{I}_{ij})) \cong \sum_{i=1}^{\nu} \sum_{j=1}^{\mu} (f \circ g)(\bar{x}_{ij})\,|J_g(\tilde{x}_{ij})|\mathfrak{I}_{ij}|.$$

If the coordinate differences $(\xi_1^i - \xi_1^{i-1})$, $(\xi_2^j - \xi_2^{j-1})$ are very small, we won't commit much of an error by replacing $\tilde{x}_{ij} \in \mathfrak{I}_{ij}$ by $\bar{x}_{ij} \in \mathfrak{I}_{ij}$, and obtain, from (111.4) and (111.7), that

$$(111.8) \qquad \sum_{i=1}^{n} \sum_{j=1}^{m} f(y_{ij})\,|\mathfrak{F}_{ij}| \cong \sum_{i=1}^{\nu} \sum_{j=1}^{\mu} (f \circ g)(\bar{x}_{ij})|J_g(\bar{x}_{ij})||\mathfrak{I}_{ij}|.$$

The left sum is an intermediate sum for $f$ on $g_*(\mathfrak{D})$ and the right sum is an intermediate sum for $(f \circ g)\,|J_g|$ on $\mathfrak{D}$. It is now clear how the transformation formula in Theorem 111.1 evolves from (111.8). The formula for the Jordan content of $g_*(\mathfrak{D})$ is obtained simply by taking $f(y) = 1$ for all $y \in g_*(\mathfrak{D})$.

One can also see from this argument how the vanishing of the Jacobian and/or the failure of $g$ to be one-to-one on a set other than of Jordan content zero, would adversely affect the validity of the transformation formula.

---

▲ *Example 3*

With $\theta$ representing the polar angle between the vector $x \in \mathbf{E}^2$ and the $\xi_1$-axis and $r = \|x\|$,

$$r = a(1 - \cos\theta)$$

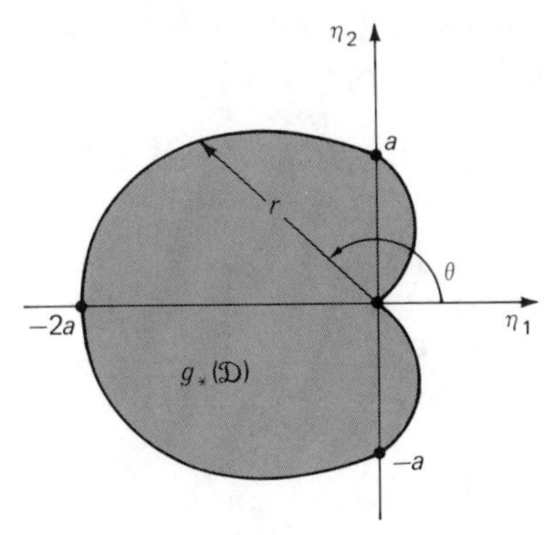


*Fig. 111.6*

represents a *cardioid* (heart-shaped curve) in the $\eta_1, \eta_2$-plane. (See Fig. 111.6.) In order to evaluate the area measure of the region that is enclosed by the cardioid, we introduce polar coordinates

$$\eta_1 = r \cos \theta,$$
$$\eta_2 = r \sin \theta,$$

with $(\theta, r)$ now playing the role of $\xi_1$, $\xi_2$ and $g(\theta, r) = (r \cos \theta,\ r \sin \theta)$, and obtain, with $\mathfrak{D}$ as in Fig. 111.7, that $g_*(\mathfrak{D})$ is the shaded region in the $(\eta_1, \eta_2)$-

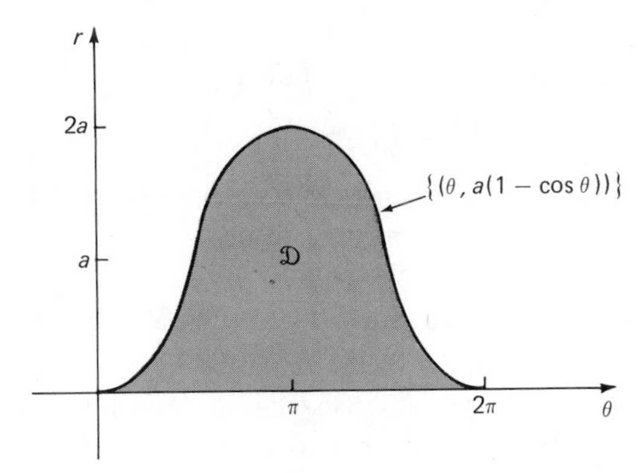


*Fig. 111.7*

plane in Fig. 111.6. Since

$$|J_g(\theta, r)| = \left| \begin{vmatrix} -r \cos \theta & \cos \theta \\ r \cos \theta & \sin \theta \end{vmatrix} \right| = r,$$

we obtain, also using Theorem 109.1,

$$J(g_*(\mathfrak{D})) = \int_{g_*(\mathfrak{D})}\int d\eta_1\, d\eta_2 = \int_{\mathfrak{D}}\int r\, dr\, d\theta = \int_{\theta=0}^{2\pi} \left( \int_{r=0}^{a(1-\cos\theta)} r\, dr \right) d\theta$$
$$= \frac{a^2}{2} \int_{\theta=0}^{2\pi} (1 - 2\cos\theta + \cos^2\theta)\, d\theta = \frac{3a^2\pi}{2}.$$

As we have pointed out already, a formula that is analogous to the one in Theorem 111.1 also holds for triple integrals. We give an illustration in the next example.

▲ *Example 4*

We shall find the volume of the solid that is cut out from the upper half of the unit ball with center at $\theta$ by a right circular cylinder of radius $\frac{1}{2}$ with the $\eta_3$-axis as axis. (See Fig. 111.8.)

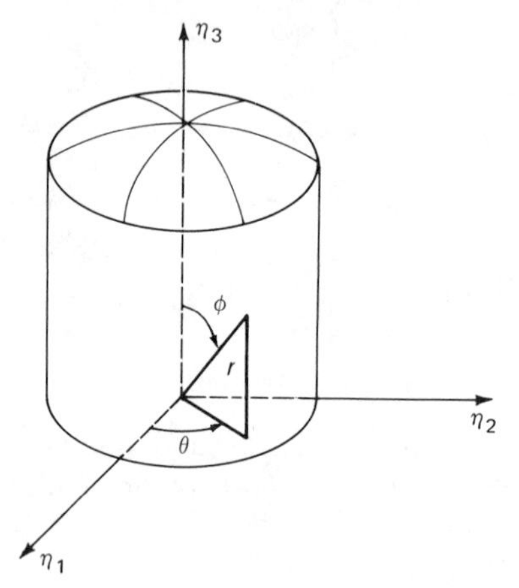


***Fig. 111.8***

We introduce spherical coordinates $\theta$, $\phi$, $r$ by

$$\begin{aligned}\eta_1 &= r\cos\theta\sin\phi,\\ \eta_2 &= r\sin\theta\sin\phi,\\ \eta_3 &= r\cos\phi.\end{aligned}$$

Then, $g(\theta, \phi, r) = (r\cos\theta\sin\phi,\ r\sin\theta\sin\phi,\ r\cos\phi)$ and

$$|J_g(\theta,\phi,r)| = \left|\begin{vmatrix} -r\sin\theta\sin\phi & r\cos\theta\cos\phi & \cos\theta\sin\phi \\ r\cos\theta\sin\phi & r\sin\theta\cos\phi & \sin\theta\sin\phi \\ 0 & r\sin\phi & \cos\phi \end{vmatrix}\right| = r^2\sin\phi.$$

With $\mathfrak{D}$ as in Fig. 111.9, $g_*(\mathfrak{D})$ is the solid under consideration and we obtain, using Theorem 109.2,

$$\begin{aligned}J(g_*(\mathfrak{D})) &= \int_{g_*(\mathfrak{D})}\iint d\eta_1\,d\eta_2\,d\eta_3 = \int_{\mathfrak{D}}\iint r^2\sin\phi\,d\theta\,d\phi\,dr\\ &= \int_{\theta=0}^{2\pi}\left(\int_{\phi=0}^{\pi/6}\left(\int_{r=0}^{1} r^2\sin\phi\,dr\right)d\phi\right)d\theta\\ &\qquad + \int_{\theta=0}^{2\pi}\left(\int_{\phi=\pi/6}^{\pi/2}\left(\int_{r=0}^{1/2\sin\phi} r^2\sin\phi\,dr\right)d\phi\right)d\theta\\ &= \frac{2\pi}{3}\left(1-\frac{\sqrt{3}}{2}\right)+\frac{\sqrt{3}\pi}{\sqrt{2}} = \pi\left(\frac{2}{3}-\frac{\sqrt{3}}{4}\right).\end{aligned}$$

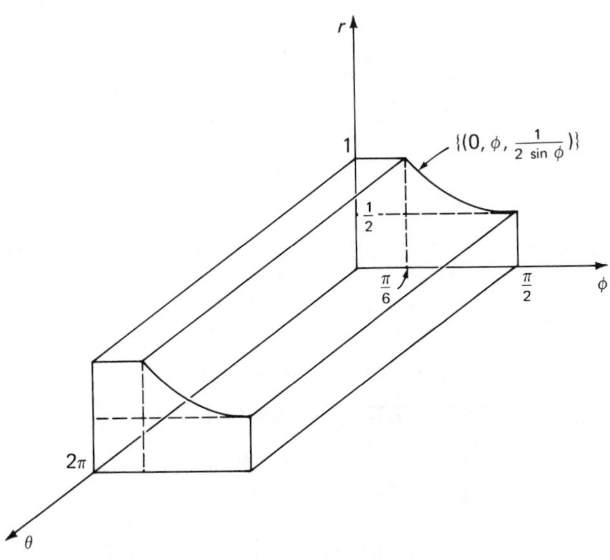


*Fig. 111.9*

## 11.111 Exercises

1. Find the area measure of the region that lies inside the cardioid $r = 1 + \cos\theta$ and outside of the circle $\eta_1^2 + \eta_2^2 = 1$.
2. The set of all points $(\eta_1, \eta_2) = (r\cos\theta,\ r\sin\theta)$ in the $(\eta_1, \eta_2)$-plane for which $r^2 = 2a^2\cos 2\theta$, is called a *lemniscate*. Make a sketch of the lemniscate, and find the area measure of the region inside the lemniscate.
3. Find the volume of a solid that is bounded below by a paraboloid $\eta_3 = \eta_1^2 + \eta_2^2$ and above by the plane $\eta_3 = \eta_2$. (*Hint*: Introduce cylindrical coordinates $\eta_1 = r\cos\theta$, $\eta_2 = r\sin\theta$, $\eta_3 = \xi_3$.)
4. Find the volume of a solid that is bounded above by the sphere $\eta_1^2 + \eta_2^2 + \eta_2^3 = 1$ and below by the paraboloid $\eta_3 = -\eta_1^2 - \eta_2^2$.

# Transformation of Integrals 12

## 12.112 Images of Jordan-Measurable Sets

If we consider a multiple integral on some domain $\mathfrak{D} \subseteq \mathbf{E}^n$ and transform the integration variable $x \in \mathfrak{D}$ into some other integration variable by means of a function $g : \mathfrak{D} \to \mathbf{E}^n$, then $\mathfrak{D}$ will be mapped onto $g_*(\mathfrak{D})$. Unless $g_*(\mathfrak{D})$ has Jordan content, the transformed integral does not make sense. It is the objective of this section to determine under what conditions on $g$ the image $g_*(\mathfrak{D})$ of a Jordan-measurable set $\mathfrak{D}$ is again Jordan-measurable.

In this chapter we shall omit the asterisk in $g_*(\mathfrak{D})$ and write $g(\mathfrak{D})$ instead. Although this results in a somewhat ambiguous notation, no misunderstanding should occur. This omission will simplify the notation when direct images under functions, that are already burdened with subscripts, are considered.

Since a set is Jordan-measurable if and only if its boundary has Jordan content zero, for the image of a Jordan-measurable set to be Jordan-measurable, it is necessary that the boundary of the image have Jordan content zero.

We shall demonstrate that under suitable assumptions about $g$, the boundary of $g(\mathfrak{D})$ is a subset of $g(\beta\mathfrak{D})$ and that $g$ maps sets of Jordan content zero (such as $\beta\mathfrak{D}$) onto sets of Jordan content zero. This will, in effect, give us a sufficient condition for the image of a Jordan-measurable set to be Jordan-measurable.

That it does not suffice for this purpose to assume mere continuity of $g$ will be shown in the following example where the continuous image of a set of two-dimensional Jordan content zero has two-dimensional Jordan content 1.

---

▲ *Example 1* (*Schoenberg's*† *Space Filling Curve*)

We consider a function $f$ from the subset

$$\mathfrak{D} = \{(x, x^2) \in \mathbf{E}^2 \mid 0 \leqslant x \leqslant 1\}$$

of $\mathbf{E}^2$ into $\mathbf{E}^2$ which is defined by

$$f = h \circ g,$$

† *Isaac J. Schoenberg*, 1903–.

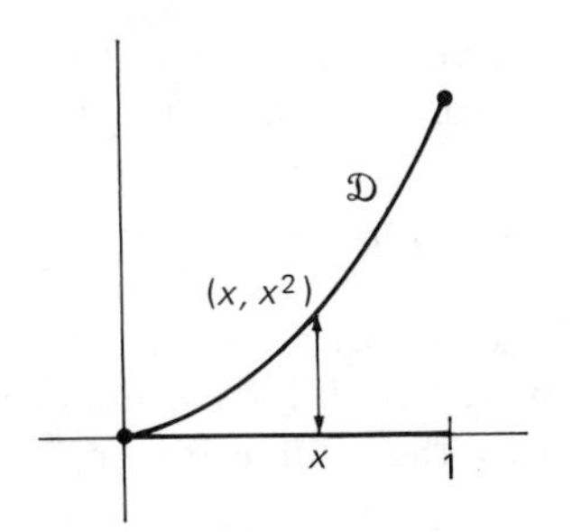


*Fig. 112.1*

where $g : \mathfrak{D} \leftrightarrow [0, 1]$ is defined by $g(x, y) = x$ and is clearly continuous (see Fig. 112.1) and where the function $h = (\chi_1, \chi_2) : [0, 1] \to \mathbf{E}^2$ is defined in terms of

$$(112.1)\quad p(x) = \begin{cases} 0 & \text{for } 0 \leqslant x \leqslant \frac{1}{3} \\ 3x - 1 & \text{for } \frac{1}{3} < x < \frac{2}{3} \\ 1 & \text{for } \frac{2}{3} \leqslant x \leqslant 1 \end{cases} \qquad p(x + 2) = p(x), \quad p(-x) = p(x),$$

(see Fig. 112.2), and

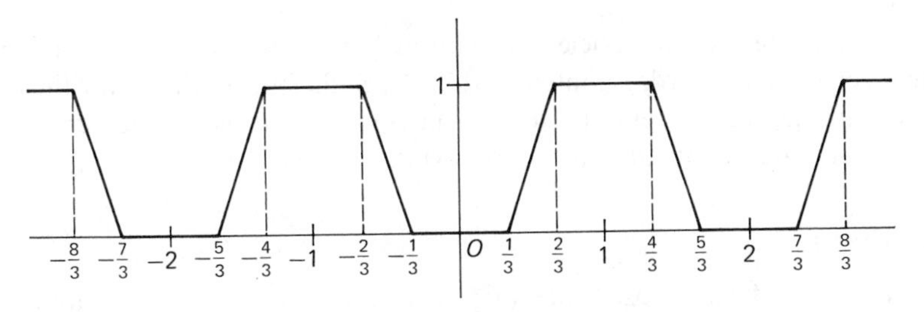


*Fig. 112.2*

$$(112.2)\quad \begin{aligned} \phi_1(x) &= \frac{1}{2} p(x), & \phi_n(x) &= \phi_{n-1}(x) + \frac{1}{2^n} p(3^{2n-2}x), \\ \psi_1(x) &= \frac{1}{2} p(3x), & \psi_n(x) &= \psi_{n-1}(x) + \frac{1}{2^n} p(3^{2n-1}x), \end{aligned}$$

by

$$\chi_1(x) = \lim_{\infty} \phi_n(x),$$
$$\chi_2(x) = \lim_{\infty} \psi_n(x).$$

Since $0 \leqslant p(x) \leqslant 1$ for all $x \in \mathbf{E}$, the sequences $\{\phi_n\}$, $\{\psi_n\}$ converge uniformly on $\mathbf{E}$ and hence, $\chi_1, \chi_2$ are continuous on $\mathbf{E}$. Therefore, $h$ is continuous on $\mathbf{E}$ and, by Theorem 67.1, $f = h \circ g$ is continuous on $\mathfrak{D}$.

$\mathfrak{D}$ has two-dimensional Jordan content zero (see example 6, section 11.105). We shall demonstrate that its continuous image under $f$, namely $f(\mathfrak{D})$, has two-dimensional Jordan content 1 by showing that

$$f(\mathfrak{D}) = [0, 1] \times [0, 1],$$

which means, in turn, that the curve $f(\mathfrak{D})$ passes through every point in $[0, 1] \times [0, 1]$. Every point $(\eta_1, \eta_2) \in [0, 1] \times [0, 1]$ has a unique representation by a pair of infinite binaries

$$\eta_1 = 0.a_1 a_2 a_3 \ldots, \eta_2 = 0.b_1 b_2 b_3 \ldots, \qquad \text{where } a_i, b_i \in \{0, 1\}.$$

(Terminating binaries such as 0.01101 are replaced by infinite binaries $0.01100\dot{1}$ and, in particular, 1 is replaced by $0.\dot{1}$.) We obtain from (112.1) and (112.2) for $x_0 \in [0, 1]$, represented by the infinite ternary

$$x_0 = 0.(2a_1)(2b_1)(2a_2)(2b_2)\ldots,$$

that $\phi_1(x_0) = 0.a_1, \phi_2(x_0) = 0.a_1 a_2 \ldots$ and $\psi_1(x_0) = 0.b_1, \psi_2(x_0) = 0.b_1 b_2 \ldots$; that is, $\phi(x_0) = \eta_1, \psi(x_0) = \eta_2$. Hence, for $(x_0, x_0^2) \in \mathfrak{D}, f(x_0, x_0^2) = h(g(x_0, x_0^2)) = h(x_0) = (\eta_1, \eta_2)$. Since there is such a point $(x_0, x_0^2) \in \mathfrak{D}$ for each $(\eta_1, \eta_2) \in [0, 1] \times [0, 1]$, we have $f(\mathfrak{D}) = [0, 1] \times [0, 1]$; that is, $J(f(\mathfrak{D})) = 1$ while $J(\mathfrak{D}) = 0$.

The graph of the function $f$ is a so-called *space filling curve*.† The first example of this kind was obtained by *Giuseppe Peano*.‡

---

In order to develop a sufficient condition for the image of a set of Jordan content zero to have Jordan content zero and a sufficient condition for the image of a Jordan-measurable set to be Jordan-measurable, we need some intermediate results which are formulated in the following lemmas.

***Lemma 112.1***

If $\mathfrak{J}_1, \mathfrak{J}_2, \ldots, \mathfrak{J}_k \subset \mathbf{E}^n$ are closed intervals, then, for every $\varepsilon > 0$, there are finitely many closed cubes $C_1, C_2, \ldots, C_l \subset \mathbf{E}^n$ with rational dimensions such that

$$\bigcup_{j=1}^{k} \mathfrak{J}_j \subseteq \bigcup_{i=1}^{l} C_i$$

and

$$0 \leqslant \sum_{i=1}^{l} |C_i| - \sum_{j=1}^{k} |\mathfrak{J}_j| < \varepsilon.$$

*Proof*

To prove this lemma, one need only note that every real number may be approximated to any given degree of accuracy by a rational number, and that an interval with rational dimensions may be represented as the union of congruent cubes with nonoverlapping interiors. The details are left to the reader (exercise 1).

† For a superbly illustrated comprehensive discussion of space filling curves we refer the reader to an article by W. Wunderlich, "Ueber Peano Kurven," Elemente der Mathematik, Vol. 28/1, 1973.

‡ *Giuseppe Peano*, 1858–1932.

***Definition 112.1***

The function $g : A \to \mathbf{E}^m$, $A \subseteq \mathbf{E}^n$, satisfies a *uniform Lipschitz condition* on $A$ if and only if there is an $L > 0$ and a $\delta > 0$ such that

$$\|g(x) - g(y)\| \leqslant L\,\|x - y\|$$

for all $x, y \in A$ for which $\|x - y\| < \delta$.

(Compare this definition with definition 75.1 and note that if $g$ satisfies a global Lipschitz condition on $A$, then it satisfies a uniform Lipschitz condition on $A$ but that the converse need not be true. Hence, to require that a function satisfy a global Lipschitz condition on a given set is *stronger* than to require that it merely satisfy a uniform Lipschitz condition on that set.)

***Lemma 112.2***

Let $\Omega \subseteq \mathbf{E}^n$ denote an open set and let $K \subset \Omega$ denote a *compact* set. If $g : \Omega \to \mathbf{E}^n$ satisfies a *uniform* Lipschitz condition on $K$ with $L > 0$, $\delta > 0$, and if for some $a > 0$,

$$\bar{J}(K) < \frac{a}{4(2L\sqrt{n})^n},$$

then

$$\bar{J}(g(K)) < a.$$

*Proof*

(Note that a compact set has outer Jordan content because it is bounded.) By definition, $\bar{J}(K) = \inf_{(P)}\bar{S}(\chi_K;P)$. Hence, there is a partition $P_a$ such that $\bar{S}(\chi_K;P_a) < \bar{J}(K) + a/4(2L\sqrt{n})^n$, i.e., there are closed intervals $\mathfrak{J}_1, \mathfrak{J}_2, \ldots, \mathfrak{J}_k$ with nonoverlapping interiors such that

$$\sum_{j=1}^{k} |\mathfrak{J}_j| < \bar{J}(K) + \frac{a}{4(2L\sqrt{n})^n} < \frac{a}{2(2L\sqrt{n})^n} \tag{112.3}$$

and $K \subseteq \bigcup_{j=1}^{k} \mathfrak{J}_j$.

By Lemma 112.1, there are $l$ cubes $C_1, C_2, \ldots, C_l$ with rational dimensions such that

$$K \subseteq \bigcup_{j=1}^{k} \mathfrak{J}_j \subseteq \bigcup_{i=1}^{l} C_i$$

and

$$0 \leqslant \sum_{i=1}^{l} |C_i| - \sum_{j=1}^{k} |\mathfrak{J}_j| < \frac{a}{2(2L\sqrt{n})^n}. \tag{112.4}$$

Since these cubes have rational dimensions, we may assume, without loss of generality, that all have the *same* dimension $s$ and that $s < (\delta/\sqrt{n})$.

If $c$ denotes a fixed point in $C_i \cap K$ and if $x \in C_i \cap K$, then $\|g(x) - g(c)\| \leqslant L\|x - c\|$ because $\|x - c\| \leqslant s\sqrt{n} < \delta$. (The length of the diagonal in a cube of dimension $s$ is $s\sqrt{n}$.) Hence $\|g(x) - g(c)\| \leqslant Ls\sqrt{n}$, and we see that $g(x)$ lies in a cube $\Gamma_i$ of dimension $2Ls\sqrt{n}$ and center at $g(c)$. Since this is true for all $x \in C_i \cap K$, we have

$$g(C_i \cap K) \subseteq \Gamma_i.$$

From (112.3) and (112.4),

$$\sum_{i=1}^{l} |C_i| < \sum_{j=1}^{k} |\mathfrak{J}_j| + \frac{a}{2(2L\sqrt{n})^n} < \frac{a}{(2L\sqrt{n})^n}.$$

Since all cubes $C_i$ have the same dimension, we have $|C_i| < a/l(2L\sqrt{n})^n, i = 1, 2, \ldots, l$, that is, $s = \sqrt[n]{|C_i|} < \sqrt[n]{a/l}/2L\sqrt{n}$. Since $\Gamma_i$ has dimension $2Ls\sqrt{n}$, we have $|\Gamma_i| = (2Ls\sqrt{n})^n < (2L\sqrt[n]{a/l}\sqrt{n}/2L\sqrt{n})^n = a/l$, and hence $\sum_{i=1}^{l} |\Gamma_i| < a$. Since $g(K) \subseteq \bigcup_{i=1}^{l} \Gamma_i$, we have $\bar{J}(g(K)) \leqslant \sum_{i=1}^{l} |\Gamma_i| < a$ and the lemma is proved. (See exercise 2 for a generalization of this lemma for $g : \Omega \to \mathbf{E}^m$, $\Omega \subseteq \mathbf{E}^n, m \geqslant n$.)

Lemma 112.2 serves as a basis for the following theorem, which, in turn, serves as basis for our investigation in sections 12.113, 12.114, and 12.115.

***Theorem 112.1***

Let $\Omega \subseteq \mathbf{E}^n$ denote an open set and let $K \subset \Omega$ denote a compact set. If $g : \Omega \to \mathbf{E}^n$ satisfies a uniform Lipschitz condition on $K$ and if $J(K) = 0$, then $J(g(K)) = 0$.

*Proof*

If $J(K) = 0$, then, for every $\varepsilon > 0$, $\bar{J}(K) < \varepsilon/[4(2L\sqrt{n})^n]$ and hence, by Lemma 112.2, $J(g(K)) < \varepsilon$; that is, $J(g(K)) = 0$. (This theorem is still true if $g : \Omega \to \mathbf{E}^m$ and $m \geqslant n$. See exercise 3.)

This theorem will take care of the boundary of a Jordan-measurable set under a mapping that is uniformly Lipschitzian because the boundary of a Jordan-measurable set is compact. (See Lemma 104.2 and note that a Jordan-measurable set is bounded.)

The following lemma provides us with a practical sufficient condition for $g$ to satisfy a uniform Lipschitz condition on a compact set.

***Lemma 112.3***

Let $\Omega \subseteq \mathbf{E}^n$ denote an open set and let $K \subset \Omega$ denote a compact set. If $g : \Omega \to \mathbf{E}^n$, $g \in C^1(\Omega)$, then $g$ satisfies a uniform Lipschitz condition on $K$.

*Proof*

Let $a \in \Omega$. Since $g'(a)$ exists, we have, from (86.5), an $L_a > 0$ and a $\delta_a > 0$ such that

$$\|g(x) - g(a)\| \leqslant L_a \|x - a\| \tag{112.5}$$

for all $x \in N_{\delta_a}(a) \cap \Omega$.

$$\{N_{\delta_a/2}(a) \mid a \in K\}$$

is an open cover of $K$. Since $K$ is compact, a finite subcover, say $N_{\delta_1}(a_1), N_{\delta_2}(a_2), \ldots, N_{\delta_l}(a_l)$ with $\delta_j = (\delta_{a_j}/2)$, will do. Let $\delta = \min(\delta_1, \delta_2, \ldots, \delta_l)$ and consider $x, y \in K$ such that $\|x - y\| < \delta$. Since $x \in K$, we have $x \in N_{\delta_j}(a_j)$ for some $j \in \{1, 2, \ldots, l\}$ and hence,

$$\|x - a_j\| < \delta_j = \frac{\delta_{a_j}}{2} < \delta_{a_j},$$

$$\|y - a_j\| \leqslant \|y - x\| + \|x - a_j\| < \delta + \frac{\delta_{a_j}}{2} \leqslant \delta_{a_j}.$$

By (112.5),

$$\|g(x) - g(y)\| \leqslant \|g(x) - g(a_j)\| + \|g(a_j) - g(y)\| \leqslant 2L_{a_j} \|x - y\|.$$

Let $L = 2\max\{L_{a_1}, L_{a_2}, \ldots, L_{a_l}\}$, and the lemma is proved.

By means of Lemma 112.3 and Theorem 112.1, we can now establish a sufficient condition for the image of a Jordan-measurable set to be Jordan-measurable.

***Theorem 112.2***

Let $\Omega \subseteq \mathbf{E}^n$ denote an open set and let $\mathfrak{D}$ denote a Jordan-measurable set where $\mathfrak{D} \cup \beta\mathfrak{D} \subset \Omega$. If $g: \Omega \to \mathbf{E}^n$, $g \in C^1(\Omega)$, and if $J_g(x) \neq 0$ for all $x \in \mathfrak{D}^\circ$, then $g(\mathfrak{D})$ is Jordan-measurable.

*Proof*

Since $\mathfrak{D} \cup \beta\mathfrak{D}$ is compact and since $g$ is continuous, $g(\mathfrak{D} \cup \beta\mathfrak{D})$ is compact and hence, $g(\mathfrak{D})$ is bounded because $g(\mathfrak{D}) \subseteq g(\mathfrak{D} \cup \beta\mathfrak{D})$. To show that $g(\mathfrak{D})$ is Jordan-measurable it suffices to show (in view of Theorem 105.2) that $\beta g(\mathfrak{D})$ has Jordan content zero.

Since $\mathfrak{D}$ is Jordan-measurable, $J(\beta\mathfrak{D}) = 0$. Since $\beta\mathfrak{D}$ is compact, we have from Lemma 112.3 and Theorem 112.1 that $J(g(\beta\mathfrak{D})) = 0$. Therefore, we need only show that $\beta g(\mathfrak{D}) \subseteq g(\beta\mathfrak{D})$.

Since $g(\mathfrak{D} \cup \beta\mathfrak{D})$ is compact, $\beta g(\mathfrak{D}) \subseteq g(\mathfrak{D} \cup \beta\mathfrak{D})$. (See exercise 5, section 11.104.) Hence, if $y \in \beta g(\mathfrak{D})$, then there is an $x \in \mathfrak{D} \cup \beta\mathfrak{D}$ such that $g(x) = y$. Suppose that $x \in \mathfrak{D}^\circ$. Since $J_g(x) \neq 0$, we have, from Theorem 97.1, that $g(x)$

is an interior point of $g(\mathfrak{D})$ contrary to our assumption. Hence, $x \in \beta\mathfrak{D}$ and $y = g(x) \in g(\beta\mathfrak{D})$. Therefore, $\beta g(\mathfrak{D}) \subseteq g(\beta\mathfrak{D})$ and $J(\beta g(\mathfrak{D})) = 0$. Hence, $g(\mathfrak{D})$ is Jordan-measurable.

This theorem will play a key role in our search for the Jordan content of the images of Jordan-measurable sets under certain maps (sections 12.113 and 12.114) and the subsequent derivation of a transformation formula for multiple integrals which contains the well-known change of integration variables into polar coordinates and spherical coordinates as special cases (section 12.115).

---

▲ *Example 2*

Let $g : \mathbf{E}^2 \to \mathbf{E}^2$ be defined by

$$\eta_1 = \xi_1 \cos \xi_2,$$
$$\eta_2 = \xi_1 \sin \xi_2.$$

Clearly, $g \in C^1(\mathbf{E}^2)$. Let $\mathfrak{D}$ denote the closed rectangle

$$\mathfrak{D} = \{(\xi_1, \xi_2) \in \mathbf{E}^2 \mid 0 \leqslant \xi_1 \leqslant 1, 0 \leqslant \xi_2 \leqslant 2\pi\}.$$

Then, $g(\mathfrak{D})$ is the closed unit disk

$$g(\mathfrak{D}) = \{(\eta_1, \eta_2) \in \mathbf{E}^2 \mid \eta_1^2 + \eta_2^2 \leqslant 1\}.$$

Since

$$J_g(x) = \begin{vmatrix} \cos \xi_2 & -\xi_1 \sin \xi_2 \\ \sin \xi_2 & \xi_1 \cos \xi_2 \end{vmatrix} = \xi_1,$$

we see that $J_g(x) \neq 0$ for all $x \in \mathfrak{D}^\circ = \{(\xi_1, \xi_2) \in \mathbf{E}^2 \mid 0 < \xi_1 < 1, 0 < \xi_2 < 2\pi\}$, and hence, $g(\mathfrak{D})$ is, by Theorem 112.2, Jordan-measurable. In Fig. 112.3 we

*Fig. 112.3*

have depicted $\beta\mathfrak{D}, \beta g(\mathfrak{D})$, and $g(\beta\mathfrak{D})$ and we see that, indeed, $\beta g(\mathfrak{D}) \subseteq g(\beta\mathfrak{D})$.

Note that the left side of $\mathfrak{D}$ is mapped onto the point $(0, 0)$, top and bottom sides of $\mathfrak{D}$ are mapped onto the line segment from $(0, 0)$ to $(1, 0)$, and the right side of $\mathfrak{D}$ is mapped onto the circumference of the unit disk $g(\mathfrak{D})$.

---

## 12.112 Exercises

1. Prove Lemma 112.1.
2. Prove Lemma 112.2 with the following modifications:

$$g : \Omega \to \mathbf{E}^m, \qquad m \geqslant n, \qquad \bar{J}(K) < \frac{\sqrt[m]{a^n}}{4(2L\sqrt{n})^n}.$$

3. Prove Theorem 112.1 for the more general case where $g : \Omega \to \mathbf{E}^m$, $m \geqslant n$. (See also exercise 2.)
4. Let $\mathfrak{D} = \{(x, y, z) \in \mathbf{E}^3 \mid x \in [0, \pi],\ y \in [0, \pi],\ z \in [0, \pi]\}$ and let $g : \mathbf{E}^3 \to \mathbf{E}^3$ be defined by

$$u = \cos x,$$
$$v = \sin x \cos y,$$
$$z = \sin x \sin y \cos z.$$

   Show that $g(\mathfrak{D})$ has Jordan content. What is $g(\mathfrak{D})$?
5. Same as in exercise 4, for $\mathfrak{D} = \{(x, y, z) \in \mathbf{E}^3 \mid 0 \leqslant x \leqslant 1, 0 \leqslant y \leqslant \pi, 0 \leqslant z \leqslant 2\pi\}$ and $g : \mathbf{E}^3 \to \mathbf{E}^3$ defined by

$$u = x \sin y \cos z,$$
$$v = x \sin y \sin z,$$
$$w = x \cos y.$$

## 12.113 Jordan Content of Linear Images of Jordan-Measurable Sets

If $l : \mathbf{E}^n \to \mathbf{E}^n$ denotes a nonsingular linear transformation, then rank $l = n$, that is, $\det l'(x) = J_l(x) \neq 0$ for all $x \in \mathbf{E}^n$. Since $l \in C^1(\mathbf{E}^n)$, the hypotheses of Theorem 112.2 are met and we see that the image of a Jordan-measurable set under $l$ is again Jordan-measurable.

It is the goal of our investigations in this section to find the Jordan content of the linear image $l(\mathfrak{D})$ of a Jordan-measurable set $\mathfrak{D}$ in terms of the Jordan content of $\mathfrak{D}$.

First, we consider the simplest Jordan-measurable sets, namely the intervals $\mathfrak{J}$ and we shall demonstrate that $J(l(\mathfrak{J})) = |\mathfrak{J}|\,|\det L|$ where $L$ is the matrix representation of $l$. This seems a fairly simple and straightforward matter. Still, the proof is somewhat tricky and we shall break it up into two parts. First, we shall investigate the effect of the elementary linear transformations on the content of an interval.

The elementary linear transformations $l_1, l_2, l_3 : \mathbf{E}^n \to \mathbf{E}^n$ may be represented in terms of the fundamental vectors $e_1, e_2, \ldots, e_n$ in $\mathbf{E}^n$ as follows:

(113.1) $\quad l_1(x) = x + (\lambda - 1)\xi_k e_k \quad$ where $\lambda \in \mathbf{R}$,

(113.2) $\quad l_2(x) = (\xi_{k_1}, \xi_{k_2}, \ldots, \xi_{k_n}) \quad$ where $\{k_1, k_2, \ldots, k_n\}$ is a permutation of $\{1, 2, \ldots, n\}$,

(113.3) $\quad l_3(x) = x + \xi_2 e_1.$

If $L_1, L_2, L_3$ are the matrix representations of $l_1, l_2, l_3$, then $L_1$ is obtained from the identity matrix by multiplying the $k$th row by $\lambda$; $L_2$ is obtained from the identity matrix by appropriate permutation of rows; and $L_3$ is obtained from the identity matrix by adding the second row to the first row. Note that a suitable permutation in (113.2) makes it possible to add any given row to any other given row by simply carrying out $l_2$ and $l_3$ in succession.

***Lemma 113.1***

If $\mathfrak{J} \subset \mathbf{E}^n$ denotes a closed interval and if $l_1, l_2, l_3 : \mathbf{E}^n \to \mathbf{E}^n$ denote the elementary linear transformations defined in (113.1), (113.2), and (113.3), then $l_1(\mathfrak{J})$, $l_2(\mathfrak{J})$, $l_3,(\mathfrak{J})$ are Jordan-measurable and

(113.4) $$J(l_j(\mathfrak{J})) = |\det L_j|\,|\mathfrak{J}|,$$

where $L_j$ is the matrix representation of $l_j$, $j = 1, 2, 3$.

*Proof*

If $\lambda = 0$, then the $k$th row of $L_1$ is the zero vector; that is, $l_1 : \mathbf{E}^n \to \mathbf{E}^m$ where $m < n$ and $l_1(\mathfrak{J})$ has $n$-dimensional Jordan content zero. This is consistent with (113.4) because $\det L_1 = 0$ for $\lambda = 0$.

If $\lambda \neq 0$, then $l_1$ is nonsingular. $l_2, l_3$ are always nonsingular. Since we also have $l_j \in C^1(\mathbf{E}^n)$, we have, from Theorem 112.2, that $l_j(\mathfrak{J})$ is Jordan-measurable for $j = 1, 2, 3$. Further,

(113.5) $\quad \det L_1 = \lambda,$

(113.6) $\quad \det L_2 = \pm 1 \quad$ (depending on whether the permutation is even or odd)

(113.7) $\quad \det L_3 = 1.$

$l_1$, for $\lambda \neq 0$, transforms $\mathfrak{J} = [a_1, b_1] \times [a_2, b_2] \times \cdots \times [a_n, b_n]$ into $l_1(\mathfrak{J}) = [a_1, b_1] \times [a_2, b_2] \times \cdots \times [\lambda a_k, \lambda b_k] \times \cdots \times [a_n, b_n]$. (If $\lambda < 0$, then $[\lambda a_k, \lambda b_k]$ is to be replaced by $[\lambda b_k, \lambda a_k]$.) Hence

(113.8) $$J(l_1(\mathfrak{J})) = |\lambda|\,|\mathfrak{J}|.$$

$l_2$ transforms $\mathfrak{J}$ into $[a_{k_1}, b_{k_1}] \times [a_{k_2}, b_{k_2}] \times \cdots \times [a_{k_n}, b_{k_n}]$ and hence,

(113.9) $$J(l_2(\mathfrak{J})) = |\mathfrak{J}|.$$

The vertices of $\mathfrak{J}$ are given by $(a_1, a_2, c_3, \ldots, c_n)$, $(b_1, a_2, c_3, \ldots, c_n)$, $(a_1, b_2, c_3, \ldots, c_n)$, $(b_1, b_2, c_3, \ldots, c_n)$ where $c_j = a_j$ or $b_j$, $j = 3, 4, \ldots, n$. Since $l_3(a_1, a_2, c_3, \ldots, c_n) = (a_1 + a_2, a_2, c_3, \ldots, c_n)$, $l_3(b_1, a_2, c_3, \ldots, c_n) = (b_1 + a_2, a_2, c_3, \ldots, c_n)$, $l_3(a_1, b_2, c_3, \ldots, c_n) = (a_1 + b_2, b_2, c_3, \ldots, c_n)$, $l_3(b_1, b_2, c_3, \ldots, c_n) = (b_1 + b_2, b_2, c_3, \ldots, c_n)$, we see that $l_3$ transforms $\mathfrak{J}$ into $P_2 \times [a_3, b_3] \times [a_4, b_4] \times \cdots \times [a_n, b_n]$, where $P_2$ is depicted in Fig. 113.1.

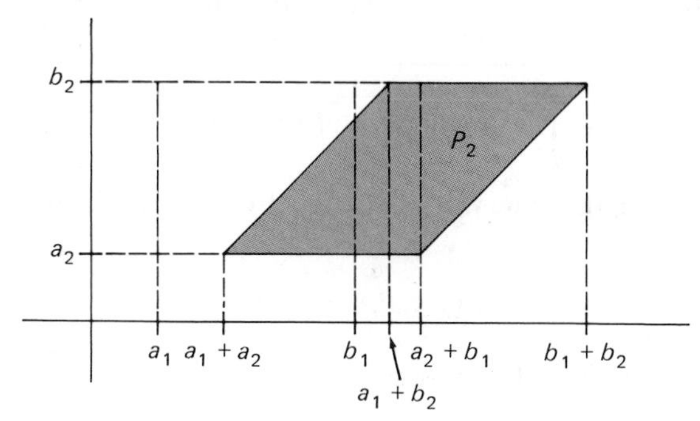


*Fig. 113.1*

As in example 2, section 11.109, we obtain

$$J(l_3(\mathfrak{J})) = |\mathfrak{J}|. \tag{113.10}$$

(113.8), (113.9), and (113.10), together with (113.5), (113.6), and (113.7), yield (113.4).

It is our goal to show that, for every linear function $l : \mathbf{E}^n \to \mathbf{E}^n$,

$$J(l(\mathfrak{J})) = |\det L|\,|\mathfrak{J}|.$$

Lemma 113.1 does not suffice for this purpose. Although we can decompose every linear function into elementary linear functions, such as $l = l_1 \circ l_2 \circ l_1 \circ l_2 \circ l_3 \circ l_1 \circ l_2$, Lemma 113.1 won't give us any information about the Jordan content of $l_1 \circ (l_3 \circ l_1 \circ l_2)(\mathfrak{J})$ because $l_3 \circ l_1 \circ l_2(\mathfrak{J})$ is *not* an interval and Lemma 113.1 *applies only to intervals*. For this purpose we need information about the effect of the elementary linear functions on parallelepipeds. It is just as simple to establish, on the basis of Lemma 113.1, the effect of the elementary linear functions on any Jordan-measurable set. This is the content of the following lemma.

***Lemma 113.2***

If $\mathfrak{D} \subset \mathbf{E}^n$ is Jordan-measurable, then $l_j(\mathfrak{D})$, where $j = 1, 2, 3$, is Jordan-measurable, and

$$J(l_j(\mathfrak{D})) = |\det L_j|\, J(\mathfrak{D}), \qquad j = 1, 2, 3, \tag{113.11}$$

where $L_j$ is the matrix representation of $l_j$.

*Proof*

If $\lambda = 0$, then $l_1$ transforms $\mathfrak{D}$ into a bounded subset of $\mathbf{E}^{n-1}$ and hence, $J(l_1(\mathfrak{D})) = 0$. Since $\det L_1 = 0$ for $\lambda = 0$, (113.11) holds.

If $\lambda \neq 0$, then $l_1$ is nonsingular. Since $l_2, l_3$ are nonsingular under any circumstances, it follows, as in the proof of Lemma 113.1, that the sets $l_j(\mathfrak{D})$ are Jordan-measurable.

Since $\mathfrak{D}$ is Jordan-measurable, we have, for any $\mathfrak{J} \supseteq \mathfrak{D}$ and every $\varepsilon > 0$, a partition $P_\varepsilon$ such that

$$(113.12) \qquad J(\mathfrak{D}) = \int_{\mathfrak{J}} \chi_{\mathfrak{D}} = \inf_{(P)} \bar{S}(\chi_{\mathfrak{D}}; P) > \bar{S}(\chi_{\mathfrak{D}}; P_\varepsilon) - \frac{\varepsilon}{|\lambda|}.$$

If $\mathfrak{J}_1, \mathfrak{J}_2, \ldots, \mathfrak{J}_k$ are the intervals generated by $P_\varepsilon$ which contain points from $\mathfrak{D}$, then

$$\bar{S}(\chi_{\mathfrak{D}}; P_\varepsilon) = \sum_{j=1}^{k} |\mathfrak{J}_j|.$$

Hence, we have, from (113.12), that

$$(113.13) \qquad \sum_{j=1}^{k} |\mathfrak{J}_j| < J(\mathfrak{D}) + \frac{\varepsilon}{|\lambda|}.$$

Since $l_1$ is nonsingular and hence, one-to-one, we have, for every $y \in l_1(\mathfrak{D})$, an $x \in \mathfrak{D}$ such that $l_1(x) = y$. Since $\mathfrak{D} \subseteq \bigcup_{j=1}^{k} \mathfrak{J}_j$, we see that $x \in \mathfrak{J}_j$ for some $j \in \{1, 2, \ldots, k\}$ and, consequently, $l_1(x) \in l_1(\mathfrak{J}_j)$. By Corollary 3 to Theorem 106.1,

$$(113.14) \qquad J(l_1(\mathfrak{D})) \leqslant J\left(\bigcup_{j=1}^{k} l_1(\mathfrak{J}_j)\right).$$

Since $l_1$ one-to-one, the intervals $l_1(\mathfrak{J}_j)$ have no interior points in common, and we have, from Corollary 1 to Theorem 106.1, that

$$(113.15) \qquad J\left(\bigcup_{j=1}^{k} l_1(\mathfrak{J}_j)\right) = \sum_{j=1}^{k} J(l_1(\mathfrak{J}_j)).$$

(See also exercise 3.) From Lemma 113.1, $J(l_1(\mathfrak{J}_j)) = |\lambda|\, |\mathfrak{J}_j|$, and we obtain from (113.14), (113.15), and (113.13), that

$$J(l_1(\mathfrak{D})) \leqslant |\lambda| \sum_{j=1}^{k} |\mathfrak{J}_j| \leqslant |\lambda|\, J(\mathfrak{D}) + \varepsilon.$$

Since this is true for every $\varepsilon > 0$, we have

$$(113.16) \qquad J(l_1(\mathfrak{D}) \leqslant |\lambda|\, J(\mathfrak{D}).$$

Since $l_1^{-1}$ is again an elementary linear function such as $l_1$ but with $(1/\lambda)$ instead of $\lambda$, we obtain, interchanging the roles of $\mathfrak{D}$ and $l_1(\mathfrak{D})$ in the above argument, that

$$(113.17) \qquad J(\mathfrak{D}) \leqslant \frac{1}{|\lambda|} J(l_1(\mathfrak{D})).$$

(113.16) and (113.17) yield, in view of (113.5),

$$J(l_1(\mathfrak{D})) = |\det L_1|\, J(\mathfrak{D}).$$

In order to prove (113.11) for $j = 2, 3$, we let $\lambda = 1$ in (113.12), and repeat the same argument. (See also exercise 5.)

***Theorem 113.1***

If $\mathfrak{D} \subset \mathbf{E}^n$ is Jordan-measurable and if $l: \mathbf{E}^n \to \mathbf{E}^n$ is a linear function, then $l(\mathfrak{D})$ is Jordan-measurable and

$$J(l(\mathfrak{D})) = |\det L|\, J(\mathfrak{D}), \tag{113.18}$$

where $L$ is the matrix representation of $l$.

*Proof*

If $l$ is singular, then $l(\mathfrak{D}) \subset \mathbf{E}^m$, $m < n$ and $\det L = 0$. Hence, (113.18) is trivially true.

If $l$ is nonsingular, then it may be represented as a composition of elementary linear functions of the type $l_1, l_2, l_3$, all of which are nonsingular:

$$l = l_{k_1} \circ l_{k_2} \circ l_{k_3} \circ \cdots \circ l_{k_m},$$

where $k_j \in \{1, 2, 3\}$, $j = 1, 2, \ldots, m$. If $L$, $L_{k_j}$ are the matrix representations of $l$, $l_{k_j}$, then $L = L_{k_1} L_{k_2} L_{k_3} \ldots L_{k_m}$. Hence, $\det L = \det L_{k_1} \det L_{k_2} \ldots \det L_{k_m}$. From Lemma 113.2,

$$\begin{aligned} J(l(\mathfrak{D})) &= J(l_{k_1} \circ l_{k_2} \circ \cdots \circ l_{k_m}(\mathfrak{D})) = |\det L_{k_1}|\, J(l_{k_2} \circ l_{k_3} \circ \cdots \circ l_{k_m}(\mathfrak{D})) \\ &= \cdots = |\det L_{k_1}|\, |\det L_{k_2}| \cdots |\det L_{k_m}|\, J(\mathfrak{D}) = |\det L| J(\mathfrak{D}). \end{aligned}$$

***Corollary to Theorem 113.1***

If $l: \mathbf{E}^n \to \mathbf{E}^n$ represents an orthogonal transformation, and if $\mathfrak{D} \subset \mathbf{E}^n$ is Jordan-measurable, then $J(l(\mathfrak{D})) = J(\mathfrak{D})$. (This may also be expressed by stating that the *Jordan content is invariant under orthogonal transformations.*)

*Proof*

If $l: \mathbf{E}^n \to \mathbf{E}^n$ is an orthogonal transformation, then $l(x) = Ax$, where $A$ is an $n \times n$ orthogonal matrix and $\det A = \pm 1$. By Theorem 113.1, $J(l(\mathfrak{D})) = |\det A|\, J(\mathfrak{D}) = J(\mathfrak{D})$.

In section 12.116 we shall show that the Jordan content is also invariant under translations. This, together with the above corollary, implies that congruent Jordan-measurable sets have the same Jordan content. (See also $(P_2)$ and $(\mu_2)$ of definition 41.1.)

---

▲ *Example 1*

The parallelepiped $P \subset \mathbf{E}^n$ which is spanned by the vectors $(a_{11}, a_{21}, \dots, a_{n1})$, $(a_{12}, a_{22}, \dots, a_{n2}), \dots, (a_{1n}, a_{2n}, \dots, a_{nn})$ is obtained from $\mathfrak{J} = [0,1] \times [0,1] \times \cdots \times [0,1]$ by a linear transformation, given by $l(x) = Ax$, where

$$A = \begin{pmatrix} a_{11} & a_{12} & \cdots & a_{1n} \\ \vdots & & & \\ a_{n1} & a_{n2} & \cdots & a_{nn} \end{pmatrix}.$$

By Theorem 113.1,

$$J(P) = J(l(\mathfrak{J})) = |\det A|\,|\mathfrak{J}| = |\det A|.$$

---

## 12.113 Exercises

1. Write down the matrix representations $L_1, L_2, L_3$ for:

   $l_1 : \mathbf{E}^4 \to \mathbf{E}^4$, defined by $l_1(x) = x + (\lambda - 1)\xi_2 e_2$;

   $l_2 : \mathbf{E}^4 \to \mathbf{E}^4$, defined by $l_2(x) = (\xi_1, \xi_4, \xi_3, \xi_2)$; and

   $l_3 : \mathbf{E}^4 \to \mathbf{E}^4$, defined by $l_3(x) = x + \xi_2 e_1$.

   Find $\det L_1$, $\det L_2$, $\det L_3$.
2. Let $l_1, l_2, l_3$ be defined as in exercise 1. Find the matrix representations of $l_1^{-1}, l_2^{-1}, l_3^{-1}$.
3. Show: If $l : \mathbf{E}^n \to \mathbf{E}^n$ is a nonsingular linear function and if $\mathfrak{J}_1, \mathfrak{J}_2, \dots, \mathfrak{J}_k$ are intervals with nonoverlapping interiors, then $l(\mathfrak{J}_1), l(\mathfrak{J}_2), \dots, l(\mathfrak{J}_k)$ have nonoverlapping interiors.
4. Let $\mathfrak{D}_1 = \{(x,y) \in \mathbf{E}^2 \mid x^2 + y^2 \leqslant 1\}$. Given that $J(\mathfrak{D}_1) = \pi$, find $J(\mathfrak{D}_2)$ where $\mathfrak{D}_2 = \{(x,y) \in \mathbf{E}^2 \mid 5x^2 + 12xy + 8y^2 \leqslant 4\}$.
5. Prove Lemma 113.2 for $l_2$ and $l_3$.

## 12.114 Jordan Content of General Images of Intervals

In the last section, a simple formula was developed for the Jordan content of the linear image of a Jordan-measurable set. In this section we shall study the image $g(\mathfrak{J})$ of an interval $\mathfrak{J}$ where $g\colon \mathbf{E}^n \to \mathbf{E}^n$ is a function that satisfies suitable differentiability conditions, preparatory to the later study of the Jordan content of the image $g(\mathfrak{D})$ of any Jordan-measurable set $\mathfrak{D}$. In the past, and in particular in Chapter 10, we have been quite successful in studying the local behavior of a function by studying the behavior of its derivative instead. The same approach will lead to results in our present investigation. Specifically, we shall find that the Jordan content of the image $g(\mathfrak{J})$ of a small interval $\mathfrak{J}$ may be approximated by the Jordan content of $g'(x)(\mathfrak{J})$ for any $x \in \mathfrak{J}$ and that this approximation is the better, the smaller the interval $\mathfrak{J}$. First we establish this result for small cubes.

***Lemma 114.1***

Let $\Omega \subseteq \mathbf{E}^n$ denote an open set and let $K \subset \Omega$ denote a *compact* set. If $g : \Omega \to \mathbf{E}^n$, $g \in C^1(\Omega)$, $J_g(x) \neq 0$ for all $x \in K$ and if $g : K^\circ \xrightarrow[1-1]{} \mathbf{E}^n$, then there is, for every $\varepsilon \in (0, 2)$, a $\delta > 0$ such that for every closed cube $C \subset K^\circ$ with side length $s < \delta$ and center $a$,

$$\left(1 - \frac{\varepsilon}{2}\right)|J_g(a)|\ |C| \leqslant J(g(C)) \leqslant \left(1 + \frac{\varepsilon}{2}\right)|J_g(a)|\ |C|. \tag{114.1}$$

(See also Fig. 114.1.)

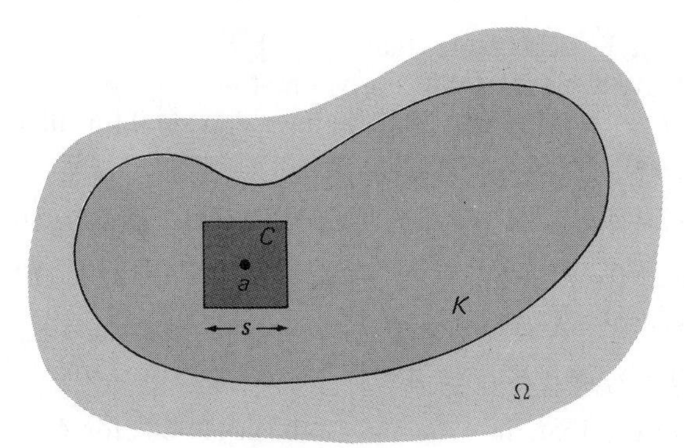


***Fig. 114.1***

*Proof*

By Theorem 112.2, $g(C)$ is Jordan-measurable. Since $J_g(x) \neq 0$ for all $x \in K$, $g'(a)$ is one-to-one for every $a \in K$. By Lemma 96.2 there is a $\mu > 0$ such that

$$\|g'(x)(u)\| \geqslant \mu \|u\| \qquad \text{for all } x \in K \text{ and all } u \in \mathbf{E}^n.$$

For any given $x \in K$ and every $v \in \mathbf{E}^n$, we let $u = [g'(x)]^{-1}(v)$. Then

$$\|v\| \geqslant \mu \|[g'(x)]^{-1}(v)\| \qquad \text{for all } v \in \mathbf{E}^n.$$

Hence, for $M = 1/\mu$,

$$\|[g'(x)]^{-1}(v)\| \leqslant M \|v\| \qquad \text{for all } x \in K \text{ and all } v \in \mathbf{E}^n. \tag{114.2}$$

We obtain, from Lemma 96.2 that, for every $\varepsilon_1 > 0$, there is a $\delta_1 > 0$ such that, for every $a \in K$,

$$\|g(x) - g(a) - g'(a)(x - a)\| \leqslant \varepsilon_1 \|x - a\| \tag{114.3}$$

for all $x \in N_{\delta_1}(a) \cap K$. For the time being, we choose

$$\varepsilon_1 < \frac{1}{M\sqrt{n}}, \tag{114.4}$$

where $M$ is defined in (114.2). A more narrow choice of $\varepsilon_1$ will become necessary later on.

Let

(114.5) $$s < \frac{2\delta_1}{\sqrt{n}}.$$

Then, if $C \subset K^\circ$ is a cube with center at $a$ and side length $s$, we have $\|x - a\| < \delta_1$ for all $x \in C$ and hence, (114.3) holds for all $x \in C$. (Note that $\sqrt{n}s$ is the length of the diagonal of $C$.)

Let $h: \Omega \to \mathbf{E}^n$ be defined by

(114.6) $$h(x) = [g'(a)]^{-1}(g(x)) \qquad \text{for all } x \in \Omega,$$

that is, $h = [g'(a)]^{-1} \circ g$. Note that $[g'(a)]^{-1}$ exists because $a \in K^\circ$. Since $g \in C^1(\Omega)$, we have $h \in C^1(\Omega)$. Since $g$ is one-to-one on $K^\circ$, $h$ is one-to-one on $K^\circ$ also. Since $J_g(x) \neq 0$ for all $x \in K$, we have $J_h(x) \neq 0$ for all $x \in K$.

From (114.6),

(114.7) $$\|h(x) - h(a) - (x - a)\| = \|[g'(a)]^{-1}(g(x) - g(a) - g'(a)(x - a))\|,$$

and we obtain, from (114.7), (114.2), (114.3), and the FIE (61.1) with $h = (\chi_1, \chi_2, \ldots, \chi_n)$, $a = (\alpha_1, \alpha_2, \ldots, \alpha_n)$, that

(114.8) $$|\chi_j(x) - \chi_j(a) - (\xi_j - \alpha_j)| \leqslant \|h(x) - h(a) - (x - a)\| \leqslant M\varepsilon_1 \|x - a\|$$

for all $x \in C$. (Note that, in order to establish (114.8) for every cube $C \subset K^\circ$ with dimension as in (114.5), we needed (114.2) for *all* $x \in K$.)

From (114.8)

$$|\chi_j(x) - \chi_j(a)| \leqslant |\xi_j - \alpha_j| + M\varepsilon_1 \|x - a\|.$$

Since $|\xi_j - \alpha_j| \leqslant (s/2)$ and $\|x - a\| \leqslant (s\sqrt{n}/2)$ for all $x \in C$, we obtain

$$|\chi_j(x) - \chi_j(a)| \leqslant \frac{s}{2}(1 + M\varepsilon_1\sqrt{n});$$

that is, $h(x)$ lies in a cube $\Gamma$ with center at $h(a)$ and dimension $s(1 + M\varepsilon_1\sqrt{n})$. This means in turn that

(114.9) $$h(C) \subseteq \Gamma.$$

As in the proof of Theorem 112.2 we see that $\beta h(C) \subseteq h(\beta C)$. Since $h$ is one-to-one on $C$ (this is why we had to assume that $C \subset K^\circ$), we also have $h(\beta C) \subseteq \beta h(C)$. (See exercise 2.) Hence, the boundary of $h(C)$ is the image of the boundary of $C$ under $h$. From (114.8),

$$|\chi_j(x) - \chi_j(a)| \geqslant |\xi_j - \alpha_j| - M\varepsilon_1 \|x - a\|.$$

For $x \in \beta C$, $|\xi_j - \alpha_j| = (s/2)$. Since $\|x - a\| \leqslant (s\sqrt{n}/2)$,

$$|\chi_j(x) - \chi_j(a)| \geqslant \frac{s}{2}(1 - M\varepsilon_1\sqrt{n});$$

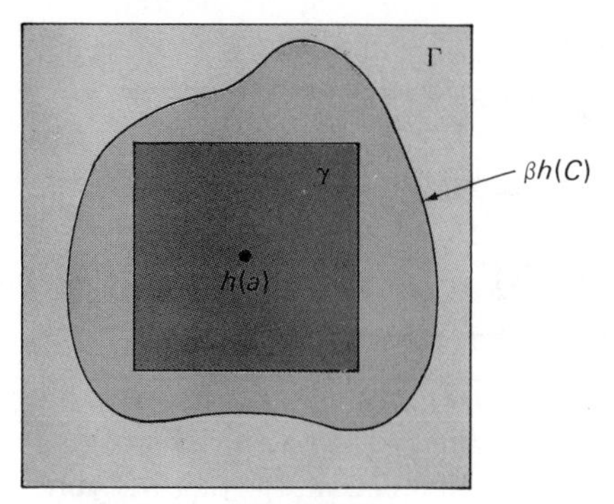


*Fig. 114.2*

i.e., all boundary points of $h(C)$ lie outside or on the boundary of a cube $\gamma$ with center at $h(a)$ and dimension $s(1 - M\varepsilon_1\sqrt{n})$. (See Fig. 114.2.) (In view of our choice of $\varepsilon_1$ in (114.4), we have $1 - M\varepsilon_1\sqrt{n} > 0$.)

We want to show that $\gamma \subseteq h(C)$. Since $h(C)$ is compact, it suffices to show that $\gamma^\circ \subseteq h(C)$. Since every $y \in \mathbf{E}^n$ lies either in $\beta h(C)$ or in $[h(C)]^\circ$ or in $c(h(C))$, and since $\gamma^\circ \cap \beta h(C) = \varnothing$ because all points on $\beta h(C)$ lie outside or on the boundary of $\gamma$, we have $([h(C)]^\circ \cap \gamma^\circ) \cup (c(h(C)) \cap \gamma^\circ) = \gamma^\circ$ and

$$([h(C)]^\circ \cap \gamma^\circ) \cap (c(h(C)) \cap \gamma^\circ) = \varnothing.$$

If $[h(C)]^\circ \cap \gamma^\circ \neq \varnothing$ and $c(h(C)) \cap \gamma^\circ \neq \varnothing$, then $[h(C)]^\circ$ and $c(h(C))$ establish a disconnection of $\gamma^\circ$ which is absurd. (See exercise 6, section 6.72.) Since $h(a) \in [h(C)]^\circ$ ($h$ maps open sets onto open sets in a neighborhood of $a$) and $h(a) \in \gamma^\circ$, we have to have $c(h(C)) \cap \gamma^\circ = \varnothing$; that is, $\gamma^\circ \subseteq h(C)$ and hence,

$$\gamma \subseteq h(C) \tag{114.10}$$

We have, from (114.9) and (114.10), that

$$\gamma \subseteq h(C) \subseteq \Gamma, \tag{114.11}$$

where $\gamma$ is a cube with center at $h(a)$ and dimension $s(1 - M\varepsilon_1\sqrt{n})$ and where $\Gamma$ is a cube with center at $h(a)$ and dimension $s(1 + M\varepsilon_1\sqrt{n})$. Since $|C| = s^n$, we have from (114.11) that

$$J(\gamma) = |C|(1 - M\varepsilon_1\sqrt{n})^n \leqslant J(h(C)) \leqslant |C|(1 + M\varepsilon_1\sqrt{n})^n = J(\Gamma). \tag{114.12}$$

Since $h(C) = [g'(a)]^{-1} \circ g(C)$, we have, from Theorem 113.1, that

$$J(h(C)) = |\det [G'(a)]^{-1}|\, J(g(C)).$$

Since

$$\det [G'(a)]^{-1} = \frac{1}{\det G'(a)} = \frac{1}{J_g(a)},$$

we have, from (114.12),

$$(1 - M\varepsilon_1\sqrt{n})^n |C| \, |J_g(a)| \leqslant J(g(C)) \leqslant (1 + M\varepsilon_1\sqrt{n})^n |C| \, |J_g(a)|.$$

We choose $\varepsilon_1$ such that (114.4) is satisfied and also

$$(1 - M\varepsilon_1\sqrt{n})^n \geqslant 1 - \frac{\varepsilon}{2}, \qquad 1 + \frac{\varepsilon}{2} \geqslant (1 + M\varepsilon_1\sqrt{n})^n.$$

This is possible (see exercise 3) and (114.1) follows readily. (Note that $g(C)$ is obtained from $h(C)$ by the linear map $g'(a)$. Hence, it follows from (114.11), that $g(C)$ contains the parallelepiped $g'(a)(\gamma)$ and is contained in the parallelepiped $g'(a)(\Gamma)$.)

If we know that for an integrable function there is a partition which generates *cubes only* and for which the Riemann condition is satisfied, we could proceed from Lemma 114.1 directly to a discussion of the Jordan content of the image of a Jordan-measurable set. This is actually true as one can show by generalizing the result of exercise 11, section 4.47 in the following sense: If $f$ is integrable on $\mathfrak{D}$, then there is, for every $\varepsilon > 0$, a $\delta > 0$ such that, for every partition $P$ for which the largest dimension of all intervals generated by $P$ is less than $\delta$, then the Riemann condition is satisfied for the given $\varepsilon$. To prove this is messy and technically difficult. We shall circumvent this difficulty by generalizing Lemma 114.1 so that it also applies to intervals and not only cubes.

***Lemma 114.2***

Let $\Omega \subseteq \mathbf{E}^n$ denote an open set and let $K \subset \Omega$ denote a compact set. If $g : \Omega \to \mathbf{E}^n$, $g \in C^1(\Omega)$, $J_g(x) \neq 0$ for all $x \in K$, and if $g : K^\circ \xrightarrow[1-1]{} \mathbf{E}^n$, then there is for every $\varepsilon \in (0, 1)$ a $\delta > 0$ such that, for every interval $\mathfrak{J} \subset K^\circ$, the largest dimension of which is less than $\delta$,

$$(1 - \varepsilon) |J_g(x)| \, |\mathfrak{J}| \leqslant J(g(\mathfrak{J})) \leqslant (1 + \varepsilon) |J_g(x)| \, |\mathfrak{J}| \qquad (114.13)$$

for all $x \in \mathfrak{J}$.

*Proof*

In order to derive (114.13) from (114.1) we shall replace $\mathfrak{J}$ by an interval with rational dimensions the content of which differs from the content of $\mathfrak{J}$ by a small amount, partition the new interval into cubes, and then apply (114.1) to each one of these cubes. For (114.13) to hold for all $x \in \mathfrak{J}$, $\mathfrak{J}$ has to be sufficiently small to begin with. How small depends on the behavior of $g$. We shall take care of this first: Since $g \in C^1(\Omega)$, we have $J_g \in C(\Omega)$, and since $J_g(x) \neq 0$ for all $x \in K$, we have a $\mu > 0$ such that

$$|J_g(x)| \geqslant \mu \qquad \text{for all } x \in K. \qquad (114.14)$$

$J_g$ is *uniformly* continuous on $K$. Hence, for every $\varepsilon_1 > 0$, there is a $\delta_1 > 0$ such that

(114.15) $\quad |J_g(x) - J_g(y)| < \varepsilon_1 \quad$ for all $x, y \in K$ for which $\|x - y\| < \delta_1$.

Hence, $|J_g(x)| < \varepsilon_1 + |J_g(y)|$ and, in view of (114.14),

(114.16) $$\left|\frac{J_g(x)}{J_g(y)}\right| < 1 + \frac{\varepsilon_1}{\mu} \qquad \text{for all } x, y \in K \text{ for which } \|x - y\| < \delta_1.$$

Also from (114.15)

(114.17) $$\left|\frac{J_g(y)}{J_g(x)}\right| > 1 - \frac{\varepsilon_1}{\mu} \qquad \text{for all } x, y \in K \text{ for which } \|x - y\| < \delta_1.$$

We choose $\varepsilon_1 > 0$ so small that

(114.18) $$\left(1 + \frac{\varepsilon}{2}\right)\left(1 + \frac{\varepsilon_1}{\mu}\right) < 1 + \varepsilon, \qquad \left(1 - \frac{\varepsilon}{2}\right)\left(1 - \frac{\varepsilon_1}{\mu}\right) > 1 - \varepsilon,$$

(see exercise 4), choose $\delta_1$ accordingly, and then choose $\mathfrak{J}$ so small that (114.16) and (114.17) hold for *all* $x, y \in \mathfrak{J}$. (This can be done by choosing $\mathfrak{J}$ so that its largest dimension is less than $\delta$ for some $\delta \in (0, \delta_1/\sqrt{n})$.)

We shall now show that (114.13) holds indeed for every interval $\mathfrak{J}$ with largest dimension less than $\delta$. For this purpose, we consider an inscribed interval $\mathfrak{J}_1$ with rational dimensions, $\mathfrak{J}_1 \subseteq \mathfrak{J}$, such that, for every given $\varepsilon_2 > 0$,

(114.19) $$|\mathfrak{J}| - |\mathfrak{J}_1| < \frac{\varepsilon_2}{4(2L\sqrt{n})^n},$$

where $L > 0$ is the Lipschitz constant of $g$ on $K$. (By Lemma 112.3, $g$ satisfies a uniform Lipschitz condition on $K$.)

$\mathfrak{J} \backslash \mathfrak{J}_1^\circ$ is compact, $|\mathfrak{J}_1| = |\mathfrak{J}_1^\circ|$, and hence, by (114.19),

$$J(\mathfrak{J} \backslash \mathfrak{J}_1^\circ) < \frac{\varepsilon_2}{4(2L\sqrt{n})^n},$$

By Lemma 112.2,

(114.20) $$\bar{J}(g(\mathfrak{J} \backslash \mathfrak{J}_1^\circ) < \varepsilon_2.$$

Since $\mathfrak{J} \subset K^\circ$ and since $g$ is one-to-one on $K^\circ$, we have, from the Corollary to Theorem 66.2, that

$$g(\mathfrak{J} \backslash \mathfrak{J}_1^\circ) = g(\mathfrak{J}) \backslash g(\mathfrak{J}_1^\circ),$$

and we see from Theorem 112.2 and Corollary 2 to Theorem 106.1 that

$$J(g(\mathfrak{J} \backslash \mathfrak{J}_1^\circ)) = J(g(\mathfrak{J}) \backslash g(\mathfrak{J}_1^\circ)) = J(g(\mathfrak{J})) - J(g(\mathfrak{J}_1^\circ))$$

exists. By (114.20),

(114.21) $$J(g(\mathfrak{J})) < \varepsilon_2 + J(g(\mathfrak{J}_1^\circ)).$$

Since $\mathfrak{J}_1^\circ \subset \mathfrak{J}_1$, we have $g(\mathfrak{J}_1^\circ) \subset g(\mathfrak{J})$ and hence $J(g(\mathfrak{J}_1^\circ)) \leqslant J(g(\mathfrak{J}_1))$. (See also Corollary 3 to Theorem 106.1.) Therefore, we have from (114.21) that

$$(114.22) \qquad J(g(\mathfrak{J})) < \varepsilon_2 + J(g(\mathfrak{J}_1)).$$

We partition $\mathfrak{J}_1$ into $k$ cubes $C_1, C_2, \ldots, C_k$ with nonoverlapping interiors and with centers at $a_1, a_2, \ldots, a_k$, which are so small that Lemma 114.1 applies to each of them. Then, by Corollary 1 to Theorem 106.1, $g(\mathfrak{J}_1) = \sum_{j=1}^k g(C_j)$. By Theorem 112.1, the Corollary to Theorem 66.2, and Theorem 106.1,

$$(114.23) \qquad J(g(\mathfrak{J}_1)) = \sum_{j=1}^{k} J(g(C_j)).$$

(See exercise 6.) Hence, we obtain from (114.22) that

$$J(g(\mathfrak{J})) < \varepsilon_2 + \sum_{j=1}^{k} J(g(C_j)).$$

From Lemma 114.1,

$$J(g(\mathfrak{J})) < \varepsilon_2 + \left(1 + \frac{\varepsilon}{2}\right) \sum_{j=1}^{k} |J_g(a_j)|\,|C_j|,$$

and from (114.16) and (114.18)

$$\left(1 + \frac{\varepsilon}{2}\right) \sum_{j=1}^{k} |J_g(a_j)|\,|C_j| < (1+\varepsilon)|J_g(x)| \sum_{j=1}^{k} |C_j| < (1+\varepsilon)|\,J_g(x)|\,|\mathfrak{J}|$$

for all $x \in \mathfrak{J}$. Hence, $J(g(\mathfrak{J})) < \varepsilon_2 + (1+\varepsilon)\,|J_g(x)|\,|\mathfrak{J}|$. Since this is true for every $\varepsilon_2 > 0$, the right side of (114.13) follows readily.

In order to demonstrate the validity of the left side of (114.13), one has to replace $\mathfrak{J}$ by a circumscribed interval $\mathfrak{J}_2$ with rational dimensions and use (114.17) instead of (114.16). (See exercise 5.)

## 12.114 Exercises

1. Show that $h'(a)$ is the identity function, where $h$ is defined in (114.6).
2. Show that $h(\beta C) \subseteq \beta h(C)$ where $h, C$ are defined as in the proof of Lemma 114.1.
3. Given $\varepsilon > 0$, $M > 0$, $n \in \mathbf{N}$. Show that one can choose an $\varepsilon_1 > 0$ such that

$$(1 - M\varepsilon_1\sqrt{n})^n \geqslant 1 - \frac{\varepsilon}{2} \qquad \text{and} \qquad 1 + \frac{\varepsilon}{2} \geqslant (1 + M\varepsilon_1\sqrt{n})^n.$$

4. Given $\varepsilon > 0$, $\mu > 0$. Show that one can choose $\varepsilon_1 > 0$ such that

$$\left(1 + \frac{\varepsilon}{2}\right)\left(1 + \frac{\varepsilon_1}{\mu}\right) < 1 + \varepsilon \qquad \text{and} \qquad \left(1 - \frac{\varepsilon}{2}\right)\left(1 - \frac{\varepsilon_1}{\mu}\right) > 1 - \varepsilon.$$

5. Under the hypotheses of Lemma 114.2, show that

$$(1 - \varepsilon)\,|J_g(x)|\,|\mathfrak{J}| \leqslant J(g(\mathfrak{J}))$$

for all $x \in \mathfrak{J}$.
6. Prove (114.23).

## 12.115 Transformation of Multiple Integrals: Jacobi's Theorem

We have done all the preparatory work for the development of a formula for the Jordan content of the image of a Jordan-measurable set. With little extra effort, we shall prove at the same time a theorem on the transformation of multiple integrals which we promised to do, in section 11.109, and which has already been stated and discussed in section 11.111 for the case of double integrals.

***Theorem 115.1*** *(Jacobi's Theorem)*

Let $\Omega \subseteq \mathbf{E}^n$ denote an open set and let $\mathfrak{D}$ denote a Jordan-measurable set such that $\mathfrak{D} \cup \beta\mathfrak{D} \subset \Omega$. If $g : \Omega \to \mathbf{E}^n$, $g \in C^1(\Omega)$, $g : \mathfrak{D}^\circ \xrightarrow[1-1]{} \mathbf{E}^n$, $J_g(x) \neq 0$ for all $x \in \mathfrak{D}^\circ$, then $g(\mathfrak{D})$ is Jordan-measurable and its Jordan content is given by

$$J(g(\mathfrak{D})) = \int_{\mathfrak{D}} |J_g|. \tag{115.1}$$

If, in addition, $f: g(\mathfrak{D}) \to \mathbf{E}$ is bounded and continuous, then

$$\int_{g(\mathfrak{D})} f = \int_{\mathfrak{D}} (f \circ g)\,|J_g|. \tag{115.2}$$

(With some extra effort—see exercises 6 to 9—one may extend the validity of this theorem to the case where $J_g(x) = 0$ on a set of Jordan content zero in $\mathfrak{D}^\circ$. With an even greater effort, one can show that the theorem still holds in the case that $f$ is discontinuous on a set of Lebesgue measure zero in $g(\mathfrak{D})$.)

*Proof*

Since $\mathfrak{D}$ is Jordan-measurable and since $(f \circ g)\,|J_g|$ is continuous, $\int_{\mathfrak{D}}(f \circ g)\,|J_g|$ exists. By Theorem 112.2, $g(\mathfrak{D})$ is Jordan-measurable. Since $f$ is continuous, also $\int_{g(\mathfrak{D})} f$ exists.

In order to show that the two integrals are equal, we may assume without loss of generality that

$$f(y) \geqslant 0 \qquad \text{for all } y \in g(\mathfrak{D}) \tag{115.3}$$

for the following reason: If $f_1, f_2 : g(\mathfrak{D}) \to \mathbf{E}$ are defined by

$$f_1(y) = \tfrac{1}{2}(|f(y)| + f(y)), \qquad f_2(y) = \tfrac{1}{2}(|f(y)| - f(y)),$$

then, $f_1(y) \geqslant 0, f_2(y) \geqslant 0, f_1, f_2$ are continuous and bounded, and

$$f(y) = f_1(y) - f_2(y).$$

Since

$$\textstyle\int_{g(\mathfrak{D})} f = \int_{g(\mathfrak{D})} f_1 - \int_{g(\mathfrak{D})} f_2$$

and

$$\textstyle\int_{\mathfrak{D}}(f \circ g)\,|J_g| = \int_{\mathfrak{D}}(f_1 \circ g)\,|J_g| - \int_{\mathfrak{D}}(f_2 \circ g)\,|J_g|,$$

it suffices to prove (115.2) under the restriction (115.3).

Let $\mathfrak{J} \subset \mathbf{E}^n$ denote a closed interval such that $\mathfrak{D} \subseteq \mathfrak{J}$. For any partition $P$ of $\mathfrak{J}$ which generates the intervals $\mathfrak{J}_1, \mathfrak{J}_2, \ldots, \mathfrak{J}_k$, let

$$\beta = \{j \in \{1, 2, \ldots, k\} \,|\, \mathfrak{J}_j \cap \beta\mathfrak{D} \neq \varnothing\},$$
$$\gamma = \{j \in \{1, 2, \ldots, k\} \,|\, \mathfrak{J}_j \subset \mathfrak{D}^\circ\}.$$

By construction, $\beta \cup \gamma \subseteq \{1, 2, \ldots, k\}$ and $\beta \cap \gamma = \varnothing$. We call an interval $\mathfrak{J}_j$ *bad* if $j \in \beta$ and *good* if $j \in \gamma$. $\bigcup_{j \in \beta} \mathfrak{J}_j = B$ is called the *bad set* and $\bigcup_{j \in \gamma} \mathfrak{J}_j = G$ the *good set*.

Clearly,

$$G \subset \mathfrak{D}^\circ \subseteq \mathfrak{D} \subseteq G \cup B.$$

(See also Fig. 115.1 where $B$ is lightly shaded and where $G$ is darkly shaded.)

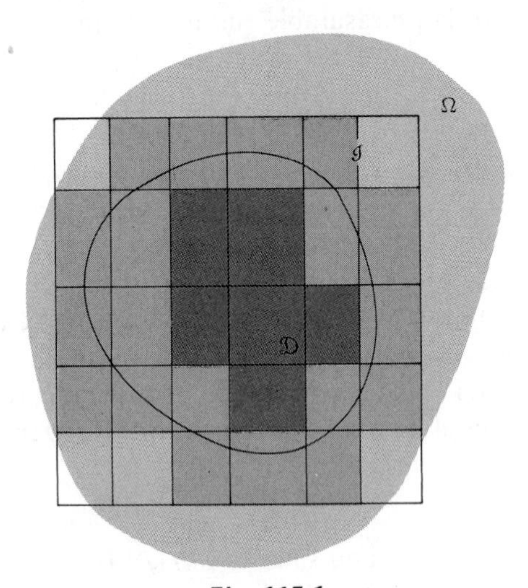


*Fig. 115.1*

For an arbitrary $\varepsilon > 0$, we construct a partition $P$ of $\mathfrak{J}$ which satisfies the following four conditions:

$$\overline{S}((f \circ g)\,|J_g|; P) - \underline{S}((f \circ g)\,|J_g|; P) < \frac{\varepsilon}{4}. \tag{115.4}$$

This is possible because $(f \circ g)\,|J_g|$ is integrable on $\mathfrak{D}$.

$$B \cup G \subset \Omega. \tag{115.5}$$

This is possible because $G \subset \mathfrak{D}^\circ \subset \Omega$ and $\beta\mathfrak{D}$ is compact. (Note that a partition $P$ that satisfies (115.4) may have to be refined to also satisfy (115.5) but that any refinement of $P$ does not affect the validity of (115.4).)

Our third requirement is that

$$J(B) = \sum_{j \in \beta} |\mathfrak{J}_j| < \min\left(\frac{\varepsilon}{16R(2L\sqrt{n})^n}, \frac{\varepsilon}{4RS}\right), \tag{115.6}$$

where $R > 0$ denotes the upper bound of $f$ on $g(\mathfrak{D})$:

$$(115.7) \qquad f(y) \leqslant R \qquad \text{for all } y \in g(\mathfrak{D}),$$

where $S > 0$ denotes a bound of $|J_g|$ on $\mathfrak{D}$:

$$(115.8) \qquad |J_g(x)| \leqslant S \qquad \text{for all } x \in \mathfrak{D},$$

and where $L > 0$ denotes the Lipschitz constant of $g$ on the compact set $B \cup G$. (That $g$ satisfies a uniform Lipschitz condition on $B \cup G$ follows from (115.5) and Lemma 112.3.) (115.6) is possible because $J(\beta\mathfrak{D}) = 0$. (See Lemma 105.1 and note that one may choose the finitely many closed intervals that cover $\beta\mathfrak{D}$ such that the boundary points of $\mathfrak{D}$ are interior points of the covering intervals.) Again, the partition satisfying (115.4) and (115.5) may have to be refined in order to satisfy (115.6) also, but such a refinement does not affect the validity of (115.4) and (115.5).

Finally, we construct $P$—if necessary by further refinement which will not affect the validity of (115.4), (115.5), and (115.6)—such that all *good* intervals are small enough for Lemma 114.2 to apply to each good interval with $(\varepsilon/4T)$ instead of $\varepsilon$ in (114.13), where $T > 0$ is chosen such that

$$(115.9) \qquad (f \circ g)(x)\,|J_g(x)| \leqslant \frac{T}{|\mathfrak{J}|} \qquad \text{for all } x \in \mathfrak{D}.$$

(Note that all good intervals lie in $\mathfrak{D}_0$.) Then

$$(115.10) \qquad \left(1 - \frac{\varepsilon}{4T}\right)|J_g(x)|\,|\mathfrak{J}_j| \leqslant J(g(\mathfrak{J}_j)) \leqslant \left(1 + \frac{\varepsilon}{4T}\right)|J_g(x)|\,|\mathfrak{J}_j|$$

for all $x \in \mathfrak{J}_j$, $j \in \gamma$.

If we define, for any $x_j \in \mathfrak{J}_j$, where $j = 1, 2, \ldots, k$,

$$S((f \circ g)\,|J_g|; P) = \sum_{j=1}^{k} (f \circ g)(x_j)\,|J_g(x_j)|\,|\mathfrak{J}_j|,$$

then the terms with $j \notin \beta \cup \gamma$ will make no contribution if we extend the definition of $f$ to $c(g(\mathfrak{D}))$ by $f(y) = 0$ for all $y \notin g(\mathfrak{D})$. Then $(f \circ g)(x)\,|J_g(x)| = 0$ for all $x \notin \mathfrak{D}$. Hence, we may just as well write

$$(115.11) \qquad S((f \circ g)\,|J_g|; P) = \sum_{j \in \beta \cup \gamma} (f \circ g)(x_j)\,|J_g(x_j)|\,|\mathfrak{J}_j|.$$

In order to show that $\int_{g(\mathfrak{D})} f - \int_{\mathfrak{D}} (f \circ g)\,|J_g| = 0$, we consider

$$(115.12) \qquad \begin{aligned} \left|\int_{\mathfrak{D}} (f \circ g)\,|J_g| - \int_{g(\mathfrak{D})} f\right| &\leqslant \left|\int_{\mathfrak{D}} (f \circ g)\,|J_g| - S((f \circ g)\,|J_g|; P)\right| \\ &+ \left|S((f \circ g)\,|J_g|; P) - \sum_{j \in \gamma} (f \circ g)(x_j)|J_g(x_j)|\,|\mathfrak{J}_j|\right| \\ &+ \left|\sum_{j \in \gamma} (f \circ g)(x_j)\,|J_g(x_j)|\,|\mathfrak{J}_j| - \sum_{j \in \gamma} (f \circ g)(x_j) J(g(\mathfrak{J}_j))\right| \\ &+ \left|\sum_{j \in \gamma} (f \circ g)(x_j) J(g(\mathfrak{J}_j)) - \int_{g(\mathfrak{D})} f\right|. \end{aligned}$$

Note that $g(\mathfrak{J}_j)$ has Jordan content according to Theorem 112.2 because $G \subset \mathfrak{D}^\circ$.

We shall now demonstrate that each of the four terms on the right of (115.12) is less than $(\varepsilon/4)$.

Since $S((f \circ g)|J_g|;P)$ as well as $\int_{\mathfrak{D}}(f \circ g)|J_g|$ lie between $\underline{S}((f \circ g)|J_g|;P)$ and $\bar{S}((f \circ g)|J_g|;P)$, we have from (115.4) that

$$\left|\int_{\mathfrak{D}} (f \circ g)|J_g| - S((f \circ g)|J_g|;P)\right| < \frac{\varepsilon}{4}. \tag{115.13}$$

From (115.11), (115.7), (115.8), and (115.6),

$$\begin{aligned}\left|\sum_{j \in \gamma} (f \circ g)(x_j)|J_g(x_j)|\,|\mathfrak{J}_j| - S((f \circ g)|J_g|;P)\right| &= \left|\sum_{j \in \beta} (f \circ g)(x_j)|J_g(x_j)|\,|\mathfrak{J}_j|\right| \\ &\leqslant RS \sum_{j \in \beta} |\mathfrak{J}_j| < \frac{\varepsilon}{4}.\end{aligned} \tag{115.14}$$

From (115.10) and in view of (115.3),

$$\begin{aligned}\left(1 - \frac{\varepsilon}{4T}\right)\sum_{j \in \gamma} (f \circ g)(x_j)|J_g(x_j)|\,|\mathfrak{J}_j| &\leqslant \sum_{j \in \gamma} (f \circ g)(x_j)\, J(g(\mathfrak{J}_j)) \\ &\leqslant \left(1 + \frac{\varepsilon}{4T}\right)\sum_{j \in \gamma} (f \circ g)(x_j)|J_g(x_j)|\,|\mathfrak{J}_j|,\end{aligned}$$

and from (115.9),

$$\sum_{j \in \gamma} (f \circ g)(x_j)|J_g(x_j)|\,|\mathfrak{J}_j| \leqslant T.$$

Hence,

$$\left|\sum_{j \in \gamma} (f \circ g)(x_j)|J_g(x_j)|\,|\mathfrak{J}_j| - \sum_{j \in \gamma} (f \circ g)(x_j) J(g(\mathfrak{J}_j))\right| < \frac{\varepsilon}{4}. \tag{115.15}$$

(Note that (115.13), (115.14), and (115.15) are valid no matter which $x_j$ we choose in the good intervals $\mathfrak{J}_j$.)

To show that the last term in (115.12) is also less then $(\varepsilon/4)$ is more difficult. The following argument will reveal our reasons for calling $\mathfrak{J}_j$ a *bad* interval when $j \in \beta$.

We have already shown that $g(\mathfrak{D})$ is Jordan-measurable. Since $G \subset \mathfrak{D}^\circ$, we have, from Theorem 112.2, that $g(G)$ is also Jordan-measurable. Since $g(G) \subseteq g(\mathfrak{D})$ it follows from Corollary 2 to Theorem 106.1 that $g(\mathfrak{D})\backslash g(G)$ is also Jordan-measurable and hence, by Theorem 107.4,

$$\int_{g(\mathfrak{D})} f - \int_{g(G)} f = \int_{g(\mathfrak{D})/g(G)} f.$$

Since $\mathfrak{D} \subseteq B \cup G$, we have $g(\mathfrak{D}) \subseteq g(B \cup G) = g(B) \cup g(G)$. Hence, $g(\mathfrak{D})\backslash g(G) \subseteq (g(B) \cup g(G))\backslash g(G) = g(B)\backslash g(G) \subseteq g(B)$. Therefore, we obtain, from (115.7), that (see also exercise 7, section 11.105):

$$\int_{g(\mathfrak{D})\setminus g(G)} f \leqslant RJ(g(\mathfrak{D})\setminus g(G)) \leqslant R\bar{J}(g(\mathfrak{D})\setminus g(G))$$
$$\leqslant R\bar{J}(g(B)).$$

(Note that $g(B)$ need not be Jordan-measurable.) From Lemma 112.2 and from (115.6),

$$\bar{J}(g(B)) < \frac{\varepsilon}{4R},$$

and hence

$$\int_{g(\mathfrak{D})\setminus g(G)} f < \frac{\varepsilon}{4}.$$

Consequently,

$$\left|\int_{g(\mathfrak{D})} f - \int_{g(G)} f\right| < \frac{\varepsilon}{4}. \tag{115.16}$$

We shall now demonstrate that we may choose $x_j \in \mathfrak{J}_j$, $j \in \gamma$ such that

$$\int_{g(G)} f = \sum_{j \in \gamma} (f \circ g)(x_j) J(g(\mathfrak{J}_j)).$$

Since $J(\mathfrak{J}_j \cap \mathfrak{J}_l) = 0$ for $j \neq l$, $j, l \in \beta \cup \gamma$, we have, from Theorem 112.1, that $J(g(\mathfrak{J}_j \cap \mathfrak{J}_l)) = 0$ for $j \neq l$, $j, l \in \beta \cup \gamma$. All good intervals lie in $\mathfrak{D}^\circ$, where $g$ is one-to-one. Hence, for $j, l \in \gamma$,

$$g(\mathfrak{J}_j \cap \mathfrak{J}_l) = g(\mathfrak{J}_j) \cap g(\mathfrak{J}_l)$$

and hence $J(g(\mathfrak{J}_j) \cap g(\mathfrak{J}_l)) = 0$ for $j, l \in \gamma$, $j \neq \gamma$. Since $g(G) = \bigcup_{j \in \gamma} g(\mathfrak{J}_j)$,

$$\int_{g(G)} f = \sum_{j \in \gamma} \int_{g(\mathfrak{J}_j)} f.$$

By Theorem 107.3 (mean-value theorem),

$$\int_{g(\mathfrak{J}_j)} f = f(y_j) J(g(\mathfrak{J}_j))$$

for some $y_j \in g(\mathfrak{J}_j)$, $j \in \gamma$. Since $g$ is one-to-one on $\mathfrak{J}_j \subseteq G$, there is a unique $x_j \in \mathfrak{J}_j$ such that $g(x_j) = y_j$. Hence, $f(y_j) = (f \circ g)(x_j)$, and we have

$$\int_{g(G)} f = \sum_{j \in \gamma} (f \circ g)(x_j) J(g(\mathfrak{J}_j)).$$

This result, together with (115.16) yields

$$\left|\int_{g(\mathfrak{D})} f - \sum_{j \in \gamma} (f \circ g)(x_j) J(g(\mathfrak{J}_j))\right| < \frac{\varepsilon}{4}. \tag{115.17}$$

In view of (115.13), (115.14), (115.15), and (115.17), we obtain from (115.12) that

$$\left|\int_{\mathfrak{D}} (f \circ g)\,|J_g| - \int_{g(\mathfrak{D})} f\right| < \varepsilon.$$

Since $\varepsilon > 0$ is arbitrary, (115.2) follows readily.

(115.1) follows from (115.2) because, by definition, $J(g(\mathfrak{D})) = \int_{g(\mathfrak{D})} \chi_{g(\mathfrak{D})}$. Since $\chi_{g(\mathfrak{D})} \circ g = \chi_{\mathfrak{D}}$, we have $\chi_{\mathfrak{D}}(x)\,|J_g(x)| = |J_g(x)|$ for all $x \in \mathfrak{D}$ and (115.1) obtains.

---

▲ *Example 1*

In example 7, section 4.53, we introduced the Gamma function

$$\Gamma(\alpha) = \int_0^\infty (x^{\alpha-1} e^{-x}\,dx), \qquad \alpha > 0.$$

(See (53.4).) It is now our objective to find $\Gamma(\frac{1}{2})$. For this purpose, we shall utilize Jacobi's theorem (115.1) and the stronger version of Fubini's theorem 108.2. We shall demonstrate that

$$\Gamma\left(\frac{1}{2}\right) = \int_0^\infty \left(\frac{1}{\sqrt{x}} e^{-x}\,dx\right) = \sqrt{\pi}. \tag{115.18}$$

A change of integration variable $\sqrt{x} = u$ yields

$$\int_0^\omega \left(\frac{1}{\sqrt{x}} e^{-x}\,dx\right) = \lim_{\varepsilon\to 0+0} \int_\varepsilon^\omega \frac{1}{\sqrt{x}} e^{-x}\,dx = \lim_{\varepsilon\to 0} 2\int_{\sqrt{\varepsilon}}^{\sqrt{\omega}} e^{-u^2}\,du = 2\int_0^{\sqrt{\omega}} e^{-u^2}\,du,$$

and hence,

$$\Gamma\left(\frac{1}{2}\right) = \int_0^\infty \left(\frac{1}{\sqrt{x}} e^{-x}\,dx\right) = 2\int_0^\infty e^{-u^2}\,du. \tag{115.19}$$

By Fubini's theorem 108.2,

$$\left(\int_0^\omega e^{-u^2}\,du\right)^2 = \int_0^\omega e^{-u^2}\,du \int_0^\omega e^{-v^2}\,dv = \int_0^\omega \left(\int_0^\omega e^{-(u^2+v^2)}\,du\right) dv \tag{115.20}$$

$$= \int_0^\omega \int_0^\omega e^{-(u^2+v^2)}\,du\,dv.$$

We have, from the Corollary to Theorem 107.4 and with reference to Fig. 115.2, that

(115.21)

$$\int_0^{\omega/\sqrt{2}} \int_0^{\omega/\sqrt{2}} e^{-(u^2+v^2)}\,du\,dv \leqslant \iint_{\mathfrak{D}} e^{-(u^2+v^2)}\,du\,dv \leqslant \int_0^\omega \int_0^\omega e^{-(u^2+v^2)}\,du\,dv.$$

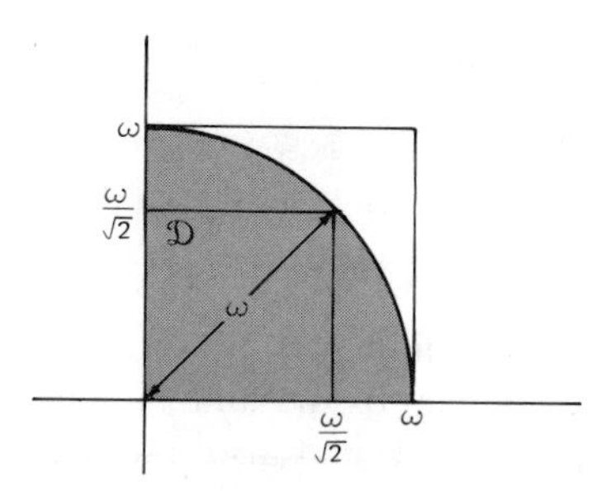


*Fig. 115.2*

We use Jacobi's theorem (115.1) to transform the integral over $\mathfrak{D}$ by means of

$$u = r\cos\theta,$$
$$v = r\sin\theta,$$

with the Jacobian

$$J(r,\theta) = \begin{vmatrix} \cos\theta & \sin\theta \\ -r\sin\theta & r\cos\theta \end{vmatrix} = r$$

into

$$\iint_{\mathfrak{D}} e^{-(u^2+v^2)}\,du\,dv = \int_0^{\pi/2}\int_0^{\omega} re^{-r^2}\,dr\,d\theta.$$

Again using Fubini's theorem and making the change-of-integration variable $r^2 = y$, we obtain

$$\int_0^{\pi/2}\int_0^{\omega} re^{-r^2}\,dr\,d\theta = \frac{\pi}{4}\int_0^{\omega^2} e^{-y}\,dy = \frac{\pi}{4}(1 - e^{-\omega^2}). \tag{115.22}$$

Since $\lim_{\omega\to\infty}\int_0^{\omega/\sqrt{2}} e^{-u^2}\,du = \lim_{\omega\to\infty}\int_0^{\omega} e^{-u^2}\,du$, we obtain from (115.20), (115.21), and (115.22), that

$$\left(\int_0^{\infty} e^{-u^2}\,du\right)^2 = \frac{\pi}{4}.$$

(115.18) follows immediately in view of (115.19). For more illustrations of Jacobi's theorem, see examples 3 and 4, section 11.111.

---

## 12.115 Exercises

1. Let

$$\mathfrak{D} = \{(\xi_1,\xi_2,\xi_3)\in\mathbf{E}^3 \,|\, 0\leqslant\xi_1\leqslant 1, 0\leqslant\xi_2\leqslant 2\pi+\alpha^2, 0\leqslant\xi_3\leqslant\pi+\beta^2, \alpha,\beta\in\mathbf{R}\}.$$

Show that, for all $\alpha,\beta\in\mathbf{R}$,

$$g(\mathfrak{D}) = \{(\eta_1,\eta_2,\eta_3)\in\mathbf{E}^3 \,|\, \eta_1^2+\eta_2^2+\eta_3^3\leqslant 1\},$$

where $g : \mathbf{E}^3 \to \mathbf{E}^3$ is defined by

$$\eta_1 = \xi_1 \cos \xi_2 \sin \xi_3,$$
$$\eta_2 = \xi_1 \sin \xi_2 \sin \xi_3,$$
$$\eta_3 = \xi_1 \cos \xi_3.$$

2. With $\mathfrak{D}$, $g$ as in exercise 1, find $J(\mathfrak{D})$ by (115.1). Note that this represents the Jordan content (volume) of the unit ball in $\mathbf{E}^3$. Find the set of all points where $J_g(\xi_1, \xi_2, \xi_3) = 0$ and also find that subset $A \subseteq \mathfrak{D}$ where $g$ is one-to-one.
3. Show that it is possible to construct a partition $P$ of $\mathfrak{J}$ such that $B \cup G \subset \Omega$, where $B, G, \Omega$ are defined in the proof of Theorem 115.1.
4. Show that it is possible to choose a partition $P$ of $\mathfrak{J}$ such that (115.6) is satisfied.
5. Let $g : \Omega \to \mathbf{E}^n$ satisfy a uniform Lipschitz condition on $A \subset \Omega$ with the constants $L_A > 0, \delta > 0$. Show: If $B \subseteq A$, then $g$ satisfies a uniform Lipschitz condition on $B$ with constant $L_B \leqslant L_A$ and $\delta > 0$.
6. Let $\mathfrak{D} \subset \mathbf{E}^n$ be Jordan-measurable and let $A \subset \mathfrak{D}$ with $J(A) = 0$. Show that $\mathfrak{D} \backslash A$ is Jordan-measurable.
7. Let $\mathfrak{D}$ denote the set in Theorem 115.1 and $A$ the set in exercise 6. If $g: \Omega \to \mathbf{E}^n$, $g \in C^1(\Omega)$, $g : \mathfrak{D}^\circ \xrightarrow[1-1]{} \mathbf{E}^n$ and $J_g(x) \neq 0$ for all $x \in \mathfrak{D}^\circ \backslash A$, show that $g(\mathfrak{D} \backslash A)$ is Jordan-measurable and that $g(\mathfrak{D})$ is Jordan-measurable.
8. Let $\mathfrak{D}' = \mathfrak{D} \backslash A$, $\mathfrak{D}, A$ as in exercise 7. If $f : g(\mathfrak{D}) \to \mathbf{E}$ is bounded and continuous, show that under the conditions stated in exercise 7, $\int_{\mathfrak{D}} (f \circ g)\, |J_g| = \int_{\mathfrak{D}'} (f \circ g)\, |J_g|$ and $\int_{g(\mathfrak{D})} f = \int_{g(\mathfrak{D}')} f$.
9. Show that Theorem 115.1 retains its validity if the hypothesis $J_g(x) \neq 0$ for all $x \in \mathfrak{D}^\circ$ is replaced by the hypothesis $J_g(x) \neq 0$ for all $x \in \mathfrak{D}^\circ \backslash A$, where $J(A) = 0$. (See exercises 6 through 8.)
10. Show that

$$\Gamma\left(\frac{n}{2} + 1\right) = \begin{cases} v! & \text{for } n = 2v \\ (v - \frac{1}{2})(v - \frac{3}{2}) \cdots \frac{1}{2}\sqrt{\pi} & \text{for } n = 2v - 1. \end{cases}$$

(For more exercises see section 11.111.)

## 12.116 Jordan Content as Length, Area, and Volume Measure*

In this section we shall summarize the properties of Jordan-measurable sets and Jordan content in order to demonstrate that the Jordan content is an acceptable length, area, and volume measure for $n = 1, 2, 3$ on the set of one-, two-, and three-dimensional Jordan-measurable sets.

For this purpose we need first the following result which is a direct consequence of Theorem 115.1:

***Corollary to Theorem 115.1***

If $t : \mathbf{E}^n \to \mathbf{E}^n$ represents a translation and if $\mathfrak{D} \subset \mathbf{E}^n$ is Jordan-measurable, then $J(\mathfrak{D}) = J(t(\mathfrak{D}))$, or, as we may put it, the Jordan content is *translation invariant*.

*Proof*

A translation $t: \mathbf{E}^n \to \mathbf{E}^n$ is given by $t(x) = x + b$, where $b \in \mathbf{E}^n$ is a constant vector. Since $J_t(x) = 1$, the corollary follows immediately from (115.1).

We have shown, in the Corollary to Theorem 113.1, that the Jordan content is invariant under orthogonal transformations. Since two sets in $\mathbf{E}^n$ are congruent if and only if they can be mapped into each other by an orthogonal transformation and a translation, we have, from the Corollary to Theorem 113.1 and the above corollary, the following important result:

***Theorem 116.1***

If $\mathfrak{D}_1 \subset \mathbf{E}^n$ is Jordan-measurable and if $\mathfrak{D}_1$ and $\mathfrak{D}_2$ are congruent, then $J(\mathfrak{D}_1) = J(\mathfrak{D}_2)$.

Theorem 116.1 establishes that the collection of Jordan-measurable sets and the Jordan content satisfy conditions $(P_2)$ and $(\mu_2)$ of definition 41.1. We have already seen in section 11.106 that $(P_1), (P_3), (P_4)$, and $(\mu_1), (\mu_3), (\mu_4)$ are satisfied. Since definition 41.1 applies, when appropriately worded, to length measure and volume measure as well, we see that the concept of Jordan content provides for $n = 1$ an acceptable length measure, for $n = 2$ an acceptable area measure, and for $n = 3$ an acceptable volume measure. Because of its importance, we summarize the pertinent properties of Jordan-measurable sets and of the Jordan content in the following theorem.

***Theorem 116.2***

Let $\mathfrak{J}$ denote the collection of all Jordan-measurable sets in $\mathbf{E}^n$.

$(\mathfrak{J}_1)$ $\mathfrak{C}_a = \{x \in \mathbf{E}^n \mid |\xi_j| \leqslant a/2,\ j = 1, 2, \ldots, n; a \geqslant 0\} \in \mathfrak{J}$,
$(\mathfrak{J}_2)$ If $\mathfrak{D}_1 \in \mathfrak{J}$ and $\mathfrak{D}_1 \sim \mathfrak{D}_2$ (congruent), then $\mathfrak{D}_2 \in \mathfrak{J}$,
$(\mathfrak{J}_3)$ If $\mathfrak{D}_1, \mathfrak{D}_2 \in \mathfrak{J}, \mathfrak{D}_1^\circ \cap \mathfrak{D}_2^\circ = \varnothing$, then $\mathfrak{D}_1 \cup \mathfrak{D}_2 \in \mathfrak{J}$,
$(\mathfrak{J}_4)$ If $\mathfrak{D}_1, \mathfrak{D}_2 \in \mathfrak{J}, \mathfrak{D}_1 \subseteq \mathfrak{D}_2$, then $\mathfrak{D}_2 \backslash \mathfrak{D}_1 \in \mathfrak{J}$;
$(\mu_1)$ $J(\mathfrak{D}) \geqslant 0$ for all $\mathfrak{D} \in \mathfrak{J}$,
$(\mu_2)$ If $\mathfrak{D}_1, \mathfrak{D}_2 \in \mathfrak{J}$ and $\mathfrak{D}_1 \sim \mathfrak{D}_2$, then $J(\mathfrak{D}_1) = J(\mathfrak{D}_2)$,
$(\mu_3)$ If $\mathfrak{D}_1, \mathfrak{D}_2 \in \mathfrak{J}$ and $\mathfrak{D}_1^\circ \cap \mathfrak{D}_2^\circ = \varnothing$, then $J(\mathfrak{D}_1 \cup \mathfrak{D}_2) = J(\mathfrak{D}_1) + J(\mathfrak{D}_2)$,
$(\mu_4)$ $J(\mathfrak{C}_1) = 1$.

---

▲ *Example 1* (*Jordan Content of the n-Dimensional Ball*)

We shall now compute the Jordan content of an $n$-dimensional ball by an application of Theorem 115.1. In view of Theorem 116.1 we may assume w.l.o.g. that the ball's center is at $\theta$.

The $n$-dimensional ball $B_\rho^n$ with center at $\theta$ and radius $\rho > 0$ is defined by

$$B_\rho^n = \{y \in \mathbf{E}^n \mid \|y\| \leqslant \rho\}.$$

We shall consider a function $g$ that maps an interval onto $B_\rho^n$ and satisfies all the conditions that were stated in Theorem 115.1, and then evaluate $J(B_\rho^n)$ by utilizing formula (115.1).

The function $g : \mathbf{E}^n \to \mathbf{E}^n$, defined by

$$\begin{aligned} \eta_1 &= \rho \cos \xi_1, \\ \eta_2 &= \rho \sin \xi_1 \cos \xi_2, \\ \eta_3 &= \rho \sin \xi_1 \sin \xi_2 \cos \xi_3, \\ &\vdots \\ \eta_n &= \rho \sin \xi_1 \sin \xi_2 \ldots \sin \xi_{n-1} \cos \xi_n, \end{aligned} \tag{116.1}$$

clearly satisfies $g \in C^1(\mathbf{E}^n)$. We have

$$\begin{aligned} \|g(x)\|^2 &= \eta_1^2 + \eta_2^2 + \cdots + \eta_n^2 \\ &= \rho^2 \cos^2 \xi_1 + \rho^2 \sin^2 \xi_1(\cos^2 \xi_2 + \sin^2 \xi_2(\cos^2 \xi_3 + \cdots \\ &\quad \cdots + \sin^2 \xi_{n-2}(\cos^2 \xi_{n-1} + \sin^2 \xi_{n-1} \cos^2 \xi_n)\ldots)). \end{aligned}$$

Hence,

$$\|g(x)\|^2 \leqslant \rho^2$$

where equality occurs only when $\xi_j = 0$ or $\pi$ for some $j \in \{1, 2, \ldots, n\}$. Therefore,

$$g : (0, \pi)^n \to (B_\rho^n)^\circ$$

where $(0, \pi)^n = (0, \pi) \times (0, \pi) \times \cdots \times (0, \pi)$. (See also exercises 1 and 2.)

Let $y \in (B_\rho^n)^\circ$; that is, $\|y\| < \rho$. Then $\eta_1^2 < \rho^2, \eta_1^2 + \eta_2^2 < \rho^2, \eta_1^2 + \eta_2^2 + \eta_3^2 < \rho^2, \ldots$. Hence, $(\eta_1/\rho) = \cos \xi_1$ has a unique solution $\xi_1 \in (0, \pi)$ and infinitely many solutions outside $(0, \pi)$. Then $\sin \xi_1 = (1/\rho)\sqrt{\rho^2 - \eta_1^2}$ and we see that $\eta_2/\rho = (1/\rho)\sqrt{\rho^2 - \eta_1^2} \cos \xi_2$ has a unique solution $\xi_2 \in (0, \pi)$—and infinitely many solutions elsewhere. We obtain

$$\sin \xi_2 = \sqrt{\frac{\rho^2 - \eta_1^2 - \eta_2^2}{\rho^2 - \eta_1^2}},$$

find a unique solution $\xi_3 \in (0, \pi)$ of

$$\frac{\eta_3}{\rho} = \frac{1}{\rho}\sqrt{\rho^2 - \eta_1^2}\sqrt{\frac{\rho^2 - \eta_1^2 - \eta_2^2}{\rho^2 - \eta_1^2}} \cos \xi_3, \ldots.$$

If we continue this process, we arrive at the result

$$g : (0, \pi)^n \leftrightarrow (B_\rho^n)^\circ.$$

Let $y \in \beta B_\rho^n$, that is, $\|y\| = \rho$. Either $\eta_j = \pm\rho$ for some $j \in \{1, 2, \ldots, n\}$ and $\eta_1 = \eta_2 = \cdots = \eta_{j-1} = \eta_{j+1} = \cdots = \eta_n = 0$, or $|\eta_j| < \rho$ for all $j = 1, 2, \ldots, n$. In the latter case, we proceed as before and obtain unique solutions $\xi_j \in [0, \pi]$, $j = 1, 2, \ldots, n$, and infinitely many solutions elsewhere. If $\eta_j = \pm\rho$ and all other $\eta_k$ vanish, then we obtain the solutions $\xi_k = \pi/2$ for $k < j$, $\xi_j = 0$ or $\pi$, and $\xi_l$, $l > j$, arbitrary. (See also exercises 3 and 4.)

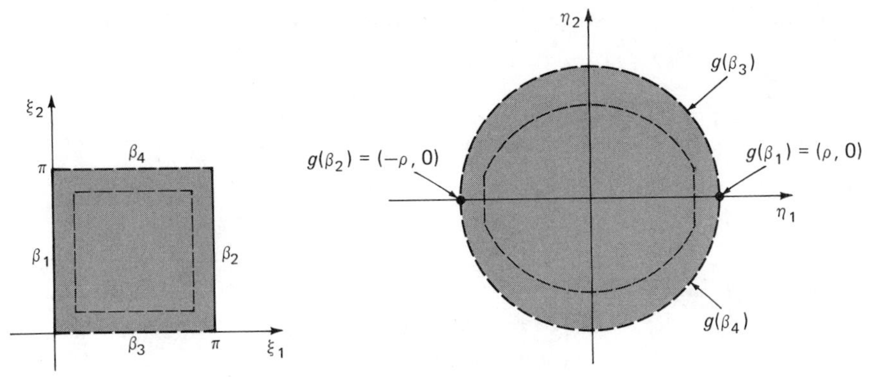


*Fig. 116.1*

Hence,

$$g : [0, \pi]^n \xrightarrow{\text{onto}} B_\rho^n,$$

where $[0, \pi]^n = [0, \pi] \times [0, \pi] \times \cdots \times [0, \pi]$.

In Fig. 116.1 we have depicted this map for the case $n = 2$. Although the boundary of $[0, \pi]^2$ is mapped onto the boundary of $B_\rho^2$, the mapping is not one-to-one. The entire left side $\beta_1$ of $[0, \pi]^2$ is mapped into the point $(\rho, 0)$, the entire right side $\beta_2$ is mapped into the point $(-\rho, 0)$, while the bottom side $\beta_3$ is mapped one-to-one onto the upper semicircle and the top side $\beta_4$ is mapped one-to-one onto the lower semicircle. The interior of $[0, \pi]^2$ is mapped one-to-one onto the interior of the disk $B_\rho^2$. For example, the dotted line in $(0, \pi)^2$ is mapped one-to-one onto the dotted line inside the disk $B_\rho^2$. (Compare Fig. 116.1 with Fig. 111.2.)

Next, we check the Jacobian: We obtain by a straightforward calculation, which is especially simple because the Jacobian of $g$ is triangular, that

$$J_g(x) = (-1)^n \rho^n \sin^n \xi_1 \sin^{n-1} \xi_2 \sin^{n-2} \xi_3 \ldots \sin^2 \xi_{n-1} \sin \xi_n, \tag{116.2}$$

and we see that

$$\begin{aligned} J_g(x) &\neq 0 \qquad \text{for all } x \in (0, \pi)^n, \\ J_g(x) &= 0 \qquad \text{if } \xi_j = 0 \text{ or } \pi \text{ for some } j \in \{1, 2, \ldots, n\}. \end{aligned}$$

Summarizing, we can state that $g \in C^1(\mathbf{E}^n)$, that $\mathfrak{D} = [0, \pi]^n$ is Jordan-measurable, $g : \mathfrak{D}^\circ \xrightarrow[1-1]{} \mathbf{E}^n$, $J_g(x) \neq 0$ for all $x \in \mathfrak{D}^\circ$, and $g([0, \pi]^n) = B_\rho^n$. Hence, (115.1) is applicable, and we obtain:

$$J(B_\rho^n) = \int_{[0, \pi]^n} |J_g(x)| = \rho^n \int_0^\pi \sin^n \xi_1 \, d\xi_1 \int_0^\pi \sin^{n-1} \xi_2 \, d\xi_2 \ldots \int_0^\pi \sin \xi_n \, d\xi_n.$$

(See also exercise 8, section 11.108.) We have

$$\int_0^\pi \sin x \, dx = 2, \qquad \int_0^\pi \sin^2 x \, dx = \frac{\pi}{2},$$

and we obtain, from two successive integrations by parts (Corollary 1 to Theorem 52.3),

$$\int_0^\pi \sin^k x\,dx = \frac{k-1}{k}\int_0^\pi \sin^{k-2} x\,dx \qquad \text{for all integers } k \geqslant 2.$$

Hence,

$$(116.3) \qquad \int_0^\pi \sin^k x\,dx = \begin{cases} \dfrac{(k-1)(k-3)\dots 4.2}{k(k-2)\dots 5.3}\,2 & \text{if } k \text{ is odd} \\[2ex] \dfrac{(k-1)(k-3)\dots 5.3}{k(k-2)\dots 4.2}\,\pi & \text{if } k \text{ is even.} \end{cases}$$

(See also exercise 9, section 4.52.) Therefore,

$$J(B_\rho^n) = \begin{cases} \rho^n \dfrac{2^{(n+1)/2}\pi^{(n-1)/2}}{1.3\dots(n-2)n} & \text{if } n \text{ is odd,} \\[2ex] \rho^n \dfrac{2^{n/2}\pi^{n/2}}{2.4\dots(n-2)n} & \text{if } n \text{ is even,} \end{cases}$$

or, as we may write it,

$$(116.4) \qquad J(B_\rho^n) = \begin{cases} \dfrac{\rho(\rho^2\pi)^{\nu-1}}{(\nu-\frac{1}{2})(\nu-\frac{3}{2})\dots(\frac{3}{2})(\frac{1}{2})} & \text{if } n = 2\nu - 1, \\[2ex] \dfrac{(\rho^2\pi)^\nu}{\nu!} & \text{if } n = 2\nu. \end{cases}$$

In terms of the *Gamma function* we may write this result in the much simpler form

$$J(B_\rho^n) = \frac{\rho^n\pi^{n/2}}{\Gamma(n/2+1)}$$

(See exercise 10, section 12.115).

(116.4) yields the well-known formulas for the length of the interval $B_\rho^1 = [-\rho, \rho]$ for $n = 1$:

$$J(B_\rho^1) = 2\rho,$$

for the area of the disk $B_\rho^2 = \{(x, y) \in \mathbf{E}^2 \mid x^2 + y^2 \leqslant \rho^2\}$ for $n = 2$:

$$J(B_\rho^2) = \rho^2\pi,$$

and for the volume of the three-dimensional ball $B_\rho^3 = \{(x, y, z) \in \mathbf{E}^3 \mid x^2 + y^2 + z^2 \leqslant 1\}$ for $n = 3$:

$$J(B_\rho^3) = \frac{4\rho^3\pi}{3}.$$

(116.4) also yields the most remarkable result that

$$\lim_{\infty} J(B_\rho^n) = 0.$$

(See also exercise 9.)

Any intuitive explanation of this phenomenon is just as good as any other and we shall not even make an attempt at confusing the issue by giving an explanation. We seek comfort in considering the Jordan contents of the inscribed cube $\gamma^n$ with dimension $(2\rho/\sqrt{n})$ and the circumscribed cube $\Gamma^n$ with dimension $2\rho$. We obtain

$$J(\gamma^n) = \left(\frac{2\rho}{\sqrt{n}}\right)^n, \qquad J(\Gamma^n) = (2\rho)^n.$$

Hence,

$$\lim_{\infty} J(\gamma^n) = 0$$

and

$$\lim_{\infty} J(\Gamma^n) = \begin{cases} 0 & \text{if } 0 < \rho < \frac{1}{2}, \\ 1 & \text{if } \rho = \frac{1}{2}, \\ \infty & \text{if } \rho > \frac{1}{2}. \end{cases}$$

---

## 12.116 Exercises

1. Show, by induction, that $\|g(x)\|^2 = \rho^2$ if $\xi_j = 0$ or $\pi$ for some $j \in \{1, 2, \ldots, n\}$, where $g$ is defined in (116.1).
2. Show that $\|g(x)\|^2 < \rho^2$ for all $x \in (0, \pi)^n$. (See also exercise 1.)
3. Show, if $\|y\| = \rho$, $|\eta_i| < \rho$, then there is *exactly one* $x \in [0, \pi]^n$ such that $g(x) = y$. $g$ is defined in (116.1).
4. Show that there is exactly one $x \in [0, \pi]^n$ such that $g(x) = (0, 0, \ldots, 0, \rho)$.
5. Check the validity of (116.2) by writing down the Jacobian in some detail.
6. Check the validity of (116.3).
7. Check the validity of (116.4).
8. Show that $\lim_\infty (a^k/k!) = 0$ where $a \in \mathbf{R}, a > 0$.
9. Show that

$$\lim_{\infty} \frac{a^k}{(k + \frac{1}{2})(k - \frac{1}{2}) \ldots (\frac{3}{2})(\frac{1}{2})} = 0.$$

10. Show that

$$\left(\frac{2\rho}{\sqrt{2v}}\right)^{2v} \leqslant \frac{(\rho^2\pi)^v}{v!}.$$

11. Consider the function $g$, as defined in (116.1) for $n = 3$, and find a representation of $g^{-1}$ on $(B_\rho^3)^\circ = \{(\eta_1, \eta_2, \eta_3) \in \mathbf{E}^3 \mid \eta_1^2 + \eta_2^2 + \eta_3^2 < \rho^2\}$.
12. Find $J(B_1^4)$, $J(B_1^5)$, $J(B_1^6)$, and $J(B_1^7)$, and show that

$$J(B_1^1) < J(B_1^2) < J(B_1^3) < J(B_1^4) < J(B_1^5) < J(B_1^6) > J(B_1^7).$$

# 13 Line and Surface Integrals

## 13.117 Introduction

In the preceding two chapters we have spent considerable time and effort to develop the theory of Jordon content and techniques for the computation of the Jordan content (length, area, volume) of objects with funny shapes. Fubini's theorem and Jacobi's theorem are the major results that were achieved pertaining to the more practical side of our endeavor.

Our results may now, in turn, be applied to simple problems in mechanics and fluid dynamics such as finding mass, moment, and centroid of given line segments, or regions, or solids of given mass density, or of computing the work to be expended by a (variable) force to push a mass-point along a straight line. In the case of a steady flow of an incompressible fluid, we may compute the amount of matter passing through a given line segment in case of a two-dimensional flow, or through a given planar region in case of a three-dimensional flow. Some of these applications will be discussed in the following three examples; others are left for the exercises.

---

▲ *Example 1* (*Mass, Moment, and Centroid of a Straight Rod*)

We consider a straight rod on the $x$-axis that extends from $a$ to $b$ and has mass density (mass per unit length) $f(x)$ at $x$, where we assume that $f\colon [a,b] \to \mathbf{E}$ is continuous. If $P = \{x_0, x_1, \ldots, x_n\}$, $x_0 = a$, $x_n = b$, is a partition of $[a,b]$, then the mass of $[x_{k-1}, x_k]$ is approximately equal to $f(\xi_k)(x_k - x_{k-1})$, where $\xi_k \in [x_{k-1}, x_k]$. Hence, the mass of the entire rod is approximately equal to $\sum_{k=1}^{n} f(\xi_k)\Delta x_k$ which is a typical intermediate sum for the function $f$ on $[a,b]$. This heuristic argument leads to

$$m = \int_a^b f(x)\,dx$$

as definition of the *mass of the rod from a to b with mass density* $f(x)$.

The *moment* of a mass-point of mass $m$ at the location $P$ with respect to a point $Q$ is defined as the product of $m$ with the distance from $P$ to $Q$. By an argument similar to the one that led to the definition of mass, we are led to

$$M = \int_a^b xf(x)\,dx$$

as definition of the *moment $M$ of the rod from $a$ to $b$ with mass density $f(x)$ with respect to the point $x = 0$.*

The $x$-coordinate of the *centroid* of the rod from $a$ to $b$, the point with respect to which the *moment of the rod is zero*, is then given by

$$\bar{x} = \frac{M}{m} = \frac{\int_a^b xf(x)\,dx}{\int_a^b f(x)\,dx}.$$

(See also exercise 1.)

▲ *Example 2* (*Work*)

If a constant force $f$ pushes a unit mass along a straight line through a distance $d$, where $f$ is applied in the direction of the motion, then $W = fd$ is called the *work* expended by $f$ to accomplish this task. If the force varies continuously from point to point in magnitude and direction, i.e., if $f$ is a continuous vector-valued function $f: \mathbf{E} \to \mathbf{E}^2$, then we arrive at the definition of the work to be expended by $f$ by a heuristic argument, as in example 1. If a unit mass is to be pushed from $x = a$ to $x = b$, we consider a partition $P$ of $[a, b]$ and argue that the work to be expended from $x_{k-1}$ to $x_k$ is approximately equal to $f(\xi_k) \cdot e_1(x_k - x_{k-1})$ where $\xi_k \in [x_{k-1}, x_k]$ and where $e_1 = (1, 0)$ is the unit vector in the direction of motion. (Note that $f(\xi_k) \cdot e_1$ is the force component acting at $\xi_k$ in the direction of motion—see Fig. 117.1.) This leads to

$$W = \int_a^b f(x) \cdot e_1\,dx$$

as the definition of *work to be expended by $f$ to push a unit mass from $x = a$ to $x = b$.*

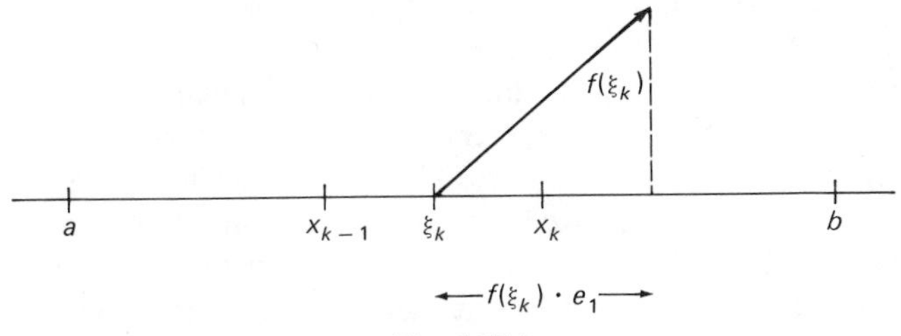


*Fig. 117.1*

▲ *Example 3* (*Steady Flow of an Incompressible Fluid*)

If a fluid is distributed over a plane or through space with constant mass density (mass per unit area or per unit volume), say 1, and moves in such a manner that the velocity at each point is independent of the time, one speaks of a *steady* two-dimensional or three-dimensional *flow of an incompressible fluid.* Such a flow is defined by the velocity vector $v(x)$ at each point $x$.

We consider a three-dimensional flow, defined by the continuous vector field $v : \mathbf{E}^3 \to \mathbf{E}^3$. To find the amount of matter that passes in unit time through a Jordan-measurable set $A$ in the $(\xi_1, \xi_2)$-plane, we consider a partition $P$ of $A$ (see Fig. 117.2), and argue that the total amount passing through $\mathfrak{J}_{ij}$ in unit time

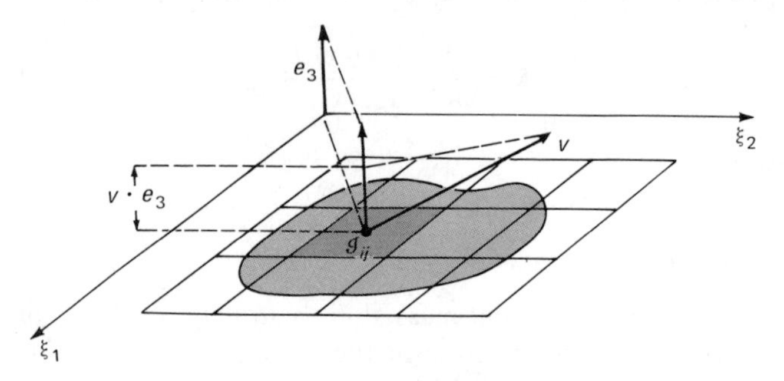


*Fig. 117.2*

in the direction of the upward normal $e_3 = (0, 0, 1)$ is approximately equal to $v(\xi_1^{(i)}, \xi_2^{(j)}, 0) \cdot e_3 \, |\mathfrak{J}_{ij}|$ where $(\xi_1^{(i)}, \xi_2^{(j)}, 0) \in \mathfrak{J}_{ij}$ and where $v(\xi_1^{(i)}, \xi_2^{(j)}, 0) \cdot e_3$ is the velocity component in the direction of $e_3$. Hence, we are led to the following definition of the *flow through A in unit time* in the direction of the upward normal:

$$F = \int_A \int v(\xi_1, \xi_2, 0) \cdot e_3 \, d\xi_1 \, d\xi_2.$$

---

These three examples raise more questions than they answer. What about the mass, moment, centroid and, for that matter, length of a bent rod (curve), or mass, moment, centroid, and area of a curved surface? What about the two-dimensional flow through a curve, or the three-dimensional flow through a curved surface? All these questions are important and legitimate questions in mechanics and fluid dynamics, but cannot be answered within the framework of the theory that we have developed up to this point.

In this chapter we shall literally bend intervals and planar regions and develop a theory of integration on such bent intervals (curves) and bent regions (surfaces), the *theory of line and surface integrals.* This, in turn, will enable us to supply answers to the questions which we have raised above, and to others. The theory we are about to discuss was principally inspired by practical problems and, at least in its early stages, developed by scientists and engineers. It has always been an important and legitimate part of mathematical analysis.

The immediately following seven sections will be devoted to a study of curves and surfaces and geometric aspects such as tangents and tangent planes, and the remaining sections will deal with integration on curves and surfaces. The theoretical background for this study is embodied in Chapters 10, 11, and 12, with the mapping theorems and the inverse function theorem of Chapter 10, Fubini's theorem of Chapter 11, and Jacobi's theorem of Chapter 12 playing a major role in the forthcoming developments. The applications are kept at a simple level and apart from the theory, so that the (hopefully rare) reader who is mystified by such mundane concepts as mass, work, flow, etc., is not unduly handicapped.

## 13.117 Exercises

1. Show that $\int_a^b (x - \bar{x})f(x)\,dx = 0$, where $\bar{x}$ is the $x$-coordinate of the centroid of the rod with mass density $f(x)$ from $a$ to $b$.
2. Find the centroid of a rod with mass density $f(x) = x$ from $x = 2$ to $x = 4$.
3. A Jordan-measurable set $A \subset \mathbf{E}^2$ has mass density $f(x)$, where $f\colon A \to \mathbf{E}$. Define the mass of $A$ in a reasonable manner.
4. The moment of a mass-point of mass $m$ in $\mathbf{E}^2$ with respect to a given line $l$ in $\mathbf{E}^2$ is defined as the product of $m$ with its perpendicular distance from $l$. Define the moment of a Jordan-measurable set $A \subset \mathbf{E}^2$ of mass density $f(x)$ with respect to $x$-axis and $y$-axis.
5. The centroid of a planar region of given mass density is the intersection of those two lines parallel to the $x$- and $y$-axes with respect to which the moments of the region are zero. Give a mathematical definition of the centroid of a planar region.
6. A two-dimensional flow of an incompressible fluid of mass density 1 is defined by $v : \mathbf{E}^2 \to \mathbf{E}^2$ where $v(x) = (\xi_1, \xi_2)$. How much matter passes in unit time through the line segment from $(1, 1)$ to $(2, 1)$?
7. How much work does the force $f(x) = (x, x^2)$ expend to push a unit mass from $x = 1$ to $x = 2$?
8. A three-dimensional flow of an incompressible fluid of constant mass density 1 is defined by $v : \mathbf{E}^3 \to \mathbf{E}^3$, where $v(x) = (\xi_1, \xi_2, \xi_3)$. How much matter passes through the disk $\xi_1^2 + \xi_2^2 \leqslant 1$, $\xi_3 = 1$, in unit time?

## 13.118 Curves

The reader has learned in calculus that

$$\left.\begin{aligned} x &= \phi(t), \\ y &= \psi(t), \end{aligned}\right\} \qquad t \in \mathfrak{J},$$

where $\mathfrak{J}$ is an interval (open or closed or neither, no matter) and where $\phi, \psi : \mathfrak{J} \to \mathbf{E}$ are continuous functions, represents a curve in the plane.

In this treatment we shall take the view that a curve exists quite apart from its representation or representations, and give the following definition:

***Definition 118.1***

$\Gamma \subset \mathbf{E}^n$ is a curve in $\mathbf{E}^n$ if and only if there exists an interval $\mathfrak{J} \subseteq \mathbf{E}$ and a continuous function $f : \mathfrak{J} \to \mathbf{E}^n$ such that $f_*(\mathfrak{J}) = \Gamma$; that is, $f$ maps $\mathfrak{J}$ onto $\Gamma$.

If $n = 2$ we speak of a *plane curve*, and if $n = 3$ we speak of a *space curve*.

$f : \mathfrak{J} \xrightarrow{\text{onto}} \Gamma$ is called a *parametrization* of $\Gamma$ and

$$x = f(t), \qquad t \in \mathfrak{J},$$

or, in components,

$$\left.\begin{aligned} \xi_1 &= \phi_1(t), \\ \xi_2 &= \phi_2(t), \\ &\vdots \\ \xi_n &= \phi_n(t), \end{aligned}\right\} \quad t \in \mathfrak{J}$$

is called a *parameter representation* of $\Gamma$. $t$ is called a *parameter*.

---

▲ *Example 1*

If $g : \mathfrak{J} \to \mathbf{E}$ is continuous, then $\{(x, g(x)) \mid x \in \mathfrak{J}\}$ is a plane curve.

$$\left.\begin{aligned} \xi_1 &= x, \\ \xi_2 &= g(x), \end{aligned}\right\} \quad x \in \mathfrak{J},$$

is a parameter representation of this curve. (See Fig. 118.1.)

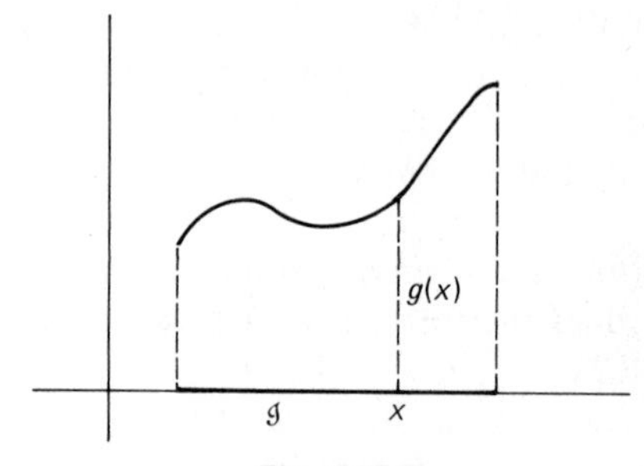


*Fig. 118.1*

▲ *Example 2*

The unit circle $C = \{x \in \mathbf{E}^2 \mid \|x\| = 1\}$ is a plane curve. If $f : \mathbf{E} \to \mathbf{E}^2$ is defined by

$$f(t) = (\cos t, \sin t), \tag{118.1}$$

then,

$$x = f(t), \qquad t \in (-\infty, \infty),$$

or, in components,

$$\left.\begin{aligned} \xi_1 &= \cos t, \\ \xi_2 &= \sin t, \end{aligned}\right\} \quad t \in (-\infty, \infty),$$

as well as

$$\left.\begin{aligned}\xi_1 &= \cos t,\\ \xi_2 &= \sin t,\end{aligned}\right\} \qquad t \in [0, 2\pi),$$

are parameter representations of $C$. (See Fig. 118.2.) If

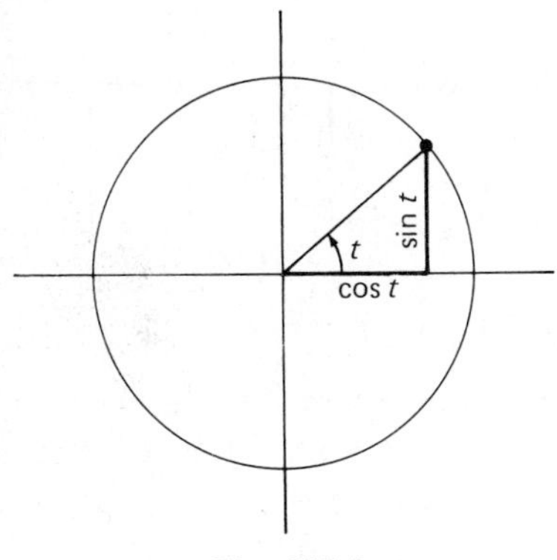


Fig. 118.2

(118.2) $$g(s) = (\sin s, \cos s),$$

then

$$\left.\begin{aligned}\xi_1 &= \sin s,\\ \xi_2 &= \cos s,\end{aligned}\right\} \qquad s \in [-\pi, \pi),$$

is also a parameter representation of $C$.

▲ *Example 3*

The boundary $\Gamma$ of the square $[0, 1] \times [0, 1]$ in Fig. 118.3 is a plane curve. In order to check this out on the basis of definition 118.1, we cut $\Gamma$ at $(0, 0)$, lay it out flat, as indicated in Fig. 118.4, and label the points that correspond to the vertices by 0, 1, 2, 3, 4.

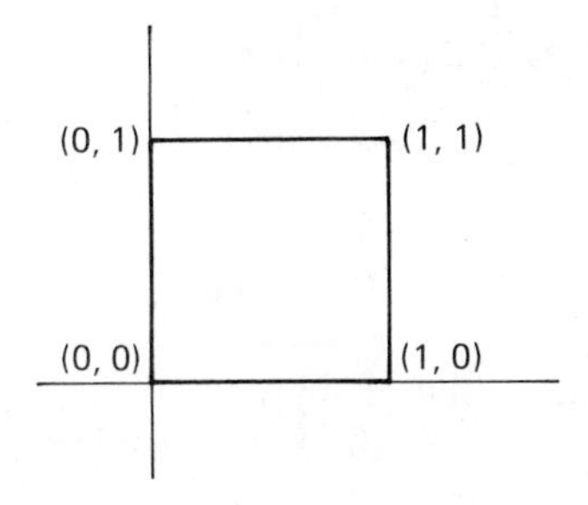


Fig. 118.3

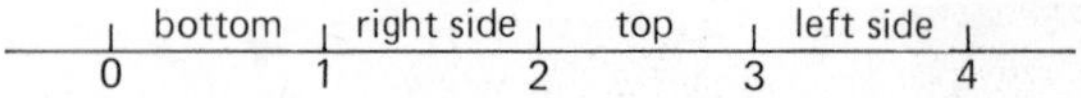

bottom | right side | top | left side
0 1 2 3 4

*Fig. 118.4*

We choose $\mathfrak{J} = [0, 4]$ and construct a parametrization of $\Gamma$ by mapping the points of $\mathfrak{J}$ into their ancestors on $\Gamma$. This yields a parametrization $f = (\phi_1, \phi_2)$ : $[0, 4] \xrightarrow{\text{onto}} \Gamma$, where

$$\phi_1(t) = \begin{Bmatrix} t & \text{for } 0 \leqslant t < 1 \\ 1 & \text{for } 1 \leqslant t < 2 \\ -t+3 & \text{for } 2 \leqslant t < 3 \\ 0 & \text{for } 3 \leqslant t \leqslant 4 \end{Bmatrix}, \qquad \phi_2(t) = \begin{Bmatrix} 0 & \text{for } 0 \leqslant t < 1 \\ t & \text{for } 1 \leqslant t < 2 \\ 1 & \text{for } 2 \leqslant t < 3 \\ -t+4 & \text{for } 3 \leqslant t \leqslant 4 \end{Bmatrix}.$$

▲ *Example 4* (*Cycloid*)

If a disk of radius $r$ rolls along a straight line, then a point $P$ that is rigidly attached to the disk at a distance $a$ from the disk's center traverses a *cycloid.* (See Fig. 118.5.)

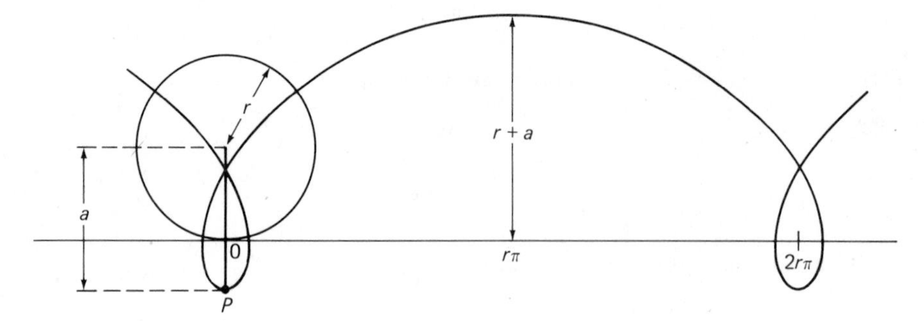


*Fig. 118.5*

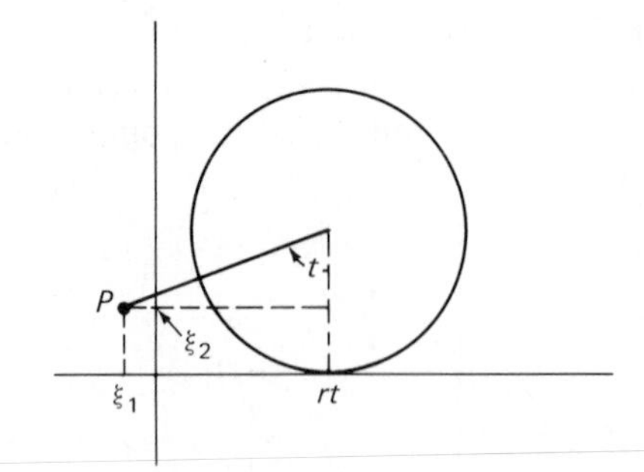


*Fig. 118.6*

If we choose the angle $t$ that is marked in Fig. 118.6 and is measured in radians as a parameter, we obtain the following parameter representation of the cycloid:

$$\left.\begin{aligned}\xi_1 &= rt - a \sin t,\\ \xi_2 &= \ \ r - a \cos t,\end{aligned}\right\} \qquad t \in (-\infty, \infty).$$

▲ *Example 5* (*Helix*)

A curve that winds itself around a right circular cylinder of radius $r$ rising at a constant rate of $a$ units per thread is called a helix. (See Fig. 118.7.)

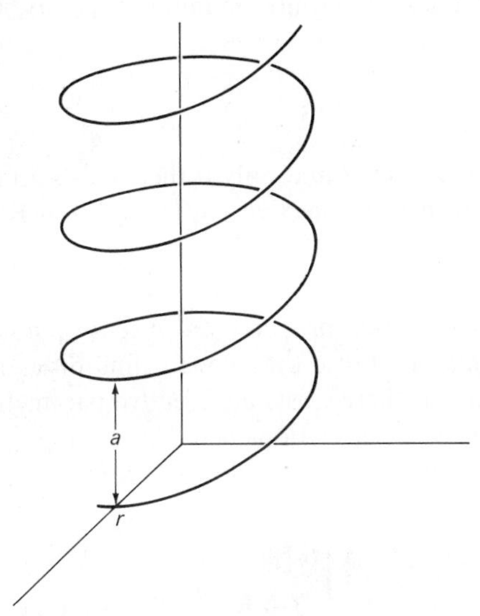


*Fig. 118.7*

Clearly,

$$\left.\begin{aligned}\xi_1 &= r \cos t,\\ \xi_2 &= r \sin t,\\ \xi_3 &= \frac{a}{2\pi} t,\end{aligned}\right\} \qquad t \in (-\infty, \infty),$$

is a parameter representation of the helix.

▲ *Example 6*

Schoenberg's space filling curve in example 1, section 12.112, is a plane curve.

---

There is nothing sacred about any particular parameter representation of a given curve. In example 2 we did find, and in the other examples we could have found, other parametrizations that would have served just as well.

***Definition 118.2***

Two parametrizations $f : \mathfrak{J} \to \mathbf{E}^n$, $g : \mathfrak{F} \to \mathbf{E}^n$ are called *equivalent* if and only if $f_*(\mathfrak{J}) = g_*(\mathfrak{F})$; i.e., they represent the same curve.

A simple relationship exists between equivalent parametrizations of *simple curves*. All curves which we have discussed in the preceding examples, with the exception of the cycloid for $a > r$ (example 4), are simple curves. If $a > r$, then the cycloid crosses itself periodically (see Fig. 118.5) in what we call *double points*. Curves without double points or multiple points, for that matter, are called simple curves.

***Definition 118.3***

$\Gamma \subset \mathbf{E}^n$ is a simple curve in $\mathbf{E}^n$ if and only if there exists an interval $\mathfrak{J} \subseteq \mathbf{E}$ and a continuous one-to-one function (injective map) $f : \mathfrak{J} \xrightarrow[1-1]{} \mathbf{E}^n$ such that $f_*(\mathfrak{J}) = \Gamma$, that is, $f : \mathfrak{J} \leftrightarrow \Gamma$.

We call a function with this property an *injective parametrization* of $\Gamma$. In this terminology we may rephrase the above definition as follows: $\Gamma \subset \mathbf{E}^n$ is a simple curve if and only if there exists an injective parametrization of $\Gamma$.

Equivalent injective parametrizations stand in a simple relationship to each other, as we shall demonstrate in the following lemma:

***Lemma 118.1***

Two injective parametrizations $f : \mathfrak{J} \to \mathbf{E}^n$, $g : \mathfrak{F} \to \mathbf{E}^n$ are equivalent if and only if there exists a unique one-to-one function (bijective map) $p : \mathfrak{J} \leftrightarrow \mathfrak{F}$ such that $f = g \circ p$ on $\mathfrak{J}$.

$p$ is called a *bijective parameter transformation.*

*Proof*

(a) Suppose $p : \mathfrak{J} \leftrightarrow \mathfrak{F}$ exists. Then, $f = g \circ p : \mathfrak{J} \xrightarrow[1-1]{} \mathbf{E}^n$ and $f_*(\mathfrak{J}) = (g \circ p)_*(\mathfrak{J}) = g_*(p_*(\mathfrak{J})) = g_*(\mathfrak{F})$; that is, $f, g$ are equivalent injective parametrizations.

(b) Suppose $f, g$ are equivalent injective parametrizations, and let $p = g^{-1} \circ f$. Then $p$ is injective because $g^{-1}$, $f$ are injective, and $p$ is surjective because $p_*(\mathfrak{J}) = g_*^{-1}(f_*(\mathfrak{J})) = g_*^{-1}(g_*(\mathfrak{F})) = \mathfrak{F}$. Hence $p : \mathfrak{J} \leftrightarrow \mathfrak{F}$ and $g \circ p = g \circ g^{-1} \circ f = f$. Suppose $g \circ p_1 = f$; then $p_1 = g^{-1} \circ f$ and $p = p_1$. Hence, $p$ is unique.

---

▲ *Example 7*

The circle $C = \{x \in \mathbf{E}^2 \mid \|x\| = 1\}$ is a simple curve and $f(t) = (\cos t, \sin t)$, $t \in [0, 2\pi)$, and $g(s) = (\sin s, \cos s)$, $s \in [-\pi, \pi)$, are injective parametrizations of $C$. We find

$$p(t) = (g^{-1} \circ f)(t) = \begin{cases} \dfrac{\pi}{2} - t & \text{for } 0 \leqslant t \leqslant \dfrac{3\pi}{2}, \\[2ex] \dfrac{5\pi}{2} - t & \text{for } \dfrac{3\pi}{2} < t < 2\pi, \end{cases}$$

and we see that $p : [0, 2\pi) \leftrightarrow [-\pi, \pi)$. (See Fig. 118.8.)

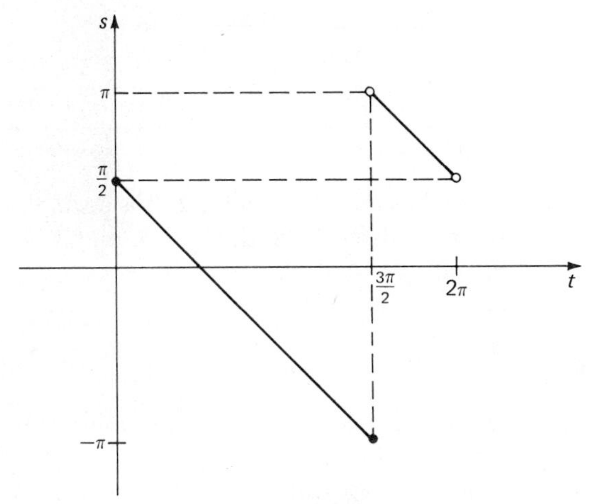


*Fig. 118.8*

---

The parametrizations which we have found in examples 2, 4, and 5 have a continuous derivative. The parametrization in example 1 may not even have a derivative, and the derivative of the parametrization in example 3 is not continuous on $[0, 4]$.

A curve that possesses a parametrization with a continuous derivative need not have a continuously turning tangent line, as the following example will show.

---

▲ *Example 8*

$\{x \in \mathbf{E}^2 \mid \xi_1^2 = \xi_2^3\}$ may be parametrized by $f(t) = (t^3, t^2)$, $t \in (-\infty, \infty)$; and we see that $f'(t) = (3t^2, 2t)$ is continuous in $(-\infty, \infty)$. Still, this curve does not have a continuously turning tangent line, as we can see from Fig. 118.9. This curve is said to have a *cusp* at $f(0) = (0, 0)$. There, $f'(0) = (0, 0)$.

---

The circle is a *simple closed curve*:

***Definition 118.4***

$\Gamma \subset \mathbf{E}^n$ is a simple closed curve if and only if it is a simple curve and there exists a parametrization $f : [a, b] \to \mathbf{E}^n$ of $\Gamma$ such that $f : [a, b) \xrightarrow[1-1]{} \mathbf{E}^n$ and $f(a) = f(b)$.

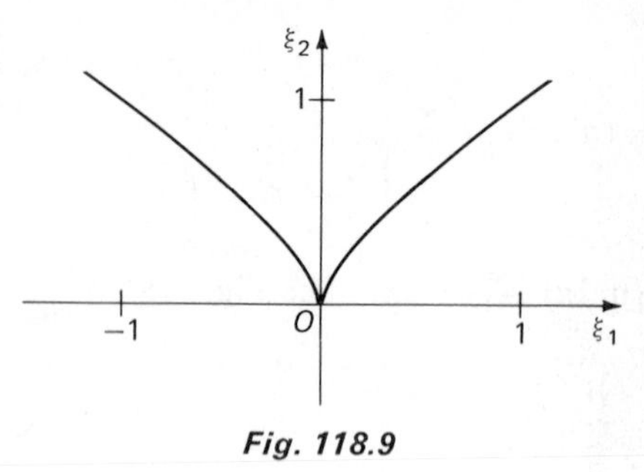


*Fig. 118.9*

---

▲ *Example 9*

We apply definition 118.4 to the circle of example 2. It is a simple curve because $f(t) = (\cos t, \sin t)$, $t \in [0, 2\pi)$, satisfies definition 118.3, and it is closed because $f(t)$ is injective on $[0, 2\pi)$ and $f(0) = f(2\pi)$.

---

## 13.118 Exercises

1. Show that the curves in examples 1, 2, 3, 5, and 8 are simple curves, by finding injective parametrizations.
2. Show that $g(s) = (r \sin s, -r \cos s, a(s - \pi/2)/2\pi)$, $s \in (-\infty, \infty)$ defines an injective parametrization, and show that it is equivalent to the parametrization of the helix in example 5. Find the bijective parameter transformation $p : (-\infty, \infty) \leftrightarrow (-\infty, \infty)$ such that $f = g \circ p$.
3. Show that the cycloid (example 4) is a simple curve for $a \leqslant r$.
4. Show that the injective parametrizations defined by $f(t) = (t, t)$, $t \in (0, 1)$, and $g(s) = (e^s, e^s)$, $s \in (-\infty, 0)$, are equivalent. Find the appropriate bijective parameter transformation from $(0, 1)$ onto $(-\infty, 0)$. Sketch the curve represented by $f : (0, 1) \to \mathbf{E}^2$.
5. Sketch the curve parametrized by $f(t) = (4t - t^2, 4t^2 - t^3)$, $t \in [-1, 5]$. Demonstrate that this curve is not simple.
6. Show that

$$f(t) = \left(\frac{1 - \sqrt{1 - t^2}}{2}, \frac{1 + \sqrt{1 - t^2}}{2}\right), \qquad t \in [-1, 1],$$

and

$$g(s) = \left(\sin^2 \frac{s}{2}, \cos^2 \frac{s}{2}\right), \qquad s \in \left[-\frac{\pi}{2}, \frac{\pi}{2}\right],$$

represent the same curve.

7. Let $f : \mathfrak{J} \to \mathbf{E}^n$, $g : \mathfrak{T} \to \mathbf{E}^n$ where $f$, $g$ are not necessarily injective. Show: If there exists a bijective parameter transformation $p : \mathfrak{J} \leftrightarrow \mathfrak{T}$ such that $f = g \circ p$ on $\mathfrak{J}$, then $f$, $g$ are equivalent.

## 13.119 Surfaces

We develop the concept of a surface in $\mathbf{E}^3$ along the same lines that we followed in developing the concept of a curve.

***Definition 119.1***

$\Sigma \subset \mathbf{E}^3$ is a *surface* in $\mathbf{E}^3$ if and only if there exists a connected set $\mathfrak{D} \subseteq \mathbf{E}^2$ with nonempty interior and a continuous function $f : \mathfrak{D} \to \mathbf{E}^3$ such that $f_*(\mathfrak{D}) = \Sigma$, that is, $f$ maps $\mathfrak{D}$ onto $\Sigma$.

$f : \mathfrak{D} \xrightarrow{\text{onto}} \Sigma$ is called a *parametrization* of $\Sigma$ and

$$x = f(t), \qquad t \in \mathfrak{D},$$

or, in components,

$$\left.\begin{aligned} \xi_1 &= \phi_1(t), \\ \xi_2 &= \phi_2(t), \\ \xi_3 &= \phi_3(t), \end{aligned}\right\} \qquad t \in \mathfrak{D},$$

is called a *parameter representation* of $\Sigma$. $t = (\tau_1, \tau_2)$ is called a *parameter*.

---

▲ *Example 1*

If $f : \mathfrak{D} \to \mathbf{E}$, $\mathfrak{D} \subseteq \mathbf{E}^2$, is continuous and $\mathfrak{D}$ has interior points, then

$$\{(\xi_1, \xi_2, f(\xi_1, \xi_2)) \mid (\xi_1, \xi_2) \in \mathfrak{D}\}$$

represents a surface;

$$\left.\begin{aligned} \xi_1 &= \tau_1, \\ \xi_2 &= \tau_2, \\ \xi_3 &= f(\tau_1, \tau_2), \end{aligned}\right\} \qquad (\tau_1, \tau_2) \in \mathfrak{D},$$

is a parameter representation of this surface. (See Fig. 119.1.)

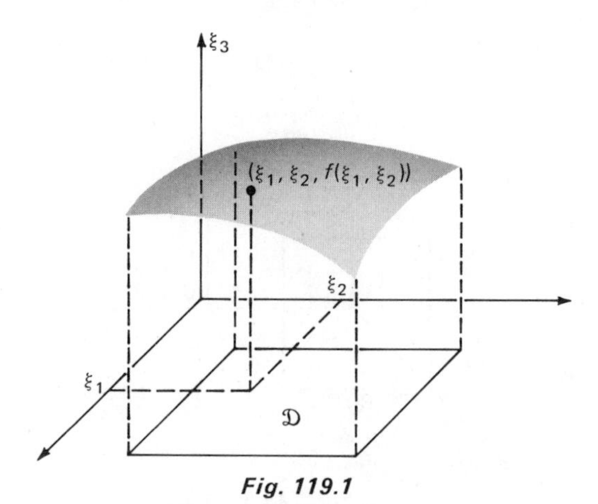


*Fig. 119.1*

▲ *Example 2* (*Sphere*)

The unit sphere $S = \{x \in \mathbf{E}^3 \mid \|x\| = 1\}$ is a surface in $\mathbf{E}^3$. If

$$f(t) = (\sin \tau_1 \cos \tau_2, \sin \tau_1 \sin \tau_2, \cos \tau_1), \tag{119.1}$$

then $x = f(\tau_1, \tau_2)$, $(\tau_1, \tau_2) \in \mathbf{E}^2$, and $x = f(\tau_1, \tau_2)$, $(\tau_1, \tau_2) \in [0, \pi] \times [0, 2\pi)$, are parameter representations of $S$. (See Fig. 119.2.)

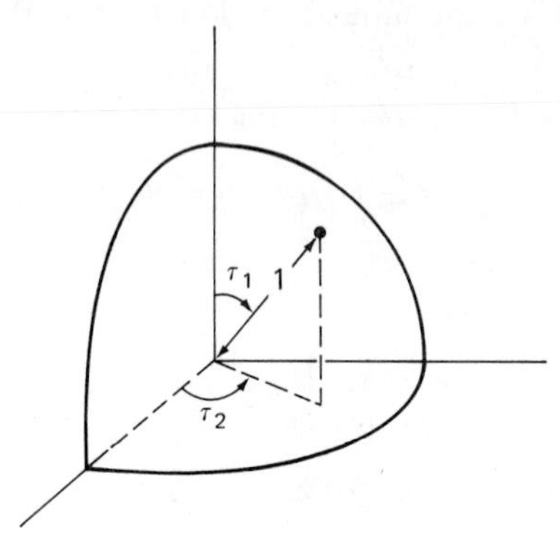


*Fig. 119.2*

If

$$g(s) = (\cos \sigma_1, \sin \sigma_1 \sin \sigma_2, \sin \sigma_1 \cos \sigma_2), \tag{119.2}$$

then $x = g(s)$, $s \in [0, \pi] \times [0, 2\pi)$, is also a parameter representation of $S$.

▲ *Example 3*

The surface $\Sigma$ of the cube $[0, 1] \times [0, 1] \times [0, 1]$ in Fig. 119.3 is a surface.

We find a parametrization of $\Sigma$ by cutting along the bold edges in Fig. 119.3 and laying the surface out flat, as indicated in Fig. 119.4.

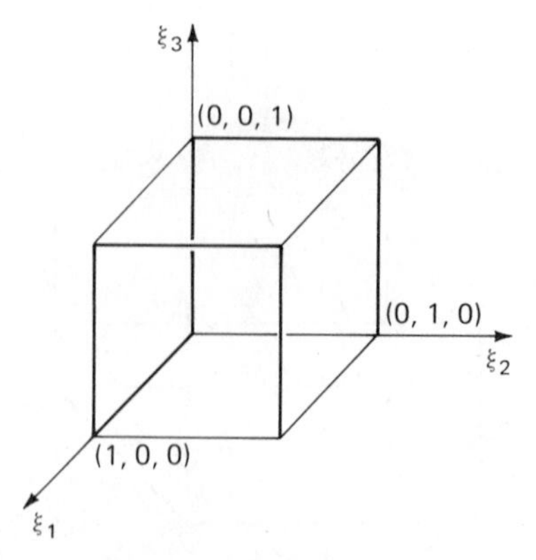


*Fig. 119.3*

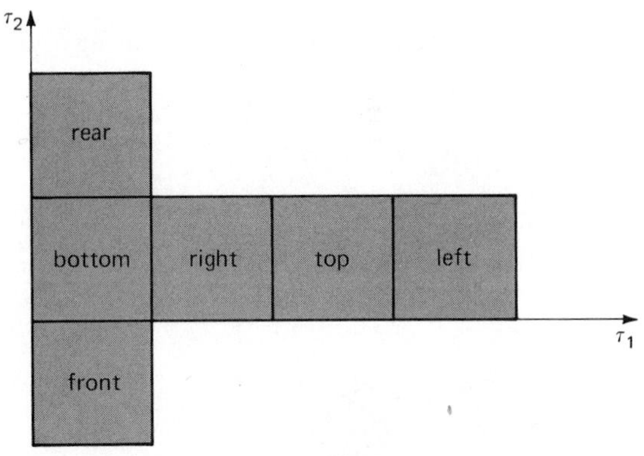


*Fig. 119.4*

We introduce a coordinate system, as indicated in Fig. 119.4, and map every point in the shaded set into its ancestor on the surface of the cube. This yields a parametrization $f$ of $\Sigma$. (See also exercise 1.)

▲ *Example 4*

The right circular cone $\{x \in \mathbf{E}^3 \mid \xi_3^2 = \xi_1^2 + \xi_2^2\}$ is a surface. $f(t) = (\tau_1 \cos \tau_2, \tau_1 \sin \tau_2, \tau_1)$, $(\tau_1, \tau_2) \in (-\infty, \infty) \times [0, 2\pi)$ defines a parametrization. (See Fig. 119.5.)

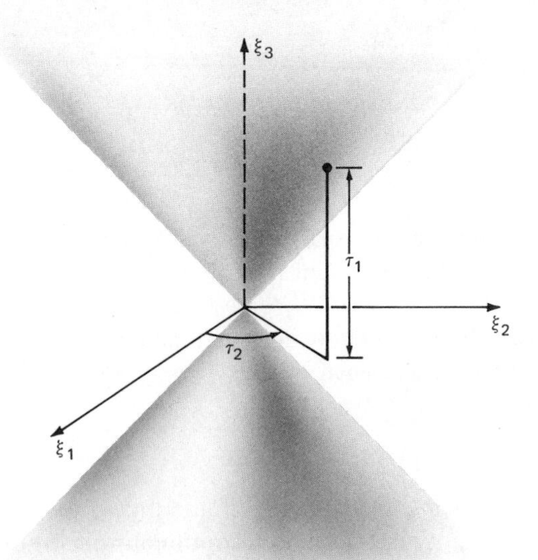


*Fig. 119.5*

▲ *Example 5*

The paraboloid generated by the parabola $\xi_3 = \xi_1^2$ rotating about the $\xi_3$-axis is a surface which may be parametrized by

(119.3) $$f(t) = (\tau_1, \tau_2, \tau_1^2 + \tau_2^2), \qquad (\tau_1, \tau_2) \in \mathbf{E}^2,$$

or, by

(119.4) $$g(s) = (\sigma_1 \cos \sigma_2, \sigma_1 \sin \sigma_2, \sigma_1^2), \qquad (\sigma_1, \sigma_2) \in [0, \infty) \times [0, 2\pi).$$

(See Fig. 119.6.)

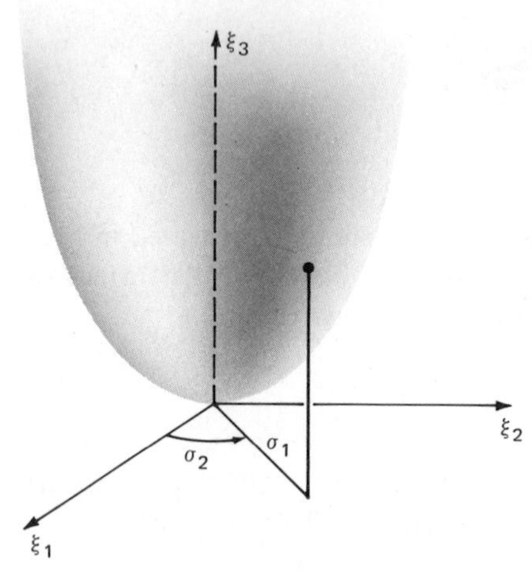


*Fig. 119.6*

---

We define equivalence of parametrizations as in section 13.118.

**Definition 119.2**

Two parametrizations $f: \mathfrak{D} \to \mathbf{E}^3$, $g: \mathcal{E} \to \mathbf{E}^3$, are called *equivalent* if and only if $f_*(\mathfrak{D}) = g_*(\mathcal{E})$; i.e., they represent the same surface.

Again, we can show that equivalent parametrizations of *simple* surfaces stand in a simple relationship to each other.

**Definition 119.3**

$\Sigma \subset \mathbf{E}^3$ is called a *simple surface* in $\mathbf{E}^3$ if and only if there exists a connected set $\mathfrak{D} \subset \mathbf{E}^2$ with nonempty interior and a continuous one-to-one function (injective map) $f: \mathfrak{D} \xrightarrow[1-1]{} \mathbf{E}^3$ such that $f_*(\mathfrak{D}) = \Sigma$; that is, $f: \mathfrak{D} \leftrightarrow \Sigma$.

Again, we call such a parametrization an *injective parametrization*.

***Lemma 119.1***

Two injective parametrizations $f : \mathfrak{D} \to \mathbf{E}^3$, $g : \mathcal{E} \to \mathbf{E}^3$, are equivalent if and only if there exists a unique bijection $p : \mathfrak{D} \leftrightarrow \mathcal{E}$ such that $f = g \circ p$.

The proof follows verbatim the proof of Lemma 118.1.

---

▲ *Example 6*

The parametrization $f$ of the paraboloid in (119.3) is injective. Hence, the paraboloid is a simple surface. The parametrization in (119.4) is *not* injective.

$$h(s) = (\sigma_2, \sigma_1, \sigma_1^2 + \sigma_2^2), \qquad (\sigma_1, \sigma_2) \in \mathbf{E}^2$$

is another injective parametrization of the paraboloid, and we see that

$$p(t) = h^{-1} \circ f(t) = h^{-1}(\tau_1, \tau_2, \tau_1^2 + \tau_2^2) = (\tau_2, \tau_1)$$

defines a bijection from $\mathbf{E}^2$ onto $\mathbf{E}^2$.

---

The parametrization (119.1) of the sphere with $\mathbf{E}^2$ as domain has continuous partial derivatives in $\mathbf{E}^2$. So does the parametrization of the cone $f(t) = (\tau_1 \cos \tau_2, \tau_1 \sin \tau_2, \tau_1)$ with domain $(-\infty, \infty) \times (-\pi, 2\pi)$. (Note that partial derivatives are defined only at interior points.) Still, as we can see from the example of the cone, this does not necessarily imply that the surface has a continuously turning tangent plane, or that it possesses a tangent plane everywhere, for that matter.

## 13.119 Exercises

1. Find an injective parametrization of the surface of the cube in example 3.
2. Show that the surface in example 1 is a simple surface.
3. Show that $f(t) = (\sin \tau_1 \cos \tau_2, \sin \tau_1 \sin \tau_2, \cos \tau_1)$, where $(\tau_1, \tau_2) \in ((0, \pi) \times (0, 2\pi)) \cup ([0, \pi] \times \{0\})$, is an injective parametrization of the sphere $\{x \in \mathbf{E}^3 \mid \|x\| = 1\}$.
4. Show that the one-sided cone $\{x \in \mathbf{E}^3 \mid \xi_3 = \sqrt{\xi_1^2 + \xi_2^2}\}$ is a simple surface.
5. Show: If $f \in C^1(\mathfrak{D})$, where $\mathfrak{D} \subseteq \mathbf{E}^2$ is an open set, and if $p : \mathcal{E} \leftrightarrow \mathfrak{D}$, where $\mathcal{E} \subseteq \mathbf{E}^2$ is an open set and $p \in C^1(\mathfrak{D})$, then $g = f \circ p \in C^1(\mathcal{E})$.
6. Prove Lemma 119.1 in detail.
7. Parametrize the right circular cylinder $\{x \in \mathbf{E}^3 \mid \xi_1^2 + \xi_2^2 = 1\}$, and show that it is a simple surface.
8. Parametrize the plane $\{x \in \mathbf{E}^3 \mid \xi_2 = 1\}$ and show that it is a simple surface.

## 13.120 Smooth Manifolds

In order to be able to discuss tangent lines to curves and tangent planes to surfaces, we have to impose certain conditions other than that the curves and surfaces not be inhabited by evil spirits. Corners such as in example 3, section

13.118, cusps as in example 8, section 13.118, edges such as in example 3, section 13.119, and vertices as in example 4, section 13.119, just won't do. We have seen, in the preceding two sections, that the requirement that a parametrization with a continuous derivative, or continuous partial derivatives, respectively, exists, is no guarantee against the appearance of cusps or vertices. To eliminate such troublesome features, we have to impose stronger conditions. It is practical, and just as easy, to deal with curves and surfaces simultaneously, and we shall do so. Curves and surfaces are special cases of *k-dimensional manifolds* in $\mathbf{E}^n$. Specifically, a curve in $\mathbf{E}^n$ is a one-dimensional manifold in $\mathbf{E}^n$ and a surface in $\mathbf{E}^3$ is a two-dimensional manifold in $\mathbf{E}^3$. What we are after are the so-called *smooth* $k$-dimensional manifolds in $\mathbf{E}^n$, which we shall eventually define in terms of (smooth) *plasters*.

***Definition 120.1***

$\Pi \subset \mathbf{E}^n$ is a *k-dimensional plaster* in $\mathbf{E}^n$ if and only if there exists an open connected set $\mathfrak{D} \subseteq \mathbf{E}^k$ and a function $f : \mathfrak{D} \to \mathbf{E}^n$ with the following properties:

1. $f : \mathfrak{D} \leftrightarrow \Pi$,
2. $f \in C^1(\mathfrak{D})$,
3. $\operatorname{rank} f'(t) = k$ for all $t \in \mathfrak{D}$.

$f$ is called a (smooth) *parametrization* of the plaster $\Pi$ and $t$ is called the *parameter*.

Clearly, every one-dimensional plaster in $\mathbf{E}^n$ is a simple curve in $\mathbf{E}^n$ and every two-dimensional plaster in $\mathbf{E}^3$ is a simple surface in $\mathbf{E}^3$.

---

▲ *Example 1*

If $\Phi: \Omega \to \mathbf{E}$ has continuous partial derivatives in the open set $\Omega \subseteq \mathbf{E}^2$, and if $\Phi(x_0) = 0$ and rank $\Phi'(x_0) = 1$ for some $x_0 = (\xi_1^0, \xi_2^0) \in \Omega$, then $\Phi(\xi_1, \xi_2) = 0$ defines a smooth one-dimensional plaster in some open set $U \subseteq \Omega$ that contains $x_0$. We may assume w.l.o.g. that $D_2\Phi(\xi_1^0, \xi_2^0) \neq 0$ and solve for $\xi_2$ in terms of $\xi_1$. By the implicit-function theorem (Theorem 99.1), there is a unique function $h : V \to \mathbf{E}$ where $V \subseteq \mathbf{E}$ is open and contains $\xi_1^0$ such that $h(\xi_1^0) = \xi_2^0, \Phi(\xi_1, h(\xi_1)) = 0$ for all $\xi_1 \in V$, and $(\xi_1, h(\xi_1)) \in U$ for all $\xi_1 \in V$. Also, $h \in C^1(V)$. Let $f(\xi_1) = (\xi_1, h(\xi_1))$. Then rank $f'(\xi_1) = \operatorname{rank}(1, h'(\xi_1)) = 1$, and we see that $(\xi_1, h(\xi_1))$, $\xi_1 \in V$ represents a smooth plaster that lies in $U$.

▲ *Example 2*

If $\Phi: \Omega \to \mathbf{E}$ has continuous partial derivatives in the open set $\Omega \subseteq \mathbf{E}^3$ and if $\Phi(x_0) = 0$ and rank $\Phi'(x_0) = 1$ for some $x_0 \in \Phi$, then we can see as in the preceding example, by invoking the implicit-function theorem, that $\Phi(\xi_1, \xi_2, \xi_3) = 0$

defines a smooth two-dimensional plaster in $\mathbf{E}^3$, namely $(\xi_1, \xi_2, h(\xi_1, \xi_2))$, $(\xi_1, \xi_2) \in V$ for some open set $V \subseteq \mathbf{E}^2$ that contains $(\xi_1^0, \xi_2^0)$, where $\Phi(\xi_1, \xi_2, h(\xi_1, \xi_2)) = 0$ for all $(\xi_1, \xi_2) \in V$, if $D_3\Phi(x_0) \neq 0$. Analogous results are obtained for the case where $D_1\Phi(x_0) \neq 0$ or $D_2\Phi(x_0) \neq 0$.

▲ *Example 3*

The circle $C = \{x \in \mathbf{E}^2 \mid \|x\| = 1\}$ without the point $(1,0)$ is a one-dimensional plaster in $\mathbf{E}^2$ because $f(t) = (\cos t, \sin t)$, $t \in (0, 2\pi)$ satisfies the requirements of definition 120.1. To wit: $(0, 2\pi)$ is open, $f: (0, 2\pi) \leftrightarrow C \backslash \{1, 0)\}$, $f \in C^1(0, 2\pi)$, and finally, rank $(-\sin t, \cos t) = 1$ since $\sin t$, $\cos t$ never vanish simultaneously.

▲ *Example 4*

The sphere $S = \{x \in \mathbf{E}^3 \mid \|x\| = 1\}$, without North and South Pole and without the meridian through $(1,0,0)$, is a two-dimensional plaster $\Pi$ in $\mathbf{E}^3$. We note that $f(t) = (\sin\tau_1 \cos\tau_2, \sin\tau_1 \sin\tau_2, \cos\tau_1)$, $(\tau_1, \tau_2) \in (0, \pi) \times (0, 2\pi)$ defines an injective parametrization of $\Pi$ and that $f \in C^1((0, \pi) \times (0, 2\pi))$. Since

$$F'(\tau_1, \tau_2) = \begin{pmatrix} \cos\tau_1 \cos\tau_2 & -\sin\tau_1 \sin\tau_2 \\ \cos\tau_1 \sin\tau_2 & \sin\tau_1 \cos\tau_2 \\ -\sin\tau_1 & 0 \end{pmatrix},$$

and since

$$\begin{vmatrix} \cos\tau_1 \sin\tau_2 & \sin\tau_1 \cos\tau_2 \\ -\sin\tau_1 & 0 \end{vmatrix} = \sin^2\tau_1 \cos\tau_2,$$

$$\begin{vmatrix} \cos\tau_1 \cos\tau_2 & -\sin\tau_1 \sin\tau_2 \\ -\sin\tau_1 & 0 \end{vmatrix} = \sin^2\tau_1 \sin\tau_2$$

do not vanish simultaneously in $(0, \pi) \times (0, 2\pi)$, we also have rank $f'(t) = 2$.

---

A *smooth k-dimensional manifold* is a plastered $k$-dimensional manifold. More precisely,

***Definition 120.2***

$M \subset \mathbf{E}^n$ is a smooth $k$-dimensional manifold in $\mathbf{E}^n$ if and only if for every point $x \in M$ there exists an open set $V \subseteq \mathbf{E}^n$ such that $M \cap V$ is a $k$-dimensional plaster.

Clearly, every $k$-dimensional plaster $\Pi$ in $\mathbf{E}^n$ is a smooth $k$-dimensional manifold in $\mathbf{E}^n$. To see this, one need only take $V = \mathbf{E}^n$. Then $\Pi \cap \mathbf{E}^n = \Pi$ is a plaster.

▲ *Example 5*

The circle $C = \{x \in \mathbf{E}^2 \mid \|x\| = 1\}$ is a smooth one-dimensional manifold (smooth curve) in $\mathbf{E}^2$. To see this, we take open sets $V_1$ and $V_2$ in $\mathbf{E}^2$ as in Fig. 120.1, and observe that the intersection of these open sets with $C$ are plasters.

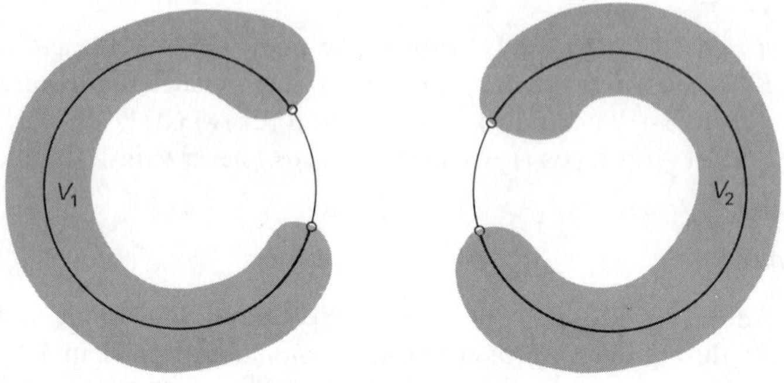


*Fig. 120.1*

▲ *Example 6*

The sphere $\{x \in \mathbf{E}^3 \mid \|x\| = 1\}$ is a smooth two-dimensional manifold (smooth surface) in $\mathbf{E}^3$ because it can be plastered, as indicated in Fig. 120.2.

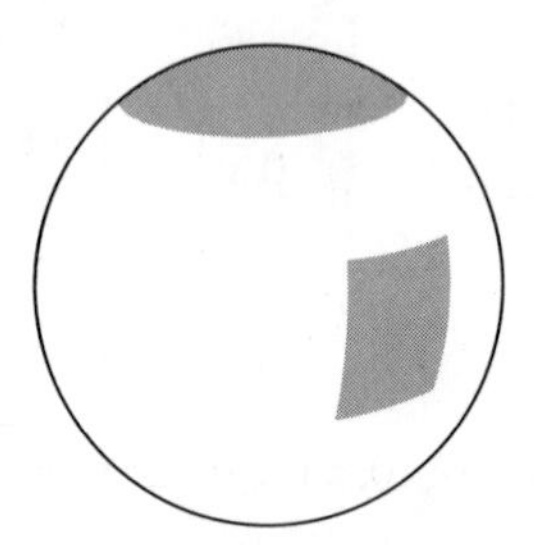

*Fig. 120.2*

To plaster the polar caps one may take the parametrization

$$f(t) = (\cos \tau_1, \sin \tau_1 \sin \tau_2, \sin \tau_1 \cos \tau_2)$$

the derivative of which has rank 2 in the polar regions. (Observe that the North Pole is obtained for $\tau_1 = (\pi/2)$, $\tau_2 = 0$, and the South Pole for $\tau_1 = -(\pi/2)$, $\tau_2 = 0$.)

▲ *Example 7*

The curve in Fig. 120.3 which is parametrized by $f(t) = (4t - t^2, 4t^2 - t^3)$, $t \in (0, 5)$, is a plaster and, *eo ipso*, a smooth manifold. Note that $\theta = f(4)$ is *not* a double point.

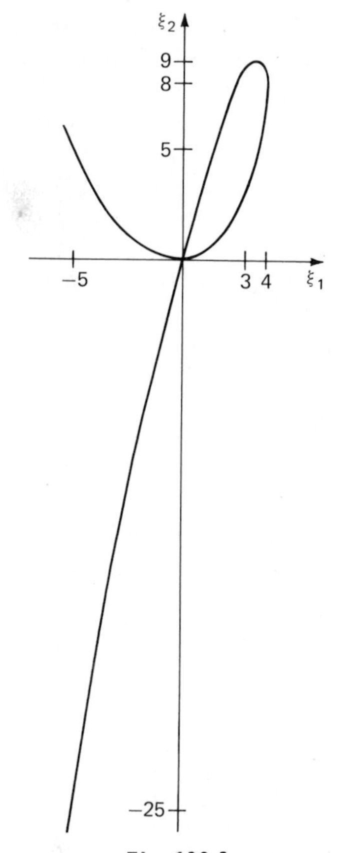


*Fig. 120.3*

▲ *Example 8*

$\Gamma = \{x \in \mathbf{E}^2 \mid \xi_1^2 = \xi_2^3\}$ is a simple curve (see example 8, section 13.118) but is neither a smooth plaster nor a smooth manifold, as we shall demonstrate. Let $V \subset \mathbf{E}^2$ denote an open set that contains $\theta$ (see Fig. 120.4). Then, $f(t) = (t^3, t^2)$, $t \in (a, b)$ for some $a < 0, b > 0$, is an injective parametrization of $\Gamma \cap V$ but does not satisfy definition 120.1 because rank $f'(0) = 0$.

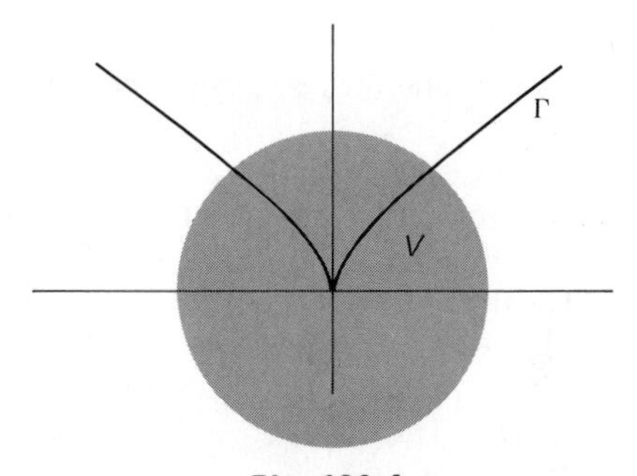


*Fig. 120.4*

If $\Gamma \cap V$ is a smooth plaster, then, there is an (injective) parametrization $g = (\psi_1, \psi_2) : (c, d) \to \mathbf{E}^2$ which satisfies definition 120.1.

We may assume w.l.o.g. that $g(0) = \theta$. Since $\psi_2$ is strictly decreasing on the left of $\theta$ and strictly increasing on the right of $\theta$, we must have $\psi_2'(0) = 0$. Hence, $\psi_1'(0) \neq 0$ and we may assume w.l.o.g. that $\psi_1'(0) > 0$. Hence, there is an $m > 0$ and a $\delta > 0$ such that $\psi_1'(s) > m$ for all $s \in N_\delta(0)$. We choose $\delta > 0$ so small that the inverse-function theorem is applicable to $\xi_1 = \psi_1(s)$ with $U = N_\delta(0)$ and that $|\psi_1(s)| < (8m^3/27)$ for all $s \in N_\delta(0)$.

We obtain, with $\xi_2 = \psi_2(\psi_1^{-1}(\xi_1)) \equiv h(\xi_1)$ from Theorem 92.3,

$$h'(\xi_1) = \psi_2'(\psi_1^{-1}(\xi_1)) \frac{1}{\psi_1'(\psi_1^{-1}(\xi_1))}.$$

From the representation of $\Gamma, \xi_2 = \xi_1^{2/3}$. Hence, $h(\xi_1) = \xi_1^{2/3}$ and therefore, $h'(\xi_1) = \frac{2}{3}\xi_1^{-1/3}$. Hence

$$\frac{\psi_2'(s)}{\psi_1'(s)} = \frac{2}{3}(\psi_1(s))^{-1/3}. \tag{120.1}$$

Let $-\delta < s < 0$. Then

$$-\frac{8m^3}{27} < \psi_1(s) < 0. \tag{120.2}$$

By (120.2), $\frac{2}{3}(\psi_1(s))^{-1/3}\psi_1'(s) < -\frac{2}{3}(3/2m)m = -1$, and consequently,

$$\psi_2'(s) < -1 \qquad \text{for } s \in (-\delta, 0).$$

If we consider $0 < s < \delta$ instead, we obtain, by symmetric reasoning, that

$$\psi_2'(s) > 1 \qquad \text{for } s \in (0, \delta).$$

Hence, the assumption that $\psi_1'(0) \neq 0$ leads to the conclusion that $\psi_2'$ is not continuous at 0, which is contrary to our assumption that $g$ satisfied definition 120.1. Therefore, $\Gamma \cap V$ is not a smooth plaster.

▲ *Example 9* (*Torus*)

The circle $(\xi_1 - 2)^2 + \xi_3^2 = 1$ generates a *torus* when it is rotated about the $\xi_3$-axis. (See Fig. 120.5.)

$$f(t) = ((2 + \sin\tau_1)\cos\tau_2, (2 + \sin\tau_1)\sin\tau_2, \cos\tau_1), \qquad (\tau_1, \tau_2) \in \mathbf{E}^2, \tag{120.3}$$

parametrizes this torus. Clearly, $f \in C^1(\mathbf{E}^2)$. Since

$$F'(t) = \begin{pmatrix} \cos\tau_1 \cos\tau_2 & -(2 + \sin\tau_1)\sin\tau_2 \\ \cos\tau_1 \sin\tau_2 & (2 + \sin\tau_1)\cos\tau_2 \\ -\sin\tau_1 & 0 \end{pmatrix}$$

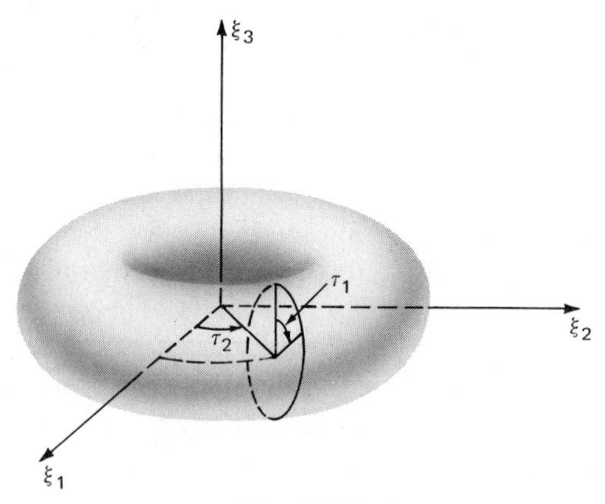


*Fig. 120.5*

and

$$\begin{vmatrix} \cos\tau_1\cos\tau_2 & -(2+\sin\tau_1)\sin\tau_2 \\ -\sin\tau_1 & 0 \end{vmatrix} = -\sin\tau_1\sin\tau_2\,(2+\sin\tau_1),$$

$$\begin{vmatrix} \cos\tau_1\sin\tau_2 & (2+\sin\tau_1)\cos\tau_2 \\ -\sin\tau_1 & 0 \end{vmatrix} = \sin\tau_1\cos\tau_2\,(2+\sin\tau_1),$$

$$\begin{vmatrix} \cos\tau_1\cos\tau_2 & -(2+\sin\tau_1)\sin\tau_2 \\ \cos\tau_1\sin\tau_2 & (2+\sin\tau_1)\cos\tau_2 \end{vmatrix} = \cos\tau_1\,(2+\sin\tau_1)$$

never vanish simultaneously anywhere, we have rank $f'(t) = 2$ for all $t \in \mathbf{E}^2$. Since $f$ is injective on $(a, a+2\pi) \times (b, b+2\pi)$ for any $a, b \in \mathbf{R}$, the torus can be plastered all over and is therefore a smooth two-dimensional manifold in $\mathbf{E}^3$.

---

## 13.120 Exercises

1. Show that $\mathbf{E}^n$ is a smooth $n$-dimensional manifold in $\mathbf{E}^n$.
2. Show that a point $x_0 \in \mathbf{E}^n$ is a smooth 0-dimensional manifold in $\mathbf{E}^n$.
3. Show that every coordinate axis in $\mathbf{E}^n$ is a smooth curve in $\mathbf{E}^n$ and that every coordinate plane is a smooth surface in $\mathbf{E}^n$.
4. Show that $\xi_1^2 - \xi_2^2 + 2\xi_1\xi_2 = 1$ defines a smooth one-dimensional plaster in some neighborhood of $(1, 2)$.
5. Show that $\{x \in \mathbf{E}^2 \mid \xi_2 = \xi_1^2\}$ is a smooth one-dimensional manifold (smooth curve) in $\mathbf{E}^2$.
6. Let $g : \mathfrak{D} \to \mathbf{E}$ where $\mathfrak{D}$ is an open set in $\mathbf{E}^2$. Impose suitable conditions on $g$ to make $\{(\xi_1, \xi_2, g(\xi_1, \xi_2)) \mid (\xi_1, \xi_2) \in \mathfrak{D}\}$ a smooth two-dimensional manifold.
7. *Prove*: For $M \subseteq \mathbf{E}^n$ to be a smooth $k$-dimensional manifold, it is sufficient that there is a parametrization $f: \mathfrak{D} \to \mathbf{E}^n$ of $M$ where $\mathfrak{D} \subseteq \mathbf{E}^k$ is open, where $f \in C^1(\mathfrak{D})$, and where rank $f'(t) = k$ for all $t \in \mathfrak{D}$.

8. Apply the result of exercise 7 to show that the torus (example 9) is a smooth two-dimensional manifold in $\mathbf{E}^3$.
9. Show: If $f : \mathfrak{D} \to \mathbf{E}^n$, $\mathfrak{D} \subseteq \mathbf{E}^k$ open, is the parametrization of a $k$-dimensional plaster $\Pi$, then $f^{-1} : \Pi \to \mathfrak{D}$ need not be continuous.
10. Let $f: \mathfrak{D} \to \mathbf{E}^n$, $\mathfrak{D} \subseteq \mathbf{E}^k$, denote a parametrization of a $k$-dimensional smooth manifold $M$ in $\mathbf{E}^n$. Show: If $t_0 \in \mathfrak{D}$, then there are open sets $U \subseteq \mathbf{E}^k \times \mathbf{E}^n$, $V \subseteq \mathbf{E}^n$, with $(t_0, f(t_0)) \in U$, $f(t_0) \in V$, and a function $h : V \to \mathbf{E}^k$ such that $x - f(h(x)) = 0$ for all $x \in V \cap M$. The process described above is called *elimination of the parameter*.
11. Use the result in exercise 10 to eliminate the parameter, whenever and wherever possible, in the following cases:
    (a) $x = a + tb$, $t \in (-\infty, \infty)$, $a, b \in \mathbf{E}^n$,
    (b) $x = (\cos t, \sin t)$, $t \in [0, 2\pi)$,
    (c) $x = (a + b \cos t, c + d \sin t)$, $t \in [0, 2\pi)$, $a, b, c, d \in \mathbf{R}$,
    (d) $x = a + b\tau_1 + c\tau_2$, $(\tau_1, \tau_2) \in \mathbf{E}^2$, $a, b, c \in \mathbf{E}^3$.

## 13.121 Diffeomorphisms and Smooth Equivalence

To cut down on the verbiage we shall henceforth refer to a smooth parametrization of a plaster simply as a parametrization of the plaster. Since plasters are simple curves or surfaces or such, Lemmas 118.1 and 119.1 apply to equivalent parametrizations of plasters. Heretofore we could only say that equivalent parametrizations (of simple curves or surfaces) are related to each other through a bijective map. In the case of plasters, we can make the much stronger statement that this bijective map is actually a *diffeomorphism*.

***Definition 121.1***

The bijective map $p : U \leftrightarrow V$ from the open set $U \subseteq \mathbf{E}^k$ onto the open set $V \subseteq \mathbf{E}^k$ is a diffeomorphism if and only if $p \in C^1(U)$ and $p^{-1} \in C^1(V)$.

We have encountered diffeomorphisms in conjunction with the inverse-function theorem (Theorem 98.1). There we have shown that whenever $p : \mathfrak{D} \to \mathbf{E}^k$, $\mathfrak{D} \subseteq \mathbf{E}^k$, $p \in C^1(\mathfrak{D})$, and, for some interior point $a \in \mathfrak{D}$, $J_p(a) \neq 0$, then there are open sets, $U, V \subseteq \mathbf{E}^k$ with $a \in U$, $p(a) \in V$ such that $p : U \leftrightarrow V$ is what we may now call a diffeomorphism.

As in the last part of the proof of Theorem 98.1, we see that:

***Lemma 121.1***

If $p : U \leftrightarrow V$ is a diffeomorphism, where $U, V \subseteq \mathbf{E}^k$ are open, then $J_p(t) \neq 0$ for all $t \in U$ and $J_{p^{-1}}(s) \neq 0$ for all $s \in V$.

For the characterization of equivalent parametrizations of plasters and for later occasions, the following property of plasters is very helpful:

***Lemma 121.2***

If $\Pi \subset \mathbf{E}^n$ is a $k$-dimensional plaster and $f: \mathfrak{D} \to \mathbf{E}^n$, $\mathfrak{D} \subseteq \mathbf{E}^k$, is a parametrization of $\Pi$, then, for every point $b \in \Pi$, there are open sets $U, V \subseteq \mathbf{E}^n$ with $(a, \theta) = (f^{-1}(b), \theta) \in U, b \in V$, such that $f$ is the restriction of a diffeomorphism $h$: $U \leftrightarrow V$ to $\mathfrak{D} \cap U$; that is, $f = (t)\,h(t, \theta)$ for all $t \in \mathfrak{D} \cap U$. (See Fig. 121.1 and 121.2.)

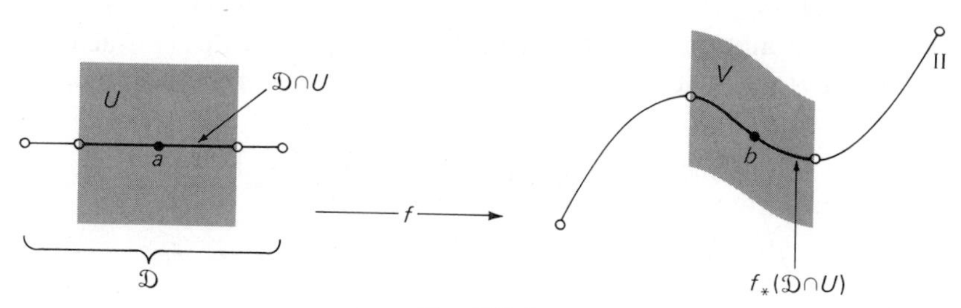


***Fig. 121.1***

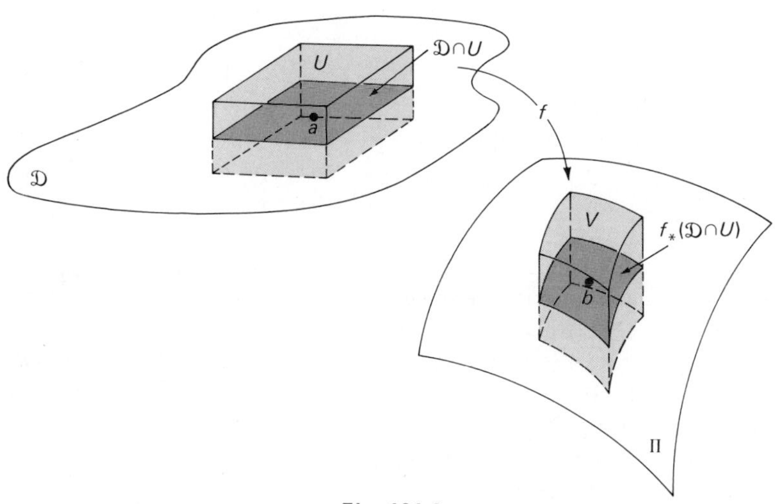


***Fig. 121.2***

*Proof*

Since rank $f'(t) = k$ for all $t \in \mathfrak{D}$, we may assume w.l.o.g. that, for $a = f^{-1}(b)$,

$$\begin{vmatrix} D_1\phi_1(a) \ldots D_k\phi_1(a) \\ \vdots \\ D_1\phi_k(a) \ldots D_k\phi_k(a) \end{vmatrix} \neq 0.$$

Now,

$$h(\tau_1, \tau_2, \ldots, \tau_k, \sigma_{k+1}, \ldots, \sigma_n) = (\phi_1(t), \ldots, \phi_k(t), \phi_{k+1}(t) + \sigma_{k+1}, \ldots, \phi_n(t) + \sigma_n)$$

defines a function $h : \mathfrak{D} \times \mathbf{E}^{n-k} \to \mathbf{E}^n$ which has the following properties:

$$h(a, \theta) = b,$$
$$h \in C^1(\mathfrak{D} \times \mathbf{E}^{n-k}),$$
$$\text{rank } h'(a, \theta) = n.$$

By the inverse-function theorem (Theorem 98.1), there are open sets $U, V \subseteq \mathbf{E}^n$ with $(a, \theta) \in U, b \in V$, such that $h : U \leftrightarrow V$ is a diffeomorphism. By construction, $h(t, \theta) = f(t)$ for all $t \in \mathfrak{D}$, and, in particular, for all $t \in \mathfrak{D} \cap U$. ($f$ is actually the restriction of $h$ to $\mathfrak{D}$ but only on $\mathfrak{D} \cap U$ are we assured that $h$ is a diffeomorphism.)

We define equivalence of parametrizations, as in sections 13.118 and 13.119.

***Definition 121.2***

Two parametrizations $f : \mathfrak{D} \to \mathbf{E}^n$, $g : \mathcal{E} \to \mathbf{E}^n$ of the $k$-dimensional plaster $\Pi \subset \mathbf{E}^n$ are called (smoothly) equivalent if and only if $f_*(\mathfrak{D}) = g_*(\mathcal{E})$.

***Lemma 121.3***

Two parametrizations $f : \mathfrak{D} \to \mathbf{E}^n$, $g : \mathcal{E} \to \mathbf{E}^n$, $\mathfrak{D}, \mathcal{E} \subseteq \mathbf{E}^k$, of the $k$-dimensional plaster $\Pi \subset \mathbf{E}^n$ are (smoothly) equivalent if and only if there exists a unique diffeomorphism $p : \mathfrak{D} \leftrightarrow \mathcal{E}$ such that $f = g \circ p$ on $\mathfrak{D}$.

*Proof*

(a) If there is such a diffeomorphism, then $f_*(\mathfrak{D}) = (g \circ p)_*(\mathfrak{D}) = g_*(p_*(\mathfrak{D})) = g_*(\mathcal{E})$. Hence, $f, g$ parametrize the same plaster. By the chain rule, $f'(t) = g'(p(t)) \circ p'(t)$. Hence, $f \in C^1(\mathfrak{D})$ if $g \in C^1(\mathcal{E})$. By Lemma 121.2, rank $p'(t) = k$ for all $t \in \mathfrak{D}$. Hence, rank $f'(t) =$ rank $g'(t)$. If rank $g'(t) = k$, then rank $f'(t) = k$ and we see that if $g$ is a (smooth) parametrization of the plaster, then so is $f$. Since $p^{-1}$ has the same properties as $p$, we may show in the same manner that $g$ is a smooth parametrization of $\Pi$ if $f$ is one.

(b) If $f, g$ are equivalent parametrizations, then $p = g^{-1} \circ f : \mathfrak{D} \leftrightarrow \mathcal{E}$ exists and is unique. This may be shown as in the proof of Lemma 118.1. To show that $p$ is a diffeomorphism, we invoke Lemma 121.2. Let $a \in \mathfrak{D}$, $f(a) = b \in \Pi$, $g^{-1}(b) = c \in \mathcal{E}$. Then there are open sets $U_1, V_1, U_2, V_2 \subseteq \mathbf{E}^n$ with $b \in V_1$, $b \in V_2$, $(a, \theta) \in U_1, (c, \theta) \in U_2$ and diffeomorphisms $h_1 : U_1 \leftrightarrow V_1, h_2 : U_2 \leftrightarrow V_2$ such that $h_1(t, \theta) = f(t)$ for all $t \in \mathfrak{D} \cap U_1, h_2(s, \theta) = g(s)$ for all $s \in \mathcal{E} \cap U_2$. (See Fig. 121.3.) Then, $P = h_2^{-1} \circ h_1 : h_1^*(V_1 \cap V_2) \leftrightarrow h_2^*(V_1 \cap V_2)$ is a diffeomorphism, and for $t \in \mathfrak{D} \cap h_1^*(V_1 \cap V_2)$, $P(t, \theta) = h_2^{-1} \circ h_1(t, \theta) = h_2^{-1} \circ f(t) = (g^{-1} \circ f(t), \theta) = (p(t), \theta)$. Since $P \in C^1(h_1^*(V_1 \cap V_2))$, we have $p \in C^1(\mathfrak{D} \cap h_1^*(V_1 \cap V_2))$. We examine $P^{-1} = h_1^{-1} \circ h_2$ in the same manner and find $p^{-1} \in C^1(\mathcal{E}) \cap h_2^*(V_1 \cap V_2))$. We can repeat these arguments for every point $a \in \mathfrak{D}$ and every point $c \in \mathcal{E}$, and find that $p \in C^1(\mathfrak{D})$, $p^{-1} \in C^1(\mathcal{E})$. Hence, $p : \mathfrak{D} \leftrightarrow \mathcal{E}$ is indeed a diffeomorphism.

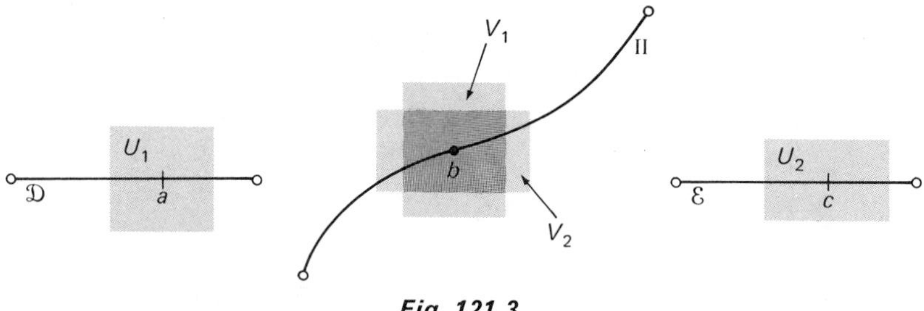


*Fig. 121.3*

Note that, in order to prove this lemma, we had to extend the definitions of $f$ and $g$ into open sets in $\mathbf{E}^n$ by means of Lemma 121.2 because we cannot differentiate $p = g^{-1} \circ f$ or $p^{-1} = f^{-1} \circ g$ by the chain rule as the derivatives of $f^{-1}, g^{-1}$ are not defined on $\Pi$, since $\Pi$ is not an open set in $\mathbf{E}^n$ for $k < n$.

Had we admitted other than open sets as domains in the definition of plasters, Lemma 121.3 would break down, as we can see from the following example.

---

▲ *Example 1*

$f(t) = (\cos t, \sin t)$, $t \in [0, 2\pi)$, and $g(s) = (\sin s, \cos s)$ $s \in [-\pi, \pi)$, both define parametrizations of the unit circle $\{x \in \mathbf{E}^2 \mid \|x\| = 1\}$. Both parametrizations satisfy definition 120.1 *except* that the domains of $f, g$ are not open. We have seen in example 7, section 12.118, that the unique bijection $p$, where $f = g \circ p$, is not even continuous and hence, cannot possibly be a diffeomorphism.

---

For the diffeomorphism $p = (\pi_1, \pi_2, \ldots, \pi_k)$ of Lemma 121.3, we have $p \in C^1(\mathfrak{D})$ and hence, $p'(t)$ exists for all $t \in \mathfrak{D}$. $\|p'(t)\| = \sqrt{\sum_{j=1}^{k} \pi_j'^2(t)}$ is continuous on $\mathfrak{D}$. However, $\|p'(t)\|$ is not necessarily bounded on $\mathfrak{D}$, as the following example shows.

---

▲ *Example 2*

$f(t) = (t, t)$, $t \in (0, 1)$, and $g(s) = (s^2, s^2)$, $s \in (0, 1)$, define equivalent parametrizations of the line segment in Fig. 121.4 which is a one-dimensional plaster. $p(t) = g^{-1} \circ f(t) = \sqrt{t}$ defines a diffeomorphism from $(0, 1)$ onto $(0, 1)$. However, $\|p'(t)\| = (1/2\sqrt{t})$ is *not* bounded on $(0, 1)$.

---

Before closing this section, we shall discuss another important application of Lemma 121.2 that will be needed in section 13.123. We consider a curve $\Gamma$ on a surface $\Sigma$ (see Fig. 121.5). Suppose that $\Sigma$ is parametrized by $f : \mathfrak{D} \to \mathbf{E}^3$ where $\mathfrak{D} \subset \mathbf{E}^2$ and that $\Gamma$ is parametrized by $g : \mathfrak{J} \to \mathbf{E}^3$, where $\mathfrak{J}$ is an interval. It seems reasonable to expect $\Gamma$ to be the image under $f$ of a curve $\gamma \subset \mathfrak{D}$, i.e., that

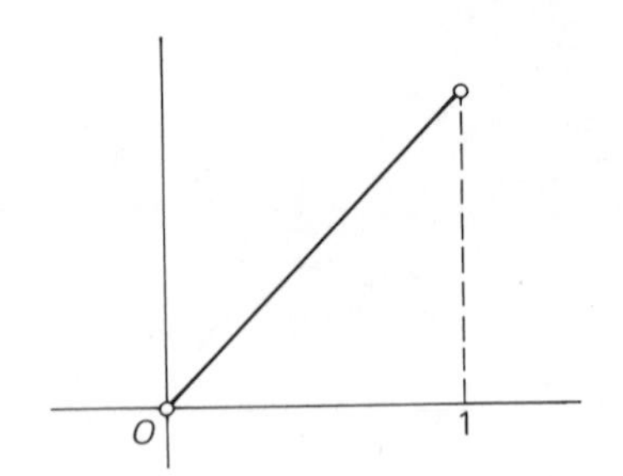


*Fig. 121.4*

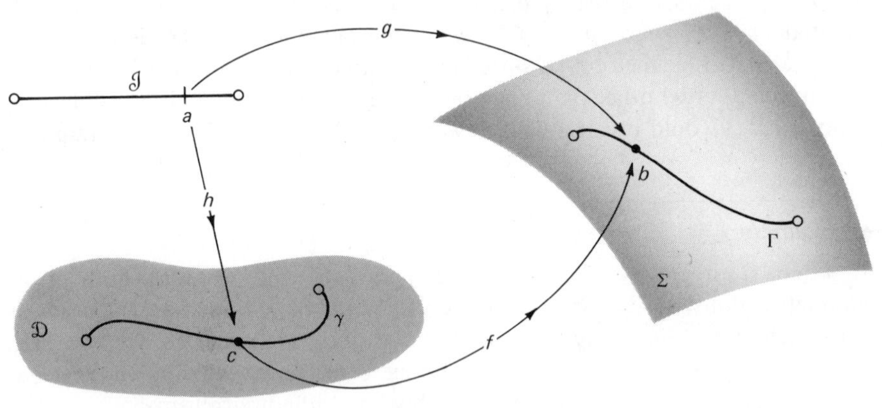


*Fig. 121.5*

there is a parametrization $h : \mathfrak{J} \to \mathfrak{D}$ such that $g = f \circ h$ on $\mathfrak{J}$. This is true if $\Sigma$ is a two-dimensional plaster and if $\Gamma$ is a one-dimensional plaster on $\Sigma$. It will follow that $\gamma \subset \mathfrak{D}$ is not just any curve but a one-dimensional plaster.

***Lemma 121.4***

Let $\Sigma$ denote a two-dimensional plaster in $\mathbf{E}^3$ and let $\Gamma$ denote a one-dimensional plaster on $\Sigma$. For any (smooth) parametrization $f : \mathfrak{D} \to \mathbf{E}^3$, $\mathfrak{D} \subseteq \mathbf{E}^2$ of $\Sigma$ there exists a one-dimensional plaster $\gamma \subseteq \mathfrak{D}$ such that $f_*(\gamma) = \Gamma$.

*Proof*

Let $g : \mathfrak{J} \to \mathbf{E}^3$ represent a (smooth) parametrization of $\Gamma$ and let

(121.1) $$h = f^{-1} \circ g : \mathfrak{J} \to \mathfrak{D}.$$

By construction, $h$ is one-to-one and $h(\mathfrak{J}) \subseteq \mathfrak{D}$. Let $h_*(\mathfrak{J}) = \gamma$. Then $f_*(\gamma) = f_*(h_*(\mathfrak{J})) = f_*(f^*(g_*(\mathfrak{J}))) = g_*(\mathfrak{J}) = \Gamma$. To show that $\gamma$ is a one-dimensional plaster we have to show, by definition 120.1, that $h \in C^1(\mathfrak{J})$ and rank $h'(t) = 1$ for all $t \in \mathfrak{J}$.

Let $a \in \mathfrak{J}$. Then $b = g(a) \in \Gamma \subset \Sigma$ and $c = f^{-1}(b) = f^{-1}(g(a)) = h(a) \in \gamma$. By Lemma 121.2, there are open sets $U, V, W_1, W_2 \subseteq \mathbf{E}^3$ with $(a, 0) \in U$, $b \in W_1$, $b \in W_2$, $(c, 0) \in V$ and diffeomorphisms $G : U \leftrightarrow W_1, F : V \leftrightarrow W_2$ such that $G(t, 0) = g(t)$ for $t \in \mathfrak{J} \cap U$ and $F(s, 0) = f(s)$ for $s \in \mathfrak{D} \cap V$.

Let $U_1 = G^*(W_1 \cap W_2)$, $V_1 = F^*(W_1 \cap W_2)$. Then, $H = F^{-1} \circ G : U_1 \leftrightarrow V_1$ is a diffeomorphism. If $t \in \mathfrak{J} \cap U_1$, then $H(t, 0) = F^{-1} \circ G(t, 0) = F^{-1} \circ g(t) = (f^{-1} \circ g(t), 0) = (h(t), 0)$. Hence, $h \in C^1(\mathfrak{J} \cap U_1)$. Since rank $H'(a, 0) = 3$ we have to have rank $h'(a) = 1$. Since $a$ is any point in $\mathfrak{J}$, $h \in C^1(\mathfrak{J})$, and rank $h'(t) = 1$ for all $t \in \mathfrak{J}$ follows.

---

▲ *Example 3*

The sphere without poles and without the meridian through $(1, 0, 0)$ is a two-dimensional plaster (see example 4, section 13.120) which may be parametrized by $f(s) = (\sin \sigma_1 \cos \sigma_2, \sin \sigma_1 \sin \sigma_2, \cos \sigma_1)$, $(\sigma_1, \sigma_2) \in (0, \pi) \times (0, 2\pi)$. The meridian through $(0, 1, 0)$ without the poles, $\Gamma$, is a one-dimensional plaster on the sphere (see Fig. 121.6) and may be parametrized by $g(t) = (0, \sin \tau_1, \cos \tau_1)$,

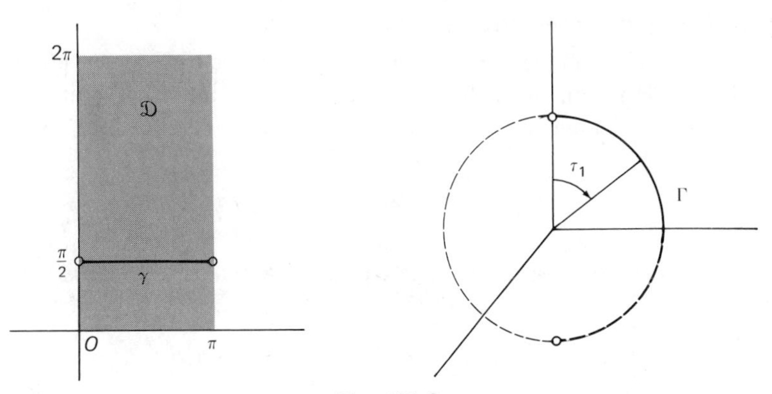


*Fig. 121.6*

$\tau_1 \in (0, \pi)$. Hence, $h(t) = f^{-1} \circ g(t) = (\tau_1, \pi/2)$, $\tau_1 \in (0, \pi)$, represents the smooth plaster $\gamma \subset \mathfrak{D}$ for which $f_*(\gamma) = \Gamma$.

---

## 13.121 Exercises

1. Prove Lemma 121.1.
2. Consider that part of the unit circle that is parametrized by $f(t) = (\cos t, \sin t)$, $t \in (-\pi/4, \pi/4)$. Construct a diffeomorphism $h$ as in Lemma 121.2 on $(-(\pi/4), (\pi/4)) \times (-1, 1)$ and sketch $h_*((-(\pi/4), (\pi/4)) \times (-1, 1))$.
3. Show that $f(t) = (t, t)$, $t \in (0, 1)$ and $g(s) = (e^s, e^s)$, $s \in (-\infty, 0)$, are equivalent parametrizations, and find the diffeomorphism $p : (0, 1) \leftrightarrow (-\infty, 0)$ for which $f = g \circ p$.

4. Show: If $f : \mathfrak{J} \to \mathbf{E}^n$ parametrizes a one-dimensional plaster in $\mathbf{E}^n$, then $f'(t) \neq \theta$ for all $t \in \mathfrak{J}$.
5. Show: If $f : \mathfrak{D} \to \mathbf{E}^3$, $\mathfrak{D} \subseteq \mathbf{E}^2$, parametrizes a two-dimensional plaster in $\mathbf{E}^3$, then $D_1 f(t)$ and $D_2 f(t)$ are linearly independent for all $t \in \mathfrak{D}$.
6. Let $\Sigma$ denote the torus of example 9, section 13.120, without top ridge and without the circle through which it intersects the $(\xi_1, \xi_3)$-plane in front.
   (a) Show that what is left is a plaster.
   (b) Consider the bottom ridge, $\Gamma$, of the torus without the point $(2, 0, -1)$ (which is a one-dimensional plaster), and find for the parametrization
   $$f(t) = ((2 + \sin \tau_1) \cos \tau_2, (2 + \sin \tau_1) \sin \tau_2, \cos \tau_1), \qquad (\tau_1, \tau_2) \in (0, 2\pi) \times (0, 2\pi),$$
   a one-dimensional plaster $\gamma$ in $(0, 2\pi) \times (0, 2\pi)$ such that $f_*(\gamma) = \Gamma$.
7. Show: If $p : U \leftrightarrow V$ is a diffeomorphism, then $\operatorname{sign} J_p(t) = \operatorname{sign} J_{p^{-1}}(p(t))$ for all $t \in U$.
8. Show: If $f, g : \mathfrak{J} \to \mathbf{E}^n$ are parametrizations of a smooth one-dimensional plaster in $\mathbf{E}^n$ where $f = g \circ p$ and if $p'(t) > 0$ for all $t \in \mathfrak{J}$, then $f'(t)/\|f'(t)\| = g'(t)/\|g'(t)\|$ for all $t \in \mathfrak{J}$.

### 13.122 Tangent Lines and Tangent Vectors

If $\Gamma \subset \mathbf{E}^n$ is a given simple curve, then the tangent line to $\Gamma$ at the point $b \in \Gamma$ is the limiting position of all lines through $b$ and some other point $y \in \Gamma$ as $y$, moving along $\Gamma$, approaches $b$. (See Fig. 122.1.)

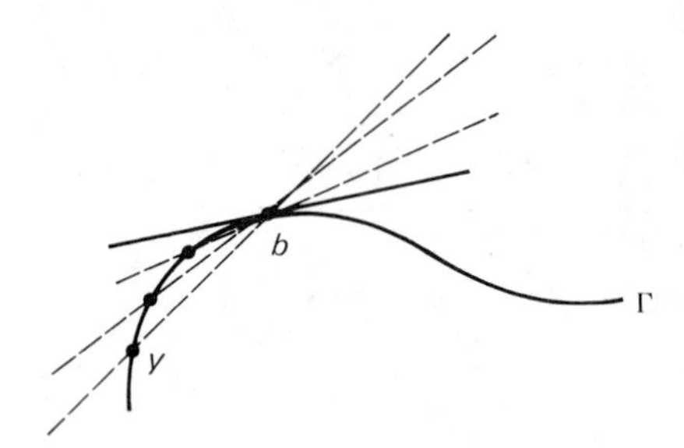


*Fig. 122.1*

$$x = b + \lambda(b - y), \qquad \lambda \in (-\infty, \infty), \tag{122.1}$$

is a parameter representation of the line through $b$ and $y$, $\lambda$ being the parameter. If $f : \mathfrak{J} \to \mathbf{E}^n$ is a parametrization of $\Gamma$ with $f^{-1}(b) = a$, $f^{-1}(y) = t$, we may write (122.1), absorbing the factor $(a - t)$ in the parameter, as

$$x = b + \lambda \frac{f(a) - f(t)}{a - t}, \qquad \lambda \in (-\infty, \infty).$$

If $t$ approaches $a$ as $y$ approaches $b$ and if $f'(a)$ exists and $f'(a) \neq \theta$, we obtain, for the limiting position of all these lines,

$$x = b + \lambda f'(a), \qquad \lambda \in (-\infty, \infty)$$

as the parameter representation of the tangent line to $\Gamma$ at $b$. This line is uniquely defined by $b$ and the direction that is given by $f'(a)$. The latter appears to depend on the parametrization of $\Gamma$ but does not, as we propose to show for smooth curves $\Gamma$.

If $f : \mathfrak{I} \to \mathbf{E}^n$, $g : \mathfrak{J} \to \mathbf{E}^n$ are parametrizations of a plaster on the smooth curve $\Gamma \subset \mathbf{E}^n$, containing the point $b \in \Gamma$, then, by Lemma 121.3, $f = g \circ p$ on $\mathfrak{I}$ where $p : \mathfrak{I} \to \mathfrak{J}$ is a diffeomorphism. With $f^{-1}(b) = a, g^{-1}(b) = c$, we have $p(a) = c, f'(a) \neq \theta, g'(c) \neq \theta$,

$$f'(a) = g'(c)p'(a), \tag{122.2}$$

where $p'(a) \neq 0$. Hence, the two parameter representations

$$x = b + \lambda f'(a) = b + \lambda p'(a)g'(c), \qquad \lambda \in (-\infty, \infty),$$
$$x = b + \lambda g'(c), \qquad \lambda \in (-\infty, \infty),$$

represent the same line.

***Theorem 122.1***

If $\Gamma \subset \mathbf{E}^n$ is a smooth curve, then the tangent line to $\Gamma$ at every point $b \in \Gamma$ exists and is unique. If $f : \mathfrak{I} \to \mathbf{E}^n$ is a (smooth) parametrization of the plaster on $\Gamma$ that contains $b$, then

$$x = b + \lambda f'(f^{-1}(b)), \qquad \lambda \in (-\infty, \infty),$$

is a parameter representation of the tangent line to $\Gamma$ at $b$.

---

▲ *Example 1*

The reader may have some misgivings about our treatment of tangent lines because our theory does not yield a tangent line to the curve $\{x \in \mathbf{E}^2 \mid \xi_1^2 = \xi_2^3\}$ at $\theta$. (See Fig. 122.2 and examples 8, section 13.118, and 8, section 13.120.) For the parametrization $f(t) = (t^3, t^2)$, $t \in (-\infty, \infty)$, or any other parametrization, for that matter, $f'(f^{-1}(\theta)) = \theta$.

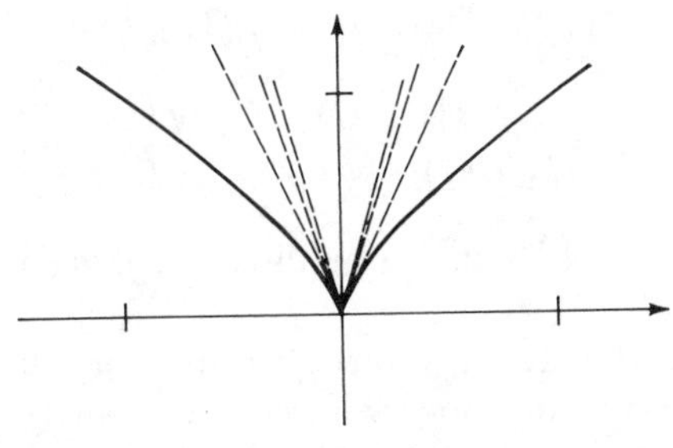

*Fig. 122.2*

Still, with a little bit of good will, one could call the $\xi_2$-axis the tangent line. The resolution of this discrepancy between theory and fact lies simply in the fact that Theorem 122.1 applies only to smooth curves, and the curve in Fig. 122.2 is not smooth in any neighborhood of $\theta$, as we have seen in example 8, section 13.120.

If we interpret $t$ as the time and $\xi_1 = t^3, \xi_2 = t^2$, as equations of motion, then, $f'(t) = (3t^2, 2t)$ represents the velocity vector at the time $t$ at the point $(t^3, t^2)$. $f'(0) = (0, 0)$ simply means that the motion comes to a momentary halt as the point $\theta$ is reached.

---

Hitherto we have viewed vectors $x = (\xi_1, \xi_2, \ldots, \xi_n)$ as position vectors, that is, vectors with their tail at the origin $\theta$ and their head at the point $(\xi_1, \xi_2, \ldots, \xi_n) \in \mathbf{E}^n$. In addition to position vectors, we shall now consider *tangent vectors*.

***Definition 122.1***

A vector $dx = (d\xi_1, d\xi_2, \ldots, d\xi_n)$ is called a tangent vector at $x \in \mathbf{E}^n$ if and only if its tail is at $x$ and its head is at $x + dx$.

(The symbolic notation $dx = (d\xi_1, d\xi_2, \ldots, d\xi_n)$ is chosen for very practical reasons, as we shall see below.)

If $x = f(t)$ is a point on a smooth one-dimensional plaster $\Gamma$ that is parametrized by $f$, then $dx = df(t, dt) = f'(t)dt$ is a tangent vector to $\Gamma$ at $x$, i.e., a vector with tail at $x = f(t)$ and head at $x + dx = f(t) + f'(t)dt$. (For the definition of the differential $df(t, dt)$, see definition 86.2.) This is evident from the discussion that preceded Theorem 122.1, wherein we saw that $f'(t)$ determines the direction of the tangent line to $\Gamma$ at $f(t) \in \Gamma$. $f'(t)$ and $f'(t)dt$ differ only by a multiplicative factor $dt$, which we shall assume to be positive, and hence they have the same direction.

If $g$ is another parametrization of $\Gamma$ and $g^{-1}(x) = s$, then, $dx = dg(s, ds) = g'(s)ds$, $ds > 0$, is also a tangent vector to $\Gamma$ at $x$. In the former case, $f'(t)dt/\|f'(t)dt\| = f'(t)/\|f'(t)\|$ is the unit vector in the direction of the tangent vector, and in the other case it is $g'(s)ds/\|g'(s)ds\| = g'(s)/\|g'(s)\|$. Since $f = g \circ p$ for some diffeomorphism $p$, we have from (122.2), that

$$\frac{f'(t)}{\|f'(t)\|} = \frac{g'(p(t))p'(t)}{\|g'(p(t))\|\,|p'(t)|} = \frac{g'(s)}{\|g'(s)\|} \operatorname{sign} p'(t); \tag{122.3}$$

i.e., the two vectors point in the same direction if $p'(t) > 0$ and in opposite directions if $p'(t) < 0$.

The reason it is possible for the tangent vectors for different, but equivalent, parametrizations to point in opposite directions is that a parametrization induces an *orientation* in a curve. We shall explain this in the following example.

▲ *Example 2*

$f(t) = (\cos t, \sin t)$, $t \in [0, 2\pi)$ and $g(s) = (\sin s, \cos s)$, $s \in [(\pi/2), (5\pi/2))$ are both parametrizations of the unit circle with center at $\theta$. In the first case, the circle is traversed in the *counterclockwise* (positive) *direction* as the parameter runs from 0 to $2\pi$, and in the second case the circle is traversed in the *clockwise* (negative) *direction*. (See Fig. 122.3.) We have $f'(t) = (-\sin t, \cos t)$ and $g'(s) = (\cos s, -\sin s)$.

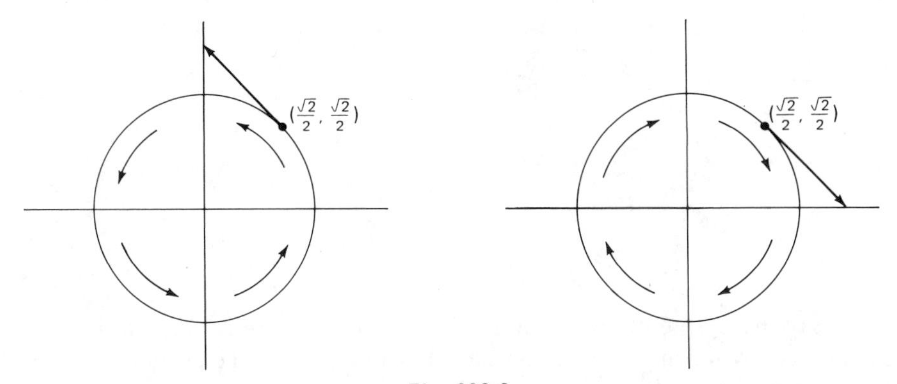


*Fig. 122.3*

We have indicated, in Fig. 122.3, the corresponding tangent vectors at $((\sqrt{2}/2), (\sqrt{2}/2))$ for both cases.

---

We see, from (122.3), that two parametrizations of the same plaster induce the same orientation if and only if $p'(t) > 0$ for all $t$ in the domain of $f$. Accordingly, we give the following definition.

**Definition 122.2**

A diffeomorphism $p : U \leftrightarrow V$, where $U, V \subseteq \mathbf{E}^k$ are open sets, is called *orientation-preserving* if and only if $J_p(t) > 0$ for all $t \in U$. Otherwise it is called *orientation-reversing*.

Two parametrizations $f : \mathfrak{D} \to \mathbf{E}^n$, $g : \mathcal{E} \to \mathbf{E}^n$, $\mathfrak{D}, \mathcal{E} \subseteq \mathbf{E}^k$ are called *consistent* if and only if they are equivalent (i.e., represent the same $k$-dimensional plaster) and if $f = g \circ p$ for some orientation-preserving diffeomorphism $p : \mathfrak{D} \leftrightarrow \mathcal{E}$.

This divides the (smooth) parametrizations of a one-dimensional plaster in $\mathbf{E}^n$ into two classes of consistent parametrizations. The parametrizations in the one class orient the plaster one way, the ones in the other class orient it the opposite way. When we speak of a curve, then the curve just sits there and it does not matter which parametrization is used. However, when we speak of an *oriented curve*, then not only the *shape* of the curve but also the *direction* in which it is traversed is prescribed and then only those consistent parametrizations $f$ will do that have $f(t)$ move along the curve in the prescribed direction as $t$ increases.

For a given parametrization $f$ of a one-dimensional plaster, the tangent vector $f'(t)dt$ could only reverse itself by passing through $\theta$ because it has to lie in the tangent line and $\|f'(t)\| = \sqrt{\phi_1'^2(t) + \phi_2'^2(t)}$ is continuous. (See Fig. 122.4.) This is not possible because rank $f'(t) = 1$.

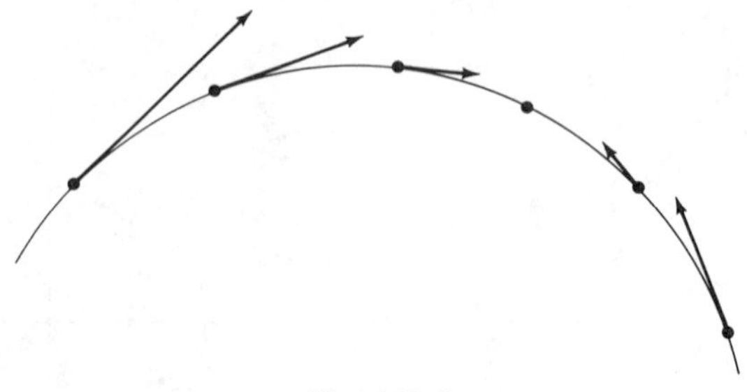

*Fig. 122.4*

We say that a smooth manifold is orientable if there are parametrizations of the plasters by which it is covered which are consistent on the portions of overlap. Clearly, *a smooth curve is orientable*.

## 13.122 Exercises

1. Find the tangent vector to the helix of example 5, section 13.118, at $((r\sqrt{2}/2), (r\sqrt{2}/2), a/8)$.
2. Show that the parametrization $f(t) = (\cos t, \sin t)$, $t \in (0, 2\pi)$, and $g(s) = (\sin s, \cos s)$, $s \in ((\pi/2), (5\pi/2))$, are equivalent but not consistent.
3. Find a unit tangent vector to the unit circle with center at $\theta$ at any point $x$ on the circle.
4. Find a vector $n$ that is orthogonal to the tangent vector to the unit circle with center at $\theta$, at the point $x = (\xi_1, \xi_2)$, and show that $n$ and $x$ either have the same direction or point into opposite directions.
5. Show: If $p : U \leftrightarrow V$ is an orientation-preserving diffeomorphism, then $p^{-1}: V \leftrightarrow U$ is also an orientation-preserving diffeomorphism.
6. Make suitable assumptions and show that $d(f \circ g)(x, u) = f'(g(x))\, dg(x, u) = f'(g(x)) \circ g'(x)(u)$. (See definition 86.2 and Theorem 92.1.)
7. Let $f: \mathfrak{D} \to \mathbf{E}^3$ parametrize a two-dimensional surface $\Sigma$ in $\mathbf{E}^3$ and let $dx$ denote the tangent vector to a curve on this surface. $(dx)^2 = dx \cdot dx$ for $x = f(t)$ is called the *first fundamental form* of $\Sigma$. Show that

$$dx \cdot dx = E(d\tau_1)^2 + 2F d\tau_1 d\tau_2 + G(d\tau_2)^2,$$

where

$$\begin{aligned} E &= (D_1\phi_1(t))^2 + (D_1\phi_2(t))^2 + (D_1\phi_3(t))^2, \\ F &= D_1\phi_1(t)D_2\phi_1(t) + D_1\phi_2(t)D_2\phi_2(t) + D_1\phi_3(t)D_2\phi_3(t), \\ G &= (D_2\phi_1(t))^2 + (D_2\phi_2(t))^2 + (D_2\phi_3(t))^2. \end{aligned}$$

8. Find the first fundamental form of:
   (a) the plane that is parametrized by

$$f(t) = (\alpha_1 + \beta_1\tau_1 + \gamma_1\tau_2, \alpha_2 + \beta_2\tau_1 + \gamma_2\tau_2, \alpha_3 + \beta_3\tau_1 + \gamma_3\tau_2),$$

   (b) the sphere that is parametrized by

$$f(t) = (\sin\tau_1 \cos\tau_2, \sin\tau_1 \sin\tau_2, \cos\tau_1),$$

   (c) the torus that is parametrized by

$$f(t) = ((2 + \sin\tau_1)\cos\tau_2, (2 + \sin\tau_1)\sin\tau_2, \cos\tau_1).$$

9. To what extent does the first fundamental form of a surface depend on the parametrization of the surface?

## 13.123 Tangent Planes and Normal Vectors

We propose to show that the tangent vectors at $b$ to all possible smooth curves that lie on a smooth surface and pass through the point $b$ on that surface lie all in one plane, the *tangent plane* to the surface at $b$.

The equation of a plane through the point $b = (\beta_1, \beta_2, \beta_3)$ is

$$\nu_1(\xi_1 - \beta_1) + \nu_2(\xi_2 - \beta_2) + \nu_3(\xi_3 - \beta_3) = 0,$$

where $n = (\nu_1, \nu_2, \nu_3)$ is a vector orthogonal to the plane or, as we shall call it, a *normal vector* to the plane. Hence, in order to find the equation of the tangent plane to a surface, we need a vector that is orthogonal to all the above-mentioned tangent vectors. It is convenient to utilize the *cross product* for that purpose. For easy reference, we review the main features of the cross product:

The cross product of two vectors $a = (\alpha_1, \alpha_2, \alpha_3)$, $b = (\beta_1, \beta_2, \beta_3) \in \mathbf{E}^3$ is defined to be the vector

$$a \times b = (\alpha_2\beta_3 - \alpha_3\beta_2, \alpha_3\beta_1 - \alpha_1\beta_3, \alpha_1\beta_2 - \alpha_2\beta_1). \tag{123.1}$$

To remember this definition more easily, we write it in the symbolic form

$$a \times b = \begin{vmatrix} e_1 & e_2 & e_3 \\ \alpha_1 & \alpha_2 & \alpha_3 \\ \beta_1 & \beta_2 & \beta_3 \end{vmatrix},$$

where $e_1, e_2, e_3$ are the fundamental vectors in $\mathbf{E}^3$. From (123.1),

$$a \times b = -b \times a. \tag{123.2}$$

By elementary computation, $a \cdot (a \times b) = 0$, $b \cdot (a \times b) = 0$; that is, $a \times b$ is orthogonal to $a$ and to $b$ and hence, to the plane spanned by $a$ and $b$.

The length $\|a \times b\|$ of $a \times b$ is equal to the area of the parallelogram that is spanned by $a$ and $b$ and the direction of $a \times b$, orthogonal to the plane spanned by $a$ and $b$, is the direction a righthanded screw would advance if turned from $a$ toward $b$ through the smaller one of the two angles between $a$ and $b$. (See Fig. 123.1.)

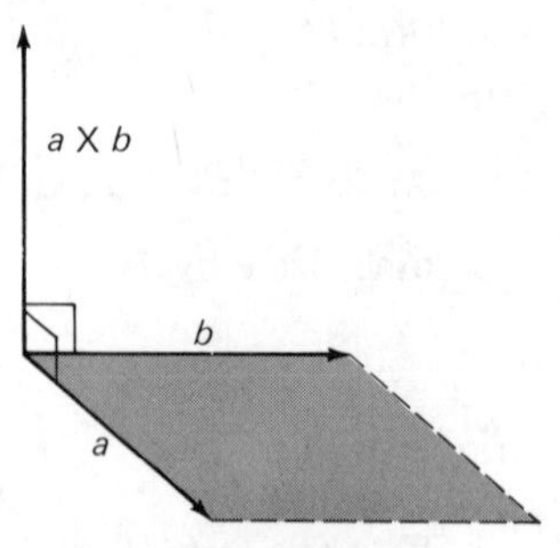


*Fig. 123.1*

If $a$, $b$ are linearly dependent, then $a \times b = \theta$. (This happens when $a,b$ point in the same direction or in opposite directions.)

Let $\Sigma \subset \mathbf{E}^3$ denote a smooth surface in $\mathbf{E}^3$ and let $b \in \Sigma$. If $f\colon \mathfrak{D} \to \mathbf{E}^3$ is a parametrization of a plaster on $\Sigma$ containing $b$ then, with $f^{-1}(b) = a = (\alpha_1, \alpha_2)$, the image under $f$ of the two lines

$$\left.\begin{aligned}\tau_1 &= \sigma_1\\ \tau_2 &= \alpha_2\end{aligned}\right\}\sigma_1 \in (\alpha_1 - \delta, \alpha_1 + \delta), \qquad \text{and} \qquad \left.\begin{aligned}\tau_1 &= \alpha_1\\ \tau_2 &= \sigma_2\end{aligned}\right\}\sigma_2 \in (\alpha_2 - \delta, \alpha_2 + \delta),$$

which lie in $N_\delta(a) \subseteq \mathfrak{D}$ for some $\delta > 0$ and cross at $a$, namely,

$$x = f(\sigma_1, \alpha_2), \sigma_1 \in (\alpha_1 - \delta, \alpha_1 + \delta), \qquad \text{and} \qquad x = f(\alpha_1, \sigma_2), \sigma_2 \in (\alpha_2 - \delta, \alpha_2 + \delta)$$

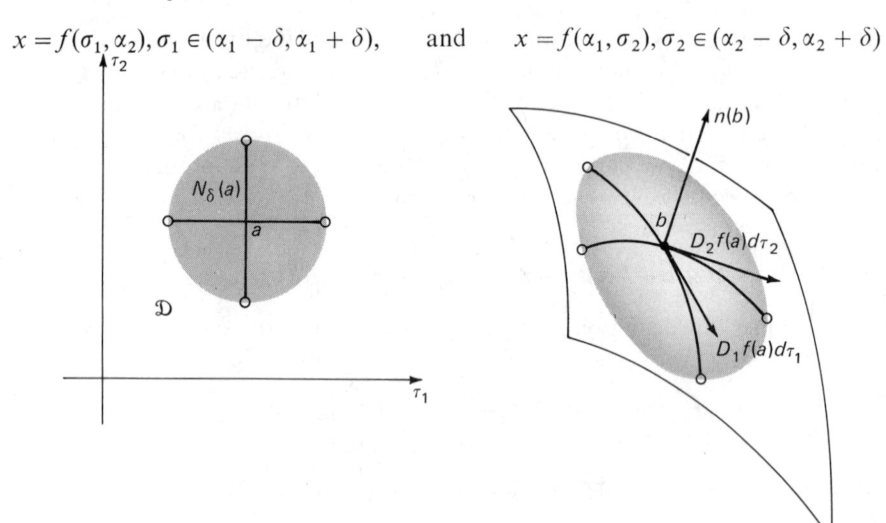


*Fig. 123.2*

are smooth one-dimensional plasters on $\Sigma$ and cross at $b$. (See Fig. 123.2.) The tangent vectors to these two curves at $b$ are $D_1 f(\alpha_1, \alpha_2) d\sigma_1$ and $D_2 f(\alpha_1, \alpha_2) d\sigma_2$ or, equivalently, $D_1 f(a) d\tau_1$ and $D_2 f(a) d\tau_2$. Hence,

(123.3) $$n(f(a)) = (D_1 f(a) \times D_2 f(a)) d\tau_1 d\tau_2$$

is a normal vector to the plane spanned by $D_1 f(a)d\tau_1$ and $D_2 f(a)d\tau_2$. Clearly, $n(f(a)) \neq \theta$ because rank $f'(t) = 2$.

***Lemma 123.1***

If $\Sigma \subset \mathbf{E}^3$ is a smooth surface, then the tangent vectors at $b \in \Sigma$, to all smooth curves that lie on $\Sigma$ and pass through $b$, all lie in the same plane.

*Proof*

Let $f: \mathfrak{D} \to \mathbf{E}^3$ parametrize a plaster of $\Sigma$ that contains $b$ and let $f^{-1}(b) = a$. If $\Gamma$ is a one-dimensional plaster on $\Sigma$ that passes through $b$, then, by Lemma 121.4, there is a one-dimensional plaster $\gamma \subset \mathfrak{D}$ such that $f_*(\gamma) = \Gamma$. Let $h = (\chi_1, \chi_2) : \mathfrak{J} \to \mathbf{E}^2$ denote a parametrization of $\gamma$ and let $h^{-1}(a) = c$. Then, $f \circ h : \mathfrak{J} \to \mathbf{E}^3$ is a parametrization of $\Gamma$ and we obtain for the tangent vector to $\Gamma$ at $b$:

$$\begin{aligned}(f \circ h)'(c)dt &= f'(a) \circ h'(c)dt \\ &= D_1 f(a)\chi_1'(c)dt + D_2 f(a)\chi_2'(c)dt,\end{aligned}$$

i.e., the tangent vector to $\Gamma$ at $b \in \Sigma$ is a linear combination of the two tangent vectors $D_1 f(a)d\tau_1$ and $D_2 f(a)d\tau_2$, i.e., lies in the same plane.

Before we can make a definite statement about the tangent plane, we have to investigate to what extent the normal vector in (123.3) depends on the parametrization.

Let $f: \mathfrak{D} \to \mathbf{E}^3$, $g : \mathcal{E} \to \mathbf{E}^3$ denote equivalent parametrizations of a plaster on $\Sigma$ containing $b$ and let $f^{-1}(b) = a$, $g^{-1}(b) = c$. Then $f = g \circ p$, where $p = (\pi_1, \pi_2) : \mathfrak{D} \leftrightarrow \mathcal{E}$ is a diffeomorphism, and $p(a) = c$. Since

$$D_k \phi_i(a) = D_1 \psi_i(c) D_k \pi_1(a) + D_2 \psi_i(c) D_k \pi_2(a), \qquad i = 1, 2, 3; k = 1, 2,$$

we have

$$D_1 f(a) \times D_2 f(a) \tag{123.4}$$

$$= \begin{vmatrix} e_1 & e_2 & e_3 \\ D_1\psi_1(c)D_1\pi_1(a) + D_2\psi_1(c)D_1\pi_2(a) & D_1\psi_2(c)D_1\pi_1(a) + D_2\psi_2(c)D_1\pi_2(a) & D_1\psi_3(c)D_1\pi_1(a) + D_2\psi_3(c)D_1\pi_2(a) \\ D_1\psi_1(c)D_2\pi_1(a) + D_2\psi_1(c)D_2\pi_2(a) & D_1\psi_2(c)D_2\pi_1(a) + D_2\psi_2(c)D_2\pi_2(a) & D_1\psi_3(c)D_2\pi_1(a) + D_2\psi_3(c)D_2\pi_2(a) \end{vmatrix}.$$

If we denote

$$D_1 f(a) \times D_2 f(a) = \begin{vmatrix} e_1 & e_2 & e_3 \\ D_1\phi_1(a) & D_1\phi_2(a) & D_1\phi_3(a) \\ D_2\phi_1(a) & D_2\phi_2(a) & D_2\phi_3(a) \end{vmatrix} = d(r_0, r_1, r_2),$$

where $r_0, r_1, r_2$ denote the rows of this determinant, we have, from (123.4), that

$$\begin{aligned}D_1 f(a) \times D_2 f(a) &= d(r_0, r_1 D_1\pi_1(a) + r_2 D_1\pi_2(a), r_1 D_2\pi_1(a) + r_2 D_2\pi_2(a)) \\ &= (D_1\pi_1(a)D_2\pi_2(a) - D_1\pi_2(a)D_2\pi_1(a))d(r_0, r_1, r_2).\end{aligned}$$

Hence,

$$D_1 f(a) \times D_2 f(a) = J_p(a)(D_1 g(c) \times D_2 g(c)), \tag{123.5}$$

and it is clear that the two normal vectors $(D_1 f(a) \times D_2 f(a))d\tau_1 d\tau_2$ and $(D_1 g(c) \times D_2 g(c))d\sigma_1 d\sigma_2$ that are obtained from equivalent parametrizations differ only by a nonzero multiplicative constant.

In view of this result and because of Lemma 123.1, we can state:

***Theorem 123.1***

If $\Sigma \subset \mathbf{E}^3$ is a smooth surface, then the tangent plane to $\Sigma$ at every point $b \in \Sigma$ exists and is unique. If $f: \mathfrak{D} \to \mathbf{E}^3$ is a (smooth) parametrization of the plaster on $\Sigma$ that contains $b$, then

$$D_1 f(f^{-1}(b)) \times D_2 f(f^{-1}(b)) \cdot (x - b) = 0$$

is an equation of the tangent plane at that point.

We see from (123.5) that the normal vectors that are obtained from consistent parametrizations (see definition 122.2) point in the same direction. Otherwise they point in opposite directions. As in the case of curves, a parametrization induces an orientation in a plaster as manifested by the direction of the normal.

---

▲ *Example 1*

We consider the plane $\xi_3 = 1$ in $\mathbf{E}^3$. For the parametrization $f(t) = (\tau_1, \tau_2, 1)$, $(\tau_1, \tau_2) \in \mathbf{E}^2$, we have

$$D_1 f(t) \times D_2 f(t) = \begin{vmatrix} e_1 & e_2 & e_3 \\ 1 & 0 & 0 \\ 0 & 1 & 0 \end{vmatrix} = (0, 0, 1),$$

and for the parametrization $g(s) = (\sigma_2, \sigma_1, 1)$, $(\sigma_1, \sigma_2) \in \mathbf{E}^2$, we have

$$D_1 g(s) \times D_2 g(s) = \begin{vmatrix} e_1 & e_2 & e_3 \\ 0 & 1 & 0 \\ 1 & 0 & 0 \end{vmatrix} = (0, 0, -1).$$

(See Fig. 123.3.) In both cases we have the same plane but not the same *oriented plane*.

---

If a surface has been oriented by means of a parametrization, we call the side facing the direction of the normal the *positive side*, and the side facing away from the normal the *negative side*. It is not at all clear that every surface has a positive side and a negative side. However, a plaster can always be parametrized such that it has a positive side and a negative side. This may be seen as follows: Since $D_1 f(t)$ and $D_2 f(t)$ have to turn, stretch, and shrink continuously as functions of $t$, $D_1 f(t) \times D_2 f(t)$ could only reverse its direction in view of the righthand screw rule if the two vectors $D_1 f(t)$ and $D_2 f(t)$ either cross each other or unfold until the angle from $D_1 f(t)$ to $D_2 f(t)$ exceeds 180 degrees. (See Fig. 123.4.)

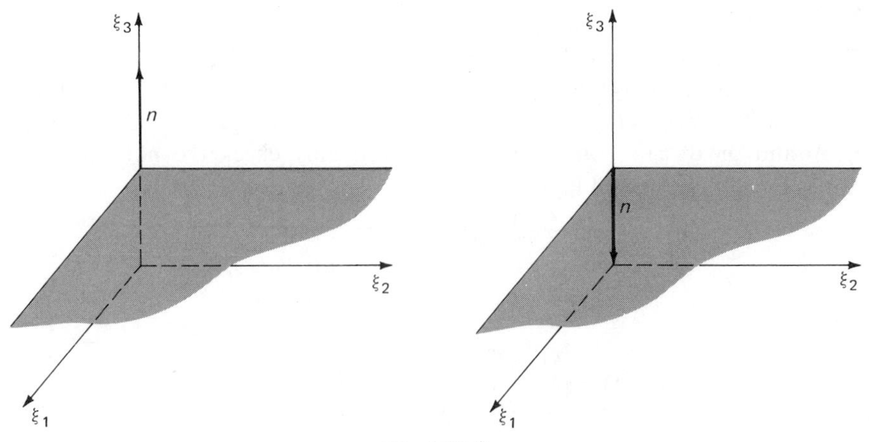


Fig. 123.3

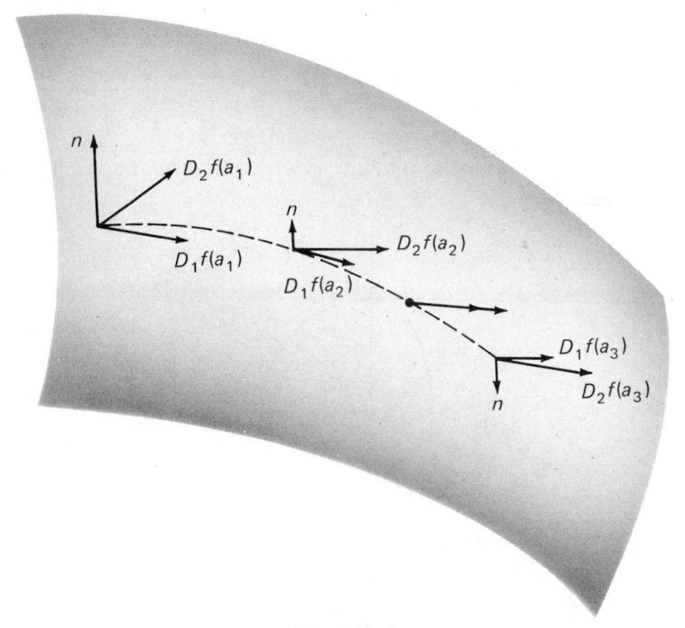


Fig. 123.4

This is impossible since rank $f'(t) = 2$ and since the crossing of $D_1 f(t)$ with $D_2 f(t)$, or the unfolding beyond 180 degrees, would mean that $D_1 f(t)$ and $D_2 f(t)$ are linearly dependent when the angle reaches 0 degrees or 180 degrees. In view of this, we can say that *a plaster is orientable.*

A smooth surface is said to be orientable if and only if it is possible to parametrize the plasters by which it is covered such that the parametrizations are consistent on the portions of overlap.

▲ *Example 2*

The sphere $\{x \in \mathbf{E}^3 \mid \|x\| = 1\}$ is orientable. We need four plasters, parametrized by $f(t) = (\sin \tau_1 \cos \tau_2, \sin \tau_1 \sin \tau_2, \cos \tau_1)$ on $(0, \pi) \times (0, 2\pi)$ and $(0, \pi) \times (-\pi, \pi)$ and by $g(s) = (-\cos \sigma_1, \sin \sigma_1 \sin \sigma_2, \sin \sigma_1 \cos \sigma_2)$ on $(0, \pi) \times (-(\pi/2), (\pi/2))$ and on $(0, \pi) \times ((\pi/2), (3\pi/2))$. We need only check the normal at one point of each plaster and find

$$n(0,0,1) = \begin{vmatrix} e_1 & e_2 & e_3 \\ 1 & 0 & 0 \\ 0 & 1 & 0 \end{vmatrix} d\sigma_1 d\sigma_2 = (0,0,1) d\sigma_1 d\sigma_2,$$

$$n(0,0,-1) = \begin{vmatrix} e_1 & e_2 & e_3 \\ 1 & 0 & 0 \\ 0 & -1 & 0 \end{vmatrix} d\sigma_1 d\sigma_2 = (0,0,-1) d\sigma_1 d\sigma_2,$$

$$n(1,0,0) = \begin{vmatrix} e_1 & e_2 & e_3 \\ 0 & 0 & -1 \\ 0 & 1 & 0 \end{vmatrix} d\tau_1 d\tau_2 = (1,0,0) d\tau_1 d\tau_2,$$

$$n(-1,0,0) = \begin{vmatrix} e_1 & e_2 & e_3 \\ 0 & 0 & -1 \\ 0 & -1 & 0 \end{vmatrix} d\tau_1 d\tau_2 = (-1,0,0) d\tau_1 d\tau_2.$$

Hence, all normals point to the outside. (*Outward normal*, see Fig. 123.5.)

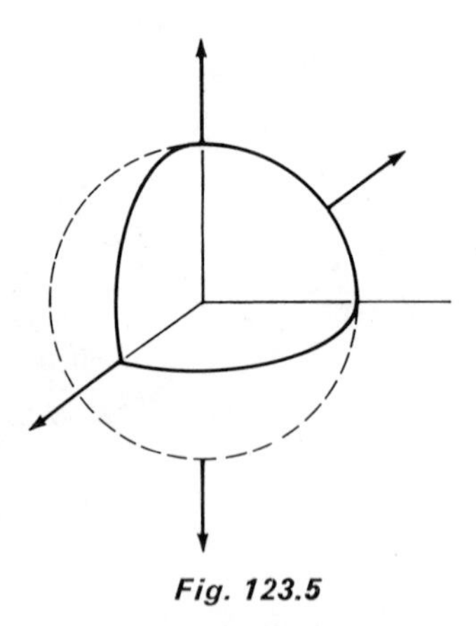

*Fig. 123.5*

▲ *Example 3* (*Möbius Strip*)

A Möbius† strip is not an orientable surface. A Möbius strip is obtained from a strip of paper (see Fig. 123.6) the ends of which are glued together after the strip has been given a half-twist.

† *A. F. Möbius*, 1790–1868.

*Fig. 123.6*

A Möbius strip as in Fig. 123.7 may be parametrized by

$$(123.6)\qquad f(t)=\left(\left(2-\tau_1\sin\frac{\tau_2}{2}\right)\cos\tau_2,\left(2-\tau_1\sin\frac{\tau_2}{2}\right)\sin\tau_2,\tau_1\cos\frac{\tau_2}{2}\right),$$
$$(\tau_1,\tau_2)\in(-1,1)\times[0,2\pi].$$

Two plasters that are parametrized by $f$ as in (123.6) on $(-1,1)\times(0,2\pi)$ and on $(-1,1)\times(-\pi,\pi)$ cover this Möbius strip. Hence, it is a smooth surface. It is, however, impossible to orient the Möbius strip because it only has one side. If one follows the normal all the way around, as indicated in Fig. 123.7, it will have reversed direction when the starting point is reached or, as one may put it, if a bug crawls along the surface, always keeping in the middle, it arrives at the starting point upside down. (Playing with Möbius strips is fun. The reader is encouraged to form a Möbius strip, cut it along the center line and cut the resulting strip again along the center line.)

---

## 13.123 Exercises

1. Find the equation of the tangent plane to the sphere $\{x\in\mathbf{E}^3 \mid \|x\|=1\}$ at the point $(1/2,1/2,(\sqrt{2}/2))$.
2. Find the equation of the tangent plane to the torus of example 9, section 13.120, at the point $(\sqrt{2}+\frac{1}{2},\sqrt{2}+\frac{1}{2},\sqrt{2}/2)$.
3. Show that the Möbius strip in Fig. 123.7 is a smooth surface.

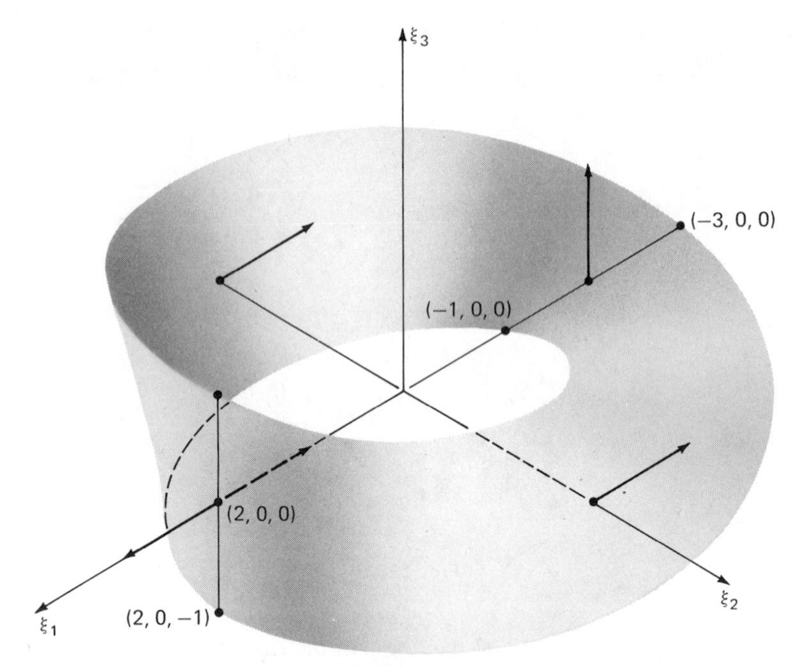


*Fig. 123.7*

4. Find the tangent plane to the Möbius strip in example 3 at the point $(\sqrt{2}, \sqrt{2}, 0)$.
5. Show that the unit vector in the direction of the normal at point $a$ to a smooth two-dimensional plaster is the same for all consistent parametrizations.
6. Let $\mathfrak{D} \subset \mathbf{E}^2$ denote an open set.
   (a) Show that $\mathfrak{D}$ is a smooth two-dimensional plaster.
   (b) Show that the normal to $\mathfrak{D}$ points upwards if the coordinate axes in the parameter space for the parametrization of $\mathfrak{D}$ ($\tau_1, \tau_2$-axes) are in standard position and that the normal points downward, if $\tau_1$- and $\tau_2$-axes in the parameter space are interchanged.
7. Show that the torus of example 9, section 13.120, is an orientable surface.

## 13.124 Patches and Quilts

If we wish to assign length and area measure to curves and surfaces, respectively, then the concept of a plaster is too general on the one hand and too restrictive on the other. $\{x \in \mathbf{E}^2 \mid \xi_2 = \sin(1/\xi_1), 0 < \xi_1 < 1\}$ is a one-dimensional plaster but it would be fairly hopeless to measure its length. (See Fig. 124.1.) On the other hand, the polygonal path in Fig. 124.2 is neither a plaster nor a smooth curve but it certainly has length.

To overcome this dilemma, we shall first consider certain subsets of plasters which we shall call *patches* and then consider manifolds that are put together from finitely many patches.

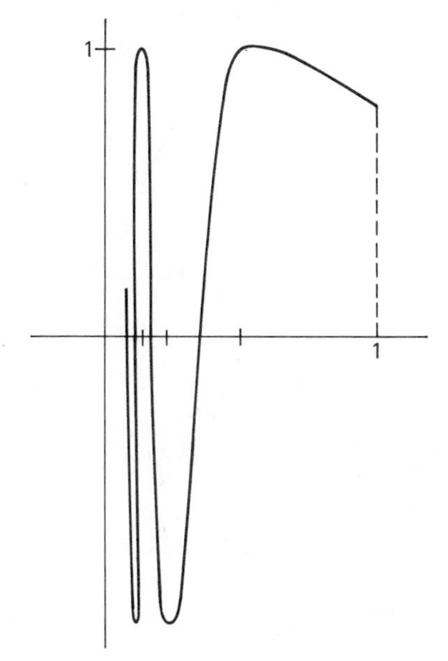


*Fig. 124.1*

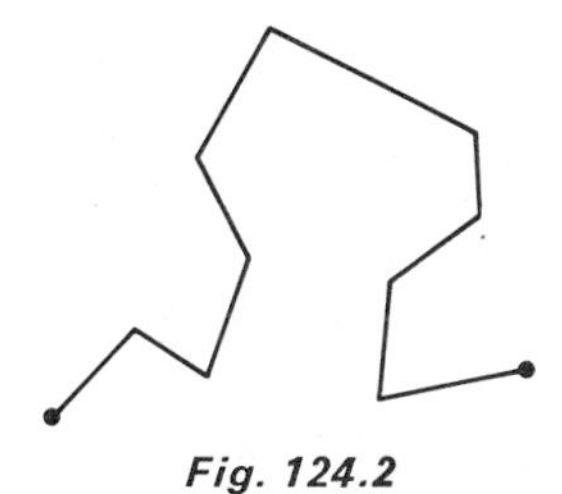

*Fig. 124.2*

***Definition 124.1***

$P \subset \mathbf{E}^n$ is a *k-dimensional patch* in $\mathbf{E}^n$ if and only if there is a $k$-dimensional plaster $\Pi \subset \mathbf{E}^n$ such that $P \subset \Pi$ and, for some parametrization $f : \mathfrak{D} \to \mathbf{E}^n$ of $\Pi$, $f^*(P)$ has the following properties:

1. $f^*(P)$ is compact,
2. $f^*(P)$ is connected,
3. $(f^*(P))^\circ \neq \varnothing$,
4. $(f^*(P))^\circ$ is connected,
5. $f^*(P)$ is Jordan-measurable,
6. $\beta(f^*(P))^\circ = \beta f^*(P)$.

$f_*[(f^*(P))^\circ] = P^\circ$ is called the *interior* of $P$ and $f_*[\beta f^*(P)] = \partial P$ is called the *boundary* of $P$. (We use the symbol $\partial$ to denote the boundary of a patch rather than the symbol $\beta$, because the boundary of a patch is not a boundary of a set

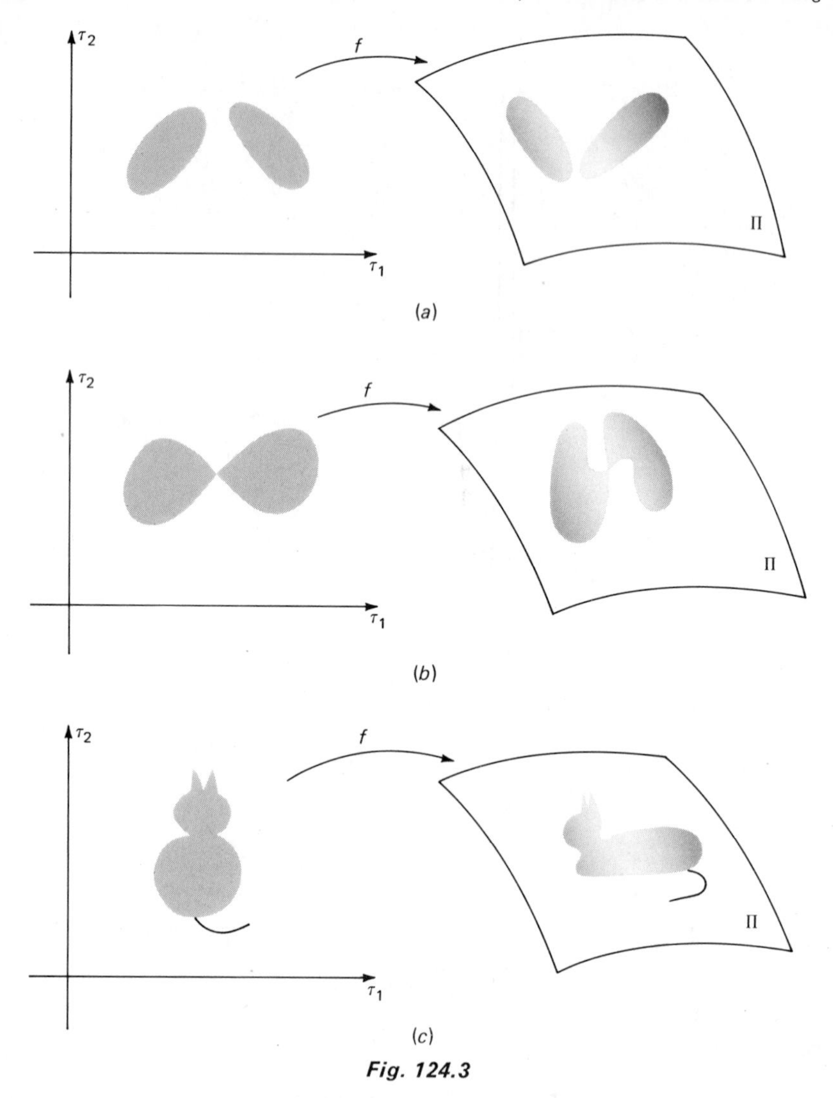


*Fig. 124.3*

by definition 104.2, since every $n$-dimensional neighborhood of a point on a $k$-dimensional patch $P$ contains points that are not in $P$ when $k < n$.)

Conditions (1), (3), and (5) in definition 124.1 are essential. Condition (2) excludes cases such as in Fig. 124.3a, condition (4) excludes cases such as in Fig. 124.3b, and condition (6) merely excludes useless appendages such as in Fig. 124.3c.

In cases such as in Fig. 124.3a and Fig. 124.3b, we simply speak of two patches —provided all other conditions are met. Cases such as the cat's tail in Fig. 124.3c are of no particular interest.

The property of being a patch is independent of the parametrization of the containing plaster $\Pi$: If $g : \mathcal{E} \to \mathbf{E}^n$ is an equivalent parametrization of $\Pi$, then

$f = g \circ p$ for some diffeomorphism $p : \mathfrak{D} \leftrightarrow \mathcal{E}$ and $g^*(P) = (p^{-1})^*(f^*(P)) = p_*(f^*(P))$. By Theorem 71.1, $g^*(P)$ is compact; by Theorem 73.1, $g^*(P)$ is connected; by Theorem 97.1, the interior of $f^*(P)$ is mapped onto the interior of $g^*(P)$ which is, therefore, not empty, and the interior of $g^*(P)$ is again connected by Theorem 73.1; by Theorem 112.2, $g^*(P)$ is Jordan-measurable; and finally, if $b$ were a boundary point of $g^*(P)$ but not a boundary point of its interior, then $p^{-1}(b)$ could not be in $f^*(P)$.

At this time we skirt the tricky question as to whether the boundary $\partial P$ of $P$ is also the "edge" of $P$. This will be discussed in section 13.131 for cases where the boundary of $f^*(P)$ has a simple structure.

When we speak of the parametrization $f$ of a patch $P$, we shall always mean the restriction of the parametrization of the containing plaster $\Pi$ to $f^*(P)$.

---

▲ *Example 1*

The shaded region $P$ on the sphere in Fig. 124.4 is part of a plaster, e.g. the right hemisphere without poles, which may be parametrized by $f(t) = (\sin \tau_1 \cos \tau_2, \sin \tau_1 \sin \tau_2, \cos \tau_1)$, $(\tau_1, \tau_2) \in (0, \pi) \times (0, \pi)$. Then, $f^*(P) = [\pi/6, \pi/3] \times [\pi/6, \pi/3]$. This set has all the properties listed in definition 124.1. Hence, $P$ is a patch.

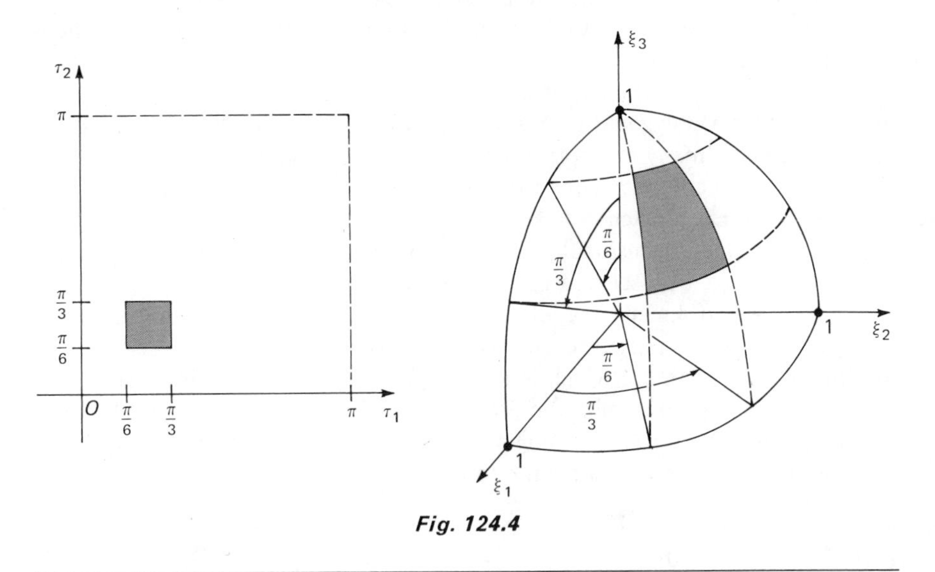


*Fig. 124.4*

---

Next we consider *quilts*. A quilt is patchwork. To wit:

***Definition 124.2***

A $k$-dimensional quilt in $\mathbf{E}^n$ is a connected set that may be represented as a union of finitely many $k$-dimensional patches with only boundary points in common.

Henceforth we shall use the terms *arc* for one-dimensional patches, *regular curve* for one-dimensional quilts without multiple points, *patch* for two-dimensional patches, and *regular surface* for two-dimensional quilts without multiple points.

Clearly, the polygonal path in Fig. 124.2 is a regular curve and the sphere and torus are regular surfaces. A stripe of width 1, for example, on the Möbius strip of example 3, section 13.123, symmetric with respect to the center line, is also a regular surface.

It will be shown in the following two sections that regular curves have length and that regular surfaces have area measure.

## 13.124 Exercises

1. Show that a patch is compact and connected and that a quilt is compact and connected.
2. Show that a sphere and a torus are regular surfaces.
3. Show that a circle is a regular curve.
4. Let $M \subset \mathbf{E}^3$ be parametrized by $f: \mathfrak{D} \to \mathbf{E}^3$, where $\mathfrak{D} \subseteq \mathbf{E}^2$ is open, where $f \in C^1(\mathfrak{D})$, and rank $f'(t) = 2$ for all $t \in \mathfrak{D}$. Show: If there is a compact Jordan-measurable subset $K$ of $\mathfrak{D}$ such that $f: K \xrightarrow{\text{onto}} M$ where $f$ is injective in the interior of $K$, then $M$ is a regular surface. Apply this theorem to the sphere.
5. Show that a line segment is an arc.
6. Show that a polygonal path consisting of finitely many line segments and with finitely many multiple points is a one-dimensional quilt.
7. Show that a compact, connected Jordan-measurable set in $\mathbf{E}^2$ with connected interior is a patch.
8. Show: A one-dimensional plaster is a simple curve and an arc is a compact and connected subset of a one-dimensional plaster.
9. Show: A two-dimensional plaster is a simple surface and a patch is a compact and connected subset of a two-dimensional plaster.
10. Show: If $\gamma$ is an arc which is parametrized by $f: [a, b] \to \mathbf{E}^n$, then $f(a)$ is the beginning point of $\gamma$ and $f(b)$ is the endpoint of $\gamma$, i.e., $\gamma$ lies between $f(a)$ and $f(b)$ on the containing plaster and $\gamma$ is traversed from $f(a)$ to $f(b)$ as the parameter increases from $a$ to $b$.

## 13.125 Arc Length

The length of an arc is to be defined in such a manner that for the case when the arc is a line segment, the familiar concept of *length* emerges. We offer the following heuristic argument to make the subsequent definition of length appear reasonable:

Let $\gamma \subset \mathbf{E}^n$ denote an arc that is parametrized by $f: [a, b] \to \mathbf{E}^n$. Let $P = \{t_0, t_1, \ldots, t_{n-1}, t_n\}$, $t_0 = a$, $t_n = b$, represent a partition of $[a, b]$ and let $f(t_j) = x_j \in \gamma$.

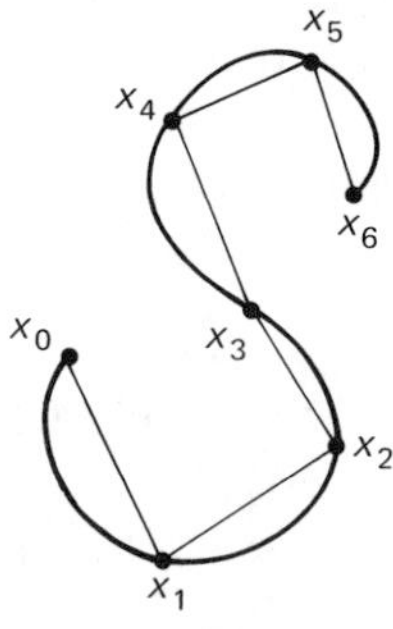


*Fig. 125.1*

We replace $\gamma$ by a polygonal path with vertices at $x_j$ (see Fig. 125.1) and argue that for a sufficiently fine partition, the length $\sum_{j=1}^{n} \|x_j - x_{j-1}\|$ of the polygonal path is a good approximation for the length of $\gamma$—if $\gamma$ has a length. From Corollary 2 to Theorem 93.1,

$$x_j - x_{j-1} = f(t_j) - f(t_{j-1}) = \begin{pmatrix} \phi_1'(t_j^{(1)}) \\ \vdots \\ \phi_n'(t_j^{(n)}) \end{pmatrix} (t_j - t_{j-1})$$

for some $t_j^{(k)} \in (t_{j-1}, t_j)$, $k = 1, 2, \ldots, n$. Hence

$$\|x_j - x_{j-1}\| = \sqrt{\sum_{k=1}^{n} \phi_k'^2(t_j^{(k)})}\,(t_j - t_{j-1}) \cong \sqrt{\sum_{k=1}^{n} \phi_k'^2(\bar{t}_j)}\,(t_j - t_{j-1})$$

if we replace $t_j^{(k)}$ by the same $\bar{t}_j$ for all $k = 1, 2, \ldots, n$. Then,

$$\sum_{j=1}^{n} \|f'(\bar{t}_j)\| (t_j - t_{j-1})$$

is an approximation for the length of the polygonal path, and as such, also an approximation for the length of $\gamma$. This is a typical intermediate sum for the function $\|f'\|$ on $[a, b]$ and we are led to the following definition:

***Definition 125.1***

The length of the arc $\gamma \subset \mathbf{E}^n$ is defined by

$$l(\gamma) = \int_a^b \|f'(t)\|\, dt, \tag{125.1}$$

where $f : [a, b] \to \mathbf{E}^n$ is a parametrization of $\gamma$.

In order for this definition to make sense, the integral in (125.1) has to exist, $l(\gamma)$ has to be independent of the parametrization of $\gamma$, and for a straight line segment $\gamma$, (125.1) has to yield the familiar result.

The integral exists, because $\|f'(t)\| = \sqrt{\sum_{k=1}^{n} \phi_k'^2(t)}$ is continuous on $[a, b]$.

If $g : [\alpha, \beta] \to \mathbf{E}^n$ is another parametrization of $\gamma$, then $f = g \circ p$ and we have, from Corollary 2 to Theorem 52.3, that

$$\int_{p(a)}^{p(b)} \|g'(s)\|\, ds = \int_a^b \|g'(p(t))\| p'(t)\, dt,$$

while

$$\int_a^b \|f'(t)\|\, dt = \int_a^b \|g'(p(t))\|\, |p'(t)|\, dt.$$

If $p'(t) > 0$ on $[a, b]$, then $p(a) = \alpha < p(b) = \beta$, and if $p'(t) < 0$, then $p(a) = \beta > p(b) = \alpha$. Hence,

$$\int_\alpha^\beta \|g'(s)\|\, ds = \int_a^b \|f'(t)\|\, dt;$$

i.e., $l(\gamma)$ does not depend on the parametrization of $\gamma$.

Finally, if $\gamma$ is a line segment from $a \in \mathbf{E}^n$ to $b \in \mathbf{E}^n$, then $f(t) = (1 - t)a + tb$, $t \in [0, 1]$ defines a parametrization of $\gamma$ and $f'(t) = -a + b$. Hence, $\|f'(t)\| = \|b - a\|$ and we obtain, as expected,

$$l(\gamma) = \int_0^1 \|b - a\|\, dt = \|b - a\|.$$

---

▲ *Example 1*

If $f \in C^1[a, b]$, then $\{(x, f(x)) \mid x \in [a, b]\}$ is an arc $\gamma$ in $\mathbf{E}^2$, and we obtain the formula

$$l(\gamma) = \int_a^b \sqrt{1 + f'^2(x)}\, dx,$$

which is familiar from calculus.

▲ *Example 2*

If $\Phi(\xi_1, \xi_2) = 0$ defines an arc $\gamma$ on some interval $[a, b]$ on the $x$-axis (see also example 1, section 13.120), then

$$l(\gamma) = \int_a^b \sqrt{(D_1\Phi(\xi_1, h(\xi_1)))^2 + (D_2\Phi(\xi_1, h(\xi_1)))^2}\, \frac{d\xi_1}{|D_2\Phi(\xi_1, h(\xi_1))|},$$

where $\Phi(\xi_1, h(\xi_1)) = 0$.

---

If $\Gamma$ is a regular curve, i.e., a finite union of arcs $\gamma_1, \gamma_2, \ldots, \gamma_l$ with only boundary points in common and without double points, which we denote symbolically by $\gamma_1 + \gamma_2 + \cdots + \gamma_l$, then we define

$$l(\Gamma) = \sum_{j=1}^{l} l(\gamma_j). \tag{125.2}$$

---

▲ *Example 3*

The circle $C = \{x \in \mathbf{E}^2 \mid \|x\| = R\}$ is a regular curve consisting of upper and lower semicircles. With $f(t) = (R\cos t, R\sin t)$, $t \in [0, 2\pi]$, we obtain

$$l(C) = \int_0^{2\pi} R dt = 2R\pi.$$

---

Our theory of length is not the most general theory that could be developed, but is more general than one might think, because it applies to cases that do not appear upon superficial examination to fit into our framework. Two such cases will be discussed in the next two examples.

---

▲ *Example 4*

The curve $\gamma$ that is defined by $\{x \in \mathbf{E}^2 \mid \xi_1^2 = \xi_2^3\}$ between $(-1, 1)$ and $(1, 1)$ (see example 8, section 13.118 and Fig. 125.2) does not appear to be a regular curve,

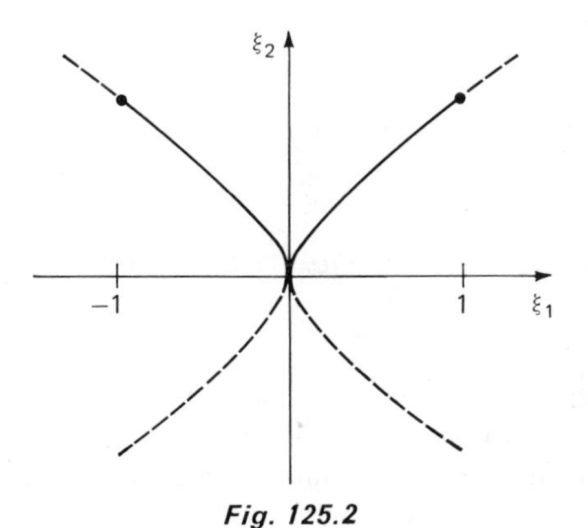


*Fig. 125.2*

as neither the left nor the right branch seems to be an arc. However, one may view the right branch as part of $\{(f(\xi_2,)\, \xi_2) \mid \xi_2 \in (-\infty, \infty)\}$ where

$$f(\xi_2) = \begin{cases} \xi_2^{3/2} & \text{for } \xi_2 \geqslant 0, \\ (-\xi_2)^{3/2} & \text{for } \xi_2 < 0, \end{cases}$$

which is a plaster as the reader can easily see. For the left branch we simply replace $f(\xi_2)$ by $-f(\xi_2)$. Hence

$$l(\gamma) = 2\int_0^1 \sqrt{\frac{9}{4}\xi_2 + 1}\, d\xi_2 = \frac{8}{27}(\sqrt{13} - 3).$$

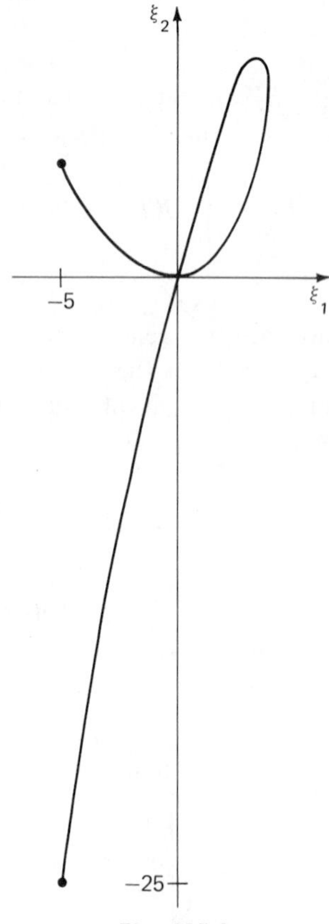


*Fig. 125.3*

▲ *Example 5*

The curve $\gamma$ in Fig. 125.3 which is parametrized by $f(t) = (4t - t^2, 4t^2 - t^3)$, $t \in [-1, 5]$, has a double point at $\theta$. Still, it may be viewed as the union of three arcs with only boundary points in common, and we obtain

$$l(\gamma) = \int_{-1}^{5} \sqrt{(4 - 2t)^2 + (8t - 3t^2)^2}\, dt.$$

This integral exists. Its evaluation is another matter. While it is very reassuring to know that regular curves have length, it is usually very difficult, if not impossible, to evaluate the pertinent integrals in a closed form.

Integrals of the above type may be expressed in terms of elementary functions and elliptic integrals.†

---

† See P. F. Byrd and M. D. Friedman, *Handbook of Elliptic Integrals for Engineers and Scientists*, 2d ed., Die Grundlehren der Mathematischen Wissenschaften, Vol. 67 (New York: Springer-Verlag Inc., 1971), p. 8.

## 13.125 Exercises

1. Find the length of that arc on the parabola $\{(x, x^2) \mid x \in \mathbf{R}\}$ that lies between $(0, 0)$ and $(1, 1)$.
2. A circle of radius $\frac{1}{3}$ rolls along the inside of the circumference of a circle of radius 1. A fixed point on the rolling circle describes a *hypocycloid.* Find the length of the hypocycloid.
3. Find the length of the helix in example 5, section 13.118, for $r = 1$, $a = 1$ between $(1, 0, 0)$ and $(1, 0, 1)$.
4. Let $\Gamma$ represent a plaster that is parametrized by $f : (a, b) \to \mathbf{E}^n$. Introduce a new parameter, call it $s$, by means of $p(t) = \int_{t_0}^{t} \|f'(\tau)\| \, d\tau$, where $t_0 \in (a, b)$.
   (a) Show that $p$ is a diffeomorphism for $t \in (a, t_0)$ and $t \in (t_0, b)$.
   (b) Show that $g = f \circ p^{-1} : (0, p(b)) \to \mathbf{E}^n$ and $(p(a), 0) \to \mathbf{E}^n$ are smooth parametrizations.
   (c) Show that $\|g'(s)\| = 1$ for all $s \in (p(a), 0)$ and $s \in (0, p(b))$.
   (d) Let $\gamma$ denote an arc on $\Gamma$ between $g(s_1)$ and $g(s_2)$ where $0 < s_1 < s_2 < p(b)$. Show that $l(\gamma) = \int_{s_1}^{s_2} ds$. Why is this result trivial?
5. A curve winds its way along the surface of the cone in example 4, section 13.119, starting at $\theta$ and rising at the constant rate of 1 unit per thread. Find the length of the first thread.
6. The mass density of an arc $\gamma \subset \mathbf{E}^2$ at $x \in \gamma$ is given by $\mu(x)$ where $\mu : \mathbf{E}^2 \to \mathbf{E}$ is a continuous function. If $f : [a, b] \to \mathbf{E}^2$ is the parametrization of $\gamma$, then

$$\int_a^b \phi_j(t)\mu(f(t)) \, \|f'(t)\| dt$$

is called the moment of $\gamma$ with respect to the $\xi_i$-axis, $i \neq j$; $i, j = 1, 2$, and

$$\xi_j = \frac{\int_a^b \phi_j(t)\mu(f(t)) \, \|f'(t)\| \, dt}{\int_a^b \mu(f(t)) \, \|f'(t)\| \, dt}, \qquad j = 1, 2,$$

are the coordinates of the centroid. Find the coordinates of the centroid of a catenary $\{(\xi_1, a\cosh(\xi_1 - b)/a) \mid 0 \leqslant \xi_1 \leqslant 2b\}$ with mass density 1.

7. Let $g : \mathfrak{J} \to \mathfrak{D}$ where $\mathfrak{J}$ is an open interval and $\mathfrak{D} \subset \mathbf{E}^2$ is open, and let $f : \mathfrak{D} \to \mathbf{E}^3$. Show: If $g, f$ are injective, $g \in C^1(\mathfrak{J})$, $f \in C^1(\mathfrak{D})$, rank $g'(s) = 1$ for all $s \in \mathfrak{J}$, and rank $f'(t) = 2$ for all $t \in \mathfrak{D}$, then $f \circ g : \mathfrak{J} \to \mathbf{E}^3$ represents a one-dimensional plaster $\gamma$ in $\mathbf{E}^3$ and, for $a, b \in \mathfrak{J}$

$$\int_a^b \sqrt{E(\psi_1')^2 + 2F\psi_1'\psi_2' + G(\psi_2')^2} \, ds$$

is the length of the arc on $\gamma$ between $f \circ g(a)$ and $f \circ g(b)$. (For the definition of $E, F, G$, see exercise 7, section 13.122.)

8. Use the formula in exercise 7 to express the length of an arc on a meridian and the length of an arc on a parallel circle on the unit sphere. (See also exercise 8(b), section 13.122.)

## 13.126 Surface Area

As in the preceding section we shall first present a heuristic argument to support the plausibility of the forthcoming definition of area measure of a (two-dimensional) patch.

Let $\sigma$ denote a patch that is parametrized by $f: A \to \mathbf{E}^3$, where $A \subset \mathbf{E}^2$, which satisfies the six conditions in definition 124.1. Let $P$ denote a partition of $A$ generating the subintervals $\mathfrak{J}_{i,j}$, $i = 1, 2, \ldots, n$; $j = 1, 2, \ldots, m$. (See Fig. 126.1.)

We approximate the area $\mu(f_*(\mathfrak{J}_{i,j}))$ of $f_*(\mathfrak{J}_{i,j})$ by the area of the parallelogram that is spanned by the two vectors in Fig. 126.2:

$$\mu(f_*(\mathfrak{J}_{i,j})) \cong \|[f(\tau_1^i, \tau_2^{j-1}) - f(\tau_1^{i-1}, \tau_2^{j-1})] \times [f(\tau_1^{i-1}, \tau_2^j) - f(\tau_1^{i-1}, \tau_2^{j-1})]\|,$$

and approximate the term on the right, in turn, by

$$\|D_1 f(t_{ij}) \times D_2 f(t_{ij})\| |\mathfrak{J}_{i,j}|$$

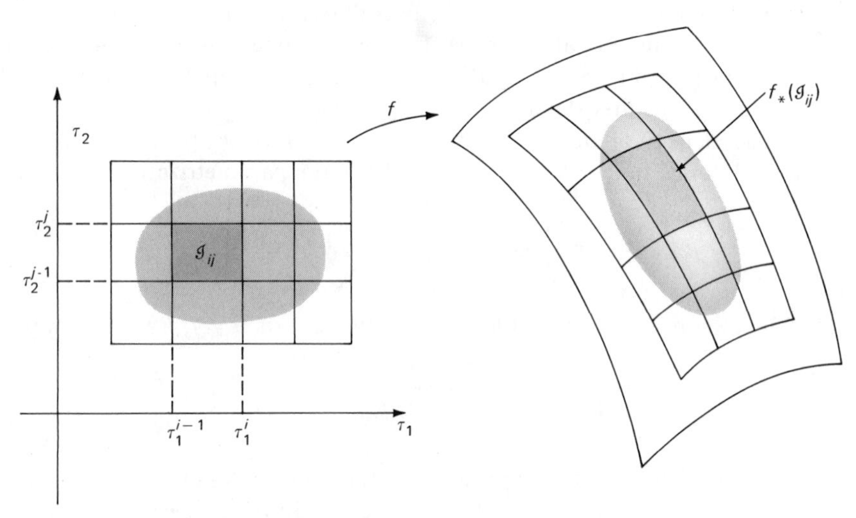


*Fig. 126.1*

*Fig. 126.2*

for some $t_{ij} \in \mathfrak{J}_{i,j}$. If we sum over all $i$ and $j$, we obtain a typical intermediate sum for the function $\| D_1 f \times D_2 f \|$ on $A$. This leads to the following definition.

***Definition 126.1***

The area measure of the patch $\sigma \subset \mathbf{E}^3$ is defined by

$$\mu(\sigma) = \iint_A \| D_1 f(t) \times D_2 f(t) \| \, d\tau_1 d\tau_2, \tag{126.1}$$

where $f : A \to \mathbf{E}^3$ is a parametrization of $\sigma$.

The integral exists because the integrand is continuous and $A$ is Jordan-measurable.

If $g : B \to \mathbf{E}^3$ is another parametrization of $\sigma$, then $f = g \circ p$ and we have, from Jacobi's theorem (Theorem 115.1),

$$\iint_{p(A)} \| D_1 g(s) \times D_2 g(s) \| \, d\sigma_1 d\sigma_2 = \iint_A \| D_1 g(p(t)) \times D_2 g(p(t)) \| \, |J_p(t)| \, d\tau_1 d\tau_2$$

while, in view of (123.5),

$$\iint_A \| D_1 f(t) \times D_2 f(t) \| \, d\tau_1 d\tau_2 = \iint_A \| (D_1 g(p(t)) \times D_2 g(p(t))) \| \, |J_p(t)| \, d\tau_1 d\tau_2.$$

Since $p(A) = B$,

$$\iint_B \| D_1 g(s) \times D_2 g(s) \| \, d\sigma_1 d\sigma_2 = \iint_A \| D_1 f(t) \times D_2 f(t) \| \, d\tau_1 d\tau_2,$$

that is, $\mu(\sigma)$ is independent of the parametrization of $\sigma$.

Finally, if $\sigma$ is a patch that lies in $\mathbf{E}^2$, i.e., a Jordan-measurable set $A$ in $\mathbf{E}^2$, we obtain, with the parametrization $f(t) = (\tau_1, \tau_2)$, $(\tau_1, \tau_2) \in A$, the expected result that

$$\mu(A) = \iint_A d\tau_1 d\tau_2 = J(A),$$

because

$$D_1 f(t) \times D_2 f(t) = \begin{vmatrix} e_1 & e_2 & e_3 \\ 1 & 0 & 0 \\ 0 & 1 & 0 \end{vmatrix} = (0, 0, 1).$$

If $\Sigma$ is a regular surface without double points, $\Sigma = \sigma_1 + \sigma_2 + \cdots + \sigma_l$, where the $\sigma_j$ are patches with only boundary points in common, we define

$$\mu(\Sigma) = \sum_{j=1}^{l} \mu(\sigma_j). \tag{126.2}$$

▲ *Example 1*

If $\{(\xi_1, \xi_2, f(\xi_1, \xi_2)) \mid (\xi_1, \xi_2) \in A\}$ defines a patch $\sigma$ (see example 1, section 13.119), we obtain

$$\begin{vmatrix} e_1 & e_2 & e_3 \\ 1 & 0 & D_1 f(\xi_1, \xi_2) \\ 0 & 1 & D_2 f(\xi_1, \xi_2) \end{vmatrix} = (-D_1 f(\xi_1, \xi_2), -D_2 f(\xi_1, \xi_2), 1),$$

and hence, from (126.1),

$$\mu(\sigma) = \iint_A \sqrt{1 + (D_1 f(\xi_1, \xi_2))^2 + (D_2 f(\xi_1, \xi_2))^2}\, d\xi_1 d\xi_2.$$

▲ *Example 2*

If $\Phi(\xi_1, \xi_2, \xi_3) = 0$ defines a patch $\sigma$ on a Jordan-measurable set $A$ in the $(\xi_1, \xi_2)$-plane, we obtain, when $D_3 \Phi \neq 0$,

$$\mu(\sigma) = \iint_A \sqrt{(D_1 \Phi)^2 + (D_2 \Phi)^2 + (D_3 \Phi)^2}\, \frac{d\xi_1 d\xi_2}{|D_3 \Phi|}, \tag{126.3}$$

where $(\xi_1, \xi_2, h(\xi_1, \xi_2))$ is the argument of the partial derivatives in the above intergrand and $\Phi(\xi_1, \xi_2, h(\xi_1, \xi_2)) = 0$ on $A$. (See also example 2, section 13.120.)

We use this formula to find the area of the cross section $E$ of the right circular cylinder $\xi_1^2 + \xi_2^2 = 1$ with the plane $\xi_1 + \xi_2 + \xi_3 = 1$.

From (126.3),

$$\mu(E) = \iint_A \sqrt{1 + 1 + 1}\, d\xi_1 d\xi_2 = \sqrt{3}\, \pi,$$

since $A$ is the unit disk with area $\pi$.

▲ *Example 3*

The sphere $S = \{x \in \mathbf{E}^2 \mid \|x\| = 1\}$ is a regular surface and to find its area, strictly following our theory, we would have to find the areas of the individual patches and then add them up. However, when everything is said and done, it amounts to applying (126.1) to the parametrization $f(t) = (\sin \tau_1 \cos \tau_2, \sin \tau_1 \sin \tau_2, \cos \tau_1)$ on $[0, \pi] \times [0, 2\pi]$. (Note that this function is injective on $(0, \pi) \times (0, 2\pi)$. See also Jacobi's theorem.) Then,

$$D_1 f(t) \times D_2 f(t) = (\sin^2 \tau_1 \cos \tau_2, -\sin^2 \tau_1 \sin \tau_2, \sin \tau_1 \cos \tau_1).$$

Hence, $\|D_1 f(t) \times D_2 f(t)\| = \sin \tau_1$, and we obtain

$$\mu(S) = \int_0^{2\pi} \left( \int_0^{\pi} \sin \tau_1 \, d\tau_1 \right) d\tau_2 = 4\pi.$$

▲ *Example 4*

We obtain, for the torus $T$ of example 9, section 13.120, after some elementary manipulations,

$$\mu(T) = \int_0^{2\pi}\int_0^{2\pi}(2+\sin\tau_1)\,d\tau_1\,d\tau_2 = 4\pi^2.$$

---

Before closing this section, let us point out the similarity between the formulas for arc length and surface area. In the one case, we have $\int_a^b \|f'(t)\|\,dt$, and in the other, $\iint_A \|D_1 f(t) \times D_2 f(t)\|\,d\tau_1 d\tau_2$. We observe that $\|f'(t)\|\,dt = \|f'(t)\,dt\|$ is the length of the tangent vector (see section 13.122) to the curve $\gamma$, and $\|D_1 f(t) \times D_2 f(t)\|\,d\tau_1 d\tau_2 = \|(D_1 f(t) \times D_2 f(t))\,d\tau_1 d\tau_2\|$ is the length of the normal to the surface $\sigma$ (see section 13.123).

## 13.126 Exercises

1. Let $g : [a,b] \to \mathbf{E}$, where $[a,b]$ is an interval on the $\xi_1$-axis, denote a continuous function with a continuous derivative. The graph of $g$ rotates about the $\xi_1$-axis and generates a surface of revolution. Show that the integral $2\pi \int_a^b g(\xi_1)\sqrt{1+g'^2(\xi_1)}\,d\xi_1$ yields the area of this surface of revolution.
2. Use the result of exercise 1 to find the area of a right circular cone of height $h$ and basis radius $r$.
3. A patch is cut out from the upper hemisphere of radius 1 with center at $\theta$ by a right circular cylinder of radius $\frac{1}{2}$ with the $\xi_3$-axis as axis. Find the area of the patch.
4. Find the area of the surface that is given by the graph of $f : \mathfrak{D} \to \mathbf{E}$ where $f(\xi_1, \xi_2) = \xi_1\xi_2$ and where $\mathfrak{D}$ is the unit disk with center at $\theta$.
5. Find the surface area of a torus that is generated by the circle $(\xi_1 - a)^2 + \xi_3^2 = 1, 1 < a$, rotating about the $\xi_3$-axis.
6. Let $f : \mathfrak{D} \to \mathbf{E}^3$ where $\mathfrak{D} \subseteq \mathbf{E}^2$ is open, let $f \in C^1(\mathfrak{D})$ and let $K \subset \mathfrak{D}$ denote a compact Jordan-measurable set with interior points. Prove: If $f : K^\circ \xrightarrow[1-1]{} \mathbf{E}^3$, rank $f'(t) = 2$ in $K^\circ$, and if $p : \mathfrak{D} \to \mathcal{E}$ where $p \in C^1(\mathfrak{D})$ and where $p$ is a diffeomorphism on $K^\circ$, then the integral in (126.1) over $K$ is invariant under $p$. (Apply this result to the computation of the surface area of a sphere by means of 126.1.)
7. The mass density of a patch $\sigma \subset \mathbf{E}^3$ at $x \in \sigma$ is given by $\mu(x)$ where $\mu : \mathbf{E}^3 \to \mathbf{E}$ is a continuous function. If $f : A \to \mathbf{E}^3$ is a parametrization of $\sigma$, then

$$\iint_A \phi_j(t)\mu(f(t))\,\|D_1 f(t) \times D_2 f(t)\|\,d\tau_1 d\tau_2$$

is the moment of $\sigma$ with respect to the $\xi_i, \xi_k$-plane $(i \neq j, k \neq j)$, and

$$\xi_j = \frac{\iint_A \phi_j(t)\mu(f(t))\,\|D_1 f(t) \times D_2 f(t)\|\,d\tau_1 d\tau_2}{\iint_A \mu(f(t))\,\|D_1 f(t) \times D_2 f(t)\|\,d\tau_1 d\tau_2}, \qquad j = 1,2,3,$$

are the coordinates of the centroid of $\sigma$. Find the centroid of the paraboloid $\{(\xi_1, \xi_2, \xi_1^2 + \xi_2^2) \mid \xi_1^2 + \xi_2^2 \leqslant 1\}$ with mass density 1.

8. Show that (126.1) may be written in the form

$$\mu(\sigma) = \iint_A \sqrt{EG - F^2}\, d\tau_1 d\tau_2,$$

where $E, F, G$ are defined in exercise 7, section 13.122.

## 13.127 Differential Forms

Differential $k$-forms in $\mathbf{E}^n$ are functions from the set of all oriented $k$-dimensional patches in $\mathbf{E}^n$ into the reals. To speak of orientation definitively, we shall always assume that the objects of our investigation are described with reference to a coordinate system in *standard position*. A coordinate system in $\mathbf{E}^1, \mathbf{E}^2$, and $\mathbf{E}^3$, respectively, is in standard position if and only if the coordinate axes are oriented as in Fig. 127.1, or if it has been obtained from such a coordinate system by a

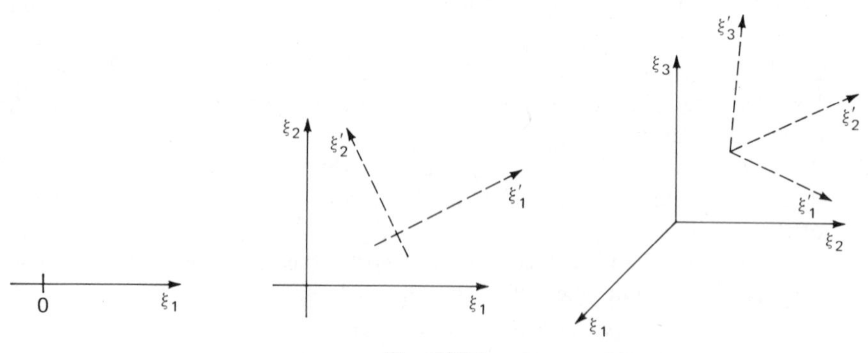


*Fig. 127.1*

translation in $\mathbf{E}^1$ in the case of $\mathbf{E}^1$, by a rotation and translation in $\mathbf{E}^2$ in the case of $\mathbf{E}^2$, and by a translation and rotation in $\mathbf{E}^3$ in the case of $\mathbf{E}^3$. (See Fig. 127.1.) (Note that the above-mentioned transformations are diffeomorphisms with Jacobian 1.)

The coordinate systems in Fig. 127.2 are *not* in standard position. They are obtained from coordinate systems in standard position by diffeomorphisms with Jacobian $-1$.

A $k$-dimensional plaster in $\mathbf{E}^k$ is called *positively oriented* through its parametrization $f: \mathfrak{D} \to \mathbf{E}^k$, $\mathfrak{D} \subseteq \mathbf{E}^k$, if and only if $J_f(t) > 0$ for all $t \in \mathfrak{D}$.

---

▲ *Example 1*

$f(t) = t$, $t \in (a, b)$, parametrizes a positively oriented one-dimensional plaster in $\mathbf{E}^1$ because $J_f(t) = f'(t) = 1 > 0$. (See Fig. 127.3.)

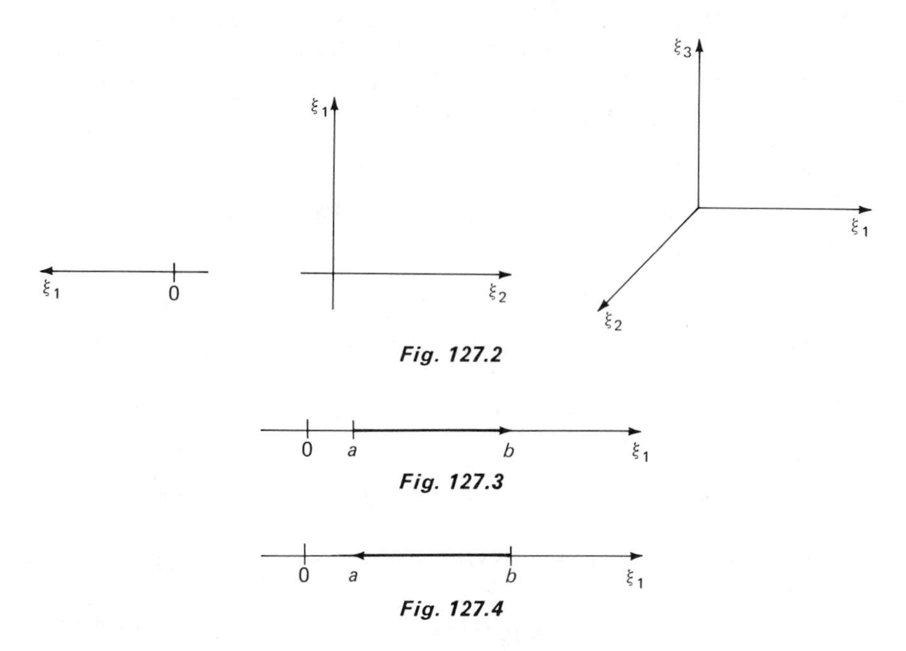


*Fig. 127.2*

*Fig. 127.3*

*Fig. 127.4*

$g(s) = -s, s \in (-b, -a)$ parametrizes the same plaster but negatively oriented because $J_g(s) = g'(s) = -1 < 0$. (See Fig. 127.4.)

▲ *Example 2*

If $\mathfrak{D} \subseteq \mathbf{E}^2$ is an open set, then $f(t) = (\tau_1, \tau_2)$, $(\tau_1, \tau_2) \in \mathfrak{D}$, parametrizes a positively oriented two-dimensional plaster in $\mathbf{E}^2$ because

$$J_f(t) = \begin{vmatrix} 1 & 0 \\ 0 & 1 \end{vmatrix} = 1 > 0;$$

and $g(s) = (\sigma_2, \sigma_1)$, $(\sigma_2, \sigma_1) \in \mathfrak{D}$, parametrizes the same plaster, but negatively oriented, because

$$J_g(s) = \begin{vmatrix} 0 & 1 \\ 1 & 0 \end{vmatrix} = -1 < 0.$$

▲ *Example 3*

If $\mathfrak{D} \subseteq \mathbf{E}^3$ is an open set, then $f_1(t) = (\tau_1, \tau_2, \tau_3)$, $(\tau_1, \tau_2, \tau_3) \in \mathfrak{D}$, $f_2(t) = (\tau_3, \tau_1, \tau_2)$, $(\tau_3, \tau_1, \tau_2) \in \mathfrak{D}$; $f_3(t) = (\tau_2, \tau_3, \tau_1)$, $(\tau_2, \tau_3, \tau_1) \in \mathfrak{D}$ parametrize the same positively oriented three-dimensional plaster in $\mathbf{E}^3$ and $g_1(s) = (\sigma_2, \sigma_1, \sigma_3)$, $(\sigma_2, \sigma_1, \sigma_3) \in \mathfrak{D}$; $g_2(s) = (\sigma_3, \sigma_2, \sigma_1)$, $(\sigma_3, \sigma_2, \sigma_1) \in \mathfrak{D}$; $g_3(s) = (\sigma_1, \sigma_3, \sigma_2)$, $(\sigma_1, \sigma_3, \sigma_2) \in \mathfrak{D}$ parametrize the same three-dimensional plaster in $\mathbf{E}^3$, but negatively oriented.

$d\xi_j$ is a differential form (a 1-form to be exact) when viewed as a function that maps the arc $\gamma \in \mathbf{E}^n$ into $d\xi_j(\gamma) = \int_a^b \phi_j'(t)dt$ where $f = (\phi_1, \phi_2, \ldots, \phi_n) : [a, b] \to \mathbf{E}^n$ is the parametrization of $\gamma$. Linear combinations of such 1-forms with scalar functions as coefficients are again 1-forms. In general:

***Definition 127.1***

Let $g : \Omega \to \mathbf{E}^n$ denote a continuous function on the open set $\Omega \subseteq \mathbf{E}^n$. The general 1-form $\omega$ in $\mathbf{E}^n$ is a function, denoted by

(127.1) $$\omega = g(x) \cdot dx = \psi_1(x)d\xi_1 + \psi_2(x)d\xi_2 + \cdots + \psi_n(x)d\xi_n,$$

which maps the oriented arc $\gamma \subset \Omega$, which is parametrized by $f : [a, b] \to \mathbf{E}^n$, into

(127.2) $$\omega(\gamma) = \int_a^b g(f(t)) \cdot f'(t)\, dt$$
$$= \int_a^b (\psi_1(f(t))\phi_1'(t) + \psi_2(f(t))\phi_2'(t) + \cdots + \psi_n(f(t))\phi_n'(t))\, dt.$$

If $\omega_1, \omega_2$ are two 1-forms in $\mathbf{E}^n$, then $(\omega_1 + \omega_2)(\gamma) = \omega_1(\gamma) + \omega_2(\gamma)$.

Clearly, the integral in (127.2) exists and is invariant under orientation-preserving diffeomorphisms, i.e., has the same value for all consistent parametrizations of $\gamma$.

We denote the oriented arc that is obtained from $\gamma$ by a reversal of the orientation by $-\gamma$ and obtain, from (127.2), that

(127.3) $$\omega(-\gamma) = -\omega(\gamma).$$

If $\Gamma = \gamma_1 + \gamma_2 + \cdots + \gamma_l$ is an oriented regular curve, $\gamma_1, \gamma_2, \ldots, \gamma_l$ being oriented arcs with only boundary points in common, we define

(127.4) $$\omega(\Gamma) = \sum_{j=1}^{l} \omega(\gamma_j).$$

The integral in (127.2) is called a *line integral* and is variously denoted by:

(127.5) $$\omega(\gamma) = \int_\gamma \omega = \int_\gamma g(x) \cdot dx = \int_\gamma (\psi_1(x)\, d\xi_1 + \psi_2(x)\, d\xi_2 + \cdots + \psi_n(x)\, d\xi_n).$$

The value $\omega(\gamma)$ is formally obtained from (127.5) be replacing $x$ by $f(t)$ and $dx$ by the differential (tangent vector) $dx = df(t, dt)$ at $x = f(t)$.

---

▲ *Example 4*

Let $\gamma$ denote the oriented arc in Fig. 127.5 and let $g(x) = (1, 1, 1)$. Then $\omega = d\xi_1 + d\xi_2 + d\xi_3$ and $\omega(\gamma) = \int_\gamma d\xi_2 = \int_a^b dt = b - a$, since $f(t) = (0, t, 0), t \in [a, b]$, is a parametrization of $\gamma$.

$-\gamma$ may be parametrized by $f(s) = -s$, $s \in [-b, -a]$, and we obtain $\omega(-\gamma) = -\int_{-b}^{-a} ds = -(b - a)$, in accord with (127.3).

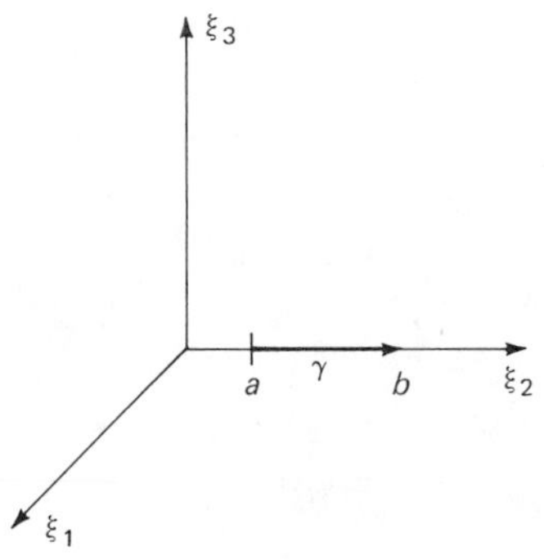


*Fig. 127.5*

▲ *Example 5*

If there exists a function $h : \Omega \to \mathbf{E}$, where $\Omega \subseteq \mathbf{E}^n$ is an open set, with continuous partial derivatives such that $g(x) = h'(x)$ for all $x \in \Omega$, then for any oriented regular curve $\gamma \in \Omega$ from $y \in \Gamma$ to $z \in \Gamma$, which is parametrized by $f : \mathfrak{J} \to \mathbf{E}^n$ with $f^{-1}(y) = a, f^{-1}(z) = b$,

$$\int_\gamma g(x) \cdot dx = \int_\gamma h'(x) \cdot dx = \int_a^b (h' \circ f)(t) \cdot f'(t)\, dt = \int_a^b (h \circ f)'(t)\, dt$$
$$= h(y) - h(z);$$

i.e., the line integral is independent of the path and depends only on beginning point and endpoint. Conversely, if the line integral is independent of the path and depends only on the beginning point and endpoint, we can find a function $h : \Omega \to \mathbf{E}^n$ such that $g(x) = h'(x)$ for all $x \in \Omega$. This may be seen as follows: Consider a fixed point $a \in \Omega$. Since $\Omega$ is open and connected, we can find a polygonal path $\gamma$ from $a$ to any point $x \in \Omega$. (See exercises 2 and 3, section 6.73.) Let

$$h(x) = \int_\gamma g(x) \cdot dx, \qquad x \in \Omega.$$

This defines a function because the line integral is independent of the path. To find its partial derivatives, we consider first a path from $a$ to $x + (\alpha_1, 0, \ldots, 0)$ via $x$ as indicated in Fig. 127.6, and obtain

$$D_1 h(x) = \lim_{\alpha_1 \to 0} \frac{h(x + (\alpha_1, 0, \ldots, 0)) - h(x)}{\alpha_1}$$
$$= \lim_{\alpha_1 \to 0} \frac{1}{\alpha_1} \int_0^{\alpha_1} g(x + (t, 0, \ldots, 0)) \cdot (1, 0, \ldots, 0)\, dt$$
$$= \lim_{\alpha_1 \to 0} \psi_1(x + (\tau_1, 0, \ldots, 0)),$$

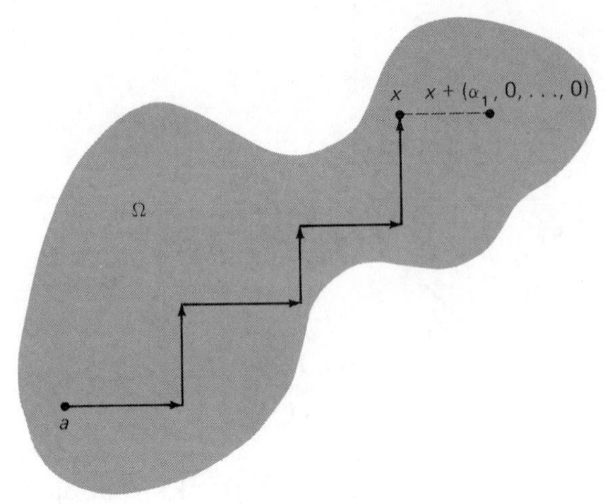


*Fig. 127.6*

where $0 \leqslant \tau_1 \leqslant \alpha_1$. Hence, $D_1 h(x) = \psi_1(x)$. Similarly, we obtain $D_j h(x) = \psi_j(x)$ for $j = 2, 3, \ldots, n$. Since the partial derivatives are continuous, $h'(x)$ exists, and $h'(x) = g(x)$.

---

*Wedge multiplication* of 1-forms leads to 2-forms. To provide some motivation for the strange definition of wedge multiplication that is to come, let us view $d\xi_1, d\xi_2, d\xi_3$ as vectors $(d\xi_1, 0, 0)$, $(0, d\xi_2, 0)$, $(0, 0, d\xi_3) \in \mathbf{E}^3$, and consider the cross products

$$(0, d\xi_2, 0) \times (0, 0, d\xi_3) = \begin{vmatrix} e_1 & e_2 & e_3 \\ 0 & d\xi_2 & 0 \\ 0 & 0 & d\xi_3 \end{vmatrix} = (d\xi_2 \, d\xi_3, 0, 0),$$

$$(0, 0, d\xi_3) \times (d\xi_1, 0, 0) = \begin{vmatrix} e_1 & e_2 & e_3 \\ 0 & 0 & d\xi_3 \\ d\xi_1 & 0 & 0 \end{vmatrix} = (0, d\xi_3 \, d\xi_1, 0),$$

$$(d\xi_1, 0, 0) \times (0, d\xi_2, 0) = \begin{vmatrix} e_1 & e_2 & e_3 \\ d\xi_1 & 0 & 0 \\ 0 & d\xi_2 & 0 \end{vmatrix} = (0, 0, d\xi_1 d\xi_2).$$

We denote these cross products by $d\xi_2 \wedge d\xi_3, d\xi_3 \wedge d\xi_1$, and $d\xi_1 \wedge d\xi_2$, and call them wedge products. From (123.2), $d\xi_i \wedge d\xi_j = -d\xi_j \wedge d\xi_i$ and, consequently, $d\xi_i \wedge d\xi_i = 0$. In general,

***Definition 127.2***

The wedge product $\omega_1 \wedge \omega_2$ of two 1-forms $\omega_1$ and $\omega_2$ is associative, distributive over addition, and

$$(127.6) \qquad \omega_1 \wedge \omega_2 = -\omega_2 \wedge \omega_1, \qquad \omega_1 \wedge \omega_1 = \omega_2 \wedge \omega_2 = 0.$$

Since $d\xi_1, d\xi_2, \ldots, d\xi_n$ are 1-forms, (127.6) implies, in particular, that

(127.7) $$d\xi_i \wedge d\xi_j = -d\xi_j \wedge d\xi_i, \qquad d\xi_i \wedge d\xi_i = 0.$$

---

▲ *Example 6*

Let $\omega_1 = d\xi_1 - d\xi_2 + \xi_1^2 d\xi_3, \omega_2 = \xi_2\, d\xi_1 + 2\xi_2\, d\xi_2 - d\xi_3$. Then,

$$\begin{aligned}\omega_1 \wedge \omega_2 &= (d\xi_1 - d\xi_2 + \xi_1^2 d\xi_3) \wedge (\xi_2\, d\xi_1 + 2\xi_2\, d\xi_2 - d\xi_3)\\ &= \xi_2\, d\xi_1 \wedge d\xi_1 + 2\xi_2\, d\xi_1 \wedge d\xi_2 - d\xi_1 \wedge d\xi_3\\ &\quad - \xi_2\, d\xi_2 \wedge d\xi_1 - 2\xi_2\, d\xi_2 \wedge d\xi_2 + d\xi_2 \wedge d\xi_3\\ &\quad + \xi_1^2 \xi_2\, d\xi_3 \wedge d\xi_1 + 2\xi_1^2 \xi_2\, d\xi_3 \wedge d\xi_2 - \xi_1^2 d\xi_3 \wedge d\xi_3\\ &= (1 - 2\xi_1^2 \xi_2) d\xi_2 \wedge d\xi_3 + (\xi_1^2 \xi_2 + 1) d\xi_3 \wedge d\xi_1 + 3\xi_2\, d\xi_1 \wedge d\xi_2.\end{aligned}$$

---

In general, if $\omega_1 = h(x) \cdot dx$ and $\omega_2 = k(x) \cdot dx$, where $h, k : \Omega \to \mathbf{E}^3$, $\Omega \subseteq \mathbf{E}^3$, then

$$\begin{aligned}\omega_1 \wedge \omega_2 &= (\chi_1(x) d\xi_1 + \chi_2(x) d\xi_2 + \chi_3(x) d\xi_3) \wedge (\kappa_1(x) d\xi_1 + \kappa_2(x) d\xi_2\\ &\qquad + \kappa_3(x) d\xi_3)\\ &= \psi_1(x) d\xi_2 \wedge d\xi_3 + \psi_2(x) d\xi_3 \wedge d\xi_1 + \psi_3(x) d\xi_1 \wedge d\xi_2 = g(x) \cdot d\sigma,\end{aligned}$$

where

(127.8) $$d\sigma = (d\xi_2 \wedge d\xi_3, d\xi_3 \wedge d\xi_1, d\xi_1 \wedge d\xi_2),$$

and where $g(x) = (\psi_1(x), \psi_2(x), \psi_3(x)) = (\chi_2(x)\kappa_3(x) - \chi_3(x)\kappa_2(x), \chi_3(x)\kappa_1(x) - \kappa_1(x)\chi_3(x), \chi_1(x)\kappa_2(x) - \chi_2(x)\kappa_1(x)) = h(x) \times k(x)$.

***Definition 127.3***

Let $g : \Omega \to \mathbf{E}^3$ denote a continuous function on the open set $\Omega \subseteq \mathbf{E}^3$. The general 2-form $\omega$ in $\mathbf{E}^3$ is a function, denoted by

(127.9) $$\omega = g(x) \cdot d\sigma = \psi_1(x) d\xi_2 \wedge d\xi_3 + \psi_2(x) d\xi_3 \wedge d\xi_1 + \psi_3(x) d\xi_1 \wedge d\xi_2,$$

which maps the oriented patch $\sigma \subset \Omega$ which is parametrized by $f : S \to \mathbf{E}^3$ into

(127.10) $$\omega(\sigma) = \iint_S g(f(t)) \cdot (D_1 f(t) \times D_2 f(t)) d\tau_1 d\tau_2,$$

where $S$ is positively oriented in $\mathbf{E}^2$. If $\omega_1, \omega_2$ are any 2-forms in $\mathbf{E}^3$, then $(\omega_1 + \omega_2)(\sigma) = \omega_1(\sigma) + \omega_2(\sigma)$. This integral exists and is invariant under orientation-preserving diffeomorphisms.

If $-\sigma$ denotes the oriented patch that is obtained from $\sigma$ by a reversal of the orientation, then we obtain, from (127.10), that

(127.11) $$\omega(-\sigma) = -\omega(\sigma).$$

If $\Sigma = \sigma_1 + \sigma_2 + \cdots + \sigma_l$ is an oriented regular surface, we define

$$\omega(\Sigma) = \sum_{j=1}^{l} \omega(\sigma_j). \tag{127.12}$$

The integral (127.10) is called a *surface integral* and is variously denoted by

$$\omega(\sigma) = \iint_\sigma \omega = \iint_\sigma g(x) \cdot d\sigma \tag{127.13}$$
$$= \iint_\sigma (\psi_1(x) d\xi_2 \wedge d\xi_3 + \psi_2(x) d\xi_3 \wedge d\xi_1 + \psi_3(x) d\xi_1 \wedge d\xi_2).$$

(127.10) is obtained from (127.13) by replacing $x$ by $f(t)$ and $d\sigma$ by the normal vector $(D_1 f(t) \times D_2 f(t)) d\tau_1 d\tau_2$ to $\sigma$ at $x = f(t)$.

The same result is also obtained if, instead of replacing $d\sigma$ by the normal vector, one replaces $d\xi_j, j = 1, 2, 3$, in (127.13) by the differentials $d\xi_j = d\phi_j(t, dt)$, $\xi_j = \phi_j(t), j = 1, 2, 3$, being the parameter representation of $\sigma$. Then,

$$d\xi_1 = d\phi_1(t, dt) = D_1\phi_1(t) d\tau_1 + D_2\phi_1(t) d\tau_2,$$
$$d\xi_2 = d\phi_2(t, dt) = D_1\phi_2(t) d\tau_1 + D_2\phi_2(t) d\tau_2,$$
$$d\xi_3 = d\phi_3(t, dt) = D_1\phi_3(t) d\tau_1 + D_2\phi_3(t) d\tau_2.$$

Noting that $D_1\phi_i(t) d\tau_1 + D_2\phi_i(t) d\tau_2$ are 1-forms, we obtain, from (127.7), that

$$d\sigma = (d\xi_2 \wedge d\xi_3, d\xi_3 \wedge d\xi_1, d\xi_1 \wedge d\xi_2) = (D_1 f(t) \times D_2 f(t)) d\tau_1 \wedge d\tau_2.$$

We may replace $d\tau_1 \wedge d\tau_2$ by $d\tau_1 d\tau_2$ because the 2-form $g(f(t)) \cdot (D_1 f(t) \times D_2 f(t)) d\tau_1 \wedge d\tau_2$ maps $S$ into $\iint_S g(f(t)) \cdot (D_1 f(t) \times D_2 f(t)) d\tau_1 d\tau_2$, with the understanding that $S$ is positively oriented in $\mathbf{E}^2$.

---

▲ *Example 7*

Let $\sigma$ denote the oriented patch in Fig. 127.7 and let $g(x) = (1, 1, 1)$. $\sigma$ is parametrized by $f(t) = (\xi_3, 0, \xi_1)$, $(\xi_3, \xi_1) \in S$, $S$ being positively oriented in the $(\xi_3, \xi_1)$-plane, and

$$D_1 f(t) \times D_2 f(t) = \begin{vmatrix} e_1 & e_2 & e_3 \\ 0 & 0 & 1 \\ 1 & 0 & 0 \end{vmatrix} = (0, 1, 0).$$

Hence,

$$\iint_\sigma \omega = \iint_\sigma (d\xi_2 \wedge d\xi_3 + d\xi_3 \wedge d\xi_1 + d\xi_1 \wedge d\xi_2) = \iint_S d\tau_3 \, d\tau_1 = J(S).$$

▲ *Example 8*

Let $\omega = d\xi_2 \wedge d\xi_3 - d\xi_1 \wedge d\xi_2$ and let $\sigma$ denote the oriented patch $\{x \in \mathbf{E}^3 \mid \xi_3 = \xi_1^2 + \xi_2^2 \leqslant 1\}$ with the normal pointing as indicated in Fig. 127.8.

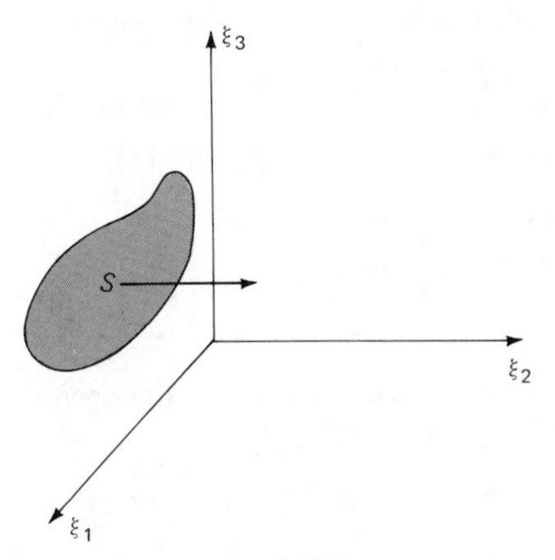


*Fig. 127.7*

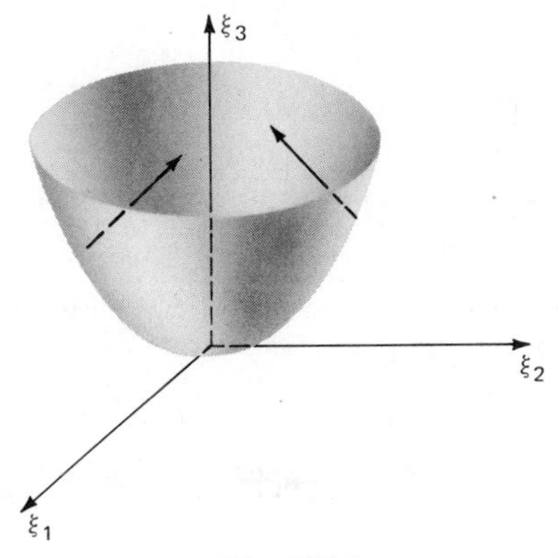


*Fig. 127.8*

We may parametrize $\sigma$ by $f(t) = (\tau_1, \tau_2, \tau_1^2 + \tau_2^2)$, $(\tau_1, \tau_2) \in \{t \in \mathbf{E}^2 \mid \|t\| \leqslant 1\}$, and obtain

$$d\xi_2 \wedge d\xi_3 = d\tau_2 \wedge (2\tau_1 d\tau_1 + 2\tau_2 d\tau_2) = -2\tau_1 d\tau_1 \wedge d\tau_2, \quad d\xi_1 \wedge d\xi_2 = d\tau_1 \wedge d\tau_2.$$

Hence,

$$\omega(\sigma) = \iint_{\|t\| \leqslant 1} (1 - 2\tau_1) d\tau_1 d\tau_2.$$

We introduce polar coordinates $\tau_1 = \rho \cos \theta$, $\tau_2 = \rho \sin \theta$, and obtain from Jacobi's theorem that

$$\omega(\sigma) = \int_0^1 \int_0^{2\pi} (1 - 2\rho \cos \theta)\rho \, d\theta \, d\rho = \pi.$$

If one takes the wedge product of a 1-form with a 2-form, one is led to a 3-form. Let $\omega_1 = h(x) \cdot dx = \chi_1(x)d\xi_1 + \chi_2(x)d\xi_2 + \chi_3(x)d\xi_3$ and $\omega_2 = k(x) \cdot d\sigma = \kappa_1(x)d\xi_2 \wedge d\xi_3 + \kappa_2(x)d\xi_3 \wedge d\xi_1 + \kappa_3(x)d\xi_1 \wedge d\xi_2$. By (127.6),

$$\begin{aligned}\omega_1 \wedge \omega_2 &= \chi_1(x)\kappa_1(x)d\xi_1 \wedge d\xi_2 \wedge d\xi_3 + \chi_2(x)\kappa_2(x)d\xi_2 \wedge d\xi_3 \wedge d\xi_1 \\ &\quad + \chi_3(x)\kappa_3(x)d\xi_3 \wedge d\xi_1 \wedge d\xi_2 \\ &= h(x) \cdot k(x)d\xi_1 \wedge d\xi_2 \wedge d\xi_3.\end{aligned}$$

***Definition 127.4***

Let $g : \Omega \to \mathbf{E}$ denote a continuous function on the open set $\Omega \subseteq \mathbf{E}^3$. The general 3-form in $\mathbf{E}^3$ is a function, denoted by

$$\omega = g(x)d\xi_1 \wedge d\xi_2 \wedge d\xi_3, \tag{127.14}$$

which maps the three-dimensional oriented patch $v \subset \Omega$ which is parametrized by $f : Y \to \mathbf{E}^3$ into

$$\omega(v) = \iiint_Y g(f(t))J_f(t)d\tau_1 d\tau_2 d\tau_3, \tag{127.15}$$

where $Y$ is positively oriented in $\mathbf{E}^3$. If $\omega_1, \omega_2$ are any 3-forms in $\mathbf{E}^3$, then $(\omega_1 + \omega_2)(v) = \omega_1(v) + \omega_2(v)$.

The integral in (127.15) exists and is invariant under orientation-preserving diffeomorphisms.

If $-v$ denotes the oriented patch obtained from $v$ by a reversal of the orientation, we have, from (127.15), that

$$\omega(-v) = -\omega(v). \tag{127.16}$$

If $\Upsilon = v_1 + v_2 + \cdots + v_l$ is a three-dimensional oriented quilt without multiple points, we define

$$\omega(\Upsilon) = \sum_{j=1}^{l} \omega(v_j). \tag{127.17}$$

The integral in (127.15) is called an *oriented volume integral* and is variously denoted by

$$\iiint_v \omega = \iiint_v g(x)dv = \iiint_v g(x)d\xi_1 \wedge d\xi_2 \wedge d\xi_3.$$

---

▲ *Example 9*

$f(t) = (\tau_1, \tau_2, \tau_3)$, $(\tau_1, \tau_2, \tau_3) \in Y$, where $Y \subset \mathbf{E}^3$ is Jordan-measurable, parametrizes a positively oriented three-dimensional patch $v$ in $\mathbf{E}^3$ (see also example 3), and we obtain, with $g(x) = 1$, that

$$\iiint_v d\xi_1 \wedge d\xi_2 \wedge d\xi_3 = \iiint_Y d\tau_1 d\tau_2 d\tau_3 = J(Y).$$

---

We have now defined 1-forms in $\mathbf{E}^n$, 2-forms in $\mathbf{E}^3$, and 3-forms in $\mathbf{E}^3$. There are also 2-forms in $\mathbf{E}^2$, and 2-forms in $\mathbf{E}^n$ for $n > 3$ and higher forms in $\mathbf{E}^n$ which are defined in an analogous manner. However, we shall not discuss these other forms in our treatment. For completeness' sake and for some practical reasons, we shall, however, define a 0-form in $\mathbf{E}^n$:

***Definition 127.5***

The general 0-form in $\mathbf{E}^n$ is a continuous function $g : \Omega \to \mathbf{E}$ on the open set $\Omega \subseteq \mathbf{E}^n$ which maps the point $x_0 \in \Omega$ (0-dimensional patch) into $g(x_0)$.

Thus a 0-form is just a real-valued function. By taking the differential of a real-valued function one obtains a 1-form. In section 13.129, we shall define the differential of 1-forms and 2-forms and see that the differential of a 1-form is a 2-form and the differential of a 2-form is a 3-form. Hence, the name *differential forms.*

Before closing this section, let us introduce some notation that is customarily used when dealing with line integrals, surface integrals, and oriented volume integrals.

If $g : \mathbf{E}^n \to \mathbf{E}$ has continuous partial derivatives, then the vector defined by

$$\text{grad } g(x) = (D_1 g(x), D_2 g(x), \ldots, D_n g(x)) \tag{127.18}$$

is called the *gradient* of $g$ at $x$. Symbolically, the gradient may be viewed as the product of the *vector operator*

$$\nabla = (D_1, D_2, \ldots, D_n)$$

with the scalar $g(x)$:

$$\text{grad } g(x) = \nabla g(x).$$

($\nabla$ is to be read as *nabla*. This name derives from the shape of $\nabla$ which resembles an ancient Hebrew string instrument of that name.)

If $g = (\psi_1, \psi_2, \ldots, \psi_n) : \mathbf{E}^n \to \mathbf{E}^n$ has continuous partial derivatives, then the scalar

$$\text{div } g(x) = D_1 \psi_1(x) + D_2 \psi_2(x) + \cdots + D_n \psi_n(x) \tag{127.19}$$

is called the *divergence* of $g$ at $x$. We may view the divergence as the dot product of $\nabla$ with $g(x)$:

$$\text{div } g(x) = \nabla \cdot g(x).$$

If $g = (\psi_1, \psi_2, \psi_3) : \mathbf{E}^3 \to \mathbf{E}^3$ has continuous partial derivatives, then the vector

(127.20)

$$\text{curl } g(x) = (D_2 \psi_3(x) - D_3 \psi_2(x), D_3 \psi_1(x) - D_1 \psi_3(x), D_1 \psi_2(x) - D_2 \psi_1(x))$$

is called the *curl* of $g$ at $x$. We may write

$$\text{curl } g(x) = \nabla \times g(x).$$

### 13.127 Exercises

1. Let $g(x) = (1, 1)$ and let $\Gamma$ denote the oriented regular curve that is composed of the arc of $\xi_1^2 + \xi_2^2 = 1$ in the first quadrant and the parabolic arc on $\xi_2 = (\xi_1 - 1)^2$ between $\xi_1 = 1$ and $\xi_1 = 2$, which is traversed from $(0, 1)$ to $(2, 1)$. Find $\int_\Gamma g(x) \cdot dx$.
2. Show that the integral in (127.2) is invariant under orientation-preserving diffeomorphisms and reverses sign under an orientation-reversing diffeomorphism.
3. Find $(\xi_1 d\xi_1 - d\xi_2 + 3d\xi_3) \wedge (d\xi_2 + d\xi_3)$, and express your answer in simplest terms.
4. Show: If $\omega_1 = h(x) \cdot dx$, $\omega_2 = k(x) \cdot dx$, then $\omega_1 \wedge \omega_2 = (h(x) \times k(x)) \cdot d\sigma$.
5. Give an interpretation of $\iiint_v d\xi_1 \wedge d\xi_2 \wedge d\xi_3 = -\iiint_v d\xi_2 \wedge d\xi_1 \wedge d\xi_3$.
6. Let $\omega = d\xi_2 \wedge d\xi_3 + d\xi_3 \wedge d\xi_1$ and let $\Sigma$ denote the oriented torus of example 9, section 13.120, with the normal pointing outwards. Find $\iint_\Sigma \omega$.
7. Let $\sigma$ denote the 2-dimensional oriented patch $\sigma$ that is cut out from the plane $\xi_1 + \xi_2 + \xi_3 = 1$ by the cylinder $\xi_1^2 + \xi_2^2 = 1$ whose normal points upwards and let $g(x) = (\xi_1, 0, \xi_3)$. Find $\iint_\sigma g(x) \cdot d\sigma$.
8. Show that the integrals in (127.10) and (127.15) are invariant under orientation-preserving diffeomorphisms.
9. $\omega = g(x) d\xi_1 \wedge d\xi_2$ is the general 2-form in $\mathbf{E}^2$ where $g : \Omega \to \mathbf{E}$, $\Omega \subseteq \mathbf{E}^2$, is a continuous function. What is $\iint_\sigma \omega$ for an oriented two-dimensional patch $\sigma \subset \mathbf{E}^2$?
10. Show that there is no 4-form in $\mathbf{E}^3$.
11. Find the general 3-form and the general 4-form in $\mathbf{E}^4$.
12. Find the general 2-form in $\mathbf{E}^n$ for $n \geqslant 2$.
13. Let $g \in C^2(\mathbf{E}^n)$ where $g : \mathbf{E}^n \to \mathbf{E}$. Express div grad $g(x)$ in components.
14. Let $g \in C^2(\mathbf{E}^3)$ where $g : \mathbf{E}^3 \to \mathbf{E}$. Express curl grad $g(x)$ in components.
15. Let $\gamma \subset \mathbf{E}^n$ denote an arc and let $\sigma \in \mathbf{E}^3$ denote a patch. Interpret $\int_\gamma \|dx\|$ and $\int_\sigma \int \|d\sigma\|$. (See also definitions 125.1 and 126.1.)
16. Let $h(x) = \xi_1 \xi_2$. By example 5, $\int_\gamma (\xi_2 d\xi_1 + \xi_1 d\xi_2) = h(y) - h(z)$, where $z, y$ are beginning point and endpoint on $\gamma$. Let $(\phi_1, \phi_2) : [a, b] \to \mathbf{E}^2$ represent $\gamma$ and derive the integration-by-parts formula.

### 13.128 Work and Steady Flow

In this section we shall discuss some applications of differential forms to mechanics and fluid dynamics.

A vector-valued function of a vector variable, $g : \mathbf{E}^n \to \mathbf{E}^n$, is said to define a *vector field* in $\mathbf{E}^n$. It associates with each point $x \in \mathbf{E}^n$ the vector $g(x)$ which may be viewed as emanating from $x$ and terminating at $x + g(x)$. (See Fig. 128.1.)

If these vectors are interpreted as representing a force acting at $x$ in the direction of $g(x)$ and of a magnitude equal to $\|g(x)\|$, one speaks of a *force field*; and if they are interpreted as velocity vectors, one speaks of a *velocity field*.

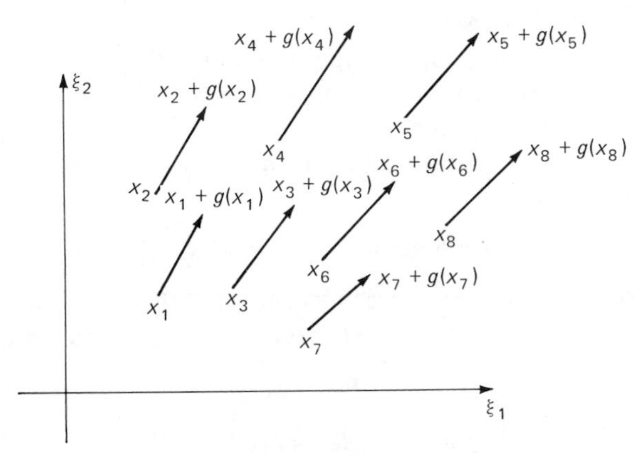


*Fig. 128.1*

---

▲ *Example 1* (*Work*)

Let $g : \mathbf{E}^3 \to \mathbf{E}^3$ define a force field, and let $\gamma$ denote an oriented regular curve in $\mathbf{E}^3$. It is our aim to give a definition of the work to be expended by $g$ to push a unit mass from the beginning point $P$ of $\gamma$ to the endpoint $Q$ of $\gamma$. At the point $x \in \gamma$, we consider the tangent vector $dx$ and the force vector $g(x)$. Then, $g(x) \cdot dx = \|g(x)\| \, \|dx\| \cos \phi$ (see Fig. 128.2), $\|g(x)\| \cos \phi$ being the component of the

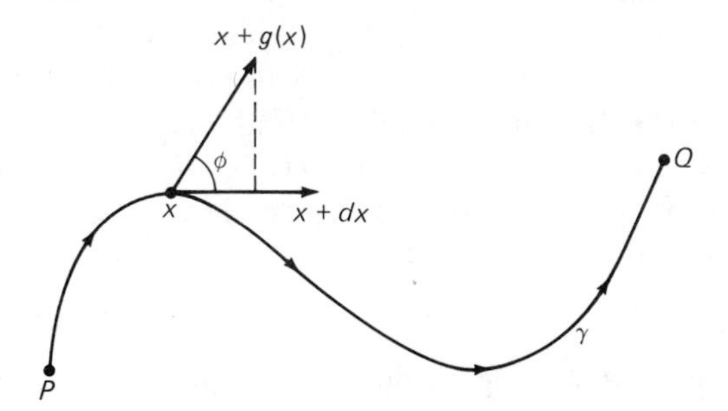


*Fig. 128.2*

force in the direction of $dx$, represents the work to be expended to push the unit mass from $x$ to $(x + dx)$ along the tangent vector, if $g(x)$ were to remain constant along the tangent vector. (See example 2, section 13.117.) Replacing $\gamma$ by a polygonal path and arguing essentially along the same lines as in the heuristic "derivation" of the formula for arc length in section 13.125, we are led to the following definition:

$$W = \int_\gamma g(x) \cdot dx$$

is called the *work* to be expended by $g$ to push a unit mass from the beginning point of the oriented curve $\gamma$ to its endpoint. Clearly, this line integral exists and is invariant under orientation-preserving diffeomorphisms when $\gamma$ is an arc or regular curve and $g$ is continuous.

In the force field is the gradient of a *potential function* $h : \mathbf{E}^3 \to \mathbf{E}$ with continuous partial derivatives:

$$g(x) = \text{grad } h(x) \qquad \text{for all } x \in \mathbf{E}^3,$$

then $\int_\gamma g(x) \cdot dx = \int_\gamma \text{grad } h(x) \cdot dx$ is independent of the path $\gamma$ and depends only on the beginning point $P$ of $\gamma$ and the endpoint $Q$ of $\gamma$, as we have seen in example 5, section 13.127:

$$\int_\gamma \text{grad } h(x) \cdot dx = h(Q) - h(P).$$

A force field defined by $g(x) = \text{grad } h(x)$ is called a *conservative* force field. Work in a conservative force field is independent of the path and depends only on the point of departure and the point of arrival. (See also example 2, section 13.130.)

▲ *Example 2* (*Two-Dimensional Steady Flow*)

If a fluid which is distributed over a plane (e.g., surface of a body of water) has constant mass density, say 1, and moves in such a manner that the velocity at each point of the plane is independent of the time, one speaks of a *steady flow of an incompressible fluid* in two dimensions. Such a flow is defined by a velocity field $v : \mathbf{E}^2 \to \mathbf{E}^2$ which assigns to each point $x$ in $\mathbf{E}^2$ a velocity vector $v(x) = (v_1(x), v_2(x))$. (See also example 3, section 13.117.) Let $\gamma$ denote an oriented regular curve in $\mathbf{E}^2$ with beginning point $P$ and endpoint $Q$. If $dx = (d\xi_1, d\xi_2)$ denotes the tangent vector to $\gamma$ at $x$, then $n(x) = (d\xi_2, -d\xi_1)$ denotes a vector perpendicular to $dx$, so that $n(x), dx$ are in standard position. (See Fig. 128.3.) If $v(x)$ remains constant along $dx$, then

$$v(x) \cdot n(x) = \|v(x)\| \, \|n(x)\| \cos \phi = \|v(x)\| \, \|dx\| \cos \phi$$

is the amount of fluid passing through $dx$ in the direction of $n(x)$ in unit time, because $\|v(x)\| \cos \phi$ is the velocity component in the direction of $n(x)$. Arguing again as in example 1 and in section 13.125, we arrive at the following

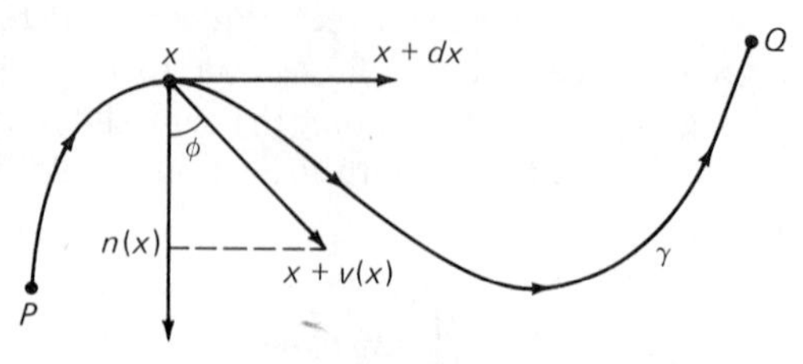


*Fig. 128.3*

definition for the total amount of fluid passing in unit time through $\gamma$ from left to right (where the right side of $\gamma$ is the side of the normal $n(x)$):

$$(128.1) \quad F = \int_\gamma (v_1(x), v_2(x)) \cdot (d\xi_2, -d\xi_1) = \int_\gamma (v_1(x)d\xi_2 - v_2(x)d\xi_1).$$

This integral exists and is invariant under orientation-preserving diffeomorphisms if $\gamma$ is an arc or regular curve and if $v$ is continuous. (Note that $F$ may actually be negative in the case where fluid flows from right to left, or *more* fluid passes from the right to the left than from the left to the right.)

If $\gamma$ is an oriented simple closed curve, i.e., beginning point and endpoint coincide, then (128.1) represents the total amount of fluid that passes in unit time through the region $A$ encompassed by $\gamma$. (See Fig. 128.4.)

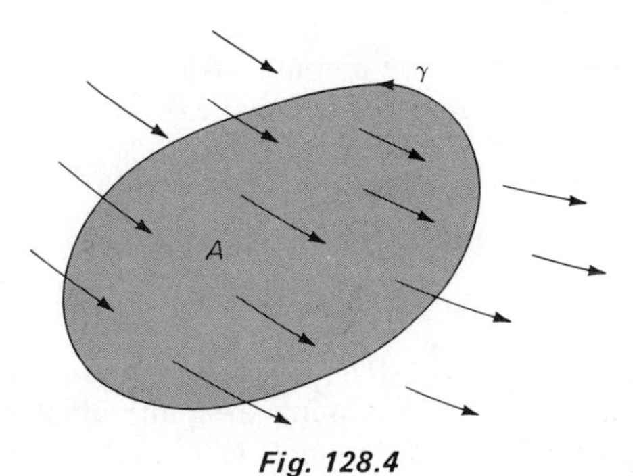


*Fig. 128.4*

We may express the latter in still another manner if $A$ is Jordan-measurable. We consider a partition of $A$ and investigate the amount of fluid passing through each individual subinterval of the partition. (See Fig. 128.5 and 128.6.)

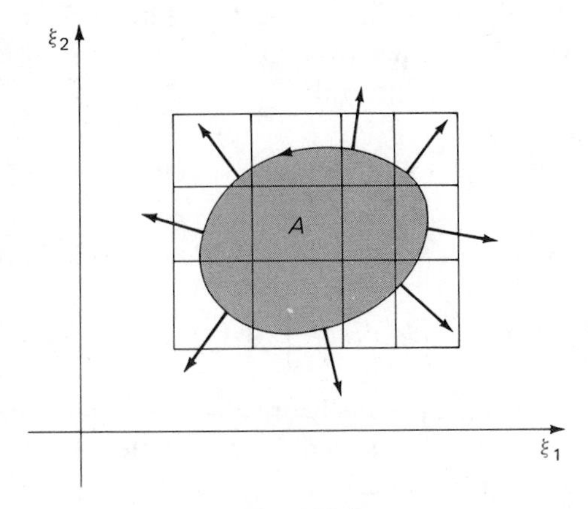


*Fig. 128.5*

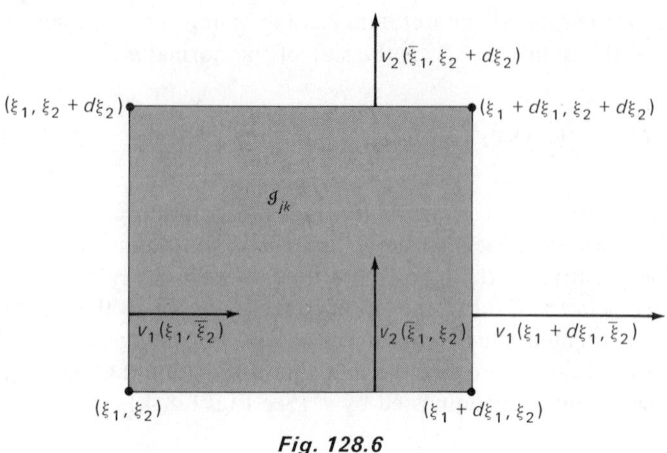


***Fig. 128.6***

With reference to Fig. 128.6, the amount of fluid passing through $\mathfrak{I}_{j,k}$ in the direction of the $\xi_1$-axis is approximately equal to

$$(v_1(\xi_1 + d\xi_1, \bar{\xi}_2) - v_1(\xi_1, \bar{\xi}_2))d\xi_2 \cong D_1 v_1(\bar{\bar{\xi}}_1, \bar{\xi}_2)d\xi_1 d\xi_2$$

for some $\bar{\bar{\xi}}_1 \in (\xi_1, \xi_1 + d\xi_1)$, and in the direction of the $\xi_2$-axis, to

$$(v_2(\bar{\xi}_1, \xi_2 + d\xi_2) - v_2(\bar{\xi}_1, \xi_2))d\xi_1 \cong D_2 v_2(\bar{\xi}_1, \bar{\bar{\xi}}_2)d\xi_1 d\xi_2.$$

Hence, $(D_1 v_1(\bar{\bar{\xi}}_1, \bar{\xi}_2) + D_2 v_2(\bar{\xi}_1, \bar{\bar{\xi}}_2))d\xi_1 d\xi_2$ is an approximation for the total flow through $\mathfrak{I}_{j,k}$ in unit time. Summing over all subintervals, we obtain a typical intermediate sum for the function $D_1 v_1(x) + D_2 v_2(x)$ on $A$, and we are led, using the notation in (127.19), to the formula

$$F = \iint_A \operatorname{div} v(x) d\xi_1 d\xi_2 \tag{128.2}$$

for the total flow through $A$ in unit time, where it is assumed that the normal to the boundary of $A$ points outwards. (See Fig. 128.5.) Since $A$ is in standard position, we may view it as a positively oriented two-dimensional patch in $\mathbf{E}^2$, and write (128.2) as a surface integral,

$$F = \iint_\sigma \operatorname{div} v(x) d\xi_1 \wedge d\xi_2 \tag{128.3}$$

where $\sigma = \{(\xi_1, \xi_2) \in A\}$. It appears from (128.1) and (128.3) that

$$\iint_\sigma \operatorname{div} v(x) d\xi_1 \wedge d\xi_2 = \int_\gamma (v_1(x) d\xi_2 - v_2(x) d\xi_1), \tag{128.4}$$

where $\sigma$ is a two-dimensional positively oriented patch in $\mathbf{E}^2$ that is encompassed by the regular curve $\gamma$ that is oriented in the counterclockwise direction. This is *Gauss's*† *Theorem* for two dimensions.

† *Carl Friedrich Gauss*, 1777–1855.

We let $v_1 = \psi_2, v_2 = -\psi_1$ and obtain, from (128.4), with $g = (\psi_1, \psi_2)$, the equivalent statement:

(128.5) $$\iint_\sigma (D_1\psi_2(x) - D_2\psi_1(x))d\xi_1 \wedge d\xi_2 = \int_\gamma g(x) \cdot dx.$$

In this form, the theorem is known as *Green's*† *Theorem.* It will be proved in section 13.130, under certain restricting assumptions on $A$.

▲ *Example 3* (*Circulation*)

Let $v : \mathbf{E}^2 \to \mathbf{E}^2$ define a velocity field for a two-dimensional steady flow of an incompressible fluid. If $\gamma$ is an arc in $\mathbf{E}^2$, then $v(x) \cdot dx = \|v(x)\| \, \|dx\| \cos \phi$ represents the velocity component in the direction of the tangent vector multiplied by the length of the tangent vector. (See Fig. 128.7.)

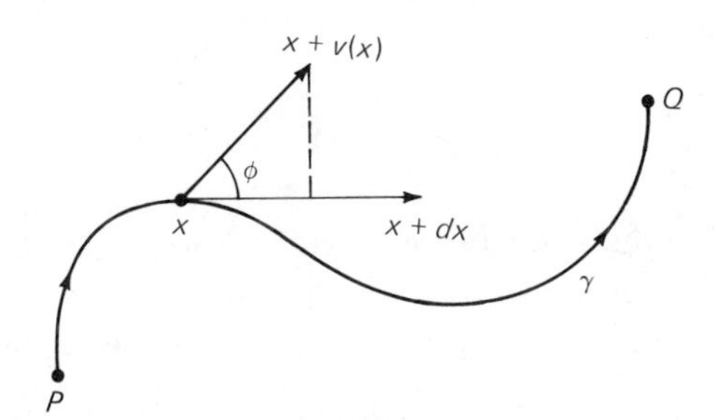


*Fig. 128.7*

$\int_\gamma v(x) \cdot dx$ is a measure for the tangential flow along $\gamma$. If $\gamma$ is an oriented regular closed curve, then

(128.6) $$C = \int_\gamma v(x) \cdot dx$$

is called the *circulation* of the fluid along $\gamma$.

By (128.5),

$$C = \int_\gamma v(x) \cdot dx = \iint_\sigma (D_1 v_2(x) - D_2 v_1(x))d\xi_1 \wedge d\xi_2,$$

where $\sigma$ is the positively oriented region encompassed by $\gamma$ and where $\gamma$ is oriented in the counterclockwise direction.

▲ *Example 4* (*Steady Flow in Three Dimensions*)

$v : \mathbf{E}^3 \to \mathbf{E}^3$ defines a velocity field for the steady flow of a space-filling incompressible fluid, of which we will assume that it has mass density 1. With reference to

† *George Green*, 1793–1841.

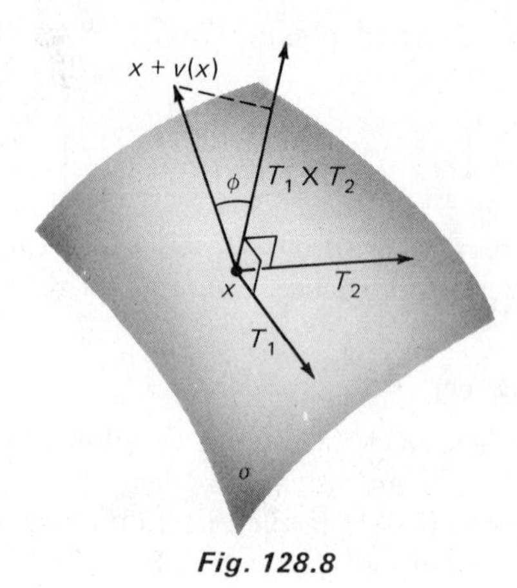


*Fig. 128.8*

Fig. 128.8, we obtain, for the flow of liquid through an oriented patch $\sigma$ in unit time, as in example 2,

$$F = \iint_{\sigma} v(x) \cdot d\sigma. \tag{128.7}$$

In Fig. 128.8, $T_1, T_2$ are two tangent vectors to $\sigma$ at $x$. Then, $v(x) \cdot (T_1(x) \times T_2(x)) = \|v(x)\| \, \|T_1(x) \times T_2(x)\| \cos \phi$ is the velocity component in the direction of the normal $T_1(x) \times T_2(x)$, times the area of the parallelogram that is spanned by $T_1(x)$ and $T_2(x)$.

If $\Upsilon$ is a positively-oriented solid that is encompassed by the regular oriented surface $\sigma$ with the normal pointing outwards, we obtain, as in example 2 and with reference to Fig. 128.9, for the total amount of liquid leaving $\Upsilon$,

$$F = \int_{\Upsilon}\iint \operatorname{div} v(x) d\xi_1 \wedge d\xi_2 \wedge d\xi_3. \tag{128.8}$$

From (128.7) and (128.8), we obtain Gauss's Theorem for three dimensions:

$$\int_{\sigma}\int v(x) \cdot d\sigma = \int_{\Upsilon}\iint \operatorname{div} v(x) d\xi_1 \wedge d\xi_2 \wedge d\xi_3, \tag{128.9}$$

where $\Upsilon$ is either positively oriented and the normal to its surface $\sigma$ points outward, or $\Upsilon$ is negatively oriented and the normal to its surface $\sigma$ points inward. We shall prove this theorem in section 13.132.

As in example 3, one obtains, for the circulation along an oriented regular closed curve $\gamma$ in $\mathbf{E}^3$,

$$C = \int_{\gamma} v(x) \cdot dx, \tag{128.10}$$

where $v : \mathbf{E}^3 \to \mathbf{E}^3$ and $dx = (d\xi_1, d\xi_2, d\xi_3)$.

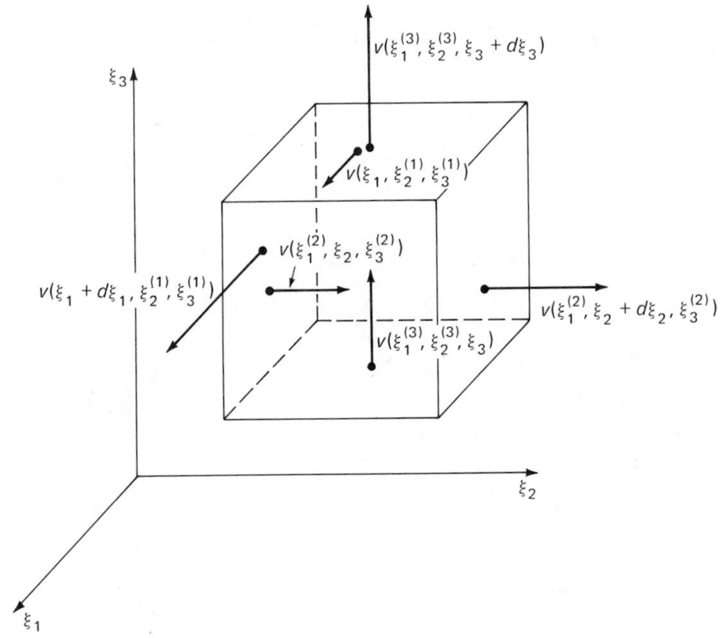


*Fig. 128.9*

---

## 13.128 Exercises

1. Given a force field by $g(x) = (1, 1, 1)$ and an oriented arc $\gamma$ by $f(t) = (\cos t, \sin t, t)$, $t \in [0, 2\pi]$. Find the work to be expended by $g$ to push a unit mass from $(1, 0, 0)$ to $(1, 0, 2\pi)$.
2. Given a velocity field by $g(x) = (1, 0, 0)$ and an oriented patch $\sigma$ by

$$f(t) = (\sin \tau_1 \cos \tau_2, \sin \tau_1 \sin \tau_2, \cos \tau_1), \qquad (\tau_1, \tau_2) \in \left[\frac{\pi}{2}, \frac{3\pi}{2}\right] \times \left[\frac{\pi}{2}, \frac{3\pi}{2}\right].$$

   Find the total amount of fluid passing through $\sigma$ in unit time.
3. Given a velocity field by $g(x) = (-\xi_2, \xi_1, 0)$ and a closed regular curve $\gamma$ by $f(t) = (\cos t, \sin t, 1)$, $t \in [0, 2\pi]$. Find the circulation along $\gamma$.

## 13.129 Differentials of $k$-forms

We have noted in section 13.127 that the differential of a 0-form is a 1-form. We shall give a definition of the differential of a form in such a manner that the differential of a $k$-form is a $(k + 1)$-form.

***Definition 129.1***

If $\omega = g(x)$ is a 0-form in $\mathbf{E}^n$, then the differential $d\omega$ of $\omega$ is the 1-form

$$(129.1)\qquad \begin{aligned} d\omega &= d(g(x)) \\ &= D_1 g(x)d\xi_1 + D_2 g(x)d\xi_2 + \cdots + D_n g(x)d\xi_n \qquad \text{in } \mathbf{E}^n. \end{aligned}$$

If $\omega = g(x) \cdot dx$ is a 1-form in $\mathbf{E}^3$, then the differential $d\omega$ of $\omega$ is the 2-form

$$(129.2)\qquad \begin{aligned} d\omega &= d(g(x) \cdot dx) \\ &= d\psi_1(x, dx) \wedge d\xi_1 + d\psi_2(x, dx) \wedge d\xi_2 + d\psi_3(x, dx) \wedge d\xi_3 \quad \text{in } \mathbf{E}^3. \end{aligned}$$

If $\omega = g(x) \cdot d\sigma$ is a 2-form in $\mathbf{E}^3$, then the differential $d\omega$ of $\omega$ is the 3-form

$$(129.3)\qquad \begin{aligned} d\omega &= d(g(x) \cdot d\sigma) \\ &= d\psi_1(x, dx) \wedge d\xi_2 \wedge d\xi_3 + d\psi_2(x, dx) \wedge d\xi_3 \wedge d\xi_1 \\ &\quad + d\psi_3(x, dx) \wedge d\xi_1 \wedge d\xi_2 \qquad \text{in } \mathbf{E}^3. \end{aligned}$$

In view of (127.18), we may write (129.1) as

$$(129.4)\qquad d(g(x)) = \operatorname{grad} g(x) \cdot dx.$$

Expansion of (129.2) yields, in view of (127.7),

$$\begin{aligned} d(g(x) \cdot dx) &= (D_1\psi_1(x)d\xi_1 + D_2\psi_1(x)d\xi_2 + D_3\psi_1(x)d\xi_3) \wedge d\xi_1 \\ &\quad + (D_1\psi_2(x)d\xi_1 + D_2\psi_2(x)d\xi_2 + D_3\psi_2(x)d\xi_3) \wedge d\xi_2 \\ &\quad + (D_1\psi_3(x)d\xi_1 + D_2\psi_3(x)d\xi_2 + D_3\psi_3(x)d\xi_3) \wedge d\xi_3 \\ &= (D_2\psi_3(x) - D_3\psi_2(x))d\xi_2 \wedge d\xi_3 + (D_3\psi_1(x) - D_1\psi_3(x))d\xi_3 \wedge d\xi_1 \\ &\quad + (D_1\psi_2(x) - D_2\psi_1(x))d\xi_1 \wedge d\xi_2. \end{aligned}$$

Using the notation (127.20), we may therefore write (129.2) in the form

$$(129.5)\qquad d(g(x) \cdot dx) = \operatorname{curl} g(x) \cdot d\sigma.$$

Expansion of (129.3) yields, in view of (127.7),

$$\begin{aligned} d(g(x) \cdot d\sigma) &= (D_1\psi_1(x)d\xi_1 + D_2\psi_1(x)d\xi_2 + D_3\psi_1(x)d\xi_3) \wedge d\xi_2 \wedge d\xi_3 \\ &\quad + (D_1\psi_2(x)d\xi_1 + D_2\psi_2(x)d\xi_2 + D_3\psi_2(x)d\xi_3) \wedge d\xi_3 \wedge d\xi_1 \\ &\quad + (D_1\psi_3(x)d\xi_1 + D_2\psi_3(x)d\xi_2 + D_3\psi_3(x)d\xi_3) \wedge d\xi_1 \wedge d\xi_2 \\ &= D_1\psi_1(x)d\xi_1 \wedge d\xi_2 \wedge d\xi_3 + D_2\psi_2(x)d\xi_2 \wedge d\xi_3 \wedge d\xi_1 \\ &\quad + D_3\psi_3(x)d\xi_3 \wedge d\xi_1 \wedge d\xi_2 \\ &= (D_1\psi_1(x) + D_2\psi_2(x) + D_3\psi_3(x))d\xi_1 \wedge d\xi_2 \wedge d\xi_3. \end{aligned}$$

Using the notation from (127.19),

$$(129.6)\qquad d(g(x) \cdot d\sigma) = \operatorname{div} g(x)d\xi_1 \wedge d\xi_2 \wedge d\xi_3.$$

We have seen in example 5, section 13.127, that

$$\int_\gamma \operatorname{grad} g(x) \cdot dx = g(Q) - g(P),$$

where $P$, $Q$ are beginning point and endpoint of $\gamma$. With the notation in (129.1), this may be written as

$$\int_\gamma d\omega = \omega \Big|_P^Q .$$

In section 13.128, we gathered strong heuristic evidence that $\iint_\sigma (D_1\psi_2(x) - D_2\psi_1(x))d\xi_1 \wedge d\xi_2 = \int_\gamma g(x)\cdot dx$ (see (128.5)). With the understanding that $\psi_3(x) = 0$ for all $x$, we may write this, in view of (129.5), as

$$\iint_\sigma d\omega = \int_\gamma \omega,$$

where $\gamma$ is the boundary of $\sigma$, suitably oriented. Finally, (128.9) may be written in the symbolic form

$$\iiint_\Upsilon d\omega = \iint_\sigma \omega,$$

where $\sigma$ is the boundary of $\Upsilon$, suitably oriented.

It is true under fairly general conditions that

$$\int_M d\omega = \int_{\partial M} \omega, \tag{129.7}$$

where $\omega$ is a $(k-1)$-form, $M$ is an oriented $k$-dimensional manifold and $\partial M$ is the boundary of $M$, suitably oriented.†

In the next section we shall prove (129.7) for what we shall call standard regions in $\mathbf{E}^2$; in section 13.131 we shall prove it for standard surfaces in $\mathbf{E}^3$, and in section 13.132 for standard solids in $\mathbf{E}^3$. What we mean by standard will be explained at the appropriate place.

(129.7) may be viewed as a generalization of the fundamental theorem of the integral calculus: The value of $\int_M d\omega$ is the value $\omega(\partial M) = \int_{\partial M} \omega$ of the "anti-derivative" $\omega$ of $d\omega$ on the boundary $\partial M$ of $M$.

## 13.129 Exercises

1. Let $\omega = \xi_1^2 d\xi_1 + \xi_1\xi_3 d\xi_2 - d\xi_3$. Find $d\omega$.
2. Same as in exercise 1, for $\omega = \xi_1\xi_2 d\xi_2 \wedge d\xi_3 - \xi_3^2 d\xi_1 \wedge d\xi_2$.
3. Let $\omega = g(x)d\xi_1 \wedge d\xi_2 \wedge d\xi_3$. Show that $d\omega = 0$.
4. Verify (129.7) for $M = [0,1] \times [0,1]$, positively oriented, with the normal to its boundary pointing outward, and $g(x) = (\xi_2, -\xi_1)$.
5. Verify (129.7) for $M = [0,1] \times [0,1] \times [0,1]$, positively oriented, with the normal to its surface pointing outward, and $g(x) = (\xi_1, \xi_2, \xi_3)$.

† See Michael Spivak, *Calculus on Manifolds* (New York: W. A. Benjamin, Inc., 1965), p. 122ff.

## 13.130 Green's Theorem

In this section we shall prove (129.7) as stated in the form (128.5) for certain patches in $\mathbf{E}^2$.

We say that $A \subset \mathbf{E}^2$ is a *region* if and only if

$$(130.1) \qquad \begin{aligned} A &= \{x \in \mathbf{E}^2 \mid a \leqslant \xi_1 \leqslant b, f_1(\xi_1) \leqslant \xi_2 \leqslant f_2(\xi_1)\} \\ &= \{x \in \mathbf{E}^2 \mid c \leqslant \xi_2 \leqslant d, h_1(\xi_2) \leqslant \xi_1 \leqslant h_2(\xi_2)\}, \end{aligned}$$

where $f_1, f_2 : [a, b] \to \mathbf{E}$ and $h_1, h_2 : [c, d] \to \mathbf{E}$ are continuous functions, where $f_1, f_2 \in C^1(a, b)$, $h_1, h_2 \in C^1(c, d)$, and $f_1(\xi_1) \leqslant f_2(\xi_1)$ for all $\xi_1 \in [a, b]$, $h_1(\xi_2) \leqslant h_2(\xi_2)$ for all $\xi_2 \in [c, d]$. (See Fig. 130.1.) Note that the two representations of

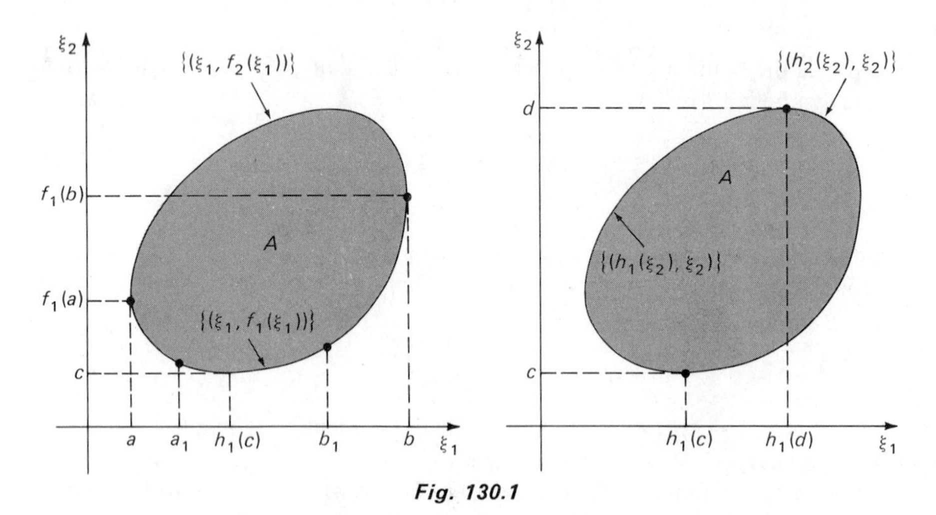


*Fig. 130.1*

$A$ have to hold simultaneously. This implies that vertical and horizontal lines intersect the boundary in at most two points.

Let $\gamma_1$ denote the lower part of the boundary of $A$ (from $(a, f_1(a))$ to $(b, f_1(b))$) and let $\gamma_2$ denote the upper part of the boundary (from $(b, f_2(b)) = (b, f_1(b))$ to $(a, f_2(a)) = (a, f_1(a))$). We choose the orientation such that $\gamma_1 + \gamma_2$ is oriented in the *positive* (counterclockwise) direction. Clearly, $\gamma_1 + \gamma_2 = \partial A$. (See Fig. 130.2.)

$\partial A$ is a regular oriented curve. We shall demonstrate only that $\gamma_1$ is a regular oriented curve, because the same argument applies to $\gamma_2$. We choose two points $a_1, b_1 \in (a, b)$ with $a_1 < h_1(c) < b_1$, and show that the parts on $\gamma_1$ from $(a, f_1(a))$ to $(a_1, f_1(a_1))$, from $(a_1, f_1(a_1))$ to $(b_1, f_1(b_1))$, and from $(b_1, f_1(b_1))$ to $(b, f_1(b))$, are oriented arcs. This is clearly the case as

$$\left.\begin{aligned} \xi_1 &= h_1(-t) \\ \xi_2 &= -t \end{aligned}\right\} \qquad t \in [-f_1(a), -f_1(a_1)],$$

$$\left.\begin{aligned} \xi_1 &= t \\ \xi_2 &= f_1(t) \end{aligned}\right\} \qquad t \in [a_1, b_1],$$

$$\left.\begin{aligned} \xi_1 &= h_2(t) \\ \xi_2 &= t \end{aligned}\right\} \qquad t \in [f_1(b_1), f_1(b)]$$

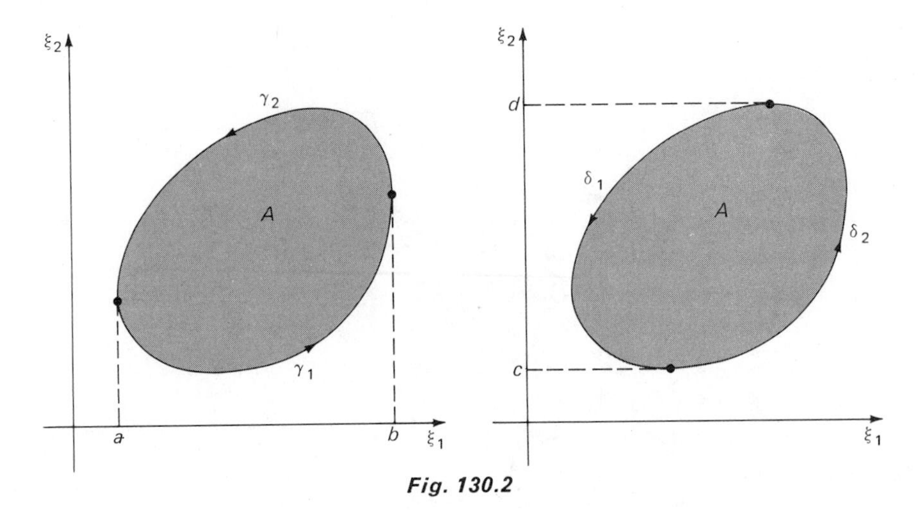


*Fig. 130.2*

are parametrizations with all the required properties. Hence, $\gamma_1$ is a regular oriented curve in $\mathbf{E}^2$. (Note that the parametrization $(\xi_1, f_1(\xi_1))$, $\xi_1 \in [a, b]$ of $\gamma_1$ cannot be used to make this point because $f_1'(a), f_1'(b)$ do not exist.) We obtain, for the 1-form $\psi_1(x)d\xi_1$, that

$$\begin{aligned}\int_{\gamma_1} \psi_1(x)d\xi_1 &= -\int_{-f_1(a)}^{-f_1(a_1)} \psi_1(h_1(-t), -t)h_1'(-t)\,dt + \int_{a_1}^{b_1} \psi_1(t, f_1(t))\,dt \\ &\quad + \int_{f_1(b_1)}^{f_1(b)} \psi_1(h_2(t), t)h_2'(t)\,dt \\ &= \int_a^{a_1} \psi_1(u, f_1(u))\,du + \int_{a_1}^{b_1} \psi_1(t, f_1(t))\,dt + \int_{b_1}^{b} \psi_1(u, f_1(u))\,du \\ &= \int_a^b \psi_1(\xi_1, f_1(\xi_1))\,d\xi_1,\end{aligned}$$

where we made the substitution $u = h_1(-t)$ in the first integral and $u = h_2(t)$ in the third integral, noting that $h_1^{-1}(u) = f_1(u)$ in the first case and $h_2^{-1}(u) = f_1(u)$ in the latter. (See Fig. 130.3.) Hence,

$$\int_{\gamma_1} \psi_1(x)\,d\xi_1 = \int_a^b \psi_1(\xi_1, f_1(\xi_1))\,d\xi_1. \tag{130.2}$$

By analogous reasoning,

$$\int_{\gamma_2} \psi_1(x)\,d\xi_1 = \int_b^a \psi_1(\xi_1, f_2(\xi_1))\,d\xi_1 = -\int_a^b \psi_1(\xi_1, f_2(\xi_1))\,d\xi_1. \tag{130.3}$$

If $\delta_1, \delta_2$ are the oriented regular curves as indicated in Fig. 130.2, we obtain, in the same manner, that

$$\begin{aligned}\int_{\delta_1} \psi_2(x)\,d\xi_2 &= -\int_c^d \psi_2(h_1(\xi_2), \xi_2)\,d\xi_2, \\ \int_{\delta_2} \psi_2(x)\,d\xi_2 &= \int_c^d \psi_2(h_2(\xi_2), \xi_2)\,d\xi_2.\end{aligned} \tag{130.4}$$

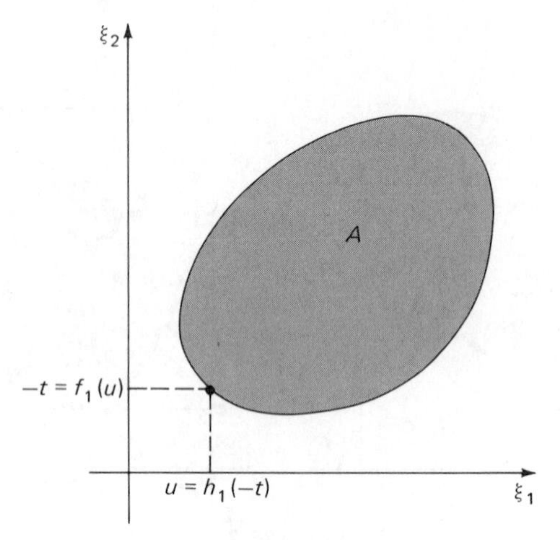


*Fig. 130.3*

It is easy to see that the formulas (130.2) to (130.4) remain valid if the derivatives of the functions $f_i, h_i$ in the representation (130.1) of $A$ have finitely many jump discontinuities in $(a, b)$ and $(c, d)$, respectively. This means that $A$ may, in effect, have a shape such as that in Fig. 130.4.

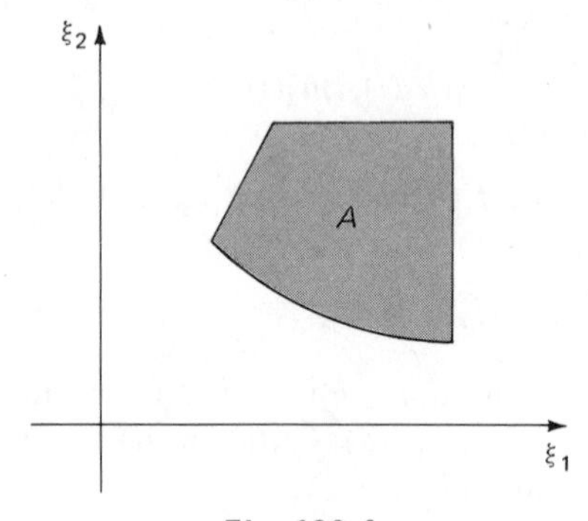


*Fig. 130.4*

***Definition 130.1***

$A \subset \mathbf{E}^2$ is a *standard region* if and only if it is represented as in (130.1), where $f_1', f_2'$ may have finitely many jump discontinuities in $(a, b)$ and $h_1', h_2'$ may have finitely many jump discontinuities in $(c, d)$.

(We note that, when the boundary of $A$ has horizontal or vertical flat spots $\beta$, then $\int_\beta \psi_2(x)d\xi_2 = 0$ along a horizontal flat spot and $\int_\beta \psi_1(x)d\xi_1 = 0$ along a vertical flat spot.)

After all these preparatory remarks, we are ready to state and prove:

***Theorem 130.1*** (*Green's Theorem*)

If $A \subset \Omega \subseteq \mathbf{E}^2$ is the union of finitely many standard regions with only boundary points in common, and if $g: \Omega \to \mathbf{E}^2$ has continuous partial derivatives in the open set $\Omega$, then

$$\int_A \int (D_1\psi_2(x) - D_2\psi_1(x))d\xi_1 \wedge d\xi_2 = \int_{\partial A} g(x) \cdot dx, \tag{130.5}$$

where either $A$ is positively oriented and its boundary $\partial A$ is oriented in the counterclockwise (positive) direction, or $A$ is negatively oriented and $\partial A$ is oriented in the negative (clockwise) direction.

*Proof*

First, we note that we need only prove (130.5) for $A$ and $\partial A$ positively oriented, because a reversal of the orientation of both changes the sign on both sides of (130.5).

Secondly, we note that we need only prove the theorem for standard regions. The validity for finite unions with only boundary points in common follows from (127.4), (127.12) and from (127.3), (127.11). (See also Fig. 130.5, and note that the line integrals along common boundaries cancel out.)

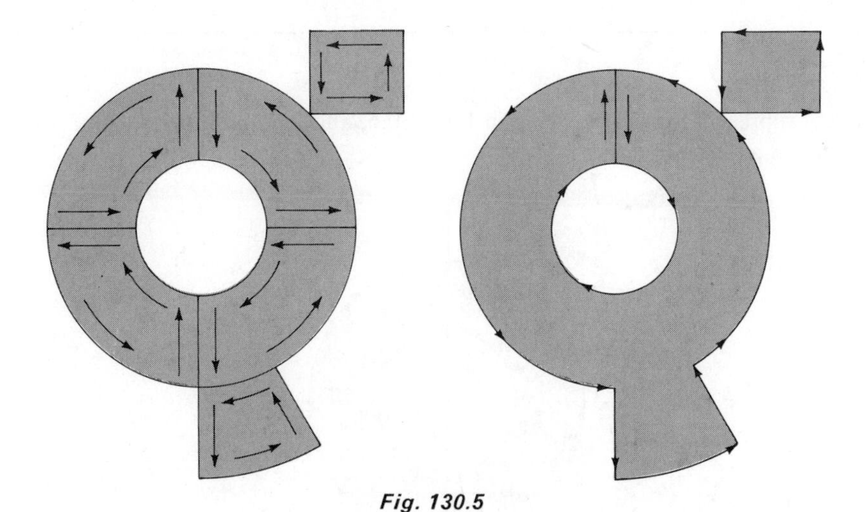

*Fig. 130.5*

By Theorem 109.1, we have, for the parametrization $f(x) = (\xi_1, \xi_2)$, $(\xi_1, \xi_2) \in A$, that

$$\int_A \int D_2\psi_1(x)\, d\xi_1 \wedge d\xi_2 = \int_A \int D_2\psi_1(x)\, d\xi_1\, d\xi_2$$
$$= \int_a^b \left( \int_{f_1(\xi_1)}^{f_2(\xi_1)} D_2\psi_1(\xi_1, \xi_2)\, d\xi_2 \right) d\xi_1.$$

For any fixed $\xi_1 \in [a, b]$,

$$\int_{f_1(\xi_1)}^{f_2(\xi_1)} D_2\psi_1(\xi_1, \xi_2)\, d\xi_2 = \psi_1(\xi_1, f_2(\xi_1)) - \psi_1(\xi_1, f_1(\xi_1)),$$

and hence, by (130.2) and (130.3),

$$\begin{aligned}(130.6)\qquad \int_A\!\!\int D_2\psi_1(x)\, d\xi_1 \wedge d\xi_2 &= \int_a^b \psi_1(\xi_1, f_2(\xi_1))\, d\xi_1 - \int_a^b \psi_1(\xi_1, f_1(\xi_1))\, d\xi_1 \\ &= -\int_{\gamma_2} \psi_1(x)\, d\xi_1 - \int_{\gamma_1} \psi_1(x)\, d\xi_1 \\ &= -\int_{\partial A} \psi_1(x)\, d\xi_1.\end{aligned}$$

From (130.4), we obtain in the same manner

$$(130.7)\qquad \int_A\!\!\int D_1\psi_2(x) d\xi_1 \wedge d\xi_2 = \int_{\partial A} \psi_2(x) d\xi_2.$$

Subtraction of (130.6) from (130.7) yields the desired result, (130.5).

---

▲ *Example 1*

Let $A$ represent the circular ring in Fig. 130.6, and let $g(x) = (\xi_2, -\xi_1)$. $\partial A$ may be taken as indicated in Fig. 130.6.

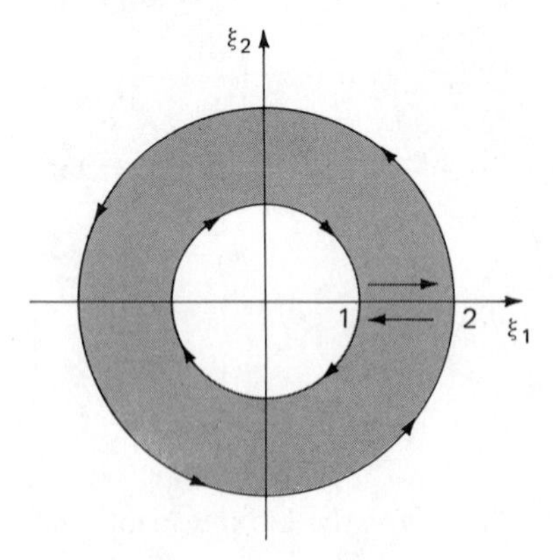


***Fig. 130.6***

Since $D_1\psi_2(x) - D_2\psi_1(x) = -2$, we have

$$\int_A\!\!\int (D_1\psi_2(x) - D_2\psi_1(x)) d\xi_1 \wedge d\xi_2 = -\int_A\!\!\int 2 d\xi_1 d\xi_2 = -6\pi.$$

We parametrize the boundary $\partial A$ by $(2\cos t, 2\sin t)$, $t \in [0, 2\pi]$; $(-t, 0)$, $t \in [-2, -1]$; $(\cos t, -\sin t)$, $t \in [0, 2\pi]$; $(t, 0)$, $t \in [1, 2]$. Then

$$\int_{\partial A} \xi_2\, d\xi_1 - \xi_1\, d\xi_2$$

$$= -4 \int_0^{2\pi} (\sin^2 t + \cos^2 t)\, dt + \int_{-2}^{-1} 0 + \int_0^{2\pi} (\sin^2 t + \cos^2 t)\, dt + \int_1^2 0 = -6\pi.$$

▲ *Example 2* (*Integrability Condition*)

If $\Omega \subseteq \mathbf{E}^2$ is open and connected, if $g \in C^1(\Omega)$, and if $\int_\gamma g(x) \cdot dx$ is independent of the path and depends only on its beginning point and endpoint, then

$$D_1\psi_2(x) = D_2\psi_1(x) \qquad \text{for all } x \in \Omega. \tag{130.8}$$

This may be deduced from the result in example 5, section 13.127. We may also deduce this result from Theorem 130.1:

Suppose $D_1\psi_2(x_0) - D_2\psi_1(x_0) > 0$ for some $x_0 \in \Omega$. Then, for some $\delta > 0$, $D_1\psi_2(x) - D_2\psi_1(x) > 0$ for all $x \in N_\delta(x_0)$, and $\int_{\mathfrak{D}} \int (D_1\psi_2(x) - D_2\psi_1(x)) d\xi_1 \wedge d\xi_2 > 0$ on the positively oriented disk $\mathfrak{D} = \{x \in \mathbf{E}^2 \mid \|x - x_0\| \leqslant (\delta/2)\}$. Hence, by Theorem 130.1, $\int_{\partial\mathfrak{D}} g(x) \cdot dx > 0$, and we see that $\int_{\gamma_1} g(x) \cdot dx \neq \int_{\gamma_2} g(x) \cdot dx$, where $\gamma_1, \gamma_2$ are the oriented arcs in Fig. 130.7. Hence (130.8) is necessary.

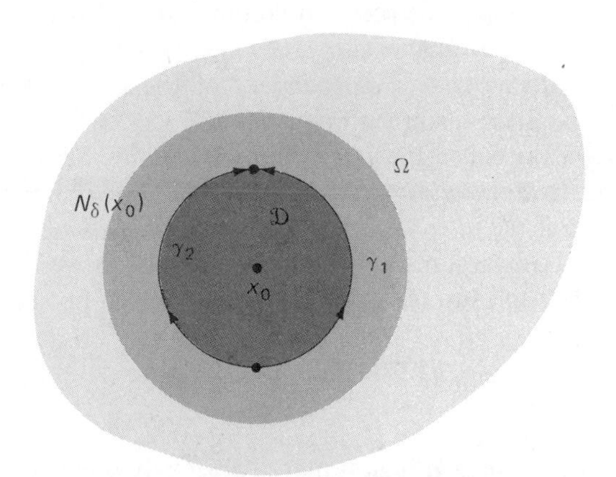


*Fig. 130.7*

If $\Omega$ is *simply connected*, i.e., *no simple closed regular curve in* $\Omega$ *has a boundary point of* $\Omega$ *in its interior*,† then (130.8) is also sufficient. To prove this, we shall demonstrate first that the line integral is independent of the path for polygonal paths, and then use this fact to construct a function $h$ as in example 5, section 13.127.

† By a theorem of Jordan, a simple closed curve divides the plane into two distinct sets with no points in common, the *interior* and the *exterior*. (See R. Courant and H. Robbins, *What Is Mathematics?* (New York: Oxford University Press, 1956), p. 267.)

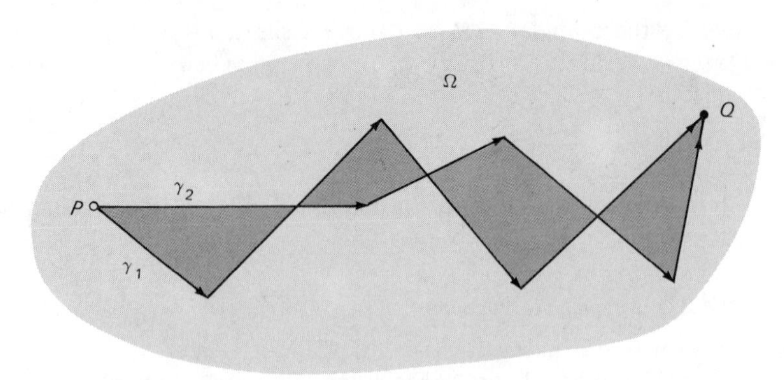


*Fig. 130.8*

Let $P, Q$ denote two points in $\Omega$, and let $\gamma_1, \gamma_2$ denote two polygonal paths joining $P$ to $Q$. (See Fig. 130.8.) The shaded region in Fig. 130.8 is the union of finitely many standard regions with only boundary points in common, and we have from Green's Theorem and (130.8), that

$$\text{(130.9)} \qquad \int_{\gamma_1 - \gamma_2} g(x) \cdot dx = \int_A \int (D_1 \psi_2(x) - D_2 \psi_1(x)) d\xi_1 \wedge d\xi_2 = 0.$$

Hence, $\int_{\gamma_1} g(x) \cdot dx = \int_{\gamma_2} g(x) \cdot dx$ for any two polygonal paths $\gamma_1, \gamma_2$ joining $P$ to $Q$. We use this fact to construct a function $h : \Omega \to \mathbf{E}$ as in example 5, section 13.127 such that $D_1 h(x) = \psi_1(x)$, $D_2 h(x) = \psi_2(x)$ for all $x \in \Omega$. Then $\int_\gamma g(x) \cdot dx = \int_\gamma \text{grad } h(x) \cdot dx$ is independent of the path, where $\gamma \subseteq \Omega$ is any regular curve, as we have also seen in example 5, section 13.127.

Condition (130.8) is called the *integrability condition*. Summarizing, we may state: *$\int_\gamma g(x) \cdot dx$ is independent of the path in the open, simply-connected set $\Omega \subseteq \mathbf{E}^2$ if and only if the integrability condition is satisfied in $\Omega$.*

Note that the above argument would not apply if $\Omega$ were not simply connected, because then the surface integral in (130.9) need not be zero. (See also exercise 8.)

▲ *Example 3*

Let $A \subset \mathbf{E}^2$ denote a standard region that is positively oriented. We have

$$J(A) = \int_A \int d\xi_1 d\xi_2.$$

For $g(x) = (0, \xi_1)$ or $g(x) = (-\xi_2, 0)$, we have $D_1 \psi_2(x) - D_2 \psi_1(x) = 1$, and hence, from Theorem 130.1,

$$\text{(130.10)} \qquad J(A) = \int_{\partial A} \xi_1 d\xi_2 = -\int_{\partial A} \xi_2 \, d\xi_1.$$

We use this formula to find the area of the disk with center at $\theta$ and radius $R$. Then $\partial A$ is parametrized by $f(t) = (R \cos t, R \sin t)$, $t \in [0, 2\pi]$, and we obtain

$$\int_{\partial A} \xi_1 \, d\xi_2 = R^2 \int_0^{2\pi} \cos^2 t \, dt = R^2 \int_0^{2\pi} \frac{1 + \cos 2t}{2} \, dt = R^2 \pi.$$

▲ *Example 4* (*Circulation Density*)

Let $v : \mathbf{E}^2 \to \mathbf{E}^2$ define a velocity field of a steady flow of an incompressible fluid as in example 3, section 13.128. We have defined $C = \int_\gamma v(x) \cdot dx$ to be taken along a regular closed curve $\gamma$ as the circulation along $\gamma$. Let $\gamma$ denote a circle with center at $x_0 \in \mathbf{E}^2$ and radius $\rho$. By Green's Theorem,

$$\int_\gamma v(x) \cdot dx = \int_D \int (D_1 v_2(x) - D_2 v_1(x)) d\xi_1 d\xi_2,$$

where $D = \{x \in \mathbf{E}^2 \mid \|x - x_0\| \leqslant \rho\}$ and its boundary $\partial D = \gamma$ are assumed to be positively oriented.

By the mean-value theorem (Theorem 107.3),

$$\int_D \int (D_1 v_2(x) - D_2 v_1(x)) d\xi_1 d\xi_2 = \rho^2 \pi (D_1 v_2(y) - D_2 v_1(y))$$

for some $y \in D$. Then

$$\lim_{\rho \to 0} \frac{1}{\rho^2 \pi} \int_\gamma v(x) \cdot dx = D_1 v_2(x_0) - D_2 v_1(x_0).$$

This gives a measure for the circulation density (circulation per unit area) at the point $x_0$. Loosely speaking, this quantity measures the "swirl" at $x_0$. Note that if we re-interpret $v : \mathbf{E}^2 \to \mathbf{E}^2$ as a velocity field in $\mathbf{E}^3$ with $v(x) = (v_1(\xi_1, \xi_2), v_2(\xi_1, \xi_2), 0)$, then $(0, 0, D_1 v_2(x) - D_2 v_1(x)) = \text{curl } v(x)$. (See (127.20).)

---

## 13.130 Exercises

1. Show that (130.2) is true if $f_1', f_2' : (a, b) \to \mathbf{E}$ and, accordingly, $h_1', h_2' : (c, d) \to \mathbf{E}$ have finitely many jump discontinuities.
2. Prove Theorem 130.1 for a two-dimensional interval without utilizing the proof in the text.
3. Use Green's Theorem to evaluate the following line integrals:
   (a) $\int_\gamma (\xi_2 \, d\xi_1 - \xi_1 d\xi_2)$ where $\gamma$ is the positively oriented boundary of $[0, 1] \times [2, 4]$.
   (b) $\int_{\gamma'} (\xi_2 \, d\xi_1 - \xi_1 d\xi_2)$ where $\gamma$ is the positively oriented boundary of $\{x \in \mathbf{E}^2 \mid 0 \leqslant \xi_1 \leqslant 1, 0 \leqslant \xi_2 \leqslant \xi_1^2\}$.
   (c) $\int_\gamma ((\xi_1 + \xi_2) d\xi_1 + \xi_2 \, d\xi_2)$ where $\gamma$ is the positively oriented boundary of $\{x \in \mathbf{E}^2 \mid -1 \leqslant \xi_1 \leqslant 1, 0 \leqslant \xi_2 \leqslant \sqrt{1 - \xi_1^2}\}$.
4. Let $A \subset \mathbf{E}^2$ denote a standard region, positively oriented. Show that

$$J(A) = \frac{1}{2} \int_{\partial A} (\xi_1 d\xi_2 - \xi_2 \, d\xi_1).$$

5. Let $A$ denote the circular ring with a circle of radius 1 and center at $\theta$ as inner boundary, and a circle of radius 5 and center at $\theta$ as the outer boundary. Let $g : \mathbf{E}^2 \to \mathbf{E}^2$ where $g \in C^1(\mathbf{E}^2)$ and $D_1\psi_2(x) = D_2\psi_1(x)$ for all $x \in \mathbf{E}^2 \setminus \{\theta\}$. Let $\gamma_1$ denote the inner boundary and let $\gamma_2$ denote the outer boundary, both positively oriented. Show that $\int_{\gamma_1} g(x) \cdot dx = -\int_{\gamma_2} g(x) \cdot dx$.
6. Use Green's Theorem to evaluate $\int_{\partial A} (\xi_1\xi_2\, d\xi_1 + \xi_1 d\xi_2)$ where $A$ is the unit disk with center at $\theta$.
7. Use the result of example 5, section 13.127, to show that if $\int_\gamma g(x) \cdot dx$ is independent of the path and depends only on the beginning point and endpoint of $\gamma$, and if $g \in C^1(\Omega)$ where $\Omega$ is open and connected, then $D_1\psi_2(x) = D_2\psi_1(x)$ for all $x \in \Omega$.
8. Let $g(x) = (\xi_2/(\xi_1^2 + \xi_2^2), -\xi_1/(\xi_1^2 + \xi_2^2))$. Show:
   (a) $D_1\psi_2(x) = D_2\psi_1(x)$ for all $x \neq \theta$,
   (b) $\int_{\gamma_1} g(x) \cdot dx \neq \int_{\gamma_2} g(x) \cdot dx$ where $\gamma_1$ and $\gamma_2$ are the left and the right halves of the unit circle with center at $\theta$, both oriented from the point $P = (-1, 0)$ to the point $Q = (0, 1)$.

## 13.131 Stokes' Theorem

In this section we shall prove (129.7), namely,

$$\int_M d\omega = \int_{\partial M} \omega$$

for certain regular surfaces in $\mathbf{E}^3$. Whenever $M$ is a $k$-dimensional manifold in $\mathbf{E}^n$ and $k < n$, as in the case of surfaces in $\mathbf{E}^3$, the theorem is customarily referred to as *Stokes' Theorem*.†

To begin with, we shall consider only oriented patches $\sigma \subset \mathbf{E}^3$ for which the containing plaster has a parametrization $f : \mathfrak{D} \to \mathbf{E}^3$ such that $f^*(\sigma)$ is a standard region in $\mathbf{E}^2$. Then, the boundary $\beta f^*(\sigma)$ is a regular simple closed curve, and the boundary of $\sigma$, $\partial\sigma = f_*(\beta f^*(\sigma))$, is, consequently, also a regular simple closed curve. $\partial\sigma$ is the "edge" of $\sigma$, i.e., all points of $\sigma$ lie on one side of $\partial\sigma$.

Otherwise, there would be two points $x, y \in \sigma^\circ = \sigma \setminus \partial\sigma$ on either side of $\partial\sigma$. (See Fig. 131.1.) Then, $f^{-1}(x), f^{-1}(y) \in (f^*(\sigma))^\circ$. Since $f^*(\sigma)$ is a standard region, we may join $f^{-1}(x)$ to $f^{-1}(y)$ by a line segment $s$, the image $f_*(s)$ of which would have to cross $\partial\sigma$, contrary to the fact that $f$ is injective on $\mathfrak{D}$.

We assign a positive orientation to a closed regular curve on an oriented plaster as follows:

***Definition 131.1***

The regular simple closed curve $\gamma \subset \Pi$ is positively oriented with respect to the oriented plaster $\Pi \subset \mathbf{E}^3$ if and only if with $h : [a, b] \to \mathbf{E}^3$, $f : \mathfrak{D} \to \mathbf{E}^3$ being the parametrizations of $\gamma$ and $\Pi$, respectively, $f^{-1} \circ h : [a, b] \to \mathfrak{D}$ represents a posi-

† *G. G. Stokes*, 1819–1903.

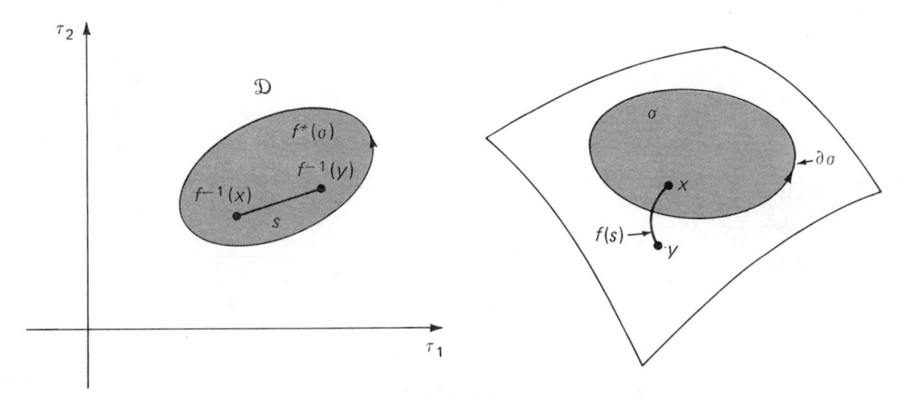


*Fig. 131.1*

tively oriented regular simple closed curve in $\mathfrak{D} \subseteq \mathbf{E}^2$. (Remember that a closed curve in the plane is positively oriented if it is oriented in the counterclockwise direction.)

---

▲ *Example 1*

The plane $\xi_3 = 1$ is a plaster in $\mathbf{E}^3$. We introduce an orientation by means of the parametrization $f(t) = (\tau_1, \tau_2, 1)$, $(\tau_1, \tau_2) \in \mathbf{E}^2$. Then, $D_1 f(t) \times D_2 f(t) = (0, 0, 1)$. (See Fig. 131.2.) If $\gamma$ is parametrized by $h(s) = (\cos s, \sin s, 1)$, $s \in [0, 2\pi]$, then $f^{-1} \circ h(s) = (\cos s, \sin s)$, $s \in [0, 2\pi]$, represents a positively oriented circle in $\mathbf{E}^2$. Hence, $\gamma$ is positively oriented with respect to the plane in Fig. 131.2.

We reverse the orientation of the plane by means of the new parametrization $g(t) = (\tau_2, \tau_1, 0), (\tau_1, \tau_2) \in \mathbf{E}^2$. Then, $D_1 g(t) \times D_2 g(t) = (0, 0, -1)$. Now, $g^{-1} \circ h(s) = (\sin s, \cos s)$, $s \in [0, 2\pi]$, parametrizes a negatively oriented circle in $\mathbf{E}^2$ and hence, $\gamma$ is not positively oriented with respect to the re-oriented plane. (See Fig. 131.3.)

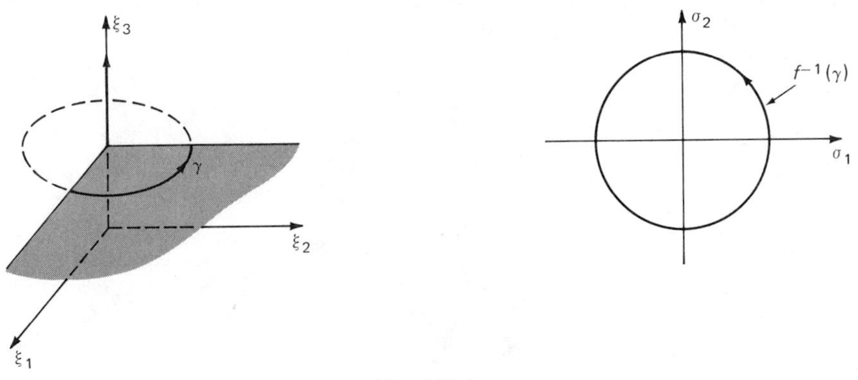


*Fig. 131.2*

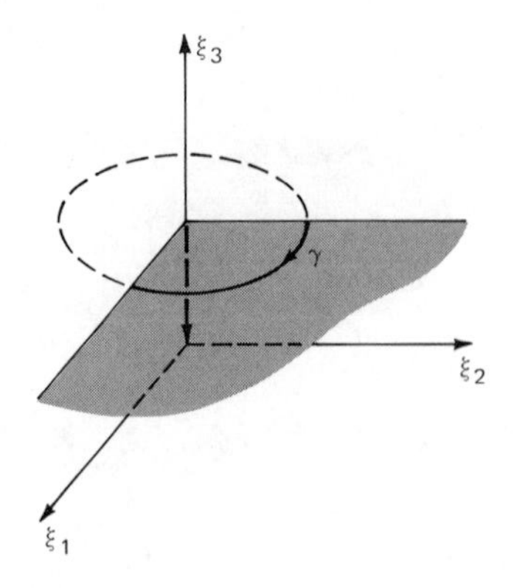


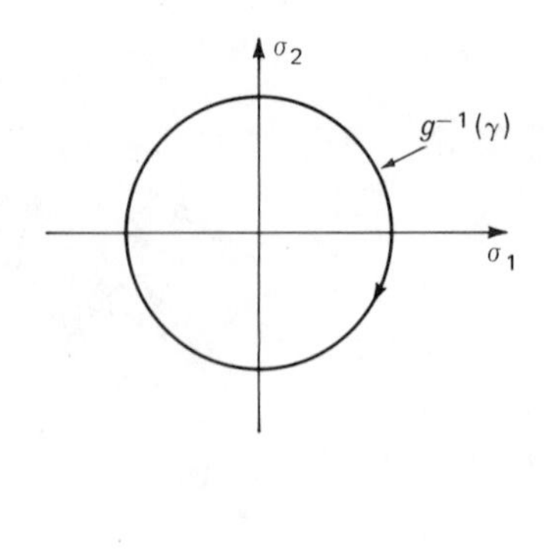


*Fig. 131.3*

---

***Definition 131.2***

$\sigma \subset \mathbf{E}^3$ is a *standard patch* if and only if there is a parametrization $f : \mathfrak{D} \to \mathbf{E}^3$ of the containing plaster such that $f \in C^2(\mathfrak{D})$ and $f^*(\sigma)$ is a standard region in $\mathbf{E}^2$.

(Note the new condition $f \in C^2(\mathfrak{D})$, which will play an important role in our proof of Stokes' Theorem.)

$\Sigma \subset \mathbf{E}^3$ is a *standard surface* if and only if it is a simple surface and is the union of finitely many standard patches with only boundary points in common.

If it is possible to orient each standard patch in a standard surface in such a manner that the orientations of the positively oriented boundaries of the patches are opposite to each other along common boundaries, we say that *the standard surface is orientable*. (See Fig. 131.4.) If a standard patch touches the rest of them in finitely many points only, then it may be oriented either way. (See Fig. 131.5.)

The positively oriented boundary $\partial\Sigma$ is the sum of the positively oriented boundaries of the individual patches. (See arrows along the boundaries of the patches in Figs. 131.4 and 131.5, and note that line integrals along arcs with opposite orientation cancel each other.)

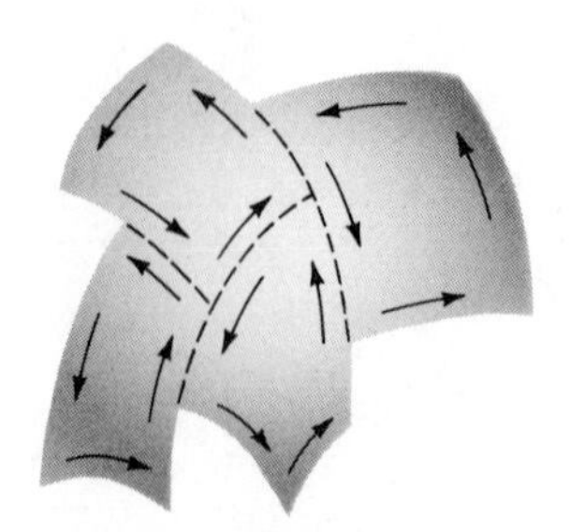

*Fig. 131.4*

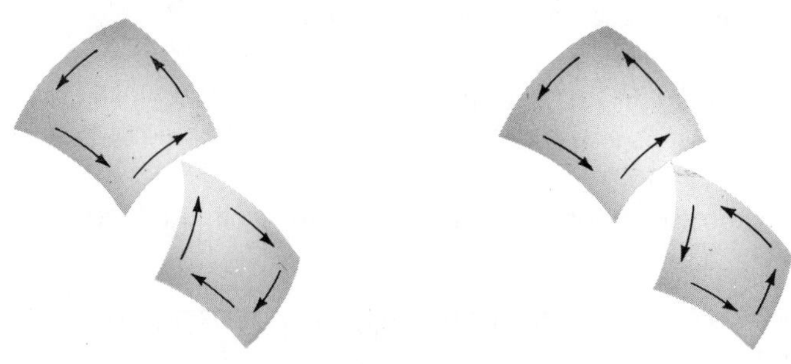

*Fig. 131.5*

The Möbius strip is a standard surface but is *not* an orientable standard surface. (See example 3, section 13.123.)

With these preliminary comments out of the way we are ready to state and prove Stokes' Theorem.

***Theorem 131.1*** (*Stokes' Theorem*)

If $\Omega \subseteq \mathbf{E}^3$ is an open set, if $\Sigma \subset \Omega$ is an oriented standard surface, and if $g : \Omega \to \mathbf{E}^3$ has continuous partial derivatives in $\Omega$, then

(131.1) $$\iint_{\Sigma} \operatorname{curl} g(x) \cdot d\sigma = \int_{\partial\Sigma} g(x) \cdot dx \qquad \text{or, equivalently,}$$

$$\iint_{\Sigma} d(g(x) \cdot dx) = \int_{\partial\Sigma} g(x) \cdot dx,$$

where $\partial\Sigma$ is positively oriented with respect to $\Sigma$.

*Proof*

We need only prove this theorem for oriented standard patches, and then obtain the result for standard surfaces by addition.

If $\sigma$ is a standard patch, it may be parametrized by $f : A \to \mathbf{E}^3$, where $A \subset \mathbf{E}^2$ is a standard region. By (127.20),

(131.2) $$\begin{aligned}\int_{\sigma}\int \operatorname{curl} g(x) \cdot d\sigma = \int_{A}\int \{&[D_2\psi_3(f(t)) - D_3\psi_2(f(t))][D_1\phi_2(t)D_2\phi_3(t) \\ &- D_1\phi_3(t)D_2\phi_2(t)] \\ &+ [D_3\psi_1(f(t)) - D_1\psi_3(f(t))][D_1\phi_3(t)D_2\phi_1(t) \\ &- D_1\phi_1(t)D_2\phi_3(t)] \\ &+ [D_1\psi_2(f(t)) - D_2\psi_1(f(t))][D_1\phi_1(t)D_2\phi_2(t) \\ &- D_1\phi_2(t)D_2\phi_1(t)]\}d\tau_1 d\tau_2.\end{aligned}$$

We collect all terms containing $\psi_1$ and obtain

$$(131.3)\quad D_3\psi_1(f(t))(D_1\phi_3(t)D_2\phi_1(t) - D_1\phi_1(t)D_2\phi_3(t)) \\ - D_2\psi_1(f(t))(D_1\phi_1(t)D_2\phi_2(t) - D_1\phi_2(t)D_2\phi_1(t)),$$

and similar expressions for the terms containing $\psi_2$ and $\psi_3$, respectively.

If we add

$$-D_1\psi_1(f(t))(D_1\phi_1(t)D_2\phi_1(t) - D_1\phi_1(t)D_2\phi_1(t)) = 0$$

to (131.3), we obtain

$$D_2\phi_1(t)(D_3\psi_1(f(t))D_1\phi_3(t) + D_2\psi_1(f(t))D_1\phi_2(t) \\ + D_1\psi_1(f(t))D_1\phi_1(t)) - D_1\phi_1(t)(D_3\psi_1(f(t))D_2\phi_3(t) \\ + D_2\psi_1(f(t))D_2\phi_2(t) + D_1\psi_1(f(t))D_2\phi_1(t) \\ = D_1(\psi_1\circ f)(t)D_2\phi_1(t) - D_2(\psi_1\circ f)(t)D_1\phi_1(t).$$

Let

$$\Psi_1 = (\psi_1\circ f)D_1\phi_1, \qquad \Psi_2 = (\psi_1\circ f)D_2\phi_1.$$

Since $f\in C^2(\mathfrak{D})$ and $A\subset\mathfrak{D}$,

$$D_1\Psi_2 - D_2\Psi_1 = D_1(\psi_1\circ f)D_2\phi_1 + (\psi_1\circ f)D_1D_2\phi_1 - D_2(\psi_1\circ f)D_1\phi_1 \\ - (\psi_1\circ f)D_2D_1\phi_1 \\ = D_1(\psi_1\circ f)D_2\phi_1 - D_2(\psi_1\circ f)D_1\phi_1.$$

Hence, we obtain from Green's Theorem, noting that $A\subset\mathbf{E}^2$ is positively oriented,

$$\int_A\int (D_1(\psi_1\circ f)(t)D_2\phi_1(t) - D_2(\psi_1\circ f)(t)D_1\phi_1(t))d\tau_1 d\tau_2 \\ = \int_A\int (D_1\Psi_2(t) - D_2\Psi_1(t))d\tau_1 d\tau_2 \\ = \int_{\partial A}(\Psi_1(t)d\tau_1 + \Psi_2(t)d\tau_2) \\ = \int_{\partial A}(\psi_1\circ f)(t)(D_1\phi_1(t)d\tau_1 + D_2\phi_1(t)d\tau_2),$$

where $\partial A$ is positively oriented. If $h:[a,b]\to\mathbf{E}^2$ is the parametrization of $\partial A$, then $f\circ h:[a,b]\to\mathbf{E}^3$ is a parametrization of $\partial\sigma$, and we have

$$\int_{\partial\sigma}\psi_1(x)\,d\xi_1 = \int_a^b(\psi_1\circ f\circ h)(s)(f\circ h)'(s)\,ds.$$

On the other hand,

$$\int_{\partial A} (\psi_1 \circ f)(t)(D_1\phi_1(t)d\tau_1 + D_2\phi_1(t)d\tau_2)$$
$$= \int_a^b (\psi_1 \circ f \circ h)(s)[D_1(\phi_1 \circ h)(s)\chi_1'(s) + D_2(\phi_1 \circ h)(s)\chi_2'(s)]ds$$
$$= \int_a^b (\psi_1 \circ f \circ h)(s)(f \circ h)'(s)ds.$$

Hence,

$$\int_{\partial A} (\psi_1 \circ f)(t)(D_1\phi_1(t)d\tau_1 + D_2\phi_1(t)d\tau_2) = \int_{\partial\sigma} \psi_1(x)d\xi_1$$

and, consequently,

$$\int_A\int (D_1(\psi_1 \circ f)(t)D_2\phi_1(t) - D_1(\psi_1 \circ f)(t)D_1\phi_1(t))d\tau_1 d\tau_2 = \int_{\partial\sigma} \psi_1(x)d\xi_1.$$

If we collect all terms containing $\psi_2$ in (131.2) we obtain, by the same reasoning, $\int_{\partial\sigma} \psi_2(x)d\xi_2$, and for all terms containing $\psi_3$ in (131.2), we obtain $\int_{\partial\sigma} \psi_3(x)d\xi_3$. Adding all these partial results yields (131.1).

---

▲ *Example 2*

Stokes' Theorem does not apply to the Möbius strip (example 3, section 13.123) because the Möbius strip is not orientable. We have seen that a Möbius strip $\mu$ may be parametrized by

$$f(t) = \left(\left(2 - \tau_1 \sin\frac{\tau_2}{2}\right)\cos\tau_2, \left(2 - \tau_1 \sin\frac{\tau_2}{2}\right)\sin\tau_2, \tau_1\cos\frac{\tau_2}{2}\right),$$
$$(\tau_1, \tau_2) \in [-1, 1] \times [0, 2\pi].$$

(See 123.6.) Its oriented edge $\eta$ may be parametrized by

$$h(\tau_2) = f(1, \tau_2), \qquad \tau_2 \in [0, 4\pi]$$

(see Fig. 131.6). We obtain, for $g(x) = (0, \xi_3^2, 0)$, that curl $g(x) = (-2\xi_3, 0, 0)$.

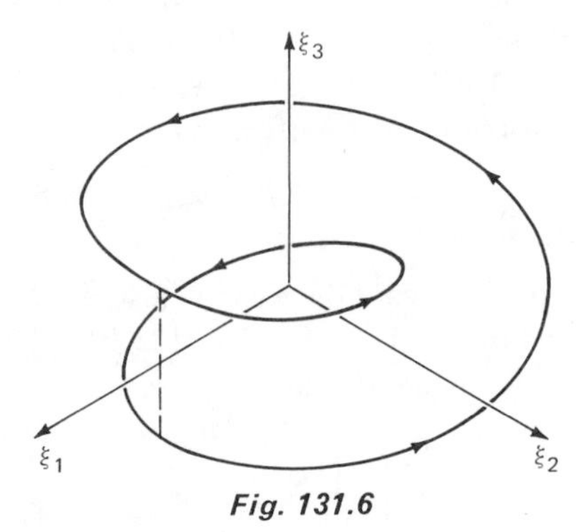


*Fig. 131.6*

Hence, we have to evaluate $\int_\mu \int 2\xi_3\, d\xi_2\, d\xi_3$ and $\int_\eta \xi_3^2\, d\xi_2$. On $\mu$,

$$d\xi_2 = -\sin\frac{\tau_2}{2}\sin\tau_2\, d\tau_1 + \left[\left(2 - \tau_1\sin\frac{\tau_2}{2}\right)\cos\tau_2 - \frac{\tau_1}{2}\cos\frac{\tau_2}{2}\sin\tau_2\right]d\tau_2,$$

$$d\xi_3 = \cos\frac{\tau_2}{2}\,d\tau_1 - \frac{\tau_1}{2}\sin\frac{\tau_2}{2}\,d\tau_2.$$

Therefore,

$$\iint_\mu \xi_3\, d\xi_2\, d\xi_3$$

$$= \int_{\tau_1=-1}^{1}\int_{\tau_2=0}^{2\pi}\left[\tau_1^2\cos\frac{\tau_2}{2}\sin^2\frac{\tau_2}{2}\sin\tau_2 - 2\tau_1^2\cos^2\frac{\tau_2}{2}\cos\tau_2\left(2-\tau_1\sin\frac{\tau_2}{2}\right)\right.$$
$$\left. + \tau_1^2\cos^3\frac{\tau_2}{2}\sin\tau_2\right]d\tau_1\, d\tau_2 = \frac{32}{9}.$$

On $\eta$,

$$\xi_3^2 = \cos^2\frac{\tau_2}{2}, \qquad d\xi_2 = \left(\left(2 - \sin\frac{\tau_2}{2}\right)\cos\tau_2 - \frac{1}{2}\sin\frac{\tau_2}{2}\sin\tau_2\right)d\tau_2.$$

Hence,

$$\int_\eta \xi_3^2\, d\xi_2 = \int_0^{4\pi}\left[\left(2-\sin\frac{\tau_2}{2}\right)\cos\tau_2\cos^2\frac{\tau_2}{2} - \frac{1}{2}\sin\frac{\tau_2}{2}\cos^2\frac{\tau_2}{2}\sin\tau_2\right]d\tau_2 = 2\pi,$$

and we see that

$$\int_\mu\int \operatorname{curl} g(x)\cdot d\sigma \neq \int_\eta g(x)\cdot dx,$$

and that a reversal of the orientation of $\eta$ would not help matters.

The fact that the Möbius strip is not orientable has, as a consequence, that, in the image of the oriented boundary of $[-1,1]\times[0,2\pi]$ under $f$, the oriented arcs at $\xi_1 = 2$, $\xi_2 = 0$ from $\xi_3 = -1$ to $\xi_3 = 1$ which are the images of $-1 \leqslant \tau_1 \leqslant 1$, $\tau_2 = 0$ and $\tau_2 = 2\pi$, do not cancel each other. (See Fig. 131.7.)

If we think of the Möbius strip as cut open along the vertical line at $\xi_1 = 2$, $\xi_2 = 0$, and let $\partial\mu = f_*(\beta([-1,1]\times[0,2\pi]))$, then Stokes' Theorem applies. (This is a case where the edge $\eta$ of the regular surface $\mu$ is not what we consider to be the boundary $\partial\mu$ of the regular surface $\mu$.)

▲ *Example 3* (*Circulation Density*)

As in the case of a two-dimensional flow we may also define a circulation density for a three-dimensional flow $v : \mathbf{E}^3 \to \mathbf{E}^3$. By (128.10), the circulation along a regular closed space curve $\gamma$ is defined as

$$C = \int_\gamma v(x)\cdot dx.$$

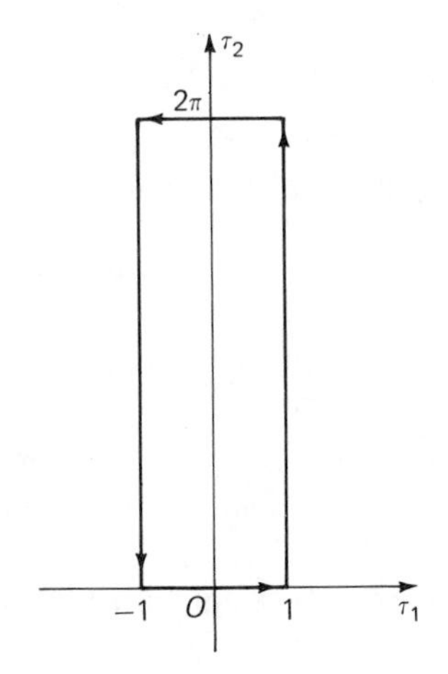


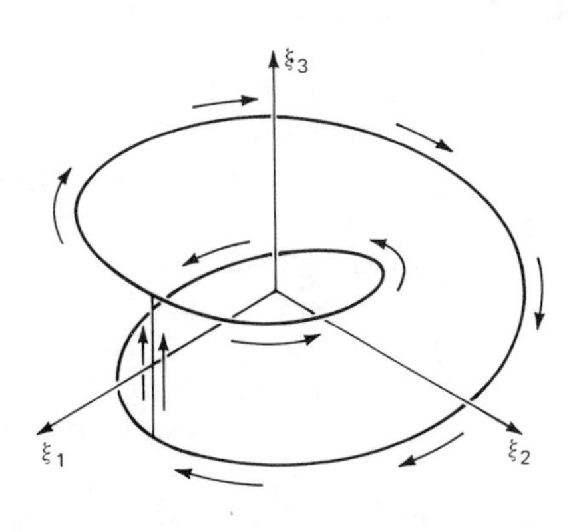


*Fig. 131.7*

We consider a plane $P$ through $x_0 \in \mathbf{E}^3$ which is parametrized by

$$f(t) = x_0 + \tau_1 u_1 + \tau_2 u_2, \qquad (\tau_1, \tau_2) \in \mathbf{E}^2,$$

where $u_1, u_2$ are orthogonal unit vectors in the plane $P: u_1 \cdot u_2 = 0$, $\|u_j\| = 1$, $j = 1, 2$. The boundary of the disk $D$ with center at $x_0$ and radius $\rho$ that lies on the plane $P$ is parametrized by

$$g(s) = x_0 + \rho u_1 \cos s + \rho u_2 \sin s, \qquad s \in [0, 2\pi].$$

Hence, $f^{-1} \circ g(s) = (\rho \cos s, \rho \sin s)$, $s \in [0, 2\pi]$, and we obtain, from Stokes' Theorem,

(131.4)

$$\int_{\partial D} v(x) \cdot dx = \iint_D \operatorname{curl} v(x) \cdot d\sigma = \int_{\tau_1^2 + \tau_2^2 \leqslant \rho^2} \int \operatorname{curl} v(f(t)) \cdot u_1 \times u_2 \, d\tau_1 \, d\tau_2.$$

We define the circulation density at $x_0$ in the plane with the normal $n = u_1 \times u_2$ as in example 4, section 13.130, as

$$\lim_{\rho \to 0} \frac{1}{\rho^2 \pi} \int_{\partial D} v(x) \cdot dx$$

and obtain, from (131.4),

$$\lim_{\rho \to 0} \frac{1}{\rho^2 \pi} \int_{\partial D} v(x) \cdot dx = \operatorname{curl} v(x_0) \cdot n. \tag{131.5}$$

Since $\|n\| = \|u_1 \times u_2\| = 1$, we see that the circulation density at $x$ in a plane perpendicular to $n$ is given by the component of curl $v(x)$ in the direction of $n$. We now realize that the result in example 4, section 13.130, is just a special case of (131.5).

### 13.131 Exercises

1. Let $\mathfrak{D} \subseteq \mathbf{E}^2$ denote an open set, and let $f : \mathfrak{D} \to \mathbf{E}^3$ denote the parametrization of a two-dimensional plaster $P$. Show: If $\gamma \subset \mathfrak{D}$ is a closed regular curve, then $f_*(\gamma) \subset P$ is also a closed regular curve.
2. $f(t) = x_0 + \tau_1 u_1 + \tau_2 u_2, (\tau_1, \tau_2) \in \mathbf{E}^2$, where $u_1 \cdot u_2 = 0, \|u_1\| = \|u_2\| = 1$, parametrizes a plane $P$ in $\mathbf{E}^3$.
   (a) Choose $x_0, u_1, u_2$ such that $P$ is the $(\xi_1, \xi_2)$-plane; the $(\xi_2, \xi_3)$-plane; the $(\xi_3, \xi_1)$-plane.
   (b) Parametrize a circle on $P$ with radius $\rho$ and center at $x_0$ such that it is positively oriented with respect to $P$; negatively oriented with respect to $P$.
3. Show that the Möbius strip in example 2 consists of two standard patches with only boundary points in common.
4. Take the standard patch of the Möbius strip that lies in front of the $(\xi_2, \xi_3)$-plane and verify Stokes' Theorem for $g(x) = (0, 0, \xi_1^2)$.
5. Show that

$$\int_\sigma\int \nabla \times \nabla h(x) \cdot d\sigma = 0,$$

   where $\sigma$ is a standard patch, and where $h : \mathbf{E}^3 \to \mathbf{E}$, $h \in C^2(\mathbf{E}^3)$.
6. Supply the details in the derivation of (131.5). (See also example 4, section 13.130.)
7. Let $v(x) = (1, 1, 1)$. Use Stokes' Theorem to find the total flux in unit time through the northern hemisphere of radius 1 with center at $\theta$.
8. Collect all terms with $\psi_2$ in the integral of (131.2), and show that the integral over $A$ of these terms is equal to $\int_{\partial\sigma} \psi_2(x) d\xi_2$.

### 13.132 The Theorem of Gauss

Finally, we get around to a proof of (129.7) for certain three-dimensional manifolds $M$ in $\mathbf{E}^3$. For this case, the theorem is, in effect, a three-dimensional version of Green's Theorem in two dimensions, and we shall give a proof that is quite similar to the proof of Green's Theorem.

We restrict ourselves to positively oriented three-dimensional manifolds $\Upsilon$ in $\mathbf{E}^3$, whose boundary may be represented simultaneously in three different ways as two shells that are fused together, as indicated in Fig. 132.1.

We assume that the functions $\alpha_i, \beta_i, \gamma_i, \delta_i$ are continuous and have continuous derivatives and partial derivatives, respectively, in the interior of their domains. It is easy to see that the boundary of such a three-dimensional manifold is a regular orientable surface.

By Theorem 109.2 for iterated triple integrals, we have, for the parametrization $(\xi_1, \xi_2, \xi_3) \in Y$ of $\Upsilon$ that

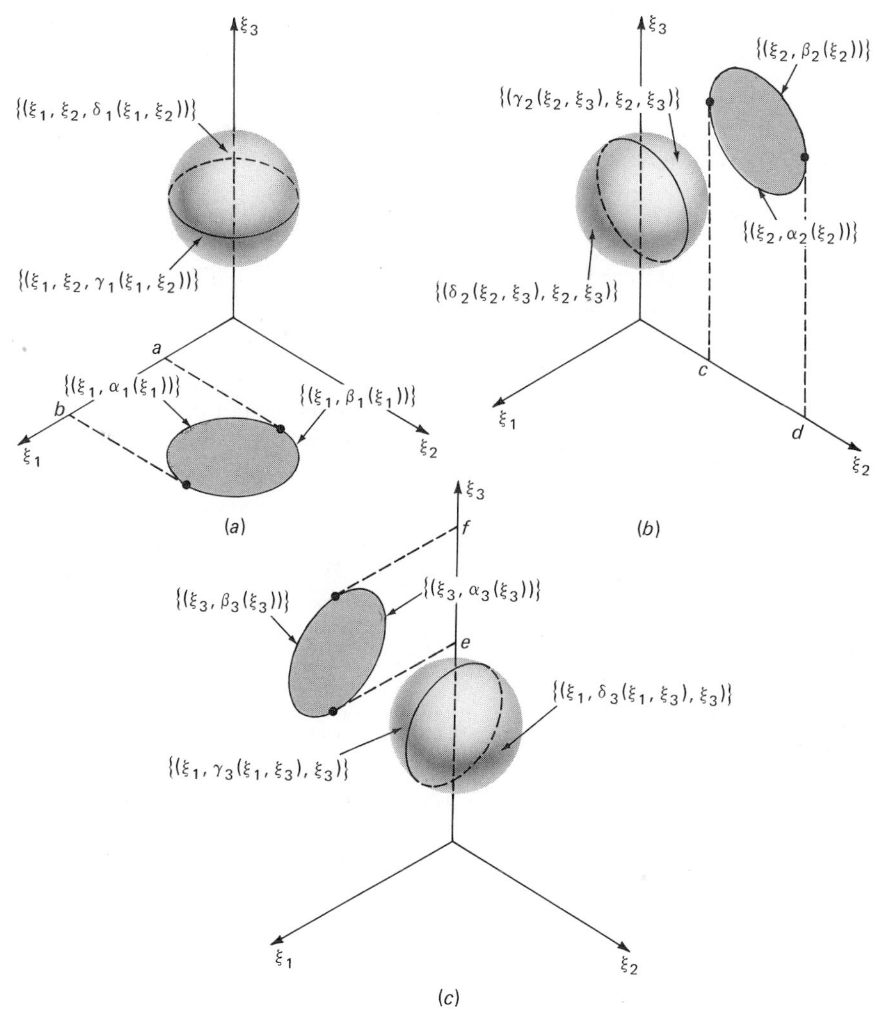


***Fig. 132.1***

$$\int_Y \iint D_3 \psi_3(x)\, d\xi_1 \wedge d\xi_2 \wedge d\xi_3 = \int_Y \iint D_3 \psi_3(x)\, d\xi_1\, d\xi_2\, d\xi_3$$
$$= \int_a^b \left( \int_{\alpha_1(\xi_1)}^{\beta_1(\xi_1)} \left( \int_{\gamma_1(\xi_1,\xi_2)}^{\delta_1(\xi_1,\xi_2)} D_3 \psi_3(x)\, d\xi_3 \right) d\xi_2 \right) d\xi_1$$
$$= \int_a^b \int_{\alpha_1(\xi_1)}^{\beta_1(\xi_1)} [\psi_3(\xi_1, \xi_2, \delta_1(\xi_1, \xi_2)) - \psi_3(\xi_1, \xi_2, \gamma_1(\xi_1, \xi_2))]\, d\xi_2\, d\xi_1.$$

If $\sigma_1$ is the lower shell in Fig. 132.1a with the normal pointing down, and $\sigma_2$ is the upper shell with the normal pointing up, then $\sigma_1 + \sigma_2$ has an outward normal, and we see, as in section 13.130, that

$$\int_a^b \int_{\alpha_1(\xi_1)}^{\beta_1(\xi_1)} \psi_3(\xi_1, \xi_2, \delta_1(\xi_1, \xi_2))\, d\xi_2\, d\xi_1 = \int_{\sigma_2}\int \psi_3(x)\, d\xi_1 \wedge d\xi_2,$$

(132.1)

$$-\int_a^b \int_{\alpha_1(\xi_1)}^{\beta_1(\xi_1)} \psi_3(\xi_1, \xi_2, \gamma_1(\xi_1, \xi_2))\, d\xi_2\, d\xi_1 = \int_{\sigma_1}\int \psi_3(x)\, d\xi_1 \wedge d\xi_2.$$

Hence,

$$\int_\Upsilon \iint D_3 \psi_3(x)\, d\xi_1 \wedge d\xi_2 \wedge d\xi_3 = \int_{\partial\Upsilon}\int \psi_3(x)\, d\xi_1 \wedge d\xi_2.$$

We derive two analogous results with reference to Figs. 132.1b and 132.1c, add up these results, and obtain Gauss' Theorem.

We convince ourselves that the above steps remain valid if one permits $\Upsilon$ to have flat spots—a term that is easier to understand than to define. We call $\Upsilon$ a *standard oriented solid* if it is oriented and if it can be represented simultaneously as in Figs. 132.1a, b, and c, with flat spots admitted. We say that $\partial\Upsilon$ is positively oriented with respect to $\Upsilon$ if the normal points outward when $\Upsilon$ is positively oriented and inward if $\Upsilon$ is negatively oriented. With this understanding we can say:

***Theorem 132.1*** *(Gauss' Theorem)*

If $\Omega \subseteq \mathbf{E}^3$ is an open set, if $\Upsilon \subset \Omega$ is a standard oriented solid, and if $g : \Omega \to \mathbf{E}^3$ has continuous partial derivatives, then

$$\int_\Upsilon \iint \operatorname{div} g(x)\, d\xi_1 \wedge d\xi_2 \wedge d\xi_3 = \int_{\partial\Upsilon}\int g(x) \cdot d\sigma,$$

or, equivalently,

$$\iiint_\Upsilon d(g(x) \cdot d\sigma) = \iint_{\partial\Upsilon} g(x) \cdot d\sigma,$$

where $\partial\Upsilon$ is positively oriented with respect to $\Upsilon$.

---

▲ *Example 1*

The cube has the flattest flat spots of them all. We apply Gauss' Theorem to the cube $[0, 1] \times [0, 1] \times [0, 1]$ with $g(x) = x$. Then, $\operatorname{div} g(x) = 3$ and

$$\int_\Upsilon \iint \operatorname{div} g(x)\, d\xi_1 \wedge d\xi_2 \wedge d\xi_3 = \int_0^1 \int_0^1 \int_0^1 3\, d\xi_1\, d\xi_2\, d\xi_3 = 3.$$

We evaluate $\int_{\partial\Upsilon}\int g(x) \cdot d\sigma = \int_{\partial\Upsilon}\int (\xi_1 d\xi_2 \wedge d\xi_3 + \xi_2\, d\xi_3 \wedge d\xi_1 + \xi_3\, d\xi_1 \wedge d\xi_2)$ one face at a time. (See Fig. 132.2.)

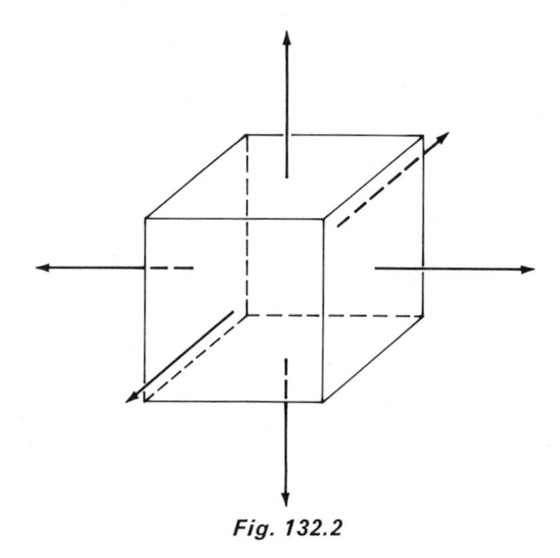

*Fig. 132.2*

With $f, t, r, b, l, ri$ denoting the properly oriented front, top, rear, bottom, left and right faces, we obtain

$$\int_f\int g(x)\cdot d\sigma = \int_0^1\int_0^1 d\xi_2\, d\xi_3 = 1, \qquad \int_t\int g(x)\cdot d\sigma = \int_0^1\int_0^1 d\xi_1\, d\xi_2 = 1,$$

$$\int_r\int g(x)\cdot d\sigma = \int_b\int g(x)\cdot d\sigma = \int_l\int g(x)\cdot d\sigma = 0,$$

$$\int_{ri}\int g(x)\cdot d\sigma = \int_0^1\int_0^1 d\xi_3\, d\xi_1 = 1.$$

Hence,

$$\int_{\partial\Upsilon}\int g(x)\cdot d\sigma = 3.$$

▲ *Example 2*

If $v : \mathbf{E}^3 \to \mathbf{E}^3$ represents a velocity field, then, by (128.7), the total flow through the surface $\partial\Upsilon$ of a standard solid in the direction of the outward normal and in unit time is given by

$$F = \int_{\partial\Upsilon}\int v(x)\cdot d\sigma.$$

If this quantity is positive, then $\partial\Upsilon$ has to enclose sources where matter is created and if $F$ is negative, there have to be sinks where matter is destroyed. Let $\Upsilon$ denote a sphere with center at $x_0$ and radius $\rho$. We define *source density* at $x_0$ as

$$\lim_{\rho\to 0}\frac{3}{4\rho^3\pi}\int_{\partial\Upsilon}\int v(x)\cdot d\sigma.$$

By Gauss' Theorem,

$$\lim_{\rho\to 0}\frac{3}{4\rho^2\pi}\int_{\partial\Upsilon}\int v(x)\cdot d\sigma = \lim_{\rho\to 0}\frac{3}{4\rho^2\pi}\int_{\Upsilon}\iint \operatorname{div} v(x)d\xi_1\wedge d\xi_2\wedge d\xi_3 = \operatorname{div} v(x_0).$$

Hence, div $v(x)$ is a measure of the source density at $x$. If div $v(x) > 0$, we say $v$ has a *source* at $x$, and if div $v(x) < 0$, we say $v$ has a *sink* at $x$.

Let $v(x) = (\xi_1, \xi_2, \xi_3)$. Then div $v(x) = 3$; that is, the source density is 3 at every point $x \in \mathbf{E}^3$. (See also example 1.)

If div $v(x) = 0$ for all $x \in \mathbf{E}^3$, then, by Gauss' Theorem,

$$\iint_{\partial\Upsilon} v(x)\cdot d\sigma = 0.$$

---

## 13.132 Exercises

1. The operator $\nabla\cdot\nabla = (D_1, D_2, \ldots, D_n)\cdot(D_1, D_2, \ldots, D_n) = D_1D_1 + D_2D_2 + \cdots + D_nD_n = \Delta$ is called the *Laplace operator*. Show that $\int_{\partial\Upsilon}\int \nabla h\cdot d\sigma = \int_\Upsilon\iint \Delta h d\xi_1\wedge d\xi_2\wedge d\xi_3$ if $\Upsilon, \partial\Upsilon$ are as in Theorem 132.1 and if $h: \mathbf{E}^3 \to \mathbf{E}$ has continuous partial derivatives of second order.
2. A function $h: \Omega \to \mathbf{E}$, $\Omega \subseteq \mathbf{E}^n$, $h \in C^2(\Omega)$, is called a *harmonic function* in $\mathbf{E}^n$ if and only if $\Delta h = 0$ for all $x \in \Omega$. Show: If $h$ is a harmonic function in $\Omega \subseteq \mathbf{E}^3$ and if $\Upsilon$ is a standard solid in $\Omega$, then $\int_{\partial\Upsilon}\int \nabla h\cdot d\sigma = 0$.
3. Show: If a velocity field is the gradient of a harmonic function, then the source density is zero everywhere.
4. Let $g(x) = (\psi_1(\xi_1), \psi_2(\xi_2), \psi_3(\xi_3))$ where $\psi_i \in C^1(\mathbf{E})$. Show that
$$\int_{\partial\Upsilon}\int \operatorname{curl} g(x)\cdot d\sigma = 0$$
for any oriented standard solid $\Upsilon \subset \mathbf{E}^3$.
5. Show that a surface as in Fig. 132.1 is a regular surface.
6. Proceed as in the proof of (130.2) to establish the validity of (132.1).
7. With $\Upsilon, \partial\Upsilon$, and $g$ as in the derivation of Theorem 132.1, show that
$$\int_\Upsilon\iint D_1\psi_1(x)d\xi_1\wedge d\xi_2\wedge d\xi_3 = \int_{\partial\Upsilon}\int \psi_1(x)d\xi_2\wedge d\xi_3.$$
8. Let $\Upsilon$ denote the positively oriented unit ball with center at $\theta$, and let $g(x) = (\xi_1 + \xi_2^2, \xi_2\xi_3 - \xi_2^2, \xi_1^2 + \xi_3)$. Evaluate $\int_{\partial\Upsilon}\int g(x)\cdot d\sigma$.
9. Let $f: [a, b] \to \mathbf{E}$ represent a continuous function where $[a, b]$ is an interval on the $\xi_3$-axis and where $f(\xi_3)$ is recorded on the $\xi_2$-axis. The graph of $f$ rotates about the $\xi_3$-axis and generates a surface $S$ of revolution. Show that $\pi\int_a^b f^2(\xi_3)d\xi_3$ is the volume of the solid bounded by $S$, and by the planes $\xi_3 = a$ and $\xi_3 = b$.

10. Let $T$ denote the oriented torus that is generated by the circle $(\xi_2 - 2)^2 + \xi_3^2 = 1$ rotating about the $\xi_3$-axis with the normal pointing outward. Let

$$g(x) = (\xi_2^2 + \xi_2 \xi_3, \xi_2 - \xi_1^2, \xi_1^2 + \xi_2^2).$$

Find $\int_T \int g(x) \cdot d\sigma$.

# 14 Infinite Series

## 14.133 Convergence of Infinite Series

Infinite series play an important role in analysis and its applications and, especially, in the analysis of complex-valued functions of a complex variable where they constitute one of the three pillars on which complex analysis is built, the others being the derivative and the integral. An important byproduct of our study of infinite series is the rigorous definition of the exponential function, the trigonometric functions and their inverse functions and the derivation of their well-known properties.

In section 2.19, we presented a brief discussion of infinite series and, in particular, the geometric series. The concept of an infinite series of real numbers may be generalized in a straightforward manner to infinite series of vectors in $\mathbf{E}^n$:

***Definition 133.1***

The infinite series $\sum_{j=1}^{\infty} a_j$, generated by the sequence $\{a_j\}$ of vectors $a_j \in \mathbf{E}^n$ is the sequence of partial sums $\{s_k\} = \{\sum_{j=1}^{k} a_j\}$.

The infinite series $\sum_{j=1}^{\infty} a_j$ converges if and only if the sequence of partial sums $\{s_k\}$ converges. $\lim_{\infty} s_k$ is called the sum of the infinite series $\sum_{j=1}^{\infty} a_j$. We write for convenience

$$\sum_{j=1}^{\infty} a_j = \lim_{\infty} s_k.$$

The generality one achieves by considering infinite series in $\mathbf{E}^n$ rather than in $\mathbf{E}$ is, by and large, an illusion. Nothing can be learned from a study of infinite series in $\mathbf{E}^n$ that cannot be learned from a study of infinite series of real numbers. Therefore we shall, as in Chapter 7, restrict our treatment to infinite series of real numbers with an occasional remark or exercise pertaining to a generalization to infinite series of vectors.

From definition 61.1 (or, equivalently, definition 15.1), we obtain immediately the following criterion for the convergence of an infinite series.

***Theorem 133.1***

The infinite series $\sum_{j=1}^{\infty} a_j, a_j \in \mathbf{E}$, converges to the sum $s \in \mathbf{E}$ if and only if, for every $\varepsilon > 0$, there is some $N(\varepsilon) \in \mathbf{N}$ such that with $s_k = \sum_{j=1}^{k} a_j$,

$$|s - s_k| < \varepsilon \qquad \text{for all } k > N(\varepsilon). \tag{133.1}$$

This theorem has a trivial generalization to the case where $a_j \in \mathbf{E}^n$.

We have, from Theorems 16.1, 16.2, and 16.3, that *the sum of an infinite series is unique*, that *the sequence of partial sums of a convergent infinite series is bounded*, and that *every subsequence of the sequence of partial sums converges to the sum of the infinite series.*

If $\{s_k\}$ does not converge, we say that $\sum_{j=1}^{\infty} a_j$ *diverges.* If $\{s_k\}$ diverges to $+\infty$ (or $-\infty$), we say that $\sum_{j=1}^{\infty} a_j$ diverges to $+\infty$ (or $-\infty$). (See definition 15.2.)

---

▲ *Example 1*

Let $a_j = (-1)^j$. Then, $s_1 = -1$, $s_2 = 0$, $s_3 = -1$, $s_4 = 0, \ldots$, and we see that $\{s_k\}$ does not converge. Hence, $\sum_{j=1}^{\infty} (-1)^j$ diverges.

---

At times we find it more convenient to start the summation with the summation index 0 rather than 1 or, for that matter, with some other number. Clearly,

$$\sum_{j=1}^{\infty} a_j = \sum_{i=0}^{\infty} a_{i+1} = \sum_{k=2}^{\infty} a_{k-1}, \ldots.$$

The fact that the sequence of partial sums of a convergent infinite series is bounded is of sufficient importance to warrant a formal statement as a lemma.

***Lemma 133.1***

If $\sum_{j=1}^{\infty} a_j$, $a_j \in \mathbf{E}$, converges, then there is an $M > 0$ such that

$$|s_k| = \left|\sum_{j=1}^{k} a_j\right| \leqslant M \qquad \text{for all } k \in \mathbf{N}.$$

The converse of this lemma is not necessarily true unless the terms $a_j$ of $\sum_{j=1}^{\infty} a_j$ are nonnegative (or nonpositive) numbers.

***Lemma 133.2***

If $a_j \geqslant 0$ (or $a_j \leqslant 0$) for all $j \in \mathbf{N}$, then $\sum_{j=1}^{\infty} a_j$ converges if and only if $|s_k| \leqslant M$ for all $k \in \mathbf{N}$ and some $M > 0$.

*Proof*

The necessity of the condition follows from Lemma 133.1.

To prove the sufficiency, we proceed as follows: Since $a_j \geqslant 0$ for all $j \in \mathbf{N}$, we have $s_{k+1} \geqslant s_k$ for all $k \in \mathbf{N}$. Application of Theorem 18.1 on the convergence of bounded increasing sequences clinches the argument.

---

▲ *Example 2*

For the infinite series $\sum_{j=1}^{\infty} (-1)^j$ of example 1, we have $|s_k| \leqslant 1$ for all $k \in \mathbf{N}$. However, the series diverges because infinitely many terms are positive and infinitely many terms are negative.

---

From the Cauchy criterion for the convergence of sequences (Theorem 62.1), we obtain a criterion for the convergence of infinite series which, in contrast to Theorem 133.1, does *not* require *a priori* knowledge of the sum of the infinite series.

Since

$$s_k - s_l = a_{l+1} + a_{l+2} + \cdots + a_k \qquad \text{for} \quad k > l,$$

we have:

***Theorem 133.2*** *(Cauchy Criterion for the Convergence of Series)*

The infinite series $\sum_{j=1}^{\infty} a_j, a_j \in \mathbf{E}$, converges if and only if for every $\varepsilon > 0$, there is an $N(\varepsilon) \in \mathbf{N}$ such that

$$(133.2) \quad |s_k - s_l| = |a_{l+1} + a_{l+2} + \cdots + a_k| < \varepsilon \qquad \text{for all } k > l > N(\varepsilon),$$

or, equivalently,

(133.3)

$$|s_{l+p} - s_l| = |a_{l+1} + a_{l+2} + \cdots + a_{l+p}| < \varepsilon \qquad \text{for all } l > N(\varepsilon) \text{ and all } p \in \mathbf{N}.$$

---

▲ *Example 3*

Let $a_j = (-1)^j(1 + 1/2^{j-1})$. Since

$$a_1 + a_2 + a_3 + a_4 + a_5 + \cdots$$

$$= -(1+1) + \left(1 + \frac{1}{2}\right) - \left(1 + \frac{1}{2^2}\right) + \left(1 + \frac{1}{2^3}\right) - \left(1 + \frac{1}{2^4}\right) + \cdots,$$

one might be tempted to write

$$a_1 + a_2 + a_3 + a_4 + a_5 + \cdots$$
$$= -(1+1) + \left(1 + \frac{1}{2} - 1 - \frac{1}{2^2}\right) + \left(1 + \frac{1}{2^3} - 1 - \frac{1}{2^4}\right) + \cdots$$
$$= -2 + \frac{1}{2^2} + \frac{1}{2^4} + \cdots,$$

and conclude that this series converges to $-\frac{5}{3}$. However,

$$|s_{l+1} - s_l| = |a_{l+1}| = 1 + \frac{1}{2^l} > 1,$$

no matter how large we choose $l$. This is in violation of (133.3), and hence, $\sum_{j=1}^{\infty} a_j$ does *not* converge. Our first and false conclusion was based on an indiscriminate use of the associative law, the validity of which is guaranteed only for finite sums. Its application to infinite series may have disastrous consequences, as we have just experienced.

---

The following lemma which is a direct consequence of Theorem 133.2 could have immediately eliminated the series in example 3 as a candidate for convergence.

***Lemma 133.3***

For $\sum_{j=1}^{\infty} a_j, a_j \in \mathbf{E}$, to converge, it is *necessary* that $\lim_{\infty} a_j = 0$.

*Proof*

By (133.3) for $p = 1$,

$$|s_{l+1} - s_l| = |a_{l+1}| < \varepsilon \qquad \text{for all } l > N(\varepsilon).$$

In example 3, $\lim_{\infty} a_j$ does *not* exist. Hence, $\sum_{j=1}^{\infty} a_j$ does *not* converge.

---

▲ *Example 4*

The series $\sum_{j=1}^{\infty} (j-1)/j$ does *not* converge because $\lim_{\infty} (j-1)/j = 1$.

---

From Lemma 133.3, we obtain in turn,

***Lemma 133.4***

If $\sum_{j=1}^{\infty} a_j, a_j \in \mathbf{E}$, converges, then $|a_j| \leqslant M$ for all $j \in \mathbf{N}$ and some $M > 0$. (See Theorem 16.2.)

Infinite series may be added, subtracted, or multiplied by a number, according to the following rules.

***Theorem 133.3*** *(Algebraic Manipulations of Series)*

If $\sum_{j=1}^{\infty} a_j, \sum_{j=1}^{\infty} b_j, a_j, b_j \in \mathbf{E}$, converge and if $\lambda \in \mathbf{R}$, then

$$\sum_{j=1}^{\infty} (a_j \pm b_j) = \sum_{j=1}^{\infty} a_j \pm \sum_{j=1}^{\infty} b_j, \tag{133.4}$$

$$\sum_{j=1}^{\infty} \lambda a_j = \lambda \sum_{j=1}^{\infty} a_j. \tag{133.5}$$

*Proof*

Let $\sum_{j=1}^{\infty} a_j = a, \sum_{j=1}^{\infty} b_j = b$.

(133.4) follows from:

$$\left|\sum_{j=1}^{k} (a_j + b_j) - (a + b)\right| \leqslant \left|\sum_{j=1}^{k} a_j - a\right| + \left|\sum_{j=1}^{k} b_j - b\right|,$$

and (133.5) follows from

$$\left|\sum_{j=1}^{k} \lambda a_j - \lambda a\right| \leqslant |\lambda| \left|\sum_{j=1}^{k} a_j - a\right|.$$

## 14.133 Exercises

1. Test for convergence or divergence:

   (a) $\sum_{j=0}^{\infty} \frac{1}{2^j}$ (b) $\sum_{j=1}^{\infty} \frac{1}{j^2}$ (c) $\sum_{j=0}^{\infty} \left(\frac{7}{6}\right)^j$ (d) $\sum_{j=0}^{\infty} (-1)^j \frac{1}{3^j}$

2. Prove Lemma 133.4.
3. Let $a_j$ be defined as in example 3. Find the numerical values of $s_1, s_2, s_3, s_4, s_5, s_6$.
4. *Prove*: $a$ is an element of the Cantor set (see definition 14.1) if and only if there is a sequence $\{a_j\}$ where $a_j \in \{0, 2\}$ such that $a = \sum_{j=1}^{\infty} a_j/3^j$.
5. Let $r \in \mathbf{R}\backslash\mathbf{Z}$. Find the sum of $\sum_{j=0}^{\infty} 1/(r + j)(r + j + 1)$.
6. *Prove*: $\sum_{j=1}^{\infty} a_j, a_j \in \mathbf{E}^n$, converges if and only if, for every $\varepsilon > 0$, there is an $N(\varepsilon) \in \mathbf{N}$ such that $\|s_k - s_l\| < \varepsilon$ for all $k > l > N(\varepsilon)$.
7. *Prove*: If $\sum_{j=1}^{\infty} a_j, a_j \in \mathbf{E}^n$, converges, then, by necessity, $\lim_{\infty} \|a_j\| = 0$ and hence, $\lim_{\infty} a_j = \theta$.
8. Let $a_j = (1/2^j, (j-1)/2j, (-1)^j/j^2) \in \mathbf{E}^3$. Does $\sum_{j=1}^{\infty} a_j$ converge?
9. *Prove*: If $a_j \in \mathbf{E}^n, c \in \mathbf{E}^n$, and if $\sum_{j=1}^{\infty} a_j$ converges, then $\sum_{j=1}^{\infty} c \cdot a_j = c \cdot (\sum_{j=1}^{\infty} a_j)$.
10. Find the sum of $1 + 1 + \frac{1}{2} - \frac{1}{3} + \frac{1}{4} + \frac{1}{9} + \frac{1}{8} - \frac{1}{27} + \frac{1}{16} + \frac{1}{81} + \frac{1}{32} - \frac{1}{243} + \cdots$.

## 14.134 The Integral Test

In this section we shall discuss a convergence test which, though limited in scope, is very practical when applicable. We shall also discuss two important applications of this test.

***Theorem 134.1*** (*Integral Test*)

Let $a_j \geqslant 0$, and let $\{a_j\}$ represent a decreasing sequence.

The infinite series $\sum_{j=1}^{\infty} a_j$ converges *if and only if* the sequence $\{\int_1^k f\}$ converges where $f: [1, \infty) \to \mathbf{R}$ may be taken to be *any* decreasing function for which $f(j) = a_j$ for all $j \in \mathbf{N}$.

*Proof*

First, let us point out that a function with the required properties always exists. The graphs of two such functions are depicted in Fig. 134.1.

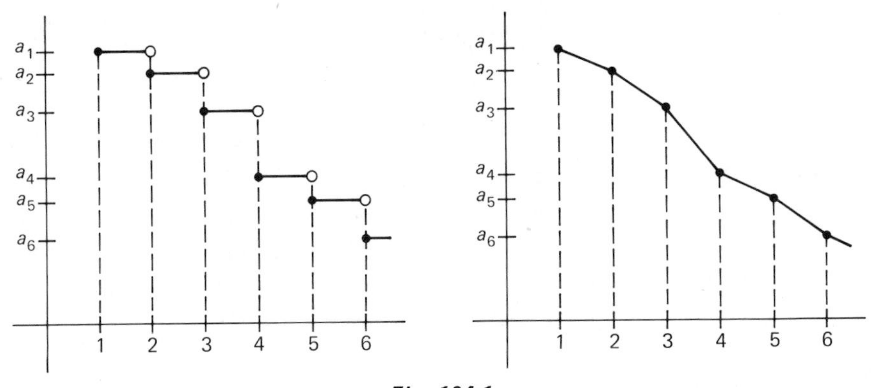


***Fig. 134.1***

By Theorem 47.2, every decreasing function $f: [1, \infty) \to \mathbf{R}$ is integrable on $[1, k]$ for all $k \in \mathbf{N}$.

Since

$$s_k - s_l \leqslant \int_l^k f \leqslant s_{k-1} - s_{l-1}, \qquad k > l > 1,$$

(see Fig. 134.2), and

$$\int_l^k f = \int_1^k f - \int_1^l f,$$

the theorem follows readily from Theorem 133.2.

The integral test is practical only when an easily integrable function with the required properties is readily available. The functions depicted in Fig. 134.1 are of no practical use other than for the purpose of demonstrating that functions, called for by Theorem 134.1, always exist.

The following two examples serve to illustrate the use of the integral test and, at the same time, lead us to an important topic which is to be discussed in the next section.

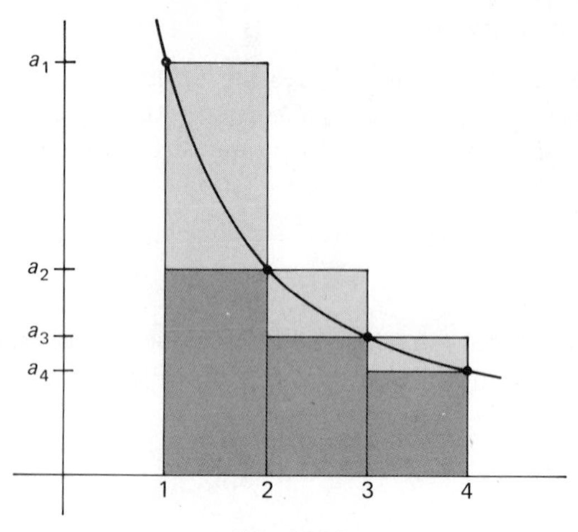


*Fig. 134.2*

---

▲ *Example 1* (*Harmonic Series*)

The *harmonic series* $\sum_{j=1}^{\infty} 1/j$ does *not* converge. We define $f: [1, \infty) \to \mathbf{R}$ by $f(x) = (1/x)$ and obtain $\int_1^k f = \log k$. (See example 5, section 4.52.) The sequence $\{\log k\}$ does *not* converge because $|\log k - \log l| = |\log k/l| = \log 2 > 0$ for all $k = 2l$, no matter how large $l$ is chosen. Hence, $\sum_{j=1}^{\infty} (1/j)$ does not converge.

▲ *Example 2*

We change the sign of every other term in the harmonic series of example 1, and obtain the *alternating series*

$$\sum_{j=1}^{\infty} (-1)^{j+1} \frac{1}{j}.$$

We use the integral test to show that the subsequence $\{s_{2k}\}$ of the sequence of partial sums converges, and then present a supplementary argument to demonstrate the convergence of $\{s_k\}$.

We have

$$\begin{aligned} s_{2k} &= \left(1 - \frac{1}{2}\right) + \left(\frac{1}{3} - \frac{1}{4}\right) + \left(\frac{1}{5} - \frac{1}{6}\right) + \cdots + \left(\frac{1}{2k-1} - \frac{1}{2k}\right) \\ &= \frac{1}{2} + \frac{1}{3 \cdot 4} + \frac{1}{5 \cdot 6} + \cdots + \frac{1}{(2k-1)(2k)} \\ &= b_1 + b_2 + b_3 + \cdots + b_k. \end{aligned}$$

Let $f : [1, \infty) \to \mathbf{R}$ be defined by

$$f(x) = \begin{cases} \frac{1}{2} & \text{for } 1 \leqslant x < 2, \\ \dfrac{1}{(2x-1)2x} & \text{for } x \geqslant 2. \end{cases}$$

Clearly, $f$ is decreasing and $f(j) = b_j$. Since $\int_1^k f = \frac{1}{2} + \frac{1}{2}\log(\frac{4}{3}(2k-1)/2k)$, since $\lim_\infty (2k-1)/2k = 1$, and since $\lim_{4/3} \log x = \log \frac{4}{3}$, we see that $\{\int_1^k f\}$ converges to $\frac{1}{2} + \frac{1}{2}\log\frac{4}{3}$ and hence, $\{s_{2k}\}$ converges.

Since $s_{2\kappa} = s_{2\kappa-1} - (1/2\kappa)$, we see that $\lim_\infty s_{2\kappa-1} = \lim_\infty s_{2\kappa}$. Hence, $\{s_k\}$ converges and, consequently $\sum_{j=1}^\infty (-1)^{j+1}(1/j)$ converges.

---

In general, an alternating series, of which $\sum_{j=1}^\infty (-1)^{j+1}(1/j)$ is a prototype, converges if the absolute values of its terms form a decreasing sequence that converges to zero.

***Theorem 134.2*** *(Convergence Test for Alternating Series)*

If $\sum_{j=1}^\infty a_j$ is an *alternating series*, i.e., if $a_1 > 0$, $a_2 < 0$, $a_3 > 0$, $a_4 < 0, \ldots$, (or: $a_1 < 0$, $a_2 > 0$, $a_3 < 0$, $a_4 > 0, \ldots$) and if $\{|a_j|\}$ decreases and converges to zero, then $\sum_{j=1}^\infty a_j$ converges.

*Proof*

Let $a_{2\kappa-1} = b_{2\kappa-1}, a_{2\kappa} = -b_{2\kappa}$. Then $a_j = (-1)^{j+1}b_j$ where we assume w.l.o.g. that $b_j > 0$ for all $j \in \mathbf{N}$. We obtain for the partial sums of odd and even order:

$$s_{2\kappa-1} = b_1 - (b_2 - b_3) - (b_4 - b_5) - \cdots - (b_{2\kappa-2} - b_{2\kappa-1}),$$
$$s_{2\kappa} = (b_1 - b_2) + (b_3 - b_4) + \cdots + (b_{2\kappa-1} - b_{2\kappa}).$$

Since $|a_{k+1}| \leqslant |a_k|$, we have $b_{k+1} \leqslant b_k$ and hence, the odd partial sums form a decreasing sequence and the even partial sums form an increasing sequence. Since

$$s_2 < s_{2\kappa} < s_{2\kappa+1} < s_1,$$

the one sequence is bounded below, and the other one is bounded above. Hence,

$$\lim_\infty s_{2\kappa-1} = \sigma_1, \qquad \lim_\infty s_{2\kappa} = \sigma_2$$

exist.

Since $s_{2\kappa} - s_{2\kappa-1} = a_{2\kappa}$ and since $\lim_\infty a_k = 0$, we have $\sigma_1 = \sigma_2$. Hence, $\lim_\infty s_k = \sigma_1$ $(=\sigma_2)$ and $\sum_{j=1}^\infty a_j$ converges.

---

▲ *Example 3*

The series $\sum_{j=1}^\infty (-1)^j(1/\sqrt{j})$ converges (slowly) because $\lim_\infty (-1)^j(1/\sqrt{j}) = 0$ and $(1/\sqrt{j+1}) < (1/\sqrt{j})$.

---

## 14.134 Exercises

1. The convergence of $\{s_{2k}\}$ does not, in general, guarantee the convergence of $\{s_k\}$. Take

$$a_j = \begin{cases} 1 & \text{if } j \text{ is odd,} \\ -\underbrace{0.99\ldots9}_{(j/2)\text{ places}} & \text{if } j \text{ is even.} \end{cases}$$

Show that $\{s_{2k}\}$ converges but that $\{s_k\}$ does not converge. (See also example 3, section 14.133 where $\{s_{2k+1}\}$ converges but where $\{s_{2k}\}$ does not converge.)

2. Give an example of convergent series $\sum_{j=1}^{\infty} a_j$, $\sum_{j=1}^{\infty} b_j$ such that $\sum_{j=1}^{\infty} a_j b_j$ does not converge.
3. Let $c_k = 1 + \frac{1}{2} + \frac{1}{3} + \cdots + (1/k) - \log k$.
   (a) Show that $c_k > 0$ for all $k \in \mathbf{N}$.
   (b) Show that $c_{k+1} \leqslant c_k$ for all $k \in \mathbf{N}$.
   (c) Show that $\lim_{\infty} c_k$ exists.
   ($\gamma = \lim_{\infty} c_k = 0.57721566\ldots$ is called *Euler's constant*. It is not known whether $\gamma$ is rational or irrational.)
4. Let $a_j = j^2$ for $j \leqslant 10^6$ and $a_j = (-1)^j(1/\sqrt[3]{j})$ for $j > 10^6$. Show that $\sum_{j=1}^{\infty} a_j$ converges.
5. Show that $\sum_{j=1}^{\infty} (1/j^{\alpha})$ converges for $\alpha > 1$ and diverges for $\alpha \leqslant 1$. ($\zeta(\alpha) = \sum_{j=1}^{\infty} 1/j^{\alpha}$ is called *Riemann's zeta function*.)
6. Let $a_j \geqslant 0$ and let $\{a_j\}$ represent a decreasing sequence. *Prove*: $\sum_{j=1}^{\infty} a_j$ converges if and only if $\sum_{j=1}^{\infty} 2^j a_{2^j}$ converges (Cauchy's condensation test).
7. Show that $\sum_{j=2}^{\infty} 1/j \log j$ does not converge.
8. Show that $\sum_{j=2}^{\infty} 1/j(\log j)(\log\log j) \cdots (\log \ldots \log j)$ does not converge.
9. Show that $\sum_{j=2}^{\infty} 1/j(\log j)^{\alpha}$ converges for $\alpha > 1$ and diverges for $\alpha \leqslant 1$.
10. Same as in exercise 9, for $\sum_{j=2}^{\infty} 1/j \log j(\log\log j) \cdots (\log \ldots \log j)^{\alpha}$.

## 14.135 Absolute Convergence and Conditional Convergence

We have seen, in examples 1 and 2, section 14.134, that the series $\sum_{j=1}^{\infty} a_j = \sum_{j=1}^{\infty} (-1)^j 1/j$ converges while the harmonic series $\sum_{j=1}^{\infty} |a_j| = \sum_{j=1}^{\infty} 1/j$ does *not* converge. A series such as this is called *conditionally convergent* as opposed to *absolutely convergent*.

***Definition 135.1***

$\sum_{j=1}^{\infty} a_j$ is called *absolutely convergent* if and only if $\sum_{j=1}^{\infty} |a_j|$ converges.

$\sum_{j=1}^{\infty} a_j$ is called *conditionally convergent* if and only if it converges but does *not* converge absolutely.

We obtain immediately from Lemma 133.2:

***Lemma 135.1***

$\sum_{j=1}^{\infty} a_j$ converges absolutely if and only if

$$|a_1| + |a_2| + \cdots + |a_k| \leqslant M$$

for all $k \in \mathbf{N}$ and some $M > 0$.

Absolute convergence implies convergence:

***Lemma 135.2***

If $\sum_{j=1}^{\infty} a_j$ converges absolutely, then $\sum_{j=1}^{\infty} a_j$ converges.

*Proof*

$|s_k - s_l| = |a_{l+1} + a_{l+2} + \cdots + a_k| \leqslant |a_{l+1}| + |a_{l+2}| + \cdots + |a_k|$.

---

▲ *Example 1*

$\sum_{j=1}^{\infty} (-1)^j(1/j^2)$ converges absolutely because $\sum_{j=1}^{\infty} |(-1)^j \, 1/j^2| = \sum_{j=1}^{\infty} 1/j^2$ converges. (See example 2, section 2.19, or exercise 5, section 14.134.)

▲ *Example 2*

The series $\sum_{j=1}^{\infty} ((-1)^j/\sqrt{j})$ converges conditionally because it converges, but $\sum_{j=1}^{\infty} |(-1)^j/\sqrt{j}| = \sum_{j=1}^{\infty} 1/\sqrt{j}$ does not converge. (See exercise 5, section 14.134.)

---

The distinction between absolute convergence and conditional convergence is not only of theoretical but also of practical importance. As we shall see, the terms in an absolutely convergent series may be rearranged any way one pleases without affecting the sum of the series while a rearrangement of terms in a conditionally convergent series will, in general, alter the sum or even destroy the convergence property of the series altogether.

---

▲ *Example 3*

We have seen in example 2, section 14.134, that $\sum_{j=1}^{\infty} (-1)^{j+1} \, 1/j$ converges. Hence, there is an $s \in \mathbf{E}$ such that

$$\sum_{j=1}^{\infty} (-1)^{j+1} \frac{1}{j} = 1 - \frac{1}{2} + \frac{1}{3} - \frac{1}{4} + \frac{1}{5} - \frac{1}{6} + \frac{1}{7} - \frac{1}{8} + \frac{1}{9} - \frac{1}{10} + \frac{1}{11} - \frac{1}{12} + \frac{1}{13} - \frac{1}{14} + \frac{1}{15} - \frac{1}{16} + \frac{1}{17} - \frac{1}{18} + \frac{1}{19} - \cdots = s. \tag{135.1}$$

(We shall see in section 14.141 that $s = \log 2$.) By Theorem 133.3,

$$\tfrac{1}{2}\sum_{j=1}^{\infty}(-1)^{j+1}\frac{1}{j} = 0 + \frac{1}{2} + 0 - \frac{1}{4} + 0 + \frac{1}{6} + 0 - \frac{1}{8} + 0 + \frac{1}{10} + 0 - \frac{1}{12} + 0 + \frac{1}{14} + 0 - \frac{1}{16} + 0 + \frac{1}{18} + 0 - \cdots = \tfrac{1}{2}s,$$

and, again by Theorem 133.3,

$$\sum_{j=1}^{\infty}(-1)^{j+1}\frac{1}{j} + \frac{1}{2}\sum_{j=1}^{\infty}(-1)^{j+1}\frac{1}{j} = 1 + \frac{1}{3} - \frac{1}{2} + \frac{1}{5} + \frac{1}{7} - \frac{1}{4} + \frac{1}{9} + \frac{1}{11} - \frac{1}{6} + \frac{1}{13} + \frac{1}{15} - \frac{1}{8} + \frac{1}{17} + \frac{1}{19} - \cdots = \frac{3s}{2}. \tag{135.2}$$

The latter series can be shown to be a rearrangement of the series in (135.1) and we see that the rearranged series converges to another sum than the original series. (See also exercises 1 and 2.)

---

We shall see in Theorem 135.2 that a conditionally convergent series of real numbers, by proper rearrangement of its terms, can be made to converge to any given number or even made to diverge.

***Definition 135.2***

$\{b_1, b_2, b_3, \ldots\}$ is a *rearrangement* of $\{a_1, a_2, a_3, \ldots\}$ if and only if there exists a function $f: \mathbf{N} \leftrightarrow \mathbf{N}$ (*rearrangement function*) such that $b_n = a_{f(n)}$ for all $n \in \mathbf{N}$.

In words: All elements of $\{a_1, a_2, a_3, \ldots\}$ are present in $\{b_1, b_2, b_3, \ldots\}$ and no others.

Absolutely convergent series are immune to rearrangements:

***Theorem 135.1***

If $\sum_{j=1}^{\infty} a_j, a_j \in \mathbf{E}$, converges *absolutely* and if $f: \mathbf{N} \leftrightarrow \mathbf{N}$ is any rearrangement function, then $\sum_{j=1}^{\infty} a_{f(j)}$ converges and

$$\sum_{j=1}^{\infty} a_{f(j)} = \sum_{j=1}^{\infty} a_j.$$

The converse is also true and follows from Theorem 135.2 which is to be taken up next.

*Proof*

By Lemma 135.2, $\sum_{j=1}^{\infty} a_j$ converges. Let $a_{f(j)} = b_j$ and

$$\sum_{j=1}^{\infty} a_j = a, \qquad \sum_{j=1}^{k} a_j = \alpha_k, \qquad \sum_{j=1}^{k} |a_j| = A_k, \qquad \sum_{j=1}^{k} b_j = \beta_k.$$

For some given $l \in \mathbf{N}$, we choose $k$ so large that

$$\{b_1, b_2, b_3, \ldots, b_l\} \subseteq \{a_1, a_2, \ldots, a_k\}. \tag{135.3}$$

Then,

$$|b_1| + |b_2| + \cdots + |b_l| \leqslant |a_1| + |a_2| + \cdots + |a_k| = A_k.$$

Since $\sum_{j=1}^{\infty} |a_j|$ converges, $\sum_{j=1}^{\infty} |b_j|$, and hence, $\sum_{j=1}^{\infty} b_j$ converge, by Lemma 133.2 and Lemma 135.2. Let

$$\sum_{j=1}^{\infty} b_j = b.$$

From Theorems 133.1 and 133.2, we have, for every $\varepsilon > 0$, an $N(\varepsilon) \in \mathbf{N}$ such that

$$|\alpha_k - a| < \frac{\varepsilon}{3} \qquad \text{for all } k > N(\varepsilon),$$

$$|\beta_l - b| < \frac{\varepsilon}{3} \qquad \text{for all } l > N(\varepsilon),$$

$$|A_k - A_j| < \frac{\varepsilon}{3} \qquad \text{for all } k > j > N(\varepsilon).$$

Hence,

$$|a - b| \leqslant |a - \alpha_k| + |\alpha_k - \beta_l| + |\beta_l - b| < \frac{2\varepsilon}{3} + |\alpha_k - \beta_l| \tag{135.4}$$

for all $k > l > N(\varepsilon)$. We choose $l > j > N(\varepsilon)$ so large that

$$\{a_1, a_2, \ldots, a_j\} \subseteq \{b_1, b_2, \ldots, b_l\},$$

and choose $k > l$ so large that (135.3) holds. Then

$$|\alpha_k - \beta_l| \leqslant |a_{j+1}| + |a_{j+2}| + \cdots + |a_k| = |A_k - A_j| < \frac{\varepsilon}{3},$$

and we obtain, from (135.4), that $a = b$; that is, $\sum_{j=1}^{\infty} a_j = \sum_{j=1}^{\infty} a_{f(j)}$.

The triangle inequality for finite sums,

$$|a_1 + a_2 + \cdots + a_k| \leqslant |a_1| + |a_2| + \cdots + |a_k|$$

can be generalized to absolutely convergent series.

***Lemma 135.3***

If $\sum_{j=1}^{\infty} a_j$ converges *absolutely*, then

$$\left|\sum_{j=1}^{\infty} a_j\right| \leqslant \sum_{j=1}^{\infty} |a_j|. \tag{135.5}$$

*Proof*

Let $s_k = \sum_{j=1}^{k} a_j$. Then,

$$|s_k| = \left|\sum_{j=1}^{k} a_j\right| \leqslant \sum_{j=1}^{k} |a_j| \leqslant \sum_{j=1}^{\infty} |a_j|. \tag{135.6}$$

Since $\sum_{j=1}^{\infty} a_j$ converges, we have

$$||s| - |s_k|| \leqslant |s - s_k| < \varepsilon \qquad \text{for all } k > N(\varepsilon).$$

Hence, $\lim_{\infty} |s_k| = |s|$, and we obtain (135.5) from (135.6). (Note that if $\sum_{j=1}^{\infty} a_j$ converges conditionally, then (135.5) reduces to the trivial statement $|\sum_{j=1}^{\infty} a_j| < \infty$.)

(An entirely different generalization of the triangle inequality will be discussed in section 14.137.)

We shall now investigate how the sum of a conditionally convergent series can be changed by rearranging its terms.

First, we note that a conditionally convergent series of real numbers has to contain infinitely many positive terms as well as infinitely many negative terms because, otherwise, it would converge absolutely. (See exercise 3.)

Let $\{p_1, p_2, p_3, \ldots\}$ denote the positive terms and $\{q_1, q_2, q_3, \ldots\}$ the negative terms in their order of appearance in $\{a_1, a_2, a_3, \ldots\}$. Then

$$\sum_{j=1}^{k} |a_j| = \sum_{j=1}^{l} p_j - \sum_{j=1}^{m} q_j, \qquad l + m = k,$$

where $l, m$ have to tend to infinity as $k$ tends to infinity. If $\sum_{j=1}^{\infty} p_j$ and $\sum_{j=1}^{\infty} q_j$ both converge, then $\sum_{j=1}^{\infty} a_j$ converges absolutely, contrary to our assumption. Since

$$\sum_{j=1}^{k} a_j = \sum_{j=1}^{l} p_j + \sum_{j=1}^{m} q_j,$$

$\sum_{j=1}^{\infty} a_j$ would not converge at all if either $\sum_{j=1}^{\infty} p_j$ or $\sum_{j=1}^{\infty} q_j$ converges and the other one does not converge. (See exercise 4.)

Hence, neither can converge. Since $p_j > 0$, $q_j < 0$ for all $j \in \mathbf{N}$, we have, for every $A > 0$, an $M \in \mathbf{N}$ such that

$$\sum_{j=1}^{k} p_j > A, \qquad \sum_{j=1}^{k} q_j < -A \qquad \text{for all } k > M. \tag{135.7}$$

(See also definition 15.2.)

Let $s \in \mathbf{R}$ and let us assume without loss of generality that $s \geqslant 0$. In view of (135.7) we can choose a $k_1 \in \mathbf{N}$ such that

$$\sum_{j=1}^{k_1 - 1} p_j \leqslant s \qquad \text{and} \qquad \sum_{j=1}^{k_1} p_j > s.$$

(If $k_1 = 1$, then the inequality on the left is to be replaced by $0 \leqslant s$.) Next, we choose $l_1 \in \mathbf{N}$ such that

$$\sum_{j=1}^{k_1} p_j + \sum_{j=1}^{l_1-1} q_j \geqslant s \quad \text{and} \quad \sum_{j=1}^{k_1} p_j + \sum_{j=1}^{l_1} q_j < s.$$

(If $l_1 = 1$, the second term on the left is to be replaced by 0.) We obtain from this choice of $k_1, l_1$ that

$$0 < \sum_{j=1}^{k_1} p_j - s \leqslant p_{k_1}, \qquad 0 < s - \sum_{j=1}^{k_1} p_j - \sum_{j=1}^{l_1} q_j < |q_{l_1}|.$$

We proceed in this manner, hereby generating two sequences $\{k_1, k_2, k_3, \ldots,\}$ and $\{l_1, l_2, l_3, \ldots\}$ such that

$$0 < \sum_{j=1}^{k_n} p_j + \sum_{j=1}^{l_{n-1}} q_j - s \leqslant p_{k_n}, \qquad 0 < s - \sum_{j=1}^{k_n} p_j - \sum_{j=1}^{l_n} q_j < |q_{l_n}|.$$

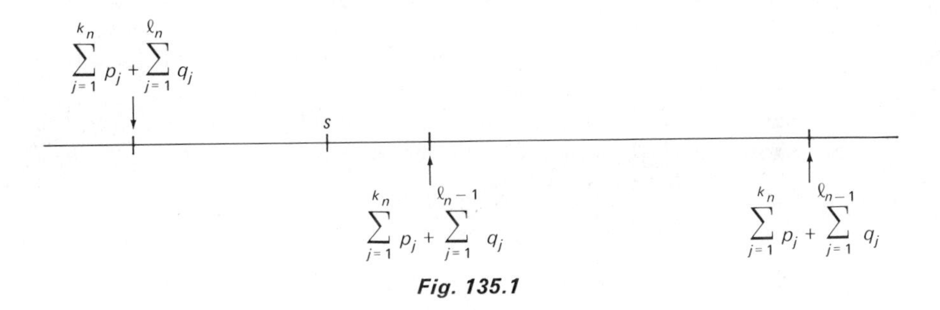


*Fig. 135.1*

(See also Fig. 135.1.) By Lemma 133.3, $\lim_\infty p_{k_n} = \lim_\infty q_{l_n} = 0$. Hence, we obtain, with the rearrangement $\{b_1, b_2, b_3, \ldots\} = \{p_1, \ldots, p_{k_1}, q_1, \ldots, q_{l_1}, p_{k_1+1}, \ldots, p_{k_2}, q_{l_1+1}, \ldots, q_{l_2}, \ldots\}$ of $\{a_1, a_2, a_3, \ldots\}$ that

$$\sum_{j=1}^{\infty} b_j = s.$$

In the same manner one can show that $\sum_{j=1}^{\infty} a_j$ can be made to diverge to $\infty$ or $-\infty$ through proper rearrangement of its terms. (See exercise 5.)

We summarize the results of our investigation in the following theorem.

***Theorem 135.2*** *(Riemann's Rearrangement Law)*

If $\sum_{j=1}^{\infty} a_j$ converges *conditionally* and if $s \in \mathbf{R}$ is any given number, then there exists a rearrangement function $f_s : \mathbf{N} \leftrightarrow \mathbf{N}$ such that $\sum_{j=1}^{\infty} a_{f_s(j)} = s$. There are also rearrangement functions $f_\infty, f_{-\infty} : \mathbf{N} \leftrightarrow \mathbf{N}$ such that $\sum_{j=1}^{\infty} a_{f_\infty(j)}$ diverges to $\infty$ and $\sum_{j=1}^{\infty} a_{f_{-\infty}(j)}$ diverges to $-\infty$.

Hence, if $\sum_{j=1}^{\infty} a_j$ does not converge absolutely, then there is a function $f : \mathbf{N} \leftrightarrow \mathbf{N}$ such that $\sum_{j=1}^{\infty} a_j \neq \sum_{j=1}^{\infty} a_{f(j)}$. The contrapositive of this statement, namely, if $\sum_{j=1}^{\infty} a_j = \sum_{j=1}^{\infty} a_{f(j)}$ for all $f : \mathbf{N} \leftrightarrow \mathbf{N}$, then $\sum_{j=1}^{\infty} a_j$ converges absolutely, is the converse to Theorem 135.1.

## 14.135 Exercises

1. Show that $\frac{1}{2} < s < \frac{5}{6}$ where $s$ is defined in (135.1).
2. Formulate the rearrangement law that yields (135.2) from (135.1).
3. *Prove*: If $\sum_{j=1}^{\infty} a_j$ converges conditionally, then there are two sequences
$$P = \{\pi_1, \pi_2, \pi_3, \ldots\}, \qquad M = \{\nu_1, \nu_2, \nu_3, \ldots\},$$
such that $P \cup M = \mathbf{N}$, $P \cap M = \emptyset$, and $a_{\pi_j} > 0$, $a_{\nu_j} < 0$, for all $j \in \mathbf{N}$.
4. Assume that $p_j$, $q_j$ are defined as in the derivation of Riemann's rearrangement law. Show: If $\sum_{j=1}^{\infty} a_j$, $\sum_{j=1}^{\infty} p_j$ converge, then $\sum_{j=1}^{\infty} q_j$ converges.
5. Show: If $\sum_{j=1}^{\infty} a_j$ converges conditionally, then there is a rearrangement function $f_\infty : \mathbf{N} \leftrightarrow \mathbf{N}$ such that $\sum_{j=1}^{\infty} a_{f_\infty(j)}$ diverges to $\infty$.
6. Devise a rearrangement of terms in the series in (135.1) such that the sum is greater than $\frac{5}{4}$.
7. $\sum_{j=1}^{\infty} a_j, a_j \in \mathbf{E}^n$, is said to converge absolutely, if and only if $\sum_{j=1}^{\infty} \|a_j\|$ converges.
   (a) Show: $\sum_{j=1}^{\infty} a_j$, $a_j \in \mathbf{E}^n$, converges absolutely if and only if there is an $M > 0$ such that $\|a_1\| + \|a_2\| + \cdots + \|a_k\| < M$ for all $k \in \mathbf{N}$.
   (b) Show: If $\sum_{j=1}^{\infty} a_j$, $a_j \in \mathbf{E}^n$, converges absolutely, then it converges.
   (c) Show: If $\sum_{j=1}^{\infty} a_j, a_j \in \mathbf{E}^n$, converges absolutely, then $\sum_{j=1}^{\infty} a_j = \sum_{j=1}^{\infty} a_{f(j)}$ for any $f : \mathbf{N} \leftrightarrow \mathbf{N}$.
8. Let
$$a_j = \left\{(-1)^j \frac{1}{\sqrt{2j}}, \qquad (-1)^{j+1} \frac{1}{\sqrt{2j}}\right\} \in \mathbf{E}^2.$$
Show that $\sum_{j=1}^{\infty} a_j$ converges conditionally.

## 14.136 Tests for Absolute Convergence

A number of practical convergence tests can be developed from the comparison of a given series with another series the convergence properties of which are already known. Such a technique is based on the following theorem:

***Theorem 136.1*** *(Comparison Test)*

If $\sum_{j=1}^{\infty} b_j$ converges, where $b_j \geqslant 0$ for all $j \in \mathbf{N}$, and if
$$|a_j| \leqslant b_j \qquad \text{for all } j \in \mathbf{N},$$
then $\sum_{j=1}^{\infty} a_j$ *converges absolutely.*

*Proof*

By Lemma 133.2, $b_1 + b_2 + \cdots + b_k \leqslant M$ for all $k \in \mathbf{N}$ and some $M > 0$. Since $|a_1| + |a_2| + \cdots + |a_k| \leqslant b_1 + b_2 + \cdots + b_k$, we have from Lemma 135.1 that $\sum_{j=1}^{\infty} a_j$ converges absolutely.

▲ *Example 1*

If $\sum_{j=1}^{\infty} a_j$ converges absolutely and if $\sum_{j=1}^{\infty} b_j$ converges, we have, from Lemma 133.4, that $|a_j b_j| \leqslant |a_j| M$ for some $M > 0$. Hence $\sum_{j=1}^{\infty} a_j b_j$ converges absolutely by Theorem 136.1.

The rewards we shall harvest from Theorem 136.1 in the form of convergence tests are rich, indeed. First, we note that if we let $b_j = \lambda q^j$ for some $q \in (0, 1)$ and some $\lambda \in \mathbf{R}^+$, then, because of the convergence of the geometric series $\sum_{j=1}^{\infty} q^j$ (example 1, section 2.19), we obtain immediately from Theorem 136.1:

***Corollary to Theorem 136.1***

If there is a $\lambda \in \mathbf{R}^+$ and a $q \in (0, 1)$ such that $|a_j| \leqslant \lambda q^j$ for all $j \in \mathbf{N}$, then $\sum_{j=1}^{\infty} a_j$ converges absolutely.

Note that in Theorem 136.1 as well as in its corollary, $|a_j| \leqslant b_j$ and $|a_j| \leqslant \lambda q^j$, respectively, need only be true for all $j \geqslant M$ for some $M \in \mathbf{N}$ because the sum of finitely many terms $a_1 + a_2 + \cdots + a_M$ is well defined anyway.

From the above corollary we obtain the following two convergence tests.

***Theorem 136.2*** (*Ratio Test and Root Test*)

Either of the following conditions is sufficient for the *absolute convergence*, and hence, convergence of $\sum_{j=1}^{\infty} a_j$:

For some $q \in (0, 1)$

$$\frac{|a_{j+1}|}{|a_j|} \leqslant q \qquad \text{for all } j \geqslant M \text{ for some } M \in \mathbf{N}. \tag{136.1}$$

(*ratio test*) or,

$$\sqrt[j]{|a_j|} \leqslant q \qquad \text{for all } j \geqslant M \text{ for some } M \in \mathbf{N}, \tag{136.2}$$

(*root test*).

*Proof*

If (136.1) is satisfied, then

$$|a_j| \leqslant q\,|a_{j-1}| \leqslant q^2\,|a_{j-2}| \leqslant \cdots \leqslant q^{j-M}\,|a_M| = \frac{|a_M|}{q^M}\,q^j$$

for all $j \geqslant M$.

If (136.2) is satisfied, then $|a_j| \leqslant q^j$ for all $j \geqslant M$.

In both cases, the absolute convergence follows from the Corollary to Theorem 136.1 with $\lambda = (|a_M|/q^M)$ and $\lambda = 1$, respectively.

▲ *Example 2*

Let $a_j = (1/j^2)$. Then $\sum_{j=1}^{\infty} (1/j^2)$ converges (see example 2, section 2.19, or exercise 5, section 14.134). Still, there is *no* $q \in (0, 1)$ such that

$$\frac{|a_{j+1}|}{|a_j|} = \frac{j^2}{(1+j)^2} \leqslant q \qquad \text{for all } j \geqslant M \text{ for some } M \in \mathbf{N}$$

or

$$\sqrt[j]{|a_j|} = \frac{1}{\sqrt[j]{j}} \leqslant q \qquad \text{for all } j \geqslant M \text{ for some } M \in \mathbf{N},$$

because $\lim_{\infty} j^2/(1+j)^2 = 1$ and $\lim_{\infty} 1/\sqrt[j]{j^2} = 1$. (See also exercise 2.)

---

If $(|a_{j+1}|/|a_j|) \geqslant 1$ for all $j \geqslant M$ for some $M \in \mathbf{N}$ or if $\sqrt[j]{|a_j|} \geqslant 1$ for infinitely many $j \in \mathbf{N}$, then $|a_{j+1}| \geqslant |a_M|$ for all $j \geqslant M$ or $|a_j| \geqslant 1$ for infinitely many $j \in \mathbf{N}$, respectively. In either case, $\lim_{\infty} |a_j| \neq 0$ if it exists at all and $\sum_{j=1}^{\infty} a_j$ does not converge.

***Theorem 136.3***

$\sum_{j=1}^{\infty} a_j$ does *not* converge if $(|a_{j+1}|/|a_j|) \geqslant 1$ for all $j \geqslant M$ for some $M \in \mathbf{N}$, or if $\sqrt[j]{|a_j|} \geqslant 1$ for infinitely many $j \in \mathbf{N}$.

In order to derive the maximum benefit from the ratio test and the root test, we introduce the notions of *superior* and *inferior limit* and then derive further tests in terms of these new concepts. For this purpose we need the concept of *limit point.*

***Definition 136.1***

The number $l \in \mathbf{E}$ is a limit point of the sequence $\{a_j\}$ if and only if, for every $\varepsilon > 0$, there is a sequence of integers $j_1, j_2, j_3, \ldots$ such that $|a_{j_k} - l| < \varepsilon$ for all $k \in \mathbf{N}$.

---

▲ *Example 3*

If $a$ is an accumulation point of $\{a_j\}$, then $a$ is also a limit point of $\{a_j\}$.

▲ *Example 4*

If $\lim_{\infty} a_j = a$, then $a$ is a limit point of $\{a_j\}$.

▲ *Example 5*

Let $a_j = (-1)^j$. Then, $-1$ and $1$ are limit points of $\{a_j\}$.

---

***Definition 136.2***

Let $\{a_j\}^l$ denote the set of limit points of the sequence $\{a_j\}$. If $\{a_j\}$ is bounded above, then

$$\overline{\lim}\, a_j = \sup\{a_j\}^l$$

is called the *superior limit* (or *lim sup*) of $\{a_j\}$. If $\{a_j\}$ is bounded below, then

$$\underline{\lim}\, a_j = \inf\{a_j\}^l$$

is called the *inferior limit* (or *lim inf*) of $\{a_j\}$.

Note that a sequence that is bounded above and below has at least one limit point. This may be proved in the same manner as the Bolzano–Weierstrass theorem. (Theorem 30.1 or Theorem 60.2.)

---

▲ *Example 6*

Let $\{a_j\} = \{1, \frac{2}{3}, -2, \frac{1}{4}, \frac{5}{6}, -5, \frac{1}{7}, \frac{8}{9}, -8, \frac{1}{10}, \frac{11}{12}, -11, \ldots\}$. We have $\{a_j\}^l = \{0, 1\}$ and $\overline{\lim}\, a_j = 1$. Although $\inf\{a_j\}^l = 0$, $\underline{\lim}\, a_j$ is *not* defined because $\{a_j\}$ is not bounded below. (This fact is sometimes expressed by the symbolic statement $\underline{\lim}\, a_j = -\infty$. Similarly, when a sequence $\{a_j\}$ is not bounded above, one says that $\overline{\lim}\, a_j = \infty$.)

---

The following lemma summarizes everything we need to know about superior and inferior limits.

***Lemma 136.1***

If $\lim_\infty a_j = a$ exists, then $\overline{\lim}\, a_j = \underline{\lim}\, a_j = a$.

$\overline{\lim}\, a_j = L$ ($\underline{\lim}\, a_j = l$) if and only if, for every $\varepsilon > 0$, there is an $N(\varepsilon) \in \mathbf{N}$ such that

(136.3) $\qquad a_j < L + \varepsilon \qquad (a_j > l - \varepsilon) \qquad$ for all $j > N(\varepsilon)$,

and there is a sequence $k_1 < k_2 < k_3 < \cdots$ of integers such that

(136.4) $\qquad a_{k_j} > L - \varepsilon \qquad (a_{k_j} < l + \varepsilon) \qquad$ for all $j \in \mathbf{N}$.

*Proof*

If $\lim_\infty a_j = a$, then $\{a_j\}$ is bounded and $\{a_j\}^l = \{a\}$. Hence, $\sup\{a_j\}^l = \inf\{a_j\}^l = a$.

If, for some $\varepsilon > 0$, $a_j \geqslant L + \varepsilon$ for infinitely many $j \in \mathbf{N}$, and if $\{a_j\}$ is bounded above, then $\{a_j\}$ has a limit point $a \geqslant L + \varepsilon$ contrary to the definition of $L$. Hence, (136.3) obtains.

If, for some $\varepsilon > 0$, $a_j > L - \varepsilon$ for finitely many elements only, then $\sup\{a_j\}^l \leqslant L - \varepsilon$. Hence, (136.4) obtains.

Conversely, if (136.3) and (136.4) hold, then $L$ is a limit point of $\{a_k\}$. If $L_1 > L$ is another limit point of $\{a_j\}$, then (136.3) is violated for $\varepsilon = (L_1 - L)/2 > 0$.

The so-called limit tests in the following theorem are obtained from Theorems 136.2 and 136.3.

***Theorem 136.4*** *(Limit Tests)*

$\sum_{j=1}^{\infty} a_j$ converges absolutely if

$$\overline{\lim}\, \frac{|a_{j+1}|}{|a_j|} < 1 \quad \text{or} \quad \overline{\lim}\, \sqrt[j]{|a_j|} < 1 \tag{136.5}$$

and does *not* converge if

$$\underline{\lim}\, \frac{|a_{j+1}|}{|a_j|} > 1 \quad \text{or} \quad \overline{\lim}\, \sqrt[j]{|a_j|} > 1. \tag{136.6}$$

*Proof*

Let $\overline{\lim}\,(|a_{j+1}|)/(|a_j|) = r < 1$. From (136.3) with $\varepsilon = (1 - r)/2 > 0$,

$$\frac{|a_{j+1}|}{|a_j|} < r + \frac{1-r}{2} = \frac{1+r}{2} < 1 \qquad \text{for all } j \geqslant N(\varepsilon),$$

and the absolute convergence follows from (136.1).

Similarly, one establishes the sufficiency for convergence, of the second condition in (136.5). (See exercise 5.)

Let $\underline{\lim}\, |a_{j+1}|/|a_j| = \rho > 1$. From (136.3), with $\varepsilon = (\rho - 1)/2 > 0$,

$$\frac{|a_{j+1}|}{|a_j|} > \rho - \frac{\rho - 1}{2} = \frac{\rho + 1}{2} > 1 \qquad \text{for all } j > N(\varepsilon),$$

and it follows from Theorem 136.3 that $\sum_{j=1}^{\infty} a_j$ does not converge.

Let $\overline{\lim}\, \sqrt[j]{|a_j|} = \rho > 1$. From (136.4) with $\varepsilon = (\rho - 1)/2 > 0$,

$$|a_{k_j}| > \left(\frac{1+\rho}{2}\right)^{k_j} > 1 \qquad \text{for all } j \in \mathbf{N},$$

and it follows again from Theorem 136.3 that $\sum_{j=1}^{\infty} a_j$ does not converge.

If we had taken the superior limit of $(|a_{j+1}|/|a_j|)$ in (136.6), we would have obtained an inconclusive result, as the following example shows.

---

▲ *Example 7*

We take the (rearranged) series (135.2) of example 3, section 14.135, of which we know that it converges:

$$1+\frac{1}{3}-\frac{1}{2}+\frac{1}{5}+\frac{1}{7}-\frac{1}{4}+\frac{1}{9}+\frac{1}{11}-\frac{1}{6}+\frac{1}{13}+\frac{1}{15}-\frac{1}{8}+\frac{1}{17}+\frac{1}{19}-\frac{1}{10}+\cdots=\frac{3s}{2},$$

where $s=\sum_{j=1}^{\infty}(-1)^{j+1}(1/j)$. We observe that

$$\frac{|a_{3j}|}{|a_{3j-1}|}=\frac{4j-1}{2j}, \qquad j\in\mathbf{N},$$

and hence, $\overline{\lim}(|a_{j+1}|/|a_j|)\geqslant 2$.

---

Theorem 136.4 remains true, of course, if superior and inferior limits in (136.5) and (136.6) are replaced by ordinary limits. This, however, would severely restrict the applicability of this test, because there are many convergent series for which

$$\lim_{\infty}\frac{|a_{j+1}|}{|a_j|} \qquad \text{or} \qquad \lim_{\infty}\sqrt[j]{|a_j|}$$

do not exist, but the superior limits do exist. Similarly, there are many divergent series where $\lim_{\infty}(|a_{j+1}|/|a_j|)$ or $\lim_{\infty}\sqrt[j]{|a_j|}$ do not exist but $\underline{\lim}(|a_{j+1}|/|a_j|)$ and/or $\overline{\lim}\sqrt[j]{|a_j|}$ exist, as the following examples show.

---

▲ *Example 8*

$$\sum_{j=1}^{\infty}a_j=1+\frac{1}{2}+\frac{1}{2\cdot 3}+\frac{1}{2^2\cdot 3}+\frac{1}{2^2\cdot 3^2}+\frac{1}{2^3\cdot 3^2}+\frac{1}{2^3\cdot 3^3}+\cdots$$

converges on account of the comparison test. (Note that $(1/2^k3^{k-1})<(1/2^{2k-1})$ and $(1/2^k3^k)<(1/2^k)$.) However, $\lim_{\infty}(|a_{j+1}|/|a_j|)$ does not exist, because

$$\left|\frac{a_{2j}}{a_{2j-1}}\right|=\frac{1}{2}, \qquad \left|\frac{a_{2j+1}}{a_{2j}}\right|=\frac{1}{3} \qquad \text{for all } j\in\mathbf{N}.$$

The superior limit does exist, and $\overline{\lim}\,|a_{j+1}/a_j|=\frac{1}{2}<1$.

▲ *Example 9*

The series

$$\sum_{j=1}^{\infty}a_j=\frac{9}{10}+\frac{1}{10^2}+\left(\frac{9}{10}\right)^3+\frac{1}{10^4}+\left(\frac{9}{10}\right)^5+\cdots$$

also converges by the comparison test ($a_j\leqslant(9/10)^j$ for all $j\in\mathbf{N}$). $\lim_{\infty}\sqrt[j]{|a_j|}$ does not exist, because

$$\sqrt[2j-1]{|a_{2j-1}|}=\frac{9}{10}, \qquad \sqrt[2j]{|a_{2j}|}=\frac{1}{10}.$$

The superior limit exists, and $\overline{\lim}\sqrt[j]{|a_j|}=\frac{9}{10}<1$.

▲ *Example 10*

The series

$$\sum_{j=1}^{\infty} a_j = 1 + \tfrac{3}{2} + 3 + \tfrac{9}{2} + 9 + \tfrac{27}{2} + 27 + \cdots$$

diverges. $\lim_\infty(|a_{j+1}|/|a_j|)$ does not exist because

$$\left|\frac{a_{2j-1}}{a_{2j}}\right| = \frac{2}{3}, \qquad \left|\frac{a_{2j+1}}{a_{2j}}\right| = 2,$$

but $\underline{\lim}\,(|a_{j+1}|/|a_j|) = (3/2) > 1$ exists.

▲ *Example 11*

The series

$$\sum_{j=1}^{\infty} a_j = \frac{1}{10} + 2^2 + \frac{1}{10^3} + 2^4 + \frac{1}{10^5} + 2^6 + \frac{1}{10^7} + 2^8 + \cdots$$

diverges; $\lim_\infty \sqrt[j]{|a_j|}$ does not exist but $\overline{\lim}\, \sqrt[j]{a_j} = 2 > 1$ exists.

---

When $\overline{\lim}(|a_{j+1}|/|a_j|) = 1$ or $\overline{\lim}\sqrt[j]{|a_j|} = 1$, then the test is not conclusive. The series may converge or diverge, as the following examples show.

---

▲ *Example 12*

The series

$$\sum_{j=1}^{\infty} a_j = 1 + 1 + \tfrac{1}{2} + \tfrac{1}{2} + \tfrac{1}{4} + \tfrac{1}{4} + \tfrac{1}{8} + \tfrac{1}{8} + \cdots = 4$$

converges, and $\overline{\lim}(|a_{j+1}|/|a_j|) = 1$, while the series $\sum_{j=1}^{\infty} b_j = 1 + \frac{1}{2} + \frac{1}{3} + \frac{1}{4} + \cdots$ diverges but

$$\overline{\lim}\, \frac{|a_{j+1}|}{|a_j|} = \lim_\infty \frac{|a_{j+1}|}{|a_j|} = 1$$

also.

▲ *Example 13*

$\sum_{j=1}^{\infty} (1/j^2)$ converges and $\overline{\lim}\, \sqrt[j]{|a_j|} = \lim \sqrt[j]{(1/j^2)} = 1$ (see exercise 2), while $\sum_{j=1}^{\infty} (1/j)$ diverges and $\overline{\lim}\, \sqrt[j]{|a_j|} = \lim_\infty \sqrt[j]{(1/j)} = 1$, also.

---

We leave it to the reader to generalize the tests of this section to the case where $\sum_{j=1}^{\infty} a_j$ is an infinite series of vectors $a_j \in \mathbf{E}^n$.

## 14.136 Exercises

1. Find a flaw in the following argument: If $|q| < 1$, then $\sum_{j=1}^{\infty} q^j$ converges because $(|a_{j+1}|/|a_j|) = |q| < 1$.
2. Show that $\lim_{\infty} \sqrt[j]{j^2} = 1$.
3. *Prove*: If $A \subset \mathbf{R}$ is bounded, then its set of limit points $A^l$ is bounded.
4. What could $\underline{\lim}\, a_j = \infty$ and $\overline{\lim}\, a_j = -\infty$ possibly mean?
5. *Prove*: If $\overline{\lim} \sqrt[j]{|a_j|} < 1$, then $\sum_{j=1}^{\infty} a_j$ converges absolutely.
6. Show: If $|a_j| \geqslant \lambda$ for all $j \geqslant M$, for some $M \in \mathbf{N}$, and some $\lambda \in \mathbf{R}^+$, then $\sum_{j=1}^{\infty} a_j$ does *not* converge.
7. Show: If $(|a_{j+1}|/|a_j|) \geqslant 1$ for all $j \geqslant M$, for some $M \in \mathbf{N}$, then $\sum_{j=1}^{\infty} a_j$ does not converge.
8. Show: If $\sqrt[j]{|a_j|} \geqslant 1$ for infinitely many $j \in \mathbf{N}$, then $\sum_{j=1}^{\infty} a_j$ does not converge.
9. Show: If $\lim_{\infty}(|a_{j+1}|/|a_j| > 1$, then $\sum_{j=1}^{\infty} a_j$ does not converge.
10. Show: $\sum_{j=1}^{\infty}(a^{j-1}/(j-1)!)$ converges for all $a \in \mathbf{E}$.
11. *Prove*: If $b_j \geqslant 0$ for all $j \in \mathbf{N}$, if $\sum_{j=1}^{\infty} b_j$ *diverges*, and if $a_j \geqslant b_j$ for all $j \in \mathbf{N}$, then $\sum_{j=1}^{\infty} a_j$ diverges.
12. *Prove*: If $a_j \geqslant 0$, $b_j > 0$ for all $j \in \mathbf{N}$ and if $\lim_{\infty}(a_j/b_j) \neq 0$, then $\sum_{j=1}^{\infty} a_j$ converges if and only if $\sum_{j=1}^{\infty} b_j$ converges.
13. Give an example of a sequence $\{a_j\}$ such that $\lim_{\infty} a_j = 0$ but for which it is *not* true that $(|a_{j+1}|/|a_j|) < q < 1$ for all $j \geqslant M$ for some $M \in \mathbf{N}$.
14. Let $a_{2j} = (1/2^j)$, $a_{2j-1} = (1/2^{j+1})$. Find $\overline{\lim}(|a_{j+1}|/|a_j|)$ and $\underline{\lim}|(a_{j+1})/a_j|$ and show that the limit tests (136.5) and (136.6) are not conclusive.
15. Let $a_j > 0$ for all $j \in \mathbf{N}$ and let

$$\alpha = \lim_{\infty} \frac{\log(1/a_j)}{\log j}.$$

*Prove*: If $\alpha > 1$ then $\sum_{j=1}^{\infty} a_j$ converges and if $\alpha < 1$, then $\sum_{j=1}^{\infty} a_j$ diverges.
16. Use the test in exercise 15 to show that $\sum_{j=2}^{\infty} (1/(\log j)^{\log j})$ converges.
17. For which $a \in \mathbf{E}$ does the ratio test guarantee convergence and divergence of $\sum_{j=1}^{\infty} (a^j/j^2)$?
18. Give examples of a convergent and a divergent series for which $\underline{\lim}$ $(|a_{j+1}|/|a_j|) = 1$.
19. *Prove*: If the sequence $\{a_k\}$ is bounded above and below, then it has at least one limit point.
20. Show that $\{a_j\}' \subseteq \{a_j\}^l$. Construct a sequence $\{a_j\}$ such that $\{a_j\}' \subset \{a_j\}^l$.
21. Let $\{b_j\}$ denote a decreasing (increasing) sequence that converges to zero and let $|a_1 + a_2 + \cdots + a_k| \leqslant M$ for some $M > 0$ and all $k \in \mathbf{N}$. Prove that $\sum_{j=1}^{\infty} a_j b_j$ converges. (This is called *Dirichlet's test.*)
22. Use Dirichlet's test (exercise 21) to show that $\sum_{j=1}^{\infty} (-1)^{j-1}(1/j)$ converges.
23. Use Dirichlet's test (exercise 21) to show that $\sum_{j=1}^{\infty} (\sin(2j-1)a)/(2j-1)$ converges for each $a \in \mathbf{R}$.
24. Show: If $\overline{\lim} \sqrt[j]{a_j} = \rho < 1$, then $\sum_{j=0}^{\infty} |a_j| < 1/(1 - (\rho + 1)/2)$.

## 14.137 The CBS Inequality and the Triangle Inequality for Infinite Series*

We consider the set of all sequences $\{\alpha_1, \alpha_2, \alpha_3, \ldots\}$ of real numbers for which $\sum_{j=1}^{\infty} \alpha_j^2$ converges (square summable sequences). This set is denoted by $l_2$:

$$(137.1) \qquad l_2 = \{(\alpha_1, \alpha_2, \alpha_3, \ldots) \mid \alpha_j \in \mathbf{R}, \quad \sum_{j=1}^{\infty} \alpha_j^2 < \infty\}.$$

In general, $l_p$ denotes the set of all sequences for which $\sum_{j=1}^{\infty} |a_j|^p$ converges.

---

▲ *Example 1*

If $\sum_{j=1}^{\infty} \alpha_j$ converges absolutely, then $(\alpha_1, \alpha_2, \alpha_3, \ldots) \in l_2$. This follows directly from example 1, section 14.136. However, if $(\beta_1, \beta_2, \beta_3, \ldots) \in l_2$, then $\sum_{j=1}^{\infty} \beta_j$ need not even converge: let $\beta_j = (1/j)$. Then $\sum_{j=1}^{\infty} (1/j^2)$ converges (i.e., $(1, \frac{1}{2}, \frac{1}{3}, \ldots) \in l_2$), but $\sum_{j=1}^{\infty} (1/j)$ does *not* converge.

If $\sum_{j=1}^{\infty} \gamma_j$ converges conditionally, then $(\gamma_1, \gamma_2, \gamma_3, \ldots)$ need not be in $l_2$: Let $\gamma_j = (-1)^{j+1}(1/\sqrt{j})$. Then $\sum_{j=1}^{\infty} \gamma_j$ converges conditionally but

$$\sum_{j=1}^{\infty} \gamma_j^2 = \sum_{j=1}^{\infty} \frac{1}{j}$$

does not converge.

---

We define addition and scalar multiplication for elements $a = (\alpha_1, \alpha_2, \alpha_3, \ldots)$, $b = (\beta_1, \beta_2, \beta_3, \ldots) \in l_2$ as follows:

$$(137.2) \qquad a + b = (\alpha_1 + \beta_1, \alpha_2 + \beta_2, \alpha_3 + \beta_3, \ldots),$$

$$(137.3) \qquad \lambda a = (\lambda\alpha_1, \lambda\alpha_2, \lambda\alpha_3, \ldots).$$

(See also (55.2) and (55.3).)

The following lemma establishes the sum of elements from $l_2$ as an element from $l_2$ and the scalar multiples of an element from $l_2$ as elements from $l_2$.

***Lemma 137.1***

If $a, b \in l_2$, $\lambda \in \mathbf{R}$, then $a + b \in l_2$, $\lambda a \in l_2$.

***Proof***

Since $\sum_{j=1}^{\infty} \alpha_j^2$ and $\sum_{j=1}^{\infty} \beta_j^2$ converge, we have, from Theorem 133.3, that $\sum_{j=1}^{\infty} (\alpha_j^2 + \beta_j^2)$ and $\sum_{j=1}^{\infty} \lambda^2\alpha_j^2$ converge. Hence, $\lambda a \in l_2$. Since $2|\alpha_j\beta_j| \leqslant \alpha_j^2 + \beta_j^2$ we see from the comparison test (Theorem 136.1) that $\sum_{j=1}^{\infty} \alpha_j\beta_j$ converges absolutely. Since $(\alpha_j + \beta_j)^2 = \alpha_j^2 + 2\alpha_j\beta_j + \beta_j^2$ we have, again from Theorem 133.3, that $\sum_{j=1}^{\infty} (\alpha_j + \beta_j)^2$ converges. Hence, $a + b \in l_2$.

By (137.2), addition in $l_2$ is commutative and associative. Also, $\theta = (0, 0, 0 \ldots) \in l_2$ and $(\lambda\mu)a = \lambda(\mu a) = \mu(\lambda a)$, $(\lambda + \mu)a = \lambda a + \mu a$, $\lambda(a + b) = \lambda a + \lambda b$, $1(a) = a$, $0(a) = \theta$, for $\lambda, \mu \in \mathbf{R}$; $a, b \in l_2$. Hence, we have, in view of definition 55.2, that:

***Lemma 137.2***

$l_2$, with addition and scalar multiplication defined as in (137.2) and (137.3), is a *real vector space.*

***Lemma 137.3***

The function $f: l_2 \times l_2 \to \mathbf{R}$ with value $f(a, b) = a \cdot b$, $a, b \in l_2$, defined by

$$a \cdot b = \sum_{j=1}^{\infty} \alpha_j \beta_j \tag{137.4}$$

is a *dot product* in $l_2$. (See definition 56.1.)

*Proof*

1. If $a \neq \theta$, then $\alpha_k \neq 0$ for some $k \in \mathbf{N}$ and $a \cdot a = \sum_{j=1}^{\infty} \alpha_j^2 \geqslant \alpha_k^2 > 0$;
2. $\lambda(a \cdot b) = \lambda \sum_{j=1}^{\infty} \alpha_j \beta_j = \sum_{j=1}^{\infty} \lambda \alpha_j \beta_j = \sum_{j=1}^{\infty} \alpha_j \lambda \beta_j = (\lambda a \cdot b) = (a \cdot \lambda b)$;
3. $a \cdot b = \sum_{j=1}^{\infty} \alpha_j \beta_j = \sum_{j=1}^{\infty} \beta_j \alpha_j = b \cdot a$;
4. $(a + b) \cdot c = \sum_{j=1}^{\infty} (\alpha_j + \beta_j)\gamma_j = \sum_{j=1}^{\infty} (\alpha_j \gamma_j + \beta_j \gamma_j) = \sum_{j=1}^{\infty} \alpha_j \gamma_j + \sum_{j=1}^{\infty} \beta_j \gamma_j$
$= a \cdot c + b \cdot c.$

Since (137.4) defines a dot product, we have, from Theorem 57.1:

***Lemma 137.4***

$$\|a\| = \sqrt{a \cdot a} = \sqrt{\sum_{j=1}^{\infty} \alpha^2} \tag{137.5}$$

is a *norm* in $l_2$.

From Lemmas 137.2, 137.3, and 137.4, we have:

***Theorem 137.1***

$l_2$ as defined in (137.1) with addition and scalar multiplication defined in (137.2) and (137.3), a dot product defined in (137.4), and a norm defined in (137.5) is a *normed linear space* with the norm defined in terms of the dot product.

(Such a space is a pre-Hilbert space. (See p. 210.) Actually, $l_2$ is a Hilbert space—but this is another story.)

Since (137.5) defines a norm, we have, from definition 57.1, part (3), that for all square summable sequences $\{\alpha_1, \alpha_2, \alpha_3, \ldots)$ and $(\beta_1, \beta_2, \beta_3, \ldots)$,

$$\sqrt{\sum_{j=1}^{\infty} (\alpha_j + \beta_j)^2} \leqslant \sqrt{\sum_{j=1}^{\infty} \alpha_j^2} + \sqrt{\sum_{j=1}^{\infty} \beta_j^2}, \tag{137.6}$$

the *triangle inequality for infinite series.*

From Theorem 56.2,

$$\left|\sum_{j=1}^{\infty} \alpha_j \beta_j\right| \leqslant \sqrt{\sum_{j=1}^{\infty} \alpha_j^2} \sqrt{\sum_{j=1}^{\infty} \beta_j^2}, \tag{137.7}$$

the *CBS inequality for infinite series.*

---

▲ *Example 2*

We have seen in example 1 that if $\sum_{j=1}^{\infty} \alpha_j$ converges absolutely, then $\{\alpha_1, \alpha_2, \alpha_3, \ldots\}$ is square summable. Hence (137.6), (137.7) hold if $\sum_{j=1}^{\infty} \alpha_j$, $\sum_{j=1}^{\infty} \beta_j$ converge absolutely. (Note that the assumption of square summability is *weaker* than the assumption that $\sum_{j=1}^{\infty} a_j$ converges absolutely.)

---

## 14.137 Exercises

1. Show that

$$\sum_{j=1}^{\infty} \frac{(-1)^j}{j 2^j} \leqslant \frac{\sqrt{3}}{3} \sqrt{\sum_{j=1}^{\infty} \frac{1}{j^2}}.$$

2. Let $l_1 = \{(\alpha_1, \alpha_2, \alpha_3, \ldots) \mid \alpha_j \in \mathbf{R},\ \sum_{j=1}^{\infty} |\alpha_j| < \infty\}$. Show: If $a \in l_1$, then $a \in l_2$.
3. Let $l_p = \{(\alpha_1, \alpha_2, \alpha_3, \ldots) \mid \alpha_j \in \mathbf{R},\ \sum_{j=1}^{\infty} |\alpha_j|^p < \infty\}$, $p \in \mathbf{N}$. Show: If $a \in l_p$, then $a \in l_q$ for all $q \geqslant p$, $q \in \mathbf{N}$.

## 14.138 Double Series

$\{s_{ij}\}$ is called a *double sequence.* We say that $\{s_{ij}\}$ converges to the limit $s$ if and only if, for every $\varepsilon > 0$, there is an $N(\varepsilon) \in \mathbf{N}$ such that $|s_{kl} - s| < \varepsilon$ for all $k, l > N(\varepsilon)$.

---

▲ *Example 1*

$\{s_{ij}\} = \{1/(i+j)\}$ is a double sequence. Since $|1/(i+j)| < \varepsilon$ for $i, j > N(\varepsilon)$ for sufficiently large $N(\varepsilon) \in \mathbf{N}$, we see that $\{1/(i+j)\}$ converges to zero.

---

We define *double series* in terms of double sequences as follows.

***Definition 138.1***

The *double series* $\sum_{i,j=1}^{\infty} a_{ij}$, generated by $\{a_{ij}\}$, is the double sequence $\{s_{kl}\}$ of partial sums $s_{kl} = \sum_{i=1}^{k} \sum_{j=1}^{l} a_{ij}$.

$\sum_{i,j=1}^{\infty} a_{ij}$ converges if and only if $\{s_{kl}\}$ converges. $\lim_{k,l\to\infty} s_{kl}$ is called the *sum of the double series*, and we write

$$\sum_{i,j=1}^{\infty} a_{ij} = \lim_{k,l\to\infty} s_{kl}.$$

In the proof of Lemma 48.2 about the Lebesgue measure of the union of denumerably many sets of Lebesgue measure zero, we had to show that the sum of the lengths $l(\mathfrak{J}_i^j)$ of the open intervals $\mathfrak{J}_i^j \subset \mathbf{R}$, $i, j \in \mathbf{N}$, is less than some $\varepsilon > 0$. For this purpose we arranged the intervals in a certain order and then found an upper bound for the thusly generated infinite series. One of the results to be derived in this section will show that we could have arranged the intervals in a much simpler manner and still come up with the same result, the reason being that $\sum_{i,j=1}^{\infty} l(\mathfrak{J}_i^j)$ converges absolutely.

***Definition 138.2***

The double series $\sum_{i,j=1}^{\infty} a_{ij}$ *converges absolutely* if and only if $\sum_{i,j=1}^{\infty} |a_{ij}|$ converges.

***Lemma 138.1***

*$\sum_{i,j=1}^{\infty} a_{ij}$ converges absolutely* if and only if

$$\sum_{i=1}^{k} \sum_{j=1}^{l} |a_{ij}|, \qquad k, l \in \mathbf{N},$$

is bounded above. (See exercise 8.)

---

▲ *Example 2*

Let $a_{ij} = (1/2^{i+j-2})$. Since $1 + \frac{1}{2} + \frac{1}{4} + \cdots + (1/2^r) \leqslant 2$ for all $r \in \mathbf{N}$, we have

$$\begin{aligned} s_{kl} &= \left(1 + \frac{1}{2} + \cdots + \frac{1}{2^{k-1}}\right) + \left(\frac{1}{2} + \frac{1}{4} + \cdots + \frac{1}{2^k}\right) + \cdots \\ &\quad + \left(\frac{1}{2^{l-1}} + \frac{1}{2^l} + \cdots + \frac{1}{2^{l+k-2}}\right) \\ &\leqslant 2 + (2-1) + \left(2 - \left(1 + \frac{1}{2}\right)\right) + \cdots + \left(2 - \left(1 + \frac{1}{2} + \cdots + \frac{1}{2^{l-2}}\right)\right) \\ &= 2 + 1 + \frac{1}{2} + \cdots + \frac{1}{2^{l-2}} \leqslant 4 \qquad \text{for all } k, l \in \mathbf{N}. \end{aligned}$$

Hence, $\sum_{i,j=1}^{\infty}(1/2^{i+j-2})$ converges absolutely. If we sum over $i$ first, we obtain $\sum_{i=1}^{\infty} 1/2^{i+j-2} = 1/2^{j-2}$ for all $j \in \mathbf{N}$. Next, we sum over $j$ and obtain the sum $\sum_{j=1}^{\infty}(\sum_{i=1}^{\infty} 1/2^{i+j-2}) = \sum_{j=1}^{\infty} 1/2^{j-2} = 2 + \sum_{j=1}^{\infty} 1/2^{j-1} = 4$. Our next major result will reveal that we would have obtained the same result, had we arranged the elements in a certain other order and that, in fact, $\sum_{i,j=1}^{\infty}(1/2^{i+j-2}) = 4$.

---

***Lemma 138.2***

The double series $\sum_{i,j=1}^{\infty} a_{ij}$ *converges absolutely* if and only if

$$\sum_{j=1}^{\infty}\left(\sum_{i=1}^{\infty}|a_{ij}|\right) \quad \text{or} \quad \sum_{i=1}^{\infty}\left(\sum_{j=1}^{\infty}|a_{ij}|\right) \tag{138.1}$$

converges. The series in (138.1) are called *iterated series.*

*Proof*

If the iterated series in (138.1) converge, then

$$\sum_{i=1}^{k}\sum_{j=1}^{l}|a_{ij}| \leqslant \begin{cases} \sum_{i=1}^{\infty}\left(\sum_{j=1}^{\infty}|a_{ij}|\right), \\ \sum_{j=1}^{\infty}\left(\sum_{i=1}^{\infty}|a_{ij}|\right), \end{cases}$$

and the absolute convergence of $\sum_{i,j=1}^{\infty} a_{ij}$ follows from Lemma 138.1.

If $\sum_{i,j=1}^{\infty}|a_{ij}|$ converges, then, by Lemma 138.1,

$$\sum_{i=1}^{k}\sum_{j=1}^{l}|a_{ij}| \leqslant M \qquad \text{for all } k, l \in \mathbf{N},$$

and some $M > 0$. For fixed $k \in \mathbf{N}$,

$$\lim_{l\to\infty}\sum_{i=1}^{k}\sum_{j=1}^{l}|a_{ij}| = \sum_{i=1}^{k}\left(\sum_{j=1}^{\infty}|a_{ij}|\right) \leqslant M.$$

Hence, $\sum_{i=1}^{\infty}(\sum_{j=1}^{\infty}|a_{ij}|)$ converges. One shows similarly that $\sum_{j=1}^{\infty}(\sum_{i=1}^{\infty}|a_{ij}|)$ converges.

Note that the convergence of the one iterated series in (138.1) follows from the convergence of the other iterated series via the absolute convergence of $\sum_{i,j=1}^{\infty} a_{ij}$.

The next theorem yields three practical methods for the evaluation of absolutely convergent double series.

***Theorem 138.1***

If $\sum_{i,j=1}^{\infty} a_{ij}$ converges absolutely, then

$$\sum_{i,j=1}^{\infty} a_{ij} = \sum_{i=1}^{\infty}\left(\sum_{j=1}^{\infty} a_{ij}\right), \tag{138.2}$$

(138.3)
$$\sum_{i,j=1}^{\infty} a_{ij} = \sum_{j=1}^{\infty} \left( \sum_{i=1}^{\infty} a_{ij} \right),$$

(138.4)
$$\sum_{i,j=1}^{\infty} a_{ij} = \sum_{p=2}^{\infty} (a_{1,p-1} + a_{2,p-2} + \cdots + a_{p-2,2} + a_{p-1,1}).$$

(Note the similarity of (138.2) and (138.3) with Fubini's Theorem, Theorem 108.2.) The series in (138.4) is called a *diagonal series* because the elements are summed along the diagonals $i + j = p$. (See below.)

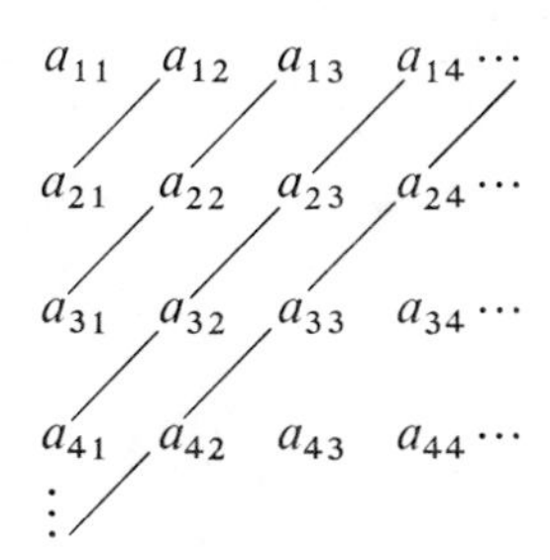

*Proof*

Since $\sum_{i,j=1}^{\infty} a_{ij}$ converges absolutely, there is an $M > 0$ such that

$$\sum_{j=1}^{l} |a_{kj}| \leqslant \sum_{i=1}^{k} \sum_{j=1}^{l} |a_{ij}| \leqslant M \qquad \text{for all } k, l \in \mathbf{N}.$$

Hence, $\sum_{j=1}^{\infty} a_{kj}$ converges absolutely for each $k \in \mathbf{N}$. Let

$$\sum_{j=1}^{\infty} a_{ij} = a_i \qquad \text{and} \qquad \sum_{i,j=1}^{\infty} a_{ij} = s.$$

Since $s_{kl} = \sum_{j=1}^{l} a_{1j} + \sum_{j=1}^{l} a_{2j} + \cdots + \sum_{j=1}^{l} a_{kj}$, we have

$$\lim_{l \to \infty} s_{kl} = a_1 + a_2 + \cdots + a_k \qquad \text{for all } k \in \mathbf{N}.$$

Since

(138.5)
$$|s_{kl} - s| < \frac{\varepsilon}{2} \qquad \text{for all } k, l > N(\varepsilon),$$

we have

$$|a_1 + a_2 + \cdots + a_k - s| \leqslant \frac{\varepsilon}{2} \qquad \text{for all } k > N(\varepsilon),$$

and hence, $\sum_{i=1}^{\infty} (\sum_{j=1}^{\infty} a_{ij}) = s$ which is (138.2). (138.3) is derived in the same manner, interchanging $i$ and $j$, $k$ and $l$.

In order to prove (138.4), we proceed as follows: Since

$$\sum_{p=2}^{q} |a_{1,p-1} + a_{2,p-2} + \cdots + a_{p-2,2} + a_{p-1,1}| \leqslant \sum_{j=1}^{q} \sum_{i=1}^{q} |a_{ij}| \leqslant M,$$

we see that the diagonal series in (138.4) converges absolutely. Let

$$A_q = \sum_{p=2}^{q+1} (a_{1,p-1} + a_{2,p-2} + \cdots + a_{p-2,2} + a_{p-1,1}).$$

We want to show that $\lim_\infty A_q = s$. Consider

$$|s - A_q| \leqslant |s - s_{kk}| + |s_{kk} - A_q| \tag{138.6}$$

for $q \geqslant 2k$. From the array below,

$$\begin{aligned} |A_q - s_{kk}| &= \left| \sum_{i=1}^{k} \sum_{j=k+1}^{q-i+1} a_{ij} + \sum_{i=k+1}^{q} \sum_{j=1}^{q-i+1} a_{ij} \right| \\ &\leqslant \sum_{j=k+1}^{q} \left( \sum_{i=1}^{\infty} |a_{ij}| \right) + \sum_{i=k+1}^{q} \left( \sum_{j=1}^{\infty} |a_{ij}| \right). \end{aligned} \tag{138.7}$$

$$\begin{array}{cccc|cccccc}
a_{11} & a_{12} & \cdots & a_{1k} & a_{1,k+1} & \cdots & a_{1,2k} & \cdots & a_{1,q} \\
a_{21} & a_{22} & \cdots & a_{2k} & a_{2,k+1} & \cdots & a_{2,2k} & \cdots & a_{2,q} \\
\vdots & & & & & & & & \\
a_{k1} & a_{k2} & \cdots & a_{kk} & a_{k,k+1} & \cdots & a_{k,2k} & \cdots & a_{k,q} \\
\hline
a_{k+1,1} & a_{k+1,2} & \cdots & a_{k+1,k} & a_{k+1,k+1} & \cdots & a_{k+1,2k} & \cdots & a_{k+1,q} \\
\vdots & & & & & & & & \\
a_{2k,1} & a_{2k,2} & \cdots & a_{2k,k} & a_{2k,k+1} & \cdots & a_{2k,2k} & \cdots & a_{2k,q} \\
\vdots & & & & & & & & \\
a_{q1} & a_{q2} & \cdots & a_{qk} & a_{q,k+1} & \cdots & a_{q,2k} & \cdots & a_{qq}
\end{array}$$

By Lemma 138.2, the iterated series (138.1) converges. The first term on the right of (138.7) is the difference between the $q$th and the $k$th partial sums of

$\sum_{j=1}^{\infty} (\sum_{i=1}^{\infty} |a_{ij}|)$ and the second term is the difference of the $q$th and the $k$th partial sums of $\sum_{i=1}^{\infty} (\sum_{j=1}^{\infty} |a_{ij}|)$. For sufficiently large $k$ we can make each less than $(\varepsilon/4)$ and we obtain from (138.5) and (138.6) that $|s - A_q| < \varepsilon$ for all $q > M(\varepsilon)$ for some $M(\varepsilon) \in \mathbf{N}$. Hence, (138.4) is valid.

---

▲ *Example 3*

In the proof of Lemma 48.2 we have seen that

$$\sum_{k=1}^{\infty} [l(\mathfrak{J}_1^k) + l(\mathfrak{J}_2^{k-1}) + \cdots + l(\mathfrak{J}_k^1)] < \varepsilon.$$

In view of Lemma 138.2 and Theorem 138.1 we also have

$$\sum_{i,j=1}^{\infty} l(\mathfrak{J}_i^j) = \sum_{i=1}^{\infty} \left( \sum_{j=1}^{\infty} l(\mathfrak{J}_i^j) \right) = \sum_{j=1}^{\infty} \left( \sum_{i=1}^{\infty} l(\mathfrak{J}_i^j) \right) < \varepsilon.$$

▲ *Example 4*

We obtain for the double series in example 2:

$$\sum_{i,j=1}^{\infty} \frac{1}{2^{i+j-2}} = \sum_{i=1}^{\infty} \left( \sum_{j=1}^{\infty} \frac{1}{2^{i+j-2}} \right) = \sum_{i=1}^{\infty} \frac{1}{2^{i-2}} = 4.$$

---

## 14.138 Exercises

1. Does the double series $\sum_{i,j=1}^{\infty} ((-1)^{ij}/ij)$ converge?
2. Let $a_{ij} = ((-1)^{i+j}/ij)$. Does the iterated series $\sum_{i=1}^{\infty} (\sum_{j=1}^{\infty} a_{ij})$ converge? Does $\sum_{i,j=1}^{\infty} a_{ij}$ converge absolutely?
3. Show that $\sum_{i,j=1}^{\infty} q^{i+j-2} = 1/(1-q)^2$ for $q \in (-1, 1)$.
4. Show that $\sum_{i,j=1}^{\infty} q^{ij} \leqslant q/(1-q)^2$ for $q \in (0, 1)$.
5. Let $s_k$ denote the $k$th partial sum of $\sum_{j=1}^{\infty} a_j$ and let $\sigma_{k,l} = |s_k - s_l|$. *Prove*: $\sum_{j=1}^{\infty} a_j$ converges if and only if the double sequence $\{\sigma_{kl}\}$ converges to zero.
6. Show that $\sum_{i,j=1}^{\infty} q^{ij-i-j+1}$ does *not* converge.
7. Show that $\sum_{i,j=1}^{\infty} (-1)^i q^{i+j-2} = 1/(1-q^2)$ for $q \in (-1, 1)$.
8. Prove Lemma 138.1.

## 14.139 Cauchy Products

In section 14.133, we have seen how to handle sums of convergent infinite series and scalar multiples of convergent infinite series. We shall now study products of infinite series. Clearly, if $\sum_{i=1}^{\infty} a_i = a$, $\sum_{j=1}^{\infty} b_j = b$, then

$$\left( \sum_{i=1}^{\infty} a_i \right) \left( \sum_{j=1}^{\infty} b_j \right) = ab.$$

How do we express the product on the left as an infinite series? We take our lead from products of finite sums. We have

$$\begin{aligned}(a_1)(b_1) &= a_1 b_1,\\ (a_1+a_2)(b_1+b_2) &= a_1 b_1 + (a_1 b_2 + a_2 b_1) + a_2 b_2,\\ (a_1+a_2+a_3)(b_1+b_2+b_3) &= a_1 b_1 + (a_1 b_2 + a_2 b_1)\\ &\quad + (a_1 b_3 + a_2 b_2 + a_3 b_1)\\ &\quad + (a_2 b_3 + a_3 b_2) + a_3 b_3.\end{aligned}$$

In general, with $a_{p+j} = b_{p+j} = 0$ for all $j \in \mathbf{N}$,

$$\begin{aligned}&(a_1 + a_2 + \cdots + a_p)(b_1 + b_2 + \cdots + b_p)\\ &\qquad = \sum_{i+j=2} a_i b_j + \sum_{i+j=3} a_i b_j + \cdots + \sum_{i+j=2p} a_i b_j.\end{aligned}$$

This suggests the following definition of the product of two infinite series:

***Definition 139.1***

The *Cauchy product* of two infinite series $\sum_{i=1}^{\infty} a_i$, $\sum_{j=1}^{\infty} b_j$ is defined by

$$\sum_{p=2}^{\infty} (a_1 b_{p-1} + a_2 b_{p-2} + \cdots + a_{p-2} b_2 + a_{p-1} b_1) = \sum_{p=2}^{\infty} \sum_{i+j=p} a_i b_j. \tag{139.1}$$

The Cauchy product of two convergent series need *not* converge, as the following example shows.

---

▲ *Example 1*

Let $a_i = b_i = ((-1)^i/\sqrt{i})$. Then, we obtain, for the $p$th term $c_p$ in (139.1),

$$\begin{aligned}c_p &= a_1 b_{p-1} + a_2 b_{p-2} + \cdots + a_{p-2} b_2 + a_{p-1} b_1\\ &= (-1)^p\left(\frac{1}{\sqrt{1}\sqrt{p-1}} + \frac{1}{\sqrt{2}\sqrt{p-2}} + \cdots + \frac{1}{\sqrt{p-2}\sqrt{2}} + \frac{1}{\sqrt{p-1}\sqrt{1}}\right).\end{aligned}$$

Since $\sqrt{i}\sqrt{p-i} \leqslant \frac{1}{2}(i + p - i) = (p/2)$, we have

$$|c_p| \geqslant \frac{2(p-1)}{p} = 2 - \frac{2}{p},$$

and hence, $\lim_{\infty} |c_p| \neq 0$. Hence, $\sum_{p=2}^{\infty} c_p$ does *not* converge, although $\sum_{i=1}^{\infty} a_i$, $\sum_{j=1}^{\infty} b_j$, both converge.

---

However, if both series converge absolutely, then the Cauchy product converges to the product of the sums of the two series.

***Theorem 139.1***

If $\sum_{i=1}^{\infty} a_i$, $\sum_{j=1}^{\infty} b_j$ converge *absolutely*, then

$$(139.2)\qquad \left(\sum_{i=1}^{\infty} a_i\right)\left(\sum_{j=1}^{\infty} b_j\right) = \sum_{p=2}^{\infty} (a_1 b_{p-1} + a_2 b_{p-2} + \cdots + a_{p-2} b_2 + a_{p-1} b_1).$$

(An analogous theorem holds for the dot product of infinite series of vectors.)

*Proof*

Since

$$\sum_{i=1}^{k} \sum_{j=1}^{l} |a_i b_j| = \sum_{i=1}^{k} \sum_{j=1}^{l} |a_i|\,|b_j| = \sum_{i=1}^{k} |a_i| \sum_{j=1}^{l} |b_j| \leqslant M$$

for some $M > 0$ and all $k, l \in \mathbf{N}$, we see that $\sum_{i,j=1}^{\infty} a_i b_j$ converges absolutely.

By Theorem 138.1, $\sum_{i=1}^{\infty} (\sum_{j=1}^{\infty} a_i b_j)$ converges. By (133.5), $\sum_{j=1}^{\infty} a_i b_j = a_i \sum_{j=1}^{\infty} b_j$. Hence, by Theorem 138.1,

$$(139.3)\qquad \sum_{i,j=1}^{\infty} a_i b_j = \left(\sum_{i=1}^{\infty} a_i\right)\left(\sum_{j=1}^{\infty} b_j\right).$$

Also from Theorem 138.1,

$$\sum_{i,j=1}^{\infty} a_i b_j = \sum_{p=2}^{\infty} (a_1 b_{p-1} + a_2 b_{p-2} + \cdots + a_{p-2} b_2 + a_{p-1} b_1).$$

This, together with (139.3) yields (139.2).

We have assumed, in Theorem 139.1, that both series $\sum_{i=1}^{\infty} a_i$ and $\sum_{j=1}^{\infty} b_j$ converge absolutely. One can show that the conclusion of Theorem 139.1 still holds true if only one of the two series converges absolutely while the other one merely converges.† The Cauchy product of two series may even converge to the product of the two sums when neither series converges absolutely. (See exercise 2 and section 14.144, exercise 6.)

## 14.139 Exercises

1. Let $a_i = b_i = (-1)^i/i$. Show that the Cauchy product of $\sum_{i=1}^{\infty} (-1)^i/i$ with itself converges. (To show that this Cauchy product converges to the product of the sums of series of which it is formed, is the subject of exercise 6, section 14.144.)
2. Let $a_i = (-1)^i/i^2$, $b_j = (-1)^j/j$. Does the Cauchy product of $\sum_{i=1}^{\infty} a_i$ and $\sum_{j=1}^{\infty} b_j$ converge?
3. Use Lemma 138.2 and Theorem 139.1 to show that

$$\sum_{i,j=1}^{\infty} q^{i+j-2} = \left(\sum_{i=0}^{\infty} q^i\right)^2 \text{ for } q \in (-1, 1).$$

† [For a proof of this stronger theorem, see T. Apostol, *Mathematical Analysis* (Reading, Mass.: Addison-Wesley Publishing Company, Inc., 1957), p. 376; or A. Devinatz, *Advanced Calculus* (New York: Holt, Reinhart and Winston, 1968), p. 168. See also exercise 1.]

4. Show that

$$\sum_{i,j=1}^{\infty} (-1)^i q^{i+j-2} = \sum_{i=0}^{\infty} (-q)^i \sum_{j=0}^{\infty} q^j.$$

5. Show that

$$\sum_{i=1}^{\infty} \frac{a^{i-1}}{(i-1)!} \sum_{j=1}^{\infty} \frac{b^{j-1}}{(j-1)!} = \sum_{k=1}^{\infty} \frac{(a+b)^{k-1}}{(k-1)!}.$$

6. Show that

$$\sum_{j=1}^{\infty} (-1)^{j-1} \frac{a^{2j-1}}{(2j-1)!} \sum_{i=1}^{\infty} (-1)^{i-1} \frac{b^{2i-2}}{(2i-2)!} + \sum_{j=1}^{\infty} (-1)^{j-1} \frac{b^{2j-1}}{(2j-1)!} \sum_{i=1}^{\infty} (-1)^{i-1} \frac{a^{2i-2}}{(2i-2)!} = \sum_{k=1}^{\infty} (-1)^{k-1} \frac{(a+b)^{2k-1}}{(2k-1)!}.$$

## 14.140 Infinite Series of Functions

Series of functions are obtained from series of numbers (or vectors) in the same manner as sequences of functions are obtained by generalization from sequences of numbers (or vectors). (See Chapter 7.)

***Definition 140.1***

Let $f_j : \mathfrak{D} \to \mathbf{E}$, $\mathfrak{D} \subseteq \mathbf{E}$, for all $j \in \mathbf{N}$. The *infinite series* $\sum_{j=1}^{\infty} f_j$ is the *sequence* $\{s_k\}$ of *partial sums* $s_k = \sum_{j=1}^{k} f_j$.

$\sum_{j=1}^{\infty} f_j$ *converges pointwise* on $A \subseteq \mathfrak{D}$ if and only if $\{s_k\}$ converges pointwise on $A$. (See definition 76.1.)

$\sum_{j=1}^{\infty} f_j$ *converges uniformly* on $A \subseteq \mathfrak{D}$ if and only if $\{s_k\}$ converges uniformly on $A$. (See definition 77.1.)

$\sum_{j=1}^{\infty} f_j$ *converges absolutely* on $A \subseteq \mathfrak{D}$ if and only if $\sum_{j=1}^{\infty} |f_j|$ converges pointwise on $A$, where $|f_j|(x) = |f_j(x)|$ for all $x \in A$. (Note that $s_k$ is a function from $\mathfrak{D}$ into $\mathbf{E}$.)

$\lim_{\infty} s_k$ is called the sum of $\sum_{j=1}^{\infty} f_j$ and we write $\sum_{j=1}^{\infty} f_j = \lim_{\infty} s_k$ on $A$ or $\sum_{j=1}^{\infty} f_j(x) = \lim_{\infty} s_k(x)$ for all $x \in A$.

We obtain from Lemma 76.1, Theorem 76.1, definition 77.1, Theorem 77.3, and Lemma 135.1:

***Theorem 140.1***

Let $f_j : \mathfrak{D} \to \mathbf{E}$, $\mathfrak{D} \subseteq \mathbf{E}$, for all $j \in \mathbf{N}$ and let $A \subseteq \mathfrak{D}$.

$\sum_{j=1}^{\infty} f_j$ *converges pointwise* on $A$ (to $s : A \to \mathbf{E}$) if and only if, for every $\varepsilon > 0$, and each $x \in A$, there is an $N(\varepsilon, x) \in \mathbf{N}$ such that

(140.1) $$|f_1(x) + f_2(x) + \cdots + f_k(x) - s(x)| < \varepsilon \qquad \text{for all } k > N(\varepsilon, x),$$

or

(140.2) $$|f_{l+1}(x) + f_{l+2}(x) + \cdots + f_k(x)| < \varepsilon \qquad \text{for all } k > l > N(\varepsilon, x).$$

$\sum_{j=1}^{\infty} f_j$ *converges uniformly* on $A$ (to $s : A \to \mathbf{E}$) if and only if, for every $\varepsilon > 0$, there is an $N(\varepsilon) \in \mathbf{N}$ such that

(140.3) $$|f_1(x) + f_2(x) + \cdots + f_k(x) - s(x)| < \varepsilon \qquad \text{for all } k > N(\varepsilon) \text{ and all } x \in A,$$

or,

(140.4) $$|f_{l+1}(x) + f_{l+2}(x) + \cdots + f_k(x)| < \varepsilon \quad \text{for all } k > l > N(\varepsilon) \text{ and all } x \in A.$$

$\sum_{j=1}^{\infty} f_j$ *converges absolutely* on $A$ if and only if there is a nonnegative function $M : A \to \mathbf{E}$ such that

(140.5) $$|f_1(x)| + |f_2(x)| + \cdots + |f_k(x)| \leqslant M(x) \qquad \text{for all } k \in \mathbf{N} \text{ and all } x \in A.$$

We observe that the partial sums of a series of continuous functions are continuous functions, and the partial sums of a series of bounded functions are bounded functions. In view of this, we obtain, from Theorems 78.1 and 78.2 on sequences, analogous results for infinite series.

***Theorem 140.2***

If $\sum_{j=1}^{\infty} f_j$ *converges uniformly* on $A \subseteq \mathfrak{D}$ to the sum $s : A \to \mathbf{E}$, then

$s$ is continuous on $A$ if all $f_j$ are continuous on $A$,
$s$ is bounded on $A$ if all $f_j$ are bounded on $A$.

---

▲ *Example 1*

If $f_j : [0, 1] \to \mathbf{E}$ is defined by $f_j(x) = x^{j-1}$, then

$$\sum_{j=1}^{\infty} x^{j-1} = \frac{1}{1 - x}$$

is continuous and bounded on $A = [0, a]$ for every $a \in (0, 1)$.

▲ *Example 2*

The function $w : \mathbf{E} \to \mathbf{E}$, which was discussed in section 7.80 and which is continuous but nowhere differentiable, may be represented by

$$w = \sum_{j=1}^{\infty} f_j,$$

where $f_j : \mathbf{E} \to \mathbf{E}$ is defined in (80.1), (80.2), and (80.4).

---

Theorems 81.1, 81.2, and 82.1 on the integration and differentiation of sequences of real-valued functions of a real variable, yield analogous results for infinite series of real-valued functions of a real variable.

***Theorem 140.3***

If the functions $f_j : \mathfrak{J} \to \mathbf{E}$ are Riemann-integrable on $[a, b] \subseteq \mathfrak{J}$ and if $\sum_{j=1}^{\infty} f_j$ converges uniformly on $[a, b]$, then $\sum_{j=1}^{\infty} f_j$ represents a Riemann-integrable function on $[a, b]$ and

(140.6) $$\int_a^b \sum_{j=1}^{\infty} f_j = \sum_{j=1}^{\infty} \int_a^b f_j.$$

If the functions $f_j : \mathfrak{J} \to \mathbf{E}$ are differentiable in the bounded interval $\mathfrak{J}$, if $\sum_{j=1}^{\infty} f_j(x_0)$ converges for some $x_0 \in \mathfrak{J}$, and if $\sum_{j=1}^{\infty} f_j'$ converges uniformly on $\mathfrak{J}$ then $\sum_{j=1}^{\infty} f_j$ represents a differentiable function on $\mathfrak{J}$ and

(140.7) $$\left(\sum_{j=1}^{\infty} f_j\right)'(x) = \sum_{j=1}^{\infty} f_j'(x) \qquad \text{for all } x \in \mathfrak{J}$$

A crude, but, whenever applicable, simple test for the uniform convergence of an infinite series, is the so-called Weierstrass test:

***Theorem 140.4*** *(Weierstrass Test)*

Let $f_j : \mathfrak{D} \to \mathbf{E}$, $\mathfrak{D} \subseteq \mathbf{E}$, and let $\sum_{j=1}^{\infty} M_j$ denote a convergent series of nonnegative numbers. If

$$|f_j(x)| \leqslant M_j \qquad \text{for all } x \in A \subseteq \mathfrak{D} \text{ and all } j \in \mathbf{N},$$

then $\sum_{j=1}^{\infty} f_j$ *converges uniformly* on $A$.

*Proof*

The uniform convergence follows directly from (140.4), and

$$|f_{l+1}(x) + f_{l+2}(x) + \cdots + f_k(x)| \leqslant M_{l+1} + M_{l+2} + \cdots + M_k.$$

---

▲ *Example 3*

Let $f_j(x) = (x^j/j!)$ for all $x \in \mathbf{E}$ and all $j \in \mathbf{N}$. $\sum_{j=0}^{\infty} (x^j/j!)$ converges pointwise for all $x \in \mathbf{E}$ (see exercise 10, section 14.136; note that we started the summation with $j = 0$) and defines a function $E : \mathbf{E} \to \mathbf{E}$:

$$E(x) = \sum_{j=0}^{\infty} \frac{x^j}{j!}, \qquad x \in \mathbf{E}.$$

$E$ is called the *exponential function.*

If $M = \max(|a|, |b|)$, then, for all $x \in [a, b]$,

$$\left|\frac{x^l}{l!} + \frac{x^{l+1}}{(l+1)!} + \cdots + \frac{x^k}{k!}\right| \leqslant \frac{M^l}{l!} + \frac{M^{l+1}}{(l+1)!} + \cdots + \frac{M^k}{k!},$$

where $\sum_{j=0}^{\infty} (M^j/j!)$ converges. By Theorem 140.4 (Weierstrass test), the series $\sum_{j=0}^{\infty} (x^j/j!)$ converges uniformly on every interval $[a, b]$. Hence, by (140.6),

$$(140.8) \qquad \int_0^x E = \int_0^x \left(\sum_{j=0}^{\infty} \frac{t^j}{j!}\right) dt = \sum_{j=0}^{\infty} \int_0^x \frac{t^j}{j!}\, dt = \sum_{j=0}^{\infty} \frac{x^{j+1}}{(j+1)!} = E(x) - 1$$

for all $x \in \mathbf{E}$.

The series of differentiated terms $\sum_{j=1}^{\infty} x^{j-1}/(j-1)! = \sum_{j=0}^{\infty} x^j/j!$ converges uniformly on every bounded interval. Hence, by (140.7),

$$(140.9) \qquad E'(x) = \left(\sum_{j=0}^{\infty} \frac{x^j}{j!}\right)' = \sum_{j=0}^{\infty} \left(\frac{x^j}{j!}\right)' = \sum_{j=0}^{\infty} \frac{x^j}{j!} = E(x) \qquad \text{for all } x \in \mathbf{E}.$$

---

Although Weierstrass' test is very effective when it works, it does not apply in more delicate situations. In such cases, the following test often proves to be useful:

***Theorem 140.5***

Let $f_j : \mathfrak{D} \to \mathbf{E}$, $\mathfrak{D} \subseteq \mathbf{E}$, and let $\phi_j : \mathfrak{D} \to \mathbf{E}$. If $\sum_{j=1}^{\infty} f_j$ converges uniformly on $A \subseteq \mathfrak{D}$ and if the sequence $\{\phi_j\}$ is increasing or decreasing on $A$, and bounded on $A$, then $\sum_{j=1}^{\infty} \phi_j f_j$ converges uniformly on $A$.

*Proof*

By hypothesis, $|\phi_j(x)| \leqslant M$ for some $M > 0$, all $j \in \mathbf{N}$, and all $x \in A$.

Since $\sum_{j=1}^{\infty} f_j$ converges uniformly on $A$ to some functions $s : A \to \mathbf{E}$, we have, for every $\varepsilon > 0$, an $N(\varepsilon) \in \mathbf{N}$ such that, by (140.3),

$$(140.10) \qquad |s(x) - f_1(x) - f_2(x) - \cdots - f_{k-1}(x)| = \left|\sum_{j=k}^{\infty} f_j(x)\right| < \frac{\varepsilon}{2M}$$

for all $k > N(\varepsilon)$ and all $x \in A$.

Let $b_k(x) = \sum_{j=k}^{\infty} f_j(x)$. By (140.10),

$$(140.11) \qquad |b_k(x)| < \frac{\varepsilon}{2M} \qquad \text{for all } k > N(\varepsilon) \text{ and all } x \in A.$$

We may assume without loss of generality that $\{\phi_j\}$ is increasing on $A$. If we note that $f_j = b_j - b_{j+1}$, we obtain, from (140.11),

$$\begin{aligned}
&|\phi_{l+1}(x)f_{l+1}(x) + \phi_{l+2}(x)f_{l+2}(x) + \cdots + \phi_k(x)f_k(x)| \\
&\quad = |\phi_{l+1}(x)b_{l+1}(x) + (\phi_{l+2}(x) - \phi_{l+1}(x))b_{l+2}(x) + \cdots \\
&\qquad + (\phi_k(x) - \phi_{k-1}(x))b_k(x) - \phi_k(x)b_{k+1}(x)| \\
&\quad \leqslant |\phi_{l+1}(x)|\,|b_{l+1}(x)| + (\phi_{l+2}(x) - \phi_{l+1}(x))\,|b_{l+2}(x)| + \cdots \\
&\qquad + (\phi_k(x) - \phi_{k-1}(x))\,|b_k(x)| + |\phi_k(x)|\,|b_{k+1}(x)| \\
&\quad < \frac{\varepsilon}{2M}(2\,|\phi_k(x)|) \leqslant \varepsilon \qquad \text{for all } k > l > N(\varepsilon) \text{ and all } x \in A.
\end{aligned}$$

By (140.4), $\sum_{j=1}^{\infty} \phi_j f_j$ converges uniformly on $A$.

---

▲ *Example 4*

We consider $\sum_{j=1}^{\infty} (-1)^{j+1}(x^j/j)$. By Theorem 134.2, this series converges pointwise for all $x \in [0, 1]$, being an alternating series, the absolute values of its terms decreasing to zero. Weierstrass' test is *not* applicable in this case because $|(-1)^{j+1} x^j/j| \leqslant 1/j$ for all $x \in [0, 1]$ and $\sum_{j=1}^{\infty} (1/j)$ does not converge. We let $f_j = (-1)^{j+1}/j$ and $\phi_j(x) = x^j$. $\sum_{j=1}^{\infty} f_j$ converges and, being a series of numbers, converges uniformly on $\mathbf{E}$ and, in particular, on $[0, 1]$. $\{x^j\}$ is decreasing on $[0, 1]$ and bounded above by 1. Hence, we have, from Theorem 140.5, that

(140.12) $$\sum_{j=1}^{\infty} (-1)^{j+1} \frac{x^j}{j} \quad \textit{converges uniformly on } [0, 1].$$

(Note that Theorem 140.5 does *not* apply to the interval $[-1, 0]$, or $(-1, 0]$, for that matter, because $\{x^j\}$ is *not* decreasing or increasing for negative $x$. In fact, the series diverges for $x = -1$.)

---

## 14.140 Exercises

1. Let $f_j(x) = (-1)^j x^{2j+1}/(2j + 1)!$, $g_j(x) = (-1)^j x^{2j}/(2j)!$. Show that
   (a) $\sum_{j=0}^{\infty} f_j$, $\sum_{j=0}^{\infty} g_j$ converge pointwise on $\mathbf{E}$,
   (b) converge uniformly on every bounded interval $\mathfrak{J}$,
   (c) converge absolutely on $\mathbf{E}$.
2. Denote $\sum_{j=0}^{\infty} f_j(x) = \sin x$, $\sum_{j=0}^{\infty} g_j(x) = \cos x$ where $f_j, g_j$ are defined in exercise 1. Show:
   (a) $\sin 0 = 0$, $\cos 0 = 1$,
   (b) sin, cos are continuous functions on $\mathbf{E}$,
   (c) $\cos(-x) = \cos x$, $\sin(-x) = -\sin x$,
   (d) $(\sin x)' = \cos x$, $(\cos x)' = -\sin x$,
   (e) $\int_0^x \sin t\, dt = -\cos x + 1$, $\int_0^x \cos t\, dt = \sin x$,
   (f) $\sin^2 x + \cos^2 x = 1$,
   (g) $|\sin x| \leqslant 1$, $|\cos x| \leqslant 1$ for all $x \in \mathbf{E}$.

3. Let $f_j(x) = x^j/(j^2(1 + x^j))$. Show that $\sum_{j=1}^{\infty} f_j$ converges uniformly on $[0, 1]$.
4. Show that $\zeta(x) = \sum_{j=1}^{\infty} (1/j^x)$ converges uniformly for $x \geqslant \alpha$ where $\alpha > 1$.
5. Show that $\sum_{j=1}^{\infty} (-1)^{j+1}(x^j/j)$ converges pointwise for $x \in (-1, 1]$.
6. Show that the series $\sum_{j=1}^{\infty} f_j$ in Theorem 140.4 converges absolutely on $A$.

## 14.141 Power Series

Power series are an important class of infinite series of functions. We have already encountered power series in section 14.140, examples 1, 3, and 4 and in exercises 1 and 5. In calculus, the reader has encountered power series in a discussion of representations of functions by Taylor series.

### *Definition 141.1*

The infinite series $\sum_{j=0}^{\infty} a_j(x - x_0)^j$, $a_j, x, x_0 \in \mathbf{R}$, is called a *power series* around $x_0$.

(For reasons of convenience, we start the summation at $j = 0$ to avoid writing the general term in the bulky form $a_{j-1}(x - x_0)^{j-1}$.)

We shall assume without loss of generality that $x_0 = 0$. We return to the case with $x_0 \neq 0$ in section 14.145.

Power series have remarkable properties which set them apart from just ordinary and common series of functions. One of these distinguishing features is dealt with in the following theorem; the others will be summarized in Theorem 141.2.

### *Theorem 141.1*

If $\sum_{j=0}^{\infty} a_j x^j$ converges for $x = c \neq 0$, then it *converges absolutely* for all $|x| < |c|$.

### *Proof*

Since $\sum_{j=0}^{\infty} a_j c^j$ converges, we have $|a_j c^j| \leqslant M$ for some $M > 0$ and all $j \in \mathbf{N}$. Let $|x| < |c|$. Then,

$$|a_j x^j| = \left|a_j c^j\left(\frac{x^j}{c^j}\right)\right| \leqslant M\left|\frac{x}{c}\right|^j.$$

Since $|x/c| < 1$, we have

$$\sum_{j=0}^{\infty} M\left|\frac{x}{c}\right|^j = \frac{M}{1 - \left|\dfrac{x}{c}\right|}$$

and hence,

$$|a_0| + |a_1 x| + \cdots + |a_k x^k| \leqslant \frac{M}{1 - \left|\dfrac{x}{c}\right|} \qquad \text{for all } |x| < |c|.$$

By (140.5), $\sum_{j=0}^{\infty} a_j x^j$ converges absolutely for all $|x| < |c|$.

Let

$$(141.1) \qquad P = \left\{ c \in \mathbf{R} \,\middle|\, \sum_{j=0}^{\infty} a_j c^j \text{ converges} \right\}.$$

If $P$ is not bounded above, then $\sum_{j=0}^{\infty} a_j x^j$ converges for all $x \in \mathbf{R}$. If it is bounded above, then it has a least upper bound $r \geqslant 0$, and $\sum_{j=0}^{\infty} a_j x^j$ converges for all $|x| < r$:

***Definition 141.2***

If $P$, as defined in (141.1), is bounded above, then

$$(141.2) \qquad r = \sup \left\{ c \in \mathbf{R} \,\middle|\, \sum_{j=0}^{\infty} a_j c^j \text{ converges} \right\}$$

is called the *radius of convergence* of $\sum_{j=0}^{\infty} a_j x^j$. If $P$ is not bounded above, we write symbolically $r = \infty$.

***Theorem 141.2***

If $r > 0$ (possibly $r = \infty$) is the radius of convergence of the power series $\sum_{j=0}^{\infty} a_j x^j$, then:

1. $\sum_{j=0}^{\infty} a_j x^j$ *converges absolutely* for all $x \in (-r, r)$,
2. $\sum_{j=0}^{\infty} a_j x^j$ *does not converge* for $x > r$ and $x < -r$,
3. $\sum_{j=0}^{\infty} a_j x^j$ converges uniformly to a continuous and bounded function on every closed interval $\mathfrak{J} \subset (-r, r)$,
4. $\int_0^x (\sum_{j=0}^{\infty} a_j t^j)\, dt = \sum_{j=0}^{\infty} \int_0^x a_j t^j \, dt$ for all $x \in (-r, r)$,
5. $(\sum_{j=0}^{\infty} a_j x^j)' = \sum_{j=1}^{\infty} j a_j x^{j-1}$ for all $x \in (-r, r)$.

(Note that by (5), a power series has derivatives of all orders.)

*Proof*

(1) and (2) follow from Theorem 141.1 and the definition of the radius of convergence.

3. Let $\mathfrak{J}$ denote a closed interval with endpoints $a, b$ and let $\rho = \max(|a|, |b|)$. If $\mathfrak{J} \subset (-r, r)$, then $\rho \in (0, r)$. We obtain, as in the proof of Theorem 141.1, that, for every $c \in (\rho, r)$,

$$|a_j x^j| \leqslant M \left| \frac{\rho}{c} \right|^j \qquad \text{for all } x \in \mathfrak{J} \text{ and all } j \in \mathbf{N}.$$

Since $|\rho/c| < 1$, the uniform convergence of $\sum_{j=0}^{\infty} a_j x^j$ on $\mathfrak{J}$ follows from Theorem 140.4. Since each term $a_j x^j$ is continuous and bounded on $\mathfrak{J}$, the continuity and boundedness of the sum follows from Theorem 140.2.

4. Follows directly from (3) and Theorem 140.3.
5. In order to apply (140.7), we have to show, first, that

$$\sum_{j=0}^{\infty} (a_j x^j)' = \sum_{j=1}^{\infty} j a_j x^{j-1}$$

converges in $(-r, r)$ which, by (3), implies uniform convergence in every closed interval $\mathfrak{J} \subset (-r, r)$.

As in the proof of Theorem 141.1, we have, for every $c \in (-r, r)$, an $M > 0$ such that $|a_j x^j| \leqslant |a_j c^j| \leqslant M$ for all $|x| < |c|$. Hence,

$$|j a_j x^{j-1}| = \left| \frac{j}{c} a_j c^j \left( \frac{x}{c} \right)^{j-1} \right| \leqslant \frac{M}{|c|} j \left| \frac{x}{c} \right|^{j-1}$$

where $|x/c| < 1$. Since

$$\lim_{\infty} \frac{(j+1)\left|\dfrac{x}{c}\right|^j}{j\left|\dfrac{x}{c}\right|^{j-1}} = \lim_{\infty} \frac{j+1}{j} \left| \frac{x}{c} \right| = \left| \frac{x}{c} \right| < 1 \qquad \text{for all } |x| < |c|,$$

we have, from (136.5) (limit test), that $\sum_{j=1}^{\infty} (M/|c|) j |x/c|^{j-1}$ converges for all $|x| < |c|$. Since $c$ is any number in $(-r, r)$, (3) and Theorem 140.3 take care of the rest.

An explicit representation of the radius of convergence of a power series is obtained from one of the limit tests in Theorem 136.4:

***Theorem 141.3*** *(Cauchy–Hadamard Rule)*†

If $\sum_{j=0}^{\infty} a_j x^j$ converges for some $x \neq 0$, then its radius of convergence is given by

$$r = \frac{1}{\overline{\lim} \sqrt[j]{|a_j|}} \qquad \text{if } \overline{\lim} \sqrt[j]{|a_j|} > 0,$$

and by

$$r = \infty \qquad \text{if } \overline{\lim} \sqrt[j]{|a_j|} = 0.$$

*Proof*

If $\sum_{j=0}^{\infty} a_j x^j$ converges for some $x \neq 0$, then $|a_j x^j|$ is bounded and $\overline{\lim} \sqrt[j]{|a_j x^j|} = |x| \overline{\lim} \sqrt[j]{|a_j|}$ exists.

We have from Theorem 136.4 that $\sum_{j=0}^{\infty} a_j x^j$ converges if $\overline{\lim} \sqrt[j]{|a_j x^j|} = |x| \overline{\lim} \sqrt[j]{|a_j|} < 1$ and does not converge if $\overline{\lim} \sqrt[j]{|a_j x^j|} = |x| \overline{\lim} \sqrt[j]{|a_j|} > 1$. Theorem 141.3 follows readily.

† *Jacques Hadamard*, 1865–1963.

▲ *Example 1*

We obtain, from the definition of the natural logarithm in (52.7), for $|x| < 1$:

$$\log(1 + x) = \int_1^{1+x} \frac{dt}{t} = \int_0^x \frac{dt}{1+t}.$$

For $|t| < 1$,

$$\frac{1}{1+t} = 1 - t + t^2 - t^3 + \cdots = \sum_{j=0}^{\infty} (-1)^j t^j. \tag{141.3}$$

This (geometric) series has the radius of convergence $r = 1/(\lim_\infty \sqrt[j]{1}) = 1$. Hence, we have, from Theorem 141.2, part (4), that, for $|x| < 1$:

$$\log(1 + x) = \int_0^x \left( \sum_{j=0}^{\infty} (-1)^j t^j \right) dt = \sum_{j=0}^{\infty} \int_0^x (-1)^j t^j \, dt$$

$$= \sum_{j=0}^{\infty} (-1)^j \frac{x^{j+1}}{j+1} = \sum_{j=1}^{\infty} (-1)^{j+1} \frac{x^j}{j}.$$

We have seen in example 4, section 14.140, that this series converges *uniformly* in $[0, 1]$. Theorem 141.2, however, gives no guarantee that it converges at $x = 1$ to $\log 2$. But, since

$$f(x) = \sum_{j=1}^{\infty} (-1)^{j+1} \frac{x^j}{j}$$

represents a continuous function on $[0, 1]$ by Theorem 140.2, since the natural logarithm is continuous for all $x \in \mathbf{R}^+$ and since

$$f(x) = \log(1 + x) \qquad \text{for all } x \in [0, 1),$$

we have

$$\log 2 = \lim_1 \log(1 + x) = \lim_1 f(x) = f(1) = \sum_{j=1}^{\infty} \frac{(-1)^{j+1}}{j}.$$

Hence,

$$\log 2 = \sum_{j=1}^{\infty} \frac{(-1)^{j+1}}{j}.$$

(See also exercise 6.)

## 14.141 Exercises

1. Find the radius of convergence of the following power series:

$$\sum_{j=0}^{\infty} x^j, \quad \sum_{j=1}^{\infty} \frac{x^j}{j}, \quad \sum_{j=1}^{\infty} \frac{x^j}{j^2}, \quad \sum_{j=0}^{\infty} j\left(\frac{x}{2}\right)^{2j}, \quad \sum_{j=0}^{\infty} j^2\left(\frac{x}{3}\right)^j.$$

2. Show: If $\lim_\infty (|a_{j+1}|/|a_j|)$ exists, then the radius of convergence of $\sum_{j=0}^\infty a_j x^j$ is given by

$$r = \frac{1}{\lim_\infty \dfrac{|a_{j+1}|}{|a_j|}} \qquad \text{if } \lim_\infty \frac{|a_{j+1}|}{|a_j|} \neq 0$$

and by $r = \infty$ if the limit is equal to zero.

3. Find the radius of convergence of

$$\sum_{j=2}^\infty \frac{x^j}{\log j}, \qquad \sum_{j=1}^\infty \frac{jx^j}{(j-1)!}, \qquad \sum_{j=0}^\infty \frac{(-1)^j\left(\dfrac{x}{2}\right)^{2j}}{(j!)^2}.$$

4. Express $\log[(1+x)/(1-x)]$ as a power series.
5. Let $y \in \mathbf{R}^+$. Show that

$$\log y = 2\left(\left(\frac{y-1}{y+1}\right) + \frac{1}{3}\left(\frac{y-1}{y+1}\right)^3 + \cdots\right).$$

6. *Prove*: If $f(x) = \sum_{j=0}^\infty a_j x^j$ for all $|x| < 1$ and if $\sum_{j=0}^\infty a_j = a$, then $\sum_{j=0}^\infty a_j x^j$ converges uniformly on $[0, 1]$ and $\lim_1 f(x) = a$.
7. Show: If $\{|a_j|\}$ is bounded, then $\sqrt[j]{|a_j|}$ is also bounded.
8. Show that the *Bessel function* of the first kind and $n$th order, $n \in \mathbf{N}$,

$$J_n(x) = \sum_{j=0}^\infty \frac{(-1)^j\left(\dfrac{x}{2}\right)^{2j+n}}{j!(n+j)!}$$

satisfies the differential equation

$$x^2 J_n''(x) + xJ_n'(x) + (x^2 - n^2)J_n(x) = 0,$$

the so-called *Bessel equation*.

## 14.142 The Exponential Function and Its Inverse Function*

It is the main purpose of this section to establish the well-known properties of the exponential function on the basis of its definition by a power series.

The *exponential function* $E : \mathbf{R} \to \mathbf{R}$ is defined by

$$E(x) = \sum_{j=0}^\infty \frac{x^j}{j!} \qquad \text{for all } x \in \mathbf{R}. \tag{142.1}$$

We have seen, in example 3, section 14.140, that this series converges for all $x \in \mathbf{R}$ and that it converges uniformly in every bounded interval. Hence, *E is continuous on* $\mathbf{R}$, and its integral and its derivative may be obtained by termwise integration and termwise differentiation, respectively. We obtained, in (140.8) and (140.9),

$$E'(x) = E(x), \qquad \int_0^x E = E(x) - 1 \qquad \text{for all } x \in \mathbf{R}. \tag{142.2}$$

In order to show that $E$ is invertible on its range, we have to find its range first, namely $\mathbf{R}^+$, as it develops, and then demonstrate that

$$E : \mathbf{R} \leftrightarrow \mathbf{R}^+. \tag{142.3}$$

Let us show first that $E : [0, \infty) \xrightarrow{\text{onto}} [1, \infty)$. We have, from (142.1), that

$$E(0) = 1, \qquad E(x) \geqslant 1 + x \qquad \text{for all } x \geqslant 0. \tag{142.4}$$

Since $E$ is continuous, we have, from the intermediate-value theorem and from (142.4), that, for every $y \in (1, \infty)$, there is an $x > 0$ such that $E(x) = y$. Hence, $E : [0, \infty) \xrightarrow{\text{onto}} [1, \infty)$.

Next we show that $E : (-\infty, 0) \xrightarrow{\text{onto}} (0, 1)$. For this purpose, we need the relation

$$E(x)E(-x) = 1. \tag{142.5}$$

From (142.2),

$$(E(x)E(-x))' = E(x)E(-x) - E(x)E(-x) = 0 \qquad \text{for all } x \in \mathbf{R}.$$

This, together with $E(0) = 1$, yields (142.5). Hence, $E(-x) = 1/E(x)$, and we obtain $E : (-\infty, 0) \xrightarrow{\text{onto}} (0, 1)$ from the previous result that $E : [0, \infty) \xrightarrow{\text{onto}} [1, \infty)$. Collecting these two partial results yields $E : \mathbf{R} \xrightarrow{\text{onto}} \mathbf{R}^+$.

For the purpose of demonstrating that $E$ is one-to-one on $\mathbf{R}$, we assume, to the contrary, that $E(x_1) = E(x_2)$ for some $x_1, x_2 \in \mathbf{R}$, $x_1 \neq x_2$. We may assume without loss of generality that $0 \leqslant x_1 < x_2$. From (142.1),

$$0 = E(x_2) - E(x_1) = \sum_{j=0}^{\infty} \frac{1}{j!} (x_2^j - x_1^j) > 0,$$

which is absurd. Since $(1/j!) > 0$, we have to have $x_1 = x_2$. Hence, $E$ is one-to-one on $\mathbf{R}$ and (142.3) is established. Hence, $E$ has an inverse function $E^{-1} : \mathbf{R}^+ \leftrightarrow \mathbf{R}$. Since $E'(x) = E(x) > 0$ for all $x \in \mathbf{R}$, we obtain from the Corollary to Lemma 37.1,

$$(E^{-1})'(y) = \frac{1}{E'(E^{-1}(y))} \qquad \text{for all } y \in \mathbf{R}^+.$$

From (142.2), $E'(E^{-1}(y)) = E(E^{-1}(y)) = y$ and hence,

$$(E^{-1})'(y) = \frac{1}{y} \qquad \text{for all } y \in \mathbf{R}^+. \tag{142.6}$$

Since $E(0) = 1$, we have $E^{-1}(1) = 0$ and hence,

$$(142.7) \qquad E^{-1}(y) = \int_1^y \frac{dt}{t} = \log y \qquad \text{for all } y \in \mathbf{R}^+,$$

i.e., *the natural logarithm is the inverse function to the exponential function.* (See also example 5, section 4.52.) From (142.6),

$$(142.8) \qquad \log'(y) = \frac{1}{y} \qquad \text{for all } y \in \mathbf{R}^+$$

and hence, $\log \in C^1(\mathbf{R}^+)$.

Since $\log(y_1 y_2) = \log y_1 + \log y_2$ for all $y_1, y_2 \in \mathbf{R}^+$ (see example 5, section 4.52), we obtain readily

$$(142.9) \qquad E(x_1)E(x_2) = E(x_1 + x_2) \qquad \text{for all } x_1, x_2 \in \mathbf{R}.$$

(The same result may be obtained by forming the Cauchy product of the power series for $E(x_1)$ and $E(x_2)$. See exercise 2.)

It is customary to denote the value of $E(1)$ by $e$ in honor of Leonhard Euler† who appears to be the first one to recognize the importance of this number. He was also the first one to designate this number by $e$. We have, from (142.1),

$$(142.10) \qquad e = E(1) = \sum_{j=0}^{\infty} \frac{1}{j!} = 2 + \frac{1}{2} + \frac{1}{6} + \frac{1}{24} + \cdots = 2.718281828459\ldots.$$

(See also exercise 5.)

From (142.9), together with (142.5), we obtain

$$E(p) = e^p \qquad \text{for all } p \in \mathbf{Z}.$$

Also from (142.9) we obtain, for any $q \in \mathbf{N}$,

$$e = E(1) = E\left(q \cdot \frac{1}{q}\right) = E\Bigg(\underbrace{\frac{1}{q} + \frac{1}{q} + \cdots + \frac{1}{q}}_{q \text{ terms}}\Bigg) = \left[E\left(\frac{1}{q}\right)\right]^q$$

and hence, $E(1/q) = e^{1/q}$. Also from (142.9), $E(p\, 1/q) = (E(1/q))^p = e^{p/q}$. Hence

$$(142.11) \qquad E(r) = e^r \qquad \text{for all } r \in \mathbf{Q}.$$

In order to show that $E(x) = e^x$ for all $x \in \mathbf{R}$, we first have to define irrational powers of a given number. If $a > 1$ and if $r_1, r_2 \in \mathbf{Q}$, we obtain, in the usual manner, that $a^{r_1}a^{r_2} = a^{r_1 + r_2}$, $(a^{r_1})^{r_2} = a^{r_1 r_2}$, and $a^{r_2} < a^{r_2}$ if and only if $r_1 < r_2$. (See exercises 6, 7, 8.) Let $x \in \mathbf{R}$. Then, $x$ may be represented as the limit of a sequence of rational numbers: $x = \lim_\infty r_k, r_k \in \mathbf{Q}$. (See Theorem 22.3.) We define

$$(142.12) \qquad a^x = \lim_\infty a^{r_k}.$$

† *Leonhard Euler*, 1707–1783.

We shall show that this limit exists and is independent of the choice of the sequence of rational numbers $\{r_k\}$ as long as $\lim_\infty r_k = x$.

If $\lim_\infty r_k = x$, then, $|r_k - r_l| < (1/n)$ for all $k > l > N(n) \in \mathbf{N}$. Hence, $-1/n < r_k - r_l < 1/n$ and, consequently, $1/\sqrt[n]{a} < a^{r_k - r_l} < \sqrt[n]{a}$. Since $\lim_\infty \sqrt[n]{a} = 1$ (see (17.2)), we have, for every $\varepsilon > 0$, an $N(\varepsilon)$ such that $1 - \varepsilon < a^{r_k - r_l} < 1 + \varepsilon$ for all $k > l > N(\varepsilon)$. Therefore, noting that $\{r_k\}$ is bounded, $|r_k| \leqslant M$ for some $M \in \mathbf{Q}$, $M > 0$, we obtain

$$|a^{r_k} - a^{r_l}| \leqslant a^M |a^{r_k - r_l} - 1| < a^M \varepsilon \qquad \text{for all } k > l > N(\varepsilon),$$

that is, $\lim_\infty a^{r_k}$ exists.

If $\lim_\infty r_k = x$, $\lim_\infty \rho_k = x$, where $\{r_k\}$, $\{\rho_k\}$ are sequences of rational numbers, we have $\lim_\infty (r_k - \rho_k) = 0$ and hence, $\lim_\infty a^{r_k} - \lim_\infty a^{\rho_k} = \lim_\infty (a^{r_k} - a^{\rho_k}) = \lim_\infty a^{r_k}(1 - a^{\rho_k - r_k}) = 0$ because $\lim_\infty a^{\rho_k - r_k} = 1$ by an argument similar to the one in the above paragraph. Hence, $\lim_\infty a^{r_k} = \lim_\infty a^{\rho_k}$.

From the usual rules for the manipulations of limits, we obtain $a^x a^y = a^{x+y}$, $(a^x)^y = a^{xy}$, $a^x < a^y$ if and only if $x < y$. (See exercise 10.)

Since $E$ is continuous, we have, from (142.12), for every $x \in \mathbf{R}$, and $\lim_\infty r_k = x$,

$$E(x) = E(\lim_\infty r_k) = \lim_\infty E(r_k) = \lim_\infty e^{r_k} = e^x;$$

that is,

$$E(x) = e^x \qquad \text{for all } x \in \mathbf{R}. \tag{142.13}$$

An important application of the above developments is the derivation of the differentiation formula for $x^\alpha$, where $\alpha$ is any real number:

Let $x > 0$ and $\alpha \in \mathbf{R}$. Then, $x = e^{\log x}$ and hence, $x^\alpha = e^{\alpha \log x}$. Therefore, by the chain rule, by (142.2) and (142.8)

$$(x^\alpha)' = (e^{\alpha \log x})' = \frac{\alpha}{x} e^{\alpha \log x} = \alpha x^{\alpha - 1}:$$

$$(x^\alpha)' = \alpha x^{\alpha - 1}, \qquad x > 0, \alpha \in \mathbf{R}. \tag{142.14}$$

The *hyperbolic sine function* and the *hyperbolic cosine function* are defined by

$$\sinh x = \tfrac{1}{2}(e^x - e^{-x}), \qquad \cosh x = \tfrac{1}{2}(e^x + e^{-x}). \tag{142.15}$$

From (142.1)

$$\sinh x = \sum_{j=0}^{\infty} \frac{x^{2j+1}}{(2j+1)!}, \qquad \cosh x = \sum_{j=0}^{\infty} \frac{x^{2j}}{(2j)!}. \tag{142.16}$$

From (142.15), (142.2), and $\sinh 0 = 0$, $\cosh 0 = 1$,

$$(\sinh x)' = \cosh x, \qquad (\cosh x)' = \sinh x,$$

$$\int_0^x \sinh t\, dt = \cosh x - 1, \qquad \int_0^x \cosh t\, dt = \sinh x.$$

We obtain

$$\sinh : \mathbf{R} \leftrightarrow \mathbf{R}, \qquad \cosh : [0, \infty) \leftrightarrow [1, \infty).$$

Hence, $\sinh^{-1}$ exists on $\mathbf{R}$, $\cosh^{-1}$ exists on $[1, \infty)$ and we obtain, from (142.15),

$$\text{(142.17)} \qquad \begin{aligned} \sinh^{-1} y &= \log(y + \sqrt{y^2 + 1}) \qquad \text{for all } y \in \mathbf{R}, \\ \cosh^{-1} y &= \log(y + \sqrt{y^2 - 1}) \qquad \text{for all } y \in [1, \infty). \end{aligned}$$

## 14.142 Exercises

1. Show in detail that $E(-x)E(x) = 1$ and $E : [0, \infty) \xrightarrow{\text{onto}} [1, \infty)$ imply that $E : (-\infty, 0) \xrightarrow{\text{onto}} (0, 1)$.
2. Show that
$$\left(\sum_{j=0}^{\infty} \frac{x_1^j}{j!}\right)\left(\sum_{j=0}^{\infty} \frac{x_2^j}{j!}\right) = \sum_{p=0}^{\infty} \frac{1}{p!}(x_1 + x_2)^p$$
and conclude from this result that (142.9) is valid.
3. Supply the details in the derivation of (142.9) from $\log(y_1 y_2) = \log y_1 + \log y_2$ for all $y_1, y_2 \in \mathbf{R}^+$.
4. Show that $\log a^b = b \log a$, $a \in \mathbf{R}^+$, $b \in \mathbf{N}$.
5. Show that $e = \lim_\infty (1 + 1/n)^n$.
6. Let $a > 0$. Define $a^{n+1} = a^n a, n \in \mathbf{N}$, and show that for $m, n \in \mathbf{N}, a^{m+n} = a^m a^n, (a^m)^n = a^{mn}$. For $a > 1$, $a^m < a^n$ if and only if $m < n$; for $0 < a < 1$, $a^m < a^n$ if and only if $m > n$.
7. Let $n \in \mathbf{N}$. Define $a^{-n} = 1/a^n$. Show that $a^{m+n} = a^m a^n, (a^m)^n = a^{mn}$ for any $m, n \in \mathbf{Z}$. For $a > 1$, $a^m < a^n$ if and only if $m < n, m, n \in \mathbf{Z}$; for $0 < a < 1$, $a^m < a^n$ if and only if $m > n, m, n \in \mathbf{Z}$.
8. Let $a > 0$. Define $a^{1/n}$, $n \in \mathbf{Z}$, $n \neq 0$, as that number for which $(a^{1/n})^n = a$. Let $p, q \in \mathbf{Q}$, $q \neq 0$. Define $a^{p/q} = (a^{1/q})^p$. Show that the properties listed in exercises 6, 7 are still valid for $p, q \in \mathbf{Q}$, $q \neq 0$.
9. Prove in detail: If $\lim_\infty r_k = \lim_\infty \rho_k$, where $r_k, \rho_k \in \mathbf{Q}$, then $\lim_\infty a^{r_k} = \lim_\infty a^{\rho_k}$ for $a > 0$.
10. Let $a > 1$ and consider $a^x, x \in \mathbf{R}$, as defined in (142.12). Show that $a^{x+y} = a^x a^y, (a^x)^y = a^{xy}, a^x < a^y$ if and only if $x < y$, for all $x, y \in \mathbf{R}$.
11. Let $C^\infty(\mathfrak{D})$ denote the set of all functions that have derivatives of *all* orders in $\mathfrak{D} \subseteq \mathbf{R}$. Show: $E$, sinh, cosh, $\sinh^{-1} \in C^\infty(\mathbf{R})$, $\log \in C^\infty(\mathbf{R}^+)$, $\cosh^{-1} \in C^\infty[1, \infty)$.
12. Show that $\cosh : (-\infty, 0] \leftrightarrow [1, \infty)$ and hence, the hyperbolic cosine has an inverse from $[1, \infty)$ onto $(-\infty, 0]$. Show that this inverse may be represented by $\log(y - \sqrt{y^2 - 1})$ for all $y \in [1, \infty)$.
13. Given a function $E : \mathbf{R} \to \mathbf{R}$ which has the properties:

    (I) $E(x)E(y) = E(x + y)$ for all $x, y \in \mathbf{R}$;

    (II) $\displaystyle\lim_{0+0} \frac{E(x) - 1}{x} = 1.$

*Prove*:

(a) $\lim_0 \dfrac{E(x)-1}{x} = 1$, (b) $E'(x) = E(x)$, (c) $E$ is continuous in $\mathbf{R}$,

(d) $E(x) > 0$ for all $x \in \mathbf{R}$.

14. Show that the function $E$ of exercise 13 is uniquely determined by the properties (I) and (II).
15. Show that the function $E$ of exercise 13 is the exponential function.

## 14.143 The Trigonometric Functions and Their Inverse Functions*

What we did in section 14.142 for the exponential function, we shall now do for the trigonometric functions. We define the *sine function* and the *cosine function* sin, cos : $\mathbf{R} \to \mathbf{R}$ by

$$\text{(143.1)} \qquad \sin x = \sum_{j=0}^{\infty} (-1)^j \frac{x^{2j+1}}{(2j+1)!}, \qquad \cos x = \sum_{j=0}^{\infty} (-1)^j \frac{x^{2j}}{(2j)!}.$$

(See also section 14.140, exercise 2.) These series converge for all $x \in \mathbf{R}$, converge uniformly on every bounded interval and hence, are continuous on $\mathbf{R}$ and are termwise integrable and differentiable. We obtain

$$\text{(143.2)} \qquad (\sin x)' = \cos x, \qquad (\cos x)' = -\sin x,$$

$$\text{(143.3)} \qquad \int_0^x \sin t\,dt = -\cos x + 1, \qquad \int_0^x \cos t\,dt = \sin x,$$

$$\text{(143.4)} \qquad \sin 0 = 0, \qquad \cos 0 = 1.$$

We obtain the fundamental trigonometric identities for $\sin(x+y)$, $\cos(x+y)$ by taking the Cauchy products

$$\sin x \cos y = \sum_{p=0}^{\infty} (-1)^p \left( \frac{x^{2p+1}}{(2p+1)!\,1!} + \frac{x^{2p-1}y}{(2p-1)!\,2!} + \cdots + \frac{xy^{2p}}{1!\,(2p)!} \right),$$

$$\cos x \sin y = \sum_{p=0}^{\infty} (-1)^p \left( \frac{y^{2p+1}}{(2p+1)!\,1!} + \frac{y^{2p-1}x}{(2p-1)!\,2!} + \cdots + \frac{yx^{2p}}{1!\,(2p)!} \right)$$

and similarly for $\sin x \sin y$, $\cos x \cos y$. We obtain

$$\sin x \cos y + \cos x \sin y = \sum_{p=0}^{\infty} (-1)^p \frac{(x+y)^{2p+1}}{(2p+1)!},$$

and similarly,

$$\cos x \cos y - \sin x \sin y = \sum_{p=0}^{\infty} (-1)^p \frac{(x+y)^{2p}}{(2p)!}.$$

If we also note in (143.1) that

$$\text{(143.5)} \qquad \sin(-x) = -\sin x, \qquad \cos(-x) = \cos x,$$

we obtain

(143.6) $$\sin(x \pm y) = \sin x \cos y \pm \cos x \sin y,$$

(143.7) $$\cos(x \pm y) = \cos x \cos y \mp \sin x \sin y$$

for all $x, y \in \mathbf{R}$.

With the ultimate objective of finding an explicit representation of *Ludolph's*† *number* $\pi$, we shall establish next, that the cosine function vanishes for some $x > 0$. Since $\cos 0 = 1$ and since the cosine is continuous, there is an $x_0 > 0$ such that $\cos x > 0$ for all $x \in [0, x_0]$. By (143.2), the sine is strictly increasing in $[0, x_0]$ and hence, $\sin x_0 > 0$. Let $\sin x_0 = a$ and let us assume that $\cos x > 0$ for all $x \in [x_0, x_1]$ for some $x_1 > x_0$. From (143.3),

$$\cos x_1 - \cos x_0 = -\int_{x_0}^{x_1} \sin t \, dt \leqslant -a(x_1 - x_0);$$

that is,

$$x_1 - x_0 \leqslant \frac{1}{a}(\cos x_0 - \cos x_1) < \frac{1}{a}\cos x_0.$$

(Note that the sine keeps increasing as long as the cosine stays positive.) Hence, the set $\{x_1 > 0 \mid \cos x > 0 \text{ for } x \in [0, x_1]\}$ is bounded above by $x_0 + (1/a)\cos x_0$ and, consequently, has a least upper bound. We define

(143.8) $$\pi = 2 \sup\{x_1 > 0 \mid \cos x > 0 \quad \text{for all } x \in [0, x_1]\}.$$

Since the cosine is continuous, we have to have

(143.9) $$\cos\frac{\pi}{2} = 0 \quad \text{and} \quad \cos x > 0 \quad \text{for all } x \in \left[0, \frac{\pi}{2}\right).$$

It is now a simple matter to establish the $2\pi$-periodicity of the sine and cosine function. Differentiation of $\sin^2 x + \cos^2 x$ yields

$$(\sin^2 x + \cos^2 x)' = 2 \sin x \cos x - 2 \cos x \sin x = 0,$$

and, in view of (143.4)

(143.10) $$\sin^2 x + \cos^2 x = 1 \quad \text{for all } x \in \mathbf{R}.$$

Consequently,

(143.11) $$|\sin x| \leqslant 1, \quad |\cos x| \leqslant 1 \quad \text{for all } x \in \mathbf{R}.$$

† *Ludolph van Ceulen's* [1540–1610] only claim to fame is that he computed $\pi$ to 35 places, using Archimedes' method of approximating the circumference of the disk by the circumferences of inscribed and circumscribed polygons—see also section 4.43. His wife was so impressed by his mathematical prowess that she had his computational result chiseled into his gravestone. *Archimedes' number* would be a more appropriate name for $\pi$, because Archimedes was the first to develop an algorithm for its systematic computation. The symbol $\pi$, incidentally, was first used by L. Euler.

From (143.9) and (143.10), $\sin^2(\pi/2) = 1$, and from (143.7),

$$\cos \pi = \cos\left(\frac{\pi}{2} + \frac{\pi}{2}\right) = \cos^2 \frac{\pi}{2} - \sin^2 \frac{\pi}{2} = -1.$$

Hence, $\sin \pi = 0$ and, again from (143.7),

$$\cos 2\pi = \cos(\pi + \pi) = \cos^2 \pi - \sin^2 \pi = 1,$$

and, therefore, $\sin 2\pi = 0$.

Consequently, we have, from (143.6) and (143.7), that

$$\sin(x + 2\pi) = \sin x, \qquad \cos(x + 2\pi) = \cos x \qquad \text{for all } x \in \mathbf{R}.$$

We leave a discussion of the inverse sine and inverse cosine for the exercises (exercises 6, 7, 8, and 9). Instead, we shall proceed right away to a discussion of the tangent function and its inverse to obtain an algorithm for the computation of $\pi$ as defined in (143.8).

From (143.5) and (143.9), $\cos x > 0$ for all $x \in (-(\pi/2), \pi/2))$. Hence, the *tangent function* $\tan : (-(\pi/2), (\pi/2)) \to \mathbf{R}$ is well defined by

$$\text{(143.12)} \qquad \tan x = \frac{\sin x}{\cos x}, \qquad x \in \left(-\frac{\pi}{2}, \frac{\pi}{2}\right).$$

By elementary differentiation rules,

$$(\tan x)' = \frac{1}{\cos^2 x} > 0 \qquad \text{for all } x \in \left(-\frac{\pi}{2}, \frac{\pi}{2}\right).$$

Hence, from the Corollary to Lemma 37.1, and since $\cos^2 x = 1/(1 + \tan^2 x)$,

$$\text{(143.13)} \qquad (\text{Tan}^{-1}\, y)' = \cos^2(\text{Tan}^{-1}\, y) = \frac{1}{1 + y^2} \qquad \text{for all } y \in \mathbf{R}.$$

Since $\tan 0 = 0$,

$$\text{(143.14)} \qquad \text{Tan}^{-1}\, y = \int_0^y \frac{dt}{1 + t^2} \qquad \text{for all } y \in \mathbf{R}.$$

(The tangent function is also one-to-one from $((\pi/2), (3\pi/2))$ onto $\mathbf{R}$, or, for that matter, from $((\pi/2) + k\pi, (\pi/2) + (k + 1)\pi)$, $k \in \mathbf{Z}$, onto $\mathbf{R}$ and, accordingly, possesses inverse functions in each such case. The one which we consider here, $\text{Tan}^{-1}$, is generally referred to as the *principal value* of the inverse tangent. This archaic terminology dates back to the time when relations that satisfied all conditions in the definition of a function, except the one that requires $(x, y_1)$, $(x, y_2) \in f$ to imply that $y_1 = y_2$, were still considered to be (multiple-valued) functions, and when a clear distinction between *function* and *value of a function* was not considered important. See also Fig. 143.1 where the dotted line represents the graph of the inverse of $\tan : ((\pi/2), (3\pi/2)) \leftrightarrow \mathbf{R}$.)

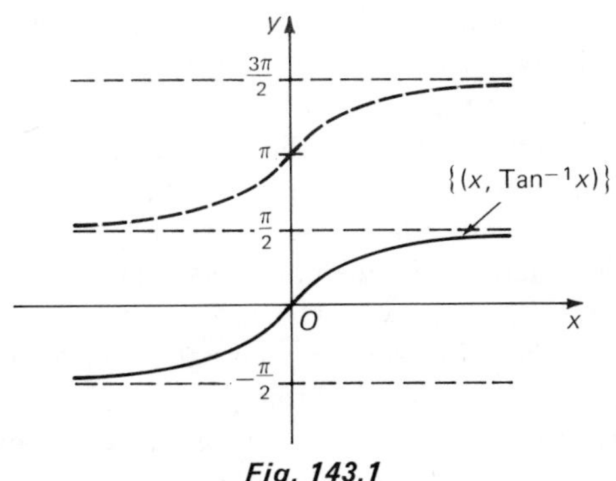


*Fig. 143.1*

From (143.7) and (143.9), $\sin^2(\pi/4) = \cos^2(\pi/4)$. Since $\sin x > 0$, $\cos x > 0$ for all $x \in (0, (\pi/2))$, we have

(143.15) $$\tan\frac{\pi}{4} = 1.$$

For $|t| < 1$,

$$\frac{1}{1+t^2} = \sum_{j=0}^{\infty}(-1)^j t^{2j}$$

and hence,

(143.16) $$\operatorname{Tan}^{-1} y = \int_0^y \sum_{j=0}^{\infty}(-1)^j t^{2j}\,dt = \sum_{j=0}^{\infty}(-1)^j \frac{y^{2j+1}}{2j+1} \qquad \text{for } |y| < 1.$$

Since the series on the right converges uniformly in [0, 1], by Theorem 140.5, and since $\operatorname{Tan}^{-1}$ is continuous, we see, as in example 1, section 14.141, that

$$\operatorname{Tan}^{-1} 1 = \sum_{j=0}^{\infty}(-1)^j \frac{1}{2j+1}.$$

This, together with (143.15) yields the following representation of $\pi$ by an infinite series

(143.17) $$\pi = 4(1 - \tfrac{1}{3} + \tfrac{1}{5} - \tfrac{1}{7} + \tfrac{1}{9} - \cdots) = 3.141592653\ldots,$$

which was discovered in 1671 by Gregory† and, quite independently, in 1674 by Leibniz.‡ This series, because of its extremely slow convergence, is of little practical value for the computation of $\pi$. Still, a fast and well-fed computer, using (143.17) and given enough time, is capable of grinding out $\pi$ to any number of decimal places. (Faster converging series for the computation of $\pi$ are developed in exercise 5 and in section 14.145, exercise 5.)

† *James Gregory*, 1638–1675.
‡ *Gottfried Wilhelm Leibniz*, 1646–1716.

In 1882, Lindemann published a theorem (the proof of which was simplified and augmented by K. Weierstrass in 1895) which implies that $E(x)$, sinh $x$, cosh $x$, sin $x$, cos $x$, tan $x$ are transcendental numbers for all rational or algebraic irrational $x$ other than zero. (See also definition 12.1.) Also, $\log x$ is transcendental for all positive rational and algebraic irrational values of $x$ other than 1; and a similar situation prevails among the inverse hyperbolic functions and the inverse trigonometric functions. If one considers that the point set $(\mathbf{Q} \cup \mathbf{I}_a) \times (\mathbf{Q} \cup \mathbf{I}_a)$ of "algebraic points" is everywhere dense in $\mathbf{R}^2$, it seems truly incredible that the graphs of these functions manage to wind their way through the plane avoiding all but one of these algebraic points. All the available proofs of the above quoted theorem of Lindemann are elementary in nature in that they do not require sophisticated methods.† Note that Lindemann's theorem implies, in particular, that $e = E(1)$ and $\pi = 4 \operatorname{Tan}^{-1} 1$ are transcendental numbers.

One calls the exponential function, the hyperbolic functions, the trigonometric functions, and their inverses, the *elementary transcendental functions*. They are called transcendental because of their above mentioned property‡ and they are called *elementary* because everybody knows them and all mathematicians have known them since the eighteenth century.

## 14.143 Exercises

1. By taking the appropriate Cauchy products, find the series representation of $\sin x \sin y$ and $\cos x \cos y$. Use this result to derive (143.7).
2. Evaluate the eighth partial sum of the series in (143.17).
3. Use (143.6) and (143.7) to show that

$$\tan 2x = \frac{2 \tan x}{1 - \tan^2 x} \qquad \text{and} \qquad \tan(x - y) = \frac{\tan x - \tan y}{1 + \tan x \tan y}.$$

4. Let $x = \operatorname{Tan}^{-1} \frac{1}{5}$. Find $\tan 2x$, $\tan 4x$, and show that $4 \operatorname{Tan}^{-1} \frac{1}{5} - \operatorname{Tan}^{-1} \frac{1}{239} = \pi/4$.
5. Use the result of exercise 4 and (143.16) to find the first four decimal places of $\pi$. (This method for the computation of $\pi$ was first discovered by John Machin§ in 1706.)
6. Show that $\sin : [-(\pi/2), (\pi/2)] \to [-1, 1]$ has a continuous inverse on $[-1, 1]$. Denote the inverse by $\operatorname{Sin}^{-1}$ and show that

$$\operatorname{Sin}^{-1} y = \int_0^y \frac{dt}{\sqrt{1 - t^2}} \qquad \text{for } |y| < 1.$$

† See, for example: I. Niven, *Irrational Numbers*, The Carus Mathematical Monographs, No. 11, The Mathematical Association of America, 1956, p. 131.

‡ For a more sophisticated definition of transcendental functions, see R. Godement, *Algebra* (Boston: Houghton Mifflin Company, 1968), p. 417.

§ *John Machin*, 1680–1751.

7. Show that $\cos : [0, \pi] \to [-1, 1]$ has a continuous inverse on $[-1, 1]$. Denote the inverse by $\text{Cos}^{-1}$ and show that

$$\text{Cos}^{-1} y = -\int_0^y \frac{dt}{\sqrt{1-t^2}} + \frac{\pi}{2} \qquad \text{for } |y| < 1.$$

8. Show that $\text{Sin}^{-1} y + \text{Cos}^{-1} y = \pi/2$ for all $y \in [-1, 1]$.
9. If the inverse function $f^{-1}$ of $f: U \leftrightarrow V$, given by $f(x) = \sum_{j=0}^{\infty} a_j x^j$, can be represented by a power series $f^{-1}(y) = \sum_{j=0}^{\infty} b_j y^j$ in $V$, then,

$$f^{-1}(f(x)) = \sum_{i=0}^{\infty} b_i \left( \sum_{j=0}^{\infty} a_j x^j \right)^i = x$$

is an identity. Use this fact to find the first two terms of the power series for $\text{Sin}^{-1} y$, if there is one.
10. Find $\lim_0 (\sin x)/x$.
11. Given the functions $S, C : \mathbf{R} \to \mathbf{R}$ which have the properties
    (I) $S(x-y) = S(x)C(y) - C(x)S(y)$,
    (II) $C(x-y) = C(x)C(y) + S(x)S(y)$,
    (III) $\lim_{0+0} \frac{S(x)}{x} = 1$.

    *Prove*:
    (a) $S(0) = 0$, $C(0) = 1$, (b) $S^2(x) + C^2(x) = 1$,
    (c) $S(-x) = -S(x)$, (d) $\lim_0 \frac{S(x)}{x} = 1$,
    (e) $S, C$ are continuous on $\mathbf{R}$, (f) $S'(x) = C(x)$, $C'(x) = -S(x)$.
12. Show that the functions $S, C$ in exercise 11 are uniquely determined by their properties (I), (II), and (III).
13. Show that $S(x) = \sin x$, $C(x) = \cos x$ where $\sin x, \cos x$ are defined as in (143.1).
14. Show: it is impossible for $E(x)$ to satisfy for all $x \in \mathbf{R}$ an equation of the type

$$p_0(x)(E(x))^n + p_1(x)(E(x))^{n-1} + \cdots + p_{n-1}(x)E(x) + p_n(x) = 0,$$

where the $p_j(x)$ are polynomials with rational coefficients which have no common factors.
15. Same as in exercise 14, for $\sin x$.
16. Show that $f: (-1, 1) \to \mathbf{R}$, defined by

$$f(x) = \frac{1}{x^2+1} \sqrt[7]{\frac{2x-3}{x^5-1} + \sqrt[3]{4 + \frac{6}{(x-1)^2}}},$$

satisfies an equation of the type in exercise 14.

## 14.144 Manipulations with Power Series

Suppose that $\sum_{j=0}^{\infty} a_j x^j$ has the radius of convergence $r > 0$ (possibly $r = \infty$). By Theorem 141.2,

$$f(x) = \sum_{j=0}^{\infty} a_j x^j$$

is continuous and bounded in every bounded interval $\mathfrak{J} \subseteq (-r, r)$.

We also have, from Theorem 141.2, that

$$f'(x) = \sum_{j=1}^{\infty} j a_j x^{j-1} \qquad \text{for all } x \in (-r, r).$$

Since the differentiated series has the same properties as the original one, we obtain

$$f''(x) = \sum_{j=2}^{\infty} j(j-1) a_j x^{j-2} \qquad \text{for all } x \in (-r, r).$$

We continue this process and obtain, after $k$ steps,

$$f^{(k)}(x) = \sum_{j=k}^{\infty} j(j-1)\cdots(j-k+1) a_j x^{j-k} \qquad \text{for all } x \in (-r, r).$$

For $x = 0$,

$$f^{(k)}(0) = k!\, a_k.$$

If we also note that $f(0) = a_0$ and denote $f(x)$ by $f^{(0)}(x)$, we may summarize these results in the following theorem.

***Theorem 144.1***

If $f(x) = \sum_{j=0}^{\infty} a_j x^j$ for all $x \in (-r, r)$, then $f$ possesses derivatives of all orders in $(-r, r) : f \in C^{\infty}((-r, r))$, and

$$a_0 = f^{(0)}(0), \qquad a_1 = f'(0), \qquad a_2 = \frac{f''(0)}{2!}, \qquad \ldots, \qquad a_k = \frac{f^{(k)}(0)}{k!}, \ldots,$$

so that, in effect,

$$f(x) = \sum_{j=0}^{\infty} f^{(j)}(0) \frac{x^j}{j!}. \tag{144.1}$$

This theorem has, within certain limitations, a converse which we shall discuss in section 14.145.

***Corollary to Theorem 144.1***

If $\sum_{j=0}^{\infty} a_j x^j = \sum_{j=0}^{\infty} b_j x^j$ for all $x \in (-r, r)$, then $a_j = b_j$ for all $j = 0, 1, 2, \ldots$.

*Proof*

Let $\sum_{j=0}^{\infty} a_j x^j = f(x)$, $\sum_{j=0}^{\infty} b_j x^j = g(x)$. By hypothesis $f(x) = g(x)$ for all $x \in (-r, r)$. Hence, $f^{(k)}(x) = g^{(k)}(x)$ for all $x \in (-r, r)$ and all $k \in \mathbf{N}$. In particular, $f^{(k)}(0) = g^{(k)}(0)$ for all $k \in \mathbf{N}$. Hence, by Theorem 144.1, $a_k = f^{(k)}(0)/k! = g^{(k)}(0)/k! = b_k$ for all $k \in \mathbf{N}$.

**Corollary to the Corollary to Theorem 144.1**

If $\sum_{j=0}^{\infty} a_j x^j = 0$ for all $x \in (-r, r)$ for some $r > 0$, then $a_j = 0$ for all $j = 0, 1, 2, 3, \ldots$.

---

▲ *Example 1*

If $f(x) = \sum_{j=0}^{\infty} a_j x^j$ for all $x \in \mathbf{R}$ and if $f(x) = E(x)$ for all $x \in (-r, r)$ for some $r > 0$, then $f(x) = E(x)$ because, for $x \in (-r, r)$, $f(x) - E(x) = \sum_{j=0}^{\infty} (a_j - 1/j!) x^j = 0$. In words: The only function that can be represented by a power series that converges everywhere and is equal to the exponential function in some interval $(-r, r)$, no matter how small $r > 0$ is, is the *exponential function* itself.

---

Noting that a power series converges absolutely for all $|x| < r$, $r$ being the radius of convergence, we obtain, from Theorem 139.1, the following theorem about the Cauchy product of two power series.

**Theorem 144.2**

If $\sum_{j=0}^{\infty} a_j x^j$ and $\sum_{j=0}^{\infty} b_j x^j$ converge in $(-r, r)$, then

$$\left(\sum_{j=0}^{\infty} a_j x^j\right)\left(\sum_{j=0}^{\infty} b_j x^j\right) = \sum_{p=0}^{\infty} (a_0 b_p + a_1 b_{p-1} + \cdots + a_{p-1} b_1 + a_p b_0) x^p \tag{144.2}$$

for all $x \in (-r, r)$.

Using (144.2) as a point of departure, we develop a process for the division of power series.

Let $\sum_{j=0}^{\infty} b_j x^j$, $\sum_{j=0}^{\infty} c_j x^j$ converge in $(-r, r)$. If $b_0 \neq 0$, then there is an interval $(-\rho, \rho) \subseteq (-r, r)$ where $(\sum_{j=0}^{\infty} c_j x^j / \sum_{j=0}^{\infty} b_j x^j)$ is defined because $\sum_{j=0}^{\infty} b_j x^j$ represents a continuous function on $(-r, r)$ which does not vanish at $x = 0$. If there is a power series $\sum_{j=0}^{\infty} a_j x^j$ which converges in $(-\rho, \rho)$ and for which

$$\left(\sum_{j=0}^{\infty} a_j x^j\right)\left(\sum_{j=0}^{\infty} b_j x^j\right) = \sum_{j=0}^{\infty} c_j x^j \tag{144.3}$$

for all $x \in (-\rho, \rho)$, then,

(144.4) $$\sum_{j=0}^{\infty} a_j x^j = \frac{\sum_{j=0}^{\infty} c_j x^j}{\sum_{j=0}^{\infty} b_j x^j} \quad \text{for all } x \in (-\rho, \rho).$$

(144.3) yields an algorithm for the determination of the coefficients $a_j$. We have, from (144.2), that

$$a_0 b_p + a_1 b_{p-1} + \cdots + a_{p-1} b_1 + a_p b_0 = c_p, \qquad p = 0, 1, 2, 3, \ldots,$$

that is,

(144.5) $$\begin{aligned} a_0 b_0 &= c_0, \\ a_0 b_1 + a_1 b_0 &= c_1, \\ a_0 b_2 + a_1 b_1 + a_2 b_0 &= c_2, \\ &\vdots \end{aligned}$$

Since $b_0 \neq 0$, we may solve these equations successively for $a_0, a_1, a_2, \ldots$. If the resulting series $\sum_{j=0}^{\infty} a_j x^j$ converges in some interval $(-\rho, \rho) \subseteq (-r, r)$, then the validity of (144.4) is assured.

---

▲ *Example 2*

Let us assume that $\tan x = (\sin x)/(\cos x)$ may be represented by a power series near $x = 0$. With the above notation, we have

$$b_0 = 1,\ b_1 = 0,\ b_2 = -\frac{1}{2!},\ b_3 = 0,\ b_4 = \frac{1}{4!},\ b_5 = 0,\ b_6 = -\frac{1}{6!},\ b_7 = 0,\ \ldots$$

$$c_0 = 0,\ c_1 = 1,\ c_2 = 0,\ c_3 = -\frac{1}{3!},\ c_4 = 0,\ c_5 = \frac{1}{5!},\ c_6 = 0,\ c_7 = -\frac{1}{7!},\ \ldots$$

Hence, we obtain from (144.5), carried up to $c_7$,

$$a_0 = 0, \quad a_1 = 1, \quad a_2 = 0, \quad a_3 = \tfrac{1}{3}, \quad a_4 = 0, \quad a_5 = \tfrac{2}{15}, \quad a_6 = 0, \quad a_7 = \tfrac{17}{315},$$

and therefore,

$$\tan x = x + \tfrac{1}{3} x^3 + \tfrac{2}{15} x^5 + \tfrac{17}{315} x^7 + \cdots.$$

Without going into the gory details, we merely wish to mention that one can show that

(144.6) $$\tan x = \sum_{j=1}^{\infty} (-1)^{j+1} \frac{2^{2j}(2^{2j} - 1) B_{2j}}{(2j)!} x^{2j-1},$$

where the $B_k$ are the *Bernoulli*† *numbers* which are defined by the recursion formula

† *Jakob Bernoulli*, 1654–1705.

(144.7)

$$B_0 = 1, \quad \frac{1}{k!}\cdot\frac{B_0}{0!} + \frac{1}{(k-1)!}\cdot\frac{B_1}{1!} + \frac{1}{(k-2)!}\cdot\frac{B_2}{2!} + \cdots + \frac{1}{1!}\cdot\frac{B_{k-1}}{(k-1)!} = 0.$$

It can be shown that the radius of convergence of the series for the tangent in (144.6) is $(\pi/2)$. (See exercise 27, section 14.145.)

---

## 14.144 Exercises

1. Find $a_0, a_1, a_2, \ldots$ such that

$$\left(\sum_{j=0}^{\infty} a_j x^j\right)\left(\sum_{j=0}^{\infty} \frac{x^j}{j!}\right) = 1 \qquad \text{for all } x \in \mathbf{R}.$$

2. Show that $\sum_{j=0}^{\infty} a_j x^j = E(-x)$ where the $a_j$ are the coefficients of exercise 1.
3. Use Theorem 144.1 and the definition of the sine function in (143.1) to show that

$$\sin^{(k)}(0) = \begin{cases} 0 & \text{for } k = 2\kappa, \\ (-1)^{\kappa} & \text{for } k = 2\kappa + 1. \end{cases}$$

4. Find the first four terms in the power series for

$$\frac{1}{1 + \dfrac{x}{2!} + \dfrac{x^2}{3!} + \dfrac{x^3}{4!} + \cdots} = a_0 + a_1 x + a_2 x^2 + a_3 x^3 + \cdots,$$

and show that $k!a_k = B_k$ for $k = 0, 1, 2, 3$, where the $B_k$ are the Bernoulli numbers which are defined in (144.7).
5. Find the first three terms of $1/(\cos x - \sin x) = \sum_{j=0}^{\infty} a_j x^j$.
6. Show that for $x \in [0, 1]$

$$(\log(1 + x))^2 = \sum_{p=2}^{\infty} (-1)^p \left(\frac{1}{p-1} + \frac{1}{(p-2)2} + \cdots + \frac{1}{2(p-2)} + \frac{1}{p-1}\right) x^p.$$

7. Verify (144.6). (*Caution*: This exercise demands time, patience, skill, and cunning.)
8. Derive $\cos^2 x = (1 + \cos 2x)/2$ from (143.6) and (143.7).
9. Assume that $1/(\cos^2 x)$ has a power series representation around $x = 0$ and find the first three terms of that power series.
10. Find the first five terms of the power series for $\tan x$ from exercise 9 and $(\tan x)' = 1/(\cos^2 x)$.
11. The hyperbolic sine and cosine are defined by

$$\sinh x = \tfrac{1}{2}(E(x) - E(-x)), \qquad \cosh x = \tfrac{1}{2}(E(x) + E(-x)).$$

Find the power series representation of $\sinh x$ and $\cosh x$ and their radius of convergence.

12. Establish the existence of the inverse of the hyperbolic sine, and find the first two terms of its power series representation by the technique of exercise 9, section 14.143.
13. Show: any function $f: \mathbf{R} \to \mathbf{R}$, which is given by $f(x) = \sum_{j=0}^{\infty} a_j x^j$ for all $x \in \mathbf{R}$ and which satisfies $f'(x) = f(x)$ for all $x \in \mathbf{R}$, $f(0) = 1$, is the *exponential function* defined in (142.1).
14. Show: any function $f: \mathbf{R} \to \mathbf{R}$, which is given by $f(x) = \sum_{j=0}^{\infty} a_j x^j$ for all $x \in \mathbf{R}$ and which satisfies $f''(x) = -f(x)$ for all $x \in \mathbf{R}$, is the *sine function* if $f(0) = 0$ and the *cosine function* if $f(0) = 1$.

## 14.145 Taylor Series

We have seen in the preceding section that a function which is represented by a power series in $(-r, r)$ for some $r > 0$ possesses derivatives of all orders in $(-r, r)$. In this section we shall discuss the converse problem of representing a function which has derivatives of all orders by a power series.

Suppose that $f$ possesses derivatives of all orders in some interval $\mathfrak{J}: f \in C^{\infty}(\mathfrak{J})$, and suppose that $a, b \in \mathfrak{J}$, $a < b$. We obtain, from Theorem 52.3, that

$$f(b) - f(a) = \int_a^b f'(x)\,dx.$$

Integration by parts (Corollary 1 to Theorem 52.3) yields, with $g(x) = -(b - x)$,

$$\int_a^b f'(x)\,dx = f(a)(b - a) + \int_a^b f''(x)(b - x)\,dx.$$

Integrating by parts again, with $g(x) = -(b - x)^2/2$, we obtain

$$\int_a^b f''(x)\,dx = f'(a)\frac{(b-a)^2}{2} + \int_a^b f'''(x)\frac{(b-x)^2}{2}\,dx, \ldots .$$

We continue in this manner and obtain, after $(n - 1)$ steps,

$$(145.1) \qquad f(b) = f(a) + \frac{f'(a)}{1!}(b-a) + \frac{f''(a)}{2!}(b-a)^2 + \cdots$$
$$+ \frac{f^{(n-1)}(a)}{(n-1)!}(b-a)^{n-1} + \int_a^b f^{(n)}(x)\frac{(b-x)^{n-1}}{(n-1)!}\,dx.$$

Since $b - x \geqslant 0$ for all $x \in [a, b]$, we have

$$(145.2) \quad \int_a^b f^{(n)}(x)\frac{(b-x)^{n-1}}{(n-1)!}\,dx = f^{(n)}(\xi_n)\int_a^b \frac{(b-x)^{n-1}}{(n-1)!}\,dx = f^{(n)}(\xi_n)\frac{(b-a)^n}{n!}$$

for some $\xi_n \in [a, b]$. (See Theorem 51.3.)

We summarize this result in the following theorem.

***Theorem 145.1*** *(Taylor's Formula)*

If $f \in C^{\infty}(\mathfrak{I})$ where $\mathfrak{I} \subseteq \mathbf{R}$ is an interval, and if $a, b \in \mathfrak{I}$, $a < b$, then there is, for every $n \in \mathbf{N}$, a $\xi_n \in [a, b]$ such that

$$f(b) = f(a) + \frac{f'(a)}{1!}(b-a) + \frac{f''(a)}{2!}(b-a)^2 + \cdots + \frac{f^{(n-1)}(a)}{(n-1)!}(b-a)^{n-1} + \frac{f^{(n)}(\xi_n)}{n!}(b-a)^n. \tag{145.3}$$

This is called *Taylor's*† *formula.*

(Note that (145.3) may be derived under the weaker hypothesis that $f \in C^{(n-1)}[a, b]$ and $f^{(n-1)} \in C'(a, b)$. (See exercise 7, section 3.37.)

$$R_n = \frac{f^{(n)}(\xi_n)}{n!}(b-a)^n \tag{145.4}$$

is called the $n$th *remainder*. Specifically, this form of the remainder is called *Lagrange's form of the remainder*. Another form of the remainder is obtained from (145.1):

$$R_n = \int_a^b f^{(n)}(x)\frac{(b-x)^{n-1}}{(n-1)!}\,dx. \tag{145.5}$$

Yet another form of the remainder, due to Cauchy, is given by

$$R_n = \frac{f^{(n)}(\xi_n)}{(n-1)!}(b-\xi_n)^{n-1}(b-a), \tag{145.6}$$

where $\xi_n \in (a, b)$ is, in general, different from the $\xi_n \in [a, b]$ in (145.4).

In order to establish the validity of (145.6), we consider a function $\phi : \mathfrak{I} \to \mathbf{R}$ which is defined by

$$\phi(x) = f(b) - f(x) - \frac{f'(x)}{1!}(b-x) - \cdots - \frac{f^{(n-1)}(x)}{(n-1)!}(b-x)^{n-1}.$$

Then,

$$\phi'(x) = -\frac{f^{(n)}(x)}{(n-1)!}(b-x)^{n-1},$$

and we obtain, from the mean-value theorem,

$$\phi(a) = \phi(a) - \phi(b) = \phi'(\xi_n)(a-b) = \frac{f^{(n)}(\xi_n)}{(n-1)!}(b-\xi_n)^{n-1}(b-a)$$

for some $\xi_n \in (a, b)$. Since $\phi(a) = R_n$, (145.6) follows readily.

† *Brook Taylor*, 1685–1731.

We replace $b$ by $x$ (145.3), note that this formula holds as well for $b < a$, observe that

$$R_n = f(x) - f(a) - \frac{f'(a)}{1!}(x-a) - \cdots - \frac{f^{(n-1)}(a)}{(n-1)!}(x-a)^{n-1},$$

and obtain, from Theorem 140.1, that $f$ may be represented by a power series around $x = a$ if $\lim_\infty R_n = 0$ for all $x \in \mathfrak{J}$.

***Theorem 145.2*** *(Taylor Series)*

If $f \in C^\infty(\mathfrak{J})$, if $a \in \mathfrak{J}$, and if $\lim_\infty R_n = 0$ for all $x \in \mathfrak{J}$, where $R_n$ is given in (145.4), (145.5), or (145.6), then

$$f(x) = \sum_{j=0}^{\infty} \frac{f^{(j)}(a)}{j!}(x-a)^j \qquad \text{for all } x \in \mathfrak{J}, \tag{145.7}$$

where $f^{(0)}(a) = f(a)$. This is called the *Taylor series* of $f$ at $a$.

A function which satisfies the hypotheses of this theorem for an *open* interval $\mathfrak{J}$ is called *analytic* at $a$.

---

▲ *Example 1* *(Binomial Series)*

Let $f : (-1, 1) \to \mathbf{R}$ be defined by

$$f(x) = (1 + x)^\alpha, \qquad \alpha \in \mathbf{R}.$$

We have, from (142.14),

$$\begin{aligned} f'(x) &= \alpha(1+x)^{\alpha-1}, \\ f''(x) &= \alpha(\alpha-1)(1+x)^{\alpha-2}, \\ &\vdots \\ f^{(n)}(x) &= \alpha(\alpha-1)\cdots(\alpha-n+1)(1+x)^{\alpha-n}. \\ &\vdots \end{aligned}$$

If $\alpha = 0$ or $\alpha \in \mathbf{N}$, then this process breaks up after finitely many steps, and (145.7) yields the binomial formula. For all other $\alpha \in \mathbf{R}$, we obtain, from (145.6), with $a = 0$:

$$R_n = \frac{\alpha(\alpha-1)\cdots(\alpha-n+1)}{(n-1)!}(1+\xi_n)^{\alpha-n}(x-\xi_n)^{n-1}x,$$

where $\xi_n$ lies between 0 and $x$.

Let $\xi_n = \theta_n x$. Then $0 < \theta_n < 1$, and we obtain

$$R_n = \frac{\alpha(\alpha-1)\cdots(\alpha-n+1)}{(n-1)!}(1+\theta_n x)^{\alpha-1}\frac{(1-\theta_n)^{n-1}}{(1+\theta_n x)^{n-1}}x^n.$$

Since $0 < (1 - \theta_n)/(1 + \theta_n x) < 1$ for all $|x| < 1$, and since

$$(1 + \theta_n x)^{\alpha - 1} < \begin{cases} (1 + |x|)^{\alpha - 1} & \text{for all } \alpha > 1, \\ (1 - |x|)^{\alpha - 1} & \text{for all } \alpha < 1, \end{cases}$$

we obtain

$$0 \leqslant |R_n| \leqslant \left| \frac{\alpha(\alpha - 1) \cdots (\alpha - n + 1)}{(n - 1)!} \right| |1 \pm |x||^{a-1} |x|^n = S_n.$$

Since $\lim_\infty |(S_{n+1})/S_n| = x$, we see, from the limit test, that $\sum_{j=1}^\infty S_j$ converges for all $|x| < 1$ and hence, $\lim_\infty S_n = 0$ for all $|x| < 1$. Therefore,

$$\lim_\infty |R_n| = 0 \qquad \text{for all } |x| < 1,$$

and we obtain, from Theorem 145.2, that, for $\alpha \neq 0$, $\alpha \notin \mathbf{N}$,

(145.8)

$$(1 + x)^\alpha = 1 + \alpha x + \frac{\alpha(\alpha - 1)}{2!} x^2 + \frac{\alpha(\alpha - 1)(\alpha - 2)}{3!} x^3 + \cdots \qquad \text{for all } |x| < 1.$$

This is called the *binomial series*. (See also exercises 16 through 20.)

We obtain from (145.8), in particular, that

$$\text{(145.9)} \qquad \frac{1}{\sqrt{1 - t^2}} = (1 - t^2)^{-1/2} = 1 + \frac{1}{2} t^2 + \frac{3}{8} t^4 + \frac{5}{16} t^6 + \cdots$$

$$= 1 + \sum_{j=1}^\infty \frac{1 \cdot 3 \cdot 5 \cdots (2j - 1)}{j! 2^j} t^{2j}.$$

---

We have pointed out, in section 14.144, that Theorem 144.1 has, within certain limitations, a converse. That a strict converse cannot possibly hold can be seen from the following example.

---

▲ *Example 2*

Let $f : \mathbf{R} \to \mathbf{R}$ be defined by

$$\text{(145.10)} \qquad f(x) = \begin{cases} E\left(-\dfrac{1}{x^2}\right) & \text{for } x \neq 0, \\ 0 & \text{for } x = 0. \end{cases}$$

(See Fig. 145.1.)

From (142.1), we obtain, for $x \neq 0$:

$$\left| x^n E\left(\frac{1}{x^2}\right) \right| = \sum_{j=0}^\infty \frac{1}{j! |x|^{2j-n}} \geqslant \begin{cases} \dfrac{1}{(v + 1)! x^2} & \text{for } n = 2v, \\ \dfrac{1}{(v + 1)! |x|} & \text{for } n = 2v + 1. \end{cases}$$

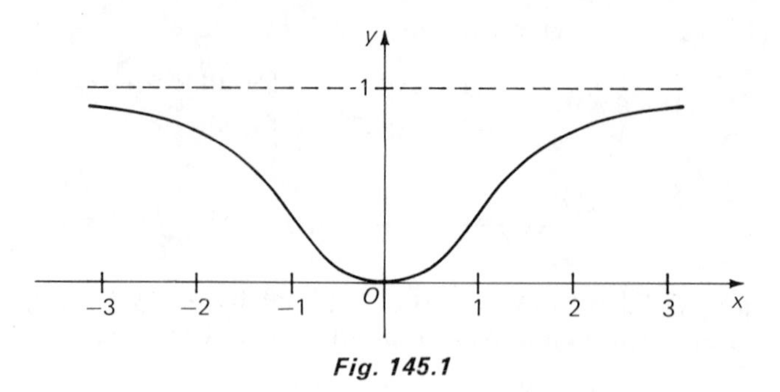


*Fig. 145.1*

Hence, $\lim_{x\to 0} 1/(x^n E(1/x^2)) = 0$ for all $n \in \mathbf{N}$, and we have

$$f^{(k)}(0) = \lim_{x\to 0} \frac{f^{(k-1)}(x)}{x} = 0 \qquad \text{for all } k \in \mathbf{N}.$$

Since $f^{(k)}$ exists for all $x \neq 0$ also, we have $f \in C^\infty(\mathbf{R})$.

If the converse of Theorem 144.1 were true, we would have, from (145.7), that

$$f(x) = 0 + 0 + 0 + \cdots \qquad \text{for all } x \in \mathbf{R},$$

which is quite absurd.

---

The preceding example demonstrates that the mere existence of all derivatives on some interval does not suffice for a given function to have a representation by a Taylor series. However, if all derivatives are uniformly bounded on $\mathfrak{J}$, that is, if, for some $M > 0$,

$$|f^{(k)}(x)| \leqslant M \qquad \text{for all } x \in \mathfrak{J} \text{ and all } k \in \mathbf{N},$$

then we obtain, from (145.4), that $\lim_\infty R_n = 0$ for all $x \in \mathfrak{J}$ and hence, by Theorem 145.2:

***Theorem 145.3***

If $f \in C^\infty(\mathfrak{J})$, if $|f^{(k)}(x)| \leqslant M$ for all $x \in \mathfrak{J}$ and all $k \in \mathbf{N}$ for some $M > 0$, and if $a \in \mathfrak{J}$, then

$$f(x) = \sum_{j=0}^{\infty} \frac{f^{(j)}(a)}{j!} (x-a)^j$$

for all $x \in \mathfrak{J}$.

(Another sufficient condition for the representation of a function by a Taylor series is developed in exercises 14 and 15.)

From Theorem 144.1, we obtain the link between power series and Taylor series.

***Theorem 145.4***

If $f(x) = \sum_{j=0}^{\infty} a_j(x-a)^j$ for all $x \in (a-r, a+r)$ where $r > 0$ is the radius of convergence, then $\sum_{j=0}^{\infty} a_j(x-a)^j$ is the Taylor series of $f$ in $(a-r, a+r)$.

We see from this theorem that the power series for $E(x)$, $\sin x$, $\cos x$, $\log(1+x)$, $\mathrm{Tan}^{-1} x$, etc., which we discussed in sections 14.142 and 14.143, are the Taylor series of these functions in the interval of their convergence.

## 14.145 Exercises

1. Find the Taylor series for the sine function around $x = (\pi/2)$.
2. Find the first four terms of the Taylor series for the tangent function around $x = 0$ and compare your result with (144.6).
3. Use the representation

$$\mathrm{Sin}^{-1} y = \int_0^y \frac{dt}{\sqrt{1-t^2}}$$

of the inverse sine (exercise 6, section 14.143), and (145.9), to find a power series representation of $\mathrm{Sin}^{-1} y$ for $|y| < 1$.

4. From (143.6) and (143.7), show that $\sin(\pi/6) = \frac{1}{2}$.
5. Show that

$$\pi = 3 + 3\sum_{j=1}^{\infty} \frac{1 \cdot 3 \cdot 5 \cdots (2j-1)}{2 \cdot 4 \cdot 6 \cdots 2j} \frac{1}{2^{2j}(2j+1)}.$$

6. Use the result in exercise 5 to compute the first two decimal places of $\pi$.
7. *Prove*: If $\sum_{j=1}^{\infty} b_j$ converges and $(b_{j+1})/b_j \geqslant (a_{j+1})/a_j$, $a_j, b_j > 0$, then $\sum_{j=1}^{\infty} a_j$ converges.
8. *Prove*: If $\lim_{\infty} j(1 - |(a_{j+1})/a_j|) > 1$, then $\sum_{j=1}^{\infty} a_j$ converges absolutely.
9. Use the test in exercise 8 to show that

$$\sum_{j=1}^{\infty} \frac{1 \cdot 3 \cdot 5 \cdots (2j-1)}{2 \cdot 4 \cdot 6 \cdots 2j} \cdot \frac{1}{2j+1}$$

converges.

10. Show that the series for $\mathrm{Sin}^{-1} y$ in exercise 3 converges uniformly in $[-1, 1]$.
11. Show that

$$\lim_{x \to 1-0} \mathrm{Sin}^{-1} y = 1 + \sum_{j=1}^{\infty} \frac{1 \cdot 3 \cdot 5 \cdots (2j-1)}{2 \cdot 4 \cdot 6 \cdots 2j(2j+1)}.$$

12. Show that

$$\pi = 2 + 2\sum_{j=1}^{\infty} \frac{1 \cdot 3 \cdot 5 \cdots (2j-1)}{2 \cdot 4 \cdot 6 \cdots 2j(2j+1)}.$$

13. Let $f : \mathbf{R} \to \mathbf{R}$ be defined by (145.10). Show that there is *no* interval containing $x = 0$ such that the derivatives $f^{(k)}$ of even order are uniformly bounded on that interval for all $k \in \mathbf{N}$.
14. Show that the remainder in Taylor's formula may be written as
$$R_n = \frac{(b-a)^n}{(n-1)!} \int_0^1 f^{(n)}((1-t)a + tb)(1-t)^{n-1}\,dt.$$
15. *Prove*: If $f \in C^\infty[0,b]$, $f(x) \geqslant 0$, $f^{(k)}(x) \geqslant 0$ for all $k \in \mathbf{N}$ and all $x \in [0,b]$, then $f$ is represented by its Taylor series in $[0,b)$.
16. Use the test in exercise 8 to show that the binomial series in (145.8) converges for $x = -1$ and $\alpha > 0$.
17. Use Theorem 140.5 to show that the binomial series in (145.8) converges uniformly on $[-1,0]$ for $\alpha > 0$.
18. Show that the representation of $(1+x)^\alpha$ in (145.8) is valid for $x = -1$ for $\alpha > 0$.
19. Show that the binomial series in (145.8) is, for $x = 1$, an alternating series from a certain term on, depending on $\alpha$, and that it converges for $\alpha > -1$.
20. Show that the representation of $(1+x)^\alpha$ in (145.8) is valid for $x = 1$ for $\alpha > -1$.
21. Prove that $e$ is irrational.
22. Prove that $\pi$ is irrational.
23. Show: If $|f(x)| < 1$ for all $x \in (-r,r)$, then $\sum_{j=0}^\infty (f(x))^j$ converges for all $x \in (-r,r)$.
24. Show: If $\left|\sum_{j=0}^\infty a_i x^j\right| < 1$ for all $x \in (-r,r)$, then $\sum_{i=0}^\infty \left(\sum_{j=0}^\infty a_i x^j\right)^i$ converges for all $x \in (-r,r)$ and $\sum_{i=0}^\infty \left(\sum_{j=0}^\infty a_j x^j\right)^i = \sum_{i=0}^\infty b_i x^i$ for all $x \in (-r,r)$ and some $b_0, b_1, b_2, \ldots$.
25. Let $f(x) = 1/\cos x$ for $-(\pi/2) < x < (\pi/2)$. Show, by induction, that $f^{(k)}(0) \geqslant 0$ for all $k \in \mathbf{N}$.
26. Show that $f(x) = 1/\cos x$ has a convergent Taylor series in $(-(\pi/2), (\pi/2))$.
27. Show that the Taylor series for $\tan x$ converges in $(-(\pi/2), (\pi/2))$.

# Appendix 1

## Cast of Characters (in Order of Their Appearance)

## The Greek Alphabet

| | | | |
|---|---|---|---|
| $\alpha, A$ | alpha | $\nu, N$ | nu |
| $\beta, B$ | beta | $\xi, \Xi$ | xi |
| $\gamma, \Gamma$ | gamma | $o, O$ | omicron |
| $\delta, \Delta$ | delta | $\pi, \Pi$ | pi |
| $\varepsilon, E$ | epsilon | $\rho, P$ | rho |
| $\zeta, Z$ | zeta | $\sigma, \varsigma, \Sigma$ | sigma |
| $\eta, H$ | eta | $\tau, T$ | tau |
| $\theta, \Theta$ | theta | $\upsilon, \Upsilon$ | upsilon |
| $\iota, I$ | iota | $\phi, \Phi$ | phi |
| $\kappa, K$ | kappa | $\chi, X$ | chi |
| $\lambda, \Lambda$ | lambda | $\psi, \Psi$ | psi |
| $\mu, M$ | mu | $\omega, \Omega$ | omega |

# Appendix 2

## Answers and Hints†

### Chapter 1

***Section 1.1,*** page 5. **1.** There is a set $A$ that does not contain the empty set. **2.** (a) If $x \in c(c(A))$, then $x \notin c(A)$ and hence $x \in A$. If $x \in A$, then $x \notin c(A)$ and hence, $x \in c(c(A))$. **4.** $\{6\}$. **6.** (c) If $x$ is an element of $A$ and $B$ or $C$, then $x$ is an element of $A$ and of $B$ or of $A$ and of $C$. **7.** If $x \in A \backslash B$, then $x \in A$ and $x \notin B$; that is, $x \in c(B)$. **8.** $(A \cup B)\backslash C = (A \cup B) \cap c(C) = (A \cap c(C)) \cup (B \cap c(C)) = (A \backslash C) \cup (B \backslash C)$. **9.** (a) If $A \cap B = \varnothing$, then, if $x \in A$ it follows that $x \notin B$; that is, $x \in A \backslash B$. If $x \in A \backslash B$, then $x \in A$. (b) If $x \in A$, then $x \in B$; that is, $A \cap c(B) = \varnothing$. (c) If $A \cap c(B) = \varnothing$, then $x \in A$ implies $x \notin c(B)$.

***Section 1.2,*** page 7. **1.** Use the result of exercise 2(a), section 1.1. **2.** $\mathbf{N}, \varnothing$. **3.** $(0, 1)$, $\varnothing$. **4.** $[0, 1]$, $\{0\}$. **5.** $\{x \in \mathbf{R} \mid x \leqslant 0 \text{ or } x \geqslant 1\}$, $\mathbf{R}$. **6.** If $x \notin B$, then $x \notin A$. **9.** $A\backslash(A\backslash B) = A \cap c(A \backslash B) = A \cap c(A \cap c(B)) = A \cap (c(A) \cup B) = (A \cap c(A)) \cup (A \cap B) = \varnothing \cup (A \cap B)$.

***Section 1.3,*** page 8. **2.** Neither. **3.** Neither.

***Section 1.4,*** page 13. **1.** (b) $\mathbf{R}$, $\{0\} \cup [1, \infty)$, (c) No, (d) No. **2.** $\{0\} \cup [1, \infty)$, $\{0\} \cup [1, \infty)$, $\{0\} \cup (1, \infty)$. **3.** $a_0 \neq 0, a_1 = a_2 = a_3 = 0$. **4.** $a_0 = a_1 = a_2 = 0$. **5.** $a \neq 0$. **6.** If $k < l$, then $f(k) = a_k < a_l = f(l)$. **7.** $[0, 1], (0, 1), (1, 4), [0, 1)$. **8.** $[0, \infty)$. **9.** If $(x, y_1), (x, y_2) \in f_{\mathfrak{D}_1}$, then $(x, y_1), (x, y_2) \in f$ and hence $y_1 = y_2$. **11.** $\{x \in \mathfrak{D} \mid f(x) \in Y_1\}, \{(x, y) \in f \mid y \in Y_1\} = \varnothing$ if $\mathfrak{R}(f) \cap Y_1 = \varnothing$.

***Section 1.5,*** page 19. **1.** $f^{-1}(y) = \sqrt{y}, y \in \mathbf{R}^+$. **2.** $f^{-1}(y) = (1/y), y \in \mathbf{R}^+$. **3.** $f^{-1}(y) = \sqrt[3]{y}, y \in \mathbf{R}$. **4.** $(-\infty, -(\sqrt{3}/3)), [-(\sqrt{3}/3), (\sqrt{3}/3)], ((\sqrt{3}/3), \infty)$. **6.** Suppose $A$ has $n$ elements and $B$ has $m < n$ elements. If $f(a_k) = b_{j_k}$, then

† Do not look up answers and/or hints unless you are absolutely stymied.

$b_{j_k} = b_{j_l}$ for some $k \neq l$. **7.** $f(x) = -x$ for all $x \in \mathbf{R}^+$. **8.** (b) If $f: A \leftrightarrow B$, then $f^{-1}: B \leftrightarrow A$, (c) If $f: A \leftrightarrow B$, $g: B \leftrightarrow C$, then $h: A \leftrightarrow C$ where $h(x) = g(f(x))$. **10.** Establish a one-to-one correspondence between $(0, 1)$ and $(-1, 1)$, between $(-1, 1)$ and all points on $\{(x, |x|) \mid x \in (-1, 1)\}$, and finally between the latter and the $x$-axis. If $f(x) = (1 - 2x)/(|2x - 1| - 1)$, then $f: (0, 1) \leftrightarrow \mathbf{R}$.

***Section 1.6,*** page 22. **1.** Suppose $\sqrt{2} = (m/n)$, $(m, n) = 1$, is rational. Then, $2 = (m^2/n^2)$. Hence, $m$ is even: $m = 2\mu$. Then $n^2 = 2\mu^2$; that is, $n$ is even and $m, n$ have factor 2 in common.

**2.** $(a + b\sqrt{2})^{-1} = \dfrac{a}{a^2 - 2b^2} - \dfrac{b}{a^2 - 2b^2}\sqrt{2}$.

**3.** $(x + iy)^{-1} = \dfrac{x}{x^2 + y^2} - i\dfrac{y}{x^2 + y^2}$.

**4.** Similar to exercise 1. **5.** Similar to exercise 2. **6.** 0 and 1 are additive and multiplicative identity, respectively. **7.** 0 and 1 are additive and multiplicative identity, respectively. **8.** (a) (4) and (2), (b) (5), (c) (6), (d) (8), (e) $a + a0 = a(1 + 0) = a$, (f) $a + (-1)a = a(1 - 1) = a0 = 0$, (g) $(-1)(a + b) = (-1)a + (-1)b$, (h) $1 + (-1) = 0$. Hence, by (6), $1 = -(-1)$. (i) $(-1)(-1) = -(-1) = 1$. **9.** $a + (-a) + x = b + (-a) = x$. If $a + y = b$, then by the same argument, $y = b + (-a)$. $(1/c)cx = (1/c)b = x$, $(1/c)cy = (1/c)b = y$.

**10.** (a) $\dfrac{1}{a}\dfrac{1}{(1/a)} = 1$. (b) If $a \neq 0$, then $(1/a)ab = b = 0$. (c) $(-a)(-b) = (-1)(a)(-1)b = ab$.
**11.** Induction: If $a^{m+1} = a^m a$ and $a^{m+n} = a^m a^n$, then $a^{m+n+1} = a^m a^n a = a^m a^{n+1}$.
**12.** Consider $a^m/a^n$ and $1/a^m a^n$.

***Section 1.7,*** page 27. **4.** (a) $a - b > 0$, $b - c > 0$. Hence, $a - b + b - c = a - c > 0$. (b) $a - b = 0$ or $> 0$ or $< 0$. (c) If $a \neq b$, then either $a > b$ or $a < b$. (d) $a - b + (c - c) > 0$. (e) $a - b + c - d > 0$. (f) $(a - b)c > 0$. (g) $-(b - a)(-c) > 0$. (h) If $(1/a) < 0$, then $a(-1/a) = -1 > 0$. (j) If $a > 0$, $b < 0$, then $a(-b) = -ab > 0$. (k) Use (7.1). **5.** $|a_1 + a_2 + a_3| \leqslant |a_1| + |a_2 + a_3| \leqslant |a_1| + |a_2| + |a_3|, \ldots$. **6.** Show that $(a/2) < a$ for $a > 0$ as follows: Since $\frac{1}{2} \in \mathbf{F}$ and $\frac{1}{2} > 0$ we have $(a/2) > 0$ and $(a/2) + (a/2) > (a/2)$. **7.** $2a < a + b < 2b$. **8.** Use result of exercise 7. **9.** (b) If $|a| \leqslant b$, then $a \leqslant b$ and $-a \leqslant b$. **10.** $(1 + a)^n = 1 + na +$ positive terms. **11.** $c^n(c - 1) > 0$ if $c > 1$; $c^n(c - 1) \leqslant 0$ if $0 < c \leqslant 1$. **12.** $0 < a^2 < ab < b^2$, etc. **13.** Let $p/q \in \mathbf{Q}$. If $p/q \leqslant 0$, then $n = 1 > p/q$. If $p/q > 0$ and $q = 1$, then $p + 1 > p/q$. If $q > 1$, then, by exercise 4(f), $(p/q)q = p > p/q$. **14.** (a) $0 \in \mathbf{Q}(x)$ for $f(x) \equiv 0$, $1 \in \mathbf{Q}(x)$ for $f(x) \equiv 1$, $g(x) \equiv 1$. (d) Let $f(x) \equiv x$. Then, for every $n \in \mathbf{N}$, $f(x) - n > 0$.

***Section 1.8,*** page 31. **2.** $0 < 4 - x^2 = (2 - x)(2 + x)$. By exercise 4(j), section 1.7, $(2 - x) > 0$ and $(2 + x) > 0$ or $(2 - x) < 0$ and $(2 + x) < 0$. The second alternative is impossible. **5.** (b) If $x = a_0 . a_1 a_2 \ldots a_n \overline{b_1 b_2 \ldots b_m}$, consider $10^{n+m}x - x$.

**6.** (a) $\frac{m}{n} = m/2^j5^k = m2^{l-j}5^{l-k}/10^l$ for $l \geqslant j, k$. (b) Because of (a), the division process $m \div n$ cannot break up. For the remainders $r_i, 0 < r_i < n$; that is, only $1, 2, \ldots, (n-1)$ can occur as remainders. Since the division process is infinite, the same remainder has to occur again eventually. **7.** Substitute $a_0 . a_1 a_2 \ldots (a_k - 1)\dot{9}$ for $a_0 . a_1 a_2 \ldots a_k, a_k > 0$. **8.** (a) 1000000001, (b) 10001.011, (c) 0.001. **9.** (a) 15, (b) 5.5, (c) 0.78125, (d) $0.\dot{3}$. **10.** See exercise 6. **11.** Substitute $a_0 . a_1 a_2 \ldots a_{k-1} 0\dot{1}$ for $a_0 . a_1 a_2 \ldots a_{k-1} 1$. **12.** (a) 102220, (b) 120.222, (c) 0.021, (d) 0.2. **13.** A rational number $m/n$ has a finite ternary expansion if and only if $n = 3^k$ for some $k \in \{0, 1, 2, \ldots\}$. **14.** See exercise 13. **15.** $0 \cdot 0 = 0, 1 \cdot 1 = 1, 2 \cdot 2 = 11$. **16.** Assume that $\sqrt{2} = m/n$ where $m, n$ do not have the common factor 3; i.e., neither has a ternary representation ending with 0. By exercise 15, $2n^2$ ends with 2 and $m^2$ ends with 1. **17.** If $x \in A$, then $x \geqslant m$. Hence, $-x \leqslant -m$.

***Section 1.9,*** page 36. **2.** If **F** is complete, then $\sup A = x$ has the required property. If $C$ is bounded above, $B$ is the set of all upper bounds of $C$ and $A = \mathbf{F} \backslash B$, then $A \cap B = \varnothing$, $A \cup B = \mathbf{F}$. Then $x = \sup C$. **4.** If $A = \{x_1\}$, then $x_1 = \sup A$. Use induction to show that if $\sup\{x_1, x_2, \ldots, x_n\} \in \{x_1, x_2, \ldots, x_n\}$, then $\sup\{x_1, x_2, \ldots, x_n, x_{n+1}\} \in \{x_1, x_2, \ldots, x_{n+1}\}$. **5.** Assume w.l.o.g. that $A$ contains the positive integer $n_1$. Either $n_1 \geqslant x$ for all $x \in A$ or $n_1 < x$ for some $x \in A$. If $n_1 < x$, show that $\sup A = \sup\{x \in A \mid x > n_1\}$ and that $\{x \in A \mid x > n_1\} \subseteq \{n_0, n_0 - 1, \ldots, 1, 0\}$. Use Theorem 9.4. **6.** Use Archimedean property. **7.** There is an $a \in A$ such that $\sup A - \varepsilon/\lambda < a \leqslant \sup A$. **8.** Modify proof of (9.4). **9.** Use the arguments of the proof of (9.3) and (9.4) both ways. **10.** $\sup A - \varepsilon < a \leqslant \sup A$ implies $-\sup A \leqslant -a < -\sup A + \varepsilon$. **11.** See exercise 13, section 1.7.

***Section 1.10,*** page 40. **1.** If $a \geqslant 0$ but $a \notin \bigcup_{k=0}^{\infty} [k, k+1)$, then either $a < 0$ or $a \geqslant (k+1)$ for all $k \in \mathbf{N}$. Use Archimedean property. **2.** If $x \in [j, j+1)$ and $x \in [k, k+1)$ for $j \neq k$, then, for $j < k, x < j + 1 \leqslant k \leqslant x$. **4.** Modify the proof of Theorem 10.1 to fit the new circumstances. **5.** Show that $\mathbf{T}^* \sim \mathbf{R}^*$ and use exercise 4.

***Section 1.11,*** page 46. **2.** If $b^* = b_0 . b_1 b_2 b_3 \ldots < a^*$, then, for some $l \in \{0, 1, 2, 3, \ldots\}$, $b_0 = a_0, b_1 = a_1, \ldots, b_{l-1} = a_{l-1}, b_l < a_l$. Hence, $\alpha^* = a_0 . a_1 a_2 \ldots a_l > b^*$. **3.** See exercise 2. **5.** For every $k \in \{0, 1, 2, 3, \ldots\}$, one can find finite decimals $\gamma^*, \delta^*$ such that $c^* \geqslant a^* > \gamma^* > a_0 . a_1 a_2 \ldots a_k$, $d^* \geqslant b^* > \delta^* > b_0 . b_1 b_2 \ldots b_k$. Hence, $\gamma^* + \delta^* > a_0 . a_1 a_2 \ldots a_k + b_0 . b_1 b_2 \ldots b_k$. Since $\gamma^* + \delta^* \in C \oplus D$ where $C = \{\gamma^* \in D^* \mid \gamma^* < c^*\}$, $D = \{\delta^* \in D^* \mid \delta^* < d^*\}$ (see exercise 3), we have $c^* + d^* > \gamma^* + \delta^*$; that is, $c^* + d^*$ is an upper bound of $\{a_0 + b_0, a_0 . a_1 + b_0 . b_1, \ldots\}$ while $a^* + b^*$ is the least upper bound of that set. **7.** $b^*c^* = \sup B \odot C$. Since $a^* \geqslant b^*$, $\beta^* \in B$ implies $\beta^* \in A$. Hence, $a^*c^*$ is an upper bound of $B \odot C$. **8.** Assume, first, that $a^* + (b^* + c^*) > (a^* + b^*) + c^*$. Show that there are finite decimals $\eta^*, \alpha^*, \delta^*, \beta^*, \gamma^*$ such that $a^* + (b^* + c^*) > \eta^* > (a^* + b^*) + c^*$, $\alpha^* < a^*$, $\delta^* < b^* + c^*$, $\eta^* < \alpha^* + \delta^*$, $\beta^* < b^*$, $\gamma^* < c^*$ such that $\delta^* < \beta^* + \gamma^*$ and hence, $\eta^* < \alpha^* + (\beta^* + \gamma^*) = (\alpha^* + \beta^*) + \gamma^*$. Find finite decimals $\alpha_1^*, \beta_1^*$ such that $\alpha^* < \alpha_1^* < a^*$, $\beta^* < \beta_1^* < b^*$ such that $\alpha^* + \beta^* < \alpha_1^* + \beta_1^* \leqslant a^* + b^*$. Then $\eta^* < (a^* + b^*) + c^*$, a contradiction. Now assume that $a^* + (b^* + c^*) < (a^* + b^*) + c^*$, and proceed in similar fashion. **9.** $a^* \leqslant 0.99\ldots98\ldots$ where the digit 8 appears in the $k$th place. Then, $a^* < 0.99\ldots99$. Choose $\delta^* = 1.00\ldots01$ where the digit 1 appears in the $k$th place. **10.** Prove by contradiction. **11.** Let $\alpha^*$ represent a finite decimal such that $\alpha^* < a^*$. Then $\alpha^*\beta^* < a^*\beta^* < 1.\dot{0}$. Hence, $1.\dot{0}$ is an upper bound of $A \odot B$; that is, $a^*b^* \leqslant 1.\dot{0}$. Suppose $a^*b^* < 1$; choose a finite decimal $\delta^* > 1.\dot{0}$—see exercise 9—such that $a^*b^*\delta^* < 1.\dot{0}$, and a finite decimal $\gamma^*$ such that $b^* < \gamma^* < b^*\delta^*$. Then, $a^*b^* < a^*\gamma^* < 1.\dot{0}$ which is a contradiction. **12.** Assume that $a^*b_1^* = 1.\dot{0}$, $a^*b_2^* = 1.\dot{0}$, and $b_1^* < b_2^*$.

***Section 1.12,*** page 48. **1.** See section 1.6, exercise 4. **2.** Rational numbers satisfy linear equations. **3.** $10^{\log_{10} 2} = 2$.

***Section 1.13,*** page 54. **1.** If $A$ is finite and $A \sim B$, then $B$ is finite. **3.** To establish the necessity, show that $A$ contains a denumerable set $\{a_1, a_2, a_3, \ldots\}$; that a denumerable set is equivalent to a proper subset such as $\{a_2, a_3, a_4, \ldots\}$; and that $f(x) = a_{n+1}$ for $x = a_n$, $f(x) = x$ for $x \neq a_n$ for all $x \in A$ is one-to-one from $A$ onto $A \backslash \{a_1\}$. To show sufficiency, modify the argument in exercise 6, section 1.5. **4.** See exercise 10, section 1.5. **5.** See exercise 4. **6.** Consider $f$: $[0, 1] \to [0, 1)$, defined by $f(x) = 1/2^{n+1}$ for $x = 1/2^n$ for all $n = 0, 1, 2, 3, \ldots$ and $f(x) = x$ for $x \neq 1/2^n$ for all $n = 0, 1, 2, 3, \ldots$. **7.** Consider $f: A \to A \cup B$ defined by $f(x) = x$ for $x \in A \backslash C$, $f(x) = b_k$ for $x = c_{2k-1}$, $f(x) = c_k$ for $x = c_{2k}$ where $B = \{b_1, b_2, b_3, \ldots\}$, $C = \{c_1, c_2, c_3, \ldots\}$. **8.** If $t^n$ satisfies an algebraic equation with integer coefficients, then $t$ is rational or algebraic irrational. **9.** By Theorem 13.5, there exists a $t \in I_t$. By exercise 8, $t, t^2, t^3, \ldots \in I_t$. Let $A = I_t$, $C = \{t, t^2, t^3, \ldots\}$, $B = Q \cup I_a$ and use exercise 7. **12.** Every positive integer may be uniquely decomposed into a product of an odd integer times a power of 2. **13.** $\{\varnothing, \{a_1\}, \{a_2\}, \{a_3\}, \{a_1, a_2\}, \{a_2, a_3\}, \{a_1, a_3\}, \{a_1, a_2, a_3\}\}$. **15.** Let $B = \{\{a\} \mid a \in A\}$. **16.** By definition 13.1, $\overline{\overline{A}} \neq \overline{\overline{2^A}}$. **19.** Let $\overline{\overline{2^A}} = 2^{\overline{\overline{A}}}$. Then, $2^{\aleph_0} > \aleph_0$, $2^{2^{\aleph_0}} > 2^{\aleph_0}$, .... **20.** Since $S_\lambda \subseteq P$ for all $\lambda \in C$, we have $\lambda \leqslant \overline{\overline{P}}$ for all cardinal numbers. What about $\overline{\overline{2^P}}$? **21.** It would have to have the largest cardinal number. Take its power set. **22.** Assume that $F \sim \mathbf{R}$. Then, there exists a function $\phi : \mathbf{R} \leftrightarrow F$, given by $\phi(\alpha) = f_\alpha \in F$ for $\alpha \in \mathbf{R}$. Consider $f: \mathbf{R} \to \mathbf{R}$, defined by $f(x) = f_\alpha(x)$ for $x \neq \alpha$ and $(x) = f_\alpha(x) + 1$ for $x = \alpha$.

***Section 1.14,*** page 57. **1.** Either $(a, b) \cap C_k = \varnothing$ for all $k \in \mathbf{N}$; that is, $(a, b) \cap C = \varnothing$, or else, $(a, b) \cap C_k \neq \varnothing$ for some $k \in \mathbf{N}$. **3.** $0.a_1a_2a_3\ldots \in [0, 1]$ but $0.a_1a_2a_3\ldots \notin D_1$ because $a_1 \neq 1, \ldots$.

## Chapter 2

*Section 2.15,* page 60. **1.** (a) 0, (b) 1, (c) does not converge, (d) $\frac{1}{2}$, (e) does not converge, (f) does not converge, (g) 0. **3.** Since $0 < c < 1$, there is an $a > 0$ such that $c = 1/(1 + a)$. By exercise 2,

$$c^n = \frac{1}{(1 + a)^n} < \frac{1}{1 + na} \qquad \text{for n} > 1.$$

Use the result of exercise $1(g)$. **5.** $|a_k - a_l| \leqslant |a_k - b| + |a_l - b|$.

*Section 2.16,* page 63. **1.** $\{a_{2k+1}\}$ is a subsequence of $\{a_k\}$. **3.** $|a_k + 3 - (a + 3)| = |a_k - a|$. **4.** No.

*Section 2.17,* page 68. **1.** $|a_k + b_k - a - b| \leqslant |a_k - a| + |b_k - b|$. **2.** $|a_k b_k - ab| \leqslant |a_k(b_k - b)| + |b(a_k - a)|$. **3.** Choose $\varepsilon = (|b|/2)$. Then, $|b| - |b_k| \leqslant |b_k - b| < |b|/2$ for $k > N(\varepsilon)$. Let $m = \min\{|b_1|, |b_2|, \ldots, |b_{N(\varepsilon)}|, |b|/2\}$. **4.** $\{\lambda, \lambda, \lambda, \ldots\}$ is a constant sequence with $\lim_\infty \lambda = \lambda$. Use Theorem 17.1(b). **5.** $c = (1 + \varepsilon_k)^k$. If $\varepsilon \leqslant 0$, then $1 + \varepsilon_k \leqslant 1$ and $(1 + \varepsilon_k)^k \leqslant 1$. **6,7,8,9,10.** Use the Corollary to Theorem 16.3. **11.** $(1 + a)^n = 1 + na + (n(n-1)/2)a^2 +$ positive terms. **12.** Let $\sqrt[k]{k} = 1 + \varepsilon_k$ and note that $\varepsilon_k > 0$ for $k \in \mathbf{N}$. By exercise 11,

$$k = (1 + \varepsilon_k)^k \geqslant 1 + \frac{k(k-1)}{2}\varepsilon_k^2$$

and hence, $0 \leqslant \varepsilon_k^2 \leqslant 2/k$. **13.** $|a_k b_k| \leqslant M|a_k| < \varepsilon$ for $k > N(\varepsilon)$ because $\lim_\infty a_k = 0$. **14.** By exercise 13: 0.

*Section 2.18,* page 70. **1.** The sequence is increasing and, because of $1 + q + q^2 + \cdots + q^{k-1} = (1 - q^k)/(1 - q) \leqslant 1/(1 - q)$, also bounded. Use Theorem 18.1. **2.** $a_1 < 2$. By induction, $a_k = \sqrt{2 + a_{k-1}} < \sqrt{2 + 2} = 2$. Hence, $\{a_k\}$ is bounded. Furthermore, $a_2 > a_1$ and, by induction, $a_k = \sqrt{2 + a_{k-1}} > \sqrt{2 + a_{k-2}} = a_{k-1}$. **3.** $\{-b_k\}$ is increasing and bounded above. Use Theorem 18.1 and exercise 10, section 1.9. **4.** Apply Theorem 18.2 to $\{-b_k\}$. **5.** $\lim_\infty b_k$ exists because $\{b_k\}$ is decreasing and bounded below by $a_1$. $\lim_\infty (b_k - a_k) = 0$ by Lemma 17.1. By Theorem 17.1($a$), $\lim_\infty ((b_k - a_k) - b_k) = -\lim_\infty a_k = 0 - \lim_\infty b_k$. **6.** $\tau = (1 + \sqrt{5})/2$.

*Section 2.19,* page 72. **2.** $2^{n+1} > n^3$ for $n \geqslant 10$. Hence, $0 < k/(2^{k+1}) < (1/k^2)$ for $k \geqslant 10$ and hence,

$$0 < s_n = s_{10} + \frac{11}{2^{12}} + \frac{12}{2^{13}} + \cdots + \frac{n}{2^{n+1}} < s_{10} + \frac{1}{11^2} + \frac{1}{12^2} + \cdots + \frac{1}{n^2}.$$

Use example 2 and Theorem 17.2.

**4.** $$\frac{9}{10}+\frac{9}{10^2}+\frac{9}{10^3}+\cdots=\frac{9}{10}\left(1+\frac{1}{10}+\frac{1}{10^2}+\cdots\right)=\frac{9}{10}\cdot\frac{1}{1-\dfrac{1}{10}}.$$

***Section 2.20,*** page 74. **1.** (c) Yes: 10/99. **2.** If $P \neq Q$, then $l[P, Q] = \delta > 0$ but $l[A_k, B_k] < \delta$ for $k > N(\delta)$.

***Section 2.21,*** page 77. **1.** If $[-m, M]$ contains infinitely many elements of $A$, then $[-m, 0]$ or $[0, M]$ or both contain infinitely many elements of $A$. Continue the argument in this manner. **2.** $a_k \leqslant x \leqslant b_k$ and $\lim_\infty |b_k - a_k| = 0$. **4.** $\mathfrak{J}_k = (0, 1/k), k \in \mathbf{N}$.

***Section 2.22,*** page 79. **1.** If $x + r = s$ is rational, then $x = s - r$ is rational. **4.** By Theorem 22.2, there is, for all $k \in \mathbf{N}$, an $i_k \in \mathbf{I}$ such that $0 < i_k < (1/k)$. By exercise 1, $x + i_k \in \mathbf{I}$. **5.** By Theorem 22.1, there is, for all $k \in \mathbf{N}$, an $r_k \in \mathbf{Q}$ such that $x < r_k < x + 1/k$. **6.** By Theorem 22.3, there are rational, as well as irrational, sequences that converge to $a$. **8.** $a < b - (1/n) \leqslant (l + 1)/n - (1/n) = (l/n) < b$.

***Section 2.23,*** page 82. **1.** If $a$ is not an accumulation point, then, for some $\varepsilon > 0, a_k \in N_\varepsilon(a)$ but $a_k \notin N'_\varepsilon(a)$ for all $k > N(\varepsilon)$. Hence, $a_k = a$ for all $k > N(\varepsilon)$. **2.** (a) $\{1\}$, (b) $\{1, -1\}$, (c) $\mathbf{N}$, (d) $\varnothing$. **3.** Let $a, b$ represent accumulation points. If $\varepsilon < |a - b|/2$, then $|a - x_k| \geqslant \varepsilon$ for infinitely many $k \in \mathbf{N}$. **4.** If $x_k = b \neq a$ for all $k > N$, then $|x_k - a| \geqslant |b - a|/2$ for all $k > N$. **6.** $\{[0, 1] \cap \mathbf{Q}\}' = [0, 1]$. **7.** $\{[0, 1] \cap \mathbf{I}\}' = [0, 1]$. **8.** If $0.a_1a_2 \ldots a_k \ldots$ is the ternary representation of an element of the Cantor set, then $0.a_1a_2 \ldots b_k \ldots$ where $b_k = 0$ if $a_k = 2$, $b_k = 2$ if $a_k = 0$, is also an element of the Cantor set. **9.** If $N'_\delta(a)$ contains finitely many elements only, then one of them is closest to $a$. **10.** If $a \in (A \cup B)'$, then $N'_\delta(a) \cap (A \cup B) = (N'_\delta(a) \cap A) \cup (N'_\delta(a) \cap B) \neq \varnothing$ for all $\delta > 0$. If $a \in A' \cup B'$, then $N'_\delta(a) \cap A \neq \varnothing$ or $N'_\delta(a) \cap B \neq \varnothing$ for all $\delta > 0$.

***Section 2.24,*** page 86. **1.** $\{1, 1/2, 1/\sqrt{3}, 1/5, 1/\sqrt{7}, 1/11, 1/\sqrt{13}, 1/17, 1/\sqrt{19}, \ldots\}$. **2.** $\{1 + (-1)^k/k\}$. **5.** Let $x_k = 1/k$. Then $f(x_k) = k$.

**6.** If $x_k = \dfrac{1}{k}$, then $\lim_\infty \dfrac{1}{1 + e^k} = 0$. If $x_k = -\dfrac{1}{k}$, then $\lim_\infty \dfrac{1}{1 + \dfrac{1}{e^k}} = 1$.

**7.** 0. **8.** Use Theorem 16.1. **9.** If $f(a - 0) = f(a + 0) = b$, then, for every nonconstant sequence $\{x_k\}$ which approaches $a$, $|f(x_k) - b| < \varepsilon$ for all $k > N(\varepsilon)$. **10.** 2, 1. **11.** If $x_k \in [0, 1]$, $x_k \neq 0$, $\lim_\infty x_k = 0$, then $x_k > 0$ for all $k \in \mathbf{N}$.

***Section 2.25,*** page 90. **2.** Use Theorem 17.1. **3.** $|f(x)+g(x)-b-c| \leqslant |f(x)-b|+|g(x)-c|, |f(x)g(x)-bc| \leqslant |f(x)(g(x)-c)|+|c(f(x)-b)|$.

**4.** $\dfrac{f(x)}{g(x)}-\dfrac{b}{c}=\dfrac{f(x)c-bg(x)+bc-bc}{cg(x)}$.

**5.** Let $x_k=\dfrac{1}{\dfrac{\pi}{2}+k\pi}$. Then $\sin\dfrac{1}{x_k}=(-1)^k$.

***Section 2.26,*** page 94. **1.** Prove by contradiction. The contrapositive of (26.1) reads: There is an $\varepsilon>0$ such that, for all $\delta>0$, there is an $x \in N_\delta(a) \cap \mathfrak{D}$ such that $|f(x)-f(a)| \geqslant \varepsilon$. **3.** Yes. **4.** If $\lim_\infty x_k=a$, then $\lim_\infty x_k^n=(\lim_\infty x_k)^n = a^n$. **6.** $|f(x)-\eta-(f(c)-\eta)|=|f(x)-f(c)|$. **7.** Let $x_k=(1/k)$. Then $f(x_k)=k$. **8.** Let $x_k=1/((\pi/2)+k\pi)$. **9.** Let $x_k>0$, $\lim_\infty x_k=a>0$. By Theorem 17.1, $\lim_\infty 1/x_k=1/a$. **10.** $\lim x\sin(1/x)=0$ by example 5, section 2.24. If $a \neq 0$,

$$\begin{aligned}\left|x\sin\frac{1}{x}-a\sin\frac{1}{a}\right| &\leqslant \left|x\left(\sin\frac{1}{x}-\sin\frac{1}{a}\right)\right|+\left|(x-a)\sin\frac{1}{a}\right| \\ &\leqslant \left|2x\cos\frac{1}{2}\left(\frac{1}{x}+\frac{1}{a}\right)\sin\frac{1}{2}\left(\frac{1}{x}-\frac{1}{a}\right)\right|+|x-a| \\ &\leqslant |x|\left|\frac{1}{x}-\frac{1}{a}\right|+|x-a|.\end{aligned}$$

See exercise 9. **11.** Let $F(a)=\lim_a f(x)$, $F(b)=\lim_b f(x)$. **12.** $||f(a)|-|f(x)|| \leqslant |f(a)-f(x)|$. **13.** $\lim_0 f(x)=0 \neq f(0)=1$. **14.** $\lim_{0-0} f(x)=0 \neq 1 = \lim_{0+0} f(x)$.

**15.** $\lim\limits_{0-0}\dfrac{1}{x}=\dot{-}\infty$, $\lim\limits_{0+0}\dfrac{1}{x}=\infty$.

**16.** Let $A=\mathbf{N}$ and $f(x)=1$ for all $x \in A$.

***Section 2.27,*** page 99. **1.** Let $x_k \in \mathfrak{D}_1$, $\lim_\infty x_k=a$. Then $\lim_a f(x)=0 \neq f(a)$. **2.** $\{\mathbf{Q} \cap [a,b]\}'=[a,b]$. See exercise 1. **3.** Let $A=f(a)+\varepsilon$, $B=f(a)-\varepsilon$. Conversely, let $\varepsilon=\max(A-f(a), f(a)-B)$. **5.** $\mathfrak{D}=(-1,0)\cup(0,1)$, $f(x) = -1$ for $x \in (-1,0)$ and $f(x)=1$ for $x \in (0,1)$. **6.** $|-f(x)-(-f(a))|=|f(x) - f(a)|$. **7.** Use Lemma 27.1.

**8.** Use exercise 7 and $\dfrac{f(x)}{g(x)}-\dfrac{f(a)}{g(a)}=\dfrac{f(x)g(a)-f(a)g(x)+f(x)g(x)-f(x)g(x)}{g(x)g(a)}$

**9.** Use exercise 4, section 2.26, and Lemma 27.3. **10.** There are points $x_l, x_r \in \mathbf{R}$ such that $f(x_l)<0$ and $f(x_r)>0$. **11.** $\cos 0>0$, $\cos 1-1<0$. **12.** Let $f(\theta)$

represent the temperature at a point of the equator with geographic longitude $\theta$. Consider $g(x) = f(x) - f(x + \pi)$, and note that $f(0) = f(2\pi)$. **13.** $f(a) < (f(a) + f(b))/2 < f(b)$, or vice versa.

***Section 2.28,*** page 103. **1.** $N_\delta(x) \subseteq A$ for some $\delta > 0$. Hence, $N'_\delta(x) \cap A \neq \varnothing$ for all $\delta > 0$. **2.** $A' = \varnothing$. **3.** If $a$ is an accumulation point of $A \cup A'$, then, for every $\delta > 0$, $N'_{\delta/2}(a)$ contains a point of $A$ or $A'$. Every deleted $(\delta/2)$-neighborhood of a point of $A'$ contains a point of $A$. Hence, every $N'_\delta(a)$ contains a point of $A$; that is, $a \in A' \subseteq A \cup A'$. **4.** Yes. **5.** $\mathbf{R} = \mathbf{R}'$. **6.** $\varnothing' = \varnothing$. **7.** Neither. **8.** Every finite set is closed. **9.** There is a $\delta > 0$ such that $N'_\delta(a_0) \cap A = \varnothing$. If $a_0 \notin A$, then $N_\delta(a_0) \cap A = \varnothing$, in contradiction to Theorem 9.1. **10.** $A, c(A)$ are open and $A \neq \varnothing$, $c(A) \neq \varnothing$. Let $B = \{x \in \mathbf{R} \mid a \leqslant x \leqslant b, a \in A, b \in c(A)\}$, $a < b$. Show that $\xi = \sup A \cap B \in B$ cannot be an interior point of $A$ nor of $c(A)$ and hence $\xi \notin \mathbf{R}$.

***Section 2.29,*** page 106. **1.** If $x \in \bigcap_{k=1}^n \Omega_k$, where $\Omega_k$ is open for all $k = 1, 2, \ldots, n$, then $x \in \Omega_k$ for all $k = 1, 2, \ldots, n$ and hence, $N_{\delta_k}(x) \subseteq \Omega_k$ for some $\delta_k$. If $\delta = \min(\delta_1, \delta_2, \ldots, \delta_n)$, then $N_\delta(x) \subseteq \bigcap_{k=1}^n \Omega_k$. To prove the part pertaining to unions of closed sets, use DeMorgan's Law. **5.** sup $S$, inf $S$ exist because $S$ is bounded. sup $S$, inf $S \in S$ because $S$ is closed. Hence $S = [\inf S, \sup S]$. **6.** Let $\{r_1, r_2, r_3, \ldots\}$ denote an enumeration of all rationals and define $f : \{\mathfrak{J}_\alpha \mid \alpha \in A\} \to \mathbf{N}$ as follows: If $r_k$ is the first rational number which is in $\mathfrak{J}_\alpha$, let $f(\mathfrak{J}_\alpha) = k$. Show that $f : \{\mathfrak{J}_\alpha \mid \alpha \in A\} \leftrightarrow B \subseteq \mathbf{N}$. **7.** For each $x \in A$, some $N_{\delta_x}(x) \subseteq A$. Show that $A = \bigcup_{x \in A} N_{\delta_x}(x)$. **8.** Note that the union of open intervals $\mathfrak{J}_\alpha$, $\alpha \in A$, for which $\bigcap_{\alpha \in A} \mathfrak{J}_\alpha \neq \varnothing$, is an open interval. Use exercises 6 and 7. **9.** Use Lemma 27.1.

***Section 2.30,*** page 108. **3.** There is at least one accumulation point that is not in the set. Use Theorem 23.1. **4.** If $S$ is finite, then $S' = \varnothing$. If $S$ is not finite, then, by Theorem 30.1, $S' \neq \varnothing$. **5.** By Theorem 30.1, $S$ has at least one accumulation point. By Theorem 28.2, $S$ contains all accumulation points.

***Section 2.31,*** page 110. **1.** Choose $\Omega$ as in example 2. **2.** Let $\xi \in (0, 1) \cap \mathbf{I}$ and consider $\Omega = \{\Omega_k \mid k \in \mathbf{N}\}$ where $\Omega_k = \{x \in \mathbf{R} \mid x > \xi + (1/k) \text{ or } x < \xi - (1/k)\}$. **3.** Let $\Omega = \bigcup_{k \in \mathbf{Z}} (k - 1, k + 1)$. **6.** Construct for each $x \in S$ an open interval $\mathfrak{J}_x$ with rational endpoints, such that $x \in \mathfrak{J}_x \subset \Omega_\alpha$ for some $\alpha \in A$.

***Section 2.32,*** page 112. **3.** $\mathbf{Q}$ is not closed. **4.** Show that $A'$ is bounded and use exercise 3, section 2.28. **6.** Since $n_k \geqslant k$, we have $a_{n_k} > A$ for all $k > N(A)$. **7.** See example 8, section 2.29.

*Section 2.33,* page 115. **2.** Use Theorem 33.2 and the fact that $\{x_k^{(1)}\}$, $\{x_k^{(2)}\}$ are bounded because $K$ is compact. By Theorem 32.2, a subsequence $\{x_{n_k}^{(1)}\}$ of $\{x_k^{(1)}\}$ converges to $a \in K$ and a subsequence

$$\{x_{n_{k_i}}^{(2)}\} \text{ of } \{x_{n_k}^{(2)}\}$$

converges to $b \in K$. Now, $a = b$ because

$$|x_{n_{k_i}}^{(1)} - x_{n_{k_i}}^{(2)}| < \frac{1}{n_{k_i}}.$$

Because $f$ is continuous at $a$,

$$\lim_{i \to \infty} f(x_{n_{k_i}}^{(1)}) = \lim_{i \to \infty} f(x_{n_{k_i}}^{(2)}) = f(a).$$

**5.** $f$ is uniformly continuous in $[-1, 2]$. For all $x_1, x_2 \in \mathbf{R}$ for which $|x_1 - x_2| < \delta < 1$, there is a $k \in \mathbf{Z}$ such that $x_1 + k, x_2 + k \in (-1, 2)$.

*Section 2.34,* page 118. **3.** $f$ is continuous for $x \in (-1, 0) \cup (0, 1)$. $f(0) - f(x) = -1 - x^2 \leqslant 0 < \varepsilon$ for all $x \in [-1, 1]$. Similar for $-1, 1$. **4.** Construct a sequence $\{x_{n_k}\}$ as in the proof of Theorem 34.1. Then $f(x_{n_k}) < 1 + f(a)$ but $f(x_{n_k}) > n_k$. **5.** By exercise 4, $\sup f(K)$ exists. By Theorems 9.1 and 32.2 there is a sequence $\{x_k\}$ such that $\lim_\infty f(x_k) = \sup f(K)$ and $\lim_\infty x_k = a \in K$. $\sup f(K) \leqslant f(a) + \varepsilon$. **8.** By Theorem 34.2, $f(x) \in [m, M]$ for all $x \in [a, b]$. By Theorem 34.3, $f(x_m) = m$, $f(x_M) = M$ for some $x_m, x_M \in [a, b]$. By Theorem 27.1, for every $y \in [m, M]$, there is an $x \in [a, b]$ such that $f(x) = y$.

## Chapter 3

*Section 3.35,* page 123. **1.** No. Take $f(x) = 1$ for $x \in (0, 1)$, $f(x) = 2$ for $x \in (1, 2)$. Then $\mathfrak{D} = (0, 1) \cup (1, 2)$ and $f'(x) = 0$ for all $x \in \mathfrak{D}$. **3.** $f(x) + g(x) - f(a) - g(a) = (f(x) - f(a)) + (g(x) - g(a))$, $f(x)g(x) - f(a)g(a) = f(x)(g(x) - g(a)) + g(a)(f(x) - f(a))$. **4.** Since $g'(a)$ exists, $g$ is continuous at $a$. Since $g(a) \neq 0$, $g(x) > 0$ (or $< 0$) for all $x \in N_\delta(a)$ for some $\delta > 0$,

$$\frac{f(x)}{g(x)} - \frac{f(a)}{g(a)} = \frac{f(x)g(a) - f(a)g(x) + f(a)g(a) - f(a)g(a)}{g(x)g(a)}.$$

**7.** $f'_+(0) = 0$, $f'(0+0) = \lim_0(2x \sin(1/x) - \cos(1/x))$ does not exist. **8.** If $f'_1(a)$, $f'_2(a)$ both are the derivative, then $f'_1(a) - f'_2(a) = \alpha_a^{(2)}(x) - \alpha_a^{(1)}(x)$ where $\lim_a(\alpha_a^{(2)}(x) - \alpha_a^{(1)}(x)) = 0$. **9.** (a) 1, (b) $2a$, (c) $na^{n-1}$, (d) $\cos a$, (e) $-\sin a$. **10.** (a) $a_0 n x^{n-1} + a_1(n-1)x^{n-2} + \cdots + a_{n-1}$, (b) $1/(\cos^2 a)$, (c) $-1/(\sin^2 a)$. **11.** $f'(0) = 0$. $f'(a)$ does not exist for $a \neq 0$.

*Section 3.36,* page 127. **2.** (c) no, (d) yes, (e) $(2x \sin(1/x) - \cos(1/x))/(2\sqrt{x^2 \sin(1/x)})$. **4.** $(3x/\sqrt{x^2+1})\cos\sqrt{x^2+1}\,(\sin\sqrt{x^2+1})^2$. **5.** $(x(1 + \cos(x^2)))/(\sqrt{x^2 + \sin(x^2)})$. **6.** 1. **7.** $1/2\sqrt{y}$. **8.** $1/(\sqrt{1-y^2})$.

***Section 3.37,*** page 135. **1.** Consider $f: [0, 1] \to \mathbf{R}$, $f(x) = x, a = 1$. **3.** $m/(m + n)$. **4.** $f(x) =$ constant for $x \in (a, b)$. Since $f$ is continuous in $[a, b]$ and $f(a) = 0$, it follows that $f(x) = 0$ for all $x \in [a, b]$. **5.** $(f(\xi) - f(x))/(\xi - x) = f'(\eta)$ for some $\eta$ between $x$ and $\xi$. Let $x$ approach $\xi$. **6.** Let $g(x) = f(b) - f(x) - f'(x)(b - x) - ((b - x)^2/(b - a)^2)(f(b) - f(a) - f'(a)(b - a))$. Then, $g(a) = g(b) = 0$ and $g \in C'(a, b)$, $g \in C[a, b]$. By Theorem 37.2, $g'(\xi) = 0$ for some $\xi \in (a, b)$. **7.** Generalize the argument in exercise 6. **8.** (a) 1 if $n = 1$, 0 if $n > 1$, (b) 0, (c) 1, (d) $\frac{1}{2}$. **9.** (a) 0, (b) $\frac{1}{2}$, (c) $\frac{1}{4}$, (d) 1. **10.** (a) 0, (b) 0. **11.** By exercise 6, $f(x) - f(c) = f''(\xi)((c - x)^2/2) > 0$ (or $< 0$) for $x$ sufficiently close to $c$. **13.** 1. **14.** $-1/(\sqrt{1 - y^2})$.

***Section 3.38,*** page 138. **3.** $f'$ has the intermediate-value property. $f'(0) = 0$, $f'(1/2k\pi) = -1$, $f'(1/(2k + 1)\pi) = 1$. **4.** If $f(x_1) = f(x_2)$, then $f'(\xi) = 0$ for some $\xi$ between $x_1$ and $x_2$.

***Section 3.39,*** page 140. **1.** If $x \neq 0$, let $x + y = xz$. Then, $h(zx) = zh(x) = ((x + y)/x)h(x) = xh(1) + yh(1) = h(x) + h(y)$. **2.** (a) $3h$, $3a^2(x - a)$, 0, (b) $(\cos x)h, h, -h$. **3.** 31° corresponds to $\pi/6 + \pi/180$ radians and $\sin(\pi/6) = \frac{1}{2}$. Consider $d \sin(\pi/6, \pi/180)$. Answer: 0.515. **4.** 2.007. **5.** 100.180.

## Chapter 4

***Section 4.40,*** page 143. **1.** $\varnothing$. **2.** $\varnothing$. **3.** $\{(x, y) \in \mathbf{R} \times \mathbf{R} \mid x^2 + y^2 < 1\}$. **4.** $\{(x, y) \in \mathbf{R} \times \mathbf{R} \mid x < 0\}$.

***Section 4.41,*** page 145. **1.** If $r = m/n$, represent $U_r$ as the union of $m^2$ squares with nonoverlapping interiors of side length $1/n$ each. **2.** If $a = m/n$, $b = \mu/\nu$, represent $\square$ as the union of $\nu n \mu m$ squares with nonoverlapping interiors and side length $(1/\nu n)$ each. **3.** If $P, P^\circ \in \mathfrak{T}$, and $P^\circ \subset P$, then $\mu(P^\circ) = \mu(P)$. **4.** $(P_2 \backslash P_1) \cap P_1 = \varnothing$, $(P_2 \backslash P_1) \cup P_1 = P_2$. Hence, $\mu(P_2 \backslash P_1) + \mu(P_1) = \mu(P_2)$.

***Section 4.42,*** page 148. **6.** Assume that for two different triangulations $T = \{T_1, T_2, \ldots, T_n\}$ and $S = (S_1, S_2, \ldots, S_m)$ of $P$, $\sum_{k=1}^{n} \mu(T_k) \neq \sum_{k=1}^{m} \mu(S_k)$. Superimpose $T$ on $S$ to obtain a decomposition of $P$ into triangles, quadrangles, pentagons, and hexagons (see exercise 3). Then use the result of exercise 1. **8.** $\phi(0) = \phi(0 + 0) = \phi(0) + \phi(0) = 2\phi(0)$. Hence $\phi(0) = 0$. $\phi(x + x) = \phi(x) + \phi(x) = 2\phi(x)$. Use induction to show that $\phi(nx) = n\phi(x)$ for $n \in \mathbf{N}$. **9.** $|\phi(x) - \phi(a)| = |\phi(x - a)| < \varepsilon$ for $0 \leqslant x - a < \delta(\varepsilon)$.

**10.**
$$\phi\left(\frac{m}{n} x\right) = m\phi\left(\frac{1}{n} x\right), \qquad n\phi\left(\frac{1}{n} x\right) = \phi(x).$$

**11.** Choose $r_k \in \mathbf{Q}$, $r_k > 0$ such that $\lim_\infty r_k = \lambda$ and note that, by exercise 10, $\phi(r_k x) = r_k \phi(x)$, and by exercise 9, $\lim_\infty \phi(r_k x) = \phi(rx)$.

***Section 4.43,*** page 151. **1.** $2^{n-1}\sin(360/2^n)$. **2.** $2^n \tan(180/2^n)$. **3.** Note that $\sin 2\alpha = 2\sin\alpha\cos\alpha$ can be established without reference to the unit circle. (See also (143.6). **4.** See exercise 2, section 2.18. **5.** $\cos(180/2^2) = \sqrt{2}/2$, $\cos^2\alpha = (1+\cos 2\alpha)/2$. Hence, $\cos^2(180/2^3) = (2+\sqrt{2})/4$, etc. Use exercise 4. **6.** Show that $\lim_\infty (\mu(a_{2^n}))/(\mu(A_{2^n})) = 1$ by using exercises 3 and 5. **8.** Make a sketch with the vertices of the inscribed and circumscribed polygon situated on the same ray emanating from the center of the circle. **9.** Let $c_{2^k 6} = a_k$, $C_{2^k 6} = b_k$. By construction, $a_0 < b_0$. By induction, $b_{k+1}^2 = 2a_k b_k b_{k+1}/(a_k + b_k) = 2b_k a_{k+1}^2/(a_k + b_k) > a_{k+1}^2, a_{k+1} = \sqrt{a_k b_{k+1}} > a_k$, and $b_{k+1} = 2a_k b_k/(a_k + b_k) < b_k$. From $\lim_\infty a_{k+1} = \sqrt{\lim_\infty a_k \lim_\infty b_{k+1}}$ follows $\lim_\infty a_k = \lim_\infty b_k$. **10.** 3.14…, 3.14….

***Section 4.44,*** page 154. **1.** Suppose that $\inf_{(B)} f(x) - \inf_{(A)} f(x) = \varepsilon > 0$. Then, there is an element $a \in A \subseteq B$ such that $f(a) < \inf_{(A)} f(x) + \varepsilon = \inf_{(B)} f(x)$, which is impossible. **2.** 3.25, 4.25.

**3.** $\dfrac{1}{2} - \dfrac{1}{2n}$, $\quad \dfrac{1}{2} + \dfrac{1}{2n}$. **4.** $\dfrac{1}{3} - \dfrac{1}{2n} + \dfrac{1}{6n^2}$, $\quad \dfrac{1}{3} + \dfrac{1}{2n} + \dfrac{1}{6n^2}$. **5.** 10.

***Section 4.45,*** page 156. **1.** Use Theorems 9.1 and 18.1. **2.** $\underline{S}(f;P) \geqslant 0$ for all $P$. **3.** $\bar{S}(f;P) \geqslant 0$ for all $P$. **4.** If $|f(x)| \leqslant M$ for all $x \in [a,b]$, then $m_k(f) \leqslant M$ for all $k \in \{1,2,\dots,n\}$. **5.** See Lemma 45.2. **7.** $m_k(f) \leqslant f(\xi) \leqslant M_k(f)$ for all $\xi \in [x_{k-1}, x_k]$. **8.** There is a $\xi' \in [x_{k-1}, x_k]$ such that $0 < M_k(f) - f(\xi_k') < (\varepsilon/n)$, $k = 1,2,\dots,n$. Similar for $\xi_k''$ and $m_k(f)$.

***Section 4.46,*** page 160. **1.** If the condition is satisfied for all refinements of $P_\varepsilon$, then it is satisfied for $P_\varepsilon$. Conversely, if it is satisfied for $P_\varepsilon$, then, by Lemma 44.2, it is satisfied for all refinements of $P_\varepsilon$. **2.** Choose

$$\xi_k \in \left[\frac{1}{k-1}, \frac{1}{k}\right] \text{ such that } f(\xi_k) = \frac{1}{2}\left(f\left(\frac{1}{k-1}\right) + f\left(\frac{1}{k}\right)\right).$$

**3.** By exercise 4, section 4.44, $\bar{S}(f;P_n) - \underline{S}(f;P_n) = 1/n$. **4.** First show, by contradiction, that $f$ has to be bounded. By exercise 8, section 4.45, $\bar{S}(f;P) - \underline{S}(f;P) \leqslant |\bar{S}(f;P) - S(f;P,\Xi')| + |S(f;P,\Xi') - I| + |I - S(f;P,\Xi'')| + |S(f;\Xi'') - \underline{S}(f;P)| < \varepsilon$. **5.** Let $I = \inf_{(P)} \bar{S}(f;P)$ and use the fact that $\underline{S}(f;P) \leqslant S(f;P,\Xi) \leqslant \bar{S}(f;P)$ for all $P$ and all $\Xi$ of $P$—see exercise 7, section 4.45. **8.** Use exercises 6 and 7. **9.** $S(f,g;P,\Xi) = f(\xi)$ where $\xi \in [a,x_1]$. **10.** Suppose, to the contrary that $|S(f,g;P_1,\Xi) - I_1| < \varepsilon/2$, $|S(f,g;P_2,\Xi) - I_2| < \varepsilon/2$ for all $P_1 \supseteq P_{1,\varepsilon}$, $P_2 \supseteq P_{2,\varepsilon}$ but $I_1 \neq I_2$.

***Section 4.47,*** page 164. **1.** $\int_0^1 f = \sup_{(P)} \underline{S}(f;P) = \frac{1}{2}$. **3.** $f(x_1) \leqslant f(x)$ for all $x \geqslant x_1$, $f(x_2) \geqslant f(x)$ for all $x \leqslant x_2$. **9.** Let $g(x) = k - 1$ for $k - 1 \leqslant x < k$, $k \in \mathbf{N}$. Then $\sum_{k=1}^{n} a_k = a_1(g(1) - g(0)) + a_2(g(2) - g(1)) + \cdots + a_n(g(n) - g(n - 1))$. **10.** Use the results of exercises 6, 7, section 4.46, and note that $P_1 \supseteq P_2$ implies $\|P_1\| \leqslant \|P_2\|$. **11.** By exercise 5, section 4.46, (*) is satisfied for all $P \supseteq P_\varepsilon$. If $m = \inf f(x) < \sup f(x) = M$ and $P_\varepsilon = \{x_0, x_1, \ldots, x_N\}$, let $\delta(\varepsilon) = \min\{\varepsilon/(3(N+1)(M-m)), (x_1 - x_0), (x_2 - x_1), \ldots, (x_N - x_{N-1})\}$. For any partition $P = \{y_0, y_1, \ldots, y_n\}$ with $\|P\| < \delta(\varepsilon)$, either $[y_{k-1}, y_k] \subset [x_{l-1}, x_l]$ for some $l \in \{1, 2, \ldots, N\}$ or $[y_{k-1}, y_k] \not\subset [x_l, x_{l-1}]$ for all $l \in \{1, 2, \ldots, N\}$. In the first case, $(\overline{M}_k(f) - \overline{m}_k(f))(y_k - y_{k-1}) \leqslant (M_l(f) - m_l(f))(x_l - x_{l-1})$ where $\overline{M}_k, \overline{m}_k$ are sup and inf, respectively, of $f$ on $[y_{k-1}, y_k]$. In the second case, there is exactly one $x_k$ such that $y_{l-1} < x_k < y_l$ and we have $(\overline{M}_l(f) - \overline{m}_l(f))(y_l - y_{l-1}) \leqslant (M - m)\delta(\varepsilon)$. Show that $P$ satisfies the Riemann condition and use exercise 5, section 4.46. **12.** See exercises 10 and 11.

**13.**
$$\text{Let } P_n = \left\{a, a + \frac{b-a}{2^n}, \ldots, a + \frac{2^n - 1}{2^n}(b - a), b\right\}.$$

Then $P_{n_1} \supseteq P_{n_2}$ if $n_1 \geqslant n_2$. Hence, $\underline{S}_n(f)$ is increasing. It is also bounded above. Hence, $\lim_\infty \underline{S}_n(f)$ exists. By exercise 8, section 4.45, there is, for every $P_n$, a $\Xi_n'$ such that $|\underline{S}_n(f) - S(f; P_n, \Xi_n')| < \varepsilon$. Use exercise 12 to show that $\lim_\infty \underline{S}_n(f) = I = \int_a^b f$.

***Section 4.48,*** page 170. **4.** There is a sequence of *open* intervals $\{\mathfrak{J}_k\}$ with the desired property. Let $\mathfrak{I}_k = \mathfrak{J}_k \cup \{\text{endpoints of } \mathfrak{J}_k\}$. **5.** Let $\mathfrak{J}_k = \mathfrak{I}_k \setminus \{\text{endpoints of } \mathfrak{I}_k\}$. Note that the set of endpoints of all $\mathfrak{I}_k$ has Lebesgue measure zero. **7.** If $a \in \mathcal{O}$, then $a$ is an *interior* point of $\mathcal{O}$ and $g(x) = x^2$ for all $x \in (a - \delta, a + \delta)$ for some $\delta > 0$. **8.** Consider a compact subset of $\mathbf{R}$ and use the Heine–Borel theorem.

***Section 4.49,*** page 175. **2.** By Theorem 47.2, $f$ is integrable and by Lemma 49.2, $f$ is continuous a.e. **3.** $\{1\}, \{1, 2\}, \{1, 2, 3\}, \{1, 2, 3\}$. **4.** $|f(\xi) - f(x)| \geqslant 1/m > 1/(m + 1)$.

***Section 4.50,*** page 177. **2.** If (50.10) is satisfied for every $\lambda, \mu \in \mathbf{R}$, then it is satisfied, in particular, for $\lambda \neq 0$, $\mu = 0$ and for $\lambda = \mu = 1$. **3.** $f(x) = 1$ for $x \in [0,1] \cap \mathbf{Q}$, $f(x) = 0$ for $x \in [0,1] \cap \mathbf{I}$, $g(x) = 1$ for $x \in [0,1] \cap \mathbf{I}$, $g(x) = 0$ for $x \in [0,1] \cap \mathbf{Q}$. $\lambda = \mu = 1$. **4.** $S(\lambda f + \mu g; P, \Xi) = \lambda S(f; P, \Xi) + \mu S(g; P, \Xi)$. $|S(f; P, \Xi) - \int_a^b f| < \varepsilon/2|\lambda|$, $|S(g; P, \Xi) - \int_a^b g| < \varepsilon/2|\mu|$ for all $P \supseteq P_\varepsilon$. **5.** Proceed as in exercise 4.

***Section 4.51,*** page 181. **2.** Take $f(x) = 0$ for $x \in [0, 1)$, $f(x) = 1$ for $x \in [1, 2]$. **3.** Take $f(x) = 1$ for $x \in [0,1] \cap \mathbf{Q}$, $f(x) = -1$ for $x \in [0,1] \cap \mathbf{I}$. Then $|f(x)| = 1$ for all $x \in [0, 1]$. **4.** The union of two sets of Lebesgue measure zero has

Lebesgue measure zero. **5.** Let $f_1(x) = f(x)$ for $x \in [a,b]\backslash\{c\}$ and $f_1(c) \neq f(c)$. Either $\underline{S}(f;P) = \underline{S}(f_1;P)$ of $\bar{S}(f;P) = \bar{S}(f_1;P)$ for all $P$. **6.** For every $\varepsilon > 0$ there is a partition $P$ such that $|\bar{S}(f_{[a,c]};P) - \bar{S}(f_1;P)| < \varepsilon$. **8.** If $g(x_0) > 0$ for some $x_0 \in [a,b]$, then $g(x) \geqslant m > 0$ for all $x \in N_\delta(x_0) \cap [a,b]$ for some $\delta > 0$ and some $m > 0$. Hence, $\int_a^b g > 0$. **9.** Let $f(x) = x$ for all $x \in [-1,1]$, $g(x) = x$ for all $x \in [-1,1]$. **10.** If $a,b,c$ are all different, then there are six possibilities: $a<b<c$, $a<c<b$, $b<a<c$, $b<c<a$, $c<a<b$, $c<b<a$. If $a<b<c$, then $\int_a^c f = \int_a^b f + \int_b^c f$. Hence, $\int_a^b f = \int_a^c f - \int_b^c f = \int_a^c f + \int_c^b f$, etc. If two of $a,b,c$, or all are equal, use (51.3).

***Section 4.52,*** page 188. **1.** The sufficiency follows from Theorem 52.3. To establish the necessity use Theorem 52.2 and note that the derivative of a constant is zero. **2.** By Theorem 52.2, $\int_a^x f$ is the antiderivative of $f$ on $[a,b]$. Note that $(\int_a^x f + C)' = f(x)$.

**3.** (a) $\dfrac{b^{n+1}}{n+1} - \dfrac{a^{n+1}}{n+1}$, (b) log 2, (c) 0, (d) 0, (e) $E(1) - 1$.

**4.** Let $Q = P \cup \Xi$ where $P$ is a partition of $[a,b]$ and $\Xi$ is a set of intermediate values of $P$. Show that $S(f,g;P,\Xi) = -S(g,f;Q,P) + f(b)g(b) - f(a)g(a)$. Note that $P$ is a set of intermediate values of $Q$. **5.** By (52.4), $\int_{x_{k-1}}^{x_k} fg' = f(x_k)g(x_k) - f(x_{k-1})g(x_{k-1}) - \int_{x_{k-1}}^{x_k} f'g$. Take sum from 1 to $n$. **6.** Use Corollary 2 to Theorem 52.3. **10.** $(x^2/2)\log x - (x^2/4) + \frac{1}{4}$ **11.** (a) $-\log(\cos x)$, (b) $\frac{1}{2}(\log x)^2$, (c) $\frac{1}{3}\log(x^3 - 3x^2 + 3x)$. **12.** $F$ is not continuous at $x = 0$, $x = 1$. **13.** Let $G(x) = F(x)$ for $x \in (a,b)$, $G(a) = \lim_a F(x)$, $G(b) = \lim_b F(x)$ and apply Theorem 52.3. **14.** $\lim_0 F(x) = 0$, $\lim_1 F(x) = 1$. $\int_0^1 dx = 1 = 1 - 0$.

***Section 4.53,*** page 194. **1.** $|x^\alpha| < \varepsilon$ for $x > \varepsilon^{1/\alpha}$. **2.** $(e^x - 1 - x)' = e^x - 1 \geqslant 0$, $e^0 - 1 = 0$. Hence, $e^x - 1 - x \geqslant 0$ for $x \geqslant 0$. **3.** Choose $n \in \mathbf{N}$ such that $\alpha \leqslant n$. Then $x^\alpha/e^{ax} \leqslant x^n/e^{ax}$ for $x \geqslant 1$. Use (53.1). **5.** $1/(1+x^2) < 1/x^2$. See example 3 and use Theorem 53.1. **6.** $2\sqrt{2}$. **7.** $L = \sup_{[a,\infty)} F(x)$ exists. By Theorem 9.1, there is an $x_0 \in [a,\infty)$ such that $0 \leqslant L - F(x_0) < \varepsilon$. Since $F$ is increasing, $0 \leqslant L - F(x) < \varepsilon$ for all $x > x_0$. **8.** $0 \leqslant \int_\varepsilon^b f \leqslant \int_a^b g$ and $\int_\varepsilon^b f$ increases as $\varepsilon \to a + 0$. **12.** Use example 3 and Theorem 53.1. **13.** Use example 6 and Theorem 53.1. **14.** $\int_0^\infty dx/(1+x^2) = \operatorname{Tan}^{-1}x\,|_0^\infty = \pi/2$. **15.** (a) $\int_{-\infty}^{\infty} \sin x\,dx = \lim_\infty(-\cos A + \cos(-A)) = 0$, but $\int_0^A \sin x\,dx = \lim_\infty(-\cos A + 1)$ does not exist. (b) $\int_{-A}^A f(x)dx = \int_{-A}^0 f(x)dx + \int_0^A f(x)dx = \int_A^0 f(x)dx + \int_0^A f(x)dx = 0$. (c) $\lim_\infty \int_0^A \cos x\,dx = \lim_\infty \sin A$ does not exist, nor does $\lim_\infty \int_{-A}^A \cos x\,dx = \lim_\infty 2\sin A$. **16.** 0, $\lim_{\varepsilon\to 0+0} \log\varepsilon$ does not exist. **17.** 0. **20.** The integral exists, by Theorem 53.1.

$$\int_0^\infty \frac{\sin^2 x}{x^2}\,dx = -\frac{1}{x}\sin^2 x\Big|_0^\infty + 2\int_0^\infty \frac{\sin x\cos x}{x}\,dx$$

$$= \int_0^\infty \frac{2\sin(u/2)\cos(u/2)}{u}\,du = \int_0^\infty \frac{\sin u}{u}\,du.$$

**21.** $\lim_0 x^2 \log x = \lim_0 \dfrac{\log x}{(1/x^2)} = 0.$ $\displaystyle\int_0^1 x \log x\, dx = \frac{x^2}{2} \log x \Big|_0^1 - \int_0^1 \frac{x}{2}\, dx = -\frac{1}{4}.$

**22.** $\alpha > 1$. (Let $u = \log x$.) **24.** 4. **25.** Consider $\lim_0 (f(1/x))/(g(1/x))$. **26.** Note that $0 \leqslant |f| - f \leqslant 2|f|$ for all $x \in [a, \infty)$ and use Theorem 53.1. **27.** (a) $1/(s-a)$, (b) $\Gamma(\alpha+1)/(s^{\alpha+1})$, (c) $\omega/(s^2-\omega^2)$, (d) $s/(s^2-\omega^2)$, (e) $\phi(s-a)$.

## Chapter 5

***Section 5.55,*** page 201. **1.** Take $\theta = (0, 0, \ldots, 0)$. **2.** Take $\lambda = \mu = 1$ and $\lambda \neq 0$, $\mu = 0$. **3.** $(2, 0, -21, 2\pi - \frac{3}{8}, 2\sqrt{7} + 81)$. **4.** $f(x) = 0$ for all $x \in [a, b]$ defines a continuous function on $[a, b]$. **7.** $e_1 = u_2 - u_3$, $e_2 = -u_1 + u_2$, $e_3 = u_1 - u_2 + u_3$, $x = (\xi_1, \xi_2, \xi_3) = (\xi_3 - \xi_2)u_1 + (\xi_1 + \xi_2 - \xi_3)u_2 + (\xi_3 - \xi_1)u_3$.

***Section 5.56,*** page 204. **1.** $x \neq \theta$ implies $x \cdot x > 0$ by 1. Hence $x \cdot x = 0$ implies $x = \theta$. Conversely, if $x = \theta$, then $x = 0y$ for $y \in V$. By (2) $x \cdot x = (0y)\cdot(0y) = 0(y \cdot 0y) = 0$. **2.** Define two Riemann-integrable functions to be equal if and only if they are equal almost everywhere. Then, $\theta$ represents all Riemann-integrable functions that are zero almost everywhere. **3.** Use (56.5) with $g(x) = 1$ for all $x \in [a, b]$. **4.** $x \cdot e_j = \xi_j$. **5.** $(x+y)\cdot(x+y) = x \cdot x + 2x \cdot y + y \cdot y \leqslant x \cdot x + 2\sqrt{(x \cdot x)(y \cdot y)} + y \cdot y$.

***Section 5.57,*** page 207. **1.** By (1), $\|x\| = 0$ implies $x = \theta$. By (2) $\|\theta\| = \|0y\| = |0|\,\|y\| = 0$ for $y \in V$. **2.** Let $a = x - y$, $b = y$ and use triangle inequality on $\|a + b\|$. **7.** $|f(x) + g(x)| \leqslant |f(x)| + |g(x)| \leqslant \max_{[a,b]} |f(x)| + \max_{[a,b]} |g(x)| = \|f\| + \|g\|$. Hence, $\max_{[a,b]} |f(x) + g(x)| \leqslant \|f\| + \|g\|$.

***Section 5.58,*** page 211. **1.** See example 4, section 5.57. **2.** See exercise 7, section 5.57. **4.** Unit disk without circumference. **5.** Unit ball without surface. **6.** $a$. **7.** $d(x, z) = td(x, y)$, $d(z, y) = (1 - t)d(x, y)$. Conversely, let $t = d(x, z)/d(x, y)$ and note that $(x - z)$, $(y - z)$ have to be linearly dependent. **8.** $\int_0^1 |f|$ is a norm in $C[0, 1]$.

***Section 5.59,*** page 215. **1.** $S = \{1, \frac{1}{2}, \frac{1}{3}, \frac{1}{4}, \ldots\}$, $S' = \{0\}$. **2.** See the Corollary to Theorem 23.1. **5.** Replace the open disks ($\delta$-neighborhoods) in the proof of Theorem 59.3 by inscribed open squares. **9.** Since $\Omega$ is open, $N_r(x)$ for $x \in \Omega$ and some $r \in \mathbf{R}$ is contained in $\Omega$. $N_{r/3}(x)$ contains some point $a$ with rational coordinates. Pick $\delta > 0$ rational and such that $r/3 < \delta < 2r/3$. Then $x \in N_\delta(a)$.

***Section 5.60,*** page 218. **1.** $\|x\| < 2$ for all $x \in A$, $a = \theta$, no. **3.** $c(\mathfrak{J}) = \{x \in \mathbf{E}^n \mid \xi_j < \alpha_j$ or $\xi_j > \beta_j$ for some $j \in \{1, 2, \ldots, n\}\}$ is open.

***Section 5.61,*** page 221. **1,2,3.** See Theorems 16.1, 16.2, and 16.3. **4.** $\|a_k - a_l\| \leqslant \|a_k - a\| + \|a - a_l\|$. **5.** Follow the pattern of the proof of Theorem 23.1. **6.** Use the FIE (61.1). **7.** If $\|x - a\| < \delta$, then $\|x - a\|_1 < n\delta$, $\|x - a\|_\infty < \delta, \ldots$.

***Section 5.62,*** page 224. **1.** $\left|\frac{1}{k} - \frac{1}{l}\right| = \frac{|l - k|}{kl} \leqslant \frac{1}{k} + \frac{1}{l}$. **2.** For $k > l$,

$$|q^k - q^l| = |q^l|\,|q^{k-l} - 1| \leqslant 2\,|q|^l.$$

**4.** Let $\{\mathfrak{I}_k\}$ denote a nested sequence of closed intervals in $\mathbf{E}$ and assume w.l.o.g. that $\lim_\infty l(\mathfrak{I}_k) = 0$. If $\mathfrak{I}_k = [a_k, b_k]$, then $\{a_k\}$ is a Cauchy sequence. Use induction for proof in $\mathbf{E}^n$. **10.** (a) Use induction. (b) Show that $|a_{k+1} - a_k| = 1/2^{k-1}$ and hence, $|a_k - a_l| \leqslant |a_k - a_{k-1}| + \cdots + |a_{l+1} - a_l| < 1/2^{l-3}$. (c) Take subsequence $\{a_{2k+1}\} = \{1, 1 + \frac{1}{2}, 1 + \frac{1}{2} + 1/2^3, \ldots, 1 + \frac{1}{2} + \cdots + 1/2^{2k-1}, \ldots\}$. Then, $\lim_\infty a_{2k-1} = 5/3$. **11.** Show that $a_{2k} - a_k > \frac{1}{2}$.

## Chapter 6

***Section 6.63,*** page 229. **1.** $\sqrt{3(\xi_1^2 + \xi_2^2)}$. **3.** $\sqrt{\xi_2^2 - 2\xi_1\xi_3^2 + \xi_2^4 + \xi_3^4}$; $x_1 = (1,1,0)$, $x_2 = (0,1,0)$.

***Section 6.64,*** page 230. **4.** $|g(\xi_2) - b| \leqslant |g(\xi_2) - f(\xi_1, \xi_2)| + |f(\xi_1, \xi_2) - b|$. **5.** No, yes. **8,9,10.** See Theorem 17.1 and exercises 1, 2, section 1.17. **11.** $\theta$.

***Section 6.65,*** page 233. **4.** Let $\varepsilon = \frac{1}{2}|f(a)|$. Then $|f(a)| - |f(x)| < \frac{1}{2}|f(a)|$ for all $x \in N_\delta(a) \cap \mathfrak{D}$. **5.** (a) $x = x_0 + (x - x_0)$, (b) $|f(x) - f(x_0)| = |f(x_0)f(x - x_0) - f(x_0)f(0)|$, (c) $E(x)$. **6.** See exercise 2, section 2.27.

***Section 6.66,*** page 239. **1.** $[0,1)$, $[0,1)$, $[0,1]$. **2.** $\varnothing$, $(-\frac{1}{2}, \frac{1}{2})$, $\{-1,0,1\}$. **3.** $\{y \in \mathbf{E}^2 \mid \eta_1^2 + \eta_2^2 < 16\}$. **4.** $\{x \in \mathbf{E}^2 \mid \xi_1^2 + \xi_2^2 < 1/16\}$. **5.** (a) $(0,1) \subset [0,1]$, (b) $\{0\} \cup (1,2) \cup (-2,-1) \subset [-2,2]$, (c) $(0,\infty)$, (d) $\mathbf{R}$, $\{0\}$. **6.** Note that $B_\mu \subseteq \bigcup_{\mu \in M} B_\mu$ for all $\mu \in M$ and use (66.10). **7.** $(-\infty, -1) \cup (1, \infty)$. **8.** $A \backslash B = A \cap c(B) = \varnothing$ implies $x \in A$ then $x \notin c(B)$. **10.** Unit circle with center at $\theta$ and the line segment from $(0,0)$ to $(1,0)$, $\theta$ included. **11.** Unit disk with center at origin. **12.** $\bigcup_{k \in \mathbf{Z}}(2k\pi, (2k+1)\pi) \subseteq \mathbf{R}\backslash\{-(\pi/2) + 2k\pi \mid k \in \mathbf{Z}\}$, $\mathbf{R}\backslash\{-(\pi/2) + 2k\pi \mid k \in \mathbf{Z}\}$, $\bigcup_{k \in \mathbf{Z}}[2k\pi, (2k+1)\pi]$, $\varnothing$. **13.** $\mathbf{E}\backslash C$ is open.

***Section 6.67,*** page 242. **1.** $\{x \in \mathbf{E}^3 \mid \xi_1^2 + \xi_2^2 + \xi_3^2 \leqslant 1\}$, $g(f(x)) = (\sqrt{1 - \xi_1^2 - \xi_2^2 - \xi_3^2}, \sqrt[4]{\xi_1^2 + \xi_2^2 + \xi_3^2})$, $(1,0)$, yes. **2.** Unit circle with center at $\theta$. **5.** 1. **6.** 1. **7.** 0. **8.** $\frac{1}{4}$. **9.** $\eta_1 = \xi_1$, $\eta_2 = \xi_2$.

***Section 6.68,*** page 247. **1.** $f^*((-1,1)) = \{0\} \cup (1,\infty) \cup (-\infty,-1)$. **2.** $f^*(\{1\}) = \{x \in \mathbf{E}^2 \mid \xi_1^2 + \xi_2^2 < 1\}$. **4.** If $G \subseteq \mathbf{E}^m$ is open, then $c(G) = H$ is closed. Hence, $f^*(H) = H_1 \cap \mathfrak{D}$ where $H_1$ is closed. Note that $f^*(H) = \mathfrak{D}\backslash f^*(G)$. **5.** $G_1 \cap \mathbf{E}^n = G_1$. **6.** $H_1 \cap \mathbf{E}^n = H_1$. **7.** See Theorem 59.3.

***Section 6.69,*** page 249. **1.** Yes. **2.** Exercise 1. **3.** Yes. **4.** The range of $f$, a parabolic cylinder, is not open in $\mathbf{E}^3$. **5.** $f_*(\Omega) = \Omega_1 \cap \mathfrak{R}(f)$ is open. **6.** Any open subset of $\mathbf{E}$ is the union of $\delta$-neighborhoods. Show that open intervals are mapped onto open sets. **7.** $f_*(\Omega) \cap \mathfrak{R}(f) = f_*(\Omega)$ is open.

***Section 6.70,*** page 251. **1.** See proof of Theorem 32.1 and Fig. 70.1. **4.** See (58.2) and Lemmas 59.2 and 59.3. **5.** Let $\Omega = \{\Omega_\alpha \mid \alpha \in \mathrm{A}\}$ denote an open cover of $K \times H$ and consider $p_{n.}(\Omega_\alpha)$ and $p_{.m}(\Omega_\alpha)$ for all $\alpha \in A$. **8.** Let $\Omega_\alpha = c(C_\alpha)$. $S \subseteq \bigcup_{\alpha \in A} \Omega_\alpha$ implies $S \subseteq \bigcup_{\alpha \in B} \Omega_\alpha$ for some finite set $B \subset A$; that is, $c(S) \supseteq \bigcap_{\alpha \in A} c(\Omega_\alpha)$ implies $c(S) \supseteq \bigcap_{\alpha \in B} c(\Omega_\alpha)$, i.e., $S \cap \bigcap_{\alpha \in A} c(\Omega_\alpha) = \varnothing$ implies $S \cap \bigcap_{\alpha \in B} c(\Omega_\alpha) = \varnothing$, i.e., $S \cap \bigcap_{\alpha \in B} c(\Omega_\alpha) \neq \varnothing$ for all $B \subset A$ implies that $S \cap \bigcap_{\alpha \in A} c(\Omega_\alpha) \neq \varnothing$.

***Section 6.71,*** page 253. **2.** $\mathfrak{D}$ is compact and $f$ is continuous. **3.** Choose $\xi_1 = \xi_2 = \cdots = \xi_{n-1} = M$, $\xi_n = (1/M^{n-1})$, $\mathfrak{D}$ is not compact.

***Section 6.72,*** page 256. **1.** $A = (-\infty, \frac{1}{2})$, $B = (\frac{1}{2}, \infty)$. **2.** $A, B$ are relatively open in $S$ if and only if there are open sets $A_1, B_1$ such that $A = S \cap A_1$, $B = S \cap B_1$. **3.** Let $A_1, B_1$ represent open sets such that $A_1 \cap S \neq \varnothing$, $B_1 \cap S \neq \varnothing$, $(A_1 \cap S) \cup (B_1 \cap S) = S$, $(A_1 \cap S) \cap (B_1 \cap S) = \varnothing$, and let $A = A_1 \cap S$, $B = B_1 \cap S$. Then $A \cap S \neq \varnothing$, $B \cap S \neq \varnothing$, $A \cup B = S$ and $A \cap B = \varnothing$. If $A' \cap B \neq \varnothing$, then $A' \cap B_1 \cap S \neq \varnothing$, hence $A' \cap B_1 \neq \varnothing$. Since $B_1$ is open, $A \cap B_1 = A_1 \cap S \cap B_1 \neq \varnothing$, a contradiction. Hence, $(A \cup A') \cap B = (A \cap B) \cup (A' \cap B) = \varnothing$. Similar for $A \cap (B \cup B') = \varnothing$. If $A \cap S \neq \varnothing$, $B \cap S \neq \varnothing$, $S \subseteq A \cup B$, $(A \cup A') \cap B = \varnothing$, $A \cap (B \cup B') = \varnothing$, then $A \cap B = \varnothing$. If $A, B$ are open, this is it. If $A$ is closed and $B$ open, then, by $A \cap B' = \varnothing$ for each $x \in A$, $N_{\delta_x}(x) \cap B = \varnothing$. Let $A_1 = \bigcup_{x \in A} N_{\delta_x}(x)$, $B_1 = B$. **5.** Proceed as in the proof of Theorem 72.2. **6.** Demonstrate convexity. **7.** If $\|x - a\| \leqslant \delta$, $\|y - a\| \leqslant \delta$, then, with $z = (1-t)x + ty$, $t \in [0,1]$, $\|z - a\| = \|(1-t)(x-a) + t(y-a)\| \leqslant \delta$.

***Section 6.73,*** page 260. **1.** $f_*([-1,1])$ is not connected. **2.** A polygonal path may be viewed as the continuous image of an interval. **3.** Let $A \subseteq S$ denote the set of all points that can be joined to some $x_0 \in S$ by a polygonal path and $B \subseteq S$ the set of all points that cannot be joined to $x_0$ by a polygonal path. $A, B$ form a disconnection for $S$. **5.** $\mathfrak{D}$ is the continuous image of $\{x \in \mathbf{E}^2 \mid 0 < \xi_2 \leqslant 1\}$ under $f: \mathbf{E}^2 \to \mathbf{E}^2$, defined by $\phi_1(\xi_1, \xi_2) = \xi_1$, $\phi_2(\xi_1, \xi_2) = \xi_1^2 \xi_2$. **7.** Use exercise 2.

*Section 6.74*, page 262. **3.** $\|f(x)-f(y)\| \leqslant \sqrt{50}\,\|x-y\|$. **4.** $\|f(x)-f(y)\| \leqslant \sqrt{\sum_{i=1}^{m}\sum_{k=1}^{n} a_{ik}^2}\,\|x-y\|$.

*Section 6.75*, page 264. **1.** $\sum_{i=1}^{n}\sum_{j=1}^{n} a_{ij}^2 < 1$. **2.** By the continuity of $f'$, there is a $\delta > 0$ such that $|f'(x)| < \frac{1}{2}$ for all $x \in N_\delta(0)$. Use the mean-value theorem to show that $|f(x)| < \delta/2$ for all $x \in N_\delta(0)$ and that $f$ satisfies a global Lipschitz condition with $L = \frac{1}{2}$ in $[-\delta, \delta]$. **3.** Show that $f(x) = x^3 + x^2 + \frac{1}{8}$ defines a contraction mapping from $[-\frac{1}{4}, \frac{1}{4}]$ into $[-\frac{1}{4}, \frac{1}{4}]$. **4.** $\|f(x)-f(y)\| \leqslant \sqrt{13/18}\,\|x-y\|$.

## Chapter 7

*Section 7.76*, page 270. **1.** $A = (-1, 1)$, $1/(1-x)$. **2.** See Theorem 61.3. **4.** $\{f_k\} \to f$ on $[0, 1]$, $f(x) = 0$. **5.** Assume that there are two limit functions $f, g$ and that $f(a) \neq g(a)$ for some $a \in A$.

*Section 7.77*, page 274. **1.** No, yes. **2.** No, no.

*Section 7.78*, page 277. **2.** $\max_K (f(x) - f_k(x)) = f(x_k) - f_k(x_k)$ exists. Assume that $f(x_k) - f_k(x_k) \geqslant \varepsilon$ for infinitely many $k \in \mathbf{N}$ and observe that $\{x_k\} \in K$ contains a subsequence that converges to some $x_0 \in K$. **5.** (a) $(0, 1, 0)$ for $x = 0$, $(0, 0, 0)$ for $0 < x < 1$, $(1, 0, 0)$ for $x = 1$, (b) $[a, b]$ for any $a, b$ with $0 < a < b < 1$. **6.** (c) If $x, y$ lie between two consecutive vertices, then $|p_\alpha(x) - p_\alpha(y)| = |p'_\alpha(\xi)|\,|x-y| \leqslant M|x-y| < \varepsilon$ for $|x-y| < \varepsilon/M$. **7.** $\mathbf{Q} \cap [a, b] = \{r_1, r_2, r_3, \ldots\}$ is denumerable. $\{f_k(r_1)\}$ is bounded. By the Bolzano–Weierstrass Theorem and by Theorem 61.2, there exists a convergent subsequence $\{f_{1k}(r_1)\}$. Consider $\{f_{1k}(r_2)\}$ and proceed as before. $\{f_{kk}(x)\}$ is the sequence we are looking for. **8.** For the $\delta(\varepsilon)$ in the definition of equicontinuity choose finitely many elements in $\mathbf{Q} \cap [a, b]$ such that each $x \in [a, b]$ is within $\delta(\varepsilon)$ of one of these elements. Then consider the sequence $\{f_{kk}(x)\}$ of exercise 7 and $|f_{kk}(x) - f_{ll}(x)| \leqslant |f_{kk}(x) - f_{kk}(r_j)| + |f_{kk}(r_j) - f_{ll}(r_j)| + |f_{ll}(r_j) - f_{ll}(x)|$ where $|x - r_j| < \delta(\varepsilon)$.

*Section 7.79*, page 281. **2.** Replace $k$ by $(k-1)$ in (79.2), multiply by $x$, replace $j$ by $(j-1)$, and note that

$$\binom{k-1}{j-1} = \frac{j}{k}\binom{k}{j}.$$

**3.** Replace $k$ by $k-2$ in (79.2) and note that

$$\binom{k-2}{j-2} = \frac{j(j-1)}{k(k-1)}\binom{k}{j}.$$

**4.** Divide the result in exercise 2 by $k$, the one in exercise 3 by $k^2$, and add both. **5.** Multiply (79.2) by $x^2$, the result of exercise 2 by $-2x$, and add both to the formula in exercise 4. **6.** $f(0)(1-x)^3 + 3f(\frac{1}{3})x(1-x)^2 + 3f(\frac{2}{3})x^2(1-x) + f(1)x^3$, $f(0)(1-x)^4 + 4f(\frac{1}{4})x(1-x)^3 + 6f(\frac{1}{2})x^2(1-x)^2 + 4f(\frac{3}{4})x^3(1-x) + f(1)x^4$. **7.** There is a $\delta(\varepsilon) > 0$ such that $|f(x) - f(y)| < \varepsilon/2$ for all $x, y \in [a, b]$ for which $|x - y| < \delta(\varepsilon)$. Choose a partition $P = \{x_0, x_1, \ldots, x_n\}$ such that $|x_k - x_{k-1}| < \delta(\varepsilon)$ for all $k = 1, 2, \ldots, n$ and put a polygonal path through $(a, f(a))$, $(x_1, f(x_1)), \ldots, (b, f(b))$.

***Section 7.80,*** page 284. **1.** For $\{\Delta_k\}$ to be a Cauchy sequence, there has to be an $N(\varepsilon) \in \mathbf{N}$ such that $|\Delta_k - \Delta_{k+1}| < \varepsilon$ for all $k > N(\varepsilon)$. **3.** $w$ is uniformly continuous on $[0, 1]$ and has period 1.

***Section 7.81,*** page 289. **1.** 0, 0. **2.** 0, 0. **5.** Convergence in $R([a, b], \mathbf{E})$ is uniform. Use Theorem 81.1. **8.** (5/18).

***Section 7.82,*** page 295. **1.** $f_k'(x) = 1 + x + (x^2/2) + \cdots + (x^{k-1})/(k-1)$. By example 5, $\lim_\infty f_k'(x) = 1 - \log(1-x)$; $f(x) = \int_0^x (1 - \log(1-t))dt + 1 = 2x + (1-x)\log(1-x) + 1$. **2.** $1/k! < 1/(k-1)^2$ for all $k \geqslant 2$. See example 2, section 2.19. **3.** $\log x$. **5.** $f_k(x) - f_k(x_0) = \int_{x_0}^x f_k'(t)dt$. Hence, $f(x) - f(x_0) = \int_{x_0}^x \lim_\infty f_k'(t)dt$. Since $\lim_\infty f_k'(t)$ is continuous, $f'(x) = \lim_\infty f_k'(x)$.

## Chapter 8

***Section 8.83,*** page 299. **1.** (a) $(2, 2, 4, 2)$, $(-1, -1, -2, -1)$, $(1, -\frac{1}{2}, -\frac{3}{2}, \frac{1}{2})$,

(b) $\begin{pmatrix} 2 & -1 & 1 \\ 2 & -1 & -\frac{1}{2} \\ 4 & -2 & -\frac{3}{2} \\ 2 & -1 & \frac{1}{2} \end{pmatrix}$, (c) $(6, \frac{3}{2}, \frac{3}{2}, \frac{9}{2})$, (d) 2.

**2.** $\lambda_1 x_1 + \cdots + \lambda_p x_p + 0x_{p+1} + \cdots + 0x_q = 0$ for $(\lambda_1, \lambda_2, \ldots, \lambda_p) \neq (0, 0, \ldots, 0)$. **3.** $n$ homogeneous equations in $(n + p)$ unknowns have a nontrivial solution. **5.** $\lambda_1 x_1 + \cdots + \lambda_p x_p + \lambda_{p+1}(x_p + x_1) = \theta$ for $(\lambda_1, \lambda_2, \ldots, \lambda_p, \lambda_{p+1}) = (1, 0, \ldots, 0, 1, -1)$.

**6.** $\begin{pmatrix} 1 & 0 \\ 0 & 0 \end{pmatrix}\begin{pmatrix} 0 \\ 1 \end{pmatrix} = \begin{pmatrix} 0 \\ 0 \end{pmatrix}$.

**9.** Contrapositive to exercise 2. **10.** $L(\xi_1, \xi_2) = \xi_1(1, 0, 1) + \xi_2(1, -1, 0)$. **11.** $l_2 \circ l_1(\lambda x + \mu y) = l_2(\lambda l_1(x) + \mu l_1(y)) = \lambda l_2(l_1(x)) + \mu l_2(l_1(y)) = \lambda l_2 \circ l_1(x) + \mu l_2 \circ l_1(y)$.

***Section 8.84,*** page 306. **1.** $y = \xi_1(1, 4, 3, 1) + \xi_2(1, 0, -1, 0) + \xi_3(3, 1, -2, 0)$ for all $\xi_1, \xi_2, \xi_3 \in \mathbf{R}$. **2.** $\sum_{i=1}^m \left(\sum_{j=1}^n a_{ij}\xi_j\right)^2$. **3.** rank $L = 3$,

$$L^{-1} = \begin{pmatrix} 0 & 0 & 1 \\ 1 & -3 & 11 \\ 0 & 1 & -4 \end{pmatrix}.$$

**4.** (a) rank $L = 2$,

(b) $L^R = \begin{pmatrix} 0 & 1 \\ 1 & -1 \\ 0 & 0 \\ 0 & 0 \end{pmatrix}$, (c) $L^N = \begin{pmatrix} 1 & 0 \\ 0 & 1 \\ -1 & -1 \\ -1 & 0 \end{pmatrix}$, (d) $\begin{pmatrix} \lambda & 1 \\ 1 & \lambda - 1 \\ -\lambda & -\lambda \\ -\lambda & 0 \end{pmatrix}$, $\lambda \in \mathbf{R}$.

**5.** Let $l(e_j) = v_j$. Every $y \in \mathfrak{R}(l)$ may be written as $y = \alpha_1 v_1 + \cdots + \alpha_n v_n$. Then $l^L(y) = (\alpha_1, \alpha_2, \ldots, \alpha_n)$. **6.** $l^L = l^L \circ (l \circ l^R) = (l^L \circ l) \circ l^R = l^R$.

***Section 8.85,*** page 310. **1.** (a) $M = 3\sqrt{3} + \sqrt{2}$, (b) rank $L = 4$, $\mu = 1/(2 + \sqrt{2} + \sqrt{3} + \sqrt{10})$, (c) $v = 2 + \sqrt{2} + \sqrt{3} + \sqrt{10}$. **2.** Let $l(x) = (\eta_1, \eta_2, \ldots, \eta_m)$. Then $|\eta_j| \leqslant |a_{j1}|\,|\xi_1| + \cdots + |a_{jn}|\,|\xi_n| \leqslant nK\|x\|$. **3.** See exercise 2, section 8.84, and note that $(\sum_{j=1}^n a_{ij}\xi_j)^2 \leqslant \sum_{j=1}^n a_{ij}^2 \sum_{j=1}^n \xi_j^2$. **4.** Use the result of exercise 3. **5.** $l^R(y) = \eta_1 u_1 + \cdots + \eta_m u_m$ where $y = (\eta_1, \ldots, \eta_m)$. Use Lemma 85.1.

**6.** $L^L = \begin{pmatrix} 1 & 1 & -1 \\ 1 & 0 & -1 \end{pmatrix}$, if $y \in \mathfrak{R}(l)$, then $y = \alpha_1(1, 1, 1) + \alpha_2(1, -1, 0)$ for $\alpha_1, \alpha_2 \in \mathbf{R}$. $l^{-1}(y) = (\alpha_1, \alpha_2)$.

**7.** $\begin{pmatrix} 1 & 0 \\ 0 & 1 \\ 0 & 0 \end{pmatrix}$.

## Chapter 9

***Section 9.86,*** page 315. **1.** See example 1. **2.** $|\xi_1^2 + \xi_2^2 - 0| \leqslant \varepsilon\|x\|$ for $\|x\| < \varepsilon$. **4.** $\xi_1^2|\xi_2| \leqslant \varepsilon\|x\|$ for $\|x\| < \sqrt{\varepsilon}$. **5.** $\|x - a - (x - a)\| = 0$. **6.** $df(a, \lambda_1 u_1 + \lambda_2 u_2) = f'(a)(\lambda_1 u_1 + \lambda_2 u_2) = \lambda_1 f'(a)(u_1) + \lambda_2 f'(a)(u_2) = \lambda_1 df(a, u_1) + \lambda_2 df(a, u_2)$. **7.** Divide (86.1) by $|x - a|$.

***Section 9.87,*** page 320. **1.** $(\omega_1, 0)$. **2.** $(\omega_2, 0, 2\omega_1)$. **3.** $D_{e_j}f(\theta) = (\partial f/\partial \xi_j)(\theta)$. **4.** 0, 0.

***Section 9.88,*** page 322.

**1.** $D_j f(\theta) = 0$, $D_j f(x) = 2\xi_j\left(\sin\dfrac{1}{\xi_1^2 + \xi_2^2} - \dfrac{1}{\xi_1^2 + \xi_2^2}\cos\dfrac{1}{\xi_1^2 + \xi_2^2}\right)$.

**2.** 0, 0. **3.** 0, 0. **4.** 0, 0. **5.** See exercise 7, section 9.86.

***Section 9.89,*** page 327.

**1.** $F'(a) = \begin{pmatrix} 2\alpha_1 & -1 \\ 1 & 2 \end{pmatrix}$, $\|f(x) - f(a) - F'(a)(x - a)\| = |\xi_1 - \alpha_1|^2$.

**2.** $\begin{pmatrix} 1 & 0 \\ 0 & 1 \\ 2 & 0 \end{pmatrix}$. **3.** $\begin{pmatrix} 0 & 0 \\ 0 & 1 \end{pmatrix}$.

***Section 9.90,*** page 333. **1.** When evaluating $D_u f(\xi_1, \xi_2)$ for $(\xi_1, \xi_2) \neq (0,0)$, apply the mean-value theorem of the differential calculus to the numerator in $\lim_{t \to 0}(f(x + tu) - f(x))/t$. **2.** No, no. **3.** $D_1 f$, $D_2 f$ exist in a neighborhood of $\theta$. $D_u f$ exists at $\theta$ only. $D_1 f$, $D_2 f$ are not continuous at $\theta$. $f'(\theta)$ does not exist, because no matter how small $\|x\|$, $|\xi_1 \xi_2|/\sqrt{\xi_1^2 + \xi_2^2} = \frac{1}{2}\sqrt{\xi_1^2 + \xi_2^2}$ for $\xi_1 = \xi_2$. **4.** As in exercise 3 except that $f'(\theta)$ exists.

***Section 9.91,*** page 338.

**5.** $\begin{pmatrix} 1 + 2\alpha_1 & -1 \\ 1 & 1 \end{pmatrix}$, $(3\alpha_1^2 - 2\alpha_1\alpha_2 + 1, 1 - \alpha_1^2)$.

***Section 9.92,*** page 343. **1.** Choose $\dfrac{\varepsilon_1}{\mu} < \min\left(\dfrac{1}{2}, \dfrac{\varepsilon}{2(M+1)}\right)$. **2.** $\begin{pmatrix} 0 & 1 \\ 1 & -1 \\ 0 & 0 \end{pmatrix}$.
**3.** $(\mathrm{Cos}^{-1})'(b) = -(1/\sqrt{1 - b^2})$. **4.** Take $f : [0,1] \to \mathbf{E}$ where $f(x) = x$, $A = [1,2]$, $a = 1$. Then $a$ is an isolated point of $f^*(A) = \{1\}$.
**5.** $\begin{pmatrix} 0 & 1 \\ \frac{1}{2} & 0 \end{pmatrix}$.
**6.** See Theorem 27.2.

***Section 9.93,*** page 346. **1.** By exercise 3, section 6.73, any two points $a, b$ in $\mathfrak{D}$ may be connected by a polygonal path from $a$ to $x_1$ to $x_2$ to ... to $x_{n-1}$ to $b$. Apply Corollary 3 to Theorem 93.1 to each two consecutive points of this path and show that $f(a) = f(b)$. **2.** Let $\mathfrak{D} = \{x \in \mathbf{E}^2 \mid \|x\| < 1\} \cup \{x \in \mathbf{E}^2 \mid \xi_1 > 1\}$ and $f(x) = 1$ for $\|x\| < 1$ and $f(x) = 2$ for $\xi_1 > 1$.

***Section 9.94,*** page 351. **1.** Note that $f(\xi_1, \xi_2) = -f(-\xi_1, \xi_2) = -f(\xi_1, -\xi_2)$ and take $\lim_0 D_1 D_2 f(0, t)$. **2.** $D_2 D_1 f(\theta) = -1$, note that $f(\xi_1, \xi_2) = -f(\xi_2, \xi_1)$. **3.** Consider $g : [0,1] \to \mathbf{E}$, defined by $g(t) = f(a + t(b - a))$, note that $g^{(j)}(t)$ exists for all $t \in [0,1]$ and $j = 1, 2, \ldots, p$, and use the result of exercise 7, section 3.37. **4.** $f(a) = f(b) + D_1 f(a)(\alpha_1 - \beta_1) + D_2 f(a)(\alpha_2 - \beta_2) + \frac{1}{2}(D_1 D_1 f(a)(\alpha_1 - \beta_1)^2 + 2D_1 D_2 f(a)(\alpha_1 - \beta_1)(\alpha_2 - \beta_2) + D_2 D_2 f(a)(\alpha_2 - \beta_2)^2) + (1/3!)(D_1 D_1 D_1 f(c)(\alpha_1 - \beta_1)^3 + 3D_1 D_1 D_2 f(c)(\alpha_1 - \beta_1)^2(\alpha_2 - \beta_2) + 3D_1 D_2 D_2 f(c)(\alpha_1 - \beta_1)(\alpha_2 - \beta_2)^2 + D_2 D_2 D_2 f(c)(\alpha_2 - \beta_2)^3)$.

***Section 9.95,*** page 358. **1.** $r = \sqrt{x^2 + y^2}$, $\theta = \mathrm{Tan}^{-1}(y/x)$ for $x > 0$, $y > 0$, $= \pi/2$ for $x = 0$, $y > 0$, $= \mathrm{Tan}^{-1}(y/x) + \pi$ for $x < 0$, $= 3\pi/2$ for $x = 0$, $y < 0$, $= \mathrm{Tan}^{-1}(y/x) + 2\pi$ for $x > 0$, $y < 0$. $U = \{(r, \theta) \in \mathbf{E}^2 \mid r > 0,\ 0 < \theta < 2\pi\}$, $V = \{(x, y) \in \mathbf{E}^2 \mid (x, y) \neq (x, 0) \text{ for } x \geqslant 0\}$. **2.** $r = \sqrt{x^2 + y^2 + z^2}$, $\theta$ as in exercise 1, $\phi = \mathrm{Cos}^{-1}(z/\sqrt{x^2 + y^2 + z^2})$, $U = \{(r, \theta, \phi) \in \mathbf{E}^3 \mid r > 0, 0 < \theta < 2\pi, 0 < \phi < \pi\}$. **4.** $\xi = -\sqrt{\eta_1^2 + \eta_2^2}$.

## Chapter 10

***Section 10.96,*** page 365.

**1.** $\eta_2 = \left(\dfrac{\eta_3 - \eta_1}{2}\right)^2$, yes, $\begin{pmatrix} 1 & -1 \\ 0 & 2\xi_2 \\ 1 & 1 \end{pmatrix}$. **2.** $\eta_1 = \eta_2^2 + \eta_3^2$, yes, $\begin{pmatrix} 2\xi_1 & 2\xi_2 \\ 1 & 0 \\ 0 & 1 \end{pmatrix}$.
**3.** See example 4. **4.** $f(1,0,0) = f(1,2\pi,2\pi)$.
**5.** $F'(\theta) = \begin{pmatrix} 0 & 2 \\ 1 & 0 \end{pmatrix}$ is one-to-one.

***Section 10.97,*** page 370. **2.** $f(-\frac{1}{2},0) = -\frac{1}{4}$, $-\frac{1}{4} + \varepsilon = f(\xi_1, \xi_2) = (\xi_1 + \frac{1}{2})^2 + \xi_2^2 - \frac{1}{4}$ does not have a solution for $\varepsilon < 0$. **3.** There is no such $\delta > 0$.

**4.**

$$\begin{pmatrix} -\dfrac{1}{\sqrt{1-\eta_1^2}} & 0 \\ -\dfrac{\eta_1\eta_2}{(1-\eta_1^2)\sqrt{1-\eta_1^2-\eta_2^2}} & -\dfrac{1}{\sqrt{1-\eta_1^2-\eta_2^2}} \end{pmatrix}.$$

**5.**

$$\begin{pmatrix} -\dfrac{1}{\sqrt{1-\eta_1^2}} & 0 & 0 \\ -\dfrac{\eta_1\eta_2}{(1-\eta_1^2)\sqrt{1-\eta_1^2-\eta_2^2-\eta_3^2}} & -\dfrac{1}{\sqrt{1-\eta_1^2-\eta_2^2}} & 0 \\ -\dfrac{\eta_1\eta_3}{(1-\eta_1^2-\eta_2^2)\sqrt{1-\eta_1^2-\eta_2^2-\eta_3^2}} & -\dfrac{\eta_2\eta_3}{(1-\eta_1^2-\eta_2^2)\sqrt{1-\eta_1^2-\eta_2^2-\eta_3^2}} & -\dfrac{1}{\sqrt{1-\eta_1^2-\eta_2^2-\eta_3^2}} \end{pmatrix}.$$

***Section 10.98,*** page 375. **1.** $a \neq (0,t)$ for all $t \in \mathbf{R}$. Then $f'(a) : \mathbf{E}^2 \leftrightarrow \mathbf{E}^2$ in $N_\rho(a)$. $f^{-1}(y) = (\eta_1, (\eta_2/\eta_1))$. $(F^{-1})'(b) = \begin{pmatrix} 1 & 0 \\ -\beta_2/\beta_1^2 & 1/\beta_1 \end{pmatrix}$.
**2.** See exercise 2, section 9.95.
**3.** $F'(\theta) = \begin{pmatrix} 0 & 0 & -1 \\ 1 & 0 & 0 \\ 0 & 1 & 0 \end{pmatrix}$, $(F^{-1})'(\theta) = \begin{pmatrix} 0 & 1 & 0 \\ 0 & 0 & 1 \\ -1 & 0 & 0 \end{pmatrix}$.
**4.** $f, g \in C^1$, $J_g(a) \neq 0$, $J_f(g(a)) \neq 0$. **5.** $f^{-1}(y) = \left(\dfrac{\eta_1 + \eta_2}{1+a}, \dfrac{\eta_2 - a\eta_1}{1+a}\right)$,

$$(F^{-1})'(y) = \frac{1}{1+a}\begin{pmatrix} 1 & 1 \\ -a & 1 \end{pmatrix}.$$

**6.** $(f^{-1} \circ f)(x) = x$ for all $x \in U$. Hence, $(f^{-1} \circ f)'(x) = (f^{-1})'(f(x)) \circ f'(x)$ is the identity function and $J_{f^{-1}}(f(x)) J_f(x) = 1$. **7.** Use $(f')^{-1}(c) \circ f$ instead of $f \circ [f'(c)]^R$.

***Section 10.99,*** page 379. **2.** It is possible to solve with respect to $\xi$ and with respect to $\eta$, but the partial derivatives of the solution for $\eta$ with respect to $\xi$ and $\zeta$ do not exist at $\eta = 1, \zeta = 2$. **3.** Yes. **4.** See also section 9.95. **5.** $\|h(y)\|^2 + \|y\|^2 < \rho^2$.

***Section 10.100,*** page 383.

**2.** $\begin{pmatrix} 0 \\ 0 \end{pmatrix}$. **4.** $\begin{pmatrix} \dfrac{\eta_1}{\sqrt{\eta_1^2+\eta_2^2}\sqrt{1-\eta_1^2-\eta_2^2}} & \dfrac{\eta_2}{\sqrt{\eta_1^2+\eta_2^2}\sqrt{1-\eta_1^2-\eta_2^2}} \\ \dfrac{-\eta_2}{\eta_1^2+\eta_2^2} & \dfrac{\eta_1}{\eta_1^2+\eta_2^2} \\ 0 & 0 \end{pmatrix}$.

**5.** In each connected portion of $V$, $f$ does not depend on $y$, but for different portions of $V$ that are not connected, $f$ may have different values for the same values of $x$. (Note that $V = \{y \in \mathbf{E}^m \mid (\theta, y) \in V_1\}$ need not be connected although $V_1 = f_*(U)$ is connected for $U = N_\rho(\theta, \theta)$.)

***Section 10.101,*** page 390. **1.** By necessity, the functions $f_j : D_j \to \mathbf{E}$ where $f_j(\xi_j) = f(\alpha_1, \alpha_2, \dots, \alpha_{j-1}, \xi_j, \alpha_{j+1}, \dots, \alpha_n)$, $a = (\alpha_1, \alpha_2, \dots, \alpha_n)$, $D_j = \{\xi_j \in \mathbf{E} \mid (\alpha_1, \alpha_2, \dots, \alpha_{j-1}, \xi_j, \alpha_{j+1}, \dots, \alpha_n) \in \mathfrak{D}\}$, must have a relative minimum at $\xi_j = \alpha_j$. **2.** $\sum_{i=1}^n \sum_{j=1}^n D_i D_j f(a)\omega_i\omega_j$ is a continuous function of $u$ and is positive for all $u \neq \theta$. Hence, it assumes a positive minimum $\mu > 0$ on the compact set $\{u \in \mathbf{E}^n \mid \|u\| = 1\}$. Since $f \in C^2(\mathfrak{D})$, there is a $\delta > 0$ such that with $v = (v_1, v_2, \dots, v_n)$, $\sum_{i=1}^n \sum_{j=1}^n D_i D_j f(c) v_i v_j \geqslant (\mu/2)$ for all $c \in N_\delta(a)$ and all $v \in \mathbf{E}^n$ for which $\|v\| = 1$. Use Taylor's formula, exercise 3, section 9.94. **4.** $D_1 D_1 f(a)\omega_1^2 + 2D_1 D_2 f(a)\omega_1\omega_2 + D_2 D_2 f(a)\omega_2^2 > 0$ for all $u \neq \theta$.
**5.** Minimum is assumed for

$$a = \frac{n\sum_{i=1}^n \xi_i\eta_i - \sum_{i=1}^n \xi_i \sum_{i=1}^n \eta_i}{n\sum_{i=1}^n \xi_i^2 - \left(\sum_{i=1}^n \xi_i\right)^2}, \qquad b = \frac{\sum_{i=1}^n \xi_i^2 \sum_{i=1}^n \eta_i - \sum_{i=1}^n \xi_i \sum_{i=1}^n \xi_i\eta_i}{n\sum_{i=1}^n \xi_i^2 - \left(\sum_{i=1}^n \xi_i\right)^2}.$$

**6.** $d = \sqrt{10}$ for $\xi_1 = 3$, $\xi_2 = \xi_3 = 0$ or $\xi_1 = 0$, $\xi_2 = 3$, $\xi_3 = 0$. See also exercise 4. **7.** Maximize the function $f(x, y) = \xi_1^2\eta_1^2 + \cdots + \xi_n^2\eta_n^2$ under the constraints $\xi_1^{2p} + \cdots + \xi_n^{2p} = 1, \eta_1^{2q} + \cdots + \eta_n^{2q} = 1$ and note that $(q/p) + 1 = q$ and $1 + (p/q) = p$. **8.** $(\xi_j + \eta_j)^p = (\xi_j + \eta_j)^{p-1}(\xi_j + \eta_j)$. Apply Hölder's inequality (exercise 7) to $(\xi_1 + \eta_1)^{p-1}\xi_1 + \cdots + (\xi_n + \eta_n)^{p-1}\xi_n + (\xi_1 + \eta_1)^{p-1}\eta_1 + \cdots + (\xi_n + \eta_n)^{p-1}\eta_n$, using $p/(p-1)$ instead of $p$ and $p$ instead of $q$.

## Chapter 11

***Section 11.102,*** page 396. **2.** If $\mathfrak{J}$ is closed and $x \notin \mathfrak{J}$, then $\xi_j < \alpha_j$ or $\xi_j > \beta_j$ for $j \in \{1, 2, \dots, n\}$. Let $\delta = (\alpha_j - \xi_j)/2$ or $\delta = (\xi_j - \beta_j)/2$. Then $N_\delta(x) \subseteq c(\mathfrak{J})$. If $\mathfrak{J}$

is open, and $x \in \mathfrak{J}$, then $\alpha_j < \xi_j < \beta_j$ for all $j = 1, 2, \ldots, n$. Let $\delta = \frac{1}{2}\min(\xi_j - \alpha_j, \beta_j - \xi_j)$. Then $N_\delta(x) \subseteq \mathfrak{J}$. **4.** $\underline{S}(\phi_2, P) = 0, \overline{S}(\phi_2, P) = 1$ for all $P$ of $\mathfrak{J}$. **5.** Adapt the proofs of Lemmas 44.2, 44.3, 45.1, and 45.2. **6.** Use Lemma 9.1.

***Section 11.103,*** page 399. **1.** Adapt the proof of Theorem 46.1. **2.** Adapt the proof of Theorem 47.1. **4.** See exercises 4, 5, section 4.48 and note that a face of an $n$-dimensional interval has $n$-dimensional Lebesgue measure zero. **5.** Adapt the proof of Lemma 48.2. **7.** Generalize the proofs of Lemmas 49.1 and 49.2. **8.** $f$ is continuous except on $c(A \cup B) \cap \mathfrak{J}$. **9.** See exercise 6, section 11.102. **10.** See section 4.50. **11.** $S = \{x \in \mathbf{E}^2 \mid 0 \leqslant \xi_1 \leqslant 1, \xi_2 = 0\}$.

***Section 11.104,*** page 403. **1.** For every $k \in \mathbf{N}$, $\{x \in \mathbf{E}^3 \mid 0 \leqslant \xi_i \leqslant 1/k, i = 1, 2, 3\}$ contains all but finitely many points from $\mathfrak{D}$. $\int_{\mathfrak{D}} f = 0$. **2.** Either there is a $\delta > 0$ such that $N'_\delta(b) \cap \mathfrak{D} = \varnothing$ or for all $\delta > 0$, $N'_\delta(b) \cap \mathfrak{D} \neq \varnothing$. **3.** Either $N_\delta(a) \cap c(\mathfrak{D}) \neq \varnothing$ for all $\delta > 0$ or there is a $\delta > 0$ such that $N_\delta(a) \cap c(\mathfrak{D}) = \varnothing$. (Note that $N'_\delta(a) \cap \mathfrak{D} \neq \varnothing$ for all $\delta > 0$). **4.** See exercise 3 and Lemma 104.2 and note that $(A \cup B)' = A' \cup B'$. **5.** If $\beta\mathfrak{D} \subseteq \mathfrak{D}$, then $\mathfrak{D} \cup \beta\mathfrak{D} \subseteq \mathfrak{D}$. See exercise 3. **6.** See exercise 2. **7.** (c) If $x \in \beta\mathfrak{D}$, by (a) and (b), $x \in c(\mathfrak{D}^\circ)$ and $x \in c((c(\mathfrak{D}))^\circ)$; that is, $x \in c(\mathfrak{D}^\circ) \cap c((c(\mathfrak{D}))^\circ) = c(\mathfrak{D}^\circ \cup (c(\mathfrak{D}))^\circ)$ and vice versa. **8.** If $x \in \beta\mathfrak{D}$, then $x \in \mathfrak{D}$ or $x \in \mathfrak{D}'$. If $x \in \mathfrak{D}\backslash\mathfrak{D}'$, then $x \in (c(\mathfrak{D}))'$ and if $x \notin \mathfrak{D}$, then $x \in c(\mathfrak{D})$. Conversely, $x \in \mathfrak{D}$ and $x \in (c(\mathfrak{D}))'$ or $x \in \mathfrak{D}'$ and $x \in c(\mathfrak{D})$. Hence $x \in \beta\mathfrak{D}$. **9.** Every point is either an interior point or a boundary point. **10.** $\mathbf{E}^3$.

***Section 11.105,*** page 408. **2.** If not, then $\mathfrak{J}_1$ contains points from $\mathfrak{D}$ as well as from $c(\mathfrak{D})$. If each point in $\mathfrak{J}_1$ is an interior point of $\mathfrak{D}$ or an interior point of $c(\mathfrak{D})$, one obtains a disconnection of $\mathfrak{J}_1$. Hence, at least one point in $\mathfrak{J}_1$ has to be a boundary point of $\mathfrak{D}$ contrary to $\mathfrak{J}_1 \cap \beta\mathfrak{D} = \varnothing$. **5.** $J(S) = 0$ implies $\lambda(S) = 0$. **6.** For $\varepsilon > 0$, there is a partition $P$ of $\mathfrak{J}$ such that $\sum_{k=1}^{\mu} (M_k(f) - m_k(f))|\mathfrak{J}_k| < \varepsilon/2m$. Let $A_1 = \{k \in \{1, 2, \ldots, \mu\} \mid \mathfrak{J}_k^\circ \cap E_m \neq \varnothing\}$, $A_2 = \{k \in \{1, 2, \ldots, \mu\} \mid \mathfrak{J}_k^\circ \cap E_m = \varnothing\}$, note that faces of $n$-dimensional intervals have $n$-dimensional Jordan content zero and proceed as in the proof of Lemma 49.2.

***Section 11.106,*** page 411. **3.** Note that $J(\beta\mathfrak{D}) = 0$ and use Theorem 106.1. **4.** $S_2 \subseteq S_1 \cup S_2$ and $(S_1 \cup S_2)\backslash S_2 = S_1\backslash S_2$. Use (106.6).

***Section 11.107,*** page 414. **1.** If $f_{\mathfrak{D}}, g_{\mathfrak{D}}$ are integrable on $\mathfrak{J} \supseteq \mathfrak{D}$, then $g_{\mathfrak{D}} - f_{\mathfrak{D}}$ is integrable on $\mathfrak{J}$. **2.** If $f$ is continuous on $\mathfrak{J} \supseteq \mathfrak{D}$ except on a set of Lebesgue measure zero, then $|f|$ is also continuous on $\mathfrak{J}$ except on a set of Lebesgue measure zero. **3.** Let $|g(x)| \leqslant M$ and choose closed intervals $\mathfrak{J}_1, \mathfrak{J}_2, \ldots, \mathfrak{J}_l$ such that $S \subseteq \bigcup_{j=1}^{l} \mathfrak{J}_j$, $\sum_{j=1}^{l} |\mathfrak{J}_j| < \varepsilon/2M$. Show that there is a partition $P$ such that $\overline{S}(h; P) < (\varepsilon/2)$, $\underline{S}(h; P) > -(\varepsilon/2)$. **6.** $\beta(\mathfrak{D}_1 \cup \mathfrak{D}_2) \subseteq \beta(\mathfrak{D}_1) \cup \beta(\mathfrak{D}_2)$ by Lemma 104.3.

***Section 11.108,*** page 419. **4.** $\psi(y) = \Psi(y) = 1$. **6.** $J(A) = \int_{\mathfrak{J}_1}(\int_{\mathfrak{J}}\int \chi_{A_t}\, dx\, dy))dt = \int_{\mathfrak{J}_1} J(A_t)\, dt$. **7.** If $D_1 D_2 f(a, b) - D_2 D_1 f(a, b) > 0$ for some $(a, b) \in \mathfrak{D}$, then there is a closed two-dimensional interval $\mathfrak{J}$ such that $D_1 D_2 f(x, y) - D_2 D_1 f(x, y) > 0$ for all $(x, y) \in \mathfrak{J}$. Use Theorem 108.2.

***Section 11.109,*** page 424. **2.** $\displaystyle\int_a^b (\beta(x) - \alpha(x))\, dx \int_c^d (\delta(z) - \gamma(z))\, dz.$

**3.** $\displaystyle\int_{-1}^{1}\left(\int_{-\sqrt{1-y^2}}^{\sqrt{1-y^2}} f(x, y)\right) dx\Bigg)\, dy.$ **4.** $\displaystyle\int_0^1\left(\int_{x^2}^{x} dy\right) dx = \frac{1}{6}.$

**6.**
$$\int_{-1}^{1}\left(\int_{-\sqrt{1-x^2}}^{\sqrt{1-x^2}}\left(\int_{-\sqrt{1-x^2-y^2}}^{\sqrt{1-x^2-y^2}} dz\right) dy\right) dx$$
$$= \int_{-1}^{1}\left(\int_{-\sqrt{1-y^2}}^{\sqrt{1-y^2}}\left(\int_{-\sqrt{1-y^2-z^2}}^{\sqrt{1-y^2-z^2}} dx\right) dz\right) dy$$
$$= \int_{-1}^{1}\left(\int_{-\sqrt{1-z^2}}^{\sqrt{1-z^2}}\left(\int_{-\sqrt{1-x^2-z^2}}^{\sqrt{1-x^2-z^2}} dy\right) dx\right) dz.$$

***Section 11.110,*** page 428. **1.** $G'(y) - G'(a) = \int_a^b (D_2 g(x, y) - D_2 g(x, a))\, dx$. **2.** Let $\tan x = t$. **3.** $(1/x)(3e^{x^3} - 2e^{x^2})$. **4.** $f'(x) + g'(x) = 0$, $f(0) + g(0) = \int_0^1 dy/(1 + y^2)$.

**6.**
$$\frac{2^{2n+1}x^{2n}(n!)^2}{(2n)!}.$$

**7.** $\int_0^1 t^{x-1}(1 - t)^{y-1} \log t\, dt$, $\int_0^1 t^{x-1}(1 - t)^{y-1} \log(1 - t)\, dt$.

***Section 11.111,*** page 437. **1.** $2 + \pi/4$. **2.** $2a^2$. **3.** $\pi/32$.

**4.**
$$\frac{2\pi}{3}\left(1 + \frac{3}{16}(\sqrt{5} - 1)^2 + \frac{\sqrt{2}}{4}(3 - \sqrt{5})^{3/2}\right).$$

## Chapter 12

***Section 12.112,*** page 445. **1.** If all dimensions of all $\mathfrak{J}_j$ are rational, the lemma is trivial. If not, replace the dimensions $s_1, s_2, \ldots, s_n$ of $\mathfrak{J}_j$ by $s_1 + \delta, \ldots, s_n + \delta$, $\delta > 0$, such that $\prod_{i=1}^{n} (s_i + \delta) - |\mathfrak{J}_j| < \varepsilon/k$ and choose rational numbers $\rho_i \in [s_i, s_i + \delta]$, $i = 1, 2, \ldots, n$. **2.** Note that $l^{(m/n)-1} \geqslant 1$. **4.** $\mathfrak{J}_g(x, y, z) = -\sin^3 x \sin^2 y \sin z \neq 0$ for $(x, y, z) \in \mathfrak{D}^\circ$. $g(\mathfrak{D})$ is the closed unit ball with center at $\theta$. **5.** $\mathfrak{J}_g(x, y, z) = x^2 \sin y \neq 0$ for $(x, y, z) \in \mathfrak{D}^\circ$. $g(\mathfrak{D})$ as in exercise 4.

***Section 12.113,*** page 450. **1.** $L_1 = (e_1, \lambda e_2, e_3, e_4)$, $L_2 = (e_1, e_4, e_3, e_2)$, $L_3 = (e_1, e_2 + e_3, e_3, e_4)$. $\det L_1 = \lambda$, $\det L_2 = -1$, $\det L_3 = 1$. **2.** $L_1^{-1} = (e_1, 1/\lambda e_2, e_3, e_4)$, $L_2^{-1} = (e_1, e_4, e_3, e_2)$, $L_3 = (e_1, e_2 - e_3, e_3, e_4)$. **3.** The interior of $\mathfrak{J}_j$ is mapped one-to-one onto the interior of $l(\mathfrak{I}_j)$. A common interior point $a$ of the images can only come from a common boundary point of the preimages. Hence, some $N_\delta(a)$ comes from a subset of a set with Jordan content zero. **4.** If $g: \mathbf{E}^2 \to \mathbf{E}^2$ is defined by $\psi_1(x, y) = 2x - 2y$, $\psi_2(x, y) = -x + 2y$, then $g(\mathfrak{D}_1) = \mathfrak{D}_2$. Hence $J(\mathfrak{D}_2) = 2\pi$.

***Section 12.114,*** page 456. **1.** $h'(x) = [g'(a)]^{-1}g'(x)$. **2.** $h(\Omega)$ is open, $h^{-1} \in C^1(h(\Omega))$ by the inverse-function theorem, $\mathfrak{I}_{h^{-1}}(y) \neq 0$ for all $y \in h(\Omega)$, $h^{-1}(h(C)) = C$ is compact. Use the argument in the proof of Theorem 112.2 with $h^{-1}$ instead of $g$, $h(\Omega)$ instead of $\Omega$, $h(C)$ instead of $\mathfrak{D} \cup \beta\mathfrak{D}$. Then $\beta C \subseteq h^{-1}(\beta h(C))$; that is, $h(\beta C) \subseteq \beta h(C)$. **3.** $\varepsilon_1 < (1/M\sqrt{n})\min(1 - \sqrt[n]{1 - \varepsilon/2}, \sqrt[n]{1 + \varepsilon/2} - 1)$. **4.** $\varepsilon_1 < (\mu\varepsilon)/(2 + \varepsilon)$. **5.** Replace $\mathfrak{I}$ by a circumscribed interval $\mathfrak{I}_2$ with rational dimensions such that $|\mathfrak{I}_2| - |\mathfrak{I}| < \varepsilon_2/4(2L\sqrt{n})^n$, use (114.17) and the left portion of (114.1). **6.** Note that $J(C_i \cap C_j) = 0$ for $i \neq j$ and $g(C_i \cap C_j) = g(C_i) \cap g(C_j)$.

***Section 12.115,*** page 463. **1.** $\eta_1^2 + \eta_2^2 + \eta_3^2 = \xi_1^2$. **2.** With $\alpha = \beta = 0$, $g$ is one-to-one and $\mathfrak{I}_g(x) = -\xi_1^2 \sin \xi_3 \neq 0$ in $\mathfrak{D}^\circ$. $J(D) = \int_{\mathfrak{D}} \iint \xi_1^2 \sin \xi_3 \, dx = 4\pi/3$. **3.** For each $x \in \beta\mathfrak{D}$ consider $N_{\delta_x}(x) \subset \Omega$. Let $C_x$ denote an open cube with center at $x$ and diagonal less than $\delta_x$. $\beta\mathfrak{D}$ is compact and finitely many $C_x$ cover $\beta\mathfrak{D}$. Consider the finite collection of closed cubes $\bar{C}_x$. **4.** Since $J(\beta\mathfrak{D}) = 0$, there are finitely many open intervals $\mathfrak{J}_1, \mathfrak{J}_2, \ldots, \mathfrak{J}_k$ such that $\beta\mathfrak{D} \subseteq \bigcup_{i=1}^k \mathfrak{J}_i$, $\sum_{j=1}^k |\mathfrak{J}_j| < \varepsilon$. Each $x \in \beta\mathfrak{D}$ is an interior point of some $\mathfrak{J}_j$. Let $\mathfrak{I}_j = \mathfrak{J}_j \cup \beta\mathfrak{J}_j$ and note that the points of $\beta\mathfrak{D}$ are interior points of the closed intervals $\mathfrak{I}_j$. **7.** $A \cup \beta A$ is compact and $J(A \cup \beta A) = 0$. Hence, $J(g(A)) = 0$. $J(g(\mathfrak{D})) = J(g(\mathfrak{D}\backslash A)) + J(g(A)) - J(g(\mathfrak{D}\backslash A) \cap g(A))$. **10.** Use (53.6) and (115.18).

***Section 12.116,*** page 469. **3.** $\cos \xi_1 = \eta_1/\rho < 1$, $\cos \xi_2 = \eta_2/\sqrt{\rho^2 - \eta_1^2} < 1$ since $\eta_1^2 + \eta_2^2 < \rho^2, \ldots$. **4.** $\xi_1 = \xi_2 = \cdots = \xi_{n-1} = \pi/2$, $\xi_n = 0$. **8.** Choose $k_0 > a$. Then

$$\frac{a^k}{k!} < \frac{a^{k_0}}{k_0!}\left(\frac{a}{k_0 + 1}\right)^{k - k_0}. \qquad \textbf{9.}\ \frac{a^k}{(k + \frac{1}{2})(k - \frac{1}{2}) \cdots (\frac{3}{2})(\frac{1}{2})} < \frac{2^{k+2}a^k k!}{(2k + 1)!} < \frac{2^{k+2}a^k}{k!}.$$

**10.** $1 \leqslant (\pi/2)^v (v^v/v!)$. **11.** $\xi_1 = \cos^{-1}(\eta_1/\rho)$, $\xi_2 = \cos^{-1}(\eta_2/\sqrt{\rho^2 - \eta_1^2})$, $\xi_3 = \cos^{-1}(\eta_3/\sqrt{\rho^2 - \eta_1^2 - \eta_2^2})$. **12.** $\pi^2/2, 8\pi^2/15, \pi^3/6, 16\pi^3/105$.

## Chapter 13

***Section 13.117,*** page 473. **2.** (28/9). **3.** $m = \int_A \int f(x, y) dx\, dy$. **4.** $m_x = \int_A \int y f(x, y) dx\, dy$, $m_y = \int_A \int x f(x, y) dx\, dy$. **5.** $m_y/m, m_x/m$. **6.** 1. **7.** $\frac{3}{2}$. **8.** $\pi$.

***Section 13.118,*** page 480. **2.** $p(t) = t + \pi/2$. **3.** If $s, t$ yield the same point on the cycloid, then $rt - a \sin t = rs - a \sin s$, $\cos t = \cos s$. Hence, $t = \pm s \pm 2k\pi$.

If $t = s \pm 2k\pi$, then $k = 0$; that is, $t = s$. If $t = -s \pm 2k\pi$, then $-s + (a/r)\sin s = \mp k\pi$. Since $a/r \leqslant 1$, only possible for $s = \pm k\pi$. Then $t = \mp k\pi \pm 2k\pi = \pm k\pi$; that is, $t = s$. **4.** $p(t) = \log t$. **5.** $f(0) = f(4)$. **6.** $p(t) = \sin^{-1} t$.

***Section 13.119,*** page 485. **1.** Proceed as in example 3, but omit the boundaries of rear, bottom, and front in Fig. 119.4, except the ones that are common to rear-bottom, bottom-front, and bottom-right. **2.** If $(\tau_1, \tau_2, f(\tau_1, \tau_2)) = (\sigma_1, \sigma_2, f(\sigma_1, \sigma_2))$, then $\tau_1 = \sigma_1, \tau_2 = \sigma_2$. **5.** $g'(t) = f'(p(t)) \circ p'(t)$. **7.** $(\cos \tau_1, \sin \tau_1, \tau_2)$, $(\tau_1, \tau_2) \in [0, 2\pi) \times \mathbf{E}$. **8.** $(\tau_1, 1, \tau_2)$, $(\tau_1, \tau_2) \in \mathbf{E}^2$.

***Section 13.120,*** page 491. **1.** $f(t) = (\tau_1, \tau_2, \ldots, \tau_n)$, $(\tau_1, \tau_2, \ldots, \tau_n) \in \mathbf{E}^n$ is a parametrization of $\mathbf{E}^n$. **2.** $f(t) = x_0$, $t \in \mathbf{E}^n$ is a parametrization of $x_0$. rank $f'(t) = 0$. **3.** $f(t) = (t, 0, \ldots, 0)$, $t \in \mathbf{E}$ is a parametrization of the $\xi_1$-axis. $f(t) = (\tau_1, \tau_2, 0, \ldots, 0)$, $t \in \mathbf{E}^2$ is a parametrization of the $\xi_1, \xi_2$-plane. **4.** $(-t + \sqrt{1 + 2t^2}, t)$, $t \in (-\infty, \infty)$. **6.** $g \in C^1(\mathfrak{D})$. **7.** Use Theorem 96.1. **9.** See example 7. **10.** Use Theorem 99.1. **11.** (a) If $\beta_j \neq 0$ where $b = (\beta_1, \beta_2, \ldots, \beta_n)$, then $t = (\xi_j - \alpha_j)/\beta_j$ and $x - a - [(\xi_j - \alpha_j)/\beta_j]b = 0$, (b) If $t \neq 0, \neq \pi$, then $\xi_2 = \sqrt{1 - \xi_1^2}$ or $\xi_2 = -\sqrt{1 - \xi_1^2}$. If $t = 0$ or $= \pi$, then $\xi_1 = \sqrt{1 - \xi_2^2}$ or $\xi_1 = -\sqrt{1 - \xi_2^2}$,

(c)
$$t = \cos^{-1}\frac{\xi_1 - a}{b} \quad \text{or} \quad t = \sin^{-1}\frac{\xi_2 - c}{d},$$

(d)
$$\xi_3 - \alpha_3 - \beta_3 \frac{[(\xi_1 - \alpha_1)\gamma_2 - (\xi_2 - \alpha_2)\gamma_1]}{\Delta} - \gamma_3 \frac{[(\xi_2 - \alpha_2)\beta_1 - (\xi_1 - \alpha_1)\beta_2]}{\Delta} = 0$$

where $\Delta = \beta_1\gamma_2 - \beta_2\gamma_1$.

***Section 13.121,*** page 497. **2.** $h(t, \sigma) = (\cos t + \sigma, \sin t)$. **3.** $p(t) = \log t$. **4.** Otherwise rank $f'(t) = 0$. **5.** Otherwise rank $f'(t) = 0$. **6.** (b) $\tau_1 = \pi$. **7.** $J_p(t)J_{p^{-1}}(p(t)) = 1$. **8.** $f'(t) = g'(p(t))p'(t)$ and

$$\frac{g'(t)p'(t)}{\|g'(t)p'(t)\|} = \frac{g'(t)}{\|g'(t)\|}.$$

***Section 13.122,*** page 502. **1.** $(-r(\sqrt{2}/2), r(\sqrt{2}/2), a/2\pi)$. **2.** $p(t) = (5\pi/2) - t$, $p'(t) = -1$. **3.** $(-\xi_2, \xi_1)$. **4.** If $(-\xi_2, \xi_1)\cdot(\alpha, \beta) = 0$ then $\dfrac{\alpha}{\beta} = \dfrac{\xi_1}{\xi_2}$.
**5.** See exercise 7, section 13.121. **8.** (a) $\sum_{i=1}^{3} (\beta_i d\tau_1 + \gamma_i d\tau_2)^2$, (b) $(d\tau_1)^2 + \sin^2 \tau_1 (d\tau_2)^2$, (c) $(5 + 4\cos\tau_1)(d\tau_1)^2 + (2 + \sin\tau_1)^2(d\tau_2)^2$. **9.** If $f = g \circ p$ where $p = (\pi_1, \pi_2)$ and $E, F, G$ are the coefficients of the form for $f$, $\bar{E}, \bar{F}, \bar{G}$ for $g$, then $E = \bar{E}(D_1\pi_1)^2 + 2\bar{F}D_1\pi_1 D_1\pi_2 + \bar{G}(D_1\pi_2)^2$, $F = \bar{E}D_1\pi_1 D_2\pi_1 + \bar{F}(D_1\pi_1 D_2\pi_2 + D_1\pi_2 D_2\pi_1) + \bar{G}D_1\pi_2 D_2\pi_2$, $G = \bar{E}(D_2\pi_1)^2 + 2\bar{F}(D_2\pi_1 D_2\pi_2) + \bar{G}(D_2\pi_2)^2$.

***Section 13.123,*** page 509. **1.** $\xi_1 + \xi_2 + \sqrt{2}\xi_3 = 2$. **2.** $\xi_1 + \xi_2 + \sqrt{2}\xi_3 = 1 + 2\sqrt{2}$. **3.** Show that $f(t)$ in (123.6) parametrizes plasters on $(-1, 1) \times (0, 2\pi)$ and on $(-1, 1) \times (-\pi, \pi)$. **4.** $\xi_1 + \xi_2 + (2 - \sqrt{2})\xi_3 = 2\sqrt{2}$.
**5.** $\dfrac{(D_1 f(a) \times D_2 f(a))}{\|D_1 f(a) \times D_2 f(a)\|} = \dfrac{D_1 g(p(a)) \times D_2 g(p(a)) J_p(a)}{\|D_1 g(p(a)) \times D_2 g(p(a)) J_p(a)\|}, \qquad J_p(a) > 0.$
**7.** (120.3) parametrizes every plaster on the torus.

***Section 13.124,*** page 514. **1.** A patch is the continuous image of a compact and connected set. **4.** For every point $t \in \mathfrak{D}$, there is an open square $S$ with center at $t$ such that $f_*(S)$ is a plaster. Finitely many such squares cover $K$. **6.** Choose the multiple points as endpoints of arcs.

***Section 13.125,*** page 519. **1.** $(\sqrt{5}/2) + \frac{1}{4}\log(2 + \sqrt{5})$. **2.** $\xi_1 = \frac{2}{3}\cos t + \frac{1}{3}\cos 2t$, $\xi_2 = \frac{2}{3}\sin t - \frac{1}{3}\sin 2t$, $0 \leqslant t < 2\pi$, $l = \frac{16}{3}$. **3.** $\sqrt{4\pi^2 + 1}$. **4.** (c) $f'(t) = g'(s)\,\|f'(t)\|$. **5.** $\tau_1 = \tau_2/2\pi$, $l = (1/2\pi)[\sqrt{2}\pi\sqrt{1 + 2\pi^2} + \sinh^{-1}(\sqrt{2}\pi)]$. **6.** $(b, (a/2)\cosh(b/a) + b/(2a^2 \sinh(b/a)))$. **8.** $\int_a^b d\tau_1$, $\int_a^b \sin\alpha\, d\tau_2$ for $\tau_1 = \alpha$.

***Section 13.126,*** page 523. **1.** $\xi_1 = \tau_2$, $\xi_2 = g(\tau_2)\cos\tau_1$, $\xi_3 = g(\tau_2)\sin\tau_1$. **2.** $\pi r\sqrt{r^2 + h^2}$. **3.** $2(1 - \sqrt{3}/2)\pi$. **4.** $(2\pi/3)(2\sqrt{2} - 1)$. **5.** $4\pi^2 a$. **7.** $(0, 0, (25\sqrt{5} + 1)/(10(5\sqrt{5} - 1)))$.

***Section 13.127,*** page 534. **1.** 2. **3.** $-4d\xi_2 \wedge d\xi_3 - \xi_1 d\xi_3 \wedge d\xi_1 + \xi_1 d\xi_1 \wedge d\xi_2$. **5.** $d\xi_1 \wedge d\xi_2 \wedge d\xi_3 + d\xi_2 \wedge d\xi_1 \wedge d\xi_3 = 0$. **6.** 0. **7.** $\frac{1}{2}$. **9.** Let $\sigma$ be parametrized by $f(t) = (\tau_1, \tau_2)$, $(\tau_1, \tau_2) \in S$. Then $\int_\sigma \int \omega = \int_S \int g(\tau_1, \tau_2) d\tau_1 d\tau_2$. **10.** $d\xi_1 \wedge d\xi_2 \wedge d\xi_3 \wedge d\xi_1 = 0, \ldots$. **11.** $\psi_1(x) d\xi_1 \wedge d\xi_2 \wedge d\xi_3 + \psi_2(x) d\xi_1 \wedge d\xi_2 \wedge d\xi_4 + \psi_3(x) d\xi_1 \wedge d\xi_3 \wedge d\xi_4 + \psi_4(x) d\xi_2 \wedge d\xi_3 \wedge d\xi_4$, $g(x) d\xi_1 \wedge d\xi_2 \wedge d\xi_3 \wedge d\xi_4$. **12.** $\sum_{i,j=1}^n \psi_{ij}(x) d\xi_i \wedge d\xi_j$, $d\xi_i \wedge d\xi_j = -d\xi_j \wedge d\xi_i$. **13.** $D_1 D_1 g(x) + D_2 D_2 g(x) + D_3 D_3 g(x)$. **14.** $\theta$.

***Section 13.128,*** page 541. **1.** $2\pi$. **2.** $-\pi$. **3.** $2\pi$.

***Section 13.129,*** page 543. **1.** $\xi_1 d\xi_3 \wedge d\xi_2 + \xi_3 d\xi_1 \wedge d\xi_2$. **2.** $(\xi_2 - 2\xi_3) d\xi_1 \wedge d\xi_2 \wedge d\xi_3$. **4.** $-2$. **5.** 3.

***Section 13.130,*** page 551. **2.** Apply Theorem 108.2 directly. **3.** (a) $-4$, (b) $-\frac{2}{3}$, (c) $-\pi/2$. **4.** Use (130.10). **5.** Introduce a cut from $(1, 0)$ to $(5, 0)$. **6.** $\pi$. **8.** $\int_{\gamma_1 - \gamma_2} g(x) \cdot dx = -2\pi$.

***Section 13.131,*** page 560. **1.** If $g : \mathfrak{J} \to \mathfrak{D}$ represents $\gamma$, then $f \circ g : \mathfrak{J} \to P$ has all the required properties. **2.** (a) $\theta$, $(1, 0, 0)$, $(0, 1, 0)$; $\theta$, $(0, 1, 0)$, $(0, 0, 1)$; $\theta$, $(0, 0, 1)$, $(1, 0, 0)$. (b) $f(s) = x_0 + u_1\cos\sigma + u_2\sin\sigma$, $f(s) = x_0 - u_1\sin\sigma + u_2\cos\sigma$. **3.** Cut at $\tau_2 = 0$, $-1 \leqslant \tau_1 \leqslant 1$ and $\tau_2 = \pi$, $-1 \leqslant \tau_1 \leqslant 1$. **7.** If $g(x) = (\xi_3, \xi_1, \xi_2)$, then curl $g(x) = (1, 1, 1)$. $\pi$.

***Section 13.132,*** page 564. **2.** See exercise 1. **3.** div grad $h = \Delta h$. **8.** $8\pi/3$. **9.** By Gauss' Theorem, $\int_\Upsilon \iint d\xi_1 \wedge d\xi_2 \wedge d\xi_3 = \int_{\partial\Upsilon} \int \xi_1 d\xi_2 \wedge d\xi_3$. Choose the parameter representation $\xi_1 = f(t)\cos\theta$, $\xi_2 = f(t)\sin\theta$, $\xi_3 = t$, $(t, \theta) \in [a, b] \times [0, 2\pi]$ of the lateral surface of $\partial\Upsilon$ and note that the surface integrals over top and bottom are zero. **10.** Note that div $g(x) = 1$ and use the formula in exercise 9 twice to find the volume $4\pi^2$ of the torus.

## Chapter 14

***Section 14.133,*** page 570. **1.** (a) see section 2.19, example 1, (b) see section 2.19, example 2, (c) diverges by Lemma 133.3, (d) $s_n = \frac{3}{4}(1 - (-1)^{n+1}/3^{n+1})$. **3.** $-2, -\frac{1}{2}, -\frac{7}{4}, -\frac{5}{8}, -\frac{27}{16}, -\frac{21}{32}$. **4.** See proof of Theorem 14.2.

**5.**
$$\sum_{j=0}^{\infty} \frac{1}{(r+j)(r+j+1)} = \sum_{j=0}^{\infty} \left(\frac{1}{r+j} - \frac{1}{r+j+1}\right)$$
$$= \frac{1}{r} - \frac{1}{r+1} + \frac{1}{r+1} - \frac{1}{r+2} + \cdots.$$

**8.** No. **9.** $\left|\sum_{j=1}^{k} c \cdot a_j - c \cdot a\right| \leqslant \|c\| \left\|\sum_{j=1}^{k} a_j - a\right\|$. **10.** 2.75.

***Section 14.134,*** page 574. **1.** $\lim_\infty s_{2k} = 1/9$, $\lim_\infty a_j \neq 0$. **2.** $a_j = b_j = (-1)^j/\sqrt{j}$.

**3.** (a) $\displaystyle\int_1^k \frac{dt}{t} < 1 + \frac{1}{2} + \cdots + \frac{1}{k+1}$, (b) $\displaystyle\int_k^{k+1} \frac{dt}{t} > \frac{1}{k+1}$.

**5.** For $\alpha \leqslant 1$, $1/j^\alpha \geqslant 1/j$. **6.** $a_1 \leqslant a_1$, $a_2 + a_3 \leqslant 2a_2$, $a_4 + \cdots + a_7 \leqslant 4a_4, \ldots,$ $a_1 \leqslant 2a_1, 2a_2 \leqslant 2a_2, 4a_4 \leqslant 2a_3 + 2a_4, \ldots$. Hence, $\frac{1}{2}(a_1 + 2a_2 + \cdots + 2^k a_{2^k}) \leqslant a_1 + a_2 + \cdots + a_{2^k} \leqslant a_1 + 2a_2 + \cdots + 2^k a_{2^k} + a_{2^k}$. **7.** $\sum_{j=2}^{\infty} 2^j a_{2^j} = \sum_{j=2}^{\infty} (1/j \log 2)$. Use exercise 6. **8,** Repeated application of Cauchy's condensation test of exercise 6 yields, ultimately,

$$\sum_{j=2}^{\infty} \frac{1}{j \log 2 \log(2 \log 2) \cdots \log(2 \log \ldots \log 2)}.$$

**9.** $2^j a_{2^j} = 1/(j \log 2)^\alpha$. Use Cauchy's condensation test. **10.** Use Cauchy's condensation test.

***Section 14.135,*** page 580. **1.** $s = \frac{1}{2}$ + positive terms and $s = \frac{1}{2} + \frac{1}{3}$ + negative terms. **2.** $f(2v) = 2v + v$, $f(2v - 1) = 2v - 1 - [v/2]$, $[x]$ = largest integer $\leqslant x$. **3.** If there are only finitely many $\pi_j$ such that $a_{\pi_j} > 0$ (or $v_j$ such that $a_{v_j} < 0$), then the series would converge absolutely. **5.** Choose $n_1$ such that $\sum_{j=1}^{n_1} p_j > 1 + |q_1|$, $n_2$ such that $\sum_{j=1}^{n_2} p_j > 1 + |q_1| + |q_2|, \ldots$. **6.** $1 + \frac{1}{3} + \frac{1}{5} + \cdots + \frac{1}{11} - \frac{1}{2} + \frac{1}{13} + \cdots$.

***Section 14.136,*** page 587. **2.** See exercise 12, section 2.17, and note that $\sqrt[j]{j^2} = \sqrt[j]{j}\sqrt[j]{j}$. **4.** $\{a_j\}$ diverges to $\infty$, $\{a_j\}$ diverges to $-\infty$. **5.** Let $\overline{\lim}\sqrt[j]{|a_j|} = r$ and $\varepsilon = (1 - r)/2$. Then $|a_j| < ((1 + r)/2)^j$. **7.** $|a_{j+1}| \geqslant |a_1|$ for all $j \in \mathbf{N}$. $\lim_\infty a_j \neq 0$. **10.** $(|a|^j/j!)/|a|^{j-1}/(j-1)! = |a|/j < 1$ for all $j > |a|$. **12.** Use exercise 11 and Theorem 136.1. **13.** $1, \frac{1}{3}, \frac{1}{2}, \frac{1}{5}, \frac{1}{4}, \frac{1}{7}, \frac{1}{6}, \ldots$. **14.** $2, \frac{1}{4}$. **15.** If $\alpha > 1$, then $\log(1/a_j)/\log j > \alpha - \varepsilon > 1$ for all $j > N(\varepsilon)$. Hence, $a_j < (1/j^{\alpha-\varepsilon})$. **17.** $a < 1$, $a > 1$. **18.** $\sum_{j=1}^{\infty} 1/j^2, \sum_{j=1}^{\infty} 1/j$. **19.** See proof of Theorem 30.1. **20.** If $a \in \{a_j\}'$ then for each $\delta > 0$, $a_{j_k} \in N'_\delta(a)$ for infinitely many $j_k$. Hence, $a_{j_k} \in N_\delta(a)$; that is, $a \in \{a_j\}^l$. $\{1, \frac{1}{2}, -1, -\frac{1}{2}, 1, \frac{1}{3}, -1, -\frac{1}{3}, 1, \frac{1}{4}, -1, -\frac{1}{4}, \ldots\}$. $\{a_k\}' = \{0\}$, $\{a_k\}^l = \{-1, 0, 1\}$. **21.** Assume w.l.o.g. that $b_j \geqslant 0$ and let $s_k = a_1 b_1 + a_2 b_2 + \cdots + a_k b_k$, $\sigma_k = a_1 + a_2 + \cdots + a_k$, $\tau_k = \sigma_1(b_1 - b_2) + \sigma_2(b_2 - b_3) + \cdots + \sigma_k(b_k - b_{k+1})$. Then, $s_k = \tau_{k-1} + \sigma_k b_k$. $\tau_k$ is the $k$th partial sum of $\sum_{j=1}^{\infty} \sigma_j(b_j - b_{j+1})$. Show that $|\tau_k|$ is bounded, and hence, $\lim_\infty \tau_{k-1}$ exists. Now, $\lim_\infty \tau_{k-1} = \lim_\infty s_k$. **22.** $\{1/j\}$ decreases to 0 and $|1 - 1 + 1 - \cdots \pm 1| \leqslant 1$. **23.** Let $b_j = 1/(2j-1)$, $a_j = \sin(2j-1)a$. From $2\sin\alpha\cos\beta = \cos(\alpha - \beta) - \cos(\alpha + \beta)$, $2\sin a \sin a = 1 - \cos 2a$, $2 \sin a \sin 3a = \cos 2a - \cos 4a, \ldots, 2\sin a \sin(2j-1)a = \cos(2j-2)a - \cos 2ja$ and hence, $2\sin a(\sin a + \sin 3a + \cdots + \sin(2j-1)a) = 1 - \cos 2ja$. For $a \neq k\pi$, $k \in \mathbf{Z}$, $|a_1 + a_2 + \cdots + a_j| \leqslant 1/|\sin a|$.

***Section 14.137,*** page 590. **2, 3.** Use Theorem 136.1. Note that $\lim_\infty |a_j| = 0$.

***Section 14.138,*** page 595. **1.** No. **2.** Yes, no. **4.** $\sum_{j=1}^{\infty}(q^i)^j = q^i/(1 - q^i) \leqslant q^i/(1 - q)$. **6.** $s_{k,l} - s_{k,l-1} \geqslant \frac{1}{4}(1 - |q|)$. **8.** See Lemma 135.1.

***Section 14.139,*** page 597. **1.** Since $p/(k(p-k)) = (1/k) + 1/(p-k)$, we have

$$\frac{1}{p-1} + \frac{1}{(p-2)2} + \cdots + \frac{1}{2(p-2)} + \frac{1}{p-1} = \frac{2}{p}\left(1 + \frac{1}{2} + \cdots + \frac{1}{p-1}\right).$$

Note that $1 + \frac{1}{2} + \cdots + 1/(p-1) < 1 + \log p$, that $\lim_\infty (\log p)/p = \lim_\infty \log \sqrt[p]{p} = 0$ and that the sign of $c_p$ is $(-1)^p$. **2.** Note that

$$\frac{1}{p-1} + \frac{1}{2^2}\frac{1}{p-2} + \cdots + \frac{1}{(p-1)^2} < \frac{1}{p-1} + \frac{1}{2(p-2)} + \cdots + \frac{1}{p-1}$$

and use the result of exercise 1. **3.** $\sum_{i,j=0}^{\infty} q^{i+j} = (\sum_{i=0}^{\infty} q_i)(\sum_{j=0}^{\infty} q_j)$.

***Section 14.140,*** page 602. **2.** (f) Use (d) and (a). **3.** Use Theorem 140.5.

***Section 14.141,*** page 606. **1.** $1, 1, 1, \frac{1}{4}, \frac{1}{3}$. **2.** Use Theorem 136.4. **3.** $1, \infty, \infty$. **4.** $\log(1 + x) - \log(1 - x) = \log[(1 + x)/(1 - x)]$. **5.** If $y = (1 + x)/(1 - x)$, then $x = (y - 1)/(y + 1)$. Use exercise 4. **6.** Use Theorem 140.5. **7.** $\lim_\infty \sqrt[k]{c} = 1$ for all $c > 0$.

***Section 14.142,*** page 611. **4.** $\int_1^{a^b} dt/t = b \int_1^a du/u$, $u^b = t$. **5.** $(\log)'(1) = \lim_0 (1/h)\log(1+h)$. Let $h = (1/n)$. By exercise 4, $n\log(1+1/n) = \log(1+1/n)^n$. Note that $\log e = \log(E(1)) = 1$. **14.** Assume that there are two such functions and take $(E_1/E_2)'$. **15.** Show that the exponential function satisfies properties (I) and (II) and use the result of exercise 14.

***Section 14.143,*** page 616. **2.** 3.01 .... **9.** $y + (1/3!)y^3 + \cdots$.
**10.** $\dfrac{\sin x}{x} = \sum_{j=0}^{\infty} (-1)^j \dfrac{x^{2j}}{(2j+1)!}$ for $x \neq 0$.
**11.** (e) From (I) and (II), $S(x) - S(y) = 2C((x+y)/2)S((x-y)/2)$ and $S$ is continuous at 0 by (d) and $|C(x)| \leqslant 1$ by (b). **12.** Differentiate $(S_1(x) - S_2(x))^2 + (C_1(x) - C_2(x))^2$. **14.** Let $x = 1$. **15.** Let $x = (\pi/2)$.

***Section 14.144,*** page 621. **1.** $a_j = ((-x)^j/j!)$. **5.** $1 + x + (3/2)x^2 + \cdots$. **6.** For $|x| < 1$,

$$\sum_{p=2}^{\infty} (-1)^p \left(\frac{1}{p-1} + \frac{1}{(p-2)2} + \cdots + \frac{1}{2(p-2)} + \frac{1}{p-1}\right) x^p$$

represents the Cauchy product of the absolutely convergent series

$$\textstyle\sum_{j=0}^{\infty} (-1)^j (x^j/j)$$

with itself. By Theorem 140.5, this series converges uniformly in $[0, 1]$. Proceed as in example 1, section 14.141. See also exercise 1, section 14.139. **9.** $1 + x^2 + \frac{2}{3}x^4 + (17/45)x^6 + \cdots$. **10.** $x + (x^3/3) + (2/15)x^5 + (17/315)x^7 + \cdots$. **12.** $y - (y^3/3!) + \cdots$. **13.** By the Corollary to Theorem 144.1, $a_{k-1} = ka_k$.

***Section 14.145,*** page 627.

**3.** $y + \sum_{j=1}^{\infty} \dfrac{1 \cdot 3 \cdot 5 \cdots (2j-1)}{2 \cdot 4 \cdot 6 \cdots (2j)} \dfrac{y^{2j+1}}{2j+1}$.

**4.** $\sin\left(\dfrac{\pi}{3} + \dfrac{\pi}{6}\right) = 1$, $\sin\dfrac{\pi}{3} = 2\sin\dfrac{\pi}{6}\cos\dfrac{\pi}{6}$, $\cos\dfrac{\pi}{3} = \cos^2\dfrac{\pi}{6} - \sin^2\dfrac{\pi}{6}$.

**5.** Use the results of exercises 3 and 4. **7.** $a_j \leqslant (a_1/b_1)b_j$; use the comparison test.

**8.** $\dfrac{1}{(j+1)^\alpha} \Big/ \dfrac{1}{j^\alpha} = \dfrac{1}{\left(1 + \dfrac{1}{j}\right)^\alpha} = 1 - \dfrac{\alpha}{j} + \alpha(\alpha+1)\left(1 + \dfrac{\theta}{j}\right)^{-\alpha-2} (1-\theta)\dfrac{1}{j^\alpha} > 1 - \dfrac{\alpha}{j}$,

$\alpha > 1$. Since $1 - \dfrac{\alpha}{j} > \left|\dfrac{a_{j+1}}{a_j}\right|$ for $j > N(\alpha)$, let $b_j = \dfrac{1}{j^\alpha}$ and use the result of exercise 7.

**9.** $\lim_\infty j(1 - |a_{j+1}/a_j|) = \frac{3}{2}$. **10.** Note $\mathrm{Sin}^{-1}\, y = -\mathrm{Sin}^{-1}(-y)$. Use exercise 9 and Theorem 140.5. **12.** Use exercise 11. **13.** $\lim_0 f''(x) = -2$, $\lim_0 f^{\mathrm{IV}}(x) = 12$, $\lim_0 f^{\mathrm{VI}}(x) = -120$, etc.... **14.** Use (145.5) and substitute $x = (1-t)a + tb$. **15.** Use the remainder in exercise 14, and note that

$$\int_0^1 f^{(n)}(tb)(1-t)^{n-1}\,dt \leqslant \frac{(n-1)!}{b^n} f(b)$$

and

$$R_n \leqslant \frac{b^n}{(n-1)!}\int_0^1 f^{(n)}(tb)(1-t)^{n-1}\,dt.$$

**17.** Absorb the factor $(-1)^j$ of $(-|x|)^j$ in the coefficients, and use exercise 16. **18.** Use the result of exercise 17. **20.** Proceed as in exercises 17 and 18. **21.** By (145.3), $e = 2 + (1/2) + (1/3!) + \cdots + (e^\xi/n!)$ where $0 \leqslant \xi \leqslant 1$. Then $2 < e < 3$. If $e$ is rational, then $e = p/q$, $q \geqslant 2$. Let $n > q$ and multiply the above formula by $(n-1)!$. Then $e^\xi/n$ is an integer which is absurd. **22.** The following simple proof is due to I. Niven.† Let $\pi = p/q$, $p, q \in \mathbf{N}$ and let $f(x) = x^n(p-qx)^n/n!$, $F(x) = f(x) - f''(x) + \cdots + (-1)^n f^{(2n)}(x)$ with $n$ so large that $p^{2n+1}/q^{n+1}n! < 1$. Then $F(0), F(\pi) \in \mathbf{Z}$. Show that $\int_0^1 f(x)\sin x\,dx = (F'(x)\sin x - F(x)\cos x)\big|_0^\pi = F(\pi) + F(0) \in \mathbf{Z}$ while $0 < \int_0^1 f(x)\sin x\,dx < 1$.

**26.**
$$\frac{1}{\cos x} = \frac{1}{2\cos^2(x/2) - 1} = \frac{\dfrac{1}{2\cos^2(x/2)}}{1 - \dfrac{1}{2\cos^2(x/2)}} = \frac{1}{2\cos^2(x/2)}\sum_{i=0}^{\infty}\left(\frac{1}{2\cos^2(x/2)}\right)^i$$

for $|x| < (\pi/2)$. Use the results of exercises 24 and 25. **27.** Use the result of exercise 26.

† *Bull. Amer. Math. Soc.*, 53 (1947), p. 509.

# Index

(A page number in parentheses signifies a reference to an exercise on that page.)

ABCDEFGHIJ–H–7987654